Low-Gravity Fluid Dynamics and Transport Phenomena

Edited by
Jean N. Koster and Robert L. Sani
University of Colorado at Boulder
Boulder, Colorado

Volume 130
PROGRESS IN
ASTRONAUTICS AND AERONAUTICS

A. Richard Seebass, Editor-in-Chief
University of Colorado at Boulder
Boulder, Colorado

Published by the American Institute of Aeronautics and Astronautics, Inc.
370 L'Enfant Promenade, SW, Washington, DC 20024-2518

American Institute of Aeronautics and Astronautics, Inc.
Washington, DC

Library of Congress Cataloging in Publication Data

Low-gravity fluid dynamics and transport phenomena/Jean N. Koster
and Robert L. Sani, eds.
  p.cm.—(Progress in astronautics and aeronautics;v.130)
  1. Fluid mechanics.   2. Liquids—Effect of reduced gravity on.
3. Capillarity.   I. Koster, Jean N.   II. Sani, Robert L.   III. Series.
TL507.P75   vol. 130   90-593
[TA357.5.R44]
629.1 s--dc20
[629.132'3]
ISBN 0-930403-74-6

Table of Contents

Chapter 2.   Transport Phenomena in Crystal Growth

Application of Energy-Stability Theory to Problems in Crystal Growth . . 73
G. P. Neitzel and D. F. Jankowski, *Arizona State University, Tempe, Arizona*

Bridgman Crystal Growth in Low Gravity: A Scaling Analysis 87
J. Iwan D. Alexander and Franz Rosenberger, *University of Alabama in Huntsville, Huntsville, Alabama*

**Steady-State Thermal-Solutal Convection and Diffusion in a
Simulated Float Zone 119**
G. W. Young, *University of Akron, Akron, Ohio*, and A. Chait, *NASA Lewis Research Center, Cleveland, Ohio*

Chapter 3. Capillary Phenomena

Chapter 4.   Gravity Modulation Effects

Chapter 5. Buoyancy, Capillary Effects, and Solidification

Chapter 6. Separation Phenomena

Separation Physics **539**
Paul Todd, *National Institute of Standards and Technology, Boulder, Colorado*

Chapter 7. Combustion

Preface

The international space-faring community is preparing to develop space stations and free-flying spacecraft into unique laboratories for fundamental physics and engineering studies. Low-gravity science, a multidisciplinary research field, can be implemented to demonstrate the character of physical processes without constraints of an accelerated environment. We need to learn how fluids behave in an acceleration-free environment before we start major commercial production efforts in such an environment.

Production of high-value materials in space for return to Earth and for use in space, two of the long-term goals of low-gravity research, requires a detailed understanding of how the low-gravity environment affects physical processes. Most materials are in the liquid phase during some period of their processing. Frequently the liquid processing phase is affected by temperature and concentration gradients. A comprehensive understanding of low-gravity fluid physics and transport phenomena is thus fundamental to exploiting the low-Earth orbit environment for possible commercial developments.

Secondary physical effects that were traditionally neglected on Earth, and were often considered nonexistent, may develop primary importance in a low-gravity environment. For centuries scientists and engineers have been trained to think in terms of a 1-g environment. A major struggle in today's space experiment preparation is just this traditional training. Scientists and engineers need to learn to think in terms of a low-gravity environment.

A focused educational effort is vital to the development of a space processing talent base and for the maximization of the return on investments (scientific and financial) in low-gravity science. A major problem encountered by scientists and engineers entering the field of low-gravity science is that of finding relevant literature dispersed over a vast number of journals and publications. There are even new journals created for microgravity science, which may drain the quality of the publications and increase their number. The result is a tendency to publish any relevant and nonrelevant details as soon as they are found, leaving ample opportunity to publish errors and errata. This abuse leaves the reader with the time-consuming task of sorting out relevant and archival information.

Fluid mechanics is often considered the foundation of space processing, and it is fundamental to the assurance of mission capability and safety. It was the idea of the editors to provide the community with a concise volume covering fluid mechanics and transport phenomena in a variety of research fields of low-gravity science. Most authors followed the request to combine an educational goal with the latest scientific insight. This special effort made many of the contributions archival ones and will make this volume a required compendium on the microgravity scientist's bookshelf.

The sequence of chapters should give the reader a rough guideline and, by no means implies that topics across chapters are not related. The contents of the papers are summarized briefly below.

Chapter 1.   Applied Fluid Mechanics and Thermodynamics

Space vehicles powered by liquid propellant and coasting interplanetary probes require that propellants or other liquids be kept at the tank drains. During boost periods sloshing of liquids constitutes a major concern for craft stability. With the advent of the Space Station Freedom these types of fluid management problems, as reviewed by F. T. Dodge, become timely again. The availability of reliable propellant management devices is crucial for the safe operation of spacecraft and the space station.

Large spacecraft require efficient thermal management systems for maximum use of the available power. Nucleate boiling is an efficient heat transfer mechanism and is especially effective in cases where small temperature differences are provided. Vapor removal from heated surfaces is a major safety concern in a reduced buoyancy environment. H. Merte Jr. provides a concise review of nucleate pool boiling primarily under reduced gravity, and includes high gravity as well. Fluids considered vary from cryogenic liquids to mercury.

Chapter 2.   Transport Phenomena in Crystal Growth

Direct numerical simulation and linear-stability theory are often used to study fluid mechanics of the Bridgman, Czochralski, and float-zoning technologies. Linear-stability theory provides a sufficient condition for instability of a given basic state. But knowledge of the linear-stability limit alone cannot guarantee that the fluid will remain stable. G. P. Neitzel and D. F. Jankowski elucidate the capabilities of energy-stability theory as a complementary tool to test the stability of basic states.

Crystal growth by directional solidification will be extensively used in space. For the Bridgman-Stockbarger technique, J. I. D. Alexander and F. Rosenberger show that the direction of the residual gravity with respect to the solidification interface must be carefully considered to optimize compositional uniformity of space-grown crystals. They also show that scaling arguments commonly employed for the prediction of low gravity results in materials processing, due to their inherently one-dimensional nature, can be rather misleading. Our understanding of transport in crystal growth can be furthered only by an approach that uses scaling, mathematical modeling, and the results of experiments in a complementary fashion.

The float-zone process has great potential for growing large, ultrapure crystals in space. In the absence of hydrostatic pressure, large diameters of molten zones may be achieved. Although buoyancy effects are reduced in low gravity, thermocapillary phenomena become of primary importance. G. W. Young and A. Chait provide an analytical/numerical model for the complete float-zone process, including all deformable interfaces and heat and mass transport by both diffusion and thermocapillary convection when the zone is subjected to the thermal profiles of an adjustable heater.

Exotic semiconductor materials, such as $Hg_{1-x}Cd_xTe$, have very attractive electronic or optical properties. The production of these semiconductors is extremely sensitive to composition, as the properties are very sensitive to stoichiometry. The fluid mechanics of such multicomponent systems is very complicated; our capabilities to accurately model the process via the Navier-Stokes equations for multiple diffusion is limited at the moment. Major efforts are needed in this field. B. N. Antar's paper is a first effort in modeling the fluid mechanics of such a multicomponent system.

An exotic material of current interest is the mercurous chloride crystal. These are grown by the Physical Vapor Transport technique. N. B. Singh reports briefly on fluid-flow experimentation in ampoules containing mercurous chloride. The ground-based efforts have shown that crystals grown at low Rayleigh number have better optical homogeneity.

Chapter 3. Capillary Phenomena

In a low-gravity setting comprising two adjacent fluid phases, such as for propellants, supply liquids, life-support liquids, or materials melts exhibiting free surfaces, fluid behavior is determined primarily by capillary phenomena. This contrasts sharply with situations on the Earth's surface, where capillarity controls behavior only in a boundary layer and is significant only for containers of small size. When gravity is removed the free surface can adopt strikingly altered configurations; the character and behavior of such free surfaces thus becomes a fundamental issue for low-gravity research. P. Concus and R. Finn have prepared a concise review of some of the salient features of equilibrium two-phase systems in microgravity, as occur for container configurations of special interest.

Adding temperature and concentration gradients to the free surface leads to the development of thermo- and/or diffusocapillary instabilities. The drastic reduction of buoyancy in the low-gravity environment proves to be an ideal environment in which to study these phenomena in depth. Legros and co-workers provide a well-balanced review of the scientific issues and current state of the art.

The problem of interface dynamics dramatically increases in complexity when capillary waves occur at the interface—an issue often neglected by the engineer. In the low-gravity environments these waves can be studied in more detail, as no gravity related hydrostatic pressure or buoyancy dampens the growth of these waves. M. G. Velarde and X.-L. Chu provide a detailed theoretical review of the problem, including nonlinear aspects and the occurrence of solitons. The review could provide a foundation for challenging and rich space experiments.

Chapter 4. Gravity Modulation Effects

The nature of space shuttle flight experiments is such that the residual gravity level in low-Earth orbit is stochastically modulated in time in both direction and amplitude. The effects of this gravity jitter are often substantial on the fluid flow, especially if the fluid exhibits interfaces or free surfaces. We currently have the capability to bound regimes of characterizable effects on different fluid situations. However, correlating experimental measurements with the real-time gravitational environment and providing numerical simulation is still an ambiguous issue. R. Monti evaluates the state of our knowledge by discussing theory and some numerical models that indicate the major distortions caused by the residual gravity and gravity jitters of the thermodiffusional fields and flow patterns established in low-gravity environments.

M. Wadih and co-workers concentrate their evaluation of gravity effects on the specific case of the differentially heated vertical cylinder. Such a system is related to Bridgman crystal growth technology. Their numerical evaluation covers large and low Prandtl-number liquids. It appears that gravity modulation effects are more important in liquid metals, especially for lower frequencies.

Chapter 5. Buoyancy, Capillary Effects, and Solidification

Many conventional material alloys are binary systems. The liquid phases are usually in stable equilibrium when isothermal. During freezing a thermal gradient is applied to the liquid layer, which may lead to double-diffusive convection. Double-diffusive convection has been investigated in some detail by the geophysical fluid mechanics community. C. F. Chen reviews the developments in double-diffusive convection and leads the reader to the equivalent problem in solidification of binary systems.

S. R. Coriell and G. B. McFadden focus on the stability analysis of binary systems that are solidified directionally. They review the basic physics related to the solidification process and the fluid flow—solid interface interactions. They discuss the development of interfacial instabilities and double-diffusive convection, and finally apply the theory to a lead-tin alloy.

A major concern during freezing is the redistribution of solute in the melt, which leads to nonuniform composition of the solid phase. This effect is known as segregation. Both micro- and macrosegregation have important consequences on the ensuing physical properties of alloys. Convective flows are largely responsible for macrosegregation so D. R. Poirier discusses models that can be used to predict the convection and macrosegregation in dendritically solidifying alloys.

Fluid flow has a definite effect on the development and progress of microstructure solidification. J. A. Dantzig and L.-S. Chao examine the interaction of macroscopic processing conditions with microstructure evolution. The presence of fluid flow changes the solute field ahead of the solidification front. The goal is to analytically separate the role of flow and diffusive transport in solidification and microstructure formation.

D. Henry presents in some detail a double-diffusive convective flow situation that exhibits thermally driven diffusion of constituents. This physical effect, known as the Soret effect, leads to a separation of constituents. In a normal-gravity environment the Soret diffusion is generally masked by other mixing effects such as buoyancy driven convection. The study, which includes linear stability and three-dimensional simulation, presents fundamental results on double-diffusive convection, separation phenomena, and convective instabilities. It is shown that a reduced-gravity environment is well suited for the study of such weaker secondary transport phenomena.

Liquid metals have a high thermal diffusivity and comparable low viscous diffusivity that corresponds to a value of the Prandtl number much smaller than 1. Experiments on low-Prandtl-number liquids are scarce; numerical modeling efforts are more common. P. Bontoux and co-workers discuss extensive numerical work on Bridgman and floating-zone model configurations using spectral methods and finite-element techniques. Their study focuses on the dynamical behavior of buoyancy-induced flows and on several steady flow regimes induced by surface tension and/or rotation. All of these cases pertain to practical crystal growth situations.

Containerless processing of materials is indicated in many cases to avoid corrosion of materials by their processing container. In a low-gravity environment containerless positioning is attractive as a processing technique for melts. In a gravity field the sample has to be levitated as well as positioned. In both cases, the melts, in the form of drops, have a free surface. Surface-tension-induced transport processes ensue from the application of temperature gradients to the freely floating drop. Transport processes become more complicated when the liquid drop, or shell, is solidified from one side. E. H. Trinh discusses the current research efforts and results of fluid dynamics and solidification of levitated drops.

Chapter 6. Separation Phenomena

Paul Todd's review on separation science in biotechnology can be considered archival. Basic physics as well as engineering fundamentals are presented. Special emphasis is on the gravity-related aspects of separation physics. Experimental results stem from laboratory studies, orbital flights, and parabolic suborbital flights.

Biotechnology typically involves handling samples of biological molecules or particulates suspended in the aqueous media necessary for sample viability. Much of the cost associated with biotechnical products is related to the processing of such mixtures, especially the analytical and preparative separation of their components. Phase partitioning is emerging as a method of choice for a number of biotechnical separations. This technique is affected by gravity as well as by forces independent of gravity. As discussed by S. Bamberger and co-workers, the low-gravity of space provides an experimental environment for improving the phase partitioning technology in space and on Earth.

Many binary metallic systems offer potential technical advantages. However, in a gravity field the two metallic components separate before solidification of the melt due to their density difference. At best, a finely dispersed mixture of one component in the other may be produced. D. Langbein discusses in great detail the separation effects that prevail when thermal convection and sedimentation of the denser component are absent. His discussions are backed by substantial low-gravity experimentation.

A special kind of separation in a variety of multiphase materials is the process of Ostwald ripening in liquids. The quantitative measurement of this effect has always been hampered by uncontrolled fluid flow originating from minute temperature or concentration inhomogeneities. A low-gravity environment provides the required reduced buoyancy environment to study Ostwald ripening in detail. L. Ratke provides a concise review of the physics of the phenomena and compares the available theories as well as microgravity experiments.

Chapter 7. Combustion

Combustion systems are usually very complicated systems in terms of fuel/oxidizer distribution, temperature fields, and chemical reactions. As these reaction rates are fast, most combustion experiments often can be designed to be performed in drop towers or aircraft. A. L. Berlad reviews the efforts to create a homogeneous particle cloud of fuel/oxidizer mixture distribution in space to study the flame propagation and extinction processes. Such an experiment lends itself ideally to the validation of theoretical and numerical results that assume perfect distributions of fuel particles. He also considers the effects of nonuniform particle distributions on observed particle cloud combustion phenomena.

For manned spacecraft, the understanding of combustion takes on new dimensions in terms of safety. Reduction of convective oxidizer transport in the low-gravity environment results in the increased importance of flames with radiative processes. For opposed-flow laminar flame spread over a solid surface, a classic combustion phenomenon, strongly radiative flames cannot be obtained in a normal environment subject to gravitationally induced flows. Such flames exhibit a characteristic size, shape, and spread rate that can only be studied in a low-gravity environment. R. A. Altenkirch and S. Bhattacharjee investigate the importance of surface and gas-phase radiative processes in these spreading flames.

The editors thank the authors for their collaboration in this fine project and those scientists who assisted in the review of the papers.

Jean N. Koster
Robert L. Sani
University of Colorado at Boulder
Boulder, Colorado
March 1990

Chapter 1. Applied Fluid Mechanics and Thermodynamics

Fluid Management in Low Gravity

F. T. Dodge*

Southwest Research Institute, San Antonio, Texas

Introduction

O N Earth, liquid in a tank settles to the bottom to minimize is gravitational potential energy. In space, where the effective gravity is small, the location of liquid in a tank is determined by the competing effects of gravity and the liquid's surface tension in order to minimize the sum of the gravitational and surface energies. Generally, the resulting position of the liquid is not over the tank outlet. Spacecraft must therefore incorporate methods to ensure that the liquid is located at the desired location in the tank even when a linear acceleration or effective gravity is directed adversely or in unknown directions. Similarly, if the tank is pressurized and must be vented periodically, methods are required to ensure that gas is positioned over the vent. One way to accomplish both of these objectives is to use an auxiliary thrusting system to provide a linear acceleration large enough to "settle" the liquid; this system will not be discussed here since its principles are similar to a tank on Earth exposed to gravity. Another method is to incorporate *passive* systems or devices that increase the effective surface tension forces sufficiently to hold the liquid at the desired position against adverse accelerations. These systems have been given the general name of propellant management devices (PMDs). (Not all of the liquids of interest are propellants, but this article will retain the conventional name.)

PMDs have been used in spacecraft for many years to control the position of nonvolatile (storable) liquids.[1] However, there is little or no flight experience with PMDs for cryogens or other easily evaporated liquids. This article will review the principles used to design and evaluate PMDs, with an emphasis on the new considerations required when the liquids are cryogenic.

*Institute Engineer, Department of Fluid Systems.

Stability Considerations

For the moment we will assume that gravity is completely absent. Then, as is well known,[2] liquid in a tank will form one or more interfaces of constant curvature (i.e., spherical), with the interface radii determined by the container geometry and size, the volume of liquid, and the contact angle θ at which the interfaces meet the wall. (The contact angle is a property of both the liquid and the tank wall material; for most liquids and tank materials of aerospace interest the contact angle is small and approximates 0 deg.) The question is: Where are the interfaces located in the tank? Reynolds and Satterlee[2] have shown that the most stable configuration of the liquid minimizes the total capillary energy $\sigma(A_i - A_w\cos\theta)$, where A_i is the liquid-gas interface area, A_w is the wetted area of the tank, and σ is the surface tension. From this relation we can prove that the liquid will collect into a single volume, rather than into several smaller, unconnected volumes, and the gas will form a single bubble that is attached to the walls at a definite location. Similar results hold true when the gravity is small but not zero, except that gravitational energy must be included in the analysis. There is a possibility that the geometric configuration can be complicated when 1) the contact angle is not zero, 2) the liquid occupies only a small part of the tank volume, and 3) the tank diameter is greater than its height. The gas will still form a single bubble, but the liquid may form an annulus around the bubble with the top and bottom of the tank being dry.[3]

It would seem, then, that for a given tank shape and size, the liquid and gas location can be predicted. Unfortunately, this predictable, stable position may become unstable when the tank is exposed to perturbing accelerations such as g jitter or when the engines are fired. The stability of a liquid-gas interface is a function of the nondimensional Bond number defined as $Bo = \rho g_d d^2/\sigma$, where g_d is the disturbance acceleration, ρ is the liquid density, and d is a relevant dimension of the interface, such as the tank diameter. The critical Bo is of order unity; thus, the g_d level that leads to instability is small for tanks of any but capillary tube size. It is likely, then, that the stable location will be disrupted periodically. From dimensional considerations the time required for the stable location to be reestablished is approximately $t = \sqrt{(\rho d^3/\sigma)}$. For a 1-m tank, this reorientation time is several minutes for typical propellants. We can conclude that it is nonconservative to assume that liquid and gas will always be in the stable position at the times when liquid flow is required or gas must be vented. Therefore, some alternative method—a PMD—must be used to ensure that gas-free liquid and liquid-free gas is available when needed.

Surface Tension Systems

The primary objective of a PMD is to keep the tank outlet covered with liquid whenever outflow is required. Secondary objectives might include control of the center of gravity of the liquid, the minimization of liquid sloshing, and the ability to vent liquid-free gas.

There are many types of propellant management devices that make use of surface tension to maintain the tank liquid at a specific position. They can be classified into one of three general types[1]: *partial communication* (sometimes called partial control), *total control*, and *total communication* systems. They can also be classified in terms of the range of acceleration levels, or effective g, for which they are able to maintain control over the liquid position.[4]

Partial communication PMDs hold only a fraction of the liquid over the outlet and leave the remaining liquid free. They are used when the spacecraft must maneuver considerably or when many small engine firings are needed. If thrusting is always sufficient to settle the liquid, the PMD can be refilled during settling operations, and the size of the PMD can be made just large enough to provide the gas-free propellant needed to start the engines each time. If the thrusting is not always sufficient to settle the liquid, the PMD must be sized to provide enough gas-free propellant for all operations when settled liquid is not available to supply the engines or pumps. In this case, the PMD need not be refillable. Both types are commonly constructed with walls make of fine-mesh screens or similar "porous" materials. Capillary forces in the pores of the screen effectively prevent external gas from entering the PMD during disturbances. Figure 1 shows a schematic of an idealized refillable PMD. Figure 2 shows a nonrefillable PMD, in which each successively lower volume of liquid is emptied in turn, with the exposed screen preventing the entrance of gas from the empty volume just above. This type of PMD is sometimes called a sponge. Because of their small size, partial communication PMDs can be made stable against large acceleration disturbances.

Total control PMDs hold all the liquid over the outlet. In effect, this type is just a nonrefillable PMD that occupies all the tank volume. Their main use is with spacecraft where slosh control is dominant.

Total communication PMDs are designed to establish a flow path from the bulk liquid to the tank outlet at all times. Since, as we discussed earlier, the liquid tends to remain bound to a wall, common forms of total communication PMDs include a liner and a series of galleries, both made of fine-mesh screens and located near the wall. Figure 3 shows a schematic of a gallery PMD. The flow channels (galleries) all are connected to a manifold at the tank outlet. As long as at least one of the channels remains

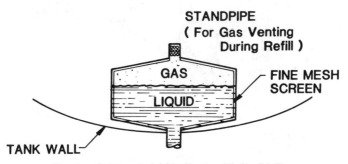

Fig. 1   Schematic of idealized refillable PMD.

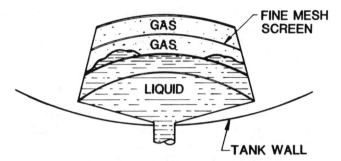

Fig. 2   Schematic of idealized nonrefillable PMD.

in contact with the bulk liquid, tank pressurization will drive the bulk liquid into the channel and then along the channel into the manifold. Capillary forces prevent gas from being driven into the other exposed channels. Because of their large size, total communication PMDs are stable against moderate accelerations but will admit gas under large accelerations.

Another type of total communication PMD uses a set of central vanes connected to a standpipe to position the liquid. If any vane is in contact with liquid, liquid is "wicked" to the standpipe. By tapering the open area of the vanes properly, a vaned PMD can also position the gas over a vent.[5] Figure 4 shows an example of the vaned PMD used in the Viking spacecraft. Because the interfaces are large, the liquid position in vaned PMDs is relatively unstable, and they can be used only when disturbing accelerations are small, such as for deep space probes and synchronous spacecraft. They can also be used when the thrusting acceleration is large

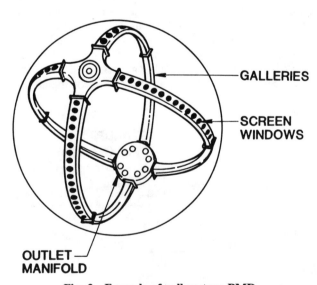

Fig. 3   Example of gallery-type PMD.

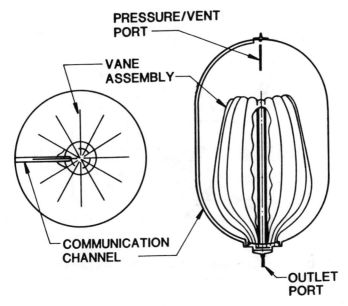

PRESSURE/VENT PORT

VANE ASSEMBLY

COMMUNICATION CHANNEL

OUTLET PORT

Fig. 4   Example of vaned PMD.

enough to settle the liquid, but after thrusting ceases, enough time must be allowed for the PMD to reacquire the liquid before any subsequent small thrust-acceleration operations occur.

Design Principles for Fine Mesh-Screen Capillary Devices

Fine-mesh screens are used extensively in the manufacture of PMDs. Some designers prefer specially manufactured perforated plates[6] or stacked, etched disks[7] rather than screens, because of their better manufacturing tolerances and reproducibility. Since the design principles are much the same for all of the screens, this discussion will consider only fine-mesh screens.

The most important characteristic of a screen is that, when wet, it can withstand a pressure differential from the gas side to the liquid side. The maximum possible differential is a function of the screen weave and liquid properties. It is commonly called the "bubble point pressure" because the test used to determine it involves increasing the pressure differential across a screen in contact with a liquid on one side and gas on the other until a gas bubble is forced through the screen (see Fig. 5). If the screen pores were exactly circular with diameter D_0, the bubble point pressure would be

$$P_{bp} = \frac{4\sigma}{D_0} \qquad (1)$$

where a small contact angle has been assumed so that $\cos\theta$ is numerically equal to 1. In general, the screen pores are not circular; hence, the

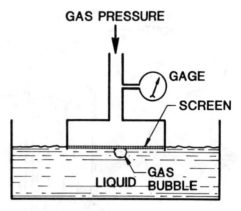

Fig. 5 Apparatus to measure bubble point pressure of a fine-mesh screen.

measured pressure P_{bp} is interpreted as defining an effective circular diameter D_{bp} for each type of screen:

$$D_{bp} = \frac{4\sigma}{P_{bp}} \qquad (2)$$

The effective diameter is related to the pore size of the screen but is not identical to it. Screens are available commercially with bubble point diameters as small as about $10\ \mu$ (Ref. 8).

It should be noted that there is no corresponding *drop point pressure*. That is, when the liquid pressure is even slightly greater than the gas pressure in Fig. 5, liquid will leak from the liquid to the gas. The reason for the absence of a drop point pressure is that liquid wets the screen and covers its entire external surface; hence, when the liquid pressure is greater than the gas pressure, a highly curved capillary-like interface cannot be formed between the screen wires to resist the pressure differential.

Knowledge of the bubble point pressure is sufficient to determine the stability of a PMD. As an example, consider the total communication channel sketched in Fig. 6a. At the intersection of the bulk liquid interface with the channel, the pressure in the channel liquid is equal to the gas pressure, since there is no capillary pressure at any point where the screen is immersed in liquid on both sides. Thus, relative to the gas pressure, the liquid at the top of the channel must support a negative hydrostatic pressure equal to $\rho g L$. To prevent gas from being drawn into the top of the channel to relieve the negative pressure, the bubble point must be greater than the following pressure:

$$P_{bp} > \rho g L \qquad (3)$$

where g is the component of the linear acceleration aligned with the channel axis. In other words, the bubble point diameter of the screen must

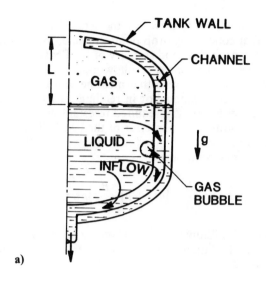

a)

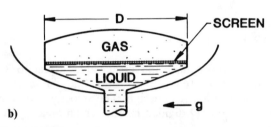

b)

Fig. 6 Stability of PMDs under adverse accelerations: a) gallery-type PMD; b) partial communication PMD.

be chosen to be smaller than

$$D_{bp} < \frac{4\sigma}{\rho g L} \tag{4}$$

Many spacecraft are launched with tanks that are not completely full. Consequently, from Eq. (4) the launch environment itself, where g is large, may impose the critical design requirement on the PMD.

As another example, the active liquid volume in the simplified partial communication PMD depicted in Fig. 6b will remain free of gas under lateral accelerations if the bubble point pressure of the bottom-most screen exposed to the gas satisfies the requirement

$$P_{bp} > \frac{4\sigma}{\rho g D} \tag{5}$$

Note that, if gas enters on the right side of the screen (where the pressure is most negative), liquid must leave on the left; in accordance with the remarks made earlier concerning drop point pressure, the pressure required to force liquid out is negligible, and stability of the PMD is a function only of the pressure differential required to allow gas in.

Another characteristic of a screen that must be considered in designing a total communication PMD channel or in refilling a partial communication PMD is the pressure difference ΔP required to establish a flow across a wet screen from the bulk liquid to the interior of the PMD. By extensive testing and correlation, Armour and Cannon[9] have shown that this ΔP has both a viscous and an inertia component:

$$\Delta P = \alpha\left(\frac{Ba^2}{\epsilon^2}\right)\mu U + \beta\left(\frac{B}{D\epsilon^2}\right)\rho U^2 \tag{6}$$

where U is the superficial liquid velocity (volume flow rate per unit surface area of screen), μ is the liquid viscosity, and the geometric parameters of the screen are defined as

  a = ratio of surface area to volume of screen
  B = thickness of screen
  D = pore diameter of screen
  ϵ = volume void fraction of screen

The parameters α and β have universal values equal to 8.61 and 0.52, respectively; other works (e.g., Blatt[8] and Cady[10]) indicate that these parameters may vary with screen design. The velocity U is computed as the total volumetric flow rate Q divided by the screen area A_w in contact with the bulk liquid.

The screen area must be large enough to yield a flow pressure drop that will not sweep any gas bubbles in the bulk liquid into the PMD. Again considering the example of the total communication PMD channel shown in Fig. 6a, the following relation must be satisfied among the various pressures:

$$P_{bp} > \rho g L + \Delta P \tag{7}$$

Equation (7) can be solved for the screen area, from which the screen width can be computed for a given channel length. If the flow rate Q is constant (independent of the tank filling level), one can see that $\Delta P + \rho g L$ will eventually exceed P_{bp}, since the screen flow area in contact with bulk liquid becomes small when the tank is nearly empty. When that occurs, gas-free liquid is no longer available. Hence, the combination of bubble point pressure and flow pressure drop, both of which are functions of the screen weave, may determine the expulsion efficiency of the tank for some missions.

A third characteristic that may need to be considered is some designs is whether or not the screen has the ability to *wick* liquid along its length. The

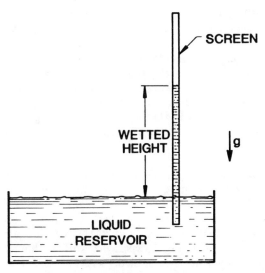

Fig. 7 Apparatus to measure wicking pressure of a fine-mesh screen.

ability to wick liquid is a function of the screen weave. A plain "square-weave" screen will not wick. More complex weaves, such as "dutch twills," which permit capillary surfaces to be established across the screen thickness, will wick liquid from wet areas to dry areas. The effective suction pressure that produces the wicking can be determined by measuring the height to which liquid will rise along a vertical sample of the screen immersed in a liquid reservoir; this test is shown schematically in Fig. 7. The maximum capillary pressure P_w that is available to cause wicking is the hydrostatic head corresponding to the height of liquid in this test; it is correlated by a modification of Eq. (2):

$$P_w = \frac{4\phi\sigma}{D_{bp}} \tag{8}$$

where ϕ is a parameter that depends on the screen weave. For a square-weave screen, $\phi = 0$, and for a dutch twill screen, it ranges from about 0.2 to slightly over 1.0, depending on the density of the weave.[11,12]

The velocity U_w of the liquid wicking to a dry area of a screen from a wet area is determined by the wicking pressure (assuming that $g = 0$ or that the screen is horizontal):

$$U_w = \frac{P_w B^2}{C_w \mu l} \tag{9}$$

where l is the length along the screen between the wicking front (wet-dry interface) and the reservoir. C_w is a nondimensional resistance that is a function of screen weave; typical values range from about 300 to several thousand.[11,12] When liquid evaporates from a wet screen (and hence must be replaced by wicking from the bulk liquid), l is interpreted as the total length of the screen. Note that Eq. (9) implies that there is a maximum possible value of evaporation rate per unit screen surface area (corresponding to the maximum possible P_w) that can be compensated by wicking.

For nonvolatile liquids wicking is not usually a quantitative consideration, as long as screens are chosen which ensure that some wicking occurs to keep wet the parts of the PMD that are exposed to gas. It becomes more important when the liquids evaporate easily, such as cryogens.

Considerations for PMDs Used with Cryogenic Liquids

Future missions will require the use of large quantities of cryogenic liquids, for which there is little or no PMD flight experience. Furthermore, many of the missions will require the transfer of liquids from tank to tank, with the implication that the tanks and PMDs will have to be refilled in space. Again, there is little flight experience with refilling large PMDs in space when the liquids are not settled. Aydelott and Rudland[13] discuss many of these issues. Entirely new devices, such as thermodynamic vents,[10] will be required for such missions, and new procedures for testing PMDs will be required.[14] It has even been speculated that the shape of a cryogenic liquid-vapor interface will be different from that of a storable liquid because of the greatly increased evaporation.[15]

The use of cryogens also affect the design of PMDs. Evaporation from screens exposed to gas or vapor is much greater than that for storable liquids; consequently, wicking considerations become more important. Furthermore, vapor may be generated within the PMDs because of unavoidable heat leaks; thus, either the PMD design must incorporate methods to remove the trapped vapor, or the liquid must be cooled periodically to condense the vapor. As another example of a new problem, if a channel-type total communication PMD is to be refilled in space, some part of the screen of each channel must be of nonwicking weave to create an open area in the channel wall for gas in the empty channels to flow out as liquid flows in, or a valve that opens to the tank interior must be incorporated in the channel to serve the same purpose.

When liquid is evaporating from a screen and being replenished by wicking, the bubble point of the screen is lowered from its normal value.[16] This phenomenon can be understood physically as a thinning or withdrawal of the liquid to the interior of the screen and a subsequent increase of the curvatures of the capillary surfaces from their minimum (the condition for maximum capillary pressure). Hence, the bubble point pressure available to prevent gas ingestion decreases as the wicking pressure, or evaporation rate, increases. To a first approximation, the available bubble point pressure P_{bp} and the available wicking pressure P_w can be related linearly[12]:

$$\frac{P_{bp}}{(4\sigma/D_{bp})} = 1 - K\left[\frac{P_w}{(4\phi\sigma/D_{bp})}\right] \tag{10}$$

where K is a parameter that must be determined experimentally as a function of screen weave (a typical value is 0.5). When the screen is not wicking, $P_w = 0$, and the equation shows that the available P_{bp} has its maximum value of $4\sigma/D_{bp}$, which is the pressure measured in a conventional bubble point test. Similarly, when the available P_w obtains its maximum value of $4\phi\sigma/D_{bp}$ corresponding to the maximum possible wicking flow, the bubble point pressure is reduced to half its maximum value (for $K = 0.5$). Because of the decrease in bubble point when wicking occurs, the ability of the PMD to keep out vapor can be substantially degraded when the tank liquid is cryogenic.

The entire subject of PMDs for use in refillable tanks with cryogenic liquids is under active investigation at NASA and elsewhere. Space-based experiments are under development to investigate design concepts and reveal potential problems. For now, the designer of PMDs for cryogenic liquids should be aware that the stability of screenlike devices will be degraded and should therefore increase safety margins accordingly.

Conclusions

Passive devices to control the location of gas and liquid in a tank in low gravity are based on the idea of augmenting interface capillary forces. For storable, nonvolatile liquids, the principles upon which the devices are designed are reasonably well understood, and the main task of the designer is to select a type of propellant management device that fits the mission. However, this task is not easy, and usually many alternatives must be examined. When the liquids are cryogenic or easily evaporated, the experience base is much more limited, and the designer should increase safety margins because of the uncertainty of performance.

References

[1]Rollins, J. R., Grove, R. K., and Jaekle, D. E., Jr., "Twenty-Three Years of Surface Tension Propellant Management System Design, Development, Manufacture, Test, and Operation," AIAA Paper 85-1199, 1985.

[2]Reynolds, W. C., and Satterlee, H. M., "Liquid Propellant Behavior at Low and Zero-G," *The Dynamic Behavior of Liquids in Moving Containers*, edited by H. N. Abramson, NASA SP-106, 1966.

[3]Concus, P., Crane, G. E., and Satterlee, H. M., "Small Amplitude Lateral Sloshing in Spheroidal Containers Under Low Gravitational Conditions," NASA CR-72500, 1969.

[4]Kleinau, W., Günther, H., Kollien, J., Schweig, H., and Würmseer, M., "Orbital Propulsion Module Study on Coni-Spherical Surface Tension Tanks," European Space Agency Rept. ESA CR(P) 2556, 1987.

[5]Dowdy, M. W., Hise, R. E., Peterson, R. G., and DeBrock, S. C., "Surface Tension Propellant Control for Viking 75 Orbiter," AIAA Paper 76-596, 1976.

[6]Tegart, J., and Wright, N. T., "Double Perforated Plate as a Capillary Barrier," AIAA Paper 83-1379, 1983.

[7]Purohit, G. P., and Loudenback, L. D., "Application of Etched Disk Stacks in Surface Propellant Management Devices," AIAA Paper 88-2919, 1988.

[8]Blatt, M. H., "Low Gravity Propellant Control Using Capillary Devices in Large Scale Cryogenic Vehicles—Design Handbook," General Dynamics Convair Rept. GDC-DDB70-006, 1970.

[9]Armour, J. C., and Cannon, J. N., "Fluid Flow Through Woven Screens," AIChE Journal, Vol. 14, No. 3, 1968, pp. 415–420.

[10]Cady, E. C., "Effect of Transient Liquid Flow on Retention Characteristics of Screen Acquisition Devices," NASA CR-135218, 1977.

[11]Symons, E. P., "Wicking of Liquids in Screens," NASA TN D-7657, 1974.

[12]Dodge, F. T., and Bowles, E. B., "Vapor Flow into a Capillary Propellant-Acquisition Device," Journal of Spacecraft and Rockets, Vol. 21, No. 3, 1984, pp. 267–273.

[13]Aydelott, J. C., and Rudland, R. S., "Technology Requirements to be Addressed by the NASA Lewis Research Center Cryogenic Fluid Management Facility Program," AIAA Paper 85-1229, 1985.

[14]Simon, E. D., "Environmental Requirements for Bubble Pressure Tests on Fine-Mesh Screens," Journal of Spacecraft and Rockets, Vol. 16, No. 4, 1979, pp. 218–222.

[15]Gershman, R., and Chu, C., "Effect of Wall Heating on Low-G Liquid-Vapor Interface Configuration," American Society of Mechanical Engineers Paper 67-WA/HT-33, 1967.

[16]Bingham, P. E., and Tegart, J. R., "Wicking in Fine Mesh Screens," AIAA Paper 77-849, 1977.

Nucleate Pool Boiling in Variable Gravity

Herman Merte Jr.*
The University of Michigan, Ann Arbor, Michigan

Introduction

NUCLEATE boiling is an important mode of heat transfer in that relatively small temperature differences can provide large rates of heat transfer, which can result in significant economic and other benefits associated with the smaller heat-transfer areas necessary to accomplish a given function.

A limitation in the development of more compact power sources using nuclear energy lies in the ability to remove the large heat-generation rates possible from the reactor core in a manner that is consistent, reliable, and predictable. Nucleate boiling would be a candidate for such an application if the fundamental mechanisms that govern the process were sufficiently well understood. Additional important applications of nucleate boiling exist, such as steam generation in convectional power plants, distillation processes in petroleum and other chemical plants, and boiling of refrigerants in cooling coils, in which the motion of the bulk liquid is generally imposed externally. This is termed forced convection boiling, and the liquid motion moves the vapor formed away from the heated surface so that the vapor may be utilized and/or further processed and the nucleate boiling process can continue.

Other applications exist in which the externally forced flow is absent, where buoyancy provides the major mechanism for vapor removal from the vicinity of the heating surface and is generally designated pool boiling. Even in circumstances where forced convection exists to some extent, the forces associated with flow acting on the vapor bubbles may be sufficiently small that buoyancy or body forces will continue to be responsible for the vapor removal process. It should then be possible to describe the behavior,

*Professor, Mechanical Engineering.

15

in terms of the basic governing mechanisms, by the pool boiling process. Devices in which pool boiling occurs are two-phase closed thermosyphons, reboilers, and heat pipes, whether gravity-assisted or not. Potentially significant applications exist in the cooling of microelectronic circuitry and the internal cooling of gas turbine blades. The latter would involve pool boiling under high-gravity fields, and its successful application would permit higher operating temperatures with attendant higher efficiencies and would also eliminate the need for the development of exotic ceramic materials with the difficulties of thermal stresses and reliability. Another important and as yet poorly understood area incorporated in the mechanisms of pool boiling is the breakdown of film boiling into the transition boiling regime. This is of concern in the loss-of-coolant accident in nuclear power plants and is encompassed in the reflooding and fuel element rewetting processes. A good understanding of this rewetting process would improve its application.

The effective and enhanced applications of both nucleate pool and forced convection boiling require a sound understanding of the mechanisms governing the processes. The vapor removal from the vicinity of the heater surface, as understood to this point, occurs primarily by buoyancy in the case of pool boiling and bulk liquid inertia in the case with forced convection. Although the variation of both gravity and forced flow are known to influence the overall heat-transfer processes, other forces or potentials are acting as well, and the relative significances of these is at present poorly understood.

The elements that constitute the nucleate boiling process—nucleation, growth, motion, collapse (if subcooled) of the vapor bubbles—are common to both pool and flow boiling. The focus of this chapter is on nucleate pool boiling. This eliminates the additional complications associated with having an external flowfield superimposed on that generated by a growing/collapsing vapor bubble and also eliminates the possibility of having other effects masked by an external flowfield, except for that which may be induced by buoyancy.

Requirements for the proper functioning of equipment and personnel in the space environment of reduced gravity and vacuum, as will be necessary in space station modules and space platforms, introduce unique problems in temperature control, power generation, energy dissipation, storage, transfer, control and conditioning of fluids (including cryogenic liquids), and liquid-vapor separation.

The temperature control in certain locations where internal heat generation takes place as a result of dissipation, from friction or Joulian heating, electronic equipment, or as a consequence of a nuclear or chemical heat source, may require that this energy be transported to other locations of the facility or stored locally for later transport and elimination. The use of the phase changes of vaporization and condensation to transport energy have the advantage of accommodating large variations in heat loads with relatively small temperature gradients and changes in temperature levels, along with the economical use of pumping power. Energy storage might be advantageous for intermittent processes or for processes where momentary surges could not be accommodated by a steady transfer of mass to a remote location and could take advantage of the latent heat associated with phase changes.

A distinction must be made between pool boiling and flow boiling when applications in the space environment of microgravity are considered, since these processes may arise in quite different specific technical applications. Pool boiling, for example, would be important for the short-term cooling of high-power electronic or other devices and for the long-term space storage of cryogens. Flow boiling, on the other hand, occurs in applications where liquid flow is imposed externally, such as in Rankine cycle vapor generation or in thermal energy managment using pumped latent heat transport.

Certain effects that can be neglected at normal Earth gravity, such as surface tension and vapor momentum, can become quite significant at microgravity conditions. Momentum imparted to the liquid by the vapor bubble during growth tends to draw the vapor bubble away from the surface, depending on the rate of growth, which in turn is governed by the temperature distribution in the liquid. On the other hand, thermophoretic forces, arising from the variation of the liquid-vapor surface tension with temperature, tend to move the vapor bubble toward the region of higher temperature. The bubble motion will be governed by which of these two effects prevail.

Variable gravity provides an opportunity to test the understanding of a phenomenon that depends on buoyancy, such as pool boiling given in this chapter, and one can look for changes in the dominant mechanisms that govern the process. One must excercise caution in the extrapolation or interpolation, from known points, of the behavior with varying gravity because of possible changes in these mechanisms. This is especially true as gravity is reduced toward very low levels. Despite the considerable activity in the study of boiling heat transfer over the past 30 years, with much of it unfortunately not well focused, the process is still not sufficiently well described at present to permit its effective applications with confidence to the environment of space. One must also exercise caution in the use of intuition when interpreting potential results with new variables. Figure 1 sets forth the results of research in which the body force normal to the heating surface was varied with saturated nucleate boiling. As will be demonstrated in detail later, the behavior is quite opposite to that observed in single-phase natural convection heat transfer.

Much of the experimental work described later in the chapter utilizes a particular simple geometry for study (a relatively large flat surface) for two reasons: 1) meaningful orientation can be defined with respect to a body force vector; and 2) surface tension effects associated with the curvature of the heating surface, which may overshadow other effects, will not be present. Many works on boiling using wires and tubes appear in the literature, and any general conclusion with such variable orientations and large curvatures must be considered very carefully. Early boiling results with spheres will be presented later. These are simple in form, convenient to use, and are sometimes useful as a particular reference state. However,

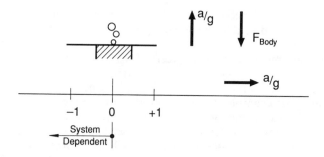

Increasing Gravity Degrades HeatTransfer
Reducing Gravity Enhances HeatTransfer

Fig. 1   Observed influence of variable buoyancy on established saturated nucleate pool boiling heat transfer.

they are rather uninteresting for learning anything fundamental about the boiling process, unless the particular application at hand involves spherical geometries.

Space limitations prohibit adequate coverage of certain important components of nucleate boiling, and some further information will be available in the references. As will become obvious, most experimental results are those with which the author and his students have been directly involved.

The basic mechanisms of nucleate pool boiling will be reviewed first, followed by a brief presentation of the status of understanding of these mechanisms. An adequate understanding implies that the behavior can be predicted in terms of the governing parameters. Such a status still remains for the future.

Basic Mechanisms of Nucleate Pool Boiling

Nucleate pool boiling may be characterized succinctly by the following:

1) A liquid-vapor phase change occurs with the formation of discrete bubbles at individual sites.

2) The bulk liquid is essentially stagnant, in contrast to the case where the liquid is set in motion by external means.

3) The energy transfer rates are large with small temperature difference driving potentials.

4) The process is inherently transient, although quasicyclic repetitions are possible with vapor-removal mechanisms such as buoyancy acting.

Before a nucleate pool boiling system can attain the steady periodic behavior normally observed in a gravity field, where buoyancy is the predominant vapor removal mechanism, the process must pass through a transient phase referred to as the nucleation, initiation, or onset of nucleate boiling. Before understanding the cyclic nature of nucleate boiling one must first understand the elements of the initial transient process.

To provide a perspective of the overall processes that constitute pool boiling, a qualitative physical description of the sequence of events that occur is presented, beginning with the transient heating of a liquid at a solid-liquid interface.

Conduction

With an initially static liquid the heat-transfer process can be described by conduction alone until buoyancy, thermophorysis, or other forces set the liquid in motion. The rate of temperature rise and the temperature distributions in this early interval depend on the nature of the heat source and the dynamic interactions with the system. The common idealizations taken as limits in analyses are step changes in either temperature or heat flux at the solid-liquid heater interface. Figure 2 illustrates the case with an imposed heat flux at the interface, in which the temperature gradient is maintained constant. The degree and extent to which the liquid becomes superheated above its saturation temperature in a given time depends on whether and how much the bulk liquid is subcooled, as also indicated in Fig. 2. This temperature distribution will be modified by the onset of natural convection or by other disturbances.

Onset of Natural Convection

Natural convection is driven by buoyancy, and its onset is described in terms of a stability in which the enervating disturbances are always present. Reducing the buoyancy by reducing the body forces, for example, to microgravity, delays the onset of the convection and reduces the resulting

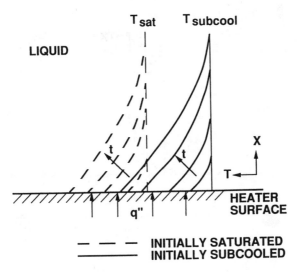

Fig. 2 Conduction heat transfer in initially stagnant liquid.

convection velocities. Both of these serve to increase the temperature levels in the liquid adjacent to the heating surface for a given heating time, regardless of whether the bulk liquid is initially saturated or subcooled. Accordingly, the opposite takes place as buoyancy is increased with increasing gravity levels: The onset of convection will take place at an earlier time and the resulting convection velocities will be increased, both of which serve to decrease the temperature levels adjacent to the heating surface for a given heating time. The liquid temperature levels and distributions adjacent to the heater surface are thus influenced by buoyancy and in turn can influence the next two elements of nucleate boiling: nucleation and bubble growth rates.

Nucleation

Vaporization can take place only at an existing liquid-vapor interface, which then constitutes the growth phase of nucleate boiling. If an interface does not exist, it must be formed. The formation of a vapor nucleus is called nucleation, and is classified as either homogeneous or heterogeneous depending on the presence of other components or species in the vicinity of the nucleation site. The circumstances under which nucleation takes place on a heated solid surface depend on the following:

1) The heater surface microgeometry. This can provide the crevices and intergranular defects that serve as preexisting interfaces. The temperature levels required to activate these preexisting nuclei have been modeled in terms of thermodynamic equilibrium at curved liquid-vapor interfaces. Assuming that the preexisting interface has the form of a hemisphere the

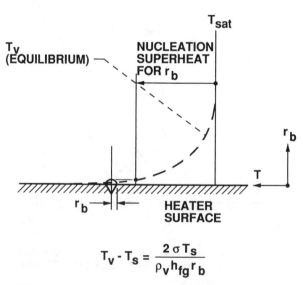

$$T_v - T_s = \frac{2\sigma T_s}{\rho_v h_{fg} r_b}$$

Fig. 3   Nucleation from preexisting surface cavities.

size of the surface defect, the liquid superheat required for subsequent bubble growth can be related to cavity size, as shown in Fig. 3. The smaller the cavity, the larger the heater surface superheat required for the onset of nucleate boiling, and the larger the bulk liquid temperatures required at the onset of the next element of boiling process.

2) The solid-fluid properties. These properties govern not only the temperature distribution in both the heater and fluids, related by their respective thermal properties, but also the surface energy relationships between the solid-liquid-vapor, often expressed in terms of a contact angle or wettability.

3) The liquid temperature distribution. This includes the solid-liquid interface temperature, since this is one spatial limit of the liquid temperature. As discussed in the subsection on onset of natural convection, the onset of natural convection governs the subsequent temperature distributions, as does the initial imposed heat flux. Once nucleation has occurred, the following bubble growth rates will be governed by the bulk liquid temperature distribution at this time.

Vapor Bubble Growth/Collapse

Vapor bubble growth requires that the liquid at the liquid-vapor interface be superheated with respect to the saturation temperature corresponding to the interfacial liquid pressure. The rate of vapor formation, and hence bubble growth, depends on this superheat and on the liquid temperature gradient at the interface, and thus on the liquid temperature distribution at the onset of bubble growth. The interfacial liquid superheat governs the internal vapor bubble pressure, which acts to move the bulk liquid away from the vicinity of the heater surface. In the dynamics of the growth process this pressure is balanced in a complex manner by the liquid inertia, liquid viscosity, buoyancy, and surface tensions. If the bulk liquid is subcooled, the pressure difference can reverse with the subsequent collapse of the vapor bubble. The various forces acting in the bubble growth/ collapse can be summarized as follows:

1) Internal bubble pressure. This is governed by the liquid temperature distribution, which in turn is influenced by buoyancy.

2) Liquid momentum. This is sometimes referred to as bulk liquid inertia.

3) Buoyancy. The pressure differences associated with the liquid-vapor density differences in a body force field act in addition to those natural convection effects that influence the liquid temperature distribution.

4) Surface tension. This includes both the tension occurring at the liquid-vapor interface and the tension at the liquid-solid-vapor interline.

5) Viscosity. This refers primarily to the liquid viscosity acting in the vicinity of the solid surface, but could include the viscous normal shear at the liquid-vapor interface away from the solid surface in circumstances where the radial growth rate is very large. Vapor viscosity could also be a factor during the very early periods when surface rates of vapor formation are large.

Since the liquid-vapor interface is deformable, the interfacial shape during growth will be governed by the net balance of the dynamic forces acting at each point on the interface, and the interface will not necessarily be spherical or hemispherical, as has been assumed in the absence of capabilities for dealing with flexible interfaces.

Departure

The subsequent motion of the vapor bubble depends on the net effect of the forces listed in the preceding subsection, plus a phenomenon associated with simultaneous evaporation and condensation across a vapor bubble, referred to as the molecular momentum effect. This is related to the molecular kinetic energy necessary for vapor molecules to escape or to be retained at a liquid-vapor interface. With thermodynamic equilibrium the net rate of evaporation and condensation is zero, but the normal nucleate boiling process is highly nonequilibrium. The net resulting molecular momentum forces are generally unobservable in the presence of the over-whelming body and other forces that usually exist. The bulk liquid momentum induced by the rapid bubble growth can act to assist in the removal of the bubble from the heater surface. In microgravity, of course, buoyancy effects are reduced significantly.

Motion Following Departure

If the circumstances of the forces acting on the vapor bubble are such that departure takes place, the subsequent motion depends on the following:

1) Buoyancy.

2) Initial velocity upon departure. This velocity induces momentum in bulk liquid, which must be conserved, and can tend to accelerate the vapor bubble if collapse takes place, or will decelerate the bubble if it grows.

3) Degree and distribution of liquid superheat and/or subcooling. The bulk liquid temperature distribution can act via the liquid-vapor surface tension or Marangoni-induced effects via the bulk liquid momentum effects associated with growth or collapse, together with liquid viscosity and via the molecular momentum effects. In variable-gravity conditions, only buoyancy will be changed, except for its indirect influence on the bulk liquid temperature distribution.

5) Subsequent steady-periodic behavior.

During the transient events described earlier, heat transfer between the solid heating surface and liquid is continuously taking place, and nucleation, growth, and departure of successive vapor bubbles will take place, either at or near the initial nucleation site. As a result of the liquid motion induced by the growing and departing bubbles, with the associated effects on the liquid temperature distribution, a seemingly chaotic and random behavior exists, and the detailed behavior of the successive bubbles may be quite variable. In spite of this, observations over a reasonable period of time reveal a statistically steady-periodic behavior of the bubbles, which can be used to describe certain features of the behavior.

An adequate understanding of the fundamental mechanisms that constitute nucleate pool boiling implies that the behavior can be predicted in terms of the governing parameters. The behavior would include the conditions for the onset of boiling, the dynamic behavior of the vapor bubbles, including both the number density of active nucleating sites and the frequency of formation, and the associated heat transfer. The governing parameters would include fluid and heater surface properties, both chemical and physical (e.g., surface tension, roughness or microgeometry, degree of wetting, and pressure), degree of initial subcooling, macrogeometry, and orientation relative to a body force.

Although a considerable amount of research has been conducted on nucleate boiling over the years and has been useful with respect to application to various technologies on Earth, the ability to predict its behavior is presently very limited, because of the involvement and interactions of the many parameters, and more so when predictive attempts are made for conditions that do not exist on Earth. One has the choice of developing either mechanistic or heuristic models for predictive purposes, but only mechanistic models can be used to confirm physical understanding. Improved physical understanding enhances opportunities for new and improved applications.

Status of Understanding

In consideration of the status of the present understanding of the nucleate boiling process, one must distinguish and separate the initial transient or boiling inception from the quasisteady periodic phase. The latter can take place only in circumstances where mechanisms act to remove the vapor bubbles from the vicinity of the heat-transfer surface. For pool boiling, in the absence of externally induced fluid motion, this will occur with the body force provided by the Earth's gravity. However, the minimum body force necessary to provide this removal mechanism is unknown.

For pool boiling in the presence of Earth gravity, the most well-known correlation for heat flux is that due to Mikic and Rohsenow[1] for saturated liquids only:

$$\frac{q_b''}{\mu_\ell h_{fg}} \left\{ \frac{\sigma}{g(\rho_\ell - \rho_g)} \right\}^{\frac{1}{2}} = B(\phi \Delta T)^{m+1} \tag{1}$$

where ϕ incorporates fluid properties given by

$$\phi^{m+1} = \frac{k_\ell^{\frac{1}{2}} \rho_\ell^{\frac{17}{8}} C_\ell^{\frac{19}{8}} h_{fg}^{(m-\frac{23}{8})} \rho_g^{(m-\frac{15}{8})}}{\mu_\ell (\rho_\ell - \rho_g)^{\frac{9}{8}} \sigma^{(m-\frac{11}{8})} T_{sat}^{(m-\frac{15}{8})}} \tag{2}$$

and B is a dimensional constant that depends on gravity and the boiling surface properties:

$$B = \left(\frac{r_{max}}{2} \right)^m \frac{2}{\pi^{\frac{1}{2}}} \frac{C_1}{g^{\frac{9}{8}}} C_2^{\frac{5}{3}} C_3^{\frac{1}{2}} \tag{3}$$

It is noted that four empirical constants are included here: m, C_1, C_2, and C_3. The variable m essentially governs the slope of the q_b'' vs ΔT_{sat} curve, with a value of 2 for most surfaces. The variable C_3 arises from an empirical correlation relating the product of the frequency of departure and departure size. The variable C_1 is an empirical constant involved with the smallest active cavity size corresponding to a given heater surface superheat. The variable C_2 is an empirical constant describing the departure size of a vapor bubble with mechanical equilibrium balance between buoyancy and surface tension.

As an illustration of the nonapplicability of Eqs. (1–3) to other circumstances, in the present instance for other levels of body force, for a given fluid, pressure, and surface, Eq. (1) can be rewritten as

$$q_b'' = Kg^{-\frac{5}{8}}\Delta T_{sat}^3 \tag{4}$$

where m is taken as 2, and all properties and constants are combined into K. For a given heat flux Eq. (4) predicts a ΔT_{sat} at $a/g = 10$ and 100 to be 1.62 and 2.61 times that at $a/g = 1$, respectively, which can be compared to measured values of 1.03 and 1.24 with water, from Ulucakli and Merte.[2]

For transient nucleate boiling occurring at the onset of boiling, one can consider individually the elements listed earlier that constitute the process. For present purposes the liquid is considered to be initially motionless and uniform in temperature, but may be saturated or subcooled, depending on the circumstances desired.

Conduction

With heat transfer to the fluid from a solid surface, the initial mechanism will be pure conduction. This process is understood quite well and can be described, given reasonably well-defined geometries and thermal properties. Transient heat-transfer measurements in fluids have agreed well with conduction calculations both with low-gravity conditions in a drop tower e.g., by Oker and Merte[4] as shown in Fig. 4, and at Earth gravity, as shown by Giarrantano.[5] Once liquid motion has occurred, temperature distributions can be determined only if the velocity distributions are known.

Onset of Natural Convection

With temperature distributions associated with conduction heat transfer, density variations and buoyant forces arise in the fluid in the presence of a body force, no matter how small. These buoyant forces give rise to motion, which will be resisted by viscous forces, and the relationship between these forces governs the stability of the process. The criteria for the onset of motion have been analytically described and verified for horizontal fine wires by Vest and Lawson[3] and for horizontal and vertical surfaces by Oker and Merte[4] all at $a/g = 1$. However, these criteria have never been tested in microgravity and in circumstances where the normal general disturbances found on Earth are no longer present. For example, the

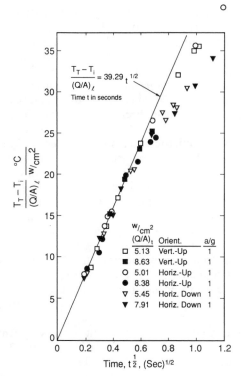

Fig. 4 **Comparison of computed and measured interface temperature, quartz and R-113 system.**

analysis in Oker and Merte[4] predicts that the onset of convection on a horizontal surface varies inversely with the square root of the body force, so that at $a/g = 10^{-5}$ the time required for the onset of convection is greater than that at $a/g = 1$ by a factor of $10^{\frac{5}{2}} = 316$. The basis for the accompanying measurement of the onset of natural convection at Earth gravity is shown in Fig. 4, where departure from the one-dimensional conduction process takes place.

Once liquid motion has begun, the process can be characterized as transient natural convection, with the degree of motion depending on the geometry, temperature distribution, body force, and fluid properties. The degree of motion, of course, will also govern the degree of thermal stratification resulting within the liquid. Natural convection processes have not been studied in low-gravity fields because of the long time constants of the process relative to the periods of time heretofore available.

Nucleation

Vapor bubble nucleation takes place when the local temperature of the liquid in the vicinity of the heating surface exceeds its saturation temperature by some amount, governed by the microgeometry of the solid surface, the solid/fluid properties, the surface temperature of the solid, and the temperature distribution in the liquid. The last two parameters depend, in turn, on the heat flux, on whether saturated or subcooled conditions exist, and on the velocity distribution in the vicinity of the heater surface as produced by buoyancy. A number of studies of nucleation and the inception of boiling under nonforced convection circumstances with reduced or simulated reduced gravity have been reported by Coeling and Merte,[6] Derewnicki,[7] Nghiem et al.,[16] and Oker and Merte.[4] A summary of incipient boiling work in microgravity was compiled by Merte and Littles,[8] and a general survey of boiling nucleation up to 1974 is available from Cole.[9] With a local hot spot or heat source on a wall, a growing single bubble may absorb sufficient liquid superheat in its vicinity in order to inhibit the formation and growth or any other vapor nuclei, in which case only this single bubble proceeds to grow. This type of behavior was observed in short-term drop tower experiments by Larkin.[10] If the heated surface is large, nucleation initiating at a single point may spread across the entire surface, being triggered by the first site. This was observed in drop tower experiments conducted by Oker and Merte,[11] where the imposed heat flux was relatively large to ensure nucleation during the short (1.4 s) fall time available. As a result, the liquid superheat was highly localized at the surface.

The onset of boiling is inherently transient in the sense that, once having begun, the dynamics of the boiling process will so change the situation that it can only be repeated by beginning anew. Results of initial nucleation site activation at Earth gravity have been reported with pool boiling by Anderson et al.[12] and Shoukri and Judd,[13] and the thickness of the thermal layer at the initiation of nucleate pool boiling has been measured by Grant and Patten.[14] In a study by Wagner and Lykoudis,[15] who used a magnetic field to reduce the effective buoyancy when heating a liquid metal, it was observed that nucleation occurred at a significantly lower level of heat flux.

Figure 5, taken from Coeling and Merte,[6] demonstrates the influence that quasisteady natural convection has on the heterogeneous nucleation process, or the onset of nucleate boiling. For $a/g = +1$ the data obtained lie on the corresponding natural convection correlation. The temperature distribution in the inverted position, $a/g = -1$, is quite different, and nucleation occurs at a considerably lower surface heat flux.

The requirement for thermodynamic homogeneous equilibrium between a spherical vapor bubble of radius r_{cr} and its pure liquid results in a liquid superheat given by Cole[9]:

$$T_v - T_s = \frac{2\sigma T_s}{\rho_v h_{fg} r_{cr}} \tag{5}$$

A liquid superheat greater than this results in growth of the vapor bubble, whereas collapse will occur with a smaller superheat. So-called cavity models have attempted to build on Eq. (5) in heterogeneous circumstances by assuming that the cavities in the surface of solid heaters are initially

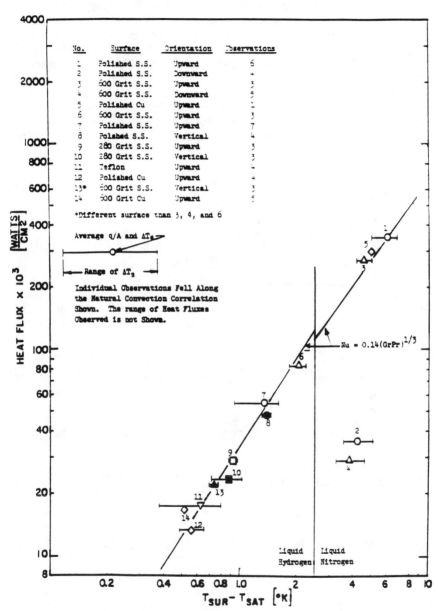

Fig. 5 Quasisteady nucleation at $a/g = +1$ and $a/g = -1$.

filled with a vapor providing a spherically shaped liquid-vapor interface. Growth occurs when the actual temperature at the top of the nucleus reaches that required for equilibrium, as shown in Fig. 3. Applications of these models require a significant degree of empiricism since the effects of microgeometry cannot be appropriately described yet, nor can the actual temperature distribution in the vicinity of the heater surfaces be described.

Attempts are being made to describe heterogeneous nucleation in terms of homogeneous nucleation theory. Some success has been achieved to date only for the transient heating case by Nghiem et al.,[16] in which a relationship between the nucleation delay time and the heater surface superheat at nucleation is given by Eq. (7) once the nucleation factor F is known, defined in Eq. (6) as the ratio of the free energy for heterogeneous nucleation to that for homogeneous nucleation:

$$F = \frac{A_{\text{het}}}{A_{\text{hom}}} \tag{6}$$

$$\frac{T_{IB}}{T_s} = \left\{ 1 - \left(\frac{\rho_\ell V_v}{\lambda} \right)_{T=T_s} \ell n \left[\left(\frac{16\pi\sigma^3 F}{P_\ell^2 3KT \, \ell n \, \dfrac{nKT\tau}{h}} \right)^{\frac{1}{2}} + 1 \right] \right\}^{-1} \tag{7}$$

Figures 6 and 7 show the experimental results that permit computation of F from Eq. (7). The F from these is plotted as a function of the dimensionless incipient boiling heater surface superheat in Fig. 8. Since this

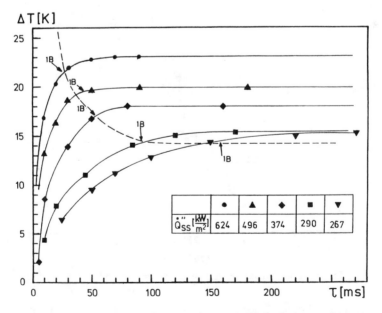

Fig. 6   Transient nucleate boiling at $a/g = +1$. H_2O-platinum wire system.

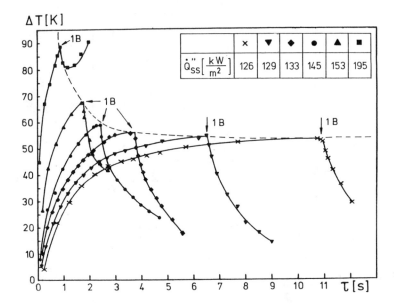

Fig. 7   Transient nucleate boiling at $a/g = +1$. R-113-thin gold film system.

work appeared, additional transient incipient boiling data have been found in the open literature and fall along the same solid lines. The systems to which these additional data apply are listed in Fig. 8.

Once boiling has initiated and reached a steady condition, the nucleation site density becomes an important parameter in the description of the boiling. A reasonable number of measurements of nucleation site density have been reported for pool boiling at Earth gravity, including those of Gaertner and Westwater[17] and Singh et al.[18] Measurements of the temperature distribution in the boundary layer, which influences the nucleation site density, where made by Marcus and Dropkin[19] and by Graham et al.[20] and the interactions taking place between adjacent nucleating sites were studied by Judd and Lavadas[21] and by Sultan and Judd.[22] The nucleation site density is also expected to be dependent on the departure size of the bubbles, and the factors that govern this are considered below later in the chapter.

Vapor Bubble Growth/Collapse
    Once a particular nucleating site has become activated, the subsequent rate of growth and later departure and/or collapse are dependent on the transient temperature distribution in the vicinity of the bubble interface, as well as on liquid inertia effects. This temperature distribution will be different depending on whether the bubble growth is the initial one at a site or is one of a succession constituting a periodic unsteady process, which can be called a steady process in a global sense. The rate of growth in this

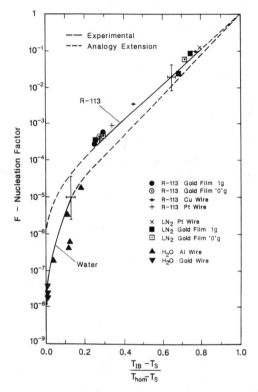

Fig. 8   Correlation of transient heterogeneous nucleation by homogeneous nucleation theory.

steady case affects the bubble frequency and together with the nucleating site density governs the relationship between heating surface superheat and flux for pool boiling as shown by Judd and Merte.[23] The rate of growth will be influenced by reduced gravity insofar as the temperature distribution is affected. Considerable work, both analytical and experimental, has been reported on the dynamics of vapor bubbles in the liquid bulk and near solid walls with pool boiling in both subcooled and saturated liquids. See, for example, Skinner,[24] Tokarev and Bogovin,[25] and Zysin and Feldberg.[26]

For the case where a bubble is in an infinite medium and the growth rate is limited by the ability of the internal vapor pressure to move the liquid surrounding the bubble away, the growth is said to be momentum or inertia controlled, and is known classically as the Rayleigh-Besant problem, which is described by Lamb.[27] The other limit involves the ability of the liquid to transport the latent heat required for evaporation to the interface, and this has been termed the thermally controlled growth. The heat transfer in the boundary layer around a growing bubble was treated originally by Fritz and Ende[28] as that of a semi-infinite solid with a step change of interface temperature. The resulting expression for the transient spherical bubble size is given by

$$R(t) = \frac{2}{\pi^{\frac{1}{2}}} J_a (a_\ell t)^{\frac{1}{2}}$$ (8)

where

$$J_a = \frac{\rho_\ell C_\ell \Delta T_s}{\rho_v h_{fg}}$$ (9)

The coefficient on the right side of Eq. (8) has a value of 1.13. Solving the transient spherical conduction problem in an approximate manner, Forster[29] obtained the same form as Eq. (8) with a coefficient of 1.77, whereas a similarity solution for the same problem by Birkhoff et al.[30] gives a value of 1.61 for the coefficient. Other analytical solutions attempted to treat the spherical growth with the combination of inertial and thermal effects, and for the asymptotic case that is thermally controlled, coefficients of 1.96 were obtained by Plesset and Zwick[31] and 2.00 by Scriven.[32] A later work of Florschuetz and Chao[33] included experimental measurements of vapor bubble collapse with uniform superheat and spherical symmetry, obtained by eliminating buoyancy in an 8-ft drop tower. Comparison with numerical solutions of both the energy and momentum equations was quite good, and the asymptotic solutions agreed with the corresponding thermally and inertially controlled cases.

For bubble growths taking place adjacent to a solid heater, surface spherical symmetry cannot occur in either shape or temperature. A number of analyses have been conducted to predict the growth rates of vapor bubbles adjacent to a solid surface, e.g., Griffith,[34] Zuber,[35] and Mikic et al.[36] In all cases a spherical shape is assumed, and empirical factors or functions are applied to compensate for the nonsymmetry in temperature distribution. In comparison with experiments, all have varying degrees of success. The discrepancies could be due to the nonspherical shape and/or to the unknown initial temperature distribution in the liquid as a result of natural convection effects.

The microconvection induced by the vapor bubble dynamics influence the heat-transfer rates and has been treated by Bahr[37] and by Fedotkin and Gulakov.[38] Other mechanisms have been proposed to account for the large heat-transfer rates with boiling, including the rapid evaporation of a microlayer confined by viscous effects beneath a rapidly growing vapor bubble[39] and the enthalpy transport arising from the action of the bubbles as effective liquid pumps, pumping superheated liquid from the boundary layer.[40] Also, the influences of surface tension may play a significant role in the heat transfer from the solid surface on which boiling is taking place. A considerable amount of literature exists on this effect with pool boiling (e.g., Refs. 41 and 42). Another factor that may become important when body forces are reduced significantly, in addition to surface tension, is the momentum effect associated with the density differences with phase change as described by Kaznovskii et al.[43] and by Blangetti and Naushahi.[44] This can influence the departure size of the vapor bubble as well as its subsequent trajectory and will be discussed later.

Bubble growth rates themselves have been found to be relatively insensitive to gravity, both at low and high gravity levels, as demonstrated by Schwartz[45] and Beckman and Merte,[46] respectively. However, as mentioned earlier, they do depend on the temperature distribution within the liquid, which in turn is governed by body forces, geometry, and heat-flux level.

Each of the mechanisms referred to earlier, which are used to explain the large heat-transfer rates associated with nucleate boiling, has its proponents backed with experimental evidence. It is possible that each of these has its distinct contribution, but the domains over which each acts to provide a major contribution are as yet very unclear. For example, as will be demonstrated later, the microlayer evaporation theory serves to qualitatively explain the influence of buoyancy on nucleate pool boiling of saturated liquids only. However, the behavior observed as both subcooling and gravity are changed requires further investigation.

Departure

Once a vapor bubble is growing as a result of liquid superheat, its subsequent motion is determined by the net effect of forces associated with six phenomena:

1) Buoyancy due to body forces or system accelerations.

2) Momentum of the liquid, either externally applied or arising from the growth of the vapor bubble itself.

3) Surface tension effects at the liquid-solid-vapor interface and at the liquid-vapor interface. The latter is referred to as the Marangoni effect when temperature differences exist over the interface, producing spacewise variations in surface tension with corresponding induced stresses.

4) Momentum effects of phase change associated with the simultaneous vaporization and condensation across a bubble.

5) Viscous effects occurring at the liquid-vapor-solid intersection region.

6) Pressure difference between the inside and outside of the vapor bubble due to surface tension and/or viscous normal shears. The pressure difference associated with hydrostatic effects is accounted for by phenomenon 1.

These phenomena are illustrated in Fig. 9. Depending on the physical circumstances, it is possible for the bubble to remain attached to the surface and grow or collapse, or to depart and then grow or collapse.

Several studies by Cochran et al.[47] and Keshock and Siegel[48] have attempted to deduce the overall relative magnitudes of phenomena 1–3 by high-speed photographs in drop tower experiments. However, the short test times available required that large heat-flux levels be used to produce a bubble in the time available, which tends to distort the results because of the high localized liquid superheats. Buoyancy effects (phenomenon 1) have been treated extensively (e.g., Ref. 49). Bubble departure sizes from a heated surface are governed by the balance between buoyant forces, surface tension forces holding the bubble to the wall, and liquid inertia. These are

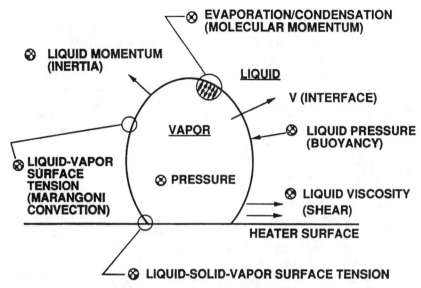

Fig. 9   Forces acting on growing/collapsing vapor bubble attached to a wall.

well understood for equilibrium cases, which would be present with low levels of heat flux or liquid superheat, and have been confirmed from Earth gravity up to $a/g = 10$ by Beckman and Merte.[46] However, the dynamic effects complicated the problem. Considerable attention has been given to the analysis and measurement of departure size because of its potential use in the prediction of the heat-transfer rates with boiling.

Using a static balance of forces between the solid-liquid-vapor surface tension and buoyancy, the earliest expression for the diameter of a spherical bubble departure is given by Eq. (10), from Fritz,[50] where the constant C includes the contact angle:

$$D_d = C\left(\frac{\sigma}{g\Delta\rho}\right)^{\frac{1}{2}} \tag{10}$$

A later work by Staniszewski[51] modified this to account for momentum imparted to the liquid during the growth process. The C_1 in Eq. (11) is an empirical constant that will depend on the surface/fluid properties, the heat flux, and bulk liquid subcooling:

$$D_d = C\left(\frac{\sigma}{g\Delta\rho}\right)^{\frac{1}{2}}\left(1 + C_1\frac{dD_d}{dt}\right) \tag{11}$$

Measurements were made by Keshock and Siegel[48] of departure diameters for both saturated distilled water vapor bubbles and water-sucrose solution vapor bubbles with various reduced-gravity body forces at times up to 1 s in a drop tower. The departure diameters were described by Eq. (10) for varying body forces only with saturated boiling distilled water. For the boiling water-sucrose solution, no differences in departure diameters were observed over the range of body forces $0.014 < a/g \leqslant 1.00$, leading the authors to conclude that the departure process was inertia dominated. Because of the short experimental time in reduced gravity, dictated by the drop tower height, it was necessary that the boiling process be initiated while still at Earth gravity, which tends to induce liquid circulation patterns that may persist after release of the test package.

In a series of works by Cochran et al.[47,52,53] measurements of vapor bubble growths were made for saturated and subcooled liquids, at Earth gravity and in free-fall conditions in a drop tower (2.2 s). It was possible to compute certain forces acting on the bubbles from the measurements: buoyancy, surface tension, pressure, and dynamic or liquid inertia by differences, as shown in Fig. 10. The results show little dynamic change between $a/g = 1$ and $a/g \simeq 0$. This is to be expected since the heat fluxes were imposed to achieve steady-state conditions at $a/g = 1$ prior to drop, and residual convective motion of the bulk liquid persisted during the 2.2-s free-fall. The equations presented in Fig. 10 are not predictive of the bubble motion unless the measured terms could be expressed in terms of parameters governing the processes.

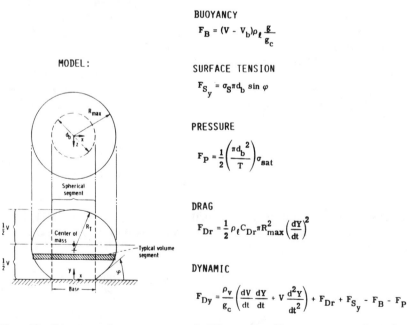

BUOYANCY

$$F_B = (V - V_b)\rho_\ell \frac{g}{g_c}$$

SURFACE TENSION

$$F_{S_y} = \sigma_s \pi d_b \sin \varphi$$

PRESSURE

$$F_P = \frac{1}{2}\left(\frac{\pi d_b^2}{T}\right)\sigma_{sat}$$

DRAG

$$F_{Dr} = \frac{1}{2}\rho_\ell C_{Dr} \pi R_{max}^2 \left(\frac{dY}{dt}\right)^2$$

DYNAMIC

$$F_{Dy} = \frac{\rho_v}{g_c}\left(\frac{dV}{dt}\frac{dY}{dt} + V\frac{d^2Y}{dt^2}\right) + F_{Dr} + F_{S_y} - F_B - F_P$$

Fig. 10 Forces acting on vapor bubbles amenable to computation from measurements.

Motion Following Departure

The trajectory of a bubble following departure depends on the dynamics of the growth process, which will be influenced by the degree and distribution of superheat or subcooling in the bulk liquid. The shapes and motion of bubbles have been studied at both Earth gravity[54] and at short-term or simulated reduced gravities.[55] In the latter case a cylindrical bubble between two closely spaced parallel plates at an angle to the horizontal was used to simulate reduced gravity. Viscous effects would appear to reduce the effectiveness of the analogy.

The description of the motion of vapor bubbles is complicated by virtual mass effects,[56] by the reactive forces of evaporating molecules,[43] and by interactions between bubbles and solid walls in the presence of temperature and velocity gradients.[57,58] It can be expected that these effects will have increasing consequences with the absence of buoyancy in microgravity.

Experimental Techniques

Practical difficulties become readily evident when attempts are made to conduct experiments with pool boiling in body force environments different from that of Earth gravity. High-gravity simulations can be conducted using either linear or centrifugal acceleration, with a reasonable period of time available only with the latter. Even here, overall size constraints may be present, the equivalent body force may vary spatially, depending on the experimental system size relative to the radius of rotation, and undesired Coriolis effects could be present. Testing at body force levels below Earth gravity present particular challenges when operation outside the Earth environment is eliminated. Many experimental studies of pool boiling under fractional and near-zero gravities have been conducted and reported using drop towers, aircraft flying parabolic trajectories, and sounding rockets. With each of these techniques the time at reduced gravity was not long enough for a reasonably sized system, for the residual convection velocities existing at $a/g = 1$ to die out at $a/g = 0$, or to achieve steady-state behavior under the new reduced-gravity conditions. In the case of aircraft the magnitude of the reduced gravity may not be as low as desired for the particular purpose because of difficulties in control. Attempts have been made to simulate reduced gravity, using the magnetic susceptibility of certain fluids or colloidal mixtures to counteract the gravity body forces. Aside from the possible disadvantage that only certain fluids can be used, the simulation is perfect only at a particular point in space because of the nonuniformity of magnetic fields.

A brief description of example experimental facilities for both high and reduced gravity is presented in the following subsections.

High Gravity

The schematic of a centrifuge used for pool boiling studies by Ulucakli and Merte[2] is shown in Fig. 11 and is designed for a maximum capacity of $a/g = 1000$ with a 50-kg test package. Provisions are incorporated for

36   H. MERTE JR.

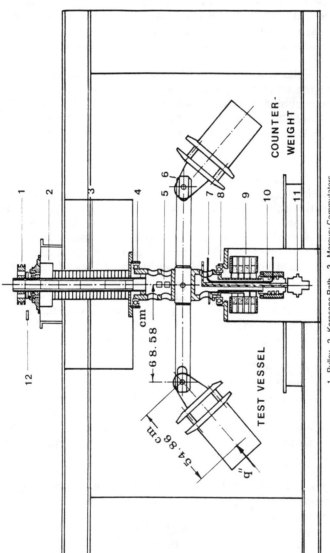

1. Pulley, 2. Kerosene Bath, 3. Mercury Commutators,
4. Upper Main Bearing, 5. Pressure Transducer,
6. Cross-Arm, 7. Cooling Water to Heat Exchangers,
8. Lower Main Bearing, 9. Copper Brushes, 10. Rotating
Fluid Coupling, 11. Mercury Commutator for AC Power,
12. Magnetic Pick-up.

Fig. 11   Centrifuge for heat-transfer studies.

transferring fluids and large electrical power while under rotation. For precision measurements at low signal levels, such as with thermocouples or thermistors, the slip-ring systems with mercury produce negligible levels of noise. When desired, the hollow vertical shaft permits the transmission of optical images to a stationary high-speed camera. By pivoting the test vessel to a horizontal cross arm, a constant orientation can be maintained between the heated surface and the effective body force vector at all rotational speeds, provided the center of gravity of the vessel is located at the heated surface. A cross section of the test vessel in Fig. 11 used with water is shown in Fig. 12. Provision is made for achieving large bulk liquid subcoolings by having submerged cooling coils around the heating surface, as well as condensing coils in the vapor space.

The cross section of the test vessel used by Clark and Merte[59,60] for the study of nucleate boiling of liquid nitrogen (LN_2) at high gravity is shown in Fig. 13. Because a rotating union to feed LN_2 to the rotating system was not practical at that time, it was necessary to operate in a batch mode, taking due account of the variation in hydrostatic head. The flow of vapor generated was directed to help isolate the inner test vessel from the warm surroundings.

A variation of such a vessel was constructed to study the effect of increased body force over the range $a/g = 1-15$ on the pool boiling of mercury at pressures up to 2.07 MPa. The high pressures, associated high temperature levels, and safety considerations provided unique design problems with a rotating system. The heater surface diameter was 50 mm, and a liquid depth of 13 mm was maintained constant.

Reduced Gravity

A drop tower designed specially for use with limited amounts of liquid hydrogen within a building is shown in Fig. 14. The free-fall distance of 9.8 m (32 ft) provided a measured body force of less than $a/g = 0.001$ for 1.4 s by the use of external fairing. Electrical connections during free-fall are made with a well-shielded drop cable. Test packages weighing up to 200 kg were accommodated within the air cushion-type deceleration assembly in Fig. 15, with a maximum deceleration of $a/g = 28$, which permitted the use of high-speed cameras. For testing under fractional gravity conditions, variable counterweights are attached to the drop package via a cable and pulley system. The provision must be made to cushion the counterweights at the end of a drop. More detailed information is given in Merte.[62]

Transient vs Steady

In certain circumstances where the times available for making boiling heat-transfer measurements are short, as in research in drop towers, it is sometimes convenient to use the heater as a transient calorimeter, as was done by Merte and Clark,[63] with a sphere in a cryogenic liquid, i.e., liquid nitrogen. In order to define the orientation better this technique was adapted to a flat surface by Merte et al.[64] Figure 16 shows one-half of the copper disk 7.5 cm in diameter, with the central portion 2.54 × 2.54 cm

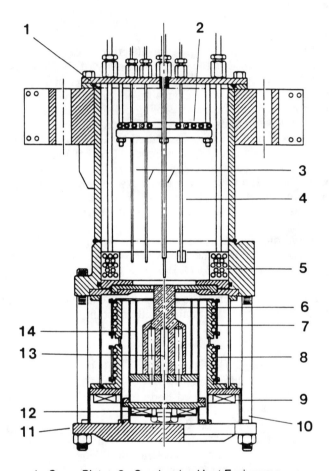

1. Cover Plate, 2. Condensing Heat Exchanger,
3. Liquid Temperature Probes, 4. Liquid Depth Control
Tube, 5. Subcooling Heat Exchanger, 6. Copper
Cylinder, 7. Upper Cylindrical Guard Heater, 8. Lower
Cylindrical Guard Heater, 9. Bottom Ring Guard Heater,
10. Support Bolts, 11. Support Plate, 12. Bottom Disc
Guard Heater, 13. Cartridge Heaters, 14. Radiation Shields

Fig. 12　Test vessel for high-gravity boiling studies with water.

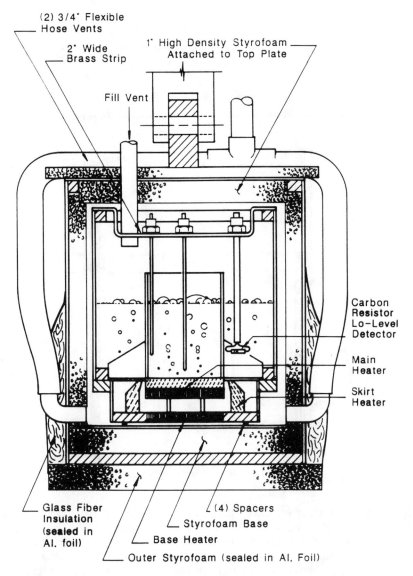

(2) 3/4" Flexible Hose Vents

2" Wide Brass Strip

1" High Density Styrofoam Attached to Top Plate

Fill Vent

Carbon Resistor Lo-Level Detector

Main Heater

Skirt Heater

Glass Fiber Insulation (sealed in Al. foil)

(4) Spacers

Styrofoam Base

Base Heater

Outer Styrofoam (sealed in Al. Foil)

Fig. 13   Test vessel for high-gravity boiling studies with liquid nitrogen.

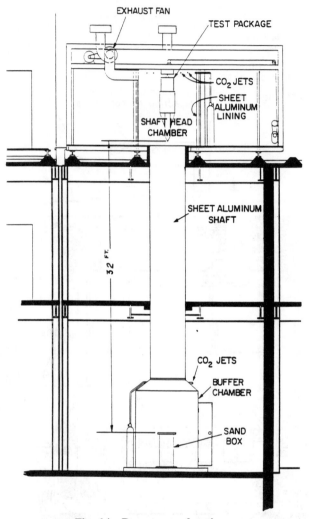

Fig. 14 Drop tower elevation.

being the test surface. A small gap exists between the two disk halves, soldered together, for thermal isolation between the two halves, and a network of grooves is machined about the central portion to within 0.7 mm of the heating surface in order to minimize edge effects. In addition, the perimeter was insulated with a tight-fitting foam insulation. The central portion of the disk thus simulates an infinite plate, insulated on one side and with boiling taking place on the other. Three thermocouples are soldered from the interior of the disk at different depths and indicate the adequacy of considering the disk to be uniform in temperature when cryogenic liquids are boiled. The criterion for justifying the use of a

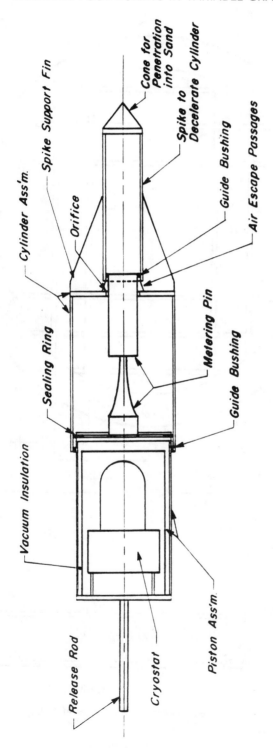

Fig. 15   Test package deceleration assembly.

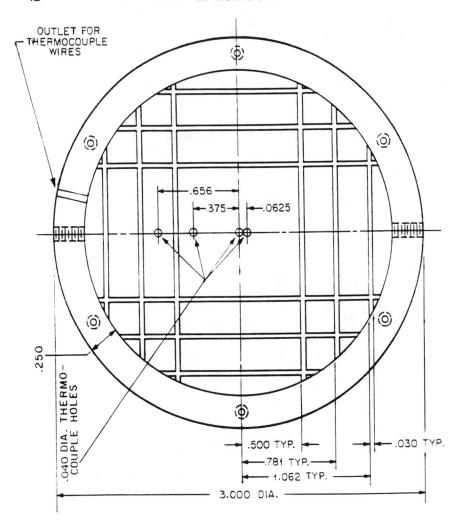

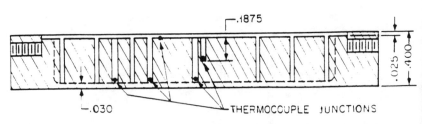

Fig. 16 Disk half as transient calorimeter for studies of boiling heat transfer.

uniform temperature in evaluating the surface heat flux is the Biot number. The orientation of the disk is varied by rotation about the supporting screws.

The disk at room temperature is immersed in the cryogenic liquid and the transient cooling curve is recorded, from which the heat flux and ΔT_{sat} are obtained. For reduced gravity the cryostat is installed in the drop package shown in Fig. 15, which is released in the drop tower such as that shown in Fig. 14 to fall 9.8 m, giving 1.4 s of free-fall. The drop package is released when the disk has reached the desired point on the cooling curve. By measurement with a sensitive accelerometer, the maximum body force due to air drag during free-fall was reduced progressively from $a/g \cong 0.008$ to $a/g \cong 10^{-3}$ by external fairing. The thermocouple and other electrical connections during free-fall are made with a drop cable.

The transient calorimeter technique for the study of boiling in reduced gravity has the disadvantage of requiring the insertion of the disk or other surface under investigation into the liquid prior to release of the drop package, producing residual motion in the liquid which persists to varying degrees during the drop. It may be desirable to be able to initiate the boiling process during free-fall, while the liquid is quiescent. This would thus include the transient onset of boiling, or nucleation, as well as the longer-term boiling. Such a process is possible using a thin, semitransparent gold film deposited on a flat quartz surface as an electrical heater. By appropriate precision measurements and resistance calibration as a function of temperature, it is possible to determine the mean boiling surface temperature and simultaneously view the boiling process from beneath, without obscuring by intervening bubbles. A disadvantage of such a surface is that the smoothness of glass results in certain differences in the boiling process when compared with metal surfaces, primarily in higher heater surface superheats for nucleation and sustained boiling. However, for purposes of qualitatively assessing the effects of orientation and buoyancy, the results should be similar. This will be seen to be the case. The heater surface used was 22×25 mm in size, and the gold was approximately 400 Å thick. The technique is described in full by Oker and Merte.[65]

By beginning the heating process after the test package is released, the liquid is initially stagnant, and the subsequent boiling behavior is unaffected by residual liquid motions. The heat-transfer mechanism up to the onset of convection or nucleation is transient conduction to the liquid and to the heater substrate, and the division of the imposed plane heat flux between the two domains is well known. After the onset of convection or boiling the transient heat flux can be computed from the measured mean surface temperature with Duhamel's superposition integral. The concept of the thin-film heater is shown in Fig. 17. Transient effects leading up to nucleation are described by Oker and Merte.[4]

Experimental Results

Pool boiling alone is considered here in order to eliminate the complicating effects of an external flowfield superimposed on that generated by

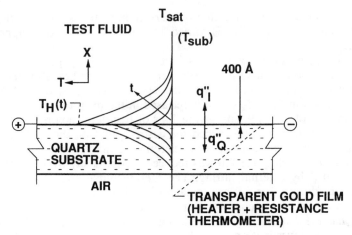

Fig. 17   Concept of thin-film heater.

growing/collapsing vapor bubbles. Also eliminated from consideration, except in a limited way, are orientations in which the body force vector is not normal to the heater surface. Bubble motions in such cases can induce liquid velocity components parallel to the heater surface, which introduce the complications corresponding to that of forced convection.

High Gravity

Water
When essentially saturated water was boiled at atmospheric pressure over the range $a/g = 1-21$, it was observed that the heat flux increased considerably with gravity at low levels of heat flux[66] because of the relatively large contribution of nonboiling convection. However, a degradation in heat transfer took place with increasing gravity as the heat-flux level was increased, manifested by an increased heater surface superheat for a given heat flux. This is attributed in part to the action of buoyancy on the growing vapor bubbles, increasing the thickness of the liquid microlayer beneath the bubble and reducing the heat transfer per bubble. For a constant heat flux, then, more nucleation sites are required, resulting in turn in a higher surface temperature. This influence of high acceleration on the microlayer is still somewhat speculative and remains to be observed directly and described analytically. As will be noted later, the behavior is consistant as buoyancy is reduced below Earth gravity. This same work included results obtained with limited amounts of liquid subcooling: At the lower levels of heat flux, increasing subcooling caused the wall superheat to increase and then to decrease. A plot of wall superheat vs liquid subcooling thus exhibited a maximum. The same trend was observed at higher levels of heat flux and acceleration, except that wall temperature increases with subcooling did not reach a maximum because of the inability to increase the subcooling further.

In a subsequent work by Ulucakli and Merte,[2] who used the test vessel in Fig. 12, liquid subcoolings of up to 90°C were attained with water near atmospheric pressure in order to explore more extensively the combined roles of subcooling and gravity on nucleate pool boiling. Results for $a/g = 1$ and 10 are presented in Figs. 18–21. The measurements of heater surface superheat are plotted as a function of bulk subcooling with various levels of heat flux as a parameter in Fig. 18 for $a/g = 1$ and in Fig. 20 for $a/g = 10$. The smoothed curves through these data points are then cross plotted in Figs. 19 and 21 corresponding to the acceleration, in the more usual boiling heat-transfer form of heat flux vs heater surface superheat using bulk subcooling as the parameter. In doing so, the competing and opposite influences of subcooling on the nonboiling and boiling contributions to the heat-transfer process become more obvious.

It is noted in Fig. 18 that the heater surface superheat at first increases and then decreases as subcooling increases from the saturation state,

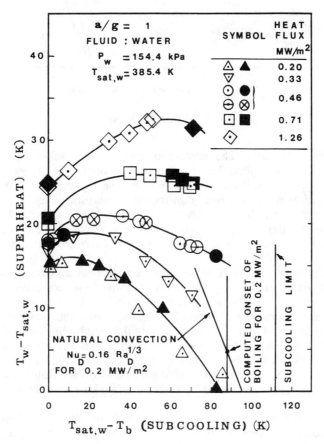

Fig. 18 Influence of large subcooling range on heater surface temperature with water; $a/g = 1$.

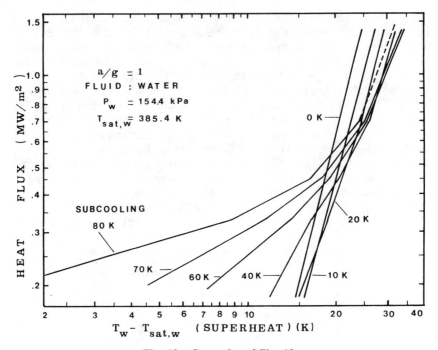

Fig. 19   Cross plot of Fig. 18.

resulting in a maximum. The increase in superheat with subcooling not only increases with heat flux, but the maximum occurs at larger levels of subcooling for larger levels of heat flux. At the highest heat flux used at $a/g = 1$, 1.26 MW/m^2, the heater surface superheat increases by 8.3 K at a subcooling of 60 K. Using the development of Zuber,[67] the critical heat flux for saturated water at $a/g = 1$ and atmospheric pressure is estimated to be 1.5 MW/m^2, somewhat above the highest level of 1.26 MW/m^2, used here at $a/g = 1$.

The relationship between heater surface superheat and bulk subcooling computed from an appropriate natural convection correlation is also included in Fig. 18 for the heat flux of 0.2 MW/m^2. The near-tangency with the data for decreasing subcooling, together with the near-zero heater surface superheat at this point demonstrate that nonboiling natural convection virtually dominates the process here. Another reasonable determination of the onset of nucleate boiling in this region can be made by incorporating the natural convection correlation into the so-called Hsu-Rohsenow-Bergles model of nucleation, presented by Rohsenow.[68] This point is also indicated in Fig. 18.

The smoothed data of Fig. 18 are cross plotted in Fig. 19 and demonstrate that nonboiling convection dominates at the low levels of heat flux, whereas as heat flux increases the heater surface superheat increases with subcooling. As will be seen, similar behavior occurs at higher acceleration levels.

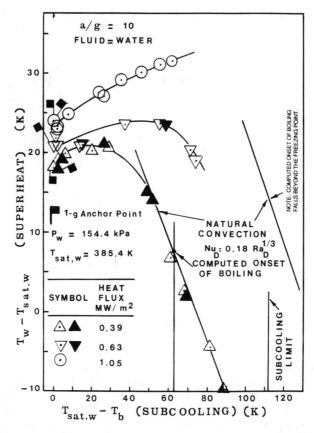

Fig. 20 Influence of large subcooling range on heater surface temperature with water; $a/g = 10$.

The next group of tests, conducted at $a/g = 10$, are shown in Fig. 20. Initial tests revealed the extreme sensitivity of the subcooling level to the cooling water flow rate, and modifications to the apparatus were made to provide for fine control. The "anchor points" indicated are obtained for each level of heat flux for the saturated state at $a/g = 1$ immediately prior to and following each test at high acceleration, as a check on reproducibility and/or hysteresis effects as a result of subjecting the boiling system to large body forces. These anchor points were found to be repeatable. At the lowest heat flux level shown, the open triangles correspond to increasing subcooling, and the dark triangles apply to decreasing subcooling. The reproducibility and lack of hysteresis are evident.

The heater surface superheat vs subcooling relationship for nonboiling natural convection from the natural convection correlation is included for the two lower heat-flux levels used. At 0.39 MW/m² the predominance of nonboiling is obvious, whereas at the heat flux of 0.63 MW/m² the

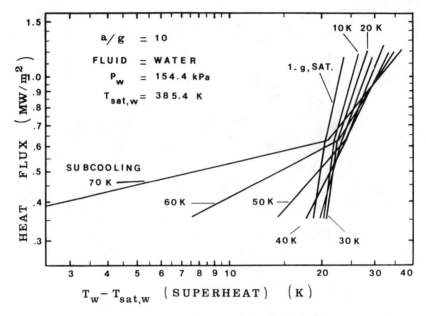

Fig. 21   Cross plot of Fig. 20.

significant departure from the nonboiling correlation indicates the presence of nucleate boiling even at this large subcooling level. The computed onset of boiling is shown for the lowest heat flux only; the subcooling for the next value used, 0.63 MW/m², falls beyond the freezing point of water. Figure 21 is the cross plot of the smoothed data of Fig. 20, with subcooling as a parameter.

Liquid Nitrogen

In Fig. 22 are plotted the nucleate pool boiling liquid nitrogen data obtained with steady-state measurements with the test vessel of Fig. 13 installed in a centrifuge and subjected to accelerations up to $a/g = 20$. Also included here are the data obtained using the transient calorimeter technique with the disk of Fig 16, as $a/g \approx 0$ and $a/g = -1$, the latter corresponding to the inverted case. These will be discussed more fully later. Results were also obtained with the transient disk at $a/g = +1$ and essentially coincide with those obtained under steady conditions with the apparatus of Fig. 13. At the higher levels of heat flux in Fig. 22, the highest of which is not far below the critical heat flux of about 200 kW/m², a degradation in the heat transfer occurs as gravity increases, which is consistant with the results obtained with water by Merte and Clark.[66] At the lower levels of heat flux, the contribution of nonboiling convection become significant, and the trend is reversed. For $a/g = 10$, 15, and 20 the measurements obtained here at the lowest heat flux with LN_2 coincide with the correlation for nonboiling natural convection obtained for H_2O in a

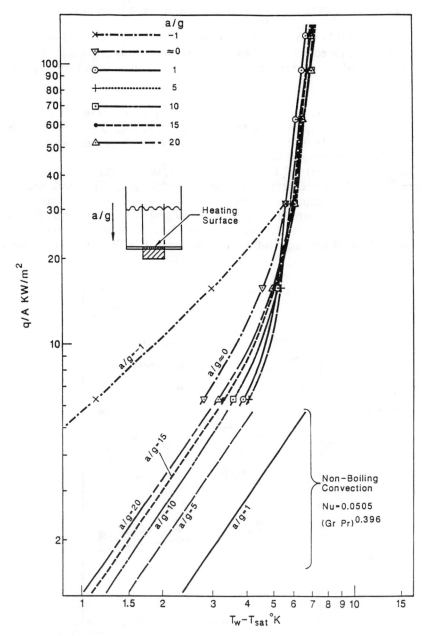

Fig. 22   Influence of body forces normal to heating surface on nucleate pool boiling of saturated LN_2.

similar configuration by Merte and Clark,[66] plotted in the lower part of Fig. 22. In comparing the trends for the disk at $a/g \cong 0$ and $a/g = -1$ with those at high gravity, it should be noted that the disk data have no significant contributions from nonboiling natural convection and therefore should be compared with the high-gravity behavior only at the higher levels of heat flux where nonboiling convection contributions become negligible.

Mercury

Representative data obtained by Merte[61] for pool boiling of mercury under high gravity up to $a/g = 15$ are presented in Figs. 23–25 in order to illustrate the similarity in behavior for a fluid whose boiling point is far above that of water and liquid nitrogen.

In the lower part of Fig. 23 the heater surface superheat and the temperature difference between the heater surface and liquid bulk nearby are plotted as a function of dimensionless equivalent gravity for a constant relatively low pressure and heat flux. $\Delta T_{sat} = T_{wall} - T_{sat}$, and T_{sat} is evaluated at the heating surface and takes into account the variation in hydrostatic head with acceleration. The decrease in both of these quantities

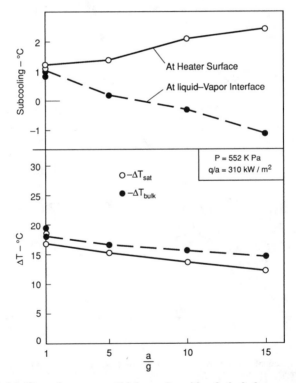

Fig. 23 Pool boiling of mercury at high gravity with relatively low pressure and heat flux.

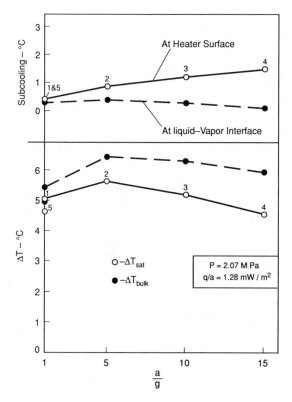

Fig. 24   Pool boiling of mercury at high gravity with relatively high pressure and head flux.

as acceleration increases is indicative of the increasing contribution of nonboiling natural convection to the total heat-transfer process. The relative behaviors of ΔT_{sat} and ΔT_{bulk} provide information as to whether boiling or nonboiling natural convection is prevalent. With nonboiling the heat-transfer coefficient remains essentially constant as the liquid subcooling at the heater surface is varied at constant acceleration, whereas it varies in a known manner as acceleration is changed. In the upper portion are plotted the respective liquid subcoolings at the heater surface and at the liquid-vapor interface. The saturation temperatures in each of these are based on the local values, i.e., the values at the heater surface and at the liquid-vapor interface, respectively, and again take into account the variations in hydrostatic pressure. The increase in subcooling of the liquid at the heater surface as acceleration increases, the upper curve, reflects the combined influence of increasing hydrostatic pressure and natural convection: The convection produces intense stirring, and the bulk liquid temperature is virtually uniform in the vertical direction. It may be noted in connection with this process that the liquid at the liquid-vapor interface becomes somewhat superheated as acceleration increases. From

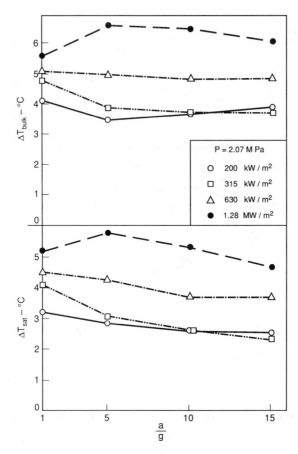

Fig. 25   Composite of influence of high gravity on ΔT for pool boiling with mercury at $P = 2.07$ MPa.

the corresponding curve in Fig. 24, for a significantly larger heat flux and pressure, the superheat is reduced due to the larger vapor bubble population at the heater surface. The lower curve in Fig. 24 represents measurements of ΔT_{sat}, which first increases and then decreases with increasing body force. This is quite similar to the behavior that Merte and Clark[66] observed with near-saturated water.

Figure 25 presents a composite of heater surface superheat and temperature difference between the heater surface and liquid bulk for a constant system pressure and four different levels of heat flux. It would appear that the heat-transfer process is relatively insensitive to wide variations in body force. However, the bulk liquid subcoolings in this chapter were deliberately maintained as small as possible in order to explore near-saturation behavior, but could not be closely controlled at that time. It is possible that relatively small changes in subcooling can produce rather exaggerated changes in boiling behavior at high levels of body force.

Reduced Gravity

Liquid Nitrogen
Figure 26 shows the boiling results obtained with the transient calorimeter technique using the disk shown in Fig. 16, with saturated liquid nitrogen at atmospheric pressure. The data cover the full range of film, transition, critical, and nucleate boiling at $a/g = 1$, for the orientations horizontal up (HU), horizontal down (HD), and vertical (V). Also included for reference purposes are results obtained with a copper sphere 2.54-cm diam, both at $a/g = 1$ and $a/g \approx 0$. No other particular significance should be given to the sphere date here. It was used initially with the transient calorimeter technique by Merte and Clark[63] because of its symmetry and simplicity, but the results depend on the size used, which covered 6- to 58-mm-diam copper spheres. The behavior of the critical heat flux and the behavior in the film boiling regions are described quite well by existing correlations for the saturated liquid case. The breakdown of film boiling to the transition region is rather sharply defined for the sphere and all orientations of the disk at a $\Delta T_{\text{sat}} = 28$ K, and the critical heat flux for the disk in the HD orientation is close to that of the sphere at $a/g \approx 0$. The relative behaviors between the three orientations follow expected patterns in the film boiling and critical heat-flux domains. It is in the nucleate boiling domain that a somewhat surprising result is noted: For a given ΔT_{sat}, the HU orientation has the lowest heat flux, the HD orientation has the highest heat flux, and the V falls between them. For $\Delta T_{\text{sat}} = 3.5$ K the

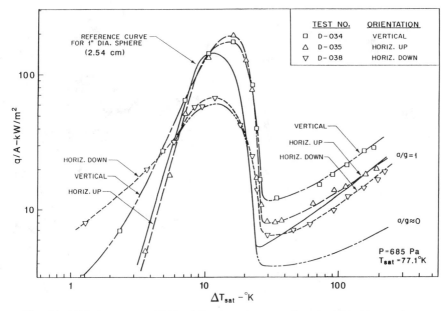

Fig. 26   Disk in saturated LN$_2$. All orientations at $a/g = 1$. All boiling regimes.

HD heat flux is five times that for HU, with the difference decreasing as heat flux increases, and reversing toward the critical heat flux. If the HU orientation is designated as $a/g = +1$, then the HD orientation is classified as $a/g = -1$, and one could conclude that the nucleate boiling heat flux increases as buoyancy is reduced for a given heater surface superheat. To test this for $a/g \approx 0$, a cryostat test vessel was installed in the drop tower, with the disk used as the test surface.

Although boiling in near-zero gravity should be expected to produce results independent of any local orientation, using the disk as a transient calorimeter requires that it be inserted into the liquid prior to release of the test package. This sets up convection patterns in the liquid that persist during the short 1.4-s free-fall time that are different, depending on the surface orientation relative to the initial $a/g = 1$ condition. This is demonstrated in Figs. 27–29 for the three orientations used, which again cover the entire boiling regime. For each test shown the test package was released when the disk has reached the desired point in the cooling curve. Consequently, the initial data point falls on the boiling curve for that particular orientation at $a/g = 1$, represented in each of Figs. 27–29 by the short dashed curves. These provide a measure of the reproducibility of the experimental technique used. Upon release of the test package, a transition in the slope of the cooling curve must take place as the boiling process accommodates itself to the new condition. In some cases the change in slope of the cooling curve took place quickly and smoothly to a single new value, remaining there. In other cases a continuous change in slope was

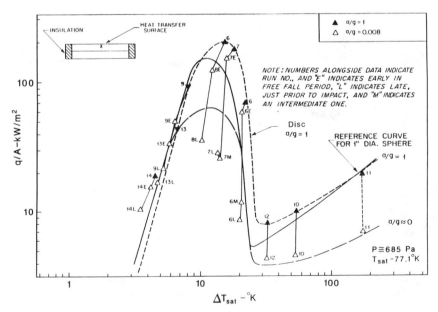

Fig. 27   Disk in saturated LN_2. Horizontal up at $a/g = 0.008$.

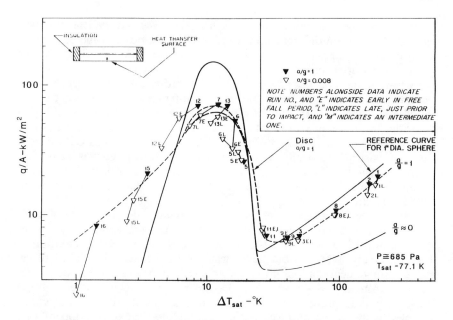

Fig. 28   Disk in saturated LN_2. Horizontal down at $a/g = 0.008$.

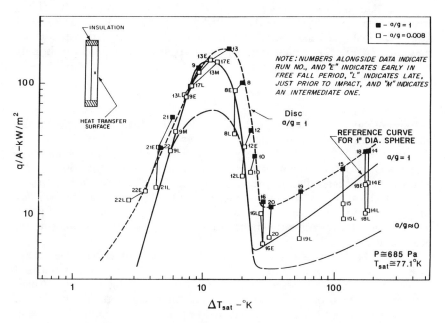

Fig. 29   Disk in saturated LN_2. Vertical at $a/g = 0.008$.

noted during the entire drop period. Where this occurred the heat flux was evaluated at several different points in the drop period, and the evaluation suggests that changes in the process are occurring continously. These points are indicated appropriately in Figs. 27–29.

For the initial HU orientation in Fig. 27, rather significant decreases in heat flux occur in the film and transition boiling regions, in contrast to the initial HD orientation in Fig. 28. This would imply that the residual liquid velocities in the vicinity of the heater persist in the latter case. Data points 7M, 7L, and 8L in Fig. 27 suggest that the phenomena of a critical heat flux can exist at drastically reduced levels of buoyancy, as was quantified approximately by Merte and Clark.[63] The larger levels of heat flux of data points 5E, 5L, 6E, and 6L in Fig. 28 reflect the influence of the residual liquid velocities.

The change in behavior in the nucleate boiling domain with reduced gravity in Fig. 27 is rather subtle, but one can note a tendency for an increase in the heat flux. This is in contrast to the initial HD orientation of Fig. 28, where a reasonable large decrease in the heat flux takes place, representing a degradation in the nucleate boiling process as the body force changes from $a/g = -1$ to $a/g \cong 0$. One might conclude here that the nucleate boiling process is enhanced if buoyancy acts to hold the vapor bubbles at the heater surface. With the initial vertical orientation of Fig. 29, negligibly small changes occur upon test package release, associated with the large residual velocity parallel to the surface persisting for some time.

The data from the HD orientation at $a/g = +1$ from Fig. 28 are plotted as those data designated $a/g = -1$ in Fig. 22, and extrapolations of the free-fall data from Fig. 27 are designated as $a/g \approx 0$ in Fig. 22. The results at $a/g = 1$ in Fig. 27 may be superimposed on the steady results at $a/g = 1$ presented in Fig. 22, which thus shows a consistency in nucleate boiling behavior as one proceeds from $a/g > 1$ to $a/g = 1$ to $a/g \approx 0$ to $a/g = -1$.

Liquid Hydrogen

Representative results for pool boiling of liquid hydrogen (LH$_2$) in the various boiling regimes obtained by Merte[62] are presented in Figs. 30–32 for $a/g = 1$, 0.235, and 0.008, respectively. The transient calorimeter technique was used with a copper sphere 2.54 cm (1 in.) for Figs. 30 and 31, and the flat disk results are presented in Fig. 32. The reference curves in Figs. 30–32 all apply to results obtained with the sphere at $a/g = 1$ for saturated liquid hydrogen at atmospheric pressure, 101.4 kPa (14.7 psia), and permit relative comparisons between the various results as parameters are changed.

Figure 30 demonstrates the effect of reduced body forces with the sphere in liquid hydrogen at atmospheric pressure. The test package was released at the desired point in the cooling curve. The film boiling and critical heat flux are reduced as buoyancy decreases and conform with the predictions of established correlations as described by Merte and Clark.[63] The decrease in the nucleate boiling regime is a distinctive departure from the behavior

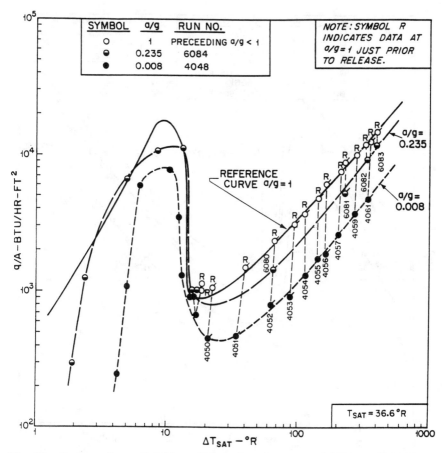

Fig. 30   Copper sphere of 2.54 cm (1 in.) diam in saturated LH₂ at 101.4 kPa (14.7 psia). Effect of reduced body force.

observed with liquid nitrogen (LN₂) described earlier and from the behavior to be described later with R-113 under reduced-gravity conditions. The drastic reduction in heat flux with buoyancy for a given ΔT in the nucleate boiling regime for the sphere is believed to be related to the relatively small latent heat of evaporation and density of liquid hydrogen. The low density results in a rather rapid dissipation of the residual velocity induced by the prior presence of buoyancy, whereas the small latent heat produces significant amounts of vapor that envelop the sphere.

The experimental data in Fig. 31 are similar in character to those of Fig. 30 except that the liquid hydrogen is now saturated at a pressure of 255 kPa (37 psia). The increased pressure results in an increased heat flux in all boiling regimes at $a/g = 1$, as is well known, except that the transition boiling is moved to a lower ΔT. Reducing the body force also results in a

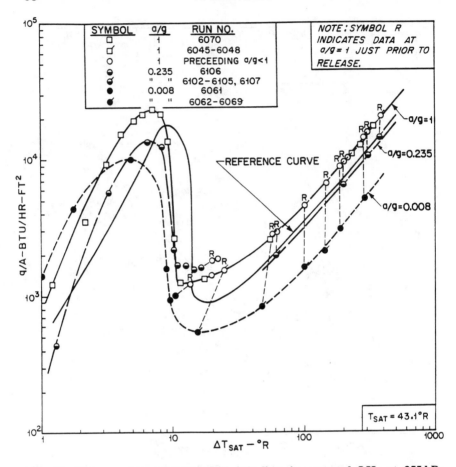

Fig. 31 Copper sphere of 2.54 cm (1 in.) diam in saturated LH₂ at 255 kPa (37 psia). Effect of reduced body force.

successive decrease in heat flux in all boiling regimes, with the exception of the lowest gravity with nucleate boiling. It is posited that here the increased vapor density reduces the tendency of the vapor to envelop the sphere and permits the continued access of liquid to the sphere, causing the increase in heat flux observed.

The boiling behavior of the flat surface with the disk in saturated liquid hydrogen for all three orientations at $a/g = 1$ is given in Fig. 32. The similarity with liquid nitrogen in Fig. 26 is to be noted, particularly the increased heat flux in the nucleate boiling domain for the horizontal-down orientation. At $\Delta T_{sat} = 1.1$ K in the nucleate boiling domain, the heat flux for the HD orientation is about five times that for the HU case. The major discrepancy in heat flux between the three orientations and the sphere appears to occur in the vicinity of the minimum heat flux.

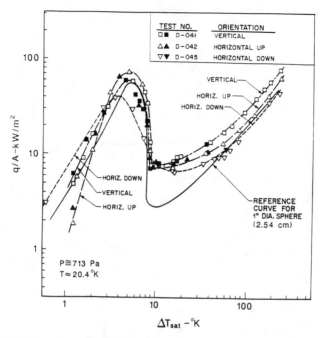

Fig. 32 Disk in saturated LH$_2$. All orientations at $a/g = 1$. All boiling regimes.

R-113

An improvement in the experimental means for examining the roles of orientation and buoyancy on nucleate pool boiling is to ensure that the disturbing influences of residual liquid motion are not present by initiating the heating process after the test package has been released in the drop tower. A thin gold film on Pyrex, described earlier, serves as the heater. In the pure transient conduction domain, the R-113 and Pyrex can be viewed as semi-infinite solids with a plane heat source between. Only 16.6% of the heat is transferred to the R-113, with the rest going into the Pyrex.

Figures 33, 34, and 35 show the transient mean surface temperatures for different levels of heat flux input, with the three orientations at $a/g = 1$ for HU, HD, and V, respectively. The open symbols represent the behavior at $a/g = 1$, and the solid ones are obtained during free-fall. Since orientation has no significance in near-zero gravity, the behavior should be the same where the heat flux levels compare. This can be seen to be approximately the case. Heater surface PCG-24 in Figs. 33 and 34 is the same surface, whereas PCG-23 in Fig. 35 is a different one, but fabricated in the same way. At the very early times in the heating process, the various curves are identical for a given heat flux regardless of buoyancy or orientation, where conduction is the only mechanism acting.

In Fig. 33 HU, nucleation, or the onset of nucleate boiling takes place at a surface temperature considerably lower with $a/g \approx 0$ and demonstrates

60   H. MERTE JR.

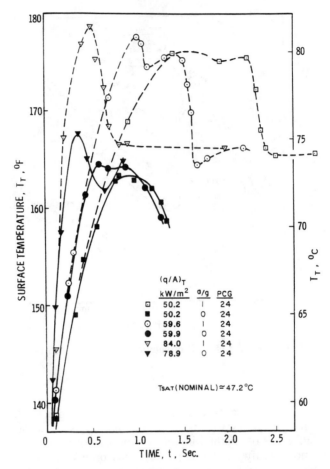

Fig. 33   Transient surface temperatures, R-113, horizontal up; $a/g = 1$ and ≈ 0.

that nucleation is influenced by the temperature distribution in the vicinity of the heating surface. Here the temperature distribution is affected by buoyancy. The boiling with the highest heat flux here goes into film boiling during free-fall, whereas those with the lower heat flux are enhanced relative to boiling at $a/g = 1$. These runs terminated at the end of the drop period, 1.4 s after release.

The onset of boiling and subsequent nucleate boiling behavior with HD in Fig. 34 are almost the same at $a/g = 1$ and $\simeq 0$, with the exception of the oscillations that take place at $a/g = 1$ as the vapor departs in slugs around the edge of the heating surface. Measurements during free-fall could be carried beyond the impact point as noted in Fig. 34 and result in the onset of film boiling at the higher power level. The mean surface temperature at $a/g = -1$ in Fig. 34 is detectably lower than that at $a/g = 1$ in Fig. 33, which is consistent with the observations made

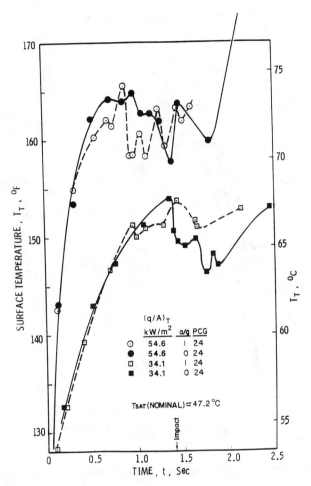

Fig. 34   Transient surface temperatures, R-113, horizontal down; $a/g = 1$ and ≈ 0.

with the disk. It is believed that no boiling took place for the lower power level in Fig. 34.

Although the onset of boiling during free-fall in Fig. 35 is about the same as that in the corresponding tests of Figs. 33 and 34, the upper two power levels proceed into film boiling prior to impact, which does not occur with the other tests. The reason for this is not known and may be related to the fact that this surface is a different one. The higher heater surface temperature prior to nucleation with the vertical surface at $a/g = 1$ in Fig. 35 is a consequence of the onset of natural convection during the heating process, which results in a decrease in bulk liquid temperature near the surface. In interpreting the results obtained with the thin film heater, one must keep in mind that the temperature measured is

62   H. MERTE JR.

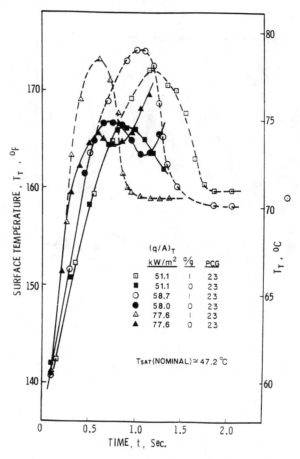

Fig. 35 **Transient surface temperatures, R-113, vertical; $a/g = 1$ and ≈ 0.**

a spatial mean and that nucleate boiling over a portion of the surface can reduce this mean.

The relatively low thermal diffusivity of Pyrex compared to copper (1 : 170), combined with the thin gold heater on its surface, provides a rapid response to changes taking place at the heater surface and permits a further comparison of the combined effects of orientation and buoyancy on nucleate pool boiling. In the next case the heating process is initiated at $a/g = 1$ to steady conditions, and the power input is maintained constant through the 1.4-s free-fall and after impact in the drop tower. Any enhancement or degradation in the boiling process should be reflected in changes in surface temperature. This is shown in Figs. 36, 37, and 38 for the initial orientation of HU, HD, and V, respectively. Since the process was steady prior to release, it is possible to compute the steady heat flux to

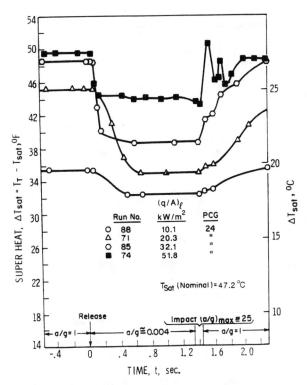

Fig. 36 Steady surface superheat, R-113, horizontal up; $a/g = 1$ and ≈ 0.

the liquid, which is indicated on each figure for each test conducted. The test package release and impact times are noted.

In Fig. 36, for the HU orientation the heater surface superheat decreases at reduced gravity, to a maximum of about 26%, and then returns to the prior existing level. The change in superheat with reduced gravity appears to increase and then decrease as the heat flux is increased. Buoyancy removes the vapor bubbles at $a/g = 1$, but the removal process is reduced as $a/g \cong 0$, and the heat-transfer process is enhanced.

With the HD orientation in Fig. 37 the opposite takes place, showing a degradation from $a/g = -1$ to $a/g \cong 0$. The oscillations taking place at $a/g = -1$ with the highest heat-flux level are also present again, as was the case with transient boiling. In this case no significant changes took place at $a/g \cong 0$, probably due to the dynamic effects associated with the periodic vapor departure process peculiar to this orientation.

With the V orientation in Fig. 38 removal of buoyancy also results in a reduction in effectiveness of the boiling process. In this case steady state at $a/g = 1$ produces a boundary-layer type of flow that dies away when buoyancy is removed. The degradation here increases with heat flux because of the increase in vapor generation rate.

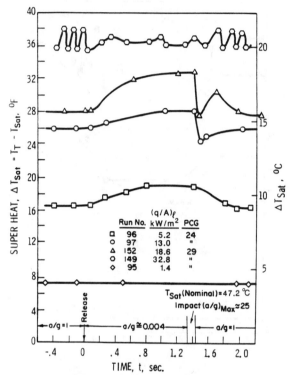

Fig. 37   Steady surface superheat, R-113, horizontal down; $a/g = 1$ and ≈ 0.

Conclusions

The results of drop tower testing presented earlier demonstrate the limitations involved in extrapolating behaviors observed in the short tests of drop towers to long-term microgravity: Insufficient time exists for transients to decay, and the long-term behavior of the vapor formed in the immediate vicinity of the surface is as yet unknown. It is also speculative to extrapolate behaviors with small or curved heating surfaces to large sizes because of surface tension effects.

At the lower levels of heat flux with nucleate boiling, the heater surface superheat first increases and then decreases as the body force is increased above $a/g = 1$, as a result of the increasing contribution of nonboiling convection. As the body force is decreased from $a/g = 1$ to $a/g \approx 0$ and $a/g = -1$, a more dramatic decrease in the heater surface superheat takes place, indicating a significant enhancement in the heat transfer process. This is also associated with buoyancy, but is proposed as being due to two different effects; buoyancy now keeps the heated fluid adjacent to the heating surface, increasing the number of nucleating sites, and negative buoyancy or its absence keeps the growing vapor bubbles closer to the heating surface, increasing the area of the microlayer and/or decreasing its

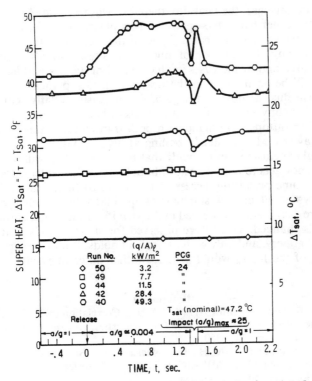

Fig. 38   Steady surface superheat, R-113, vertical; $a/g = 1$ and ≈ 0.

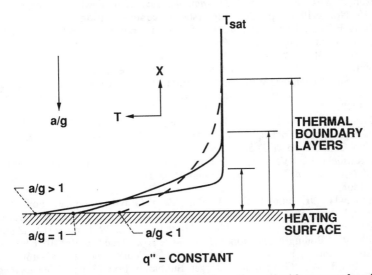

Fig. 39   Qualitative liquid temperature distributions observed with saturated nucleate pool boiling in various gravity fields.

thickness. In any case, the heat flux therefore will increase. At high gravity levels, on the other hand, the opposite takes place; the increased buoyancy reduces the extent of superheated liquid in the vicinity of the heat transfer surface, reducing the number of nucleating sites, and also reduces the microlayer area and/or increases its thickness. For an imposed heat flux the net effect will be an increase in the heater surface superheat at the higher levels of heat flux, as observed. Again, the increasing relative contribution on nonboiling convection at the lower levels of heat flux reverses this trend.

These observations were made with both transient and steady boiling at reduced gravity and with steady boiling at high gravity.

A general conclusion can be made that nucleate boiling will be enhanced if the buoyancy acts to hold the vapor bubbles near the heater surface, while at the same time permitting access of the liquid to the surface in order to prevent dryout. Figure 39 summarizes qualitatively the changes in the temperature distributions observed in the vicinity of the heater surface as the buoyancy is changed. These were observed for short periods of time only, for reduced gravity, and only with saturated liquids. It may be expected that subcooling of the liquid would produce rather different results.

References

[1] Mikic, B. B., and Rohsenow, W. M., "A New Correlation of Pool-Boiling Data Including the Effect of Heating Surface Characteristics," *Journal of Heat Transfer*, Vol. 91, May 1969, pp. 245–350.

[2] Ulucakli, M. E., and Merte, H., Jr., "Nucleate Pool Boiling with High Gravity and Large Subcooling," ASME/AIChE National Transfer Conference, Vol. 2, Paper HTD96, 1988, pp. 415–422.

[3] Vest, C. M., and Lawson, M. L., "Onset of Convection Near a Suddenly Heated Horizontal Wire (in English)," *International Journal of Heat and Mass Transfer*, Vol. 15, No. 6, 1972, pp. 1281–1283 (shorter communication).

[4] Oker, E., and Merte, H., Jr., "A Study of Transient Effects Leading up to Inception of Nucleate Boiling," *Proceedings of the 6th International Heat Transfer Conference*, Vol. 1, Paper PB-5, 139–144.

[5] Giarratano, P. J., "Transient Boiling Heat Transfer From Two Different Heat Sources: Small Diameter Wire and Thin Film Flat Surface on a Quartz Substrate," *International Journal of Heat and Mass Transfer*, Vol. 27, No. 8, 1984, pp. 1311–1318.

[6] Coeling, K. J., and Merte, H., Jr., 1969, "Incipient and Nucleate Boiling of Liquid Hydrogen," *Transactions of ASME, Journal of Engineering for Industry*, Vol. 91B, No. 2, 1969, pp. 513–521.

[7] Derewnicki, K., "Vapor Bubble Formation During Fast Transient Boiling on a Wire," *International Journal of Heat and Mass Transfer*, Vol. 26, No. 9, 1983, pp. 1405–1408.

[8] Merte, H., Jr., and Littles, J. W.; "Zero Gravity Incipient Boiling Heat Transfer," *Proceedings of the Space Transportation System Propulsion Technology Conference*, Vol. 4 (Cryogens), NASA, Huntsville, AL, 1971, pp. 1312–1348.

[9] Cole, R., "Boiling Nucleation," *Advances in Heat Transfer*, Vol. 10, edited by J. P. Hartnett and T. F. Irvine, Jr. Academic, New York, 1974, pp. 85–166.

[10] Larkin, B. K., "Growth of Hydorgen Bubbles in Low Gravity," *Proceedings of the Conference on Long-Term Cryo-Propellant Storage in Space*, NASA George C. Marshall Space Flight Center, Huntsville, AL, 1966.

[11]Oker, E., and Merte, H., Jr., "Transient Boiling Heat Transfer in Saturated Liquid Nitrogen and F113 at Standard and Zero Gravity," Heat Transfer Lab., Dept. of Mechanical Engineering, Univ. of Michigan, Ann Arbor, Final Rept., ORA Rept. 074610-52-F, Contract NAS8-20225, Oct. 1973.

[12]Anderson, D. L. J., Judd, R. L., and Merte, H., Jr., "Site Activation Phenomena in Saturated Nucleate Boiling," American Society of Mechanical Engineers, Paper 70-HT-14, May 1970.

[13]Shoukri, M., and Judd, R. L., "Nucleation Site Activation in Saturated Boiling," Journal of Heat Transfer, Vol. 1, Pt. 97C, Feb. 1975, pp. 93–98.

[14]Grant, I. D. R., and Patten, T. D., "Thickness of the Thermal Layer at the Initiation of Nucleate Pool Boiling," Proceedings of the Institute of Mechanical Engineers, Vol. 180, Pt. 3C, 1965, pp. 124–134.

[15]Wagner, L. Y., and Lykoudis, P. S., "Mercury Pool Boiling Under the Influence of a Horizontal Magnetic Field," International Journal of Heat and Mass Transfer, Vol. 24, No. 4, 1981, pp. 635–644.

[16]Nghiem, L., Merte, H., Jr., Winter, E. R. F., and Beer, H., "Prediction of Transient Inception of Boiling in Terms of a Heterogeneous Nucleation Theory," Journal of Heat Transfer, Vol. 103, No. 1, 1981, pp. 69–73.

[17]Gaertner, R. F., and Westwater, J. W., "Populations of Active Sites in Nucleate Boiling Heat Transfer," CEP Symposium, Vol. 46, No. 30, 1960, p. 39.

[18]Singh, A., Mikic, B. B., and Rohsenow, W. M., "Active Sites in Boiling," Journal of Heat Transfer, Vol. 98, Ser. C, No. 3, 1976, pp. 401–406.

[19]Marcus, B. D., and Dropkin, D., "Measured Temperature Profiles Within the Superheated Boundary Layer Above a Horizontal Surface in Saturated Nucleate Pool Boiling of Water," Journal of Heat Transfer, Vol. 87C, No. 3, 1965, pp. 333–341.

[20]Graham, R. W., Hendricks, R. C., and Ehlers, R. C., "An Experimental Study of the Pool Heating of Liquid Hydrogen in the Subcritical and Supercritical Pressure Regimes Over a Range of Accelerations," NASA TM-X-52039, 1964.

[21]Judd, R. L., and Lavadas, C. H., "The Nature of Nucleation Site Interaction," Journal of Heat Transfer, Vol. 102, No. 3, 1980, pp. 461–464.

[22]Sultan, M., and Judd, R. L., "Interaction of the Nucleation Phenomena at Adjacent Sites in Nucleate Boiling," Journal of Heat Transfer, Vol. 105, No. 1, 1983, pp. 3–9.

[23]Judd, R. L., and Merte, H., Jr., "Influence of Acceleration on Subcooled Nucleate Pool Boiling," Proceedings of the 4th International Heat Transfer Conference, Vol. VI, Paper B.8.7, Elsevier, Amsterdam, 1970.

[24]Skinner, L. A., "Vapor Bubble Growth on a Semi-Infinite Solid," Zeitschrift fuer Angewandte Mathematik und Physik, Vol. 19, No. 6, 1968, pp. 833–843.

[25]Tokarev, V. M., and Bogovin, A. A., "Experimental Study of Dynamic Effects in Bubble Boiling of Liquids," Journal of Engineering Physics, Vol. 37, No. 6, 1980, pp. 1420–1423.

[26]Zysin, L. V., and Feldberg, L. A., "Experimental and Analytic Study of the Shapes of Vapor Bubbles in Subcooled Boiling," Heat Transfer-Soviet Research, Vol. 14, No. 3, 1982, pp. 64–70.

[27]Lamb, H., Hydrodynamics, 6th ed., Dover, 1945, p. 122.

[28]Fritz, W., and Ende, W., "On the Vaporization Process of Vapor Bubbles with Kinematic Photography," Physikalische Zeitschrift, Vol. 37, 1936, pp. 391–396.

[29]Forster, H. K., "On the Conduction of Heat into a Growing Vapor Bubble," Journal of Applied Physics, Vol. 25, 1954, p. 1067.

68   H. MERTE JR.

[30] Birkhoff, G., Margulies, R. S., and Horning, W. A., "Spherical Bubble Growth," *Physics of Fluids*, Vol. 1, 1958, p. 201.

[31] Plesset, M. S., and Zwick, S. A., "A Non-Steady Heat Diffusion Problem with Spherical Symmetry," *Journal of Applied Physics*, Vol. 23, 1952, p. 95.

[32] Scriven, L. E., "On the Dynamics of Phase Growth," *Chemical Engineering Science*, Vol. 10, 1959, p. 1.

[33] Florschuetz, L. W., and Chao, B. T., "On the Mechanics of Vapor Bubble Collapse," *Journal of Heat Transfer*, Vol. 87C, 1965, pp. 209–220.

[34] Griffith, P., "Bubble Growth Rates in Boiling," *Transactions of the ASME*, Vol. 80, 1959, p. 721.

[35] Zuber, N., "The Dynamics of Vapor Bubbles in Non-Uniform Temperature Fields," *International Journal of Heat and Mass Transfer*, Vol. 2, 1961, pp. 83–105.

[36] Mikic, B. B., Rohsenow, W. M., and Griffith, P., "On Bubble Growth Rates," *International Journal of Heat and Mass Transfer*, Vol. 13, 1970, pp. 657–666.

[37] Bahr, A., "Significance of Microconvection Induced by Vapor Bubbles on Heat Transfer With Boiling," *Chem. Ing. Techn.*, Vol. 38, No. 9, 1966, pp. 922–925.

[38] Fedotkin, I. M., and Gulakov, A. V., "Kinetics of Boiling and Bubbling," *Fluid Mechanics-Soviet Research*, Vol. 2, Pt. II, 1982, pp. 96–101.

[39] Moore, F. D., and Mesler, R. B., "The Measurement of Rapid Surface Temperature Fluctuations During Nucleate Boiling of Water," *AIChE Journal*, Vol. 7, 1961, pp. 620–629.

[40] Han, C.-Y., and Griffith, P., "The Mechanism of Heat Transfer in Nucleate Pool Boiling," *International Journal of Heat and Mass Transfer*, Vol. 8, 1965, pp. 887–914.

[41] Kao, Y. S., and Kenning, D. B. R., "Thermocapillary Flow Near a Hemispherical Bubble on a Heated Wall," *Journal of Fluid Mechanics*, Vol. 53, No. 4, 1972, pp. 715–735.

[42] Baranenko, V. I., and Chichkan, L. A., "Thermocapillary Convection in the Boiling of Various Fluids," *Heat Transfer-Soviet Research*, Vol. 12, No. 2, 1980, pp. 40–44.

[43] Kaznovskii, S. P., et al., "Reactive Forces Exerted on an Expanding Vapor Bubble by Evaporation Molecules," *High Temperature*, Vol. 14, No. 5, 1977, pp. 894–900.

[44] Blangetti, F., and Naushahi, M. K., "Influence of Mass Transfer on the Momentum Transfer in Condensation and Evaporation Phenomena," *International Journal of Heat and Mass Transfer*, Vol. 23, No. 12, 1980, pp. 1694–1695.

[45] Schwartz, S., "Saturated Pool Boiling of Water in a Reduced Gravity," PhD Dissertation, University of Southern California, Los Angeles, CA, 1966.

[46] Beckman, W. A., and Merte, H., Jr., "A Photographic Study of Boiling in an Accelerating System," *Journal of Heat Transfer*, Vol. 87C, Aug. 1965, pp. 374–380.

[47] Cochran, T. H., et al., "Experimantal Investigation of Nucleate Boiling Bubble Dynamics in Normal and Zero Gravities," NASA TN D-4301, Feb. 1968.

[48] Keshock, E. G., and Siegel, R., "Forces Acting on Bubbles in Nucleate Boiling Under Normal and Reduced Gravity Conditions," NASA TN D-2299, 1964.

[49] Haggard, J. B., Jr., "Unstable Bubble Motion Under Low Gravitational Conditions," NASA TN D-5809, 1970.

[50] Fritz, W., "Calculation of the Maximum Volumes of Vapor Bubbles," *Physikische Zeitschrift*, Vol. 36, 1935, pp. 379–384.

[51] Staniszewski, B. E., "Nucleate Boiling Bubble Growth and Departure," Massachusets Institute of Technology, Cambridge, MA, TR-16, 1959.

[52] Cochran, T. H., and Aydelott, J. C., "Effects of Subcooling and Gravity Level on Boiling in the Discrete Bubble Region," NASA TN-D-3449, Sept. 1966.

[53] Cochran, T. H., and Aydelott, J. C., "Effects of Fluids Properties and Gravity Level on Boiling in a Discrete Bubble Region," NASA TN-D-4070, 1967.

[54] Miksis, M. J., Vanden-Broeck, J. M., and Keller, J. B., "Rising Bubbles," *Journal of Fluid Mechanics*, Vol. 123, Jan. 1982, pp. 31–42.

[55] Siekmann, J., Eck, W., and Johann, W., "On Bubble Motion Through Liquid Under Reduced Gravity," *Archives of Mechanics*, Vol. 28, No. 5/6, 1976, pp. 795–801.

[56] Drew, D., Cheng, L., et al, "The Analysis of Virtual Mass Effects in Two-Phase Flow," *International Journal of Multiphase Flow*, Vol. 5, No. 4, 1979, pp. 233–242.

[57] Bratukhin, Y. K., and Evdokimova, O. A., "Motion of Gas Bubbles in a Nonuniformly Heated Liquid," *Fluid Dynamics*, Vol. 14, No. 5, 1980, pp. 679–681 (translated by Consultants Bureau).

[58] Lee, S. H., and Le Al, L. G., "Motion of a Sphere in the Presence of a Plane Interface. Part Two: An Exact Solution in Bipolar Coordinates," *Journal of Fluid Mechanics*, Vol. 98, No. 1, 1980, pp. 193–224.

[59] Clark, J. A., and Merte, H., "Boiling Heat Transfer to a Cryogenic Fluid in Both Low and High Gravity Fields," *Proceedings of XIth International Congress of Refrigeration*, Munich, FGR, Aug. 27–Sept. 4, 1963.

[60] Clark, J. A., Merte, H., et al, "Pressurization of Liquid Oxygen Containers," Dept. of Mechanical Engineering, University of Michigan Contract NAS-8-825 with NASA George C. Marshall Space Flight Center, Huntsville, AL, Progress Rept. 6, Rept. 04268-8-P, April 1964.

[61] Merte, H., Jr., "Liquid Metal Boiling in Agravic Fields," *Investigation of Liquid Metal Boiling Heat Transfer*, Air Force Aero Propulsion Lab., Wright-Patterson AFB, OH, AFAPL-TR-66-85, Jan. 1967.

[62] Merte, H., Jr., "Incipient and Steady Boiling of Liquid Nitrogen and Liquid Hydrogen Under Reduced Gravity," Heat Transfer Lab., Dept. of Mechanical Engineering, University of Michigan, Ann Arbor, TR-7, ORA Rept. 07461-51-T, Contract NAS8-220228, Nov. 1970.

[63] Merte, H., Jr., and Clark, J. A., "Boiling Heat Transfer with Cryogenic Fluids at Standard, Fractional and Near-Zero Gravity," *Journal of Heat Transfer*, Vol. 86C, Aug. 1964, pp. 315–319.

[64] Merte, H., Jr., Littles, J. W., and Oker, E., "Boiling Heat Transfer to LN_2 and LH_2: Influence of Surface Orientation and Reduced Body Forces," *Proceedings of the XIIIth International Congress of Refrigeration*, Commission 1, Paper 1.62, 1971.

[65] Oker, E., and Merte, H., "Semi-Transparent Gold Film as Simultaneous Surface Heater and Resistance Thermometer for Nucleate Boiling Studies," *Journal of Heat Transfer*, Vol. 103, No. 1, 1981, pp. 65–68.

[66] Merte, H., Jr., and Clark, J. A., "Pool Boiling in an Accelerating System," *Journal of Heat Transfer*, Vol. 83C, Aug. 1961, pp. 233–242.

[67] Zuber, N., "Hydrodynamic Aspects of Boiling Heat Transfer," U.S. Atomic Energy Commission Rept. AECU-4439, 1959.

[68] Rohsenow, W. M., "Boiling," *Handbook of Heat Transfer, Fundamentals*, McGraw-Hill, New York, 1985, Chap. 12, pp. 12-1–12-94.

Chapter 2. Transport Phenomena in Crystal Growth

Application of Energy Stability Theory to Problems in Crystal Growth

G. P. Neitzel* and D. F. Jankowski†
Arizona State University, Tempe, Arizona

Introduction

TECHNIQUES for manufacturing semiconductor materials to be used as substrates for microelectronic devices are numerous and varied in their approaches. Bulk crystal growth methods such as Bridgman-Stock-barger, Czochralski, and float zoning produce ingots of material that are subsequently sliced and processed into wafers; more recent techniques such as metal-organic chemical vapor deposition (MOCVD), liquid phase epitaxy (LPE), and molecular beam epitaxy (MBE) are concerned with the direct deposition of substrate layers on underlying material. However, most microelectronic material that is presently available is still being produced by decades-old crystal growth methods that rely on the solidification of bulk melts. These methods are capable of producing large quantities of material at low cost. Recent advances in microelectronics technology, particularly in the area of miniaturization, are imposing stricter requirements on the microstructure of substrate materials, and these require concomitant advances in the technology that produces them. Such advances in the bulk crystal growth methods mentioned earlier are unlikely to result simply from further trial-and-error engineering approaches, since these have been pushed to their limits in many years of usage. Rather, significant improvements will require increased knowledge of the transport processes associated with these methods.

*Professor, Department of Mechanical and Aerospace Engineering; currently at The George W. Woodruff School of Mechanical Engineering, Georgia Institute of Technology, Atlanta, GA.
†Professor, Department of Mechanical and Aerospace Engineering.

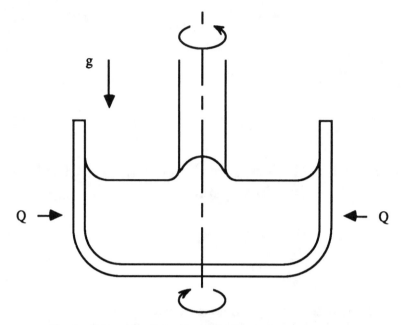

Fig. 1　Schematic of the Czochralski crystal growth process.

There is already a significant amount of literature on various fluid dynamic aspects of several crystal growth processes.[1] The flow, thermal, and solute environments associated with these can be quite complex. Consider, for example, the Czochralski process depicted schematically in Fig. 1. A crucible containing molten material, e.g., silicon, is heated from the outside. A seed crystal of known crystallographic orientation is placed in contact with the free surface of the melt and withdrawn at a variable rate until the diameter of the solidified crystal increases to the desired size. The process continues until the supply of material in the crucible is exhausted. In order to minimize the effect of unavoidable asymmetry in the furnace, both the crystal and crucible are rotated, usually in opposite directions. For the purposes of this paper we shall ignore convective mechanisms associated with the solidification process,[2] although it is acknowledged that these are of paramount importance since they most directly impact the constitution of the material. Instead, we concentrate on larger-scale effects that can also influence material quality, although the type of analysis we shall later describe is not restricted to these effects.[3]

　The flow in the bulk melt within a Czochralski crucible is the result of several forms of convection. Associated with the counter-rotation of the crystal and crucible is a form of forced convection driven by an imbalance between the local centrifugal force and the prevailing pressure gradient force. Such convection is also observed in the case of a disk rotating in still fluid. It is well known that the flow due to a rotating disk can become unstable, resulting in the appearance of waves at the surface of the disk (see, for example, the photograph by Gregory et al.[4] appearing as the frontispiece to the book edited by Rosenhead[5]). Buoyancy forces also drive convection due to the existence of a radial temperature gradient that is perpendicular to the axial orientation of the gravitational field. An instability of this type of convection can appear as an onset of unsteadiness in the flowfield. This oscillatory buoyant convection has been studied extensively in recent years.[6] Finally, the existence of a free surface provides yet another avenue for convection. It is known that the surface tension of liquids is a function of temperature, in most cases, decreasing with increasing surface temperature. Since there is a surface temperature gradient on the free surface in a Czochralski melt, there is a corresponding surface tension gradient that causes motion of the free surface from hot to cold regions. Because the underlying fluid is viscous, this surface motion induces a bulk motion known as "thermocapillary convection." Recent experiments[7] for a model of the float-zone crystal growth process have demonstrated that steady thermocapillary convection can also undergo a transition to an oscillatory mode of convection. The appearance of any of these flow instabilities can provide a nonuniform flow environment that, when coupled with the solidification process, may lead to the appearance of undesirable nonuniformities in the final material.

Therefore, the study of the stability properties of various types of convection associated with crystal growth processes is a worthwhile endeavor with regard to the refinement of the processes and the production of better quality material. Several such analyses and experiments have been performed. Most theoretical efforts have employed either direct numerical simulation of the flowfields or linear stability theory. Linear stability theory provides a sufficient condition for instability of a given basic state. That is, a value of a relevant, adjustable dimensionless parameter is obtained, above which the flow is guaranteed to be unstable to ever-present infinitesimal disturbances. It is well known, however, that some flows experience subcritical instability; i.e., they can become unstable to finite-amplitude disturbances at lower values of the same parameter. Therefore, knowledge of the linear stability limit alone cannot guarantee that a flow will remain stable. For this reason, we have chosen to employ an alternate stability theory for the analysis of a basic state of relevance to crystal growth melts. Energy stability theory adopts a different approach to testing the stability of a basic state and results in a sufficient condition for stability to disturbances of arbitrary amplitude. If this condition, in terms of an adjustable dimensionless parameter, is such that a process can be operated at parametric values below this limit, then the absence of instability can be guaranteed. Such a result could be of technological value in several crystal growth processes.

In subsequent sections we shall outline the use of energy stability theory and justify reasons for its use in terms of the convection mechanisms described earlier. We then present some results of calculations of our own for a particular basic state of relevance to the float zone process. Finally, we discuss extensions that are necessary for further study.

Stability Theory

Hydrodynamic stability theory is concerned with determining the response of a given basic state, which, in the most general case, involves velocity, pressure, temperature, and concentration fields, to disturbances. Such disturbances are always present due to the effects of surface roughness, imperfect thermal control, vibrations, and other sources. The mathematical approach to all stability problems begins with the determination of the basic state, which is a solution to the appropriate governing equations. For some situations, such as viscous fluid at rest between a pair of differentially heated horizontal planes, this may be trivial. For others it may be possible to obtain an analytical representation of the basic state. However, for more complex flows in finite geometries, it is necessary to resort to numerical methods for this purpose. It is only in recent years that sufficient progress in computational algorithms and computer hardware has been made so that these may be performed[8,9] with reasonable effort. An example of a numerically generated basic state[10] for a model suggested by the float-zone process is shown in Fig. 2. One of the complicating features of this computation is the presence of an unknown, deformable free surface.

Following the determination of the basic state, the next step is the derivation of a set of disturbance equations, performed by assuming that any scalar or vector dependent variable q can be represented as

$$q(x,t) = Q(x) + q'(x,t)$$

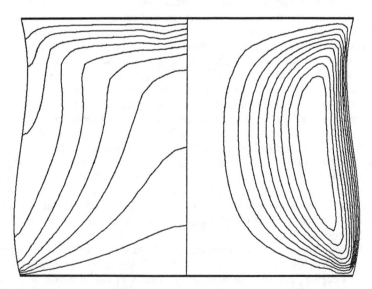

Fig. 2 Isotherms and meridional plane streamlines for a model half-zone with a deformable free surface.

where Q is the basic state quantity, q' is a disturbance to it, x is the spatial (Eulerian) position, and t is time. Although it is not necessary, we have assumed that the basic state quantities are steady. This assumed solution is substituted into the governing nonlinear equations, resulting in a set of exact disturbance equations, the solution of which will describe the behavior of the disturbances as functions of time and space. For a problem governed in part by the Navier-Stokes or simplified Boussinesq equations, these disturbance equations will be nonlinear due to the presence of the convective terms or free surfaces. The two primary means for analyzing these equations differ in the way with which these nonlinear terms are treated.

Linear Stability Theory

Linear stability theory assumes that the magnitude of any disturbance quantity is small in comparison to its basic state counterpart, i.e., $|q'| \ll |Q|$, so that products of disturbances are small in comparison to disturbances alone and may be neglected. Hence, the disturbance equations are linearized. For steady (or time-periodic) basic state, the Floquet theory then allows the decomposition of the solution into a product of two terms: one is independent of time (or periodic in time with the same period as the basic state), whereas the other exhibits an exponential time dependence of the form $\exp(ct)$. The real part of the growth rate c thus determines whether the assumed infinitesimal disturbances will grow (instability) or decay (stability) with time. Further assumptions (e.g., normal modes) usually result in the formulation of a boundary-value problem, the solution of which yields a value R_L of an adjustable parameter R above which disturbance growth is guaranteed. In some cases for which a loss of stability occurs, the imaginary part of c does not vanish when $R = R_L$. This type of transition, a so-called Hopf bifurcation (see, for example, Ref. 11), is observed in flows associated with crystal growth melts. Since linear theory has restricted its attention to perturbation of the basic state by infinitesimal disturbances, the growth of finite-amplitude disturbances at $R < R_L$ cannot be precluded. Therefore, the net result of a linear theory analysis of a given basic state is a sufficient condition for instability. A recent text by Drazin and Reid[12] provides several examples of the application of linear stability theory.

Energy Stability Theory

Energy stability theory dates to the work of Reynolds[13] and Orr,[14] but its modern revival is primarily due to Joseph.[15] The elegance of its mathematics has served as an attraction. The fact that the symmetry associated with energy stability theory carries over to its discrete counterparts is very convenient. More importantly, it will be argued that it is capable of yielding useful information. An energy theory investigation of a given basic state also begins with the exact disturbance equations, but then takes a different path. Instead of assuming the disturbance quantities to be infinitesimal, an exact evolution equation for a generalized energy functional is derived. For simplicity, let us assume that the governing equations are the momentum balance equations and an equation of energy. The energy identity is derived by taking the inner product of the disturbance velocity u' with the disturbance momentum equation, multiplying the disturbance energy equation by the disturbance temperature T', combining these using a coupling parameter λ, and integrating the result over a relevant volume V. Defining the (quadratic) energy functional E as

$$E = \int_V (u' \cdot u' + \lambda T'^2) \, dV$$

the equation is one for dE/dt, with the right-hand side being composed of various energy production and dissipation terms, depending on the problem being treated. The free parameter $\lambda > 0$ can be chosen for convenience; its role will be described in more detail later. The beauty of the energy stability theory is that the nonlinear convective terms $u' \cdot \nabla u'$ and $u' \cdot \nabla T'$ vanish exactly upon performing the integration, irrespective of their amplitudes. Physically, the reason for this is that, although these nonlinear terms are responsible for redistributing disturbance energy between various disturbance components, they do not contribute to the net increase or decrease of disturbance energy within V. Hence, this nonlinearity is eliminated without the necessity of restricting the amplitudes of disturbance quantities.

We are thus led to making a decision on the stability or instability of the basic state depending on whether the global quantity E decreases or increases with time. Unfortunately, the evaluation of the right-hand side of the energy identity requires knowledge of the disturbance quantities themselves, which is not available. However, one can usually construct an upper bound for the right-hand side and therefore for dE/dt. Stability can thus be guaranteed if this upper bound is, at most, zero.[16] The variational problem associated with this upper bound is usually treated by expressing it as a set of equivalent Euler-Lagrange equations that constitute an eigenvalue problem for R_E, a value of the adjustable parameter below which stability to finite-amplitude disturbances is guaranteed. However, the formulation as an Euler-Lagrange problem is not necessary. It is possible to attack the variational problem *directly*, and this is the approach that was taken for the example to be discussed soon.

Energy stability theory therefore provides a sufficient condition for stability, which is complementary to the sufficient condition for instability obtained from linear stability theory. These concepts are demonstrated in Fig. 3. Here R is the relevant adjustable parameter. The size of the region $R_E < R < R_L$ is a measure of the usefulness of these theories in yielding practical information. The global stability limit, R_G, which separates stability from instability and for which no general theory exists, lies in this region. It can be argued that $R_L - R_E$ is dependent on the physical mechanism responsible for the instability. Some specific results have been collected in Ref. 17. They suggest that, for buoyancy, thermocapillary, and centrifugal instabilities, which can be expected to be present in various

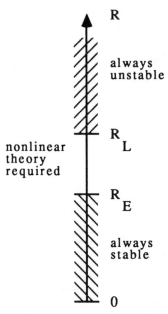

Fig. 3   Results of energy and linear stability calculations.

crystal growth melts, the results of energy theory can be useful in the sense of $R_L - R_E$ being of moderate size.

The next section discusses some of the details of the application of energy theory to a problem that is motivated by the float-zone crystal growth process. The actual float-zone process is too complicated to attempt an all-encompassing calculation at the present time. However, model experiments have identified an instability whose presence in the actual process could be expected to be detrimental to crystal quality. Thus, a study of the so-called half-zone model is a natural preliminary to further analysis of the complete float-zone process.

Model Half-Zone

Basic State

The half-zone model is intended to isolate thermocapillarity as the primary convection mechanism in a cylindrical geometry and, in addition, approximate the lower half of a float-zone melt. A schematic diagram is shown in Fig. 4. Two cylindrical rods of radius R with planar ends are oriented a distance H apart with their axes in the direction of gravity. A liquid zone is created between the ends and the rods are heated differentially, with the upper rod being at a higher temperature (T_H) than the lower (T_C). The temperature gradient along the free surface drives a bulk

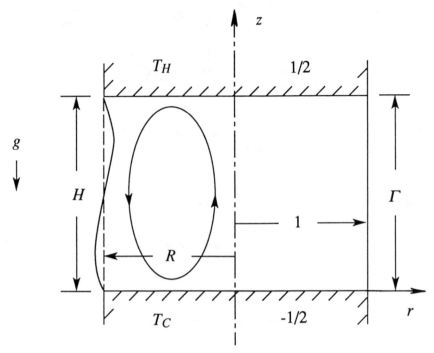

Fig. 4   Definition sketch for a model half-zone.

thermocapillary convection. Since the axial temperature gradient is vertically upward, the fluid is stably stratified in the axial direction with respect to buoyancy. In general, there will be flow- and heat-transfer-induced radial temperature gradients, that can cause buoyant convection. Several experiments have been performed both in full and microgravity environments[7,18-20] that have identified the potentially troublesome onset of oscillatory convection. The first theoretical stability studies[21,22] of this convection in a cylindrical geometry employed linear stability theory and assumed the aspect (length-to-radius) ratio of the zone to be asymptotically large, so that an analytical solution for the basic state could be found. The results of these analyses are linear limits that appear to be a couple of orders of magnitude below the experimentally obtained limits, and it was speculated by Xu and Davis that part of the reason for the disagreement may be the assumption of a long zone, which is subject to a larger class of disturbances than one of finite length. The analysis to be presented here employs a numerically determined basic state in a finite geometry.

In general, as shown in the left half of Fig. 4, the half-zone experiences free-surface deformation under the influence of gravity. For these first energy stability calculations for this basic state, surface deformation was neglected and the free surface is assumed to be the undeformed cylinder shown on the right half of the figure. For microgravity applications basic state calculations[10] show that there is little flow-induced surface deformation. However, energy stability calculations of buoyancy thermocapillary instability in planar layers[23] have shown that surface deformation may be stabilizing; thus, these computations for the half-zone are being pursued.

The liquid in the half-zone is assumed to be a Newtonian, Boussinesq fluid, and scales for length, velocity, and pressure are chosen to be R, $\gamma(T_H - T_C)/\mu$ and $\gamma(T_H - T_C)/R$, respectively, where γ is the (assumed constant) rate of decrease of surface tension σ with temperature, and μ is the coefficient of dynamic viscosity. A dimensionless temperature Θ is defined by

$$\Theta = (T - T_m)/(T_H - T_C)$$

where $T_m = (T_H + T_C)/2$ is the mean temperature. The dimensionless geometry and thermal end conditions are indicated on the right-hand side of Fig. 4, where $\Gamma = H/R$ is the aspect ratio of the zone.

The usual kinematic and no-slip conditions are applied and the temperature is assumed constant at the fluid/solid interfaces. On the free surface, a shear stress balance is required, together with a heat-transfer boundary condition; this is taken to be a convective condition of the form

$$\Theta_r = -Bi[\Theta - \Theta_a(z)]$$

where the subscript r denotes partial differentiation, Bi is a Biot number, and $\Theta_a(z)$ is an assumed external temperature distribution. This condition is convenient from the point of view of the application of energy stability theory, but clearly must be extended to the case of a nonlinear radiation condition. The difficulty associated with this will be discussed later.

The details of the finite-difference, basic state calculations will not be provided here, but may be found in Ref. 24. Plots of the meridional plane streamlines show the presence of a single, toroidal cell and distortion of the isotherms from a pure conduction state similar to that seen in Fig. 2, but without free-surface deformation. This distortion depends primarily on the Prandtl number $Pr = \nu/\kappa$, where ν and κ are the kinematic viscosity and thermal diffusivity, respectively, and the strength of the motion through the Marangoni number

$$Ma = \gamma(T_H - T_C)R/\mu\kappa$$

or, equivalently, by the Reynolds number $Re = Ma/Pr$.

Energy Stability Limit

The derivation of the energy identity follows the usual procedure with a single exception: Only axisymmetric disturbances are considered; i.e., all disturbance quantities are independent of the azimuthal coordinate and there is no disturbance velocity component in this direction. This restriction, which was originally necessary in light of available computational resources, is also relevant to some experimental observations. Further implications of this assumption will be discussed later.

The energy identity is

$$E^{-1}\, dE/dt = E^{-1}(-PrD - MaI + PrJ)$$

where the energy functional (modified by the inclusion of the Prandtl number) is

$$E = \int_V (u' \cdot u' + \lambda Pr T'^2)\, dV$$

where $D \geqslant 0$ represents energy dissipation, and I and J represent volume and surface energy production, respectively. The definitions of these terms may be found in Ref. 24. The integral I is dependent on the basic state, and this complicates the calculation of the energy limit, which, in the present formulation, is based upon the Marangoni number.

The right-hand side of the energy identity, called F, can be bounded from above over a space of kinematically admissible functions[15] that includes the physical disturbances as a subset. If we require this upper bound to be, at most, zero, this will ensure that $E \rightarrow 0$ as $t \rightarrow \infty$. Thus, we seek Ma_E as the minimum positive value of Ma that accomplishes this. The coupling parameter λ is a free parameter and can be chosen to maximize Ma_E. It should be pointed out that, although we are referring to this value as Ma_E, it is really an *upper bound* to the true stability limit, because of the assumption of axisymmetric disturbances and the possibility that non-axisymmetric ones could grow for Ma less than our calculated Ma_E.

The actual calculation of Ma_E is based on a discrete version F_h of the continuous, quadratic functional F, obtained by the application of finite-difference concepts. The condition that F_h be a stationary leads to a generalized algebraic eigenvalue problem

$$(A - MaB)x = 0$$

where A and B are sparse, symmetric, indefinite matrices with B depending upon the basic state. Because of the specific structure of A and B, the extraction of Ma_E from this algebraic system is a matter of some computational difficulty; the various details, as well as checks on the numerical procedure, are discussed in Ref. 24.

Results

One of the most interesting and unexpected aspects of the numerical results is the important influence of the coupling parameter λ. Typically, in energy stability analyses a value of this parameter is assigned a priori, with the understanding that the result obtained will be a lower bound to the true

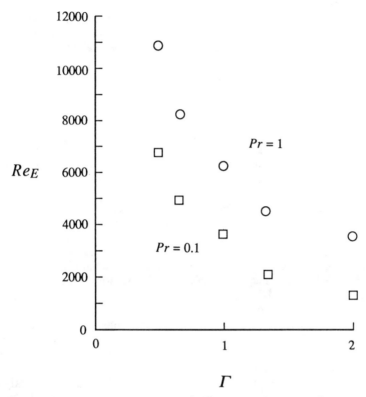

Fig. 5   Axisymmetric energy limits vs Prandtl number and aspect ratio.

energy limit. Here a maximization procedure was used to obtain the optimum value of λ. This was found to be a worthwhile endeavor, with order-of-magnitude improvements being made. The optimizing values of λ were usually very small. Because of the way that the coupling parameter enters the definition of the energy functional E, this probably means that the hydrodynamic disturbances are more important than the temperature disturbances.

Some specific results are shown in Fig. 5. These are calculated for a microgravity environment and show the variation of $Re_E = Ma_E/Pr$ with the half-zone aspect ratio for two different values of the Prandtl number. The smaller value of Pr is very roughly representative of the values for microelectronic materials, whereas the larger value is closer to that appropriate to some model experiments. Since it is not yet possible to perform the computations at $Pr = 7$, a rough extrapolation of the present results was necessary to make comparisons with the experiments of Ref. 7. This extrapolation implies that the computed axisymmetric energy limits would lie above the experimental results. In turn, this demonstrates the necessity of computing stability limits for general disturbances. However, the fact that these computed limits are as large as they are is very promising, since even a significant reduction because of three dimensionality could still yield stability limits in close agreement with these model experiments. If this desirable situation occurs, then the results of the calculations may be useful in bridging the gap in Prandtl number between real and model liquids.

Discussion

The results of an energy stability analysis of a model half-zone are encouraging. There is significant potential for these concepts to yield technologically useful information when applied to convective basic states associated with crystal growth processes. Most progress has been made with regard to the float-zone process, but even here there are substantial, nontrivial extensions that should still be implemented. It can be expected that the following topics have important implications with regard to the stability characteristics of crystal-growing processes in general: 1) three-dimensional disturbances, 2) radiation heat transfer to or from the free surface, 3) deformable free surfaces, 4) melting and solidification interfaces, 5) three-dimensional basic states induced by rotation and/or nonuniform heating, and 6) free-surface contamination. Extensions to a real processes are desirable in the hopes of providing the crystal grower with specific recommendations for choosing operating conditions that will yield striation-free material.

Our current research is concerned with the first three of these topics as applied to a model half-zone. The extension to include three-dimensional disturbances does not have any additional conceptual or analytical difficulties, but does require increased computer resources and changes in the numerical algorithms to reflect the presence of Hermitian, rather than real, symmetric matrices. Nonlinear radiation boundary conditions *do* represent an analytical challange, because the quadratic nature of the functional F is destroyed. One anticipates, in this case, that the energy stability limit will be *conditional*,[15] i.e., dependent on the amplitude of the disturbances. Finally, the general case of a deformable free surface can be approached in approximate ways. Most simply, the disturbances can be assumed to not further deform the basic-state free surface. Conceptually, this problem is identical to the half-zone problem discussed earlier. A more difficult problem allows the disturbances to deform the free surface. It is for this case that Davis and Homsy[23] found increased energy limits for planar layers. Like the radiation problem, this also leads to a conditional stability limit.

Considerable effort is currently being directed toward computation of what we have called basic states of relevance to various crystal growth processes. Linear stability calculations for these states are also receiving increasing attention due, in part, to the maturation of numerical bifurcation theory. Both finite-difference[25] and finite-element[11] approaches have been successfully implemented. There is no doubt that such information can be useful, but in a somewhat limited way. The possible presence of a subcritical instability requires additional information if a guarantee of stability is desired, as it clearly is in the crystal growth processes. Energy stability theory is currently the only possible *rigorous* source for this complementary information.

Acknowledgments
We wish to acknowledge the collaboration of our colleagues Y. Shen, H. D. Mittelmann, and J. R. Hyer in the research which served as the basis for this paper. Support for this research has been provided by the Microgravity Science and Applications Division of the National Aeronautics and Space Administration and by the National Science Foundation (GPN).

References
[1]Ostrach, S., "Fluid Mechanics in Crystal Growth—the 1982 Freeman Scholar Lecture." *ASME Journal of Fluids Engineering*, Vol. 105. 1983, pp. 5–20.

[2]Davis, S. H., "Hydrodynamic Interactions in Directional Solidificaiton," *Journal of Fluid Mechancis*, Vol. 212, 1990, pp. 241–262.

[3]Braun, R. J., and Davis, S. H., "Energy Stability for Directional Solidification," *Bulletin of the American Physical Society*, Vol. 34, 1989, p. 2310.

[4]Gregory, N., Stuart, J. T., and Walker, W. S., "On the Stability of Three-Dimensional Boundary Layers with Application to the Flow due to a Rotating Disk," *Philosophical Transactions of the Royal Society of London, Series A*, Vol. 248, 1955, pp. 155–199.

[5]Rosenhead, L. (ed.), *Laminar Boundary Layers*, Oxford Univ. Press, Oxford, UK,1963.

[6]Bottaro, A., and Zebib, A., "Bifurcation in Axisymmetric Czochralski Natural Convection," *Physics of Fluids*, Vol. 31, 1988, pp. 495–501.

[7]Preisser, F., Schwabe, D., and Scharmann, A., "Steady and Oscillatory Thermocapillary Convection in Liquid Columns with Free Cylindrical Surface," *Journal of Fluid Mechanics*, Vol. 126, 1983, pp. 545–567.

[8]Cuvelier, C., and Driessen, J. M., "Thermocapillary Free Boundaries in Crystal Growth," *Journal of Fluid Mechanics*, Vol. 169, 1986, pp. 1–26.

[9]Sackinger, P. A., Brown, R. A., and Derby, J. J., "A Finite Element Method for Analysis of Fluid Flow, Heat Transfer and Free Interfaces in Czochralski Crystal Growth," *International Journal for Numerical Methods in Fluids*, Vol. 9, 1989, pp. 453–492.

[10]Hyer, J. R., Jankowski, D. F., and Neitzel, G. P., "Thermocapillary Convection in a Model Float-Zone," *Journal of Thermophysics and Heat Transfer*, 1991 (to be published).

[11]Winters, K. H., "Oscillatory Convection in Liquid Metals in a Horizontal Temperature Gradient," *International Journal for Numerical Methods in Engineering*, Vol. 25, 1988, pp. 401–414.

[12]Drazin, P. G., and Reid, W. H., *Hydrodynamic Stability*, Cambridge Univ. Press, Cambridge, UK, 1981.

[13]Reynolds, O., "On the Dynamical Theory of Incompressible Viscous Fluids and the Determination of the Criterion," *Philosophical Transactions of the Royal Society of London, Series A*, Vol. 186, 1895, pp. 123–164.

[14]Orr, W. M., "The Stability or Instability of the Steady Motions of a Liquid. Part II: A Viscous Liquid," *Proceedings of the Royal Irish Academy, Series A*, Vol. 27, 1907, pp. 9–138.

86   G. P. NEITZEL AND D. F. JANKOWSKI

[15]Joseph, D. D., *Stability of Fluid Motions*, Vols. I and II, Springer-Verlag, Berlin, 1976.

[16]Davis, S. H., and von Kerczek, C., "A Reformulation of Energy Stability Theory," *Archive for Rational Mechanics and Analysis*, Vol. 52, 1973, pp. 112–117.

[17]Davis, S. H., "Energy Stability of Unsteady Flows," *Proceedings of the IUTAM Symposium on Unsteady Boundary Layers*, Laval Univ., Quebec, 1971.

[18]Chun, C.-H., "Experiments on Steady and Oscillatory Temperature Distribution in a Floating Zone due to the Marangoni Convection," *Acta Astronautica*, Vol. 7, 1980, pp. 479–488.

[19]Schwabe, D., Preisser, F., and Scharmann, A., "Verification of the Oscillatory State of Thermocapillary Convection in a Floating Zone Under Low Gravity," *Acta Astronautica*, Vol. 9, 1982, pp. 265–273.

[20]Kamotani, Y., Ostrach, S., and Vargas, M., "Oscillatory Thermocapillary Convection in a Simulated Float-Zone Configuration," *Journal of Crystal Growth*, Vol. 66, 1984, pp. 83–90.

[21]Xu, J.-J., and Davis, S. H., "Convective Thermocapillary Instabilities in Liquid Bridges," *Physics of Fluids*, Vol. 27, 1984, pp. 1102–1107.

[22]Xu, J.-J., and Davis, S. H., "Instability of Capillary Jets with Thermocapillarity," *Journal of Fluid Mechanics*, Vol. 161, 1985, pp. 1–25.

[23]Davis, S. H., and Homsy, G. M., "Energy Stability for Free-Surface Problems: Buoyancy-Thermocapillary Layers," *Journal of Fluid Mechanics*, Vol. 98, 1980, p. 527–553.

[24]Shen, Y., Neitzel, G. P., Jankowski, D. F., and Mittelmann, H. D., "Energy Stability of Thermocapillary Convection in a Model of the Float-Zone, Crystal-Growth Process," *Journal of Fluid Mechanics*, Vol. 217, 1990, pp. 639–660.

[25]Dijkstra, H. A., and van de Vooren, A. I., "Multiplicity and Stability of Steady Solutions for Marangoni Convection in a Two-Dimensional Rectangular Container with Rigid Sidewalls," *Numerical Heat Transfer, Part A*, Vol. 16, 1989, pp. 59–75.

Bridgman Crystal Growth in Low Gravity: A Scaling Analysis

J. Iwan D. Alexander* and Franz Rosenberger†
University of Alabama in Huntsville, Huntsville, Alabama

Nomenclature

c_m = solute concentration (melt)
c_m^0 = interfacial solute concentration (melt)
c_s^0 = interfacial solute concentration (crystal)
c_{smax} = maximum interfacial solute concentration (crystal)
c_{smin} = minimum interfacial solute concentration (crystal)
c_{sav} = laterally averaged interfacial solute concentration (crystal)
c_{av} = laterally averaged interfacial solute concentration (melt)
c_m^∞ = solute concentration far from the crystal-melt interface (melt)
$c(0)$ = maximum interfacial solute concentration (melt)
D = solute diffusivity (melt)
e_w = unit vector normal to ampoule wall
Gr = Grashof number
g = dimensionless effective gravity vector
g = gravitational acceleration magnitude
g_0 = 980 cm-s^{-2}, terrestrial gravitational acceleration
j = solute flux
k = distribution coefficient
k_{eff} = effective distribution coefficient
L = characteristic system dimension
L_a = adiabatic zone length
L_D = characteristic distance at which $Pe_g = 1$
N = unit vector normal to the crystal-melt interface

*Senior Research Scientist, Center for Microgravity and Materials Research.
†Professor and Director, Center for Microgravity and Materials Research.

Pe = solutal Peclet number
Pe_g = growth Peclet number
Pr = Prandtl number
R = ampoule radius
Re = Reynolds number
Sc = Schmidt number
s = dimensionless solute concentration (melt)
T = melt temperature
T_m = melting temperature of crystal
T_h = temperature of hot zone
ΔT = characteristic temperature difference
ΔT_r = radial temperature difference
t = dimensionless time
$\boldsymbol{u}$ = dimensionless velocity vector (melt)
u,v = dimensionless velocity components (melt)
V = rescaled velocity component $V = vPe^{\frac{1}{3}}$
V^* = characteristic speed (melt)
V_s = characteristic diffusive speed (melt)
V_0 = characteristic speed due to buoyancy (melt)
V_δ = characteristic speed in "steep gradient zone" due to buoyancy (melt)
V_g = crystal growth velocity
V_ω = characteristic rotational flow velocity (melt)
x_0 = parameter relating $Re_{\max}$ and Gr
x_c^* = dimensionless thickness of solute boundary layer
β = coefficient of thermal expansion (melt)
δ_v = momentum boundary-layer thickness
δ_c = solute boundary-layer thickness
ζ = stretched coordinate, Eq. (5)
κ = thermal diffusivity (melt)
ν = kinematic viscosity (melt)
ξ = measure of nonuniformity in interfacial composition
ξ_m = alternative measure of nonuniformity in interfacial composition
ρ_m = density (melt)
ρ_s = density (crystal)
σ = ratio of melt and crystal densities
θ = dimensionless temperature

Introduction

IN this paper we examine the role of dimensionless or scaling analyses as an interpretive and predictive tool for practitioners of crystal growth under low-gravity conditions. The motivation for the study of transport conditions in melts subject to low gravity arises from the current interest of crystal growers in taking advantage of the reduction in effective gravity, and thus, buoyancy-driven convection, afforded by laboratories in low-Earth orbit. It has been recognized for some time that convective effects in doped semiconductor melts can result in undesirable compositional variations in crystals grown from these melts.[1-5] In many of these cases, for example, growth by the Bridgman-Stockbarger technique, the convection is buoyancy driven.[6,7] Recent numerical calculations[8] indicate that, depending on the orientation of the interface in the residual gravity field, convective effects in Bridgman-type systems can still cause significant lateral compositional variations in the crystal even at steady effective gravity levels 10^{-6} times that experienced on Earth. Since numerical simulations are generally complex, order-of-magnitude scaling analyses have been popular tools for estimating the effects of residual accelerations characteristic of a space laboratory. Therefore, in this paper we examine the reliability of such analyses when applied to steady weakly convecting buoyancy flows, in comparison with our and other numerical results.

In practically all technologically interesting systems, solidification is associated with a change in composition. Components or dopants are either partly rejected or preferentially incorporated into the solid. This redistribution, or segregation, of dopants is of great importance for many aspects of materials preparation,[9-11] including, for example, the purification of materials, or the predetermination of the composition of the nutrient phase in order to achieve the desired dopant distribution in the solid. The compositional change at the interface can be described in terms of the interfacial segregation coefficient of a component, $k = c_s^0/c_m^0$, where c_s^0 and c_m^0 represent, respectively, the concentrations in the solid and liquid at the interface. The value of k reflects whether the solute is preferentially rejected ($k < 1$) or incorporated ($k > 1$) upon solidification. For cases in which local equilibrium prevails at the phase boundary, k is equal to the equilibrium distribution coefficient and is thus independent of the mass transfer kinetics.[11]

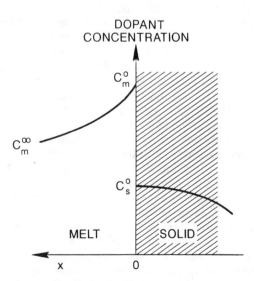

Fig. 1 Concentration profiles in the liquid and solid caused by partial rejection of a melt component upon solidification.

Mass-transfer rates in the liquid are typically slower than the solidification rates. Consequently, during growth of the solid the liquid cannot adjust its composition uniformly throughout, and compositional gradients will be established in both the solid and liquid phases (see Fig. 1). In addition, depending on the specific mass-transfer mode in the liquid, concentration gradients can develop parallel to the interface. This then leads to lateral compositional uniformities in the growing crystal. Similarly, any time dependence or nonuniformity in the growth rate (i.e., in the solute incorporation rate) will also result in compositional nonuniformity in the crystal.

In order to predict the operating conditions under which desirable dopant uniformities can be obtained, for example, for doped semiconductor crystals, it is necessary to understand the interplay between the various processes affecting mass transfer and their effect on the composition of the growing solid. We shall examine the results of numerical simulations of the effect of steady residual acceleration on the transport of solute (dopant) in a gallium-doped germanium (Ge : Ga) melt during directional solidification under low-gravity conditions. We shall interpret these results in terms of the relevant dimensionless groups associated with the process and evaluate scaling techniques by comparing their predictions with the numerical results.

We will demonstrate that, when convective transport is comparable with diffusive transport, some specific knowledge of the behavior of the system is required before scaling arguments can be used to make reasonable predictions. Preferably, such knowledge can be obtained from practical experience, or otherwise from numerical simulations. Even then, we emphasize that scaling cannot be expected to provide better than an order-of-magnitude accuracy. Scaling analyses, like all simplified models (analytical or numerical), will often suffer from the fact that they will fail to capture all the essential physics. This is especially true when the inherently one-dimensional scaling techniques are applied to intrinsically multidimensional problems. On the other hand, for essentially one-dimensional problems (for example, within boundary layers associated with high-Reynolds-number flows), scaling techniques can provide reasonable estimates.

Previous scaling analyses of transport conditions in Bridgman-type systems[12] have used the concepts of solute and momentum boundary layers in the development of their arguments. If a momentum boundary layer exists, then there are two basic flow regions present. One region extends from a rigid boundary some distance δ_v into the fluid. The second region meets the first at δ_v and extends into the *bulk* of the fluid. Prandtl[13] first recognized and rigorously justified the existence of such a situation. He used the term "boundary layer" for the region immediately adjacent to the rigid boundary. A boundary layer is thin in comparison to the characteristic dimension of the system,[14] and the velocity gradient across the layer is large. The dominant mechanism of momentum transfer within the layer is by diffusion. In other words, viscous forces are larger than inertial forces within the boundary layer. Outside this region, in the bulk of the fluid, inertial forces dominate and momentum transfer occurs mainly by convection; that is, the characteristic velocity of the flow, V^*, greatly exceeds the rate at which momentum is able to diffuse through the system. The momentum diffusion rate can be characterized as L/v. That is, in order for a momentum boundary layer to form, the Reynolds number, $Re = V*L/v$, which characterizes the bulk flow, must be much greater than one. For low Reynolds numbers, momentum transfer due to diffusion is significant throughout the *bulk* flow; i.e., it is everywhere at least as large as convective transfer (viscous forces are at least as large as inertial forces).

In order to evaluate the convective contribution to solute transfer, it is advantageous, in analogy to the Reynolds number for momentum transfer, to define a solutal Peclet number as the ratio of the convective and diffusive velocities.[9,11,15,16] That is, $Pe = V*/V_s = V*L/D$ (see also Table 1). Here $V*$ represents a characteristic velocity that may result from buoyancy-driven motion or merely from the translation of the ampoule through the furnace temperature gradient, and $V_s = D/L$ represents the characteristic diffusive speed.

In order for a solutal boundary layer to form in the system, the value of the bulk Peclet number must be much greater than one. For low-Peclet-number systems, solute transport due to diffusion is significant throughout the bulk of the fluid; i.e., it is everywhere at least as large as convective solute transport. In systems with a bulk $Pe \gg 1$, a solutal boundary layer is thus understood as the (interfacial) region of width δ_s in which the local Peclet number $Pe = V*\delta_s D \ll 1$. Note that, for planar interfaces, non-uniformities in the interfacial solute concentration can only occur if there is a component of velocity in the bulk that is parallel to the interface and convective solute transport there is comparable or larger than the

Table 1   Dimensionless groups associated with transport processes in the melt

Group	Physical interpretation	Definition
Solutal Pe	Convective solute transport rate / Solute diffusion rate	$Pe = \dfrac{V_0}{V_s} = \dfrac{V_0 L}{D} = ReSc$
Re	Inertial momentum transport rate / Viscous momentum transport rate	$Re = \dfrac{V_0 L}{v}$
Gr	Buoyancy force / Viscous force	$Gr = \dfrac{g\beta \Delta T L^3}{v^2}$
Sc	Momentum diffusivity / Solute diffusivity	$Sc = \dfrac{v}{D}$
Pr	Momentum diffusivity / Thermal diffusivity	$Pr = \dfrac{v}{\kappa}$
Growth Pe	Growth speed / Solute diffusion speed	$Pe_g = \dfrac{V_g L}{D}$

diffusive transport which would otherwise lead to homogeneous interfacial composition. In addition to Pe, it is also useful to define a Peclet number $Pe_g = V_g L/D$ that represents the ratio of the growth velocity to the diffusive velocities.

Conclusions that can be drawn from previous work related to transport in Bridgman systems will be described briefly in the next section. In the section on numerical analysis of crystal growth under low-gravity conditions, we shall summarize the results of our numerical analysis of the sensitivity of the Bridgman-Stockbarger technique to steady residual accelerations. The application of scaling techniques to this situation is discussed in the section on order-of-magnitude analysis. Two approaches are used: one simply expresses the results of the section on numerical analysis of crystal growth under low-gravity conditions, in terms of the relevant dimensionless parameters; the other approach follows the procedure outlined by Camel and Favier.[12]

Previous Work Related to Transport in Bridgman-Type Systems

Brown[6] discussed solute transport in a variety of melt growth systems. He presented a simple asymptotic theory (outlined later) which predicts that, when $Pe \gg 1$ and $Pe_g \sim \mathcal{O}(1)$, the melt composition takes the form of a solute boundary layer near a no-slip boundary and that the spatial variations in concentration at the melt-crystal interface should scale with $Pe^{-\frac{1}{3}}$.

To illustrate this scaling we refer to the work of Wilson,[17,18] Burton et al.,[19] and Levich,[15] and follow the arguments used by Brown.[6] Burton et al.[19] examined the segregation of one component of a two-component melt during directional solidification of an infinitely large uniformly rotating crystal. The melt velocity was taken to be the sum of the velocities due to the uniform growth rate and the stagnation flow caused by the disk rotation. Since, in this case the axial velocity is assumed to be independent of the radial coordinate,[20] it was assumed that the solute field depended only on the axial coordinate. These assumptions admit a closed-form solution of the equations. The solute concentration c_m^0 at the crystal-melt interface is then given by

$$\frac{c_m^0}{c_m^\infty} = k_{\text{eff}} = \frac{k}{k + (1 - k)\,\exp(-\Delta)} \tag{1}$$

where c_m^∞ is the concentration in the melt far from the interface, k is the interfacial distribution coefficient introduced earlier, and Δ is defined by

$$\Delta \equiv -\ell n\left[1 - \int_0^\infty \exp-\left(z + \frac{V_\omega}{3V_g}z^3\right)dz \right] \tag{2}$$

As pointed out by Wilcox[9] and re-emphasized by Wilson,[17,18] the quantity Δ, which is sometimes misidentified as the diffusion boundary-layer thickness, can only be identified with this thickness when $V_\omega \gg V_g$. Under such conditions,

$$\Delta \approx \frac{\Gamma(1/3)}{3} \left(\frac{V_\omega}{3V_g}\right)^{-\frac{1}{3}}$$ (3)

where $\Gamma(-)$ is the gamma function.[21] This result is analogous to that obtained by Levich,[15] and the scaling corresponds to that for a solute boundary layer adjacent to a no-slip wall.[6,15,22] Whereas the preceding situation is idealized, a generalized form of the Burton et al.[19] solution can be applied to regions where an intense stagnation flow impinges on a melt-crystal interface and creates a solute boundary layer. Brown[6] demonstrated the robust nature of the Burton et al.[19] solution as follows. Consider the steady form of the dimensionless solute transport equation

$$u\frac{\partial s}{\partial x} + v\frac{\partial s}{\partial y} = = \frac{1}{Pe}\left(\frac{\partial^2 s}{\partial x^2} + \frac{\partial^2 s}{\partial y^2}\right)$$ (4)

where x is parallel to the crystal interface, and y lies along the growth direction. Now assume that $Pe \gg 1$ and that $Pe_g \sim \mathcal{O}(1)$, and let

$$V = vPe^{\frac{2}{3}}, \qquad \zeta = yPe^{\frac{1}{3}}$$ (5)

Then Eq. (4) becomes

$$-\int_{-L/2}^{L/2} \frac{\partial V}{\partial \zeta}\,dx\,\frac{\partial s}{\partial x} + V\frac{\partial s}{\partial \zeta} = \frac{\partial^2 s}{\partial \zeta^2} + \mathcal{O}(Pe^{-\frac{2}{3}})$$ (6)

where u has been related to V through the integration of div$u = 0$. The boundary condition for solute conservation at the crystal-melt interface is then

$$\frac{\partial s}{\partial \zeta}(x,0) = Pe_g Pe^{-\frac{1}{3}}(k-1)s(x,0)$$ (7)

Equations (7) and (8) have solutions of the form[6]

$$s = s_0 + s_1(x,\zeta)Pe^{-\frac{1}{3}} + \mathcal{O}(Pe^{-\frac{2}{3}})$$ (8)

and the solute field in the boundary layer is seen to scale with $Pe^{-\frac{1}{3}}$. Solute fields that exhibit such scaling have been found in numerical computations when the laminar convection in the melt is intense.

There are several other instances in melt growth where the use of boundary-layer techniques has provided insight into transport conditions in intensely convecting melts.[24,25] However, for many crystal growth situations the multidimensional nature of the associated convective flows can cause difficulties for the application of boundary-layer analyses. For instance, in Czochralski crystal growth, Riley[26] notes that oxygen transport cannot be adequately described by boundary-layer analyses. Likewise, for

Bridgman-Stockbarger crystal growth, Brown[6] notes that the multicellular character of convection in high-aspect-ratio (L/R) Bridgman-type systems complicates the interpretation of boundary-layer analyses because the nearly constant level of solute in each convection cell is not known a priori. This would similarly hamper the application of order-of-magnitude scaling analyses.

Perhaps the most thorough combination of numerical modeling and scaling of a Bridgman-Stockbarger system to date has been undertaken by Chang and Brown[27] and Adornato and Brown.[23] Chang and Brown used a three-dimensional axisymmetric model of the growth of gallium-doped germanium (Ge : Ga) by the Bridgman-Stockbarger process. Their model is similar to that described in the next section, except that it included heat transfer in the crystal and allowed for curvature of the crystal-melt interface, and solidification took place in a circular cylinder. This work led to the important result that the transition from diffusion-controlled solute transport to weak convection is accompanied by a maximum in radial compositional nonuniformity at the crystal-melt interface. As the intensity of convection in the bulk is increased, the interfacial nonuniformity decreases as a boundary layer forms along the solidification interface. Adornato and Brown[23] extended these calculations. They also found a maximum in interfacial compositional nonuniformity at the transition between the diffusion-controlled and convection-controlled transport regimes. Furthermore, when the velocity scale was chosen to be $V_0 = [g\beta(\Delta T)_r R]^{\frac{1}{2}}$, which is that for a vertical buoyant layer,[28] they showed that, in the regime dominated by convective transport, the magnitude of the compositional nonuniformity scaled as $Pe^{-\frac{1}{3}} = (GrSc)^{-\frac{1}{3}}$. This is the scaling predicted by the simple asymptotic theory discussed earlier. These results have important consequences for crystal growth under low-gravity conditions: Unless the effective gravity level is reduced sufficiently, transport conditions could result in greater nonuniformity for the low-gravity experiments than under terrestrial conditions.

Camel and Favier[12] have outlined an elegant procedure for obtaining estimates of the transport regimes and lateral segregation in vertical Bridgman systems. The procedure is based on a comparison of the various dimensionless groups listed in Table 1 and centers upon the determination of the length scales for convection and diffusion.

For convection Camel and Favier identify flow regimes based on a comparison of the relative magnitudes of the dimensionless maximum flow velocity in terms of the Reynolds number $Re_{max} = V_{max}L/\nu$ and the Grashof number Gr for a known flow. For $Re_{max} \ll 1$ the maximum velocity and Gr were assumed to be linearly related by a factor x_0^2, where x_0 is interpreted as a length scale on which viscous forces and buoyancy forces are of the same order of magnitude. This length scale x_0 must be determined from numerical simulations or analytical approximations. The flow regimes are defined in Table 2.

Camel and Favier argue that, whenever convection is present, diffusion will still be the dominant mode of solute transport in a region close to the crystal-melt interface. The width of this region will vary locally according

Table 2 Classification of thermal buoyancy-driven flow regimes predicted by Camel and Favier[12]

Regime	Domain	Re
Stokes flow	$Gr < x_0^{-4}$	$x_0^2 Gr$
Boundary-layer flow	$Gr > x_0^{-4}$	$Gr^{\frac{1}{2}}$

Reynolds number represents the maximum characteristic velocity. x_0 is a parameter that must be determined empirically for a given system.

to the strength and direction of the convective flow. They define an interfacial region of thickness δ_c, in which the convective and diffusive contributions to solute transport are of the same order of magnitude. The nondimensional size of this region is $x_c^* = \delta_c / L$. Following the introduction of this length scale into the transport equations, the criteria for the determination of x_c^* are obtained and the transport regimes are defined by an order-of-magnitude comparison. The transport regimes depicted in Fig. 2 represent the different relationships between the solutal Peclet number $Pe = GrSc$ and the growth Peclet number Pe_g which arise from the order-of-magnitude analysis. An estimate of the amount of lateral composition variation in the grown crystals was then obtained upon consideration of the boundary condition (7) at the crystal-melt interface. A composition $c(0)$ was defined at the interface. For values of the distribution coefficient $k < 1$ (> 1), $c(0)$ should be the maximum (minimum) composition at the interface. Consequently, in their model the lateral nonuniformity cannot exceed $|c(0) - c_m^\infty|$, where c_m^∞ is again the far-field composition. Now the value of $\partial s / \partial x$ at the location of $c(0)$ is at least $|c(0) - c_m^\infty| / x_c^*$. The compositional nonuniformity may then be obtained from

$$\xi_m = \frac{|c(0) - c_m^\infty|}{c(0)} = (1 - k)x_c^* Pe_g \qquad (9)$$

where the value of x_c^* is determined by one of the transport regimes shown in Fig. 2. In the diffusive regime the nonuniformity is proportional to the ratio of convective velocities and the growth rate.

The predictions of solute transport conditions using the scaling techniques previously given have been compared with experimental and numerical results.[12,29] In some cases the estimates are in good agreement with numerical calculations, whereas for others the combination of $GrSc$ and Pe that leads to maximum compositional nonuniformity tends to be overestimated.[12] This is not surprising when one considers that the criteria for defining the transport regimes neglect the effects of lateral boundaries, corner flows,[30] and the details of the thermal profile that drives convection.

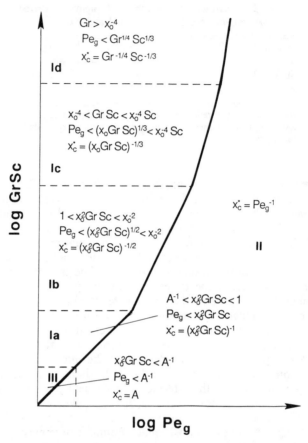

log GrSc

Id

$Gr > x_o^{-4}$
$Pe_g < Gr^{1/4} Sc^{1/3}$
$x_c^* = Gr^{-1/4} Sc^{-1/3}$

Ic

$x_o^{-4} < Gr\,Sc < x_o^{-4}\,Sc$
$Pe_g < (x_o Gr\,Sc)^{1/3} < x_o^{-4}\,Sc$
$x_c^* = (x_o Gr\,Sc)^{-1/3}$

$x_c^* = Pe_g^{-1}$

II

Ib

$1 < x_o^2 Gr\,Sc < x_o^{-2}$
$Pe_g < (x_o^2 Gr\,Sc)^{1/2} < x_o^{-2}$
$x_c^* = (x_o^2 Gr\,Sc)^{-1/2}$

$A^{-1} < x_o^2 Gr\,Sc < 1$
$Pe_g < x_o^2 Gr\,Sc$
$x_c^* = (x_o^2 Gr\,Sc)^{-1}$

Ia

$x_o^2 Gr\,Sc < A^{-1}$
$Pe_g < A^{-1}$
$x_c^* = A$

III

log Pe$_g$

Fig. 2   Transport regimes defined by *GrSc* vs *Pe$_g$* following Camel and Favier.[12] Ia–Id: convective regime; II: advective-diffusive regime; III: fast diffusion regime. The variable x_0 is the parameter relating *Gr* to *Re*, x_c^* is the dimensionless length for the solute boundary layer, and *A* is the system aspect ratio.

Numerical Analysis of Crystal Growth Under Low-Gravity Conditions

Formulation

The formulation and solution of the model described in this section have been discussed in detail in a previous work.[8] The basic model system is shown in Fig. 3. The physical situation under investigation involves an ampoule containing a dilute two-component melt. Directional solidification takes place as the ampoule is translated through fixed "hot" and "cold" zones. The zones are separated by a thermal barrier[31] that is modeled using adiabatic sidewalls. Translation of the ampoule is modeled by supplying a doped melt of dilute bulk composition c_∞ at a constant velocity V_g at the

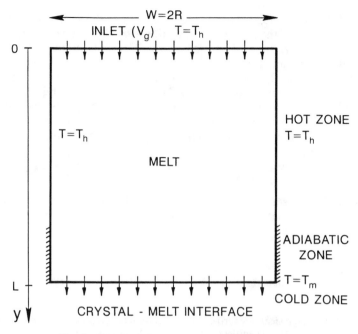

Fig. 3 Idealized Bridgman-Stockbarger system.

top of the computational space (inlet) and withdrawing a solid of composition $c_s = c_s(x,t)$ from the bottom. The crystal-melt interface is located at a distance L from the inlet. The temperature at the interface is taken to be T_m, the melting temperature of the crystal, and the upper boundary is held at a higher temperature T_h. Since we wish to confine our attention to compositional nonuniformities caused by buoyancy convection, rather than variations resulting from nonplanar crystal-melt interfaces,[32,33] the interface is held flat in our model.

The governing equations are cast in dimensionless form using L, κ/L, $\rho_m \kappa^2 / L^2$, $T_h - T_m$, and c_∞ to scale the lengths, velocity, pressure, temperature, and solute concentration. The dimensionless equations governing momentum, heat, and solute transfer in the melt are then

$$\frac{\partial u}{\partial t} + (\text{grad}u)u = Pr\,\Delta u + GrPr^2 g \tag{10}$$

$$\text{div}u = 0 \tag{11}$$

$$\frac{\partial \theta}{\partial t} + u \cdot \text{grad}\theta = \Delta \theta \tag{12}$$

$$\frac{Sc}{Pr}\left(\frac{\partial s}{\partial t} + u \cdot \text{grad}s\right) = \Delta s \tag{13}$$

where, $u(x)$ is the velocity, $\theta = [T(x) - T_m]/(T_h - T_m)$ is the temperature, and $s = c_m(x)/c_m^\infty$ is the solute concentration. Pr, Gr, and Sc are defined in Table 1. The magnitude of g used in the Grashof number will be the magnitude of the actual acceleration. For the calculations presented here, we have used the thermophysical properties corresponding to gallium-doped germanium melts (see Table 3). The term g in Eq. (1) has a magnitude of unity and specifies the orientation of the gravity vector. The magnitude of the acceleration is then represented by the Grashof number. The following boundary conditions apply at the crystal-melt interface:

$$\theta = 0, \quad u \cdot N = \frac{Pe_g Sc}{Pr\sigma}, \quad u \wedge N = 0, \quad \frac{\partial s}{\partial y} = Pe_g(1 - k)s \quad (14)$$

We define the measure of compositional nonuniformity in the crystal at the interface to be the lateral range in concentration given by

$$\xi = \frac{c_{smax} - c_{smin}}{c_{sav}}\% \quad (15)$$

where c_s is the (dimensional) solute concentration in the crystal, and c_{av} is

Table 3   **Thermophysical properties of gallium-doped germanium[27] and operating conditions used in the calculations**

		Thermal conductivity of melt	0.17 W-K^{-1}-cm^{-1}
		Heat capacity of melt	0.39 J-g^{-1}-K^{-1}
		Density of melt (ρ_m)	5.6 g-cm^{-3}
		Density of solid (ρ_s)	5.6 g-cm^{-3}
Property		Kinematic viscosity of melt (v)	1.3×10^{-3} cm^2-s^{-1}
		Melting temperature (T_m)	1231 K
		Solute diffusivity (D)	1.3×10^{-4} cm^2-s^{-1}
		Thermal diffusivity of the melt (κ)	1.3×10^{-1} cm^2-s^{-1}
		Segregation coefficient (k)	0.1
		Thermal expansion coefficient (β)	2.5×10^{-4} K^{-1}
Associated		Prandtl number $Pr = v/\kappa$	0.01
dimensionless		Schmidt number $Sc = v/D$	10
parameters		Density ratio $\sigma = \rho_m/\rho_s$	1.0
	A)	Hot zone temperature (T_h)	1331 K
		Distance between inlet and interface (L)	1.0 cm
		Height of adiabatic zone (L_a)	2.5 mm
		Ampoule width/diam	1.0 cm
Operating	B)	Hot zone temperature (T_h)	1251 K
conditions	C)	Ampoule width	2.0 cm
	D)	Ampoule width	0.5 cm
	V1)	Translation (supply) rate (V_g)	6.5 μm-s^{-1}
	V2)	Translation (supply) rate (V_g)	3.25 μm-s^{-1}
	V3)	Translation (supply) rate (V_g)	0.65 μm-s^{-1}

the average concentration. The following boundary conditions are applied at the "inlet" ($y = 0$):

$$\theta = 1, \quad u \cdot N = \frac{Pe_g Sc}{Pr}\sigma, \quad u \wedge N = 0, \quad \frac{\partial s}{\partial y} = \frac{Pe_g Sc}{Pr}(s - 1) \quad (16)$$

At the side walls the conditions are

$$u \cdot N = \frac{Pe_g Sc}{Pr}\sigma, \quad u \cdot e_w = 0, \quad \text{grad} s \cdot e_w = 0 \quad (17)$$

with $\theta = 1$ in the isothermal zone and $\text{grad}\theta \cdot e_w = 0$ in the adiabatic zone.

In an actual experiment, because of the finite length of the ampoule, there is a gradual decrease in length of the melt zone during growth. In this model transient effects related to this change are ignored. This assumption is referred to as the quasisteady assumption and is frequently used in melt-growth modeling.[8,23,27] It is thus assumed that the ampoule is sufficiently long for these effects to be negligible, a condition that should be checked from case to case.

The equations were solved using a pseudospectral Chebyshev collocation method[8] for the operating conditions and thermophysical properties of the Ge : Ga system listed in Table 3.

Results and Discussion

Because of the small magnitude of both the Prandtl number and the residual accelerations considered, the temperature field is relatively

INLET

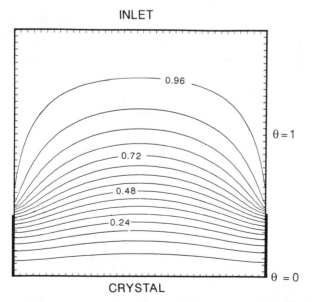

CRYSTAL

Fig. 4 **Dimensionless temperature field θ. Note that there is no difference between this and a purely conductive profile (i.e., $g = 0$).**

insensitive to convection and is characterized by the profile shown in Fig. 4. The presence of lateral as well as axial gradients in temperature leads to convective motion regardless of the orientation of the gravity vector. Velocity and concentration distributions have been obtained for various magnitudes and orientations of the acceleration vector and different growth rates and ampoule widths. The results for the ξ are given in Table 4.

Figures 5–7 illustrate two examples of computed velocity fields and consequent solute fields. If there had been no buoyancy-driven convection, the velocity vectors would be parallel, they would all have the same length equal to the V_g, and they would be oriented perpendicular to the crystal-melt interface. Furthermore, the isoconcentrates would be parallel to the interface.

In contrast to the single-roll flow depicted in Fig. 5a, Fig. 7 shows that, when the same magnitude gravity vector is oriented perpendicular to the interface, no rolls are present. The associated compositional nonuniformity is symmetric about the centerline of the ampoule and is significantly smaller than that for asymmetric case.

Figures 8a–8c depict the lateral variation in composition of the crystal for gravity orientations parallel, at 45 deg, and perpendicular to the interface, further illustrating the importance of ampoule orientation with respect to the residual acceleration vector. For a fixed acceleration magnitude, as the ampoule axis and acceleration vector are brought into

Table 4 Compositional nonuniformity $\xi(\%)$ for Ge : Ga

Ampoule width, cm: 1, 0.5, 2.0. Growth rate, $\mu\text{m-s}^{-1}$.

Residual acceleration magnitude (g/g_0)	Orientation N	e_g	1 / 6.5	1 / 0.65	0.5 / 6.5	2.0 / 6.5
A) 10^{-4}	↑	←	80			
10^{-5}		←	92.7		11.9	12.0
		↗	70.9		11.3	
		↓	6.4		0.95	
$5(10)^{-6}$		↓	3.2			
		↗	39			
		←	54.2			
10^{-6}		←	11.3		2.0	
		↗	8.0			
		↓	0.7		0.0	
B) 10^{-5}		←	22.6			64.5
10^{-6}		←	2.3			

e_g is the unit vector parallel to g, $L = 1$ cm for all cases.
Values in parentheses indicate three-dimensional results. A) and B) refer to the operating conditions listed in Table 3.

$\overset{g}{\longleftarrow}$

INLET

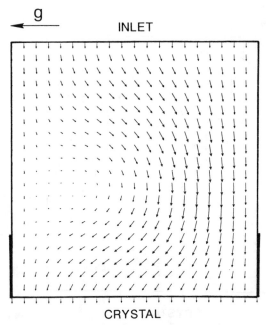

CRYSTAL

Fig. 5a   Steady flowfield produced by a residual acceleration with a magnitude 10^{-5} g_0 acting parallel to the crystal-melt interface. The maximum speeds are approximately twice the growth speed.

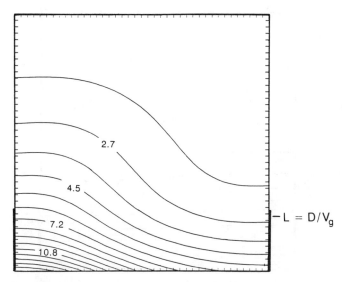

Fig. 5b   Dimensionless solute field s associated with the flow depicted in Fig. 5a. The contour interval is 0.9. For this case $\xi = 92.7\%$.

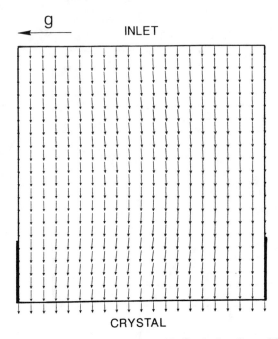

Fig. 6a   Steady flowfield produced by a residual acceleration with a magnitude $10^{-6}g_0$ acting parallel to the crystal-melt interface. The maximum speeds are slightly greater than the growth speed.

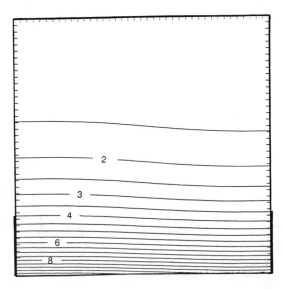

Fig. 6b   Dimensionless solute field associated with the flow depicted in Fig. 6a. For this case $\xi = 11.3\%$.

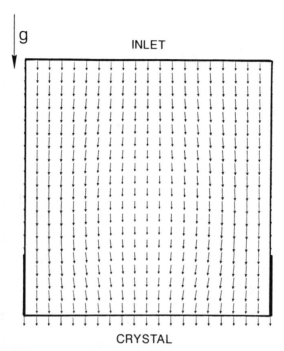

Fig. 7 Steady flowfield produced by a residual acceleration with a magnitude $10^{-5}g_0$ acting perpendicular to the crystal-melt interface; otherwise, the operating conditions are the same as those for Figs. 3–5. The maximum velocity shown is approximately equal to the growth speed. For this case $\zeta = 6.4\%$.

alignment, the transition form asymmetric to symmetric compositional nonuniformity is accompanied by a reduction in the magnitude of the nonuniformity.

The characterization of melt grown crystals often includes radially or laterally averaged longitudinal or axial crystal composition profiles.[34–36] Laterally averaged longitudinal *melt* composition profiles are shown in Fig. 9 for a selection of residual accelerations. Note that only the $10^{-6}\,g_0$ profile in Fig. 9a deviates "significantly" from the purely diffusive case (dashed curve). In fact, one might have expected the 10^{-5} profile to have exhibited a more pronounced departure from the diffusion profile, since the two-dimensional concentration field exhibits more distortion than does the lower-gravity case (cf. Figs. 5b and 6b). Thus, the reduction to a one-dimensional description (via averaging) results in a considerable loss of information. Obviously, in these cases the laterally averaged profiles cannot be used to assess the convective contribution to solute transfer.

The consequences of reducing the growth rate are illustrated by Fig. 10, which was obtained for $10^{-5}\,g_0$ parallel to the interface and $V_g = 0.65\,\mu\text{m-s}^{-1}$. A comparison of Figs. 5a and 10a shows that, as a result of reduced melt translation rate, the buoyancy-driven recirculation

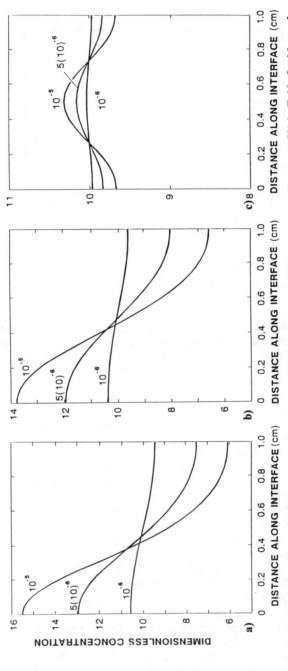

Fig. 8 Lateral variations in melt composition at the interface for operating conditions corresponding to A–V1 in Table 3 with acceleration magnitudes of 10^{-5}, 5×10^{-6}, and $10^{-6} g_0$ with a) parallel, b) 45-deg orientations c) perpendicular with respect to the crystal-melt interface.

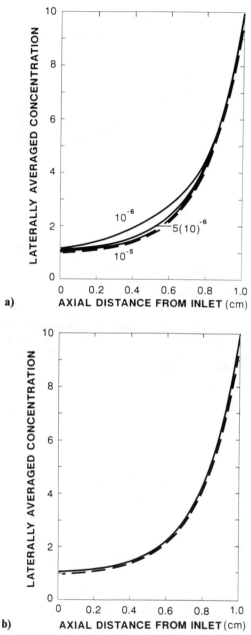

a)   AXIAL DISTANCE FROM INLET (cm)

b)   AXIAL DISTANCE FROM INLET (cm)

Fig. 9 Laterally averaged, axial composition profiles in the melt for operating conditions corresponding to A–V1 in Table 3 for accelerations with magnitudes of 10^{-5}, 5×10^{-6}, and $10^{-6} g_0$ oriented a) parallel and b) perpendicular to the crystal-melt interface. Note that the latter are indistinguishable from the pure diffusion profile (dashed line).

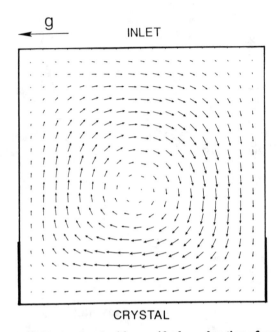

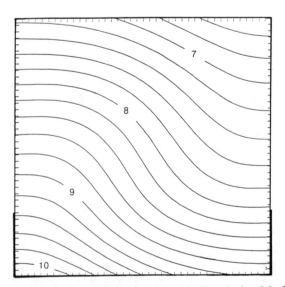

Fig. 10a   Velocity field associated with a residual acceleration of magnitude $10^{-5}g_0$ acting parallel to the crystal-melt interface. The operating conditions correspond to A–V3 in Table 3; i.e., the growth rate is 0.65 μm-s^{-1}. Maximum velocities are approximately 20 times the growth speed.

Fig. 10b   Dimensionless solute field associated with flow depicted in Fig. 10a. The contour interval is 0.2. For this case $\xi = 11.9\%$.

now dominates. Thus, the axial concentration gradient is reduced (Figs. 5b and 10b), and a smaller concentration gradient along the interface results (in Fig. 10b $\xi = 11.9\%$, compared to 92.7% in Fig. 5b).

Table 4 also illustrates the effect of changing the temperature gradient and the ampoule dimensions. For an acceleration of 10^{-5} g_0 oriented parallel to the interface, a reduction of the temperature difference between the hot zone and the crystal-melt interface from 100 K to 20 K changes ξ from 92.7% to 22.6%. An increase in the width of the ampoule from 1 to 2 cm with the 20 K temperature difference yields $\xi = 64.5\%$. Clearly, for a given steady acceleration the value of ξ is sensitive to the particular operating conditions employed, and it is obvious that the adiabatic zone used in this model is not the optimum choice for a low-gravity environment.

In addition to the calculations undertaken for Ge : Ga melts, we also varied the D of the solute. We have examined cases with $Sc = 10, 20, 30,$ and 50 for operating conditions otherwise corresponding to those listed as A-V1 in Table 3. (Recall that in the relative weighting of convective and diffusive fluxes, an increase in Sc corresponds to a reduction in diffusive transport.) The results are displayed in Figs. 11 and 12. The relationship between Sc and ξ is not linear. In particular, for acceleration magnitudes of 10^{-6} and $10^{-5}g_0$ oriented parallel to the interface, the maximum value of ξ is attained for $Sc = 20$, whereas at $10^{-5}g_0$ oriented perpendicular to the interface, there is little difference in the degree of nonuniformity between $Sc = 20$ and 50. Laterally averaged composition profiles in the melt for $10^{-4}g_0$ acting parallel to the interface are depicted for $Sc = 10, 20, 30,$ and 50 in Fig. 12b. In contrast to Fig. 9, these composition profiles exhibit a flat section in the bulk of the melt as a result of the increase in convective transport at $10^{-4}g_0$.

Order-of-Magnitude Analysis and Discussion

To illustrate the reliability of scaling analyses, we will attempt to obtain some of the results from the previous section from the relevant scales associated with momentum and solute transport and also apply the scaling techniques of Camel and Favier.[12] We are interested in comparing the relative rates of solute transport by convection and diffusion in the vicinity of the crystal-melt interface. At a planar interface, in the absence of buoyancy-driven convection, diffusion of the rejected solute into the bulk melt causes a deviation of the melt concentration from its initial bulk value up to a distance L_D from the interface. Note that $L_D = D/V_g$ represents the characteristic distance for which $Pe_g = 1$, i.e., the distance over which diffusion can propagate the concentration perturbation against the advective inflow of bulk melt due to the growth of the crystal.[16] When buoyancy-driven convection occurs, the spatial extent, $\bar{\delta}$, of this concentration perturbation can be larger or smaller than L_D, depending on whether the flow locally augments or diminishes the rate at which solute is transported away from the interface compared to the purely advective-diffusive case.

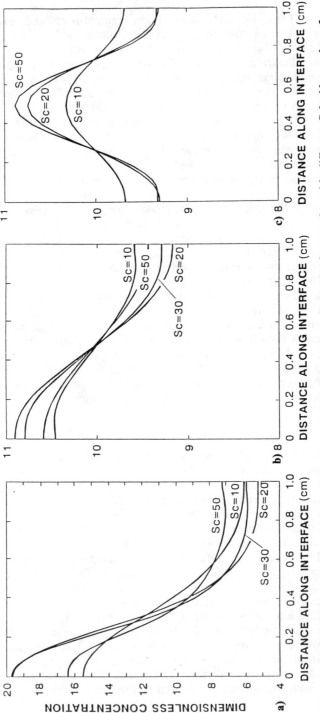

Fig. 11 Lateral variations in melt composition at the interface in crystals grown from melts with different Schmidt numbers for accelerations with the following magnitudes and orientations with respect to the crystal-melt interface: a) $10^{-5}g_0$, parallel; b) $10^{-6}g_0$, parallel; and c) $10^{-5}g_0$, perpendicular.

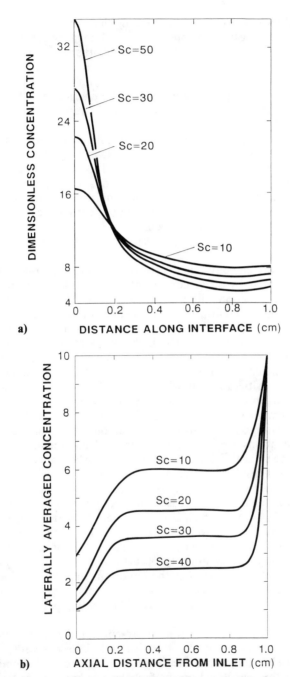

Fig. 12   a) Lateral variations in melt composition at the interface and b) laterally averaged axial composition profiles for melts with different Schmidt numbers, in crystals grown from melts with different Schmidt numbers, for a $10^{-4}g_0$ acceleration oriented parallel to the crystal-melt interface.

We have shown in the previous section that lateral distortion of the isoconcentrates can occur as a consequence of buoyancy-driven convection. In general, for steady transport the following situations can occur:

1) Buoyancy-driven convection is weak, and the concentration gradients are shallow everywhere, even at the crystal-melt interface. Thus, the lateral nonuniformity in interfacial solute distribution is limited. This situation will arise for relatively slow growth rates and for rapid diffusive transport, i.e., Gr, Sc, $Pe/Pe_g \ll 1$.

2) Buoyancy-driven flow penetrates the whole melt. Return flow at side walls is such that locally $\bar{\delta} > L_{D'}$. In contrast, at other locations (see, for example, Fig. 5) buoyancy-driven flow directed toward the interface results in $\bar{\delta} < L_{D'}$. This results in large lateral variations in crystal composition as the convective flow distributes solute nonuniformly along the interface. This situation will arise when growth rates, convective velocities, and diffusive speeds are of the same order, i.e., Pe, $Pe_g \sim \mathcal{O}(1)$. This regime represents a transition between convective and advective-diffusive transport.

3) Buoyancy-driven convection is strong. The distance $\bar{\delta} < L_D$ is narrow and is characterized by steep concentration gradients. The bulk melt is well mixed. This situation will arise when $Pe \gg 1$. The melt concentration profiles of Fig. 12b present a transition between cases 2 and 3.

4) Buoyancy-driven convection is intense and forms a momentum boundary layer of thickness δ_v. The highly convecting flow creates a thin solute boundary layer of thickness $\bar{\delta} \ll L_D$. When $\bar{\delta}$ is small compared to the system diameter, variation of solute along the interface should scale with $(Pe)^{-\frac{1}{3}}$. This situation will arise when $PeSc^{-1} = Gr \gg 1$, $Pe_g \sim \mathcal{O}(1)$.

We shall deal only with the first three cases that span the conditions covered by our numerical simulations. The fourth case has been discussed earlier in the section on previous works based on the arguments of Brown.[6]

In our simulations we found that, for a given orientation of g, the maximum velocity scales linearly with the Grashof number. Figure 13 depicts numerical results for ζ as a function of V_0/V_g, with triangles and squares indicating two different growth rates, $V_g = 6.5$ and 0.65μm-s^{-1}, respectively. For the 6.5 μm-s^{-1} cases, the maximum compositional uniformity occurs at $V_0/V_g \cong 1$, which is in agreement with trends demonstrated by previous works,[23,27] in which maximum nonuniformity occurs at an intermediate value of the Peclet number. However, for the lower growth rate case the higher nonuniformity occurs for $Pe/Pe_g = V_0/V_g \cong 10$, but $Pe = V_0/V_s \cong 1$. It is evident that the condition that $Pe \sim 1$ does not, as might be expected from linear scaling arguments, necessarily lead to a larger nonuniformity than, say, $Pe = 10$. This finding is further supported by Figs. 14 and 15.

The original intent of Fig. 14 was to present a unified description of the dependence of ζ on the ratio of convective and diffusive speeds, Pe, which was varied by changing either the Grashof or the Schmidt numbers. The motivation for this was based on the fact that we knew, from the material presented in the previous section and the work of Chang and Brown[23] and Adornato and Brown[27] that maxima in compositional nonuniformity could

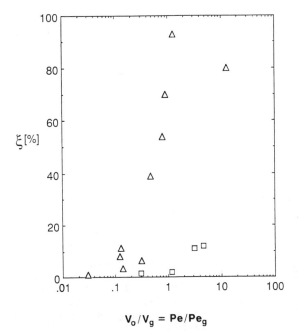

$$V_0/V_g = Pe/Pe_g$$

Fig. 13   Compositional nonuniformity ξ vs V_0/V_g for $Sc = 10$. Triangles correspond to cases for which $V_g = 6.5\ \mu\text{m-s}^{-1}$; squares correspond to $V_g = 0.65\ \mu\text{m-s}^{-1}$. For each case V_0 is the maximum velocity obtained numerically.

be obtained at a fixed Sc by increasing the Grashof number and that for fixed Gr a maximum nonuniformity could be found at some intermediate value of Sc. It can be seen that such a ξ vs Pe presentation does not lead to a single curve for the cases we considered. The reason for this is that all of the details of the interaction of the flow with the solute field cannot be folded into a scalar parameter. However, all is not lost.

Curve 3 in Fig. 14 corresponds to the case where there is a well-developed flow with maximum velocities an order of magnitude higher than the growth rate and the diffusive rates. The points on this curve were obtained by examining the effect of a fixed level of convection on melts with different solute diffusivities (see also Fig. 12); i.e., the Gr is the same for all points, and Pe was increased through a decrease in the solute diffusivity (increases in Sc). The increase in ξ with increasing Sc can be explained as follows. For the advective-diffusive case a decrease in diffusivity at fixed V_g leads to a decrease in L_D. This means that the bulk melt composition is closer to the interface. In the presence of convection the local variations in $\bar{\delta}$ (discussed in case 2 of this section) will thus result in larger compositional nonuniformities. However, eventually the value of L_D will be small enough that convective velocities which decrease with distance from the interface will be too small to disturb the solute distribution resulting from advective-diffusive transport. Consequently, as discussed

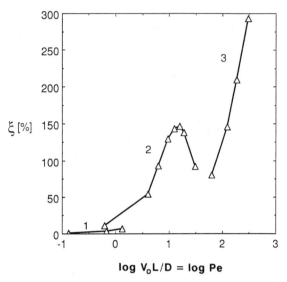

Fig. 14   Compositional nonuniformity ξ vs $V_0 L/D$. **Curves 2 and 3 correspond to g oriented parallel to the crystal-melt interface; curve 1 corresponds to a perpendicular orientation. For each case V_0 is the maximum velocity obtained numerically, and V_g is 6.5 μm-s^{-1}.**

earlier, ξ vs Sc can be expected to have a maximum at an intermediate value of Sc. This is borne out by curve 2 in Fig. 14 and is shown explicitly by Figs. 11a–11c.

For the lower values of Pe it is interesting to note the difference between curves 1 and 2, which have been calculated for the same values of Gr but with different orientations of the gravity vector. It appears that the change in flow structure from a single roll (parallel to two symmetric rolls; see Figs. 5a and 10a) is responsible for this difference. This is another example of the limitations of scalar arguments.

To further emphasize the problem of neglecting the multidimensional nature of the system, we compare our predicted nonuniformities with those predicted by the approach of Camel and Favier.[12] First it is necessary to convert ξ into the same measure of nonuniformity they used. This is simply

$$\xi_m = \frac{c_{sav}}{c_{smax}} \xi = \frac{c_{smax} - c_{smin}}{c_{smax}} \tag{18}$$

where ξ_m is given by Eq. (9).

In order to apply their technique it was necessary to evaluate a proportionality constant x_0^2 (see the section on previous works) relating the maximum convective velocity, obtained in the numerical simulations, with the Grashof number. We found the value of this constant depended on the orientation of the acceleration and on the chosen length scale. We found

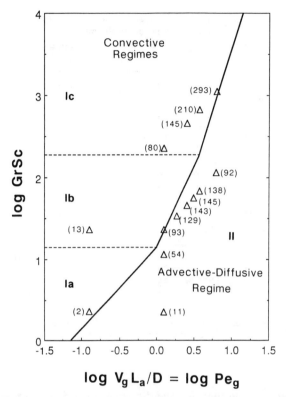

$$\log V_g L_a/D = \log Pe_g$$

Fig. 15   A plot of $GrSc$ vs Pe_g for the results of numerical simulations for steady accelerations parallel to the crystal-melt interface. Solid and dashed lines delineate the transport regimes defined in Fig. 2. Note that the adiabatic zone length has been used to calculate the Grashof and growth Peclet numbers and that $x_0 = 0.27$. Numbers in parentheses indicate the value of ξ associated with each of the numerical results.

the adiabatic zone length L_a to yield results that best showed the trends predicted by the Camel-Favier approach. Table 5 lists values of x_0 associated with the choice of L_a as a reference length. Having calculated x_0, we then determined the transport regime as defined by Fig. 2 by graphing $GrSc$ vs Pe_g in Fig. 15 for the case of the residual acceleration parallel to the interface. The nonuniformities ξ obtained from our numerical simulations are also given in Fig. 15. In Table 5 a comparison is made between values of ξ_m obtained numerically and those from the Camel-Favier recipe. The latter values are calculated by first determining the transport regime (see Fig. 2) based on the values of $GrSc$, Pe, and x_0. If the case under consideration lies in the convective regime, the ξ_m are calculated using Eq. (9) and the value of x_c^* corresponding to the associated transport regime. According to Camel and Favier, if advection-diffusion is dominant, ξ_m is proportional to the ratio of the convective velocity within the solute layer of extent x_c^* and the growth velocity. The nonuniformity isolines in Fig. 8

Table 5   Comparison of numerical (ξ_m^*) and order-of-magnitude (ξ_m) estimates of compositional nonuniformity

g Orientation	x_0	Gr	Sc	ξ_m^*, %	ξ_m, %	Pe_g	Regime
Axial	0.12	0.227	10	0.6	2.4	1.25	II (a)
Axial	0.12	1.135	10	3.2	12	1.25	II (a)
Axial	0.12	2.27	10	6.2	24	1.25	II (a)
45 deg	0.25	0.227	10	7.6	9.9	1.25	II (a)
45 deg	0.25	1.135	10	33	29	1.25	II (a)
45 deg	0.25	2.27	10	47	90	1.25	II (b)
Parallel	0.27	0.227	10	11	12	1.25	II (a)
Parallel	0.27	1.135	10	54	59	1.25	II (a)
Parallel	0.27	2.27	10	61	73	1.25	I (b)
Parallel	0.27	22.7	10	48	28	1.25	I (c)
Parallel	0.27	22.7	20	65	45	2.5	I (c)
Parallel	0.27	22.7	30	72	68	3.75	I (c)
Parallel	0.27	22.7	50	82	72	6.25	I (c)
Parallel	0.27	2.27	15	71	40	1.875	II (b)
Parallel	0.27	2.27	20	73	29	2.5	II (b)
Parallel	0.27	2.27	25	73	21	3.125	II (b)
Parallel	0.27	2.27	30	70	16	3.725	II (b)
Parallel	0.27	2.27	50	56	7	6.25	II (b)

Note that for these cases Gr and Pe have been computed using the adiabatic zone length L_a as a reference length.

of their paper indicate that, for the diffusive subregimes complementary to the convective regimes Ia–Id, this ratio is respectively given in terms of the ratio of $x_0^2 GrSc$, $(x_0^2 GrSc)^{\frac{1}{2}}$, $(x_0 GrSc)^{\frac{1}{3}}$, and $Gr^{\frac{1}{4}} Sc^{\frac{1}{3}}$ to Pe_g. In this sense the quantity $x_0^2 GrSc$ may be considered in terms of a local solutal Peclet number.

As Table 5 shows, at high Schmidt numbers the Camel-Favier approach meets with some success in predicting the nonuniformity but is less faithful at low values of Gr. Figure 15 also contains two cases for which Pe_g is reduced at fixed $GrSc$. According to the Camel-Favier estimates, the lower $GrSc$ case should have yielded a higher compositional nonuniformity as the growth rate was reduced; i.e., the estimates predict that the system moves closer to the convective diffusive transition. This was not reflected in the nonuniformity obtained from the simulations. Evaluation of our numerical results shows that the increase in L_D (associated with reduction in growth rate) led to an order-of-magnitude decrease in the solute gradient at the interface (see Fig. 10b). Hence, since the systemwide variation in composition is small, the interfacial nonuniformity is correspondingly small.

Finally, in order to explore the sensitivity of the choice of reference length when using the dimensionless groups Gr and Pe_g, we compare estimated nonuniformities based on the adiabatic zone length L_a and system length L. Either one may be a reasonable choice. The results are presented in Table 6. Although it is true that large nonuniformities

Table 6   Comparison of numerical (ξ^*_m) and order-of-magnitude (ξ_m) of compositional nonuniformity for different choices of length scale

g Orientation	L, cm	x_0	Gr	Sc	ξ^*_m, %	ξ_m, %	Pe_g	Regime
Parallel	1.0	0.066	145	10	61	10	5	II (b)
Parallel	0.25	0.27	2.27	10	61	73	1.25	I (b)
Parallel	1.0	0.066	145	15	71	5	7.5	II (b)
Parallel	0.25	0.27	2.27	15	71	40	1.875	II (b)
Parallel	1.0	0.066	145	25	73	2.5	12.5	II (b)
Parallel	0.25	0.27	2.27	25	73	21	3.125	II (b)
Parallel	1.0	0.066	145	10	12	18	0.5	I (b)
Parallel	0.25	0.27	2.27	10	12	8	0.125	I (b)

appear to occur when Pe is $\mathcal{O}(1)$ and $Pe_g \sim \mathcal{O}(1)$, no clear distinction between individual cases can be made. This is hardly surprising. An order-of-magnitude estimate is no more than its name suggests. Indeed, we should only expect our best estimates to agree within an order of magnitude with reality (or our numerical simulations). Different physical situations may yield estimates that have the same order of magnitude. Any attempt to distinguish between these cases based on the estimates alone is futile.

Conclusions

We have compared the results of an order-of-magnitude or scaling analysis with those of numerical simulations of the effects of steady low gravity on compositional nonuniformity in crystals grown by the Bridgman-Stockbarger technique. The comparison indicates that scaling arguments alone can at best be expected to yield order-of-magnitude accuracy. This is hardly surprising if one considers the neglect of the multidimensionality (boundary conditions, flow structure, etc.) of the physical situation which is inherent in such scalar descriptions. Of course, prior knowledge of the system behavior can be used to *locally* improve the accuracy of the scaling (e.g., Ref. 30). Similarly, if a system locally exhibits one-dimensional behavior, such as boundary-layer flow, then the appropriate scaling for the formal reduction of the transport equations can be used effectively to estimate the local system response. Our understanding of transport during crystal growth can be furthered only by an approach that uses scaling, mathematical modeling, and, naturally, the results of experiments in a complementary fashion.

Acknowledgment

This work was supported by NASA Grants NAG8-684 and NAG8-724, by the State of Alabama through the Center for Microgravity and Materials Research at the University of Alabama in Huntsville, and through the Alabama Supercomputer Network.

References

[1] Hurle, D. T. J., Jakeman, E., and Pike, E. R., "Striated Solute Distributions Produced by Temperature Oscillations During Crystal Growth from the Melt," *Journal of Crystal Growth*, Vol. 3/4, 1968, pp. 633–640.

[2] Pimputkar, S. M., and Ostrach, S., "Convective Effects in Crystals Grown from Melt," *Journal of Crystal Growth*, Vol. 55, 1981, pp. 614–646.

[3] Müller, G., "Convection in Melts and Crystal Growth," *Convective Transport and Instability Phenomena*, edited by J. Zierep and H. Oertel, Braun Verlag, 1982, pp. 441–468.

[4] Müller, G., "Convection and Inhomogeneities in Crystal Growth from the Melt," *Crystals: Growth, Properties and Applications*. Vol. 12, Springer-Verlag, Berlin, 1988.

[5] Müller, G., Neumann, G., and Weber, W., "Natural Convection in Vertical Bridgman Configurations," *Journal of Crystal Growth*, Vol. 70, 1984, pp. 78–93.

[6] Brown, R. A., "Theory of Transport Processes in Single Crystal Growth from the Melt," *AIChE Journal*, Vol. 34, 1988, pp. 881–911.

[7] Langlois, W. E., "Buoyancy Driven Flows in Crystal Growth Melts," *Annual Review of Fluid Mechanics*, Vol. 17, 1985, pp. 191–215.

[8] Alexander, J. I. D., Ouazzani, J., and Rosenberger, F., "Analysis of the Low Gravity Tolerance of Bridgman-Stockbarger Crystal Growth: I. Steady and Impulse Accelerations," *Journal of Crystal Growth*, Vol. 97, 1989, pp. 285–302.

[9] Wilcox, W. A., "The Role of Mass Transfer in Crystallization Processes," *Preparation and Properties of Solid State Materials*, edited by R. A. Lefever, Dekker, New York, 1971.

[10] Flemings, M. C., *Solidification Processing*, McGraw-Hill, New York, 1974.

[11] Rosenberger, F., *Fundamentals of Crystal Growth*, Springer-Verlag, New York, 1979.

[12] Camel, D., and Favier, J. J., "Scaling Analysis of Convective Solute Transport and Segregation in Bridgman Crystal Growth from the Doped Melt," *Journal de Physique (Paris)*, Vol. 47, 1986, pp. 1001–1014.

[13] Prandtl, L., *Verhandlungen des Dritten Internationalen Mathematiker-Kongresses, Heidelberg 1904*, Leipzig, 1905, p. 484.

[14] Goldstein, S., *Modern Developments in Fluid Dynamics*, Dover, New York, 1965.

[15] Levich, V. G., *Physicochemical Hydrodynamics*, Prentice-Hall, Englewood Cliffs, NJ, 1962.

[16] Rosenberger, F., and Müller, G., "Interfacial Transport in Crystal Growth, a Parametric Comparison of Convective Effects," *Journal of Crystal Growth*, Vol. 65, 1983, pp. 91–104.

[17] Wilson, L. O., "On Interpreting a Quantity in the Burton, Prim and Slichter Model Equation as a Diffusion Boundary Layer Thickness," *Journal of Crystal Growth*, Vol. 44, 1978, pp. 247–250.

[18] Wilson, L. O., "A New Look at the Burton, Prim and Slichter Model of Segregation during Crystal Growth from the Melt," *Journal of Crystal Growth*, Vol. 44, 1978, pp. 371–376.

[19] Burton, J. A., Prim, R. C., and Slichter, W. P., "The Distribution of Solute in Crystals Grown from the Melt. I. Theoretical," *Journal of Chemical Physics*, Vol. 21, 1953, pp. 1987–1990.

[20] Von Kármán, T., "Über Laminare und Turbulente Reibung," *Zeitschrift fuer Angewandte Mathematik und Mechanik*, Vol. 1, 1921, pp. 244–247.

[21] Abramowitz, M., and Stegun, E. A., *Handbook of Mathematical Functions*, Vol. 55, National Bureau of Standards Applied Mathematics Series, Gaithersburg, MD, 1964.

[22] Pan, Y. F., and Acrivos, A., "Heat Transfer at High Peclet Number in Regions of Closed Streamlines," *International Journal of Heat and Mass Transfer*, Vol. 11, 1968, pp. 439–444.

[23] Adornato, P. M., and Brown, R. A., "Convection and Segregation During Vertical Bridgman Growth of Dilute and Non-Dilute Binary Alloys: Effects of Ampoule Design," *Journal of Crystal Growth*, Vol. 80, 1987, pp. 155–190.

[24] Hjellming, L. N., and Walker, J. S., "Mass Transport in Czochralski Puller with a Strong Magnetic Field," *Journal of Crystal Growth*, Vol. 85, 1987, pp. 25–31.

[25] Wheeler, A., "Boundary Layer Models in Czochralski Crystal Growth," *Journal of Crystal Growth*, Vol. 97, 1989, pp. 64–75.

[26] Riley, N., "Species Transport in Magnetic Field Czochralski Growth," *Journal of Crystal Growth*, Vol. 97, 1989, pp. 76–84.

[27] Chang, C. J., and Brown, R. A., "Radial Segregation Induced by Natural Convection and Melt/Solid in Vertical Bridgman Growth," *Journal of Crystal Growth*, Vol. 63, 1983, pp. 343–364.

[28] Acrivos, A., "On the Combined Effect of Forced and Free Convection Heat Transfer in Laminar Boundary Layer Flows," *Chemical Engineering Science*, Vol. 21, 1966, pp. 343–352.

[29] Rouzaud, A., Camel, D., and Favier, J. J., "A Comparative Study of Convective Solute Transport and Segregation in Bridgman Crystal Growth from the Doped Melt," *Journal of Crystal Growth*, Vol. 73, 1985, pp. 149–166.

[30] Kimura, S., and Bejan, A., "Natural Convection in a Differentially Heated Corner Region," *Physics of Fluids*, Vol. 28, 1985, pp. 2980–2989.

[31] Dahkoul, Y. M., Farmer, R., Lehoczky, S. L., and Szofran, F. R., "Numerical Simulation of Heat Transfer During the Crystal Growth of HgCdTe Alloys," *Journal of Crystal Growth*, Vol. 86, 1988, pp. 49–55.

[32] Coriell, S. R., and Sekerka, R. F., "Lateral Solute Segregation During Unidirectional Solidification of a Binary Alloy with a Curved Solid-Liquid Interface," *Journal of Crystal Growth*, Vol. 46, 1979, pp. 479–482.

[33] Coriell, S. R., Boisvert, R. F., Rehm, R. G., and Sekerka, R. F., "Lateral Solute Segregation and Interface Instabilities During Unidirectional Solidification," *Journal of Crystal Growth*, Vol. 54, 1981, 167–175.

[34] Gatos, H. C., "Semiconductor Crystal Growth and Segregation Problems on Earth and in Space," *Materials Processing in the Reduced Gravity Environment of Space*, edited by G. Rindone, Elsevier, New York, 1982, pp. 355–371.

[35] Witt, A. F., Gatos, H. C., Lichtensteiger, M., Lavine, M. C., and Herman, C. J., "Crystal Growth and Steady State Segregation Under Zero Gravity," *Journal of the Electrochemical Society*, Vol. 125, 1975, pp. 276–283.

[36] Tiller, W. A., Jackson, K. A., Rutter, J. W., and Chalmers, B., "The Redistribution of Solute Atoms During the Solidification of Metals," *Acta Metallurgica*, Vol. 1, 1953, pp. 428–437.

Steady-State Thermal-Solutal Convection and Diffusion in a Simulated Float Zone

G. W. Young*
University of Akron, Akron, Ohio
and
A. Chait†
NASA Lewis Research Center, Cleveland, Ohio

Nomenclature

a = width of the heater temperature profile, cm
B = Bond number $\rho_L g L^3 / \gamma_S R$
B_L = Biot number in liquid phase, $h_L R / k_L$
$\overline{B_L}$ = scaled Biot number, $B_L = \epsilon^2 \overline{B_L}$
B_S = Biot number in solid phase, $h_S R / k_S$
$\overline{B_S}$ = scaled Biot number, $B_S = \epsilon^2 \overline{B_S}$
C = capillary number, $\mu w^* / \gamma_S$
C_0 = concentration of solute in the feed rod (assumed uniform), wt %
$\overline{C}$ = scaled capillary number $C = \epsilon^4 \overline{C}$
C_S = concentration of solute in the solidifying crystal
$c^S(x,\eta)$ = concentration of solute in the melt (freezing interface boundary layer)
$c^F(x,\rho)$ = concentration of solute in the melt (melting interface boundary layer)
c_p = heat capacity in liquid phase, J/gK
$c(x,z)$ = scaled concentration of solute in the melt (core)
$c(x,\eta)$ = concentration introduced in Eq. (47c)
D = solute diffusivity, cm²/s
g = acceleration of gravity, cm/s²

*Professor, Mathematical Sciences.
†Research Scientist, Materials Division.

119

g_0 = terrestrial acceleration of gravity, cm/s^2
$h(x)$ = position of the solidifying interface
$h_m(x)$ = position of the melting interface
$h_{L,S}$ = heat-transfer coefficient, W/cm$^2 \cdot$ K
K = conductivity ratio, k_S/k_L
k = segregation coefficient
$k_{S,L}$ = thermal conductivity, W/cm $\cdot$ K
L = mean length of the float zone, cm
Le = Lewis number, κ/D
$-L_S$ = mean position of the solidifying interface, cm
L_F = mean position of the melting interface, cm
L_1 = position on feed crystal where boundary conditions at infinity are applied, cm
L_2 = position on product crystal where boundary conditions at infinity are applied, cm
$\ell_F = L_F/L$
$\ell_S = L_S/L$, latent heat, J/cm^3
M = morphological number, mC_0/T_M
Ma = Marangoni number, $\gamma R\Delta T/\mu\kappa$
m = liquidus slope, K/wt %
P_D = solute Peclet number, VR/D
Pe = thermal Peclet number (liquid), VL/κ
Pr = Prandtl number, kinematic viscosity/thermal diffusivity
P_S = thermal Peclet number (solid), VL/κ_S
$p(z)$ = pressure
p' = hydrostatic pressure scale, $\gamma_S R/L^2$
p^* = flow pressure scale, $\mu w^* L/R^2$
R = half-width of the crystal sheet, cm
Re = surface tension Reynolds number, $\gamma R\Delta T/\mu\nu$
R_L = radiation number in the liquid phase, $\sigma\epsilon_m RT_M^3/k_L$
$\overline{R}_L$ = scaled radiation number, $R_L = \epsilon^2\overline{R}_L$
R_S = radiation number in the solid phases, $\sigma\epsilon_S RT_M^3/k_S$
$\overline{R}_S$ = scaled radiation number, $R_S = \epsilon^2\overline{R}_S$
R_V = translation Reynolds number, LV/ν
$r(z)$ = position of the melt/gas interface
Sc = Schmidt number, kinematic viscosity/solute diffusivity
St = Stefan number, $\mathscr{L}/T_M\rho_L c_p$
T_a = scaled ambient temperature
$\overline{T}(x,n)$ = temperature of melt introduced in Eq. (47a)
$T(x,z)$ = scaled temperature of the melt (core)
$\hat{T}(x,\rho)$ = temperature of melt introduced in Eq. (60a)
$T_F(x,z)$ = scaled temperature of the feed crystal (core)
$T^F(x,\rho)$ = temperature of the melt (melting boundary layer)
T_M = melting point of the pure substance, K
T_h = scaled maximum heater temperature
$T^S(x,\eta)$ = temperature of melt (solidifying boundary layer)
$T_S(x,z)$ = scaled temperature of the product crystal (core)
ΔT = temperature difference, $T_h - T_M$

u = melt velocity in the x direction
V = translation speed of the heater, cm/s
w = melt velocity in the z direction
w^* = melt velocity scale, $\gamma \Delta TR/L\mu$
x = dimensionless coordinate across the width of the crystal sheet measured from the centerline
z = dimensionless coordinate along the axis of the sheet measured from the plane of the heater
α = contact angle between the solidifying interface and the product crystal/gas interface
Γ = surface free energy number, $\gamma'/\mathscr{L}R$
γ = surface tension gradient with respect to temperature, $|d\gamma_S/dT|$
γ' = surface free energy, J/cm^2
γ_S = surface tension, dyne/cm
δ_j = exponent defined in Eq. (53c)
ϵ = aspect ratio, R/L
$\epsilon_{m,S}$ = emmisivity of the melt and solid surfaces
η = solidifying boundary-layer coordinate, $(z + \ell_S)/\epsilon$
$\bar{\theta}(x,\eta)$ = temperature of product crystal introduced in Eq. (47b)
$\hat{\theta}(x,\rho)$ = temperature of feed crystal introduced in Eq. (60b)
$\theta_F(x,\rho)$ = temperature of feed crystal (boundary layer)
$\theta_S(x,\eta)$ = temperature of product crystal (boundary layer)
$\theta(z)$ = scaled temperature profile of the heater
κ = thermal diffusivity in the liquid phase, cm^2/s
κ_S = thermal diffusivity in the solid phase, cm^2/s
μ = viscosity, g/cm · sec
ρ = melting boundary-layer coordinate, $(z - \ell_F)/\epsilon$
ρ_L = density of the liquid phase, g/cm^3
σ = Stefan-Boltzman constant, $W/cm^2 \cdot K^4$
ϕ = contact angle between the melt/gas interface and the product crystal
$\bar{\phi}$ = scaled contact angle, $\phi = \epsilon^2 \bar{\phi}$
ψ' = contact angle number $\frac{1}{2}\bar{\phi}P_D(k - 1)/k$
ψ = stream function, $u = -\psi_z$, $w = \psi_x$

Introduction

THE floating-zone technique is one of the methods suggested for crystal growth of electronic materials in space. The process consists of translating an energy source (heater, laser, or electron beam) along the axis of a cylindrical feed stock for the growth of bulk crystals or along the long dimension of a crystal sheet for ribbon-to-ribbon growth. As shown in Fig. 1, the traveling energy source establishes a molten zone in the feed crystal. The molten zone is shaped by the melting interface of the feed crystal, the solidifying interface of the product crystal, and the liquid-gas interface of the melt. In the electronic materials arena the silicon-based device technology is rapidly approaching a maturity level in terms of both scale and performance. Future advances for a wide variety of device applications

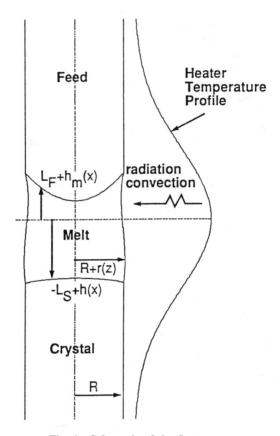

Fig. 1   Schematic of the float zone.

may come from new classes of materials, such as the III–V and II–VI families. Some advantages offered by these two classes are higher carrier mobilities and the possible tailoring of the available wide range of bandgap energies to specific applications. However, several major difficulties in processing these materials are present, making device production very limited. These difficulties lie in the large number of dislocations, the presence of twins, and their growth-related defects hindering both the growth of large high-quality crystals and the subsequent production of the wide variety of devices potentially offered by these classes of materials. The technique of float zoning may uniquely offer some answers to these problems in a single processing stage.

The primary advantages of this containerless technique are no contamination or expansion (contraction) problems from a crucible; for ribbon-to-ribbon growth the technique eliminates the need for wafering.

The primary disadvantages are that Earth-established float zones require small melt volumes in order to maintain a stable melt configuration. This limits the applicability of the technique to material systems such as silicon, which have a low density to surface tension (Bond number) ratio. Further-more, such short melt columns make it difficult to achieve a controlled thermal profile, which is a necessity for producing crystals of high quality. Finally, the presence of gravity may cause buoyancy-driven flows, leading to crystal striations. However, such shortcomings may be allieviated in the microgravity environment of space. Longer melt columns may be established for a wider variety of material systems. This allows better control of the imposed heater temperature profile. Unfortunately, though, surface-tension-driven flows remain significant in microgravity, and these, too, may lead to striations. For obvious reasons the small amount of available microgravity data necessitate the use of ground-based efforts, both experimental and theoretical, to gain an understanding of the physical mechanisms underlying the float-zone system. The heat, momentum, and mass-transfer processes all couple together to establish the zone configuration.

Hence, a complete model of the float-zone process must account for 1) the heat transfer between the heater, ambient, and solid and liquid phases; 2) the fluid mechanics of the melt; 3) the distribution of solute in determining the shape and stability of the zone; and 4) the effects of the deformable liquid-gas free surface and surface tension.

As discussed by Duranceau and Brown,[1] the theoretical studies that have considered the float-zone technique have mainly concentrated on one of the preceding points, thus neglecting the interplay of the remaining three. Additionally, the analyses have either neglected solidification (melting) and/or assumed planar interfaces for some (all) of the free surfaces.

For example, Kuiken and Roksnoer[3] analytically investigate only the heat transfer by a procedure similar to that which we shall propose, but do not calculate the interfacial shapes. Fu and Ostrach[4] numerically solve the heat-transfer problem with flat interfaces. Harriot and Brown[5] examine the solute transport in the melt due to rotation of the feed rod with flat interfaces and no heat transfer. Coriell et al.[6] and Surek and Coriell[7] investigate the effects of surface tension and contact angle on the shape and stability of the zone but for an isothermal system with flat interfaces. Harriot and Brown[8,9] analytically and numerically solve for the flowfields due to rotation of the feed rod for a zone of fixed size. Chang and Wilcox[10,11] Clark and Wilcox,[12] Sen and Davis,[13] and Lai et al.[14] examine flows induced by surface tension gradients (Marangoni convection) in zones of fixed size with flat surfaces and no solidification.

As a step toward the development of a model incorporating the inter-actions of the physical mechanisms 1-4 outlined earlier, Duranceau and Brown[1] consider a conduction-dominated heat-transfer model coupling 1 and 4 and more recently a flow model[2] coupling requirements 1, 2, and 4. The positions of the melting, solidifying, and liquid-gas interfaces are determined as a result of their numerical solution of the governing equa-tions. Hence, the shape and location of the zone relevant to the imposed heater temperature profile are defined as a consistent response to the heat transfer, melt convection, and surface tension.

The purpose of this chapter is to extend the model of Duranceau and Brown[1] to include steady-state diffusion of a solute and also to examine surface-tension-driven flows. Our approach is to reduce the highly nonlinear coupled set of partial differential equations, which governs the float-zone system, to a set of nonlinear coupled ordinary differential equations through asymptotic methods. Under conditions where the zone half-width R is small compared to the zone length L and the heat transfer due to convection and radiation are small, we show that it is possible to obtain analytical solutions for the temperature and concentration profiles, as well as for the melting, solidifying, and melt/gas interfacial shapes. The procedure we follow is similar to the calculation of the temperature distribution in floating-zone silicon crystals considered by Kuiken and Roksnoer[3] and similar to the analysis of Brattkus and Davis[22] concerning the effects of side-wall heat losses in a small-aspect-ratio directional solidification geometry. The flow model is an extension of Sen and Davis[13] to include interface morphology.

In the next section we formulate the model for a float zone in a thin sheet. We choose this geometry over the cylindrical version since smaller aspect ratios ($R/L = \epsilon \sim 0.1$) have been reported in experiments.[18] However, we anticipate that our results can qualitatively be extended to the cylindrical geometry. Calculations for the cylindrical case are currently being performed. In both cases the models include convective and radiation heat transfer as well as the definition of the material-dependent contact angle ϕ shown in Fig. 2. The contact angle α does not enter the calculation since we set the surface free energy equal to zero.

In the section on scaling we discuss the validity of the scalings we choose. We define the convective and radiation heat transfer, as well as the ϕ, to be $O(\epsilon^2)$ quantities. Furthermore, we examine the small ϵ limit in conjunction with static instabilities of the melt column as predicted by Plateau,[19] Rayleigh,[20] and Heywang and Ziegler.[21] The result of these scalings and limits leads to the matched asymptotic analysis briefly described in the section on asymptotic analysis. In the results section numerical solutions of the energy balances at the melting and solidifying fronts give the mean

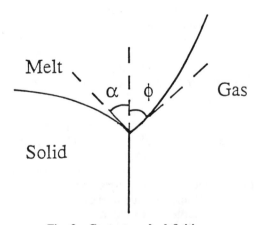

Fig. 2   Contact angle definition.

position of these interfaces. From these solutions we construct float zones given the translation rate of the energy source, the heater and ambient temperature profiles, and the material properties. We conduct a parametric investigation of the effects of these input parameters to identify the mechanisms involved in establishing the zone shape. Additionally, we discuss the use of our results as a design tool for system predictions. In particular, we examine the concentration profiles at the solidifying interface due to various heater temperature profiles.

Model Formulation

The governing equations for the float zone shown in Fig. 1 are non-dimensionalized by scaling the vertical coordinate z with the L of the liquid zone,

$$L = L_S + L_F \qquad (1)$$

where $-L_S$ and L_F denote the mean positions of the solidifying and melting interfaces, respectively. All other lengths are scaled on R, where $2R$ is the width of the feed and product crystals. The concentration of solute c in the melt is scaled with C_0, the concentration of solute in the feed crystal, which is assumed to be uniform. The temperatures T in the melt, T_F in the polycrystalline feed crystal, and T_S in the solidifying crystal are scaled with the equilibrium melting temperature T_M of the pure substance. The melt velocity, $u = (u,w)$, is scaled with the Marangoni velocity scale

$$w^* = \frac{\gamma \Delta T R}{L \mu} \qquad (2a)$$

where $\gamma = |d\gamma_s/dT|$, γ_s is the surface tension of the liquid/gas interface, and μ is the melt viscosity. The temperature difference is

$$\Delta T \approx (T_h - T_M) \qquad (2b)$$

Two pressure scales are introduced. The hydrostatic scale is given by

$$p' = \frac{\gamma_s R}{L^2} \qquad (2c)$$

whereas for the system including surface-tension-driven flow we use

$$p^* = \frac{\mu w^* L}{R^2} \qquad (2d)$$

We define a coordinate system translating with the interfaces and heater at a dimensional speed V and defined such that $z = 0$ denotes the maximum temperature position of the heater. The diffusion of solute in both the feed and product crystals is neglected because of the low diffusivity in the solids vs the melt, and we assume that there is no density change upon a phase transformation. The following equations describe the steady two-dimensional heat and solute transport in the melt and diffusion of heat in the feed and solidifying crystals.

Melt $(-\ell_S < z < \ell_F)$:

$$\epsilon^2 Ma[-\psi_z T_x + \psi_x T_z] - \epsilon^2 Pe\, T_z = T_{xx} + \epsilon^2 T_{zz} \tag{3a}$$

$$\epsilon^2 MaSc/Pr[-\psi_z c_x + \psi_x c_z] - \epsilon P_D c_z = c_{xx} + \epsilon^2 c_{zz} \tag{3b}$$

$$\epsilon^2 Re[\psi_x \psi_{xxz} - \psi_z \psi_{xxx} + \epsilon^2(\psi_x \psi_{zzz} - \psi_z \psi_{zzx})] - \epsilon^2 R_V(\psi_{xxz} + \epsilon^2 \psi_{zzz})$$

$$= \psi_{xxxx} + 2\epsilon^2 \psi_{xxzz} + \epsilon^4 \psi_{zzzz} \tag{3c}$$

Feed crystal $(\ell_F < z < \infty)$:

$$T_{Fxx} + \epsilon^2 T_{Fzz} = -\epsilon^2 P_S T_{Fz} \tag{4}$$

Solidifying crystal $(-\infty < z < -\ell_S)$:

$$T_{Sxx} + \epsilon^2 T_{Szz} = -\epsilon^2 P_S T_{Sz} \tag{5}$$

In the definitions of the melt and the feed and solidifying crystals we have introduced the nondimensional mean positions

$$\ell_S = \frac{L_S}{L} \tag{6a}$$

$$\ell_F = \frac{L_F}{L} \tag{6b}$$

of the solidifying the melting interfaces, the aspect ratio,

$$\epsilon = \frac{R}{L} \tag{7}$$

of the size of the liquid zone, and a scaled diffusional distance,

$$P_D = \frac{VR}{D} \tag{8}$$

for the diffusion of solute in the melt. The Peclet numbers Pe and P_S are defined by

$$Pe = \frac{VL}{\kappa} \tag{9a}$$

$$P_S = \frac{VL}{\kappa_S} \tag{9b}$$

where κ and κ_S denote the thermal diffusivities of the liquid and solid phases, respectively.

The Marangoni number, which is a measure of the heat transport by thermocapillary convection relative to thermal conduction, is defined by

$$Ma = \frac{\gamma R \Delta T}{\mu \kappa} \tag{10a}$$

the surface tension Reynolds number, which is a measure of the forces arising from thermocapillarity and viscosity, is given by

$$Re = \frac{\gamma R \Delta T}{\mu \nu} \tag{10b}$$

As a result of the moving coordinate system,

$$R_V = \frac{LV}{\nu} \tag{10c}$$

The Schmidt number, Sc, measures the transport of solute by convection relative to diffusion, and is given by

$$Sc = \frac{\gamma}{D} \tag{10d}$$

From Eqs. (8) and (9) we see that solutal Peclet numbers were defined using the crystal width, whereas thermal Peclet numbers were defined using the zone length. Under the assumption that ϵ is small, these scalings are consistant with systems for which the Lewis number, κ/D, is large. Using the material properties for GeGa given in Table 1, we find this to be the case. Thus, we have different length scales over which the diffusion of heat and solute occur. This disparity in length scales is the motivation for the use of asymptotic methods presented in the section on asymptotic analysis.

Letting $h(x)$ denote deviations from the planar solidifying interface, the governing system [Eqs. (3–5)] is subject to the following boundary conditions at the solidification front, $z = -\ell_S + \epsilon h(x)$:

$$T = T_S \tag{11a}$$

$$T = Mc + 1 + \frac{\Gamma h_{xx}}{[1 + (h_x)^2]^{\frac{3}{2}}} \tag{11b}$$

$$\epsilon PeSt = K[\epsilon T_{Sz} - T_{Sx}h_x] - [\epsilon T_z - T_x h_x] \tag{11c}$$

$$P_D c(k - 1) = \epsilon c_z - c_x h_x \tag{11d}$$

$$\psi = \psi_z = 0 \tag{11e}$$

Table 1 Material properties and operating parameters for Ge doped with Ga

Material: Ge 1% wt Ga

$R = 0.5$ cm
$V = 0.0003$ cm/s
$\phi = 10$ deg
$g_0 = 981$ cm/s^2
$\gamma = 700$ dyne/cm
$k_L = 0.17$ W/K $\cdot$ cm
$k_S = 0.17$ W/K $\cdot$ cm
$\rho_L = 5.6$ g/cm^3
$\rho_{\text{solid}} = 5.6$ g/cm^3
$c_p = 0.39$ J/g $\cdot$ K
$c_{p,\text{solid}} = 0.39$ J/g $\cdot$ K
$C_0 = 1$
$\pounds = 2833$ J/cm^3
$T_M = 1231$ K
$T_a = 340$ K
$T_h = 2250$ K
$a = 1.0$
$B_L = 0.005$
$B_S = 0.005$
$R_L = 0.03$
$R_S = 0.05$
$m = -4$ K/% wt
$k = 0.1$
$D = 1.3 \times 10^{-4}$ cm^2/s
$\gamma = 0.26$ dyne/cm $\cdot$ K

The nondimensional groups appearing in the Eqs. (11) are a morphological number

$$M = \frac{mC_0}{T_M} \qquad (12)$$

where m is the slope of the liquidus line, the nondimensional surface free energy

$$\Gamma = \frac{\gamma'}{\mathscr{L}R} \qquad (13)$$

where γ' is the surface free energy, and $\mathscr{L}$ is the latent heat per unit volume, the Stephan number

$$St = \frac{\mathscr{L}}{T_M \rho_L c_p} \qquad (14)$$

where ρ_L is the density of the melt, c_p is the heat capacity, and the ratio of solid and liquid conductivities is

$$K = \frac{k_S}{k_L} \tag{15}$$

Additionally, the parameter k denotes the segregation coefficient.

Letting $h_m(x)$ denote deviations from the planar melting interface, the following boundary conditions are applied at the melting front, $z = \ell_F + \epsilon h_m(x)$:

$$T = T_F \tag{16a}$$

$$T = Mc + 1 - \frac{\Gamma h_{mxx}}{[1 + (h_{mx})^2]^{\frac{3}{2}}} \tag{16b}$$

$$\epsilon PeSt = K[\epsilon T_{Fz} - T_{Fx}h_{mx}] - [\epsilon T_z - T_x h_{mx}] \tag{16c}$$

$$P_D(1 - c) = \epsilon c_z - c_x h_{mx} \tag{16d}$$

$$\psi = \psi_z = 0 \tag{16e}$$

For simplicity only we assume symmetry with respect to the centerline, $x = 0$. Here

$$T_x = T_{Fx} = T_{Sx} = c_x = h_x = h_{mx} = \psi = \psi_{xx} = 0 \tag{17}$$

Along the liquid/gas interface, $x = 1 + r(z)$, $-\ell_S < z < \ell_F$, where $r(z)$ denotes deviations from the planar, we have

$$c_x - \epsilon^2 c_z r_z = 0 \tag{18a}$$

$$\frac{-[T_x - \epsilon^2 T_z r_z]}{\sqrt{1 + \epsilon^2 (r_z)^2}} = B_L[T - \theta] + R_L[T^4 - \theta^4] \tag{18b}$$

$$-p + Bz = \frac{r_{zz}}{[1 + \epsilon^2 (r_z)^2]^{\frac{3}{2}}} \tag{18c}$$

$$\left(-\frac{Rv}{Re} + \psi_x\right)r_z + \psi_z = 0 \tag{18d}$$

$$-p + \frac{2\epsilon^2[-\psi_{xz} - r_z\psi_{xx} + \epsilon^2 r_z(\psi_{zz} + r_z\psi_{xz})]}{1 + \epsilon^2 r_z^2} = [\epsilon^3 C^{-1} - \epsilon^2 T]\frac{r_{zz}}{[1 + \epsilon^2 r_z^2]^{\frac{3}{2}}} \tag{18e}$$

$$(1 - \epsilon^2 r_z^2)(\psi_{xx} - \epsilon^2 \psi_{zz}) - 4\epsilon^2 r_z\psi_{xz} = -(1 + \epsilon^2 r_z^2)^{\frac{1}{2}}(T_z + r_z T_x) \tag{18f}$$

Equation (18a) denotes zero solute flux at the liquid/gas interface. Equation (18b) represents an energy balance for the heat transfer between the melt and the surrounding gas. The temperature profile established by the heater is given by $\theta(z)$. For the results presented in the results section we take

$$\theta(z) = T_a + (T_h - T_a) \exp\left(-\left[\frac{Lz}{a}\right]^2\right) \qquad (19)$$

where a is a scale factor to measure the width of the heater profile. Equation (18c) represents the normal force balance across the liquid/gas interface assuming no melt motion and using the hydrostatic pressure scale in Eq. (2c). Equation (18d) is the kinematic condition, and Eqs. (18e) and (18f) are the normal and tangential force balances in the presence of flow and using the pressure scale in Eq. (2d). The nondimensional groups appearing in Eq. (18) are the Biot number

$$B_L = \frac{h_L R}{k_L} \qquad (20)$$

where h_L is the heat transfer coefficient, the radiation number

$$R_L = \frac{\sigma \epsilon_m R (T_M)^3}{k_L} \qquad (21)$$

where σ is the Stefan-Boltzmann constant and ϵ_m is the emissivity, the Bond number

$$B = \frac{\rho_L g L^2}{\gamma_s \epsilon} \qquad (22)$$

which measures the strength of the gravitational force, and the capillary number

$$C = \frac{\mu w^*}{\gamma_s} \qquad (23)$$

which measures the viscous forces relative to the surface tension force.

Along the solid boundaries of the feed and product crystals, the energy balances are

Feed ($x = 1, z > \ell_F$):

$$-T_{Fx} = B_S[T_F - \theta] + R_S[T_F^4 - \theta^4] \qquad (24)$$

Product ($x = 1, z < -\ell_S$):

$$-T_{Sx} = B_S[T_S - \theta] + R_S[T_S^4 - \theta^4] \qquad (25)$$

The Biot number B_S and the radiation number R_S are defined by replacing h_L with h_S and k_L with k_S in Eqs. (20) and (21).

Far from the melting and solidifying interfaces, we assume that the temperature of the crystals equals the ambient temperature,

$$T_F = T_a, \qquad z \to \infty \tag{26a}$$

$$T_S = T_a, \qquad z \to -\infty \tag{26b}$$

This condition can be replaced by a flux constraint.

Finally, where the liquid/gas interface meets the melting and solidifying interfaces, we have the following contact conditions:

$$r(z) = 0, \qquad z = \ell_F + \epsilon h_m(1) \tag{27a}$$

$$r(z) = 0, \qquad z = -\ell_S + \epsilon h(1) \tag{27b}$$

To set the value of the pressure, we impose an additional contact angle condition. As shown in Fig. 2, there exists two material-dependent contact angles α and ϕ. By setting $\Gamma = 0$ in Eqs. (11b) and (16b), the contact angle α is consequently free to take on any value consistent with the determination of h and h_m. In the case that $\Gamma \neq 0$, then α is a material property and the solid/liquid interfaces adjust to this value. However, it has been shown that the value of ϕ at the solidifying front is necessarily a material property determining the stability of the crystal shape. Hence, we assume

$$\phi = \epsilon^2 \bar{\phi} \tag{28}$$

where $\bar{\phi}$ is a constant and require

$$\epsilon r_z = \tan(\epsilon^2 \bar{\phi}), \qquad z = -\ell_S + \epsilon h(1) \tag{29}$$

For binary systems we require an additional condition guaranteeing conservation of mass. Since the feed crystal is assumed to have a scaled uniform concentration of 1, then the solidifying crystal must have a concentration C_S such that

$$\int_0^1 C_S(x,z) \, dx = 1 \tag{30}$$

for all z. At the interface, $z = -\ell_S + \epsilon h(x)$, we require the segregation condition

$$C_S = kc \tag{31}$$

Hence, integrating across the crystal width and using Eq. (30) we obtain

$$\int_0^1 c[x,z = -\ell_S + \epsilon h(x)] \, dx = 1/k \tag{32}$$

We also require zero net mass flow in the melt, which leads to the condition

$$\psi = 0 \qquad \text{at} \qquad x = 1 + r(z) \tag{33}$$

Summarizing, for the float zone shown in Fig. 1, the governing equations are Eqs. (3–5) subject to the boundary conditions (11) and (16) at the solidifying and melting interfaces, respectively. The symmetry conditions [Eq. (17)] apply along the centerline. At the liquid/gas interface, we require Eq. (18), and along the solid boundaries Eqs. (24) and (25) apply. Far away from interfaces, Eq. (26) applies, and finally Eqs. (27) and (29) establish the trijunction and set the value of the pressure. For binary systems Eq. (32) guarantees conservation of solute, whereas Eq. (33) guarantees zero net mass flow.

We shall consider two limiting cases of this system. First, we examine a conduction-dominated heat-transfer model for which $u \equiv 0$. We consider both pure and binary systems in this regime. Second, we investigate the effects of surface-tension-driven flow, $u \neq 0$, and $g = 0$ for pure systems ($c = 0$) only. For the limits discussed in the next section an examination of Eqs. (3a) and (3c) shows that the flow model for small ϵ will characterize conduction-dominated heat transfer in a viscous melt. Thus, the results will be limited to Marangoni numbers less than $0(10^2)$ in magnitude.

Scaling

Like Brattkus and Davis,[22] we seek an approximate solution to the governing system in the limit of small aspect ratio ϵ. Furthermore, we assume the following asymptotic limits:

$$B_L = \bar{B}_L \epsilon^2, \qquad B_S = \bar{B}_S \epsilon^2, \qquad R_L = \bar{R}_L \epsilon^2, \qquad R_S = \bar{R}_S \epsilon^2 \tag{34}$$

$$C = \epsilon^4 \bar{C}, \tag{35}$$

whereas all remaining nondimensional groups defined in the previous section are taken to be order one quantities. In particular, we note that the surface tension γ_S is typically quite large for the materials under consideration so that the Bond number defined in Eq. (22) can be taken as order one, while the capillary number is quite small ($0(10^{-3})$ for silicon). The choice of ϵ^4 as discussed in Ref. 13 is to preserve conservation of mass.

The result of the small ϵ limit is that the solution to the governing system separates into core and boundary-layer regions as shown in Fig. 3. Equation (34) scales the heat transfer so that the core temperature fields are not identical to the heater profile. Larger heat-transfer results in the latter. The magnitudes established by Eq. (34) are also reasonable considering the size of these numbers, $0(10^{-2})$, as listed by Duranceau and Brown[1] for small-scale silicon floating zones.

Additionally, small ϵ results in the consideration of small contact angles ϕ as given by Eq. (28). For silicon, ϕ is approximately 0.2 rad; thus, this

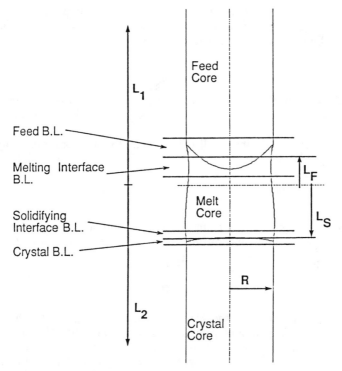

Fig. 3   Boundary-layer representation.

assumption does not seem too restrictive. The factor ϵ^2 in Eq. (28) is chosen for convenience of the asymptotic analysis to follow so that the heat transfer and nonplanar liquid/gas interfacial effects appear at the same order.

Finally, we must consider the consequences of small ϵ on the static stability of the solutions obtained. It is known that for a cylindrical liquid zone trapped between two rigid plates that the maximum zone length L must not exceed $2\pi R$. Thus, clearly ϵ has a lower bound. Experiments for thin vertical sheets have reported stable zone configurations for which $\epsilon \sim 1/2$ (Baghdadi and Gurtler[17]). Smaller aspect ratios ($\epsilon \sim 1/10$) (Yablonovitch and Gmitter[18]) have been observed when oxide films are present or when the zone is established only in the central portion of the sheet with the edges remaining solid. Additionally, Murphy et al.[16] report cylindrical melt zones surpassing the $2\pi R$ limit.

In our procedure the value of R is given and the value of L is determined through an asymptotic expansion that is valid for $\epsilon \ll 1$. The value of ϵ is therefore determined only when the solution is completed and is not assumed a priori. The validity of the results we obtain in the following sections can be tested either through a stability analysis, by comparison with a numerical solution of the full governing system, or by comparison with experiments. The stability analysis is not presented here. Rather, we consider a qualitative comparison with the axisymmetric computational results of Duranceau and Brown[1] to justify the small ϵ limits.

Asymptotic Analysis

In this section the solution procedure for the conduction-dominated, no-flow, heat-transfer model is described, together with results for the flow model.

As shown in Fig. 3, the small ϵ limit leads to core solutions away from the melting and solidifying interfaces and boundary-layer solutions near these interfaces.

Core

In the core regions away from the melting and solidifying interfaces we assume a power series expansion for all dependent variables $\psi(x,z)$ such that

$$\psi(x,z) = \psi_0 + \epsilon\psi_1 + \cdots \qquad (36)$$

Substituting Eq. (36) into the governing system, using the scalings in Eq. (34), and neglecting boundary conditions [Eqs. (11) and (16)], we obtain a series of systems of equations for each order of ϵ. The details of the solution of these equations is presented in Young and Chait.[23] We find that core temperature, concentration, and pressure fields are given by

$$T = T_0(z) + \epsilon T_1(z) + \epsilon^2\{T_2(z) - \tfrac{1}{2}x^2[T_{0zz} + PeT_{0z}]\} + \cdots \qquad (37a)$$

$$T_F = T_{F0}(z) + \epsilon T_{F1}(z) + \epsilon^2\{T_{F2}(z) - \tfrac{1}{2}x^2[T_{F0zz} + P_S T_{F0z}]\} + \cdots \qquad (37b)$$

$$T_S = T_{S0}(z) + \epsilon T_{S1}(z) + \epsilon^2\{T_{S2}(z) - \tfrac{1}{2}x^2[T_{S0zz} + P_S T_{S0z}]\} + \cdots \qquad (37c)$$

$$c = c_0 + \epsilon c_1 + \epsilon^2 c_2 + \cdots \qquad (37d)$$

$$P = 2\{B[-\tfrac{1}{2}\ell_S^2 + \tfrac{1}{6}(\ell_S^2 - \ell_S\ell_F + \ell_F^2)]\} + \epsilon 2\bar{\phi} + 0(\epsilon^2) \qquad (37e)$$

and the position of the liquid/gas interface is

$$r(z) = \tfrac{1}{6}B[z^3 + z^2(2\ell_S - \ell_F) + z(\ell_S^2 - 2\ell_S\ell_F) - \ell_S^2\ell_F]$$
$$+ \epsilon\bar{\phi}[-z^2 + (\ell_F - \ell_S)z + \ell_S\ell_F] + 0(\epsilon^2) \qquad (37f)$$

In the above ℓ_S and ℓ_F have yet to be determined. The c_i are all constants to be determined through matching conditions at the interfaces, and the $T_i(z)$, $T_{Fi}(z)$, and $T_{Si}(z)$ are all functions determined from the heat-transfer conditions, Eqs. (18b), (24), and (25). At leading order, these conditions give

$$(1 + r_0)(T_{0zz} + PeT_{0z}) + T_{0z}r_{0z} - \bar{B}_L T_0 - \bar{R}_L T_0^4$$

$$= -\bar{B}_L \theta - \bar{R}_L \theta^4 (-\ell_S < z < \ell_F) \tag{38a}$$

$$T_{F0zz} + P_S T_{F0z} - \bar{B}_S T_{F0} - \bar{R}_S T_{F0}^4 = -\bar{B}_S \theta - \bar{R}_S \theta^4 (z > \ell_F) \tag{38b}$$

$$T_{S0zz} + P_S T_{S0z} - \bar{B}_S T_{S0} - \bar{R}_S T_{S0}^4 = -\bar{B}_S \theta - \bar{R}_S \theta^4 (z < -\ell_S) \tag{38c}$$

The differential Eqs. (38a–38c) determining the leading-order core temperature profile are similar to those derived by Brattkus and Davis.[22] However, ours are nonlinear due to the radiation term and have nonconstant coefficients due to the liquid/gas interfacial shape, $r_0(z)$. Additionally, $\theta(z)$ allows a spatially varying heater temperature. Boundary conditions for Eq. (38a) are determined from matching criteria at both the solidifying and melting interfaces. We use Eq. (26a) and a matching condition at the melting interface as boundary conditions for Eq. (38b), whereas Eq. (26b) and matching at the solidifying interface give the boundary conditions for Eq. (38c).

Equations similar to Eqs. (38) are derived for the temperature functions T_1, T_2, etc. Again the boundary conditions for these differential equations are determined from higher-order matching. Additionally, corrections to the liquid/gas interfacial shape, $r(z)$, are determined as functions of the solidifying and melting interfaces.

For the flow model we find

$$T = T_0(z) + \epsilon T_1(z) + \epsilon^2 \left\{ T_2(z) - \frac{1}{2} x^2 [PeT_{0z} + T_{0zz}] \right.$$

$$\left. + \frac{1}{12} MaT_{0z}^2 \left[x^2 - \frac{x^4}{2} \right] \right\} + 0(\epsilon^3) \tag{39a}$$

$$\psi = \tfrac{1}{6} T_0'(z)[x - x^3] + \epsilon \{ \tfrac{1}{6} T_1'(z)[x - x^3] + \tfrac{1}{6} T_0'(z)r_1(z)[x + x^3] \} + 0(\epsilon^2) \tag{39b}$$

$$p = \bar{C}^{-1} r_1'' + 0(\epsilon) \tag{39c}$$

$$r = r(z)[\text{hydrostatic Eq. (37f)}] + \epsilon r_1(z) + 0(\epsilon^2) \tag{39d}$$

with Eqs. (37b) and (37c) still valid in the solid phases. Here

$$r_1''' = \bar{C} T_0' \tag{40}$$

subject to the contact conditions

$$r_1(-\ell_S) = 0 \tag{41a}$$

$$r_1(\ell_F) = 0 \tag{41b}$$

$$r_1'(-\ell_S) = 0 \tag{41c}$$

and T_0 satisfies Eq. (38a). This agrees with our earlier comment that the heat transfer is conduction dominated. A comparison of Eqs. (37a) and (39a) reveals that the convective heat transfer (term proportional to Ma) does not appear until $0(\epsilon^2)$. Also, note that the presence of flow alters the pressure profile so that the liquid/gas interfacial shape is no longer hydrostatic. However, for the systems of interest the capillary number is so small that the term r_1 is negligible, and the liquid/gas interface is essentially the hydrostatic solution.

Boundary Layer: Solidifying Interface

The asymptotic representations for the core region do not satisfy the boundary conditions [Eq. (11)] at the solidifying front. Here, solutions no longer vary slowly in the z direction so that appropriate balances between horizontal and lateral diffusion are needed. These are obtained by rescaling the z coordinate as

$$\eta = \frac{z + \ell_S}{\epsilon} \tag{42}$$

In this region the temperature field of the solid feed, T_F, is not needed, and we write

$$(c, T, T_S) = \sum_{j=0}^{\infty} (c_j^S, T_j^S, \theta_{Sj}) \epsilon^j \tag{43}$$

Also, since the heat transfer is small according to Eq. (34), the solidifying interface will be flat to leading order in ϵ; thus, we expand $h(x)$ as

$$h = \sum_{j=1}^{\infty} h_j(x) \epsilon^j \tag{44}$$

For the same reason the melting interface $h_m(x)$ will also be flat to leading order. We find the following solutions valid in the solidifying boundary layer:

$$T^S = \left[T_0(-\ell_S) = \frac{Mc_0}{k} + 1 \right] + \epsilon[T_{0z}(-\ell_S)\eta] + 0(\epsilon^2) \tag{45a}$$

$$c^S = \left\{ c_0 \left[1 + \frac{1-k}{k} \exp(-P_D\eta) \right] \right\} + 0(\epsilon^2) \tag{45b}$$

$$\theta_S = \left[T_{S0}(-\ell_S) = \frac{Mc_0}{k} + 1 \right] + \epsilon[T_{S0z}(-\ell_S)\eta] + 0(\epsilon^2) \tag{45c}$$

$$h_1 = 0 \tag{45d}$$

with $c_0 = 1$ due to Eq. (32). The leading-order expressions for T^S and θ_S give additional boundary conditions for Eqs. (38a) and (38c). From Eq. (11c) we find a compatibility constraint for the heat balance at the solidifying interface:

$$KT_{S0z}(-\ell_s) - T_{0z}(-\ell_S) = PeSt \tag{46}$$

In the next section we show that this equation couples to a similar version derived at the melting boundary layer, and together the solution of the two determines the mean positions $-\ell_S$ and ℓ_F of the solidifying and melting interfaces.

The $0(\epsilon^2)$ problem is the first system in which heat transfer in the boundary layer occurs at the liquid/gas interface. It is also the first order in which the effects of a nonplanar liquid/gas interface are felt. We find the following $0(\epsilon^2)$ corrections by making use of the definition of $-\ell_S$ as the mean position of the solidifying interface:

$$T_2^S = \bar{T} + T_2(-\ell_S) - \tfrac{1}{2}x^2[T_{0zz}(-\ell_S) + PeT_{0z}(-\ell_S)]$$

$$+ T_{1z}(-\ell_S)\eta + \tfrac{1}{2}\eta^2 T_{0zz}(-\ell_S) \tag{47a}$$

$$\theta_{S2} = \bar{\theta} + T_{S2}(-\ell_S) - \tfrac{1}{2}x^2[T_{S0zz}(-\ell_S) + P_S T_{S0z}(-\ell_S)]$$

$$+ T_{S1z}(-\ell_S)\eta + \tfrac{1}{2}\eta^2 T_{S0zz}(-\ell_S) \tag{47b}$$

$$c_2^S = \bar{c} + c_2 + \exp[-P_D\eta]\{\psi_1\eta + \psi_2 x^2 + \psi_3 x^2\eta + \psi_4\eta^2 + \psi_5 x^4\} \tag{47c}$$

where

$$\left.\begin{array}{c} \psi' = \dfrac{1}{2}\bar{\phi}P_D\left(\dfrac{k-1}{k}\right) \\[2mm] \psi_1 = \dfrac{2\psi_2}{P_D} + \dfrac{2\psi_3}{P_D^2} \\[2mm] \psi_2 = \tfrac{1}{2}\psi - \tfrac{1}{6}\psi_3 P_D \\[2mm] \psi_3 = \dfrac{1}{6}BP_D\dfrac{1-k}{k} \\[2mm] \psi_4 = \dfrac{\psi_3}{P_D} \\[2mm] \psi_5 = \dfrac{1}{12}\psi_3 P_D \end{array}\right\} \tag{48}$$

and

$$\bar{\theta} = \frac{\left\{(\bar{B}_S - \bar{B}_L)\left[\frac{M}{k} + 1 - \theta(-\ell_S)\right] + (\bar{R}_S - \bar{R}_L)\left[\left(\frac{M}{k} + 1\right)^4 - \theta^4\right]\right\}}{2(1 + K)}$$

$$\times \sum_{j=1}^{\infty} \frac{4(-1)^j}{(j\pi)^2} \exp(j\pi\eta)\cos(j\pi x) \tag{49}$$

$$\bar{T} = -K\bar{\theta}(x, -\eta) \tag{50}$$

$$\bar{c} = \left(\frac{-1}{kP_D}\right)\exp(-P_D\eta)\left[-\psi_1 - \frac{1}{3}\psi_3 + kP_D\left(\frac{1}{3}\psi_2 + \frac{1}{5}\psi_5\right)\right.$$

$$\left. - P_D c_2(1 - k)\right]\frac{-kP_D}{2M(1 + K)}$$

$$\times \left\{(\bar{B}_L + K\bar{B}_S)\left[\frac{M}{k} + 1 - \theta(-\ell_S)\right] + (\bar{R}_L + K\bar{R}_S)\left[\left(\frac{M}{k} + 1\right)^4 - \theta^4\right]\right\}$$

$$\times \sum_{j=1}^{\infty} \frac{4(-1)^j \exp(\delta_j\eta)\cos(j\pi x)}{(j\pi)^2(P_D + \delta_j)} - \psi_3 \sum_{j=1}^{\infty} \frac{4(-1)^j \exp(\delta_j\eta)\cos(j\pi x)}{(j\pi)^2(P_D + \delta_j)} \tag{51}$$

$$h_2 = \frac{k}{(1 - k)P_D}\psi_2\left\{\left(x^2 - \frac{1}{3}\right) + \psi_5\left(x^4 - \frac{1}{5}\right)\right\}$$

$$+ \frac{k(\bar{B}_L + K\bar{B}_S)\left[\frac{M}{k} + 1 - \theta\right] + (\bar{R}_L + K\bar{R}_S)\left[\left(\frac{M}{k} + 1\right)^4 - \theta^4\right]}{(1 - k)2MP_D(1 + K)}$$

$$\times \left\{\left(x^2 - \frac{1}{3}\right) - \frac{4kP_D}{\pi^2}\sum_{j=1}^{\infty} \frac{(-1)^j\cos(j\pi x)}{j^2(P_D + \delta_j)}\right\}$$

$$- \frac{k\psi_3}{(1 - k)P_D}\sum_{j=1}^{\infty} \frac{4(-1)^j\cos(j\pi x)}{(j\pi^2)(P_D + \delta_j)} \tag{52}$$

and

$$\delta_j = -\frac{1}{2}[P_D^2 + \sqrt{P_D^2 + 4j^2\pi^2}] \tag{53}$$

Also, we find

$$c_2 = -\frac{(\psi_1 + \frac{1}{3}\psi_3)}{P_D} = \left(\frac{1 - k}{kP_D}\right)\left[\bar{\phi} - \frac{B}{3P_D}\right] \tag{54}$$

Considering Eq. (52), one finds that the nonplanar correction h_2 to the solidifying interface is influenced by a scale factor inversely proportional to the leading-order solutal gradient which multiplies the sum of two terms reflecting curved isotherms resulting from the heat transfer across the liquid/gas interface and the local solute distribution and, finally, by the nonplanar liquid/gas interface through the contact angle ϕ.

Boundary Layer: Melting Interface

Similar to the analysis at the solidifying interface, one finds that the asymptotic conditions for the core region do not satisfy *all* of the boundary conditions [Eq. (16)] at the melting front. A balance between horizontal and lateral thermal diffusion is required. This is obtained by rescaling the z coordinate as

$$\rho = \frac{z - \ell_F}{\epsilon} \tag{55}$$

In this region the temperature field of the solid product crystal is not needed, and we write

$$(c, T, T_F) = \sum_{j=0}^{\infty} (c^F, T_j^F, \theta_{Fj}) \epsilon^j \tag{56}$$

As mentioned earlier, the small heat-transfer results in the melting interface being flat to leading order in ϵ. Thus, we expand $h_m(x)$ as

$$h_m = \sum_{j=1}^{\infty} h_{mj}(x) \epsilon^j \tag{57}$$

Similar to the analysis at the solidifying interface, we find the following solutions to be valid in the melting boundary layer:

$$T^F = [T_0(\ell_F) = M + 1] + \epsilon[T_{0z}(\ell_F)\rho] + 0(\epsilon^2) \tag{58a}$$

$$c^F = 1 + 0(\epsilon^2) \tag{58b}$$

$$\theta_F[T_F(\ell_F) = M + 1] + \epsilon[T_{F0z}(\ell_F)\rho] + 0(\epsilon^2) \tag{58c}$$

The constant concentration profile at the melting interface causes this free surface to distort more than the solidifying front. Hence, $h_{m1} \neq 0$ for this case. Its determination is postponed until the $0(\epsilon^2)$ system. The leading-order expressions for T^F and θ_F give additional boundary conditions for Eqs. (38a) and (38b). Also, Eq. (16c) leads to the compatibility constraint

$$KT_{F0z}(\ell_F) - T_{0z}(\ell_F) = PeSt \tag{59}$$

This equation couples to Eq. (46). A discussion of their significance is given in the next section.

At the next order of approximation, heat transfer in the boundary layer occurs at the liquid/gas interface. Making use of the definition of ℓ_F as the mean position of the melting interface, we find

$$T_2^F = \hat{T} + T_2(\ell_F) - \tfrac{1}{2}x^2[T_{0zz}(\ell_F) + PeT_{0z}(\ell_F)]$$

$$+ T_{1z}(\ell_F)\rho + \tfrac{1}{2}\rho^2 T_{0zz}(\ell_F) \tag{60a}$$

$$\theta_{F2} = \hat{\theta} + T_{F2}(\ell_F) - \tfrac{1}{2}x^2[T_{F0zz}(\ell_F) + P_S T_{F0z}(\ell_F)]$$

$$+ T_{F1z}(\ell_F)\rho + \tfrac{1}{2}\rho^2 T_{F0zz}(\ell_F) \tag{60b}$$

where

$$\hat{\theta} = \left(\left\{\left[\bar{B}_S\left(K - \frac{PeST}{T_{F0z}(\ell_F)}\right) - \bar{B}_L\right][M + 1 - \theta(\ell_F)]\right.\right.$$

$$\left.\left. + \frac{\left[\bar{R}_S\left(K - \dfrac{PeSt}{T_{F0z}(\ell_F)}\right) - \bar{R}_L\right][(M + 1)^4 - \theta^4]}{2\left(2K - \dfrac{PeSt}{T_{F0z}(\ell_F)}\right)}\right\}\right)$$

$$\times \sum_{j=1}^{\infty} \frac{4(-1)^j}{(j\pi)^2} \exp(-j\pi\rho)\cos(j\pi x) \tag{61a}$$

$$\hat{T} = -K\hat{\theta}(x, -\rho) \tag{61b}$$

$$h_{m1} = \frac{(x^2 - \tfrac{1}{3})}{2T_{0z}(\ell_F)}\left(\left\{\bar{B}_L + \frac{K\left[\bar{B}_S\left(K - \dfrac{PeSt}{T_{F0z}(\ell_F)}\right) - \bar{B}_L\right]}{\left(2K - \dfrac{PeSt}{T_{F0z}(\ell_F)}\right)}\right\}[M + 1 - \theta(\ell_F)]\right.$$

$$\left. + \left\{\bar{R}_L + \frac{K\left[\bar{R}_S\left(K - \dfrac{PeSt}{T_{F0z}(\ell_F)}\right) - \bar{R}_L\right]}{\left(2K - \dfrac{PeSt}{T_{F0z}(\ell_F)}\right)}\right\}[(M + 1)^4 - \theta^4]\right) \tag{61c}$$

Considering Eq. (61c), one finds that the nonplanar correction, h_{m1}, to the melting interface is influenced by curved isotherms resulting from heat transfer across the liquid/gas interface and an overall scale factor that is *inversely proportional to the leading-order temperature gradient at the melting interface*. Because of the constant concentration profile, the melting interface is an order of magnitude larger in deformation than the solidify-ing interface and is strongly influenced by the core temperature profile.

In the limit $M \to 0$ (pure system limit) the same will occur at the solidifying interface. As discussed by Brattkus and Davis,[22] $h_1 \neq 0$ in this limit, and the interface undergoes a greater deformation since it is an isotherm. In the case where $M \neq 0$, the leading-order solutal gradient determines the magnitude of the deformation through an inverse propor-tionality. Hence, the interface (melting or solidifying) possessing the shal-lowest concentration gradient will undergo the largest deflection from the planar.

The boundary-layer analyses for the flow model are very similar to those for the conduction-dominated, no-flow system. The leading-order boundary-layer solutions for the flowfield at the solidifying and melting interfaces, respectively, are

$$\psi_0^S = \frac{1}{6} T_0'(-\ell_S)[x - x^3] + 2T_0'(-\ell_S) \sum_{j=1}^{\infty} \frac{(-1)^j}{(j\pi)^3} [1 + j\pi\eta] \, e^{-j\pi\eta} \sin(j\pi x)$$

$$(62a)$$

$$\psi_0^F = \frac{1}{6} T_0'(\ell_F)[x - x^3] + 2T_0'(\ell_F) \sum_{j=1}^{\infty} \frac{(-1)^j}{(j\pi)^3} [1 + j\pi\rho] \, e^{j\pi\rho} \sin(j\pi x) \quad (62b)$$

The higher-order boundary-layer corrections require the solution of a biharmonic equation forced by the nonlinear convective terms. The effects of nonplanar liquid/gas, melting, and solidifying interfacial shapes are included in the boundary conditions for this system. However, we shall only present composite solutions based on the leading-order expressions (39b), (62a), and (62b). These are valid in a rectangular melt bounded by $-1 \leqslant x \leqslant 1$ and $-\ell_S \leqslant z \leqslant \ell_F$. It is this flow, though, that alters the temperature profile of the melt and leads to the solidifying and melting interfacial shapes:

$$h_1^* = h_{m1}(x; -\ell_S, T_{S0}) + \frac{Ma}{48} \left[\frac{PeSt - KT_{S0}'(-\ell_S)}{-PeSt + 2KT_{S0}'(-\ell_S)} \right]$$

$$\times \left(x^2 - \frac{1}{2} x^4 - \frac{7}{30} \right) T_{S0}'(-\ell_S) \qquad (63a)$$

$$h_{m1}^* = h_{m1}(x; \ell_F, T_{F0}) + \frac{Ma}{48} \left[\frac{PeSt - KT_{F0}'(\ell_F)}{-PeSt + 2KT_{F0}'(\ell_F)} \right]$$

$$\times \left(x^2 - \frac{1}{2} x^4 - \frac{7}{30} \right) T_{F0}'(\ell_F) \qquad (63b)$$

where h_{m1} is defined as in Eq. (61c), setting $M = 0$ and replacing ℓ_F with $-\ell_S$ and T_{F0} with T_{S0} for the solidification front. Thus, the interfaces are composed of a combination of the conduction heat-transfer solution with corrections due to the convective heat transfer. In these corrections it can be shown that the term in brackets is always negative; thus, the sign of the temperature gradients, $T'_{S0}(-\ell_S)$ and $T'_{F0}(\ell_F)$, determines the concavity of the convective corrections. Note that the convective corrections are also *linearly proportional to the leading-order temperature gradients*, whereas the conduction pieces are inversely proportional. We shall discuss this further in the next section.

Results

Equations (47–52) and (60–63) display a dependence on the mean positions ℓ_F and $-\ell_S$ of the melting and solidifying interfaces. These values have yet to be determined. They are calculated as the solution to the system of Eqs. (46) and (59), which represents energy balances at the respective interfaces. These equations are coupled through the gradient of the core melt temperature. Thus, the mean positions of the interfaces will be influenced by the latent heat and the heat transfer across the liquid/gas interface in the core regions.

To determine ℓ_F and ℓ_S, one must calculate the leading-order core temperature fields, T_0, T_{F0}, and T_{S0}. T_0 is found by solving Eq. (38a) subject to Eqs. (45a) and (58a). T_{F0} is found by solving Eq. (38b) subject to Eqs. (26a) and (58c). Finally, T_{S0} is found by solving Eq. (38c) subject to Eqs. (26b) and (45c). In practice, the boundary conditions at $\pm\infty$ are applied at arbitrary but large values of L_1 and L_2 (see Fig. 3). In general, for an order of unity Bond number, nonzero radiation parameters $\bar{R}_L$ and $\bar{R}_S$, and an arbitrary heater temperature profile $\theta(z)$ (such as the Gaussian profile used by Duranceau and Brown[1,2]) the preceding task will require a numerical solution of the differential equations. The results are input to a Newton iteration of Eqs. (46) and (59), with convergence being usually obtained in less than 0.1 CPU seconds on a Cray XMP, allowing rapid examination of the effects of varying system and material parameters.

We now consider a parametric study of the system using the data for silicon as listed in Table 2. These properties were taken from Duranceau and Brown.[1] We shall qualitatively compare our sheet results with the axisymmetric results of Duranceau and Brown.[1,2] Recall that the solidifying interface is given by

$$z = -L_S + \epsilon^2 Rh_2\left(\frac{x}{R}\right), \qquad -R \leqslant x \leqslant R \tag{64}$$

where Eq. (64) is in dimensional terms, in the presence of a solute with no flow, and for h_2 defined as in Eq. (52). For a pure system with no flow, the solidifying interface is given by

$$z = -L_S + \epsilon Rh_{m1}\left(\frac{x}{R}; -\ell_S\right) \tag{65}$$

where h_{m1} is defined in Eq. (61c) setting $M = 0$ and replacing ℓ_F with $-\ell_S$.

Table 2   Material properties and operating parameters for Si

Material: Si
$R = 0.5$ cm
$V = 0.00417$ cm/s
$\phi = 11$ deg
$g_0 = 981$ cm/s^2
$\gamma = 720$ dyne/cm
$k_L = 0.64$ W/K · cm
$k_S = 0.22$ W/K · cm
$\rho_L = 2.55$ g/cm^3
$\rho_S = 2.33$ g/cm^3
$c_p = 1.059$ J/g · K
$c_{p,\text{solid}} = 1.038$ J/g · K
$\pounds = 4598$ J/cm^3
$T_M = 1695$ K
$T_a = 340$ K
$T_h = 2542$ K
$a = 1.0$
$B_L = 0.005$
$B_S = 0.005$
$R_L = 0.018$
$R_S = 0.043$

The melting interface is given by

$$z = L_F + \epsilon R h_{m1}\left(\frac{x}{R} ; \ell_F\right) \tag{66}$$

for both pure ($M = 0$) and binary systems ($M \neq 0$) with h_{m1} defined as in Eq. (61c). For pure systems with surface-tension-driven flow, the solidifying and melting interfaces are given by

$$z = -L_S + \epsilon R h_1^*\left(\frac{x}{R}\right) \tag{67}$$

$$z = L_F + \epsilon R h_{m1}^*\left(\frac{x}{R}\right) \tag{68}$$

with h_1^* and h_{m1}^* given by Eqs. (63a) and (63b).

The liquid/gas interfaces are given by

$$\pm x = R[1 + r_0(z) + \epsilon r_1(z)] \tag{69}$$

where Eq. (37f) gives $r_0(z)$ and $r_1(z)$ for the no-flow system, and Eq. (39d) gives these values in the presence of a melt flow. Finally, after the Newton

Table 3   Mean positions of melting and solidifying
interfaces

	L_F	L_S
Duranceau and Brown[1]	0.271	0.614
Present study	0.300	0.675

iteration is completed, L is determined using Eq. (1), the value R is input,
and the magnitude of $\epsilon = R/L$ is determined. Typically we calculate values
for ϵ in the range 0.5–0.7. Table 3 lists a quantitative comparison with
Duranceau and Brown.[1]

We have determined values of L_F and $-L_S$ for the work of Duranceau
and Brown[1] by averaging their centerline and edge values of the respective
interfacial shapes. In Fig. 4 we display our calculated zone shape with the
comment that Duranceau and Brown also find that the melting interface
undergoes greater deformation than the solidifying interface for these
conditions as predicted by our analysis. The agreement we have with their
results for the mean positions suggests that the higher-order corrections,
which we have not presented, may be negligible.

In the following parametric study all input data are held fixed as given
in Table 2, whereas the parameter under investigation is varied. We note
that the following results are qualitatively equivalent to those of Duranceau
and Brown[1] and are applicable to the no-flow model, unless otherwise
stated.

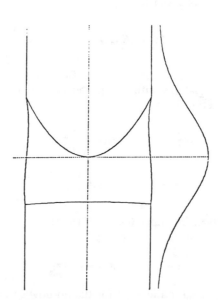

Fig. 4   Float zone with parameters due to Duranceau and Brown.[1,2]

g/g_0:   0   1   4

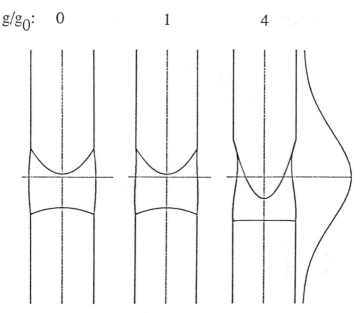

Fig. 5   Effects of gravity variation.

Gravity

In Fig. 5 we examine the results for varying gravity. As g is increased from zero, the pressure changes from constant to hydrostatic. The liquid/gas interface correspondingly changes from parabolic to a necked cubic profile. This change in interfacial shape results in the maximum value of the imposed temperature profile being applied at the necked region. From the left-hand side of Eq. (38a) one sees that changes in the interfacial shape r_0 (surface area) alters the heat transfer. In particular, there is less heat transfer due to less surface area for the necked profile. Hence, the liquid zone decreases in size as gravity increases.

Furthermore, notice that as gravity increases the contact angle at the melting interface is also decreasing. Should this angle decrease below the equilibrium receding value, the contact line may move to dynamically dewet the feed rod, causing failure of the zone. Additionally, short-range forces, such as Van der Waal's forces, may be predominant in this thin melt region. Our model does not account for such physics; and thus, we do not find a solution under conditions of large g or small surface tension, i.e., large Bond number.

Contact Angle

In Fig. 6 we examine the results for varying contact angle. Clearly, all interfacial shapes change with the contact angle; however, the magnitude of the changes is quite small. Thus, the calculated zones are similar even when

φ (deg) :   11   22

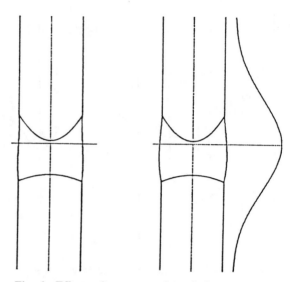

Fig. 6   Effects of contact angle variation.

the contact angle is doubled. This illustration, coupled with the next discussion on the effects of the emissivity, point out the potential use of our model for the purpose of identifying the critical material parameters defining the float zone and the desired accuracy of their measurement.

Emissivity

In Fig. 7 we examine the results for varying the liquid phase emissivity. Since radiation occurs as the fourth power of temperature, it is the predominant mode of heat transfer. The zone height increases as the emissivity increases due to the greater heat transfer. We find that the zone shape is most sensitive to changes in the emissivity. As an example, consider Eq. (52), which defines the solidifying front. For a pure system this expression defines an isotherm at the melting point, T_M. The term $(M = 0)$

$$T_M^4[1 - \theta^4(-\ell_S)] \tag{70}$$

in this expression determines the concavity of the solidifying front. For the low emissivity case, the heater temperature at the interface on the outer wall between the solidifying front and the gas/melt interface is higher than the melting point. Thus, heat transfers in at the interface, leading to a convex toward the melt interface profile. On the other hand, for the high-emissivity case, more heat transfers into the bulk melt, causing a

ε_m: 0.6   0.3

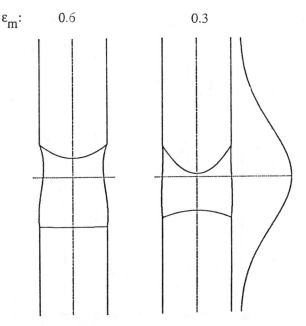

Fig. 7   **Effects of surface emissivity variation.**

larger zone. Here the heater temperature at the outer wall interface is lower than the melting point. Hence, heat transfers out at the interface, leading to a concave toward the melt interface profile. This scenario suggests the possibility of using the model to design the temperature profile of the heater with the criterion of producing a flat interface.

Velocity

In Fig. 8 we examine the results for varying V, the translation rate of the heater. Only when $V \equiv 0$ do we obtain a liquid zone that is symmetric with respect to the x axis. For $V \neq 0$ the maximum in the heater profile is always nearer to the melting interface. From Eqs. (46) and (59) we see that the latent heat requirements ($PeSt$) increase as the velocity increases. Finite thermal conductivities lead to melting of only the edges of the feed rod at large velocities so that, if the velocity is sufficiently high, no liquid zone could be established. For this situation the asymptotic expansions are no longer uniformly valid, and we do not obtain a feasible solution as shown in Fig. 9.

Heater Parameters

In Fig. 10 we examine the results for varying the maximum value T_h and width of the heater profile a. Increasing T_h has qualitatively a similar effect as increasing a (increasing the width of the heater profile); hence, we

V (cm/s) :   0   0.004   0.009

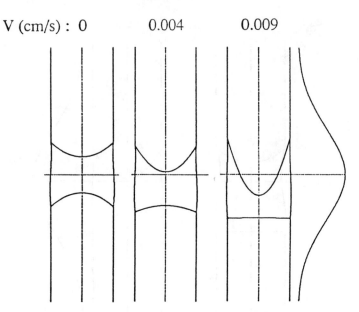

Fig. 8   Effects of growth velocity variation.

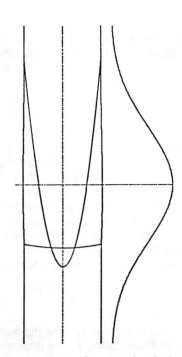

Fig. 9   Incomplete zone due to fast growth velocity ($V = 0.012$ cm/s).

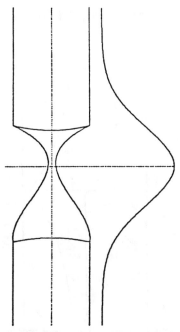

Fig. 10   Unstable zone configuration due to increasing aspect ratio ($\epsilon = 0.33$).

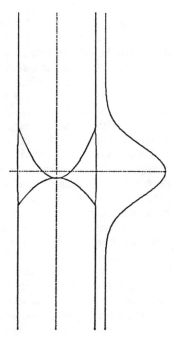

Fig. 11   Marginal zone due to narrow heater profile.

gaussian   exponential   step

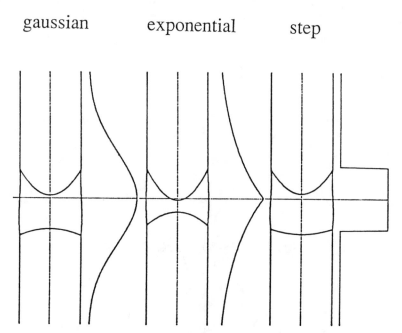

Fig. 12   Effects of heater temperature profile variation.

restrict ourselves to a discussion of the latter. The aspect ratio of the zone in Fig. 10 is $\epsilon \sim 0.36$. We have increased the heater width to cause the longer zone. The necking of the liquid gas interface indicates that we are approaching a nonfeasible solution. This may correspond to a capillary instability (Rayleigh and Plateau type) in a stability analysis. The solution in Fig. 11 is approaching a nonfeasible state since the heater profile is so narrow that there is insufficient heat to melt through the feed rod.

Another important built-in capability of our model is the ability to test different heater configurations, since the heater profile is an arbitrary function in z. Figure 12 shows a comparative study of three heater profiles. Gaussian profiles are usually established by radiative coil heaters, exponential profiles are possible for laser heating,[17] and step profiles could be approached by placing insulating zones at desired locations. Each of the preceding profiles has been selected by practitioners of crystal growth for some assumed unique qualifications. For example, concentrated localized hot spots generated by surface laser heating (in an opaque media) were supposed to reduce the zone size and flatten its interfaces, whereas step heaters are thought to force the melting and solidifying interfaces to be flat. To compare the three configurations, the maximum temperature of the heaters was kept constant, and the width parameter a, although it had a different interpretation for each profile, was also kept the same for comparison. It is evident that merely changing a profile does not guarantee the desired results. This is not to say that, for example, a flat solidifying interface is not possible. This comparison simply points out that the entire system should be taken into consideration when experimental parameters are being changed.

Concentration Field

The preceding study is for the pure silicon system. However, we find similar behavior for binary systems. In Table 1 we list input data for Ge doped with Ga. The float zones that we obtain are similar in shape to those shown for silicon. The additional feature is the concentration profile of the solute. The concentration at the solidification front is obtained by evaluating the boundary-layer concentration profile [composed of Eqs. (37d), (45b), and (47c)] at the solidifying interface. We find

$$
c^S(\eta = \epsilon^2 h_2) = \frac{1}{k} + \frac{\epsilon^2}{2M(1+K)} \left\{ \left(\frac{1}{3} - x^2 \right)(\bar{B}_L + K\overline{B_S}) \left[\frac{M}{k} + 1 - \theta(-\ell_S) \right] \right.
$$
$$
\left. + \left(\frac{1}{3} - x^2 \right)(\bar{R}_L + K\bar{R}_S) \left[\left(\frac{M}{k} + 1 \right)^4 - \theta^4(-\ell_S) \right] \right\} \qquad (71)
$$

Note that the curvature of the concentration profile is opposite to that of the interfacial shape as given by Eq. (52). This is in agreement with the experimental results of Carlberg.[15] Figure 13 shows the concentration profile at the interface (normalized with respect to C_0) for two different

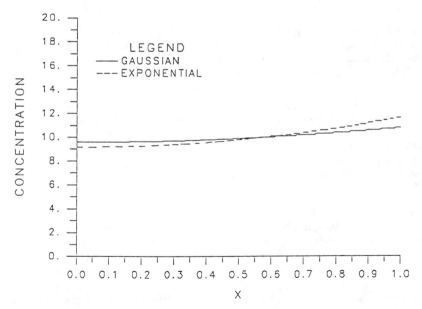

Fig. 13   **Lateral concentration profile along the solidifying interface for Ge doped with Ga.**

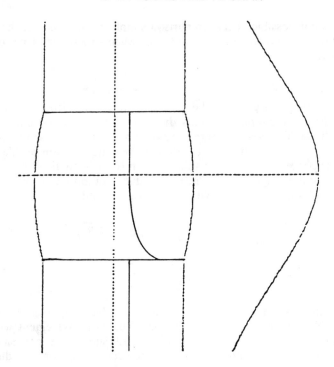

Fig. 14   Leading-order longitudinal concentration profile for Ge doped with Ga.

heater profiles for the same material and operating conditions. This suggests the possibility of optimizing the temperature profile to minimize the lateral segregation.

The concentration profile for fixed x across the height of the liquid zone is obtained from a composite expansion of c, c^S, and c^F. The profile in the solid crystal, C_S, is given by k times Eq. (71) due to steady-state conditions. Figure 14 displays the typical exponentially decaying profile.

An interesting consequence of Eq. (71) compared with Eq. (52) is that a simultaneous flat interface and flat concentration profile may not occur since the interface is no longer an isotherm for binary systems. This agrees with the results of Brattkus and Davis.[22]

Surface-Tension-Driven Flow

The results given in the previous subsection are particular to the no-flow model. We have examined the effects of surface-tension-driven flow for a steady, laminar, viscous regime. Having obtained the same leading-order temperature field as the conduction-dominated no-flow model, we anticipate that the following results are valid for $Ma < 100$. Here the numerical results of Kobayashi[24] and Duranceau and Brown[1] suggest that the conduction temperature field is not affected much by the flow. However, as we shall see, the $0(\epsilon^2)$ presence of convective heat transfer significantly alters the interfacial shapes. Such low Marangoni systems are desirable because this regime is not susceptible to the oscillatory flows present in higher Ma number systems. Additionally, proposed space experiments for both II–VI and III–V systems rely on hot-wall furnace techniques, which may result in lower Ma numbers.

We examine the flow features using Eqs. (39b), (62a), and (62b). These equations define the streamlines shown in Fig. 15, representing the leading-order flowfield valid in a rectangular melt. We note that the flow is linearly proportional to the temperature gradient in the melt, which has been established by the heater. We find that the surface flows are roughly twice as large in magnitude as the centerline return flows. Figure 15a, valid for a stationary heater, shows cells that are symmetrical top to bottom and left to right. The cell cores are located nearest to the outer one-third of the melt zone, left to right, and nearest to the interfacial corners where the temperature gradients are the highest. For the situation of a translating heater, Fig. 15b shows that the cells are no longer symmetrical top to bottom. The

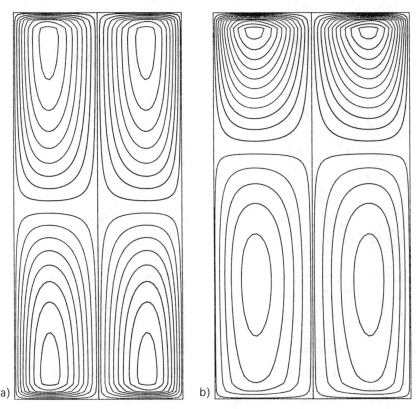

a)   b)

Fig. 15 Leading-order flow streamlines at a) $Ma = 27$ and $V = 0$ and b) $V = 0.004$ cm/s.

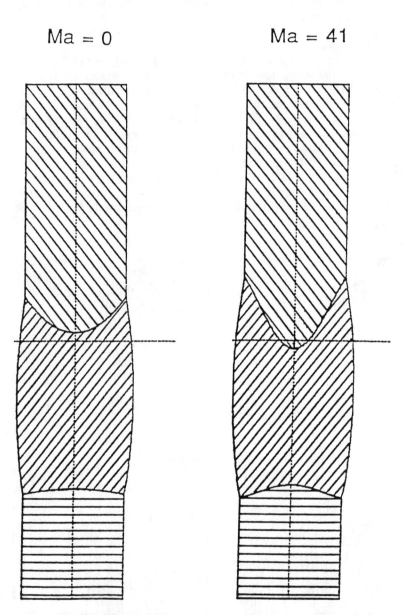

Fig. 16 **Effects of Marangoni convection on zone shape and size.**

upper cells have been compressed nearer the melting interface since the maximum in the heater profile is closer to this front.

In Fig. 16 we examine the effects of this flow on the interfacial shapes. First, as previously mentioned, the liquid/gas interface is nearly identical for both since the capillary number is so small. Examining the solidification front, we find that the flow-induced shape is no longer parabolic as the conduction front. The flow shape is a fourth-order polynominal with an inflection point. The hot fluid is dragged toward the solidifying edge by surface tension gradients. This causes more heat to transfer into the solidifying crystal, leading to melt back of the edges due to the larger temperature gradients. This heat loss to the crystal coupled with the convective heat transfer in the melt also leads to a colder temperature field at the centerline. Hence, the interface grows in comparison to the conduction, no-flow model. Overall these effects in the boundary layer are fed back into the core by the return flows, causing an $0(\epsilon^2)$ decrease in the maximum value of the melt temperature at $z = 0$, the heater location.

Finally, since the conduction heat-transfer contribution to the interfacial shape is inversely proportional to the temperature gradient, whereas the convective heat-transfer contributions are directly proportional, there is an interesting compromise for control over the solidifying shape. Large gradients give better control over the conduction profile but lead to increased flow, whereas low gradients lead to weaker flows but increased parabolic deflection of the interfacial shape.

Summary and Conclusions

We present models describing the steady-state thermal diffusion in a pure system, the thermal-solutal diffusion in a binary system, and heat and momentum transfer in a pure system. The geometry of the model is described by a two-dimensional Cartesian coordinate system that is applicable for crystal sheets. Heat transfer between the ambient and system takes place via both convective and radiation heat transfer. The melting, solidifying, and melt/gas interfacial shapes, as well as the thermal, flow, and solutal profiles, are analytically evaluated as functions of the heater and ambient temperature profiles and material properties.

The solution procedure involves a coupled asymptotic/numerical approach. The asymptotic expansions are based on the assumption that the aspect ratio $\epsilon = R/L$ is small and that the convective and radiation heat transfer are $0(\epsilon^2)$. These scalings lead to boundary-layer solutions around the melting and solidifying interfaces. The melt/gas interface is determined by the surface tension and melt pressure. The pressure is set by the melt flow and contact angle conditions. The core temperature profiles and the mean positions of the melting and solidifying interfaces are calculated numerically as functions of the latent heat and heater parameters. Lateral corrections to the interfacial shapes are then determined.

The advantage of this solution procedure is that it reduces the coupled set of partial differential equations to ordinary type. These equations are more easily integrated; thus, it is possible to readily investigate the response of the zone to parametric changes. The major restriction of the model is the small-aspect-ratio assumption. Quantitative comparisons with full numerical and experimental studies show that this restriction is not severe, since zones with $\epsilon \sim 0.7$ were routinely computed with excellent agreement to the studies previously mentioned.

Our results should be applicable in situations where melt flows are not intense enough ($Ma < 100$) to change the thermal field in pure systems, or where the physical properties of the melt are such that the convective field is decoupled from the thermal field, the latter being established primarily by diffusion. Low-Pr-number materials, such as semiconductor and most metals with low Marangoni convection, fall into this class. This analysis is expected to provide correct zone sizes, shapes, and responses to material properties and thermal conditions variations for many material systems of interest. However, in binary (or doped) systems any melt motion is expected to be significant in redistributing the solute (dopant). For most materials systems of importance the Schmidt number Sc, which measures the relative importance of convection to diffusion in setting the solutal field, is relatively large; hence, the results obtained here concerning solute distribution are limited at best to either low Sc numbers or to float zones established in microgravity conditions with an oxide or encapsulant layer where both natural and surface tension driven convection are hopefully of less importance in setting the solutal field.

Finally, it is proposed that the methodology presented in this paper is very attractive both as an analysis and design tool, in part because of the rapid succession with which parameters can be changed and new solutions can be obtained.

Acknowledgment
This work is supported by the Microgravity Science and Applications Program of NASA, under cooperative agreement NCC 3-104.

References
[1] Duranceau, J. L., and Brown, R. A., "Thermal-Capillary Analysis of Small-Scale Floating Zones: Steady-State Calculations," *Journal of Crystal Growth*, Vol. 75, 1986, pp. 367–389.

[2] Duranceau, J. L., and Brown, R. A., "Finite Element Analysis of Melt Convection and Interface Morphology in Earthbound and Microgravity Floating Zones," *Proceedings of the Third International Colloquium on Drops and Bubbles* (submitted for publication).

[3] Kuiken, H. K., and Roksnoer, P. J., *Journal of Crystal Growth*, Vol. 47, 1979, pp. 29–34.

[4] Fu, B. I., and Ostrach, S., "Numerical Solutions of Floating Zone Thermocapillary Flows," *Proceedings of the 4th European Symposium on Materials Sciences Under Microgravity*, Madrid, Spain, 1983, pp. 239–245.

[5] Harriot, G. M., and Brown, R. A., "Steady Solute Fields Induced by Differential Rotation in a Small Floating Zone," *Journal of Crystal Growth*, Vol. 69, 1984, pp. 589–604.

[6]Coriell, S. R., Hardy, S. C., and Cordes, M. R., "Stability of Liquid Zones," *Journal of Colloid and Interface Science*, Vol. 60, 1977, pp. 126–136.

[7]Surek, T., and Coriell, S. R., "Shape Stability in Float Zoning of Silicon Crystals," *Journal of Crystal Growth*, Vol. 37, 1977, pp. 253–271.

[8]Harriot, G. M., and Brown, R. A., "Flow in a Differentially Rotated Cylindrical Drop at Low Reynolds Number," *Journal of Fluid Mechanics*, Vol. 126, 1983, pp. 269–285.

[9]Harriot, G. M., and Brown, R. A., "Flow in a Differentially Rotated Cylindrical Drop at Moderate Reynolds Number," *Journal of Fluid Mechanics*, Vol. 144, 1984, pp. 403–418.

[10]Chang, C. E., and Wilcox, W. R., "Inhomogeneities Due to Thermo-Capillary Flow in Floating Zone," *Journal of Crystal Growth*, Vol. 28, 1975, pp. 8–12.

[11]Chang, C. E., and Wilcox, W. R., "Analysis of Surface Tension Driven Flow in Floating Zone Melting," *International Journal of Heat and Mass Transfer*, Vol. 19, 1976, pp. 355–366.

[12]Clark, P. A., and Wilcox, W. R., "Influence of Gravity on Thermo-Capillary Convection in Floating Zone Melting of Silicon," *Journal of Crystal Growth*, Vol. 50, 1980, pp. 461–469.

[13]Sen, A. K., and Davis, S. H., "Steady Thermo-Capillary Flows in Two-Dimensional Slots," *Journal of Fluid Mechanics*, Vol. 121, 1982, pp. 163–186.

[14]Lai, C. L., Ostrach, S., and Kamotani, Y., *1985 U.S. Japan Heat Transfer Joint Seminar*, San Diego, CA, 1985.

[15]Carlberg, T., *Journal of Crystal Growth*, Vol. 79, 1986, pp. 71–76.

[16]Murphy, S. P., Hendrik, S. I., Martin, M. J., Grant, T. W., and Lind, M. D., *MRS Symposium Proceedings*, Vol. 87, 1987, pp. 391–397.

[17]Baghdadi, A., and Gurtler, R. W., "Recent Advances in Ribbon-to-Ribbon Crystal Growth," *Journal of Crystal Growth*, Vol. 50, 1980, pp. 236–246.

[18]Yablonovitch, E., and Gmitter, T., "Ribbon-to-Ribbon Float Zone Single Crystal Growth Stabilized by a Thin Silicon Dioxide Skin," *Applied Physics Letters*, Vol. 45, 1984, pp. 63–65.

[19]Plateau, J. A. F., *Statique expérimentale et théorique des liquides soumis aux seules forces moléculaires*, Gauthier-Villars, Paris, 1873.

[20]Rayleigh, J. W. S., *Proceedings of the Royal Society of London*, Vol. 29, 1879, pp. 71–79.

[21]Heywang, W., and Ziegler, G., "Stability of Vertical Melting Zones," *Zeitschrift fuer Naturforschung*, Vol. 11a, 1956, pp. 238–243.

[22]Brattkus, K., and Davis, S. H., "Directional Solidification with Heat Losses," *Journal of Crystal Growth*, Vol. 91, 1988, pp. 538–556.

[23]Young, G. W., and Chait, A., "Steady-State Thermal-Solutal Diffusion in a Float Zone," *Journal of Crystal Growth*, Vol. 96, 1989, pp. 65–95.

[24]Kobayashi, N., "Computer Simulation of the Steady Flow in a Cylindrical Floating Zone under Low Gravity," *Journal of Crystal Growth*, Vol. 66, 1984, pp. 63–72.

Thermosolutal Convection in Liquid HgCdTe
Near the Liquidus Temperature

Basil N. Antar*
University of Tennessee Space Institute, Tullahoma, Tennessee

Introduction

THE pseudobinary alloy material $Hg_{1-x}Cd_x Te$ has received a great deal of attention recently due to its utility as both a semiconductor and a photoconductor. Its versatility lies in the fact that this material exhibits semiconducting properties over much of the composition range. The forbidden energy gap is dependent on the composition variable x, ranging from a wide-gap semiconductor for $x = 1$ to a semimetal at $x = 0$. $Hg_{1-x}Cd_x Te$ is one of the most studied semiconductors, outranked only by Si and GaAs. Presently, $Hg_{1-x}Cd_x Te$ is considered to be the most promising narrow-gap semiconductor for future infrared detector arrays. The demonstration of long-wavelength photoconductivity in these alloys has led the way for the subsequent development of infrared detectors. These materials are widely used today in photoconductive and photovoltaic infrared detectors. The present generation of $Hg_{1-x}Cd_x Te$ detectors serve as the eyes of practical and widely used infrared systems involved in thermal imaging, surveillance, and other military, space, and commercial applications.

Because of the sensitivity of its properties to both composition and its uniformity of distribution, the manufacturing of $Hg_{1-x}Cd_x Te$ requires extra care. Traditionally $Hg_{1-x}Cd_x Te$ has been produced via the Bridgman-Stockbarger process. In that technique the elements are sealed within a quartz ampoule that is initially heated above the liquidus temperature for the appropriate composition and then lowered through a temperature gradient. The quality of crystals produced in this method depends on

*Professor, Engineering Science and Mechanics.

159

both the thermodynamic processes taking place during solidification and on the speed of motion of the ampoule. The amount of convective motion occurring near the solid-liquid interface determines, to a large extent, the quality of the crystal. Normally, it is desirable to avoid convection altogether and to maintain a planar solid-liquid interface with gradients of temperature and concentration parallel to the growth direction. It can be difficult in practice to maintain such an interface shape under the normal Bridgman-type solidification conditions. Past experience with this type of process has shown that horizontal temperature gradients in the melt are unavoidable due to the discontinuity in the conductivities of the ampoule, crystal, and melt at the point where the interface meets the ampoule. Such horizontal gradients give rise to thermal convection in the melt.

Normally, under all solidification conditions the sample will possess both temperature and solute gradients in the axial direction. Convection may arise in the liquid phase depending on the magnitude and direction of these gradients. Although $Hg_{1-x}Cd_xTe$ is a pseudobinary alloy, the melt is considered to consist of the mixture of HgTe and CdTe. Thus, there are basically two mechanisms influencing convective instabilities in the melt in the vicinity of the solid-liquid interface. When convection occurs in a liquid under these conditions, it will be of a double diffusive nature. In circumstances where the crystal is below the melt, the sign of the temperature gradient is such as to inhibit convection. Furthermore, it is known that in the case of $Hg_{1-x}Cd_xTe$ the heavier constituent is usually rejected at the solid-liquid interface. Casual observation then reveals that both the solute and temperature gradients are stabilizing. It will be shown that, because of the peculiar nature of the thermodynamic properties of this binary alloy, it may be thermally unstable regardless of the direction and sign of the temperature gradients.

In order to diminish the likelihood of the onset of buoyancy-driven convection in the melt during solidification, $Hg_{1-x}Cd_xTe$ has been a prime candidate for exploiting microgravity environment to produce the material. This study is primarily focused on understanding the criteria leading to convective instabilities in the melt of $Hg_{1-x}Cd_xTe$ under thermodynamic conditions representative of the Bridgman-Stockbarger solidification process.

The chemical and physical properties of $Hg_{1-x}Cd_xTe$ have been under intensive investigation for the past 30 years.[1,2] The most significant properties that are relevant to the present study are the phase diagram properties of the mixture and the variation of the density with both temperature and mole fraction x. Chandra and Holland[3] have measured the density variation of liquid $Hg_{1-x}Cd_xTe$ for four values of x in the neighborhood of the solidification temperature. Their measurements reveal that the density possesses a maximum in the interior of the liquid close to the solid-liquid interface as shown in Fig. 1. Such density variation indicates that convective instabilities may always be present regardless of the magnitude and direction of the temperature gradient.

In many naturally occurring phenomena, thermal convection occurs in an unstably stratified fluid layer that is bounded from above and/or below

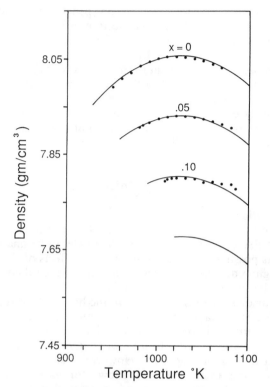

Fig. 1   Density variation of $Hg_{1-x}Cd_xTe$ with temperature and concentration. Circles, measured data of Ref. 3; solid lines, Eq. (3).

by stably stratified regions. For such configurations the convective motion may extend a substantial distance into the adjacent stable zones. If the convective mixing is efficient enough, the stable thermal stratification may be weakened sufficiently to allow the motion to penetrate far deeper into the stable zones than might be predicted by analyzing the initial state. Such states of penetrative convection are the subject of the present study.

Penetrative convection has been the subject of considerable research since the pioneering work of Veronis.[4] Regime diagrams have been established for a wide range of fluid properties through linear stability analyses.[5-7] In addition, there are several very accurate finite-amplitude analyses including full numerical simulation of the convection environment.[8-11] Most of the pertinent stability results of these analyses have been confirmed by experiments on penetrative convection; in most of these experiments water was the working fluid. Among the earliest experimental confirmations of the theoretical results of Veronis were the experiments of Malkus.[12] Recently, Walden and Ahlers[13] conducted detailed experiments on liquid helium I in which they studied, among other effects, penetrative convection.

All of the studies previously cited on penetrative convection were concerned with the Benard type of convection in which heat was the only diffusing component. In this study the linear stability calculations for penetrative convection are extended to fluids possessing two diffusing components in order to apply them to the case of solidifying $Hg_{1-x}Cd_x Te$. Until recently there did not exist in the literature any stability diagram on double diffusive convection in which the density possesses a maximum. This work is an extension of the recent study by the author[14] of penetrative double diffusive convection in which the parameter space is further explored in order to cover a wide range of fluids.

The Model

Governing Equations

In order to establish the most basic physical criteria affecting the onset of penetrative double diffusive convection, the model to be analyzed here will be as simple as possible. Thus, for this study the effects of the ampoule side walls will be ignored. In this study we consider a heated fluid layer of depth d that is infinite in the horizontal direction. Let the coordinate system be rectangular Cartesian with the x,y axes in the horizontal direction and z in the vertical. The fluid layer is assumed to support both temperature and concentration gradients in the vertical direction. To make the analysis more amenable to comparison with previous work, these gradients are assumed to be constant throughout the layer. However, it should be noted that the method of solution used here also allows for variable gradients. The motionless (basic state) temperature and concentration distributions appropriate for energy and species conservation with boundary conditions of constant temperature and concentration are given by

$$\bar{T} = T_0 - (d_0 - z)\,\Delta T / d_0 \qquad (1)$$

$$\bar{S} = S_0 + (d_0 - z)\,\Delta S / d \qquad (2)$$

where T_0 and S_0 are the temperature and solute concentration values, respectively, at the specific vertical location d_0. A sketch of the coordinate system used and typical temperature, concentration, and density profiles are shown in Fig. 2.

Before we proceed with the model further it is instructive to specify the response of the fluid density to small variations in both temperature and concentration. Customary analytical approaches to double diffusive convection (e.g., Turner[15]) restrict the density variation of the fluid to be linear with both temperature and concentration changes. This is our point of departure from previous studies of double diffusive convection. This study will be concerned only with fluids in which the density variation with temperature is quadratic. This is because the primary motivation for this study is to investigate the fluid response in a solidifying binary alloy that

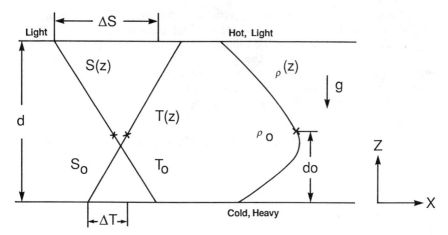

Fig. 2 A schematic diagram of the model.

belongs to the class of mixtures characterized by this specific density variation within the fluid layer. A good representative sample of this class of materials is the pseudobinary system $Hg_{1-x}Cd_xTe$, whose density variation with both temperature and concentration is shown in Fig. 1. The measured density values appear to be well approximated in the range of mole fraction, $x : 0 < x < 0.10$, by the following equation:

$$\rho = 7.931\{1 - 1.3515 \times 10^{-6}(T - 1028)^2 - 3.1819 \times 10^{-3}(S - 5)\} \quad (3)$$

where ρ is in grams per square centimeters, T is in degrees Kelvin, and S is in percent mole fraction (i.e., $S = 100x$).

The intent of this study is to investigate the influence of such density variation on the onset of double diffusive convection. Thus, in the following analysis the density dependence on temperature and concentration will be assumed to obey the equation

$$\rho = \rho_0[1 - \alpha(T - T_0)^2 + \beta(S - S_0)] \quad (4)$$

where

$$\alpha = -\frac{1}{2\rho T}\frac{\partial \rho}{\partial T}, \qquad \beta = \frac{1}{\rho}\frac{\partial \rho}{\partial S}$$

and ρ_0 is the density at temperature T_0 and concentration S_0. In this case the height d_0 is the position of maximum density when $\Delta S = 0$. Note that, when $(\beta\Delta S/d) \geqslant [2\alpha(\Delta T)^2/d_0]$, the density distribution as given by Eq. (4) will vary monotonically with height z. However, the following discussion will only concern those conditions in which the density possesses a maximum in the interior of the fluid layer. It should be emphasized that expression (4) applies only to a specific fluid within specific ranges and is not to be viewed as a general law for all fluid mixtures.

In order to establish the conditions leading to the onset of double diffusive convection in a fluid, the governing equations (mass, momentum, and energy) are first linearized about the basic state given by Eqs. (1), (2), and (4).[16] The linearized Boussinesq perturbation equations with the density representation (4) for such a configuration are given by

$$\left(\frac{1}{\sigma}\frac{\partial}{\partial t} - \nabla^2\right)\nabla^2 w = (-1 + \lambda_0 z)\nabla_1^2 T - Rs\nabla_1^2 S \tag{5}$$

$$\left(\frac{\partial}{\partial t} - \nabla^2\right)T = -Rw \tag{6}$$

$$\left(\frac{\partial}{\partial t} - L\nabla^2\right)S = w \tag{7}$$

where

$$\nabla^2 = \frac{\partial^2}{\partial x^2} + \frac{\partial^2}{\partial y^2} + \frac{\partial^2}{\partial z^2}; \qquad \nabla_1^2 = \frac{\partial^2}{\partial x^2} + \frac{\partial^2}{\partial y^2}$$

In the preceding equations w, T, and S are the perturbation vertical velocity component, the temperature, and the concentration, respectively. Equations (5–7) are written in a nondimensional form where the velocity, space, and time scales are κ/d, d, and d^2/κ, respectively. Temperature and concentration were made dimensionless using $\nu\kappa/(2\alpha g d^3 \Delta T)$ and ΔS for respective scales. The variables α, g, ν, and κ are, respectively, the coefficients of thermal expansion, gravity, kinematic viscosity, and thermal diffusivity. Equations (5–7) possess the following nondimensional numbers σ, L, λ_0, R, and Rs, which are the Prandtl number, the Lewis number, the inverse of the maximum density position for zero concentration gradient, the Rayleigh number, and the solutal Rayleigh number, respectively. These numbers are defined in the following manner:

$$\alpha = \nu/\kappa; \qquad L = \kappa_s/\kappa; \qquad \lambda_0 = d/d_0$$

$$R = \frac{2\lambda_0^4 \alpha g (\Delta T)^2 d_0^3}{\kappa\nu}; \qquad Rs = \frac{\beta g \Delta S d^3}{\kappa\nu}$$

where κ_s is the solute diffusion coefficient.

Equations (5–7) are solved subject to two different sets of boundary conditions. For the free-free surface conditions the following constraints are applied:

$$w = \frac{\partial^2 w}{\partial z^2} = S = T = 0 \quad \text{at} \quad z = 0,1 \quad (8a)$$

and for the rigid-rigid surface conditions, the following constraints are applied:

$$w = \frac{\partial w}{\partial z} = S = T = 0 \quad \text{at} \quad z = 0,1 \quad (8b)$$

The solution to the perturbation Eqs. (5–7) satisfying the preceding boundary conditions takes the following form:

$$(w,T,S) = \{w'(z),T'(z),S'(z)\}\{\exp(pt)\}f(x,y)$$

where

$$\frac{\partial^2 f}{\partial x^2} + \frac{\partial^2 f}{\partial y^2} = -a^2 f$$

Upon substituting the preceding solution form into Eqs. (5–7) and dropping the primes, the following set of ordinary differential equations for the perturbation functions, w, S, and T is obtained:

$$\left[\frac{p}{\sigma} - (D^2 - a^2)\right](D^2 - a^2)w = a^2(1 - \lambda_0 z)T + a^2 SRs \quad (9)$$

$$[p - (D^2 - a^2)]T = -Rw \quad (10)$$

$$[p - L(D^2 - a^2)]S = w \quad (11)$$

where $D = d/dz$. Similarly, the boundary conditions take the following form:

$$w = D^2 w = S = T = 0 \quad \text{at} \quad z = 0,1 \quad (12a)$$

or

$$Dw = S = T = 0 \quad \text{at} \quad z = 0,1 \quad (12b)$$

Note that in the solution representation above, a, which is the horizontal wave number, is taken as a purely real number. On the other hand, p represents a complex frequency, i.e., $p = p_r + ip_i$, where p_r is the perturbation growth rate, and p_i is the perturbation frequency.

The homogeneous sets of Eqs. (9–11) with either of the homogeneous boundary conditions [Eqs. (12)] form an eigenvalue problem which may be represented in the following functional form:

$$p = p(a,\lambda_0,R,R_s,\sigma,L) \qquad (13)$$

This eigenvalue problem consists of a set of coupled linear, ordinary differential equations with six parameters. Such problems are solved by determining the eigenvalue p (in this case both p_r and p_i) for specific values of the parameters involved in the equations. If, for a specific set of values of all the parameters in Eq. (13), p_r is found to be positive, then the perturbations are unstable; otherwise they are stable. A positive value for p_r, $p_r > 0$, implies that the motionless basic state will evolve into convective motion in the form of rolls or cells. Whenever several simultaneous eigenvalues exist for the same values of the parameters, the one with the largest value for p_r is called the mode of maximum growth rate. The monotone modes are those for which $p_i = 0$ for any p, and the oscillatory modes are those for which $p_i \neq 0$.

Solution Method

For the solution of the present eigenvalue problem a computer code was developed with an eighth-order, variable-step, Runge-Kutta-Fehlberg[17] initial-value integrator. A Newton-Raphson method was used for the iteration procedure, and an orthonormalization procedure was implemented at each integration step. All eigenvalues presented in the results and discussion section were produced with a maximum relative tolerance of 10^{-4} in the iteration process (i.e., all eigenvalues presented were converged to four significant figures). Normally the shooting method, in the local form, requires an initial guess for the eigenvalues, and the whole process for obtaining the results is started by first trying several guesses. When convergence on a single eigenvalue is obtained, that eigenvalue is then used as the initial estimate to obtain a neighboring one, and the process is repeated to obtain all desired eigenvalues. For an existing eigenvalue convergence here was achieved typically in under 10 iterations. The eigenvalues reported in the next section were produced on a VAX 11/780, and it took approximately 1 min of CPU time per iteration for a moderate number of integration steps. The number of steps required in the integration process increased with increasing values of the Rayleigh and the solutal Rayleigh numbers.

Results and Discussion

It is customary to present the results of the linear stability analysis in the form of regime diagrams in which the various possible stability regions are identified. A typical regime diagram is shown in Fig. 3. The figure shows the various instability and stability regions in the first quadrant of the R-Rs plane for $\sigma = 10$, $L = 0.01$, and $\lambda_0 = 2$. The results shown are for both the

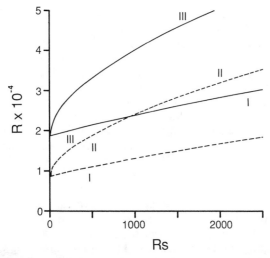

Fig. 3   Regions of stability and instability in the first quadrant of the R-Rs plane for $\lambda_0 = 2$, $\sigma = 10$, and $L = 0.01$. Dashed lines, free-free boundary conditions; solid lines, rigid-rigid boundary conditions.

free-free (dashed lines) and the rigid-rigid (solid lines) boundary conditions. The stability discussion below is limited to the first quadrant of the R-Rs plane because, on one hand, according to previous work,[14,16] both the monotone and the oscillatory modes of instability were found to possess maximum growth rates in this quadrant only. However, on the other hand, as discussed in the Introduction, the original physical configuration motivating the present analysis is that for the solidification of $Hg_{1-x}Cd_xTe$ in which the temperature gradient of the basic state is destabilizing, $R > 0$, whereas the concentration gradient is stabilizing, $Rs > 0$. Figure 3 shows that there are two curves belonging to each type of boundary condition in that segment of the plane. Each set of two curves originates from a different position on the ordinate. These are the points of critical instability belonging to the case of penetrative convection in a single-component fluid as identified by Veronis[4] and verified in Ref. 14. The two curves belonging to each boundary condition divide the first quadrant of the R-Rs plane into three distinct regions. The three regions are labeled I, II, and III and are defined in the following manner: In region I all instability modes are damped and the basic state is stable to infinitesimal disturbances. In each of the regions II and III there exists at least one unstable mode. The mode of maximum growth rate in region II is oscillatory ($p_r > 0$, $p_i \neq 0$), whereas in region III it is monotone ($p_r > 0$, $p_i = 0$). This specific division of the quadrant into the stable and unstable zones as shown in Fig. 3 resembles the regime diagram for the classical double diffusive convection case discussed in Turner.[16] Note that the instability regimes for the free-free and the rigid-rigid boundary conditions are qualitatively similar but differ in their exact location.

The analysis in the previous work[14] revealed that the division of the first quadrant of the plane R-Rs plane into the three regions shown in Fig. 3 is characteristic of the problem and does not vary qualitatively with variations in λ_0. The variation of the stability and instability zones with λ_0 is shown in Fig. 4 for the same values of σ and L as in Fig. 3 and for $\lambda_0 = 2.0$, 1.6, and 1.0 for the free-free boundary conditions. It is seen that, as the value of λ_0 is increased, the position of the critical point of instability at $\Delta S = 0$ is shifted to higher values of the Rayleigh number R, whereas both regions I and II appeared to increase substantially with increasing λ_0.

The results of the present study are discussed next. The intent is to generalize the results of Ref. 14 to a wider range of fluids that are characteristic of $Hg_{1-x}Cd_xTe$ as discussed in the previous section. To achieve this, the basic parameters entering the eigenvalue problem [Eq. (13)] and reflecting the different physical characteristics of the fluid are varied. These are the Prandtl and the Lewis numbers, σ and L, respectively. Figure 5 illustrates the modification of the basic stability criteria brought about by changing the Lewis number. This figure shows the three stability and instability regions in the first quadrant of the R-Rs plane for three values of the Lewis number and for fixed values of σ and λ_0 at 10.0 and 2.0, respectively. It is seen that, as L is increased, the area of the plane where

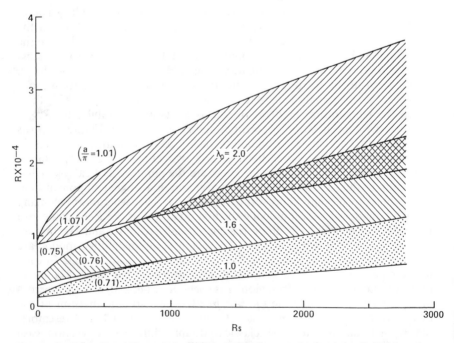

Fig. 4   Regions of stability and instability in the first quadrant of the R-Rs plane for $\lambda_0 = 1.0$, 1.6, and 2.0, $\sigma = 10$, and $L = 0.01$. Shaded areas are respective regions in which oscillatory modes are most unstable.

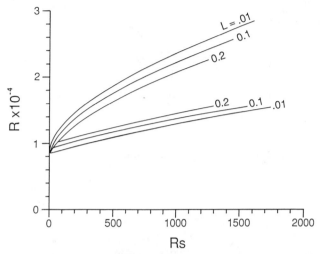

Fig. 5   Regions of stability and instability in the first quadrant of the R-Rs plane for $L = 0.005$, 0.01, 0.1, and 0.2, $\lambda_0 = 2.0$, and $\sigma = 10$.

oscillatory modes of instability are dominant decreases, whereas the stability zone grows. Also, for a fixed value of σ the regions of dominant oscillatory unstable modes appear to be embedded within each other for successively higher values of L in such a way that the upper bound for this area is the one delineated for $L = 0.01$. Figure 5 suggests further that in the limit, as $L \rightarrow \infty$, while σ is held constant, the width of region II will shrink to zero. Consequently, it may be concluded that, for this limiting case, only monotone modes possess maximum growth rates. Further analysis demonstrates that the three stability and instability zones that are delineated by the curves for $L = 0.01$ in Fig. 5 are very close to their asymptotic values in the limit as $L \rightarrow 0$. This implies that, as L approaches zero, the three characteristic stability and instability zones for this problem are not altered substantially from the respective zones shown for $L = 0.01$.

Figure 6 is similar to Fig. 5, except in this case a different value of the Prandtl number is used. The variation of the three stability and instability zones with L is shown for $\lambda_0 = 2.0$ and $\sigma = 0.1$. Again, it is observed that the oscillatory unstable region diminishes in size with increasing L whereas the stable region grows. The point at which the oscillatory region originates, labeled T in the figure, shifts to larger values of for both R and Rs with increasing L. This point is also seen in Fig. 5 for both cases $L = 0.2$ and $L = 0.1$. This point will be called the triple point, since it is the intersection of the three curves that divide the first quadrant of the R-Rs plane into the different stability and instability regions. It is observed that to the left of the triple point there exists only a single zone of instability that is separated from the stable region by the marginal stability curve. The stable region in this case is below the curve, whereas the unstable region is above it. Furthermore, the unstable region to the left of the triple point is

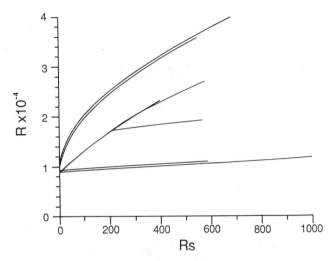

Fig. 6   Same as Fig. 5 but for $\sigma = 0.1$.

characterized by monotone modes of maximum growth rates. To the right of the triple point there exist the three customary regions of stability and instability. It should be pointed out that a triple point at vanishingly small values of Rs exists for each of the cases discussed in both this chapter and in Ref. 14. However, some of these points evidently lie very close to the ordinate and are not visually discernible in the figures.

Figure 7 illustrates the modification of the stability and instability regime diagram, again in the same region of the R-Rs plane, as a result of variations in the value of the Prandtl number while L and λ_0 are held fixed. It shows a plot of the three regions for $\lambda_0 = 2.0$, $L = 0.1$, and $\sigma = 10.0$, 1.0, 0.1, and 0.05. Several interesting features are observed. First, it can be seen that the area in which the oscillatory modes are the most unstable shifts up diagonally with decreasing Prandtl numbers. This is manifested in the shifting of the triple point to larger values of both R and Rs. The trend in this figure suggests that, in the limit as $\sigma \to 0$, the triple point occurs at very large values of both R and Rs, leaving a substantial portion (if not all) of the first quadrant of the R-Rs plane devoid of a region in which the modes of maximum growth rates are oscillatory. Thus, in the limit as $\sigma \to 0$, there will remain only one region of instability separated from the stable zone by the marginal stability curve. The modes of maximum growth rate in that unstable region are monotone. It is also observed that the zones of stability, region I, become larger with decreasing σ.

The second feature observed in Fig. 7 is that all of the regions where the most unstable modes are oscillatory appear to originate from a single smooth curve for different values of σ. This is the solid line in the figure. Further analysis shows that this curve can be constructed by plotting, in the R-Rs plane, the location of a second marginal stability mode, $p_r = 0$, in addition to the one making up the lower boundary of region II, for the

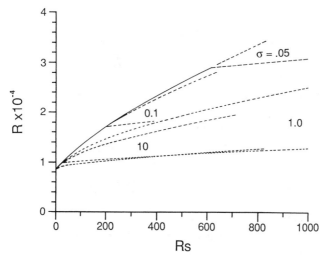

Fig. 7   Regions of stability and instability in the first quadrant of the R-Rs plane for $\sigma = 10$, 1, 0.1, and 0.05, $\lambda_0 = 2.0$, and $L = 0.1$.

same values of σ, λ_0, L, and Rs. Depending on the value of σ, it is found that there exist two simultaneous modes in the first quadrant of the R-Rs plane possessing $p_r = 0$. One of these modes has $p_i = 0$, whereas the other has $p_i \neq 0$. The marginal stability modes making up the solid line in Fig. 7 are neutral stability modes (i.e., $p_r = p_i = 0$). Whether two marginal stability modes exist for a single value of Rs, or only one, depends on the value of σ. Of course, whenever both modes exist, the one for which $p_i \neq 0$ occurs for a lower value of R and hence is the relevant one. This is the reason why the neutral stability curve shown in Fig. 7 was not discussed earlier. However, since the location of the neutral stability modes coincides with the locus of the triple points for all σ, the identification of their position in the R-Rs plane is of value. As shown in Fig. 7, the curve on which these modes lie forms an envelope for the oscillatory region, region II, in addition to the stable one, region I, for all values of σ. Thus, this curve defines an absolute upper boundary for both the region of stable modes and the region of oscillatory most unstable modes in the first quadrant of the plane.

The qualitative modification of the three stability zones, as well as the presence of the envelope curve discussed earlier, were found to be independent of the specific value of λ_0. Figure 8 shows the various regions for $L = 0.1$, $\lambda_0 = 1.6$, and $\sigma = 1.0$, 0.1, and 0.05. This figure exhibits an almost identical trend to that in Fig. 7, except for the quantitative differences brought about by the values of the parameters. Note, however, that the curve delineating the upper boundary of region II and the locus of the triple points in Figs. 7 and 8 are not identical. There exists a small region between these two curves in which the most unstable modes are oscillatory. The width of this region appears to diminish with decreasing σ and

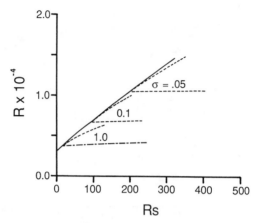

Fig. 8 Same as Fig. 7 but for $\lambda_0 = 1.6$.

thus may not be discernible in Figs. 7 and 8 for lower values of σ, but nevertheless it exists.

The position of the solid curves in both Figs. 7 and 8 represents an absolute lower bound for the instability region in the first quadrant of the R-Rs plane. In other words, any basic state configuration for which both R and Rs fall above that curve is linearly unstable. Thus, the location of this curve is valuable as a general stability criteria for double diffusive convection.

Figure 9 shows the various envelope curves formed by the loci of the triple points for four values of the Lewis number. Two significant points

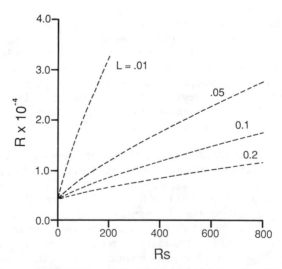

Fig. 9 The locus of the position of the triple points for $\lambda_0 = 2$ and for $L = 0.2, 0.1, 0.5,$ and 0.01.

are to be observed in this figure. The first is that all curves shown originate from a single point on the ordinate. This is the location of the critical point for $\Delta S = 0$. The second is that the inclination of each curve to the abscissa appears to increase with decreasing values of L. It may be concluded on the basis of this figure that the region in which the monotone unstable modes are dominant diminishes with decreasing L for any one value of Rs. However, this does not imply an equivalent increase in the stability region, since, as was shown earlier, the size of that region is a function of σ. Depending on the value of σ, there could be a substantial region below each of these curves in which the most unstable modes are oscillatory. It was also found that all curves shown in Fig. 9 collapse onto a single curve in the first quadrant of the R-Rs plane when plotted on a R-(Rs/L) scale (i.e., on a redefined Rs scale to $Rs' = \beta g \Delta S d^3 / \kappa_s \nu$).

Conclusions

This chapter examined in detail the influence of the Prandtl and Lewis numbers on the onset of penetrative double diffusive convection. Their effect on the ensuing motions was analyzed by studying different regime diagrams in which only one of these parameters was varied while the other was kept constant. These two parameters contain all of the thermodynamic properties necessary to define a specific fluid in this problem. This work has, in essence, examined the response of different fluids due to conditions appropriate to the onset of penetrative double diffusive convection.

The analysis of the previous section revealed that the three stability and instability regions in the first quadrant of the R-Rs plane may be altered substantially by varying either σ or L. It was found that the region in which the oscillatory modes are found to possess maximum growth rates, in that portion of the plane, may diminish in size with increasing L. In fact, the results indicated that in the limit $L \to \infty$ only monotone modes possess maximum growth rates. This is a situation in which the instability is manifested by steady convective motion. This is also the case where mass diffusion has the dominant influence. Furthermore, it was found that, for fluids possessing vanishingly small Prandtl numbers, again, only monotone modes have maximum growth rates for moderate values of both R and Rs. In this case, again, the instability is manifested by steady convection.

The preceding analysis also leads to the establishment of a simple general criterion for the onset of penetrative double diffusive convection. This was manifested by a single curve in the first quadrant of the R-Rs plane for all values of σ and L. The area above the curve comprises a region of instability through monotone modes, whereas below it the basic state configuration is either unstable through oscillatory modes or stable. There exists one such curve for every value of λ_0.

References

[1] Rogalski, A., and Piotrowski, J., "Intrinsic Infra Red Detectors," *Progress in Quantum Electronics*, Vol. 12, 1988, pp. 87–289.

[2] Willardson, R. K., and Beer, A. C., "Mercury Cadmium Telluride," *Semiconductors and Semimetals*, Vol. 18, 1981, pp. 1–385.

[3]Chandra, D., and Holland, L. R., "Density of Liquid $Hg_{1-x}Cd_x Te$," *Journal of Vacuum Science and Technology*, Vol. A1, 1983, pp. 1620–1624.

[4]Veronis, G., "Penetrative Convection," *Astrophysical Journal*, Vol. 137, 1963, pp. 641–663.

[5]Sun, Z.-S., Tien, C., and Yen, Y.-C., "Thermal Instability of a Horizontal Layer of Liquid with Maximum Density," *AIChE Journal*, Vol. 15, 1969, pp. 910–915.

[6]Merker, G. P., Waas, P., and Grigull, U., "Onset of Convection in a Horizontal Water Layer with Maximum Density Effects," *International Journal of Heat and Mass Transfer*, Vol. 22, 1979, pp. 505–515.

[7]Straughan, B., "Finite Amplitude Instability Thresholds in Penetrative Convection," *Geophysical and Astrophysical Fluid Dynamics*, Vol. 34, 1985, pp. 227–242.

[8]Moore, D. R., and Weiss, N. O., "Nonlinear Penetrative Convection," *Journal of Fluid Mechanics*, Vol. 61, 1973, pp. 553–581.

[9]Musman, S., "Penetrative Convection," *Journal of Fluid Mechanics*, Vol. 31, 1968, pp. 343–360.

[10]Blake, K. R., Poulikakos, D., and Bejan, A., "Natural Convection Near 4°C in a Horizontal Water Layer Heated from Below," *Physics of Fluids*, Vol. 27, 1984, pp. 2608–2616.

[11]Nansteel, M. W., Medjani, K., and Lin, D. S., "Natural Convection of Water Near its Density Maximum in a Rectangular Enclosure: Low Rayleigh Number Calculations," *Physics of Fluids*, Vol. 30, 1987, pp. 312–317.

[12]Malkus, W. V. R., "A Laboratory Example of Penetrative Convection," *Geofisica Internacional*, Vol. 5, 1965, pp. 89–95.

[13]Walden, W., and Ahlers, G., "Non-Boussinesq and Penetrative Convection in a Cylindrical Cell," *Journal of Fluid Mechanics*, Vol. 109, 1981, pp. 89–114.

[14]Antar, B. N., "Penetrative Double Diffusive Convection," *Physics of Fluids*, Vol. 30, 1987, pp. 322–330.

[15]Turner, J. S., "Multicomponent Convection," *Annual Review of Fluid Mechanics*, Vol. 17, 1985, pp. 11–43.

[16]Turner, J. S., *Buoyancy Effects in Fluids*, Cambridge Univ. Press, Cambridge, UK, 1973.

[17]Fehlberg, E., "Classical Fifth-, Sixth-, Seventh-, and Eighth-Order Runge Kutta Formulas with Stepsize Control," NASA TR-R-287, 1968.

Transport Phenomena During Vapor Growth of Optoelectronic Material: A Mercurous Chloride System

N. B. Singh*

Westinghouse Science & Technology Center, Pittsburgh, Pennsylvania

Introduction

IN spite of exciting and desirable properties of mercurous chloride, relatively little attention has been paid to the purification process, vaporization, its congruency, and fabrication of devices. A phase diagram in the literature[1] shows the occurrence of a mercury salt-rich eutectic in equilibrium and the existence of a syntactic line extending well beyond the 90 mole % of mercury. The mercurous chloride phase exists at a 1:1 composition of mercury and mercuric chloride with a large miscibility gap. Thus, when mercurous chloride is melted, it separates into mercury and mercuric chloride; mercury settles down in the growth tube, leaving mercuric chloride on the top. Thus, it is not possible to grow crystals by solidification from the melt. Another disadvantage of the melt growth is the increasing vapor pressure of mercurous chloride, which can lead to explosion near the melting point. Considering those points, we have been growing crystals by the vapor phase transport method.

During the last five years we have extensively studied the purification process, growth mechanism and anisotropy, transport behavior, fabrication, and testing of devices. In this chapter results of mass flow and growth rate are reported for different convective conditions.

*Advisory Engineer, Electro-Optical Material.

175

Experimental Methods

Purification of Source Materials

Source material was listed as 99 + % pure. A repeated vacuum sublimation was carried out in a multichamber glass apparatus. Purified material used in the present study has fewer than 10 ppm metallic impurities.

Crystal Growth Velocity Measurements

Crystal growth velocity measurements were made in a two-zone transparent furnace. The temperature of each zone could be controlled independently. The growth velocity was measured by the relaxation method described in Ref. 2. Crystals were grown only in a [110] direction for growth velocity measurements. Orientation was controlled by preoriented seed. The growth velocity measurements were carried out as a function of the aspect ratio a/L, where a is the radius of the growth tube and L is the length between source and crystal.

Results and Discussion

The present study is part of our ongoing attempt to produce large and optically homogeneous mercurous chloride crystals suitable for acousto-optical devices. In this regard the influence of convection is very complex. It produces a variety of spatial and temporal transport rate patterns that become the cause of undesirable micro and macrosegregation. Figure 1 illustrates pieces from two crystals polished on opposite faces. A crystal shown in Fig. 1a shows the presence of bands and striations indicating more inhomogeneity in the refractive index than the crystal shown in Fig. 1b. Convective flows due to thermal and/or solutal flow might be the cause of this type of internal structure. By changing the aspect ratio, attempts have been made to study the variation of growth velocity due to thermal convection in this experiment.

The furnace used in the present study is shown in Fig. 2a along with the temperature profile in Fig. 2b. The temperature profile was measured by using a platinum probe that was embedded in a thin glass. The insulated region produced a temperature hump between the top and bottom furnaces that prevented spontaneous nucleation of crystallites away from the seed. Because the insulation was transparent, we could easily examine the crystal-vapor interface. Mercurous chloride is extremely anisotropic, and we observed[3] that the growth velocity in the [001] direction was approximately two times higher than that in the [110] direction. In order to minimize this effect, all measurements were carried out in a [110] orientation.

For the geometry used in our system the Rayleigh number Ra can be defined[4] as

$$Ra = \frac{g \cdot \beta \cdot a^4 \cdot C_p \cdot d^2 \cdot \Delta T}{k \cdot \eta \cdot L} \tag{1}$$

a)

5 mm

b)

Fig. 1   Mercurous chloride crystal: a) with striations; b) without striations.

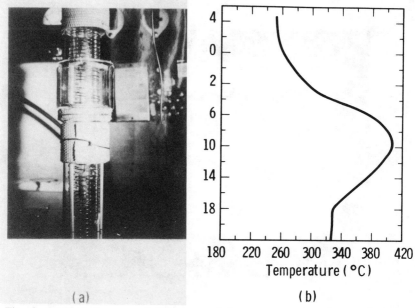

(a)   (b)

Fig. 2   Vertical transparent furnace used in present study: a) growth furnace;
b) temperature distribution.

where g is the gravitational constant, β the volume expansion coefficient, ΔT the temperature difference between hot and cold zones, a the radius of the growth tube, C_p the heat capacity of fluid, d the density, k the thermal conductivity, η the viscosity, and L the transport length. From Eq. (1) it is clear that thermal convection conditions can be changed by varying g, ΔT, and/or a/L. It is strongly dependent on the geometry factor a^4/L. The dependence of mercurous chloride growth velocity on Ra is shown in Fig. 3. As the Ra increases, the growth velocity increases to a certain value. At a critical value, growth velocity drops to a much lower value. This again confirms that convection increases the growth velocity up to certain value; after that, convection destabilizes flow in the system. Contrary to the present observations, the existing theories predict that growth velocity should increase with increasing a/L.

Charlson and Sani[5] have developed stability curves for high-aspect-ratio geometries. These studies are basically for the high-Prandtl-number fluids. For conditions representing the practical crystal growth range and aspect ratios, the stability curve for mercurous chloride is much more complex. This curve is shown in Fig. 4 for the quartz container. It shows a transition point at an aspect ratio of 0.06–0.08. A direct comparison of this data with the theoretical stability curves is not possible because this system has very low a/L values. However, from Ref. 5 it does appear that stability curves for the conducting side walls show a transition point at a low aspect ratio.

Growth Velocity x 10^3 (cm / hr)

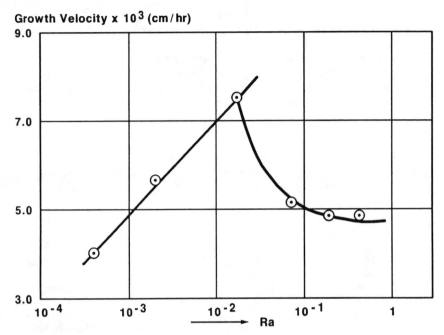

Fig. 3 Variation of crystal growth velocity with Rayleigh number.

$Ra^{1/4}$ x a/L x 10^3

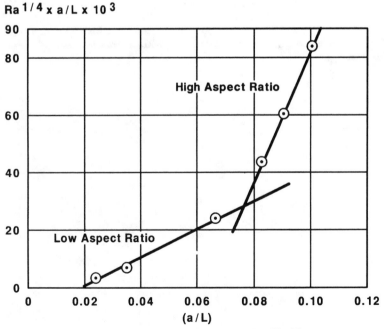

Fig. 4 Stability curve for mercurous chloride.

Summary

Crystal growth velocity was measured in a mercurous chloride system as a function of the Rayleigh number by varying a/L. Growth velocity data showed different trends at low and high aspect ratios. This does not support the velocity-aspect ratio trend predicted by theories. The system cannot be scaled on the basis of measurements done at a low aspect ratio. Some change in fluid flow behavior occurs in the growth tube as the aspect ratio increases.

Acknowledgment

The financial support by NASA Microgravity Science and Application Division Code EN, through NASA Lewis Research Center, is sincerely acknowledged.

References

[1]Yosim, S. J., and Mayer, S. W., "The Mercury-Mercuric Chloride System," *Journal of the American Chemical Society*, Vol. 64, 1960, pp. 909–911.

[2]Singh, N. B., Hopkins, R. H., Mazelsky, R., and Conroy, J. J., "Purification and Growth of Mercurous Chloride Single Crystals," *Journal of Crystal Growth*, Vol. 75, 1986, pp. 173–180.

[3]Singh, N. B., Hopkins, R. H., Mazelsky, R., and Gottlieb, M., "Phase-Relations and Crystal Growth of Mercurous Iodine Crystals," *Journal of Crystal Growth*, Vol. 85, 1987, pp. 240–248.

[4]Carruthers, J. R., *Preparation and Properties of Solid State Materials*, Vol. 4, edited by W. R. Wilcox, Dekker, New York, 1979.

[5]Charlson, G. S., and Sani, R. L., "On Thermoconective Instability in a Bounded Cylindrical Fluid Layer," *Journal of Heat and Mass Transfer*, Vol. 14, 1971, pp. 2157–2160.

Chapter 3. Capillary Phenomena

Capillary Surfaces in Microgravity

Paul Concus*
University of California, Berkeley, California
and
Robert Finn†
Stanford University, Stanford, California

IN this chapter we present a selection of older and newer results on capillary surface interfaces, which complement each other and form elements of a new and developing theory. Although much of the material applies in general gravity fields, we have tried to emphasize the particular interest related to a microgravity setting and the applications peculiar to microgravity. We have made no attempt at completeness. Our attention is directed naturally to matters on which we feel a special competence because we discovered them ourselves; beyond that, we have included work by others that has stimulated our interest and to which we could respond. Even in this limited context, much has been omitted in the interest of exhibiting clear lines of conceptual development. We have included almost no proofs. For readers seeking more complete understanding of the field or detailed information on particular points, we trust that we have included enough references to provide a basis for a fruitful literature search. For those with a more casual interest as well as (and especially) for new initiates to the wealth of exciting and challenging problems that have appeared in the last decades and continue to appear, we hope to have provided a useful overview of the current state of the art.

For the general notion of capillary surface and associated variational characterization, we refer the reader to Ref. 1, Chapter 1. We restrict ourselves here to particular configurations.

---

*Senior Scientist, Lawrence Berkeley Laboratory and Adjunct Professor, Department of Mathematics.
†Professor, Department of Mathematics.

Capillary Tube

We consider a semi-infinite cylindrical tube of homogeneous material and general section, closed at one end by a base Ω, in a uniform gravity field g either zero or directed toward the base. We attempt to cover Ω by a volume V of fluid, making contact angle γ with the walls, and to characterize the resultant free surface S (see Fig. 1). At first we seek the solution as a "graph" $z = u(x,y)$ over Ω and are led by the principle of virtual work to look for a stationary "point" for the energy functional

$$E[u;\gamma] \equiv \int_{\Omega} \sqrt{1 + |\nabla u|^2}\, \mathrm{d}x\, \mathrm{d}y - \beta \oint_{\Sigma} u\, \mathrm{d}s + \frac{1}{2}\kappa \int_{\Omega} u^2\, \mathrm{d}x\, \mathrm{d}y \qquad (1)$$

under the constraint

$$V = \int_{\Omega} u\, \mathrm{d}x\, \mathrm{d}y \qquad (2)$$

Here Σ is the boundary of Ω; $\beta = \cos\gamma$ is the relative adhesion coefficient (wetting energy density) of the fluid on the walls, constant for homogeneous materials; $\kappa = \rho g/\sigma$, where ρ is the density change across S and σ the surface tension. The indicated direction for the gravity field corresponds to $g > 0$. The coordinates are chosen so that $z = 0$ corresponds to the base Ω. For details of the derivation, see, e.g., Ref. 1, Chapter 1.

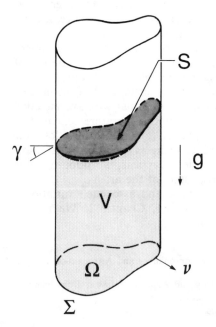

Fig. 1 Partly filled cylindrical tube with base Ω.

The Euler-Lagrange equation for $E[u;\gamma]$ becomes

$$\operatorname{div} Tu = \kappa u + \lambda, \qquad Tu = \frac{\nabla u}{\sqrt{1 + |\nabla u|^2}} \tag{3}$$

in Ω, with

$$v \cdot Tu = \beta \tag{4}$$

on Σ, where v is the exterior unit normal on Σ, and λ is a Lagrange multiplier corresponding to the constraint (2).

To determine the constant λ, we integrate Eq. (3) over Ω, obtaining by the divergence theorem and Eq. (2)

$$\lambda |\Omega| = -\kappa V + \oint_{\Sigma} v \cdot Tu \, ds$$

which by Eq. (4) yields

$$\lambda = \frac{-\kappa V + \beta |\Sigma|}{|\Omega|} \tag{5}$$

The symbols $|\Sigma|$, $|\Omega|$, ... denote the length or area of the indicated geometric quantity. Here we note that if V is replaced by $V + C$, where C is a constant, then addition of $C/|\Omega|$ to u yields again a solution of Eqs. (3) and (4) under the (new) constraint (2); that is, once a solution for some V is known, then solutions for any V are obtained by rigid vertical translation. It is necessary to show that by choosing V large enough we will have $u > 0$, so that Ω is covered by S. Clearly this can be achieved (by the indicated family) if and only if u is bounded below in Ω. We shall see that such a bound cannot always be achieved, even for relatively simple geometries, and thus the problem is in many interesting cases mathematically ill-posed.

We note that the left side of Eq. (3) can be identified as twice the mean curvature H of the surface S. Also, Eq. (4) states that S meets the cylinder walls Z over Σ in the constant angle γ. Thus, the problem posed by Eqs. (3–5) can be interpreted geometrically: We are to find a surface of prescribed mean curvature $H(z)$, which bounds a prescribed volume V over the cylinder base Ω and which meets Z in a prescribed angle γ.

Circular Tube

Here we seek solutions with rotational symmetry about an axis. Even in this case, the question of existence of a solution with prescribed data is not a triviality; the first proof in the literature appears in Johnson and Perko,[2] over 150 years after the initial estimates of Young and of Laplace for the center and meniscus heights of the (presumed) solution surface.

In 1806 Laplace presented his celebrated estimate for the height u_0 of the surface on the axis of a narrow tube of radius a. The estimate is equivalent to approximation of the surface by a spherical cap. In the volume-constrained case considered here we find, in terms of nondimensional variables $\mathcal{U}_0 = (1/a)u_0$ and $\mathcal{V} = (1/a^3)V$ and contact angle γ,

$$\mathcal{U}_0 \approx \frac{1}{\pi}\mathcal{V} - \frac{1}{\cos\gamma} + \frac{2}{3}\frac{1 - \sin^3\gamma}{\cos^3\gamma} \equiv \mathcal{L}[\mathcal{V};\gamma] \tag{6}$$

Laplace offered no proof for the asymptotic validity of Eq. (6) and no indication of how small a must be for prescribed accuracy. The first formal proof of asymptotic correctness of Eq. (6) was given by Siegel.[3] In Ref. 4 the explicit bounds in terms of Bond number $B = \kappa a^2$,

$$\mathcal{L}[\mathcal{V};\gamma] < \mathcal{U}_0 \quad \text{for all} \quad B, \quad 0 \leqslant \gamma < \frac{\pi}{2} \tag{7}$$

$$\mathcal{U}_0 < \mathcal{L}[\mathcal{V};\gamma] + \frac{\cos\gamma(1 + 2\sin\gamma)}{6(1 + \sin\gamma)^4}B, \quad B < 6, \quad 0 \leqslant \gamma < \frac{\pi}{2} \tag{8}$$

are established. Both inequalities reverse if $\pi/2 < \gamma \leqslant \pi$, with $\mathcal{V}$ replaced by $-\mathcal{V}$. It should be noted that these bonds are strict, not merely asymptotic. It can be shown that Eqs. (7) and (8) are best possible, in the sense that

$$\mathcal{U}_0 = \mathcal{L}[\mathcal{V};\gamma] + \frac{\cos\gamma(1 + 2\sin\gamma)}{6(1 + \sin\gamma)^4}B + O(B^2) \tag{9}$$

uniformly in γ, as $B \to 0$.

Similar bounds are established in Refs. 5 and 6 for the height $\mathcal{U}_1$ on the contact line and for the "meniscus height" $\mathcal{U}_1 - \mathcal{U}_0$. Siegel[7] has given bounds that hold throughout the trajectory $0 \leqslant r \leqslant a$, although they are somewhat less exact at the endpoints.

General Sections

If the system (3–5) is to reflect reality, one should expect solutions to exist when containers have reasonable sections, other than that of the disk. This topic has been studied mathematically in considerable generality (see Ref. 1 and the references therein). For simplicity we restrict our attention here to sections Ω whose boundaries Σ consist of a finite number n of smooth curves that meet in well-defined corners $P_1, ..., P_n$, with interior half-angles $\alpha_1, ..., \alpha_n$. We note that the condition (4) cannot be prescribed at the P_j, since v is not defined at these points. Nevertheless, if $g > 0$, we obtain the general result that *there is exactly one surface $z = u(x,y)$ over Ω, such that* Eq. (3) *holds strictly in Ω, the given volume V is achieved, and* Eq. (4) *holds strictly on Σ except at the $\{P_j\}$*. This result was first proved by Emmer[8] under some restrictions. For greater generality see Ref. 9 and Ref. 1, Chapter 7.

Observe that uniqueness holds without growth conditions at the $\{P_j\}$. In this respect the behavior of solutions of Eq. (3) differs strikingly from that typically encountered for solutions of linear problems, for which failure to prescribe boundary conditions at even a single point can lead to nonuniqueness. The nonlinearity in the present problem constrains the solution near the exceptional boundary points, even though no data are prescribed there.

The uniqueness follows from a general comparison principle, which is central to the material that follows. We define $Nu \equiv \operatorname{div} Tu - \kappa u$.

COMPARISON PRINCIPLE (CP). *Suppose* $Nu \geqslant Nv$ *in* Ω, $v \cdot Tv$ $\geqslant v \cdot Tu$ *on* $\Sigma \backslash \{P_j\}$. *There follows* i) *if* $\kappa > 0$, *then* $v \geqslant u$ *in* Ω; *equality holds at any point if and only if* $v \equiv u$ *in* Ω; *and* ii) *if* $\kappa = 0$, *then* $v(x,y) \equiv u(x,y) + \text{const}$ *in* Ω.

This statement is adequate for the applications to be made here. More general statements (and results) can be found in Ref. 10. The crucial point is that the result holds without growth hypotheses at the exceptional points $\{P_j\}$.

Application I

Suppose Ω contains a corner of interior angle 2α, as indicated in Fig. 2. Let $v(x,y)$ denote a lower hemisphere, whose equatorial circle lies over Γ.

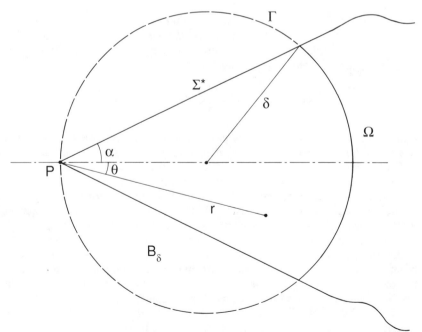

Fig. 2 Domain Ω containing a corner. The disk B_δ.

Since $\mathrm{div}Tv$ is twice the mean curvature of the surface $v(x,y)$, we have

$$\mathrm{div}Tv = \frac{2}{\delta} \tag{10}$$

in the disk B_δ bounded by Γ; also, v meets the vertical (planar) walls Z over Σ^* in the constant angle $\gamma_0 = (\pi/2) - \alpha$. Let $u(x,y)$ satisfy

$$\mathrm{div}Tu = \kappa u, \qquad \kappa > 0 \tag{11}$$

in Ω, with boundary angle γ such that $\gamma_0 \leqslant \gamma \leqslant \pi/2$. Then $v \cdot Tv \geqslant v \cdot Tu$ on $\Sigma^* \cap B_\delta$, and $v \cdot Tv = 1$ on $\Gamma \cap \Omega$. We observe that $|v \cdot Tu| < 1$ on $\Gamma \cap \Omega$ by virtue of the finiteness of $|\nabla u|$, and we choose $\Sigma = \Sigma^* \cup (\Gamma \cap \Omega)$, $\{P_i\}$ to be the three intersection points $\Gamma \cap \Sigma^*$. We adjust v by an additive constant, so that its minimum is $v_0 = 2/(\kappa\delta)$. We then have by Eq. (11)

$$\mathrm{div}Tv = \frac{2}{\delta} = \kappa v_0 \leqslant \kappa v \tag{12}$$

Thus, $Nv \leqslant Nu$ in $\Omega \cap B_\delta$. From **CP** we find immediately $u < v$ in $\Omega \cap B_\delta$. Similarly, $u > -v$. But $v \leqslant v_0 + \delta$ and *there follows*

$$|u| < \frac{2}{\kappa\delta} + \delta \tag{13}$$

throughout $\Omega \cap B_\delta$. This inequality holds for any configuration for which γ lies in the interval $|(\pi/2) - \gamma| \leqslant \alpha$, including the endpoints. But outside that interval a very different behavior prevails. *In fact,*[11,12] *if $|(\pi/2) - \gamma| > \alpha$, there exist positive constants C, ϵ such that*

$$\left| u - \frac{\cos\theta - \sqrt{k^2 - \sin^2\theta}}{k\kappa r} \right| < Cr^\epsilon, \qquad k = \frac{\sin\alpha}{\cos\gamma} \tag{14}$$

near P (see Fig. 2). Thus, in this case $|u| \to \infty$ at P, and we see that there is a discontinuous change in behavior as $|(\pi/2) - \gamma|$ passes through α.

The discontinuity can be used effectively for accurate measurement of contact angle, at least when $|\pi/2 - \gamma|$ is not too large. Figure 3 shows results of a "kitchen sink" experiment carried out by Tim Coburn in the Medical School of Stanford University, using distilled water between plates of acrylic plastic. A change (effected by hand) of about 2 or 3 deg in the half-angle between plates yielded the change in observed peak height at P, from slightly under the bound of Eq. (13) when $|(\pi/2) - \gamma| \leqslant \alpha$, to over 10 times that bound when $|(\pi/2) - \gamma| > \alpha$, thus leading to a preliminary estimate of between 78 and 81 deg for the contact angle γ_{wa} between water and acrylic plastic. Coburn's procedure was repeated recently under more controlled conditions by Mark Weislogel at NASA Lewis Research Center. For his materials he obtained accurately repeatable results of $\gamma_{wa} = 80 \pm 2$ deg. By comparison, he found the reproducibility of other, currently standard

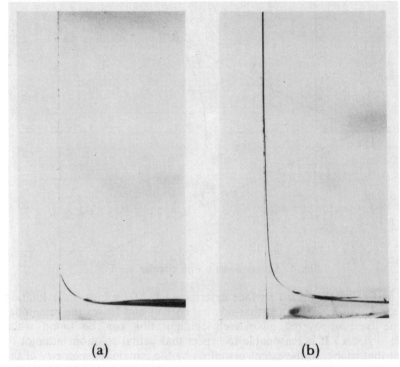

(a)   (b)

Fig. 3   Discontinuous dependence on data: $g > 0$: a) $\alpha \approx 12$ deg; b) $\alpha \approx 9$ deg.

methods to be more strongly affected by hysteresis, with uncertainty several times as large.

The accuracy of the method just described decreases rapidly as $|(\pi/2) - \gamma|$ increases, since the critical opening angle then becomes large and the height changes become restricted to a small neighborhood of the vertex, where observations are difficult. This point is addressed in the following considerations.

Application II

The above results are for $g > 0$. If $g = 0$, the corner effect becomes still more striking. In fact, *if $g = 0$ and Ω contains a corner at which $|(\pi/2) - \gamma| > \alpha$, then there is no solution of Eqs. (3) and (4) over Ω for any λ*. This can be proved very simply by applying the divergence theorem to Eq. (3) over a domain cut off by a segment Γ at the corner, using the boundary condition (4) on Σ^* and the bound $|Tu| < 1$ on Γ. If $|(\pi/2) - \gamma| > \alpha$, a contradiction is immediately obtained by letting $\Gamma \to P$ (see Ref. 13).

What does such a result mean physically? To get a feeling for what happens, observe that, when $|(\pi/2) - \gamma| > \alpha$ and the boundary segments at the corner are long enough, a circular arc Γ of radius $1/\lambda$ can be positioned in the corner to meet Σ^* in angle γ at both intersection points (see Fig. 4).

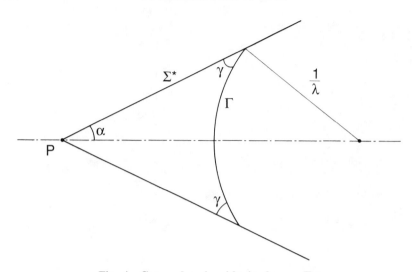

Fig. 4   Corner domain with circular arc Γ.

The vertical (cylindrical) surface determined by Γ defines, in a limiting sense, a solution of Eq. (3) that fills the corner, and leaves the remainder of the base uncovered. (No such configuration can be found when $|(\pi/2) - \gamma| \leqslant \alpha$.) It is reasonable to expect that actual solutions attempt to adopt that shape, to the extent permitted by the remaining geometry of the container. This view is supported by experiments made by W. Masica at the NASA Lewis Research Center Zero Gravity Facility, using a cylindrical container with hexagonal cross section. In this case, if $|(\pi/2) - \gamma| \leqslant \alpha$, a solution surface exists and can even be given explicitly as the spherical cap

$$u(x,y;\delta) = u_0 \mp \sqrt{\frac{\delta^2}{\cos^2\gamma} - r^2} \tag{15}$$

where δ is the radius of the inscribed circle, and u_0 is determined by the constraint

$$\int_\Omega u \, \mathrm{d}x \, \mathrm{d}y = V \tag{16}$$

Figure 5 shows the results of drop tower experiments, in which identical acrylic plastic containers and two different liquids are used. In Fig. 5a the liquid is a 20% ethanol in water solution, for which $\gamma \approx 48$ deg, and there holds $|(\pi/2) - \gamma| < \alpha$; here the expected spherical cap is obtained. In Fig. 5b the liquid is a 30% ethanol in water solution, for which $\gamma \approx 25$ deg, and there holds $|(\pi/2) - \gamma| > \alpha$; the fluid now attempts to fill the corners and climbs to the top of the container, meeting the top and the upper boundary walls in the prescribed angle. The fluid height over the remainder of the base is significantly lowered.

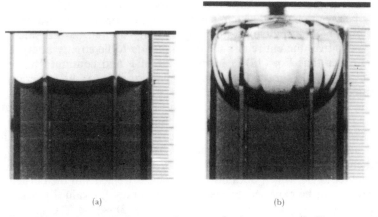

(a) (b)

Fig. 5 **Discontinuous dependence on data: $g = 0$: a) $\alpha + \gamma \geqslant \pi/2$; b) $\alpha + \gamma < \pi/2$.**

This behavior is the key to the following observations.[14] Consider the domain of Fig. 6, bounded by two line segments and a circular arc of radius δ. We suppose that $g > 0$ and $|(\pi/2) - \gamma| \leqslant \alpha$. It can then be shown[9] that a solution of Eq. (3) exists, with data γ on Σ. We note that a hemisphere of radius $\delta/\cos\gamma$ and center at O meets the bounding cylinder Z in the constant angle γ (except at the vertex, where the angle is undefined). Applying **CP** in ways similar to that used above, we are led to the inequality

$$\frac{V}{|\Omega|} - \frac{\delta}{\cos\gamma}\left(1 - \frac{\sqrt{k^2 - 1}}{k}\right) < u(x,y;g) < \frac{V}{|\Omega|} + \frac{\delta}{\cos\gamma}\left(1 - \frac{\sqrt{k^2 - 1}}{k}\right) \quad (17)$$

throughout Ω when $0 \leqslant \gamma < \pi/2$, and the same inequality with $u \to -u$, $\gamma \to (\pi - \gamma)$ when $\pi/2 < \gamma \leqslant \pi$. To fix the ideas, consider the case $0 \leqslant \gamma <$

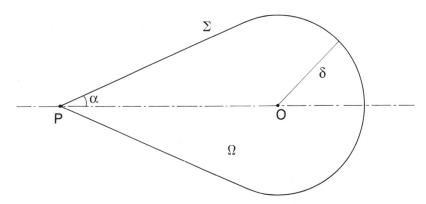

Fig. 6 **"Ice cream cone" domain.**

$\pi/2$. If V is prescribed in advance, the formal solution $u(x,y;g)$ might be partly below the base Ω, which is physically unacceptable. However, we see from Eq. (17) that by choosing V large enough so that the left side of Eq. (17) is positive, the entire surface can be made to lie strictly above Ω. We note further that Eq. (17) is independent of g and continues to hold as stated as $g \searrow 0$, the surfaces converging in fact to the spherical cap indicated earlier.

Now we suppose $|(\pi/2) - \gamma| > \alpha$. In this case the fluid tries to fill out the corner (as indicated earlier), as $g \searrow 0$; thus, if the cylindrical container is sufficiently high, then the bottom Ω must become uncovered, except for a small neighborhood of P. In particular, $u(0,0;g)$ will violate Eq. (17) for small enough g. Thus, we have found a procedure for the determination of the contact angle, which does not require measurements in the corner. This procedure can be expected to yield great accuracy for values γ reasonably distant from 0 and π, perhaps in the range $|(\pi/2) - \gamma| < 70$ deg.

Application III

As we have seen, when $g = 0$, surfaces $u(x,y)$ satisfying Eqs. (3) and (4) need not always exist. Explicit geometric criteria have been given only under restrictive conditions (cf. Theorem 8 in Ref. 15, and Ref. 16). However, the following useful (indirect) criterion appears in Ref. 15 (see also Ref. 1, Chapters 6 and 7).

Let

$$R_\gamma = \frac{|\Omega|}{|\Sigma| \cos\gamma}$$

A solution of Eqs. (3) and (4) in Ω exists if and only if, for every strict subarc Γ (in Ω) of a semicircle of radius $|R_\gamma|$ that meets Σ with angle γ at both intersection points, as indicated in Fig. 7, there holds

$$\Phi[\Omega^*] \equiv |\Gamma| - |\Sigma^*| \cos\gamma + \frac{1}{R_\gamma} |\Omega^*| > 0 \qquad (18)$$

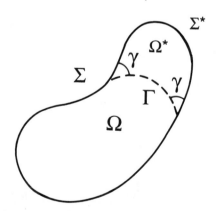

Fig. 7   **Domain partitioned by circular arc Γ of radius R_γ.**

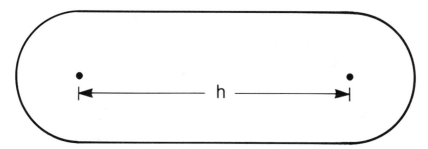

Fig. 8   Domain with parallel sides and semicircular ends.

In many cases of interest there can be found (up to inessential displacements) only a finite number of curves Γ with the required properties; thus, the matter can be settled by examining a finite number of cases. Of particular interest are those situations in which Eq. (18) holds vacuously, in the sense that there is no Γ satisfying the conditions. For example, that is the case, regardless of γ, in the configuration of Fig. 8, bounded by two parallel lines and semicircles, and we are thus assured of existence, regardless of h, for that configuration.

This situation changes dramatically if we consider instead the configuration of Fig. 9, bounded by two nonparallel lines and smoothly joined circular arcs. We observe from Eq. (18) that to any domain Ω there corresponds a critical value $\gamma_0(\Omega)$ of γ (and the supplementary critical value $\pi - \gamma_0$), such that a solution exists if $|(\pi/2) - \gamma| < |(\pi/2) - \gamma_0|$ and fails to exist if $|(\pi/2) - \gamma| > |(\pi/2) - \gamma_0|$. For the Ω of Fig. 8 we have $\gamma_0 = 0$ (or π); nevertheless, if we let $\alpha \to 0$ in Fig. 9, we find

$$\lim_{\alpha \to 0} \gamma_0(\Omega) = \pi/2$$

That is, *when the sides are parallel, a solution exists for any γ; but if they are nonparallel, then the closer they become to being parallel, the more solutions are excluded.* This nonuniformity in behavior is illustrated by the curves in Fig. 10, relating the value of γ_0 in $[0,\pi/2]$ to ρ for differing values

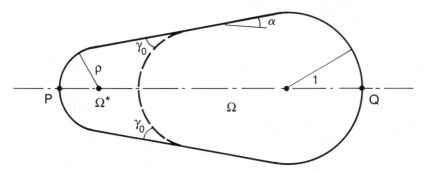

Fig. 9   Domain with nonparallel sides and circular arc ends.

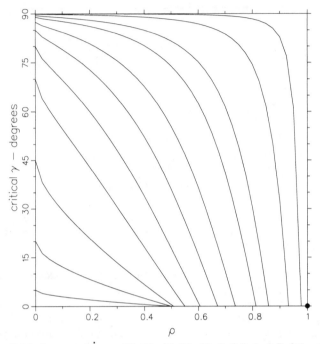

Fig. 10   **Critical** γ **vs** ρ **for** $\alpha = 85, 70, 45, 20, 10, 5, 2.5, 1, 0.5, 0.1,$ **and** 0.01 **deg, from lower left to upper right. For** $\alpha = 0$ **deg the curves degenerate to the single point** $\rho = 1, \gamma_0 = 0$ **deg.**

of α. For each α-curve, solutions fail to exist in the region below and to the left of the curve. As $\alpha \to 0$, the curves tend to the entire upper and right-hand boundary segments, but $\alpha = 0$ yields only the indicated single point $(1,0)$. (The depicted curves join piecewise linearly the tabulated values calculated for increment 0.025 in ρ.)

Application IV

     The domains Ω^* that appear when $\Phi = 0$ for some Γ, as above, can be interpreted physically. As mentioned above, solution surfaces over Ω exist when $|(\pi/2) - \gamma| < |(\pi/2) - \gamma_0|$. Letting $\gamma \to \gamma_0$ from within the range of solvability, we obtain a sequence $u^{(n)}$ of surfaces satisfying Eq. (3) with $\kappa = 0$; *this sequence can be normalized to converge throughout Ω^* to $\pm \infty$ (according as $\gamma_0 \lessgtr \pi/2$) and throughout $\Omega \backslash \bar{\Omega}^*$ to a solution $u^0(x,y)$.* Thus, we can consider $u^0(x,y)$ to determine a "generalized solution" over Ω that is identically positive infinite over Ω^*. Such "solutions" were first introduced in a formal mathematical way by Miranda[17] and are basic for the general existence theory (Ref. 1, Chapters 6 and 7).

     Alternatively, we could have normalized the $u^{(n)}$ to converge to a solution in Ω^* and to $\mp \infty$ in $\Omega \backslash \Omega^*$. These properties can be exploited in conjunction with particular geometries (chiefly modifications of that in Fig. 11) to obtain procedures for accurate measurement of small contact angles. For example,

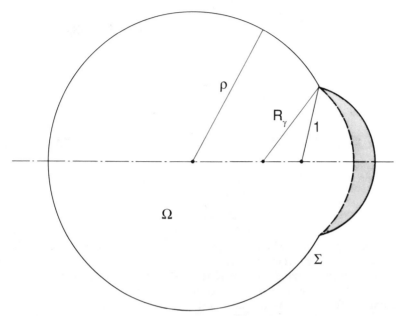

Fig. 11 Two-circle domain.

in Fig. 11 the fluid height becomes infinite in the shaded region when $|(\pi/2) - \gamma| \geqslant |(\pi/2) - \gamma_0|$. This work is currently in progress, and preliminary results are encouraging.

Application V

What happens if we fix γ and let $g \searrow 0$ (or, equivalently, $B \searrow 0$)? This question is of considerable importance for application II. It is known that for reasonably smooth Ω a solution $u^B \equiv u(x,y;B)$ always exists when $B > 0$. Siegel[18] proved that *whenever a solution $u^0 = u(x,y;0)$ exists, then (under some restrictions) there exists a constant C such that $|u^B - u^0| < CB$ uniformly in Ω.* Tam[19] later removed the restrictions and also to some extent the requirement that u^0 exist. He showed that, *whenever Ω^* is uniquely determined, there exist functions $C_1(B), C_2(B)$ such that*

$$\lim_{B \to 0} [u^B + C_1(B)] = \begin{cases} \pm \infty & in & \Omega^*, \text{ according as } \gamma_0 \lessgtr \pi/2 \\ solution\ in & \Omega \backslash \bar{\Omega}^* \end{cases}$$

$$\lim_{B \to 0} [u^B + C_2(B)] = \begin{cases} solution\ in & \Omega^* \\ \mp \infty & in & \Omega \backslash \Omega^*, \text{ according as } \gamma_0 \lessgtr \pi/2 \end{cases}$$

This is the sense in which transition to zero gravity must be interpreted in general (practical) situations.

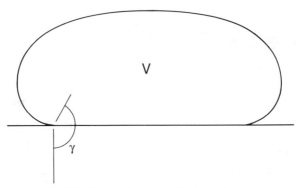

Fig. 12   Sessile drop on a horizontal plane.

Uniqueness and Nonuniqueness

Surfaces $u(x,y)$ that satisfy Eqs. (3) and (4) in capillary tubes are uniquely determined when $g \geqslant 0$ and provide an absolute minimum for the mechanical energy. This follows easily from **CP**. Miranda[20] showed that any energy minimizing surface that covers Ω must in fact have the form $u(x,y)$ (i.e., the surface cannot bend over itself if it minimizes). More recently, Vogel[21] showed that there can be no stationary (equilibrium) configuration distinct from the minimizer. Thus, *capillary surfaces in cylindrical tubes are strongly stable configurations that cover the base simply.*

A drop of liquid volume V resting on a horizontal plane (Fig. 12) is also unique and energy minimizing (Ref. 1, Chapter 3, Note 3). Now imagine the plane continously deformed, through a family of convex surfaces, into the cylinder (Fig. 13). Does the liquid surface remain unique? In general, certainly not! Consider the configuration of Fig. 14, in which a conical base meets the right-circular-cylindrical wall in a 45-deg angle, and suppose that $\gamma = 45$ deg. If the cone is filled from the bottom until just below the juncture point, a horizontal surface at that height provides a solution, with the volume V of the cone. Now observe that any volume $V^+ > V$ yields a capillary surface meeting the cylinder walls. This surface cannot be horizontal in view of the 45-deg contact angle. If fluid is removed until the contact line is just above the juncture point, a congruent surface is obtained with volume $V^- < V$. By considerations of continuity, there must be an intermediate configuration distinct from the horizontal one, and with the same volume.

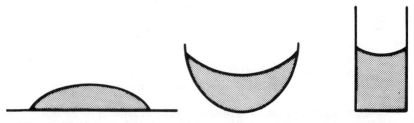

Fig. 13   Fluid on a plane, in an "intermediate" container, and in a cylinder.

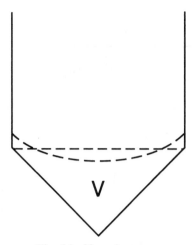

Fig. 14 Nonuniqueness.

This discussion can be modified in order to yield an entire continuum of equilibrium interfaces, all of which bound the same volume and have identical mechanical energy.[22,23] In this continuum the horizontal surface has been shown to be unstable[23,24]; thus, it is unlikely to be observed physically. Using these considerations, *one can design an axially symmetric container, as close to a circular cylinder as desired, which when half filled with liquid admits a continuum of symmetric solutions, but for which no interface of minimizing energy can be symmetric.*[23] In Fig. 15 the radial section is shown of such a container admitting a continuum of symmetric solutions for $\gamma = 60$ deg and zero gravity, along with some of the solution surfaces. The procedure works in any gravity field; however, the size of the container becomes small and the effects can be difficult to observe, except under microgravity conditions.

Stability

If $g > 0$, then the free surface in a capillary tube yields a global energy minimum and hence is stable. Under reasonable conditions this will also be so when $g = 0$, but in some senses there can be exceptions. For example, in the configuration of Fig. 6, when $|(\pi/2) - \gamma| = \alpha$, the free surface is clearly unstable under small perturbations of the contact angle. See Langbein[25] for a detailed discussion of the instability.

Similarly, a liquid drop on a horizontal surface (sessile drop) is stable for any $g \geqslant 0$. As shown above, if the support surface is curved, stability cannot in general be expected. For a horizontal support surface Π with $g < 0$ (or, equivalently, with $g > 0$ and the drop contacting Π from below) one obtains the "pendent drop," for which stability criteria have a long history (see, e.g. Ref. 1, Sec. 4.15). We indicate here recent results of Wente.[26] To

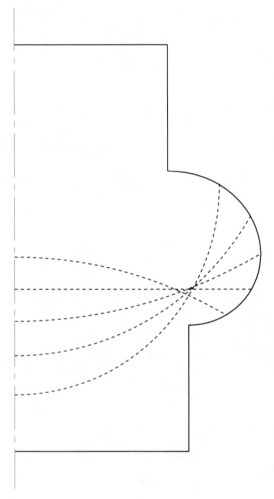

Fig. 15   Radial section of a container (solid curve) with selected free surfaces (dashed curves) from the continuum having the same energy, contact angle, and enclosing the same liquid volume with the container ($\gamma = 60$ deg, $g = 0$).

describe them, we normalize the equation to the form

$$\mathrm{div}\,Tu = -u \qquad (19)$$

in terms of nondimensional variables ($\sqrt{|\kappa|}\,u \to u$, $\sqrt{|\kappa|}\,r \to r$). It can be shown[27] that, for any prescribed $u_0 < 0$, there exist global axisymmetric solutions for which $u(0) = u_0$; these solutions are bounded in volume, up to the horizontal contact plane with positive u, and (for large $|u_0|$) exhibit a succession of "bubblelike" shapes (see Fig. 16). Wente[26] shows the following:

Suppose that $0 < \gamma < \pi$. *Then, for any sufficiently small volume V, there exist stable drops that are convex and resemble spherical caps. These drops are generated by profile curves whose tip is at* $|u_0|$, *where* $|u_0|$ *is large. As V increases, so does* u_0, *until an inflection point is reached. This drop is stable. With further increase of V,* u_0 *decreases; the drop profile contains an inflection but continues in some interval to remain stable. Instability occurs prior to the occurrence of a second inflection (see Fig. 16).*

If $\gamma = 0$, *all profile curves contain an inflection. The curves are generated by starting with a solution* $u \equiv 0$, *then letting* u_0 *decrease and considering the portion of the curve up to the first maximum. The limit of stability is reached prior to the appearance of a vertical point, i.e., before* $|u_0| \approx 2.5678$ (cf. Ref. 27).

For any angle of contact the drop height increases monotonically with volume throughout the range of stability.

Wente also establishes stability criteria for two other configurations, in which the fluid drop hangs from a fixed circular aperture, under conditions of either constant pressure (the nearly empty medicine dropper) or of constant volume (the filled medicine dropper). In the latter case he shows that *stable configurations can occur in which both a bulge and a neck appear.* Figure 17 shows a verification of this behavior in the latter author's kitchen sink, with a drop of water (coloured with soluble ink) suspended in ricinoleic acid, which has a density of about 0.95, so that a near neutrally buoyant (small Bond number) condition arises.

If $g = 0$, then all sessile (or pendent) drops on a horizontal homogeneous support plane are spherical caps. That is also the case for the "medicine dropper" problem with circular orifice. If the orifice is not circular, one is led to a generalization of the classical Plateau problem for minimal surfaces: *among all surfaces through a given closed curve $\mathscr{C}$ and bounding with a given surface through $\mathscr{C}$ a prescribed volume V, to find one that minimizes area.* The solution (soap bubble) is a surface of constant mean curvature H. An existence proof was given by Wente.[29] Wente also showed that for sufficiently small Bond number a pendent drop through the orifice and supported by $\mathscr{C}$ must exist.[30]

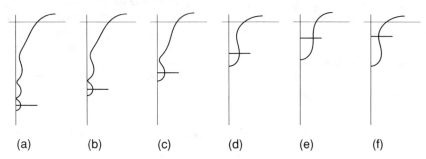

| (a) | (b) | (c) | (d) | (e) | (f) |

Fig. 16 Pendent drop formation with increasing volume, $\gamma = 90$ deg. The line segments indicate the plane of support. Configurations a–e are always stable; f will be stable if the increase of volume from e is sufficiently small.

Fig. 17   Stable pendent drop with neck and bulge.

Another stability problem that has attracted much attention is that of a liquid column (liquid bridge) of volume V joining two parallel planes of distance h in the absence of gravity. The problem can be traced back to Rayleigh,[31] who studied the linearized stability of infinite liquid columns whose sections are circular and found energy-decreasing perturbations of period equal to the circumference. The column joining parallel planes, with contact angle $\pi/2$, was studied independently by Athanassenas[32] and Vogel.[33] These authors obtained, by different procedures, the result that *any stable configuration is a circular cylinder. If $V > (1/\pi)h^3$, the cylinder is stable; if $V < (1/\pi)h^3$, it is unstable.* We note that the onset of instability appears at half the cylinder length at which it occurs for the classical Rayleigh instability. That is because of the differing boundary condition, which allows a larger choice of perturbations in the present case.

Vogel continued his study with a series of papers[34–36] investigating stability criteria for contact angle $\gamma \neq \pi/2$, and also for differing angles γ_1, γ_2 on the two planes. He found a remarkable diversity in the kinds of behavior that can occur. We summarize some of the results.

The drop is necessarily rotationally symmetric and described by its meridional distance $f(x)$ from an axis, with $x \in [0,h]$. The function $f(x)$ must satisfy the equation

$$ff'' - (1 + f'^2) = 2H(1 + f'^2)^{\frac{3}{2}}f \tag{20}$$

for some constant H, and the boundary conditions

$$f'(0) = -\cot\gamma_1, \qquad f'(h) = \cot\gamma_2 \tag{21}$$

The following conditions suffice for stability:

1) *The Sturm-Liouville problem*

$$L(\psi) \equiv -\left(\frac{f\psi'}{(1+f'^2)^{\frac{3}{2}}}\right)' - \frac{\psi}{f(1+f'^2)^{\frac{1}{2}}} = \lambda\psi \tag{22a}$$

$$\psi'(0) = \psi'(h) = 0 \tag{22b}$$

has exactly one negative eigenvalue.

2) *There is a smoothly parametrized family $f(x;\epsilon)$ with $f(x;0) \equiv f(x)$, such that $\gamma_1'(\epsilon) = \gamma_2'(\epsilon) \equiv 0$ and such that*

$$H'(0)V'(0) > 0$$

where $H(\epsilon)$, $V(\epsilon)$ are the mean curvature and volume of the drops in the family. Physically this means that volume increases with increasing pressure in the drop.

If $L(\psi)$ has two or more negative eigenvalues, or if $H'(0)V'(0) < 0$, then the drop is unstable.

For sufficiently large V a stable drop exists whose profile is uniformly close to an arc of a circle.

If $\gamma_1 + \gamma_2 \neq \pi$, then the family of all profile curves without inflections can be parametrized by H, with condition 1) above holding throughout the family.

Every convex drop is stable.

There exist stable drops with inflection points in the profile curve (and which are thus not convex).

If $|f''(x)| > a_0 > 0$ for a solution of Eqs. (20) and (21), then the first two eigenvalues λ_0, λ_1 of Eq. (22) satisfy $\lambda_0 < 0 < \lambda_1$. Thus, the only way an instability can develop is for condition 2) to fail. If that occurs, Vogel notes that for given γ_1, γ_2 each profile determines a pair of values H, V; he defines the point (H, V) corresponding to a profile on which $V'(H)$ changes sign to be a point of instability of type 2.

Another way for a family of profiles to become unstable is for condition 1) to fail; if λ_1 passes through zero while condition 2) continues to hold, then we refer to a point of instability of type 1.

In the following we consider the case $\gamma_1 = \gamma_2$. Let Π be the plane parallel to the two given ones and midway between them. Vogel[35] shows that, *at a type 1 point, the instability manifests itself through a perturbation that is asymmetric with respect to Π; at a type 2 point, symmetric perturbations give rise to the instability.*

Vogel found numerical evidence to support the view that there is an angle $\gamma_0 \approx 31.14$ deg, such that, if $\gamma_1 = \gamma_2 > \gamma_0$, the family of inflectionless profiles will become type 1 unstable, corresponding to a bifurcation that occurs when an inflection appears at the boundary (he proves the existence of the bifurcations), whereas, if $\gamma_1 = \gamma_2 < \gamma_0$, the family of inflectionless profiles will exhibit a type 2 instability. For $\gamma_1 = \gamma_2$ no stable configurations with interior inflections were found. For $\gamma_1 \neq \gamma_2$, no type 1 instability was observed.

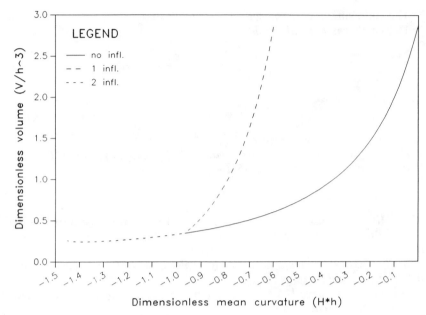

Fig. 18 V vs H for $\gamma_1 = \gamma_2 = 60$ deg.

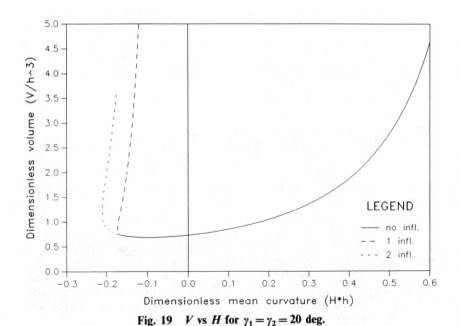

Fig. 19 V vs H for $\gamma_1 = \gamma_2 = 20$ deg.

In Refs. 34 and 36 Vogel presents a systematic numerical investigation of stability criteria and modes of breakdown for varying combinations of angles γ_1, γ_2. In Figs. 18 and 19 (taken from Ref. 35), the two modes of breakdown are exhibited in two specific cases. Figure 18 corresponds to $\gamma_1 = \gamma_2 = 60$ deg, and we see that, with decreasing $H, V'(H) > 0$ until a bifurcation occurs. In Fig. 19, $\gamma_1 = \gamma_2 = 20$ deg; in this case $V'(H)$ changes sign prior to the bifurcation.

We remark that, since no breakdown by bifurcation was observed when $\gamma_1 \neq \gamma_2$, the former of the two cases just mentioned is presumably an isolated event; a slight change in γ_1 or γ_2 should thus lead to a splitting into two distinct branches, with type 2 instability replacing the bifurcation.

Convexity

When is a capillary surface convex? For a given constant contact angle $\gamma \neq 0, \pi$ in a capillary tube, it does *not* suffice that the base domain Ω be convex. To see that, consider Ω as indicated in Fig. 9, with $0 < \gamma_0 < \pi/2$, where γ_0 is the critical angle. Take a sequence $\gamma \searrow \gamma_0$. Then the corresponding zero-gravity solutions $u(x,y;\gamma)$ of Eqs. (3) and (4) exist and can be normalized so that $u \to u^{(0)}(x,y)$ (a solution of Eq. (3) with $\kappa = 0$) uniformly in $\Omega \backslash \bar{\Omega}^*$, $u \to \infty$ uniformly in Ω^*. The solutions are symmetric with respect to the line PQ; thus,

$$\left. \frac{\partial u}{\partial x} \right|_P = -\tan\gamma \to -\tan\gamma_0, \qquad \left. \frac{\partial u}{\partial x} \right|_Q = \tan\gamma \to \tan\gamma_0$$

Hence, if γ is close enough to γ_0 that $u(P;\gamma) - u(Q;\gamma) > |P - Q| \tan\gamma$, an inflection must appear in the curve lying directly above PQ in the surface, and thus convexity will fail. A similar construction can also be effected when $g > 0$ (see Korevaar[37]).

Chen and Huang[38] proved that, *if $g = 0$ and $\gamma = 0$ or π, then convexity of Ω implies convexity of the solution surface.* Under some restrictions, Korevaar[37] found by a different method the same result when $g > 0$.

Although the surface itself need not be convex when Ω is convex, Chen[39] proved that *there can be only one minimal point if $0 < \gamma < \pi/2$.* Chen's proof (and result) were improved in some respects by Siegel.[40] Huang[41] proved that, *if $g = 0$ and Ω is convex, then the (unique) minimal point must have a distance of at least $|\Omega|/|\Sigma|$ from the boundary Σ of Ω. If, in addition, Ω admits a line of symmetry, then the level curves of the solution surface are convex.*

In Ref. 10 it is shown by example that a drop resting on a horizontal plane (with variable contact angle so that it need not be symmetric) can wet a convex region and nevertheless not itself be convex. We have been informed that A. N. Wang has now found an example for which also the level curves are not all convex.

We close by mentioning a striking result of Vogel[21] on sessile drops with variable contact angle: *If $0 < \gamma < \pi/2$, then the entire surface projects simply onto the support plane Π. For any distribution of γ, let z_0 be the* *maximum height of the drop above* Π. *Then the portion of the drop with height exceeding* $\frac{1}{2}z_0$ *projects simply onto* Π.

Acknowledgment

This work was supported in part by the Applied Mathematical Sciences subprogram of the Office of Energy Research of the U.S. Department of Energy under Contract DE-AC03-76SF000098 and in part by the National Science Foundation.

References

[1] Finn, R., *Equilibrium Capillary Surfaces*, Springer-Verlag, New York, 1986.

[2] Johnson, W. E., and Perko, L. M., "Interior and Exterior Boundary Value Problems from the Theory of the Capillary Tube," *Archive for Rational Mechanics and Analysis*, Vol. 29, 1968, pp. 125–143.

[3] Siegel, D., "Height Estimates for Capillary Surfaces," *Pacific Journal of Mathematics*, Vol. 88, 1980, pp. 471–516.

[4] Finn, R., *Equilibrium Capillary Surfaces*, Mir, Moscow, 1989 (Russian Translation, with Appendix by H. C. Wente).

[5] Finn, R., "On the Laplace Formula and the Meniscus Height for a Capillary Surface," *Zeitschrift fuer Angewandte Mathematik und Mechanik*, Vol. 61, 1981, pp. 165–173.

[6] Finn, R., "Addenda to my Paper 'On the Laplace Formula and the Meniscus Height for a Capillary Surface'," *Zeitschrift fuer Angewandte Mathematik und Mechanik*, Vol. 61, 1981, pp. 175–177.

[7] Siegel, D., "Explicit Estimates of a Symmetric Capillary Surface for Small Bond Number" (to be published).

[8] Emmer, M., "Esistenza, Unicità e Regolarità nelle Superfici di Equilibrio nei Capillari," *Annali dell'Università di Ferrara, Sezione 7*, Vol. 18, 1973, pp. 79–94.

[9] Finn, R., and Gerhardt, C., "The Internal Sphere Condition and the Capillary Problem," *Ann. Mat. Pura Appl.*, Vol. 112, 1977, pp. 13–31.

[10] Finn, R., "Comparison Principles in Capillarity," *Partial Differential Equations and Calculus of Variations*, edited by S. Hildebrandt and R. Leis, Springer-Verlag, 1988, pp. 156–197 (Springer Lecture Notes in Mathematics No. 1357).

[11] Concus, P., and Finn, R., "On Capillary Free Surfaces in a Gravitational Field," *Acta Mathematica*, Vol. 132, 1974, pp. 207–223.

[12] Miersemann, E., "On the Behavior of Capillaries at a Corner," *Pacific Journal of Mathematics*, Vol. 140, 1989, pp. 149–154.

[13] Concus, P., and Finn, R., "On Capillary Free Surfaces in the Absence of Gravity," *Acta Mathematica*, Vol. 132, 1974, pp. 177–198.

[14] Concus, P., Finn, R., and Langbein, D., "The Contact Angle in Capillarity" (unpublished observations).

[15] Finn, R., "A Subsidiary Variational Problem and Existence Criteria for Capillary Surfaces," *Journal für die Reine und Angewandte Mathematik*, Vol. 353, 1984, pp. 196–214.

[16] Concus, P., and Finn, R., "Continuous and Discontinuous Disappearance of Capillary Surfaces," *Variational Methods for Free Surface Interfaces*, edited by P. Concus and R. Finn, Springer-Verlag, Berlin, 1987, pp. 197–204.

[17] Miranda, M., "Superfici Minime Illimitate," *Ann. Scuola Norm. Sup. Pisa*, Vol. 4, 1977, pp. 313–322.

[18] Siegel, D., "The Behavior of a Capillary Surface for Small Bond Number," *Variational Methods for Free Surface Interfaces*, edited by P. Concus and R. Finn, Springer-Verlag, Berlin, 1987, pp. 109–113.

[19] Tam, L.-F., "The Behavior of Capillary Surfaces as Gravity Tends to Zero," *Comm. P.D.E.*, Vol. 11, 1986, pp. 851–901.

[20] Miranda, M., "Superfici Cartesiane Generalizzate ed Insiemi di Perimetro Localmente Finito sui Prodotti Cartesiani," *Ann. Scuola Norm. Sup. Pisa*, Vol. 3, 1976, pp. 501–548.

[21] Vogel, T. I., "Uniqueness for Certain Surfaces of Prescribed Mean Curvature," *Pacific Journal of Mathematics*, Vol. 134, 1988, pp. 197–207.

[22] Gulliver, R., and Hildebrandt, S., "Boundary Configurations Spanning Continua of Minimal Surfaces," *Manuscr. Math.*, Vol. 54, 1986, pp. 323–347.

[23] Finn, R., "Nonuniqueness and Uniqueness of Capillary Surfaces," *Manuscr. Math.*, Vol. 61, 1988, pp. 347–372.

[24] Concus, P., and Finn, R., "Instability of Certain Capillary Surfaces," *Manuscr. Math.*, Vol. 63, 1989, pp. 209–213.

[25] Langbein, D., "The Shape and Stability of Liquid Menisci in Solid Edges" (to be published).

[26] Wente, H. C., "The Stability of the Axially Symmetric Pendent Drop," *Pacific Journal of Mathematics*, Vol. 88, 1980, pp. 421–470.

[27] Concus, P., and Finn, R., "The Shape of a Pendent Liquid Drop," *Philosophical Transactions of the Royal Society of London Series A*, Vol. 292, 1979, pp. 307–340.

[28] Finn, R., "The Pendent Drop Revisited" (unpublished observations).

[29] Wente, H. C., "A General Existence Theorem for Surfaces of Constant Mean Curvature," *Mathematische Zeitschrift*, Vol. 120, 1971, pp. 277–288.

[30] Wente, H. C., "An Existence Theorem for Surfaces in Equilibrium Satisfying a Volume Constraint," *Archive for Rational Mechanics and Anal.*, Vol. 50, 1973, pp. 139–158.

[31] Rayleigh, J. W. S., "On the Capillary Phenomena of Jets," *Scientific Papers*, Vol. 1, Cambridge Univ. Press, Cambridge, UK, 1899, pp. 377–401.

[32] Athanassenas, M., "A Variational Problem for Constant Mean Curvature Surfaces with Free Boundary," *Journal Reine Angewandte Math.*, Vol. 377, 1987, pp. 97–107.

[33] Vogel, T. I., "Stability of a Liquid Drop Trapped Between Two Parallel Planes," *SIAM Journal of Applied Mathematics*, Vol. 47, 1987, pp. 516–525.

[34] Vogel, T. I., "Stability of a Liquid Drop Trapped Between Two Parallel Planes II: General Contact Angles," *SIAM Journal of Applied Mathematics*, Vol. 49, 1989, pp. 1009–1028.

[35] Vogel, T. I., "Types of Instability for the Trapped Drop Problem with Equal Contact Angles," *Geometric Analysis and Computer Graphics*, edited by P. Concus, R. Finn, and D. Hoffman, Springer-Verlag, Berlin (to be published).

[36] Vogel, T. I., "Numerical Results on the Stability of a Drop Trapped Between Parallel Planes," Preprint, Mathematics Dept., Texas A&M Univ., 1989.

[37] Korevaar, N. J., "Capillary Surface Convexity Above Convex Domains," *Indiana University Mathematics Journal*, Vol. 32, 1983, pp. 73–81.

[38] Chen, J.-T., and Huang, W. S., "Convexity of Capillary Surfaces in the Outer Space," *Inventiones Mathematicae*, Vol. 67, 1982, pp. 253–259.

[39] Chen, J.-T., "Uniqueness of Minimal Point and its Location of Capillary Free Surfaces over Convex Domains," *Asterisque*, Vol. 118, 1984, pp. 137–143.

[40] Siegel, D., "Uniqueness of the Minimum Point for Capillary Surfaces over Convex Domains" (to be published).

[41] Huang, W.-H., "Level Curves and Minimal Points of Capillary Surfaces over Convex Domains," *Bulletin of the Institute Math. Acad. Sin.*, Vol. 2, 1983, pp. 390–399.

Thermohydrodynamic Instabilities
and Capillary Flows

J. C. Legros,* O. Dupont,† P. Queeckers,† and S. Van Vaerenbergh†
Université Libre de Bruxelles, Brussels, Belgium
and
D. Schwabe‡
Giessen University, Giessen, Federal Republic of Germany

Nomenclature

ρ = density of the liquid
k = coefficient of heat conduction in the liquid
K = heat diffusivity of the liquid
Q = rate of heat loss per unit area from the upper free surface (function of x and z)
Q_0 = rate of heat loss per unit area from the upper free surface in the unperturbed system
t = time
T = temperature distribution in the liquid (function of $x;y;z$)
$T' = T - T_0$ is thus the temperature perturbation (supposed small); T'_s is its value at the surface
T_{0B} = steady-state temperature of the lower surface in the unperturbed system
v = velocity
σ = surface tension of the liquid
v = kinematic viscosity of the liquid

*Professor, Department of Chemical Physics.
†Doctor, Department of Chemical Physics.
‡Professor, Physics Institute.

General Introduction

Surface Tension

INSIDE a homogeneous fluid (e.g., pure liquid phase), a molecule is surrounded by other molecules, and statistically on the average, it is attracted equally in all directions. This is different at the surface of the liquid, because the particle density is much larger in the liquid phase than in the gas. As a result, the interface tends to contract as a consequence of this inward pull.

In order to increase the area of the interface, work has to be performed to bring molecules toward this surface against the inward attraction force.

The properties inside the superficial layer and in the two bulk phases are very different. Nevertheless, in 1805 Young explained that, from a mechanical point of view, it appears as if the two fluids (in the present case, liquid and gas) were separated by a membrane without thickness and in a state of tension.

In order to define the surface tension in one point of the interface, one can imagine that this surface is divided into two regions by a line. If, through the length element δl of this line, one of the regions is pulling with an intensity $\sigma \delta l$, oriented tangentially to the surface, σ is the surface tension at this point and along this line. It has the dimensions of a force per unit length (N/m).

The interface is homogeneously stretched (under tension) when it has the following properties:

1) In each point σ is perpendicular to the separating line and has the same amplitude independently of the separation direction.

2) The σ has the same value in each point of the interface.

Regarding the complex description of the two phases separated by a nonuniform superficial layer, the Young concept is a very simplified model in which the two bulk phases remain uniform up to the contact of a geometric surface of tension. This model gives a description that is extremely useful for all of the mechanical phenomena of capillarity that we shall consider in this paper.

If we could evaluate exactly the forces existing in the superficial layer, we could be able to calculate the exact position where σ should be applied so that the simplified model could be mechanically equivalent to the real system. But the surface tension is experimentally measured using macroscopic phenomena that allow one to know only with a very low accuracy the position of the surface on which it is applied. Nevertheless, from a theoretical point of view the position of this surface (called surface of tension) is perfectly determined and is not at all arbitrary.

Interface Curvature

When the interface is not flat, at equilibrium the pressure into two bulk phases is not the same on each side of the interface. This pressure difference is described by the Laplace law, which can be written as

$$p'' - p' = \sigma\left(\frac{1}{R_1} + \frac{1}{R_2}\right) \qquad (1)$$

where R_1 and R_2 are the main radii of the surface at the point of interest. As a consequence of this property, in weightlessness the pressure p'' has the same value everywhere in the second phase and the pressure p' is homogeneous in the first phase. From the previous relation this means that under these particular conditions only systems with a constant mean curvature have to be considered.

Capillary Motions

In the case of a pure compound the surface tension depends only on temperatures and generally decreases with T, except for some liquid crystals and some metals such as cadmium, iron, and copper. However, there is no really well-established theory to describe this phenomenon.

Independently of the force normal to the curved interface, which induces p and is described by the Laplace equation, a tangential force may be exerted on the surface of the liquid if the surface tension changes from point to point (e.g., if the temperature is not uniform along the interface). The magnitude of this force is equal to the surface tension gradient,

$$p_T = \text{grad}\,\sigma \tag{2}$$

and communicates to the surface of the liquid a motion from lower to higher surface tension regions. In weightlessness this mechanism can induce rapid motions, which are generally called "Marangoni convection."

Let us examine this fluid motion in a thin layer of depth d and laterally bounded by two walls at temperatures T_1 and T_2 (e.g., with $T_2 > T_1$). The x axis is taken parallel to the temperature gradient, and the y axis is perpendicular to the interface. The liquid/gas interface is at $y = 0$, and the bottom plate is at $y = d$. The distance between the walls is l. In this simplified model the Prandtl number will be sufficiently low so that the induced velocity field will not disturb the linear temperature distribution.

The gradient of surface tension is given by

$$\text{grad}\,\sigma = \delta\sigma/\delta T\ \text{grad}\,T \tag{3}$$

The surface tension is taken, as in the more general case, as decreasing when the temperature increases. Thus,

$$\sigma = \sigma_{T_1} + \frac{\delta\sigma}{\delta T}\frac{T_2 - T_1}{l}\,x \tag{4}$$

This describes the fact that the surface tension changes from point to point along the liquid/gas interface; it is maximum at the cold wall and decreases to minimum at the hot wall.

The tangential force per 1 cm^2 of surface area is given according to Eqs. (3) and (4):

$$p_T = \delta\sigma/\delta x = (\delta\sigma/\delta T)\ \text{grad}\,T \tag{5}$$

To estimate the velocity amplitude produced by this force, we shall use the previously defined assumptions in order to simplify the hydrodynamic equations.

The problem will be considered as two dimensional because there is no temperature variation in the transverse z direction, and we shall study the velocity distribution sufficiently far from the walls in order to avoid any y-velocity component. Thus, the Navier-Stokes equation can be written as

$$\frac{\delta p}{\delta x} = \mu\left(\frac{\delta^2 v_x}{\delta x^2} + \frac{\delta^2 v_x}{\delta y^2}\right) \tag{6}$$

Because the layer is thin ($d \ll 1$), the depth is constant, and there is no expansion of the liquid:

$$v = v_x(y)l_x$$

The term $\delta v_x/\delta x$ is thus vanishing, and the first term of the right-hand side of Eq. (6) is omitted as a higher-order term. Equation (6) can thus be approximated by

$$\frac{\delta p}{\delta x} = \mu\frac{\delta^2 v_x}{\delta y^2} \tag{7}$$

In weightlessness the pressure component in the direction y is

$$\delta p/\delta y = 0 \tag{8}$$

Thus, the pressure p will be considered as a function of x only.

The continuity equation expressing that the total liquid flow through a transverse section of the layer is vanishing becomes

$$\int_0^d v_x \, dy = 0 \tag{9}$$

To solve the system of Eqs. (7–9) describing the behavior of the liquid, boundary conditions need to be defined.

The rigid wall at the bottom of the layer gives the equation

$$v_x\big|_{y=d} = 0 \tag{10}$$

At the interface the viscous stress and the surface force per unit surface area must be equal, leading to the following conditions:

$$\mu\left(\frac{\delta v_x}{\delta y}\right)_{y=0} = p_T = \frac{\delta\sigma}{\delta T}\,\mathrm{grad}\,T \tag{11}$$

Integration of Eq. (7) yields

$$v_x = \frac{1}{\mu}\frac{\delta\sigma}{\delta x}d - \frac{1}{2\mu}\frac{\delta p}{\delta x}d^2 - \frac{1}{\mu}\frac{\delta\sigma}{\delta x}y + \frac{1}{2\mu}\frac{\delta p}{\delta x}y^2$$

$$= \frac{1}{\mu}\frac{\delta\sigma}{\delta x}(d - y) - \frac{1}{2\mu}\frac{\delta p}{\delta x}(d^2 - y^2) \tag{12}$$

The substitution of the velocity in the continuity equation allows us to write

$$\frac{\delta p}{\delta x} = \frac{3}{2d}\frac{\delta\sigma}{\delta x} \tag{13}$$

Integration of Eq. (13) along the x axis from the first to the second wall yields

$$p = p_0 + \frac{3}{2d}[\sigma(x) - \sigma(0)] \tag{14}$$

which defines the pressure gradient (where p_0 is an undetermined constant). This allows us to define the velocity distribution as

$$v_x = \frac{1}{4d\mu}\frac{\delta\sigma}{\delta T}(3y^2 - 4dy + d^2)\frac{dT}{dx} \tag{15}$$

The calculation of the maximum velocity at the interface shows that this velocity is increasing with the temperature gradient and with the thickness of the layer (in the limit of validity of the thin-layer assumption). To show that these induced velocities are rather large, let us estimate the velocity at the interface between water and air in a thin layer 0.001 m depth at 20°C ($\mu = 0.001$ N $\cdot$ s $\cdot$ m^{-2}) submitted to a temperature gradient of 100 K $\cdot$ m^{-1} with $\delta\sigma/\delta T = -0.5\ 10^{-3}$ N $\cdot$ m^{-1} $\cdot$ K^{-1}. From Eq. (15),

$$v_x\big|_{y=0} \approx -12.5\ 10^{-3}\ \text{m} \cdot \text{s}^{-1} = -12.5\ \text{mm} \cdot \text{s}^{-1}$$

Heat Flux Perpendicular to the Liquid-Gas Interface: The Marangoni-Bénard Problem

Introduction
In 1900 Henri Bénard[1] began scientific investigations of fluid motions in a shallow fluid layer heated uniformly from below. He observed cellular flows produced on the free surface of the liquid film and cellular deformations associated with the motions. These studies were made on thin liquid layers (0.5- to 1.0-mm depths). The most notable observation of Bénard was the hexagonal convection cells, now commonly referred to as Bénard cells.

In 1916 Lord Rayleigh's pioneering theoretical analysis[2] of the stability of a layer of fluid heated from below seemed to provide a basis for the understanding of the Bénard cells. The fundamental result found by Lord Rayleigh was the existence of a minimum temperature gradient required for the onset of convection. The model of Lord Rayleigh was a thin layer of fluid confined between two flat, rigid, horizontal plates, so that there is no free surface.[3] Rayleigh's conclusion was that the critical temperature gradient necessary to induce convection was 10^4–10^5 times larger than the critical gradient found experimentally by Bénard (this was recognized only recently).

In 1956 Block[4] introduced a new aspect into the explanation of the Bénard cells. He carried out the following experiment in a hydrocarbon liquid in which Bénard cells were established: A needle point was wetted with a silicone fluid that was insoluble in the hydrocarbon. The needle point was then made to touch the surface of the hydrocarbon film exhibiting Bénard cells. A thin film was thereby floated off, and the effects were observed as it spread over the surface. Whenever and as soon as the layer passed over the Bénard cell, the surface deformation disappeared and the flow stopped.

Clearly, the variation of vertical density could not be affected so rapidly by the covering of the layer. Block concluded from his experiments that surface tension should play a significant role in the formation of the hexagonal cells (as recognized intuitively by Bénard himself). He also pointed out that convective motions occurred at temperature differences of a fraction of the critical value predicted by Lord Rayleigh.

Linear Stability Analysis[7]

Surface-Tension-Induced Motions

Two years after Block's conclusions, the same features were established theoretically by Pearson.[5] The essence of the phenomenon comes from the fact that surface tension in most of the fluids is a monotonically decreasing function of temperature, and in the case of two constituents it is a function of relative concentration. Thus, if the free surface of a fluid layer is not at uniform temperature or at uniform relative concentration, effective surface tractions are present and motions within the field must be expected to take place.

If we consider the case of a homogeneous liquid layer cooled from the upper free surface or heated at the lower rigid surface, a simple qualitative explanation of the existence of steady cellular motions can be given.

If a microscopic disturbance increases the temperature locally at the interface, the surface tension will decrease and will induce motions starting at this place. Because of the law of conservation of matter this will induce a motion from the bulk phase toward the interface. The upcoming liquid is warmer than the liquid/gas interface. This mechanism can be maintained if it overcomes the dissipation mechanisms due to viscosity and heat diffusivity. In that case it has the form of cellular convection.

In the center of the cells warm fluid is drawn toward the surface; this spreads across the surface, cooling as it does so, until it reaches the edges of the cell, where it descends toward the lower surface of the layer and is warmed there. The decrease in temperature across the surface from the center to the edge of cell is accompanied by an increase in surface tension. The amplitude of the motions will, of course, be determined by the physical parameters of the fluid and by the temperature gradients involved. The driving force for the motion is provided by the flow of heat from the heated lower surface toward the cooled upper surface.

In order to evaluate the parameters a small-disturbance analysis is performed in the particular case of an infinite homogeneous liquid layer of uniform thickness whose lower surface is in contact with a rigid plate and whose upper surface is free.

The only physical quantities that are assumed to vary within the fluid are the temperature, the surface tension, which is a function of temperature only, and the rate of heat loss from the surface, also a function of temperature only. The liquid is confined between two planes: $y = 0$ [the bottom (rigid) surface] and $y = d$ [the upper (free) surface].

We take a linear undisturbed temperature distribution T_0 in the liquid:

$$T_0 = T_{0B} - \beta y \qquad (16)$$

where

$$Q_0 = k\beta \qquad (17)$$

The Navier-Stokes and heat equations in the linearized disturbance state are

$$\left(\frac{\delta}{\delta t} - \nu \nabla^2\right)\nabla^2 v = 0 \qquad (18)$$

$$\left(\frac{\delta}{\delta t} - K\nabla^2\right)\nabla^2 T' = \beta v \qquad (19)$$

We also define the surface tension variation of the liquid in which $\sigma_T = -(\delta\sigma/\delta T)_{T=T0s}$ represents the rate of change of surface tension with temperature, evaluated at $\sigma = \sigma_{0s}$ with $\sigma_0 = \sigma(T_{0s})$:

$$\sigma = \sigma_0 - \sigma_T T'_s \qquad (20)$$

$$Q = Q_0 + q T'_s \qquad (21)$$

where $q = (\delta Q/\delta T)_{T=T0s}$ represents the rate of change with temperature of the rate of loss of heat per unit area from the upper surface to its upper environment.

The boundary conditions for the velocity are

$$v = \frac{\delta v}{\delta y} = 0 \quad \text{at} \quad y = 0 \tag{22}$$

$$v = 0; \quad \rho v \frac{\delta^2 v}{\delta y^2} = \sigma \nabla_1^2 T', \quad \text{at} \quad y = d \tag{23a}$$

$$\nabla_1^2 = \nabla^2 - \frac{\delta^2}{\delta z^2} \tag{23b}$$

Equation (23a) equates the change in surface tension due to temperature variations across the surface to the shear stress experienced by the fluid at the surface.

Boundary conditions for the temperature can be formulated as followed:

1) If the boundary consists of a good conductor of large extent, the temperature T_{0B} can be considered constant and in the perturbed case $T' = 0$ at the lower boundary.

2) If the bottom boundary of the liquid consists of a layer of a bad conductor in contact at its lower surface with a good conductor, then the temperature gradient in the materials bounding the liquid may be large compared with the temperature gradient in the liquid itself; T_{0B} will, in the steady unperturbed state, be a constant.

A small change in the bottom surface temperature will not affect the rate of heat conduction through the bad conductor by any appreciable amount. Hence, the lower boundary condition on the perturbation temperature will be approximately $\delta T'/\delta y = 0$.

The preceding boundary conditions are limiting cases of the general mixed boundary condition

$$T' = Y \delta T'/\delta y \tag{24}$$

Where Y is a constant that depends on the thermal properties of boundary and liquid.

The second boundary condition for the temperature is

$$-k \frac{\delta T'}{\delta y} = q T' \tag{25}$$

We introduce classically the dimensionless variables

$$(\xi; \eta; \epsilon) = \left(\frac{x}{d}; \frac{y}{d}; \frac{z}{d} \right); \qquad \tau = \frac{tK}{d^2} \tag{26}$$

with the assumption that

$$v = -\frac{K}{d} F(\xi; \epsilon) f(\eta) \, e^{p\tau} \tag{27}$$

$$T' = \beta d F(\xi; \epsilon) f(\eta) \, e^{p\tau} \tag{28}$$

Because the motions have a periodic structure,

$$\frac{\delta^2 F}{\delta \zeta^2} + \frac{\delta^2 F}{\delta \epsilon^2} + \alpha^2 F = 0 \tag{29}$$

where α is a dimensionless constant arising from the separation of variables. We introduce Eqs. (27–29) in Eqs. (18) and (19) with $D \equiv d/dy$ and $Pr = v/K$ (Prandlt number):

$$[p - Pr(D^2 - \alpha^2)](D^2 - \alpha^2)f = 0 \tag{30}$$

$$[p - (D^2 - \alpha^2)]g = -f \tag{31}$$

When the boundary conditions are expressed in dimensionless variables, Eqs. (22), (23), and (25) become

$$f(0) = f'(0) = 0$$

$$f(1) = 0; \quad f''(1) = \alpha^2 Ma\, g(1), \quad g'(1) = -Bi\, g(1) \tag{32}$$

where Ma is the Marangoni number,

$$Ma = \sigma_T \frac{\beta d^2}{\rho v K}$$

and $Bi = qd/k$ is the Biot number.

Equation (24) becomes for the conducting case ($Y = 0$), $g(0) = 0$ and for the insulating case ($Y^{-1} = 0$), $g'(0) = 0$. The general relation becomes $g'(0) = g(0)Y/d$. The solution in the case of neutral stability ($p = 0$) of

$$(D^2 - \alpha^2)(D^2 - \alpha^2)f = 0 \tag{33}$$

$$(D^2 - \alpha^2)g = f \tag{34}$$

for the conducting and insulating case given in Figs. 1 and 2 for various Biot numbers.

All of the curves converge asymptotically toward $Ma = 8\alpha^2$ for large values of α, while each of them displays a critical value of Ma corresponding to a particular value for α for which particular disturbances are first growing, causing the rest state to become unstable. The case $Bi = 0$ for the insulating boundary condition has a particular behavior; the critical value of α is zero and $Ma = 48$.

Except for this special case, the shapes of the neutral stability curves are very similar in the conducting and insulating cases; in general, larger positive values of L lead to a greater stability.

Inverted Bifurcation Existence

Now let us discuss the nature of the onset of convection. We demonstrated that motions appear when the dissipative effects of viscosity and thermal conductivity are unable to maintain the no-flow state; any pertur-

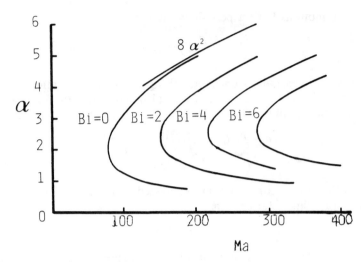

Fig. 1   Neutral stability curves, conducting case.

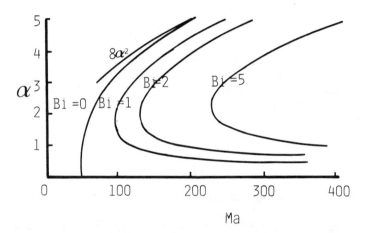

Fig. 2   Neutral stability curves, insulating case.

bations grow into a steady convective cellular flow with a number of cells depending on the width-to-height ratio of the cavity. We have a bifurcation[13] at this point. This bifurcation can be of two kinds: a normal bifurcation (also call "pitchfork" or "supercritical bifurcation") or an inverted bifurcation (also call "transcritical" or "subcritical bifurcation"). With a normal bifurcation the velocity grows continuously, but in the case of an inverted bifurcation the velocity grows abruptly to a nonzero value and presents a hysteresis loop behavior (see Fig. 3). The type of bifurcation present in the Marangoni-Bénard instability (MBI) has to be found by the nonlinear and bifurcation theories.[14,15]

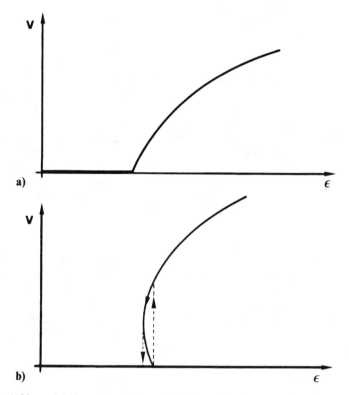

Fig. 3   a) Normal bifurcation; b) inverted bifurcation (arrows show the hysteresis loop).

Using symmetry arguments we can establish the nature of the bifurcation.[16] If we have identical boundary conditions for the velocity at the upper and lower sides (for example, two solid plates), the problem has symmetry with respect to the horizontal and the vertical axes. The equations are invariant under reflection about the horizontal midplane and the vertical midplane. When the convection appears, it breaks at least one symmetry. Hence, we have a normal bifurcation in this case.

For the MBI, at the free surface, we do not have the same boundary condition as we had at the lower solid surface. Here we have only a symmetry with respect to the vertical axis. The bifurcation to an odd number of cells breaks the vertical symmetry and thus is a normal bifurcation. The bifurcation to an even number of cells breaks no symmetry and must be an inverted bifurcation that does not break a symmetry.[17]

Winters et al.[18] have verified these results by numerical simulations and have determined that the nature of the bifurcation depends on the aspect ratio of the container. For a small aspect ratio (e.g., 2.2) it is a normal bifurcation and for larger aspect ratios it is an inverted bifurcation. Winters and co-workers obtained values of critical Marangoni numbers for an infinite layer in excellent agreement with the classical results obtained by linear stability theory.

An important characteristic of the inverted bifurcation is the hysteresis loop present at the beginning of the instability. Several experimenters have tried to observe this hysteresis behavior. For example, Gerbaud[19] measured the horizontal velocities at the free surface for a layer of silicone oil (see Fig. 4). Another example is in an experiment[20] with an n-heptanol aqueous solution heated from the upper side in a temperature region where the surface tension increases with the temperature. The gravity in these cases is stabilizing and convection is observed, only due to the surface tension stresses. Figure 5 shows the temperature difference at the liquid layer boundaries as a function of the heat flux. We can see a decrease of the temperature difference at the onset of the convection. We suppose that this decrease is due to an inverted bifurcation.

Surface Tension and Gravity-Induced Motions

In 1964 Nield[6] combined the two forces acting on a fluid layer heated from below: the buoyancy forces studied by Rayleigh and the surface tension forces studied by Pearson. He used the linear stability analysis techniques and the Fourier series method to obtain the eigenvalue equation for the case where the lower boundary surface is a rigid conductor and the upper free surface is subjected to a general thermal condition. It was found that these two contributions to instabilities reinforce one another and are tightly coupled. Nield's work is summarized in Fig. 6, where a plot of

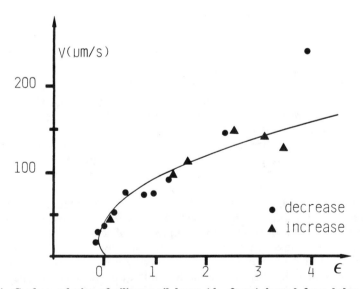

Fig. 4   Surface velocity of silicone oil layer ($d = 2$ mm) heated from below as a function of $\epsilon = (Ra - Ra^c)/Ra^c$. The curve is a parabolic regression.[19]

N−HEPTANOL

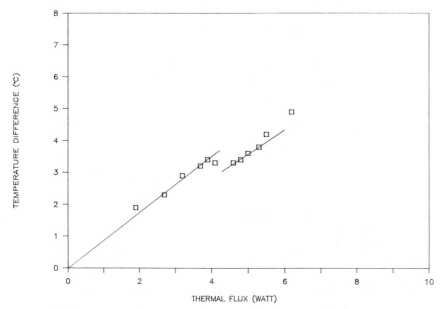

Fig. 5   Induced vertical temperature difference as a function of the imposed heat flux. We heat from above a *n*-heptanol solution ($d = 1$ mm).[20]

Marangoni and Rayleigh numbers limits the stable and the unstable regions. The points below the curve represent stable states.

In a first, but good, approximation the contributions of the two forces can be calculated by

$$\frac{Ra_c}{Ra_c^0} + \frac{Ma_c}{Ma_c^0} \approx 1 \tag{35}$$

where Ma_c^0 is the critical Marangoni number ($Ma_c^0 = 79$), when $Ra = 0$, corresponding to, e.g., $g = 0$.

Ra^0 is the Rayleigh number when no surface forces are acting:

$$Ra = \frac{g\alpha\Delta T d^3}{\nu K} \tag{36}$$

Table 1   Critical Rayleigh and Marangoni numbers for two common mechanical boundary conditions in the conductive cases

Boundary conditions	Ra_c^0	Ma_c^0
Both rigid	1704	——
Lower rigid, upper free	1100	79

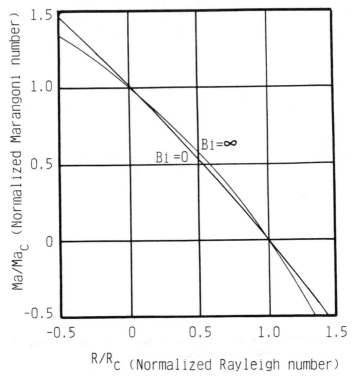

Fig. 6   Stability diagram. Plot of reduced critical Marangoni and Rayleigh numbers of marginal stability. $Bi = 0$ refers to insulating free surface. $Bi = \infty$ refers to conducting free surface.

Table 2   Temperature difference and nondimensional numbers of the pattern formation

	d	$T1$, °C	$T2$, °C	Ra_1^c	Ma_1^c	Ra_2^c	Ma_2^c
	2.57	6.68	5.98	——	——	91.5	76.8
Silicone oil,	1.81	6.68	6.68	35.9	61.0	35.9	61.0
100 cS	1.31	4.12	7.92	8.21	26.4	16.4	52.9
	0.93	3.86	11.3	2.73	17.5	8.73	56.2
	0.72	2.09	12.4	0.67	7.17	4.49	48.2
Silicone oil,	1.34	2.88	5.16	13.8	37.0	25.4	68.0
50 cS	0.93	2.74	6.82	4.4	24.5	11.5	63.8

$T1$ corresponds to the first pattern, $T2$ to the second one

The independent critical Marangoni number (Ma_c^0) and Rayleigh number (Ra_c^0) are given in Table 1 for two couples of mechanical boundary conditions.

Following Nield's approach, it should be possible to simulate the microgravity conditions by the use of very thin fluid layers. In 1986 Koschmieder and Biggerstaff[10] investigated the onset of surface-tension-driven Bénard convection. They performed experimental investigations of the onset of convection in shallow fluid layers and found that, for fluid layers smaller than 2 mm, the onset of convection takes place in two stages. First, a weak pattern occurs that is characterized by its appearance at even smaller temperature gradients as the depth of the fluid is decreased. When the temperature difference across the fluid is increased, a second strong pattern forms near the predicted critical Marangoni number. As shown by Koschmieder, the first pattern is not caused by the variation of surface tension with temperature. The values deduced by Koschmieder and Biggerstaff are given in Table 2.

Microgravity Experiments on Marangoni-Bénard Instability

Previous Works

The first experiment performed under microgravity conditions on the onset of surface-tension-driven Bénard convection was realized onboard Apollo 14 and 17.

In Apollo 14 (Ref. 8), the experiment was conducted during the lunar flyback on February 7, 1971. The purpose of the experiment was to heat four different fluids (Krytox, carbon dioxide, gas, and a water and sucrose solution) under a gravity level less than $10^{-6} g$. Each fluid was in a container of distinctive geometry, and resultant fluid flow or temperature changes were noted. During this microgravity experiment it was impossible to create a flat liquid/gas interface. Because of these experimental difficulties, no quantitative data were obtained during these Apollo 14 experiments; nevertheless, important aspects were pointed out:

1) Surface tension alone could activate cellular convective flows of macroscopic amplitude.

Table 3 Critical temperature difference, Rayleigh and Marangoni numbers during Apollo 17 and ground experiments

	Nominal oil depth, mm	Calculated cell diam, cm	Theoretical cell diam, cm	Onset time, s	T at onset, °C	Ra at onset	Ma at onset
Apollo 17	2	0.7 ± 0.2	0.73	18–21	≈ 2.9	3×10^{-7}	400
	4	1.4 ± 0.4	1.46	46–60	≈ 5.3	3×10^{-6}	1320
Ground	2			120	≈ 6.1	695	927
tests	4			120	≈ 8.6	7710	2580

2) A critical value of the temperature gradient must be exceeded before cellular convection can be initiated.

During Apollo 17 (Ref. 9) two experiments with convex oil layers were performed with 2- and 4-mm depths, respectively. The results of the flight were very surprising: The critical Marangoni numbers were much larger than the predicted value of 80. Furthermore, the critical temperature differences were smaller for the onset of convection in low-gravity conditions than in 1-g tests, the opposite of the theoretical predictions. Table 3 summarizes the results.

Spacelab D2 Mission and TEXUS 21

The main purpose of the MBI experiment is to determine the onset of convection by the measurement of the critical transverse temperature difference at given fluid layer depths. The associated critical wave numbers will also be investigated.

To prepare the Spacelab D2 Mission scheduled for 1992, a short microgravity time experiment was performed in a TEXUS sounding rocket. TEXUS 21, in which the experiment was planned, was launched in spring 1989 at Kiruna, Sweden. The objectives of the rocket flight were directed toward two directions: a technical and a scientific point of view. The technical objective was to create a flat liquid/gas interface in order to study the onset of convection with a temperature gradient perpendicular to the interface and to study the filling of the cell, which has to be performed without jet or liquid drop formation. The scientific objectives of the flight consisted of the following:

1) Verify the Marangoni instability under microgravity in a well-defined experiment (flat, free surface and accurate temperature control).

2) Measure the critical Marangoni number Ma^c under microgravity conditions as accurately as possible due to limited microgravity time.

3) Measure the critical wavelength (e.g., the size of the convection cells) under microgravity and the convective pattern (e.g., upstream in the center of the hexagons, regularity of hexagons, effect of lateral walls).

4) Develop special microgravity-relevant experiment techniques for experiments on the Marangoni effects.

Because of the limited time (6 min) available during the flight of a TEXUS sounding rocket, the experimental boundary conditions are away from the optimal quasi-steady-state ones that are normally required for measuring the transition from no flow to the hexagonal convection pattern. We optimized the experiments and the experimental time-temperature profile to cope with this limitation in microgravity time and developed special measurement techniques and a parameter evaluation to gain significant results.

The TEXUS 21 module includes two different cells: one circular, the other rectangular with two independent experimental volumes. In both types of cells, the heat conductivity of the gaseous phase was chosen as high as possible by filling the experiment volume with helium. The detailed description of the hardware can be found elsewhere.[11,12] The circular convection cell was chosen (by D. Schwabe from Giessen University) because it is the classical geometry used by most other experimenters in 1-*g* investigations and thus allows comparison with these experiments. The circular geometry is best suited to the expected hexagonal cell pattern because the possible extension (radius) of the experiment cell is limited due to the restrictions in experiment time, experiment mass, and power available in TEXUS. In this experiment we specially address the question of side-wall effects, which are more severe in time-dependent experiments than under quasisteady experiment conditions, and we address how to reduce them adequately under microgravity conditions. The rectangular shape was chosen by the ULB team because this geometry is better suited to measure convective patterns and stream velocity through the side walls as it is forseen on the following experiments in the Advanced Fluid Physics Module (AFPM).

TEXUS 21 Results

From a technical point of view one of the main results of this flight was that it demonstrates that it is possible to create a flat, stable liquid/gas interface under microgravity conditions into both cells characterized by small aspect ratios. During the TEXUS 21 flight we succeeded in establishing flat gas/oil interfaces without any menisci both in the rectangular and the circular cells.

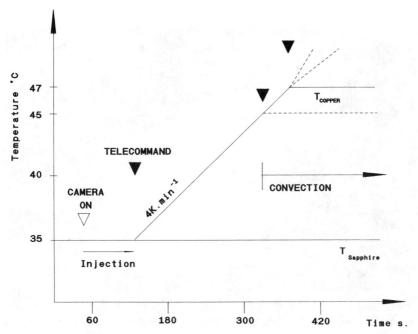

Fig. 7   Temperature-time profile for the TEXUS flight of the circular cell (dashed lines are optional by telecommand).

Circular Cell. This cell using the so-called liquid side-wall technique had a free surface of 75-mm diam. The operation scenario for the flight experiment of the circular cell is given on Fig. 7. After the injection of the predetermined amount of oil (50 cS of kinematic viscosity), the copper plate is heated with a constant temperature ramping of a 4 K $\cdot$ min^{-1}. This was imposed in order to reach a ΔT (Cu-sapphire) $= 12$ K after the forseen 330 s, taking into account that the temperature of the liquid/gas interface is increasing during the experiment. This gives rise to approximately ΔT (Cu-oil interface) $= 6$ K corresponding to $\approx 4 \, \Delta T^c$ (*Ma*). The estimated value (for steady state) for the onset of the purely surface-tension-driven Marangoni instability was 1.4 K for AK 50 silicone oil).

In Fig. 8 we show the cellular convection pattern shortly after the threshold in the circular cell. The pattern is still faint and shows cells of different sizes, but one can observe one central convection cell (large and shifted somewhat to the right) surrounded by a ring of six convection cells (two smaller ones are on the right of the central one), surrounded by a ring of 12 convection cells adjacent to the rim of the plastic foil. Since the available microgravity time lasted 452 s, this pattern has almost 2 min to be established. It was observed that, as the convection pattern increased its contrast, the cell size became more regular and hexagon-like.

We note that the convection pattern was established earlier under 1 g and that the size of the convection cells under 1 g was much smaller than those under microgravity. Both effects can be directly understood as being

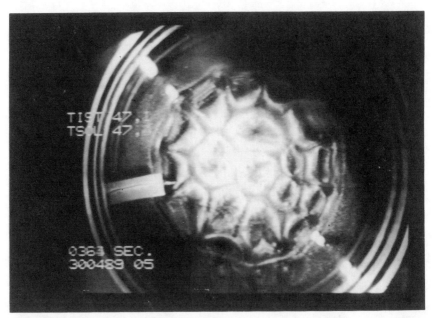

Fig. 8 Photograph from video monitor of convection pattern under $10^{-4} g$ during TEXUS flight.

due to gravity, which allows the instability to penetrate faster from the heated copper plate into the fluid layer in the time-dependent 1-g reference experiment. Under microgravity conditions the motor for the instability is located at the free liquid surface only, resulting in a slower development of the convection pattern and in a larger convection cell size.

Rectangular Cell. The rectangular cell has side walls made from Lexan, and its lateral dimensions are 40×60 mm. The injected fluid was silicone oil with 150 cS viscosity. As for circular cell, the gas phase consisted of helium at 1.5 bar pressure.

The heating of the copper plate starts directly after the end of the oil injection. The heating rate was 20 W during the first 20 s; this was done to inject enough heat into the copper plate to bring the temperature difference between the copper and sapphire plate ΔT (Cu-S) to 5.0°C, giving rise to a $\Delta T = 2.5$°C across the fluid layer. This was smaller than the critical temperature difference estimated at $\Delta T^c = 4.96$°C for a quasisteady experiment under microgravity conditions. This 20-W heat-injection period was followed by a 9-W heat injection to pass the transition to convection with a lower heating rate.

Four temperature sensors were placed in the copper plate, in the liquid very near the free surface of the oil, in the He gap just above the interface, and in the He gap just below the sapphire plate. They allowed the heat flux and its variation due to the onset of convection to be measured. The sensors in the oil and helium are extremely thin in order to give disturbances as small as possible (wire = 0.025 mm; head = 0.4 mm).

During the flight it was possible to uplink commands to change the heating power injected in the copper block; the downlink data were the temperatures of the various heat sensors and video pictures. The temperatures are taken each 0.5 s during flight at the level of the copper block (lower rigid surface), near the sapphire window (the top of the helium layer), and near the fluid gas interface in the gas phase.

In Fig. 9 the temperature drop in the liquid is plotted vs the temperature in the gas phase. The reported points are taken between $t = 405$ s and the end of the flight ($t = 0$ corresponds to the beginning of the fluid injection). The break in the curve for ΔT liquid ≈ 8.8°C corresponds to the transition from the rest state toward the convective state. Because of the nonlinearity of the temperature distribution in the liquid, this value is higher than predicted by a linear stability analysis performed with a linear temperature distribution.

The values of interest in these nonsteady microgravity experiments are the thermal gradients at the level of the liquid/gas interface (ΔT surface) because the surface tension forces are located at this level. With the assumption verified by numerical simulation[12] that the temperature distribution in the gas layer is nearly linear, it is possible to evaluate the temperature gradient at the level of the interface.

The ΔT gas corresponding at the break in the curve is around 4.7°C. The distance between the temperature sensors in the gas layer is 3.8 mm; this leads to a temperature gradient of about 1.24 K $\cdot$ mm^{-1}. That can be compared with the theoretical value (4.96 K/5 mm) of 0.992 K $\cdot$ mm^{-1}.

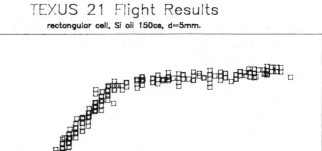

TEXUS 21 Flight Results
rectangular cell, Si oil 150cs, d=5mm.

Fig. 9   Schmidt-Milverton plotting.

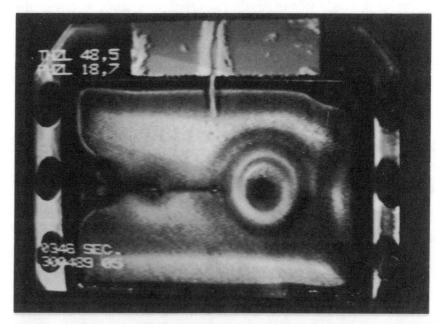

Fig. 10   Photograph from video monitor of the perturbation resulting from the gas bubble.

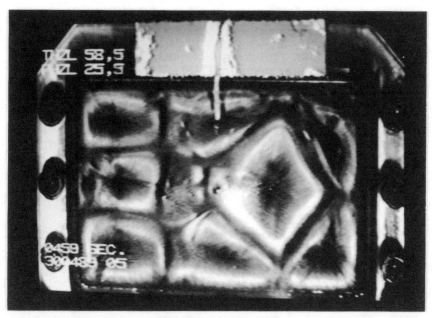

Fig. 11 Photograph from video monitor of convection pattern under 10^{-4} g during TEXUS flight (rectangular cell).

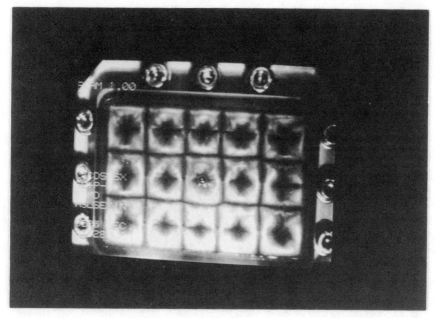

Fig. 12 Corresponding convection pattern under 1 g (ground test).

J. C. LEGROS ET AL.

During the filling a bubble of gas was injected. In Fig. 10 the perturbation due to this gas bubble is clearly observed as the ring around it, which appeared during the development phase of the motions.

Figure 11 shows the convective pattern under microgravity conditions (just before the end of the experiment), 450 s after the liftoff of the rocket. In the right side of the cell, the bubble has disturbed the organization, but on the left side, three rectangular cells are created.

With respect to the ground experiment test, in Fig. 12 we can see that these convective cells have roughly the same dimensions. The estimated nondimensional wavelength in the 1g test realized in ground is $k_x \approx 2.8$ along the longer side and $k_y \approx 3.4$ along the shorter side. This has to be compared with the microgravity wavelengths, $k_x \approx 3.0$ and along $k_y \approx 3.4$.

Marangoni-Bénard Thermosolutal Instabilities

Introduction

Until now we have considered pure substances for which Marangoni-Bénard instabilities are due to the temperature dependence of the surface tension only. For liquid mixtures the mass fraction of each compound may deeply affect the stability and convective behavior of the liquid layer. We shall only consider binary mixtures with one free surface perpendicular to the steady-state temperature gradient (see Fig. 13).

The surface tension of the free interface also depends on the surface mass fraction N chosen by convention as the mass fraction of the denser component. Thus, at the free boundary, we have

$$-p + \mu \frac{\delta v_x}{\delta x} = \frac{\delta \sigma}{\delta N} \frac{\delta N}{\delta x} + \frac{\delta \sigma}{\delta T} \frac{\delta T}{\delta x} \qquad (37)$$

If we disregard Gibbs absorption (see Refs. 21 and 22 for a detailed analysis) and chemical reactions, one additional equation will describe the

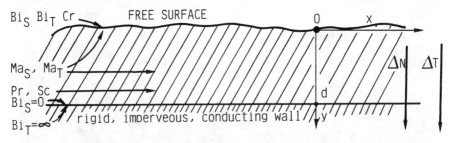

Fig. 13 Geometry, convention of signs and "location" of adimensional numbers used in Marangoni-Bénard in binary mixtures.

conservation of species. If the Soret effect is neglected, three additional numbers will then be required to perform the stability analysis of the layer: the Schmidt number,

$$Sc = v/D \tag{38}$$

where D is the isothermal diffusion coefficient, the solutal Marangoni number,

$$Ma_s = - \frac{\frac{\delta\sigma}{\delta N} d\Delta N}{D\mu}$$

and the solutal Biot number,

$$Bi = q_s d/D$$

where q_s is the mass transfer coefficient.

The previously defined Marangoni and Biot numbers will be qualified "thermal" and noted Ma and Bi, respectively. The list of adimensional numbers used is given in Table 4.

Numbers related only to bulk characteristics are Ra, Ra_s, Sc, and Pr numbers. The Biot and Marangoni numbers related surface to bulk properties. However, it is not always clear if the "dissipative" coefficients (K, D, and μ) used in the later ones are bulk or surface characteristics,[20] but we shall not analyze this question in detail in this chapter.

Stability is analyzed by looking at what conditions a fluctuation will be amplified. It is assumed that fluctuations obey macroscopic conservation laws. Thus, potential energy supplied at interface by surface tension forces and in the bulk by buoyancy forces will have to overpass dissipation for a convective motion to be sustained. As in the classical approach, we assume that dissipation happens only in the bulk. Instability will then arise only after critical values of surface/bulk or bulk/bulk ratios.

When a stability analysis of a fluid initially at rest is performed with linear theory, the threshold value corresponds to the onset of convection induced by infinitesimal fluctuations. The system will then be said to be "unconditionally unstable," "unstable," or "spontaneously unstable." The normal mode analysis[23] will give the critical numbers as a function of the wavelength and frequency by the dispersion equation, with frequency as a function of wavelength.

The main factor predicted by linear stability analysis for binary mixtures is the possiblity of having oscillatory time-dependent motions at the threshold (overstability). But if linear stability gives the wavelengths and frequencies of convection, it does not give any information on the amplitude of convection (as, for example, the mean value of velocity over half-wave number and half period). In this respect we have to pass the threshold and take into account nonlinear terms by numerical simulation or by bifurcation theory. We may have two situations shown in Fig. 14.

Table 4   Adimensional numbers

Thermal		
Ma	Thermal Marangoni number	$= -\dfrac{\delta\sigma/\delta T \Delta T d}{K\mu}$
Pr	Prandtl number	$= \dfrac{\nu}{K}$
Bi	Biot number	$= \dfrac{qd}{k_T}$
Ra	Rayleigh number	$= \dfrac{\alpha\rho g \Delta T d^3}{K\nu}$

Solutal		
Ma_s	Solutal Marangoni number	$= -\dfrac{\delta\sigma/\delta N \Delta N d}{D\mu}$
Sc	Schmidt number	$= \dfrac{\nu}{D}$
Bi_s	Solutal Biot number	$= \dfrac{q_s d}{D}$
Ra_s	Rayleigh number	$= \dfrac{\beta\rho g \Delta T d^3}{D\nu}$

Thermosolutal numbers		
S	Soret separation number	$= -\dfrac{\alpha N_0(1-N_0)D'/D}{\beta}$
S'		$= -N_0(1-N_0)\Delta T D'/D$
Le	Lewis number	$= \dfrac{Pr}{Sc} = \dfrac{D}{K}$

Other		
Cr	Crispation number	$= \dfrac{\mu k}{\sigma d}$

The variable q is the thermal transfer coefficients appearing in boundary conditions, k_T the thermal conductivity, and k the wavelength of convective pattern.

The "amplitude" of convection may increase monotonically near the threshold when the control parameter is increased, or it may pass directly to a finite value. In the latter situation (subcritical bifurcation), there will be a hysteresis behavior in the sense that convection will be sustained when the control parameter is decreased below the linear theory threshold.

There may be competition between supercritical and subcritical instabilities. When two control parameters are involved (as temperature gradient and concentration gradient), the situation may become quite intricate. A useful concept is, then, the motion of codimension, which, in the vicinity of a critical point, may be assimilated with the minimal number of parameters

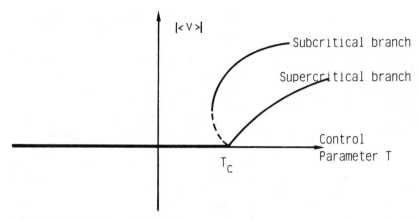

Fig. 14   Sketch of possible bifurcations when one control prameter is modified.

needed to cross the threshold. Although bifurcations of codimension two has already been studied experimentally in binary mixtures for Rayleigh-Bénard convection,[24] here we will deal only with bifurcations of codimension one. The bifurcations will generally be supercritical so that it will be difficult to observe their exact threshold values. Indeed, if finite-amplitude fluctuations could be avoided, we would have to slightly overpass the threshold to observe any nonvanishing amplitude.

In a more practical vision it is clear that some external imponderable will disturb the system. It is also intuitive that these induced finite-amplitude fluctuations may quench convection for lower values of the critical parameter. The appropriate way to delimit this domain of conditional stability is to answer the following question: For which value of the control parameter(s) will the system always return to its previous configuration, whatever the amplitude of fluctuation? The answer is given by the energy theory. It is of great practical relevance for experiments (to control the quality of the experiment and the resolution obtained), for material processing, and for understanding the mechanisms of instability to consider both linear and energy theories. The energy theory and linear theory limits behave differently when there are two or more control parameters (as in binary mixtures).

Binary Isothermal Fluid: Solutocapillary Instability
The stability problem of an isothermal binary layer with one free surface is formally analogous to the problem of pure nonisothermal liquid layer. Ma, Bi, and Pr numbers are replaced by Ma_s, Bi_s, Sc, and a time constant:

$$\frac{d^2}{K} \quad \text{by} \quad \frac{d^2}{D}$$

A spontaneous steady convection will appear if Ma_s overpasses a threshold value. For example, for $Sc = 10^3$ this value is $Ma_s = 54$.[25] Suppose that $\delta\sigma/\delta N$ is positive and ΔN is negative (free surface enriched in denser component). Then a negative concentration fluctuation on the surface will cause a local decrease in surface tension. Because of this, the pressure drops below this point and liquid will rise. The fluid coming up will enhance the motions because its surface tension is lower than in surface. Thus, the system is unstable, provided the concentration difference is high, the diffusivity is poor, and the ratio $(\delta\sigma/\delta N)/\mu$ is high enough.

When there is mass transfer of the denser component from liquid to gas, there will be a stabilizing effect. Mass transfer from gas to liquid would exist by heating a pure fluid from the free interface. The main difference between the problems of nonisothermal pure fluid and isothermal binary mixture is that $\delta\sigma/\delta N$ can be positive as well as negative, whereas often the surface tension decreases when the temperature is increased. The initial concentration gradient may happen for various reasons: impurities diffusing from the free interface, salts diffusing from a rigid boundary, a Soret concentration gradient once the temperature gradient is suppressed (solutal diffusion is low with respect to thermal diffusion), surface evaporation, etc.

Such a solutocapillary experiment where the concentration difference was controlled by surface evaporation has been performed during a D1 mission by Lichtenbelt et al.[26] The concentration gradient of acetone diluted in water was created by absorbing the acetone for the gaseous phase in active carbon on the top of the gaseous layer.

Although a solutal Marangoni unconditional instability has not been observed, probably because of surface pollution and side-wall effects,[27] convection easily developed when the interface was curved due to the fact that this curvature induces a concentration gradient along the interface via mass transfer in gas phase. Experiments performed on the ground presented unconditional instabilities, but buoyancy forces have to be taken into account in that case. Numerical simulations performed on a square cavity taking into account nonlinear terms of the momentum conservation equations show that, in the absence of buoyancy forces, the system evolves, soon after the threshold value evaluated by a linear theory, toward a steady convective state. The center of the cells is located near the interface because of mass transfer throughout interface and finite diffusivity of acetone in water. By the reached steady convection, the efficiency of the acetone transfer is increased from two or three times with respect to the steady state.

Thermosolutocapillary Instabilities Without Soret Effect

Now we consider a nonisothermal binary layer initially at rest and in weightlessness. Thermal properties are characterized by Ma, Bi, and Pr numbers and solutal properties by Ma_s, Bi_s, and Sc numbers. Surface deformability is characterized by the crispation number.

Choosing peculiar boundary conditions will greatly simplify the stability analysis. Boundaries will be supposed impervious (thus excluding evaporation or other mass transfer). It will be assumed that the free surface is insulating, whereas the lower boundary will be assumed to be perfectly conducting (this is possible for liquid mixtures whose thermal diffusivity is smaller compared to the rigid boundary material but much higher than the gaseous phase). Furthermore, we consider in a first approach that the surface is undeformable ($Cr = 0$), since it gives only second-order corrections. Thus, the stability may be analyzed in the (Ma, Ma_s) control parameters plane parametrized by the constants Pr and Sc. Critical thermal and solutal Marangoni numbers given by the linear theory (Ma_0^c and Ma_{s0}^c) correspond to the limit situations already analyzed.

The results obtained for small Lewis numbers are summarized in Fig. 15 (after Velarde and Castillo[28]). Systems in quadrant III are unconditionally stable because both thermal and solutal effects are stabilizing the layer. When destabilizing (quadrant I), steady convection will appear unconditionally when

$$\frac{Ma}{Ma_0^c} + \frac{Ma_s}{Ma_{s0}^c} = 1 \tag{39}$$

Since concentration and temperature fluctuations are not coupled at steady state, they will act additively. Finite-amplitude fluctuations may lead to

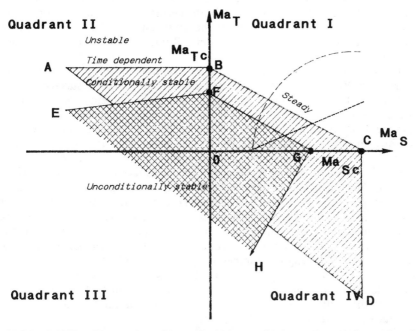

Fig. 15 Stability diagram for a binary liquid layer with impervious surfaces: dotted line, example of isothermal increase; dashed line, example of thickness increase (assuming T and N are kept constant).

convection at lower values of Marangoni numbers. The lower limit for them, as computed by the energy theory,[29] is given by the segment FG on Fig. 15.

Note that for a given system Eq. (39) may be written as

$$\frac{dN}{(dN)^c} + \frac{dT}{(dT)^c} = \text{cste} \tag{40}$$

Thus, the linear theory threshold may be reached by increasing ΔT or ΔN, but care has to be taken that $\delta^2\sigma/\delta T\delta N$ could be large. For example, Ma_s may increase when the upper temperature is increased because $\delta\sigma/\delta N$ depends on temperature. Since in a practical situation it is difficult to increase ΔT keeping the free surface temperature constant, the trajectory taken by the system in the (Ma, Ma_s) plane when ΔT is increased will typically be the one shown by dotted curve on Fig. 15.

Another way to approach the threshold would be to increase the depth d. However, it is difficult in such a situation to maintain both ΔT and ΔN constant except if the concentration gradient is thermally driven (Soret effect).[30]

In quadrants II and IV, defined by Marangoni numbers of opposite signs, there appears to be no stabilizing effect of the negative Marangoni numbers (again it is the absence of coupling between temperature and concentration infinitesimal fluctuations at steady state that is responsible for this). The competition between both thermal and solutal tension forces will be sensible soon after the linear theory threshold. The situation is then very similar to that of time-dependent Rayleigh-Bénard convection for binary mixtures.

Let us suppose that there is a local decrease in surface tension. The consequent pressure drop will drive fluid from bulk to surface. If the decrease in surface tension is induced by the faster diffusing agent (temperature or concentration), the fluid coming up will tend to restore the initial surface tension value. Because of this difference in diffusion times, the surface tension will become larger at this point than in the surrounding area so that motion will be inverted after some time. If the restoring capability is small, this inversion will take a longer time, but oscillations will be sustained because the inversion will cause fluid to come up at another point (only in the limit case where restoring capability vanishes will the frequency of oscillation vanish). In counterpart the destabilizing effect should be high enough so that potential energy released by the surface forces could balance bulk viscous dissipation. This defines the value of the critical Marangoni number in quadrant I. On the other quadrant (IV) the destabilizing agent will be the slower diffusing one. If a local increase appears in the surface tension, some fluid will go from the surface toward the bulk. The slower diffusing agent will finally dominate and produce a destabilization at this point.

The convection will be established with a pattern characterized by the wavelength of the most unstable mode. Approximately this later depends nearly only on the depth of the cell (about twice the depth). The frequency of the oscillations increases from zero to a constant value when the absolute value of the negative Marangoni number is increased.

The energy theory approach predicts that the system is conditionally less stable when the negative Marangoni number increases. This may be explained by the fact that finite-amplitude fluctuations produce more and more overshooting of surface tension when the "stabilizing" effect increases.

Nonisothermal Binary Mixture with Soret Effect

When the Soret effect is acting, the stationary temperature field will determine the stationary concentration distribution. At steady state, the mass flux of each species is constant, and for a temperature gradient perpendicular to imperveous walls it will thus vanish everywhere:

$$\frac{\delta N}{\delta y} + N_0(1 - N_0)\frac{D'}{D}\frac{\delta T}{\delta y} = 0 \qquad (41)$$

The sign of ΔN will be the same as the sign of ΔT when the Soret coefficient D'/D is negative. If N is chosen to be the mass of the denser component, this means that the denser component migrates toward the hot plate.

The steady state is only determined by the applied temperature gradient. Thus, in the model previously described, the stability of the layer may be controlled only by the applied temperature difference and by the layer thickness. This is an interesting property for the study of the capillary Marangoni-Bénard instabilities in mixtures during space experiments without buoyancy forces. The control of both temperature and concentration fields by one control parameter is an unique situation in this respect.

However, the value of the Soret coefficient has to be known with precision. Experimental results present great discrepancies. Legros and Goemaere[31] have shown with stability considerations that those discrepancies should be induced by buoyancy force effects. Numerical simulations[32] performed on closed cylindrical cells confirm this analysis. A fraction of the steady nonconvective separation may be obtained even in convective regime. But the amplitude of convention in Soret measurement cells is nearly unpredictable due to their three-dimensional, complicated geometry.

Accurate measurements of Soret coefficients will be performed under the very good microgravity conditions ($10^{-5} g$) of the EURECA automatic platform. The launch and the recovery of the platform by the Space Shuttle is planned for 1991. In this Soret coefficient measurement (SCM) experiment (see Fig. 16), 20 organic and electrolytic aqueous solutions will be studied.[33] Those systems have a typical Lewis number of 10^{-2}. The Soret coefficient is of the order of 10^{-2} to $10^{-3} K^{-1}$. Thus, the mass fraction separation may be of the order of 10^{-1} to 10^{-2} for a temperature difference of 10 K. Even for small values of the Soret coefficients, the thermodiffusion should not be neglected in the stability analysis.We may compare solutal

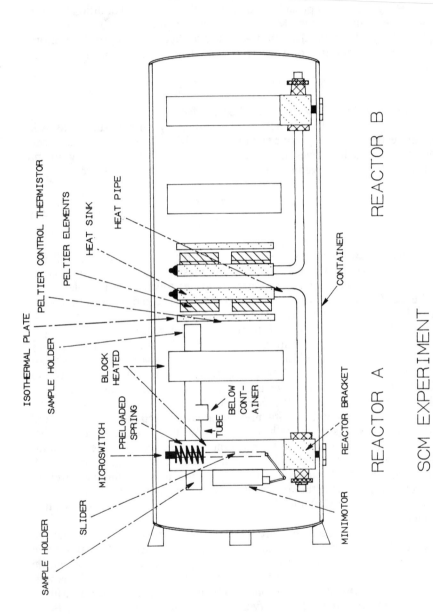

Fig. 16 Schematic of the Soret coefficient measurement reactor. Only 1 of the 20 tubes is represented.

and thermal influence on Marangoni instabilities by the ratio

$$\frac{Ma_s}{Ma} = \left(\frac{\delta\sigma}{\delta N}\Delta NK\right)\bigg/\left(\frac{\delta\sigma}{\delta T}\Delta TD\right) \tag{42}$$

because critical values of solution and thermal Marangoni numbers are of the same order of magnitude.

For Soret-driven concentration gradients Eq. (43) may also be written as

$$\frac{Ma_s}{Ma} = \left(\frac{\delta\sigma}{\delta N}\bigg/\frac{\delta\sigma}{\delta T}\right)N_0(1-N_0)\frac{D'}{D}\frac{1}{Le} \tag{43}$$

Because of the high concentration dependence of the surface tension, the solutal effect induced by the temperature gradient may be as important as the pure thermal effect.

According to Castill and Velarde,[25] we have only two cases in the linear approach. When the solutal effect is destablizing, the criticality is given by

$$\frac{Ma}{Ma_0^c} = 1 \tag{44}$$

whereas when $Ma_s > 0$ we have

$$\frac{Ma_s}{Ma_{s0}^c} = 1 \tag{45}$$

Of course, Ma_s is not an independent parameter here. If we consider that $(\delta\sigma/\delta N)$ and $(\delta\sigma/\delta T)$ and other physical parameters are constant, Eq. (45) may be written as

$$\Delta T\left[1 + \frac{\Delta N}{\Delta T}\bigg|_{\text{Soret}}\frac{\Delta T_c(\Delta N = 0)}{\Delta N_c(\Delta T = 0)}\right] = \Delta T_c(\Delta N = 0) \tag{46}$$

It is clear from Eq. (46) that, if the Soret effect is large enough and negative (the denser component migrates toward the hot wall), we may have an instability when the temperature gradient is stabilizing. This corresponds to the situation in quadrant IV that because of Soret coupling is an "extension" of quadrant I. In particular, there will not be oscillatory motions, i.e., a concentration fluctuation will finally be thermally driven.

In quadrant I the situation is similar to that of a mixture without a Soret effect (here $Le \leqslant 1$). The wavelength will decrease with increasing Ma_s. In contradiction with the corresponding Rayleigh-Bénard problem, no experimental results for this situation presently exist.

Acknowledgments
This investigation has been made possible due to the European Space Agency's support for microgravity experiments and the financial support of

the Belgian Science Policy Office (SPPS) by an Actions de Recherches Concertées contract, of Fonds National de la Recherche Scientifique (FNRS), by an FRFC contract. The collaboration between the Université Libre de Bruxelles and Giessen University was supported by NATO Contract 0916/83.

References

[1]Bénard, H., "Les tourbillons cellulaires dans une nappe liquide," *Revue Générale des Sciences Pure et Appliquées*, 1901, pp 1261–1271, 1309–1328.

[2]Rayleigh, J. W. "On the Convention Currents in a Horizontal Layer of Fluid when the Higher Temperature is on the Underside," *Philosophical Magazine*, Vol. 32, 1919, pp. 529–535.

[3]Chandraseckhar, S., *Hydrodynamic and Hydromagnetic Stability*, Clarendon, Oxford, UK, 1961.

[4]Block, J. M., "Surface Tension as the Cause of Bénard Cells and Surface Deformation in a Liquid Film," *Nature (London)*, Vol. 178, 1956, pp. 560–562.

[5]Pearson, J. R. A., "On Convection Cells Induced by Surface Tension," *Journal of Fluid Mecahnics*, Vol. 4, 1958, pp. 489–493.

[6]Nield, D. A., "Surface Tension and Buoyancy Effects in Cellular Convection," *Journal of Fluid Mechanics*, Vol. 19, 1964, pp. 341–345.

[7]Platten, J. K., and Legros, J. C., *Convection in Liquids*, Springer-Verlag, Berlin 1984.

[8]Grodzka, P. G., and Bannister, T. C., "Heat Flow and Convection Demonstration Experiments Aboard Apollo 14," *Science (Washington DC)*, Vol. 176, May 5, 1972.

[9]Grodzka, P. G., and Bannister, T. C., "Heat Flow and Convection Demonstration Experiments Aboard Apollo 14," *Science (Washington DC)*, Vol. 187, Jan. 17, 1975.

[10]Koschmieder, E. L., and Biggerstaff, M. I., *Journal of Fluid Mechanics*, Vol. 167, 1986, pp. 49–55.

[11]Schwabe, D., Dupont, O., Queeckers, P., and Legros, J. C., "Experiments on Marangoni-Bénard Instability Problems Under Normal and Microgravity Conditions," *Oxford 89-Proceedings of the VIIth European Symposium on Materials and Fluid Sciences in Microgravity*, European Space Agency SP.

[12]Dupont, O., Queeckers, P., Legros, J. C., and Schwabe, D., "Thermo-Hydrodynamical Instabilities Under Microgravity Conditions: The Marangoni-Bénard Problem. 1. Reference Ground Tests of Texus 21 Flight" (unpublished observations).

[13]Bergé P., Pomeau, Y., and Vidal C., *L'ordre dans le chaos*, Hermann, 1984.

[14]Norman, C., Pomeau, Y., Velarde, M. C., "Convective Instability: A Physicist's Approch," *Review of Modern Physics*, Vol. 46, 1977, pp. 581–586.

[15]Davis, S. H., *Physicochemical Hydrodynamics Interfacial Phenomena*, Vol. 174, edited by M. G. Velarde, NATO ASI Series B: Physics, Plenum, New York, 1988.

[16]Cliffe, K. A., and Winters, K. H., "The Use of Symmetry in Bifurcation Calculations and its Application to the Bénard Problem." *Journal of Computational Physics*, Vol. 67, 1986, pp. 310–316.

[17]Benjamin, T. B., "Bifurcation Phenomena in Steady Flows of a Viscous Fluid," *Proceedings of the Royal Society of London, Series A*, Vol. A59, 1978, pp. 1–6.

[18]Winters, K. H., Plesser, T., Cliffe, K. A., "The Onset of Convection in a Finite Container due to Surface Tension and Buoyancy," *Physica D*, Vol. 29, 1988, pp. 387–389.

[19]Gerbaud, C., "Etude des instabilités convectives dans les phénomènes de Rayleigh-Bénard-Marangoni," PhD Dissertation, Université de Provence, Aix-en-Provence, France, 1981.

[20]Queeckers, P., Dupin, J. C., and Legros, J. C., "Influence of the Surface Viscosity on the Marangoni-Bénard Instabilities," *Heat Transfer in Convective Flows*, Vol. 107, American Society of Mechanical Engineers HTD, New York, 1989, pp. 405–409.

[21]Brian, P. L. T., and Smith, K. A., "Influence of Gibbs Adsorption on Oscillatory Marangoni Instability," *AIChE Journal*, Vol. 18, 1972, pp. 231–237.

[22]Chun, X. L., and Verlarde, M., "Bénard-Marangoni Convection in Liquid Layers with a Deformable Open Surface: Note on the Role of Solute Accumulation at the Air-Liquid Interface," *Journal of Colloid and Interface Science* (to be published).

[23]Platten, J. K., and Legros, J. C., *Convection in Liquids*, Springer-Verlag, Berlin, 1984.

[24]Kolodner P., Williams, H., and Moc, C., "Optical Measurement of Soret Coefficient of Ethanol/Water Solutions," *Journal of Chemical Physics*, Vol. 88, 1988, pp. 6512–6516.

[25]Castill, J. C., and Velarde, M. G., "Thermal Diffusion and the Marangoni-Bénard Instability of a Two-Component Fluid Layer Heated from Below," *Physics Letters*, Vol. 66A, 1978, pp. 489–496.

[26]Lichtenbelt, J. H., Drinkenburg, A. A. H., and Dijkstra, H. A., "Marangoni Convection and Mass Transfer from the Liuqid to the Gas Phase," *Scientific Results of the German Spacelab Mission D1*, Vol. 127, Aug. 1986.

[27]Dijkstra, H. A., "Analysis of Flow Development due to Marangoni Convection in a Mass Transfer System," *Physiochemical Hydrodynamics Interfacial Phenomena*, edited by M. G. Verlarde, Plenum, New York, 1988.

[28]Velarde, M. G., and Castillo, J. L., *Convective Transport & Instability Phenomena*, edited by J. Zierp and H. Oerlof, Jr., 1982, p. 235.

[29]Joseph, D. O., *Stability of Fluid Motions, I and II*, Springer-Verlag, Berlin, 1976.

[30]D2 Proposal by J. C. Legros, O. Dupont, P. Queeckers, S. Van Vaerenbergh.

[31]Legros, J. C., and Goemaere, P., "Soret Coefficient and the Two-Component Bénard Convection in the Benzene-Methanol System," *Physical Review A*, Vol. 32, 1985, pp. 1903–1908.

[32]Henry, D., and Roux, B., "Three-Dimensional Numerical Study of Convection in a Cylindrical Thermal Diffusion Cell: Inclination Effect," *Physics of Fluids*, Vol. 30, 1986, pp. 1656–1659.

[33]Van Vaerenbergh, S. R., and Legros, J. C., "Kinetics of the Soret Effect and its Measurement Under Microgravity Conditions," *Progress in Astronautics and Aeronautics*, Vol. 110, AIAA, New York, 1987, pp. 222–228.

Interfacial Oscillators

Manuel G. Velarde* and Xiao-Lin Chu†
Universidad Nacional de Educacion a Distania Madrid, Spain

Introduction

CAPILLARY waves, or ripples, gravity waves, and solitons have been well studied since the works of Laplace, Kelvin, Stokes, Boussinesq, Rayleigh, and Korteweg and de Vries.[1-7] Their properties were shown to be determined mainly by the stress condition at liquid surfaces. The major influence of the boundary condition for normal stress to the surface has been especially emphasized. Although these waves are rather transverse motions due to the deformation of the surface, there is yet another type of wave discovered not too long ago by Lucassen.[8] (See also Refs. 9 and 10.) It refers to mostly longitudinal motion along the surface, in the limit along a flat surface. The existence of this motion is not surprising, considering that a strong analogy is expected between a monolayer-covered surface and a stretched elastic membrane. The coverage with a surfactant monolayer, either by adsorption from solution or by spreading, indeed gives elastic properties to a surface so that it tends to resist the periodic surface expansion and compression that appears as wave motion. Normally, any wave motion of the surface has both transverse and longitudinal components, and they are not separable. Only in the case of small-amplitude wave motion (corresponding to linear theory) can transverse waves and longitudinal waves be considered as two genuinely different modes of oscillation. Longitudinal waves have received much less attention than the others since generally the motion is damped out much more rapidly than the transverse one. However, Lucassen showed under which circumstances in the elastic surface of a

*Professor of Physics, Facultad de Ciencias.
†Postdoctoral Fellow, Facultad de Ciencias.

highly viscous fluid longitudinal waves could be observed easier than transverse waves.

Longitudinal waves are, to a major extent, related to the boundary conditional for tangential stress with a frequency that depends on the viscosity and surface elasticity modulus (or just the Marangoni number). However, gravity-capillary waves have a frequency that depends on gravity and on surface tension (Laplace overpressure) but not on viscosity. The latter only appears in the damping factor and frequency deviation in the dispersion relation.

Generally these waves, these oscillatory motions, are damped, albeit differently by viscosity. However, if a nonequilibrium distribution of surfactant/temperature is imposed in the liquid, with mass/energy transfer across the surface, the Marangoni effect may transform the thermal or chemical energy into convective motion, overtaking the viscous dissipation and thus sustaining the wave motions.[11-15]

Sternling and Scriven[16] were the first authors to provide a mathematical description of spontaneous convection at the interface of two fluids. In their now-classical paper, published in 1959, they clearly considered the Marangoni effect as the "interfacial engine" that converts chemical energy into convection. Marangoni stresses due to surfactant concentration and/or temperature gradients along the interface generate convective motions that are transmitted to the bulk. If there is high enough surfactant or temperature gradient in the bulk, convection may promote the surfactant and/or thermal distribution at the liquid-liquid interface, thus permitting disturbances to develop in the form of steady cellular convection or oscillatory convection.[17-20] The latter case may correspond to waves that appear as either longitudinal or transverse capillary gravity waves or a combination of both.

Although Sternling and Scriven[16] were aware of transverse motions in experiments, and referred to them in their paper as "localized stirring with rippling and twitching of the interface," they treated the interface as flat in order to simplify the mathematical problem. This excluded the transverse waves in their analysis. Another approximation in the model of Sternling and Scriven was the elimination of surface adsorption, although they considered a surface-active solute. However, solute adsorption and eventual accumulation at the surface takes place in liquids with surface-active solutes and drastically affects stability.[21]

The Sternling and Scriven theory has had subsequent developments by several authors.[22-23] Sanfeld and collaborators studied a liquid-liquid interface model with due account of the deformable interface and solute adsorption at the interface. However, they only discussed one of the various possible modes, the long-wave longitudinal waves. It was also shown in their results that the deformability of the interface has a negligible influence on the stability of these waves. An extension of the Scriven and Sternling theory and of the theory of Sanfeld and Linde and their collaborators has recently been provided by the present authors.[11-15,34-43] Thus, the rest of this chapter is a succinct account of our findings with recipes for experimental tests in both ground-based conditions and in the microgravity environment of a spacelab. Note that most of our findings for surfactant gradients can be transferred verbatim to the case of temperature gradients and vice versa.

Case of a Liquid Open to Air and the Role of Surfactants (Marangoni Effect)

Disturbance Equations
For simplicity let us consider a two-dimensional liquid layer with infinite depth. The motionless undisturbed surface is located at $z = 0$. The linearized equations that disturbances upon the quiescent state obey are

$$\nabla^* v = 0 \tag{1}$$

$$\rho \frac{\partial v}{\partial t} = -\nabla p + \eta \nabla^2 v \tag{2}$$

$$\frac{\partial c}{\partial t} - \beta^c w = D\nabla^2 c \tag{3}$$

where $v = (u, w)$, p is the velocity pressure, c the concentration of the surfactant, ρ the density of the liquid, $\eta = \rho v$ the dynamic viscosity, v the kinematic viscosity, D the mass diffusivity, and $\beta^c = (\partial c/\partial z)_0$ the gradient of surfactant at the quiescent state. The disturbances obey the following (linearized) boundary conditions at the deformable surface:

$$\frac{\partial \zeta}{\partial t} = w \tag{4}$$

$$-T_0\nabla_\Sigma^2\zeta + g\rho\zeta - p - 2\eta\frac{\partial w}{\partial z} = 0 \tag{5}$$

$$\left(\frac{\partial T}{\partial \Gamma}\right)_0 \nabla_\Sigma\gamma - \eta\left(\nabla_\Sigma w + \frac{\partial u}{\partial z}\right) = 0 \tag{6}$$

$$\frac{\partial \gamma}{\partial t} + \Gamma_0 \nabla_\Sigma * u_\Sigma - D_\Sigma \nabla_\Sigma^2\gamma + D\frac{\partial c}{\partial z} = 0 \tag{7}$$

$$\gamma = k^1(c - \beta^c\zeta)_\Sigma \tag{8}$$

where ζ is the surface deviation from the $z = 0$ level, Γ the excess surfactant concentration at the surface, the subscript 0 indicates a value in a reference state, γ is the disturbance on Γ_0, g the gravitational acceleration, T the surface tension, and the subscript Σ accounts for either a value taken on the surface or a derivative along the surface. Note that, although under microgravity conditions or with shallow layers we may neglect buoyancy, we cannot neglect the role of gravity in the boundary condition (bc). For, if we set g to zero, the capillary length [Eq. (9)], which is the natural length scale in our analysis, diverges to infinity, thus leading to an irrelevant instability in the long wavelength limit, i.e., at a vanishing Fourier wave number.

For universality in presentation we introduce new units to rescale the quantities in the equations. The capillary length

$$l = \sqrt{\frac{T_0}{g\rho}} \tag{9}$$

is chosen as our length scale; v/l, l^2/v, $v^2\rho/l^2$, $\beta^c l$, and Γ_0 are used as units for velocity, time, pressure, surfactant concentration and excess surface concentration, respectively. Thus, Eqs. (1–8) become

$$\nabla * v = 0 \tag{10}$$

$$\frac{\partial v}{\partial t} = -\nabla p + \nabla^2 v \tag{11}$$

$$\frac{\partial c}{\partial t} - w = S^{-1} \nabla^2 c \tag{12}$$

with boundary conditions at $z = \zeta$,

$$\frac{\partial \zeta}{\partial t} = w \tag{13}$$

$$-\frac{1}{SC} \nabla^2_\Sigma \zeta + \frac{Bo}{SC} \zeta - p - 2\frac{\partial w}{\partial z} = 0 \tag{14}$$

$$\frac{HE}{SH_z} \nabla_\Sigma v + \left(\nabla_\Sigma v + \frac{\partial u}{\partial z}\right) = 0 \tag{15}$$

$$HS\left(\frac{\partial \gamma}{\partial t} + \nabla_\Sigma * u_\Sigma - S_\Sigma^{-1} \nabla_\Sigma^{-2} \gamma\right) + \frac{\partial c}{\partial z} = 0 \tag{16}$$

$$\gamma = \frac{H_z}{H}(c - \zeta)_\Sigma \tag{17}$$

Note that, although we have used the same notation for v, p, c, γ, and ζ in these equations, as in the dimensional case, all of the units have changed, and they are dimensionless now. The following dimensionless parameters have also been used: $S = v/D$, Schmidt number; $S_\Sigma = v/D_\Sigma$, surface Schmidt number; $C = \eta D/lT_0$, capillary number; $Bo = \rho g l^2/T_0$, Bond number (for simplicity it is taken equal to unity); $E = -(\partial c/\partial \Gamma)_0 k'\beta l^2/(\eta D)$, surfactant (elasticity) Marangoni number; $H = \Gamma/(\beta^c l^2)$, surface excess surfactant number; and $H_z = k'/l$, Langmuir adsorption number. Then Eqs. (10–17) have solutions

$$w = (A\,e^{az} + B\,e^{mz})e^{iax + \lambda t} \tag{18}$$

$$u = (iA\,e^{az} + imB\,e^{mz})\,e^{iax + \lambda t} \tag{19}$$

$$p = -\left(\frac{\lambda A\,e^{az}}{a}\right) e^{iax + \lambda t} \tag{20}$$

$$c = \left(\frac{A}{\lambda}\,e^{az} + \frac{SB}{\lambda(S - l)}\,e^{mz} + F\,e^{qz}\right) e^{iax + \lambda t} \tag{21}$$

where

$$m = \sqrt{a^2 + \lambda} \quad \text{and} \quad q = \sqrt{a^2 + S\lambda}$$

Here a is the Fourier wave number, and λ denotes the time constant. The real part of λ determines the stability of the system, whereas the imaginary part determines the convective mode. An oscillatory motion requires the imaginary part of λ, $\omega = \text{Im}(\lambda)$, not being zero. In this case we have a Hopf bifurcation from the motionless state to a mode of oscillatory convection that shows up as a wave at the air-liquid interface. However, if $\omega = 0$, the transition is expected from a steady state of rest to a steady cellular convective pattern. A, B, and F are arbitrary constants to be obtained using the boundary conditions.

Based on these conditions we can study the evolution of the system by either traditional hydrodynamic stability theory or by a intuitively more appealing approach. It is known that a simple linear oscillatory system can be described by the harmonic equation

$$\frac{\partial^2 f}{\partial t^2} + \Phi_f = -\Psi \frac{\partial f}{\partial t} \tag{22}$$

where f is the unknown, and Φ and Ψ are factors specific to each case. The terms in the left-hand side of this equation, $\partial^2 f/\partial t^2$ and Φf, account for the inertia and restoring forces of the system. They are ideal, nondissipative parts. The right-hand side of the equation, $-\Psi(\partial f/\partial t)$, represents the dissipation in the system and determines the stability of the oscillator. When $\Psi > 0$, one has a damped oscillation, and the motionless state is asymptotically stable, whereas, if $\Psi < 0$, we have an oscillatory explosion eventually saturated by nonlinear terms in the Navier-Stokes equations left out in the linear approximation. At $\Psi = 0$, we may speak of the threshold for oscillation. Two modes of oscillation of the air-liquid interface will be described using the harmonic oscillator approach [Eq. (22)], the transverse and longitudinal waves earlier mentioned. Later on we shall give the nonlinear counterpart of Eq. (22) for nonlinear limit cycle oscillation and in the final section for soliton excitation.

Transverse Gravity-Capillary Waves

Assuming, for simplicity, that surface adsorption and accumulation have a negligible effect on the transverse suface waves, Eqs. (15-17) reduce to

$$\frac{Ea^2}{S}(c - \zeta) + \left(\frac{\partial^2}{\partial z^2} + a^2\right)w = 0 \tag{23}$$

$$\frac{\partial c}{\partial z} = 0 \tag{24}$$

Equation (11) gives the dynamic evolution equation of the liquid layer. It is valid in the volume as well as at the surface, a part of the liquid. On the other hand, we have a kinematic relation [Eq. (13)] on the surface. Substituting w in the left side of Eq. (11) by Eq. (13), a relation on the surface $z = \zeta$ is written as

$$\frac{\partial^2 \zeta}{\partial t^2} = -\frac{\partial p}{\partial z} + \nabla^2 w \tag{25}$$

From Eq. (20) one has $\partial p/\partial z = ap$. Using Eqs. (14) and (23), Eq. (25) becomes

$$\frac{\partial^2 \zeta}{\partial t^2} + \frac{Bo + a^2}{SC}a\zeta = -2a\frac{\partial w}{\partial z} - 2a^2 w - \frac{Ea^2}{S}(c - \zeta) \tag{26}$$

To proceed further, we estimate the coefficients A, B, and F. From Eq. (14) one has

$$B = \frac{\lambda f(Bo) + 2a}{\lambda f(Bo) - \lambda/a - 2p}A \tag{27}$$

where

$$f(Bo) = \frac{Bo + a^2}{SC\lambda^2} + \frac{1}{a} \tag{28}$$

For transverse waves the Laplace equation $f(Bo) = 0$ holds, and as the interfacial disturbance penetrates little in the liquid, $\omega \gg a^2$. Then from Eq. (27) it follows that

$$B \approx -\frac{2a^2}{\lambda}A \ll A \tag{29}$$

Thus, we take, in fact, the potential part as the major ingredient in the velocity field, which is consistent with the fact that transverse waves exist in ideal nonviscous fluids. Taking advantage of this simplification, Eq. (26) becomes

$$\frac{d^2\zeta}{dt^2} + \frac{Bo + a^2}{SC}a\zeta = -4a^2\frac{dz}{dt} - \frac{Ea^2}{S}(c - \zeta) \tag{30}$$

The coefficient F can be obtained from Eq. (24):

$$F = -\frac{1}{q}\left(\frac{a}{\lambda}A + \frac{mS}{\lambda(S-1)}B\right) \approx -\frac{a}{q\lambda}A \qquad (31)$$

which leads to

$$c - \zeta = \left(F + \frac{B}{\lambda(S-1)}\right)e^{iax+\lambda t} \approx -\frac{aA}{q\lambda}e^{iax+\lambda t} \approx a\frac{d\zeta/dt - \omega\zeta}{\omega\sqrt{2S\omega}} \qquad (32)$$

Thus, Eq. (30) becomes

$$\frac{d^2\zeta}{dt^2} + \left(\frac{Bo + a^2}{SC}a - \frac{Ea^3}{S\sqrt{2s\omega}}\right)\zeta = -\left(4a^2 + \frac{Ea^3}{S\omega\sqrt{2S\omega}}\right)\frac{d\zeta}{dt} \qquad (33)$$

Assuming now that the surfactant Marangoni number E is not exceedingly high, albeit of order $\mathcal{O}(C^{-1})$, it follows that

$$(B + a^2)/(SC)a \gg (Ea^3)/(S\sqrt{2s\omega})$$

Therefore, Eq. (33) can be reduced to

$$\frac{d^2\zeta}{dt^2} + \frac{Bo + a^2}{SC}a\zeta = -\left(4a^2 + \frac{Ea^3}{S\omega\sqrt{2S\omega}}\right)\frac{d\zeta}{dt} \qquad (34)$$

Equation (34) describes the oscillatory convective motion of the surface and takes the form of a simple oscillator, Eq. (22). One can directly judge that the left-hand-side terms of Eq. (34), which is the potential part, is the leading part of the equation. In the absence of dissipation and Marangoni effect, the equation appears as an ideal oscillator governed by the Laplace equation. Thus, potential motion is the main character of a transverse wave. The right-hand side of Eq. (34) reflects the competition between viscous dissipation and Marangoni stresses. According to our earlier discussion, when

$$4a^2 + \frac{Ea^3}{s\omega\sqrt{2S\omega}} > 0\left(\text{or} \quad -E < \frac{4S\omega\sqrt{2S\omega}}{a}\right) \qquad (35)$$

the oscillatory convection will be damped out. In contrast, when

$$4a^2 + \frac{Ea^3}{s\omega\sqrt{2S\omega}} < 0\left(\text{or} \quad -E > \frac{4S\omega\sqrt{2S\omega}}{a}\right) \qquad (36)$$

the oscillation grows exponentially. At the neutral state,

$$-E = \frac{4S\omega\sqrt{2S\omega}}{a} \qquad (37)$$

The kinetic energy dissipated by viscosity and that produced by surface tension work just compensate each other, giving a sustained oscillator. Thus, the oscillation frequency is given by the dispersion relation

$$\frac{Bo + a^2}{SC} a - \omega^2 = 0 \tag{38}$$

By taking $dE[(a,w(a)]/da = 0$, the necessary condition for minimum yields the neutral curve, i.e., the critical values for sustained transverse waves:

$$E_c^T = 4\sqrt{10}\left(\frac{6S}{5\sqrt{5}\,C}\right)^{\frac{3}{4}} \approx -7.931\left(\frac{S}{C}\right)^{\frac{3}{4}} \tag{39}$$

$$\omega_c^T = \sqrt{\frac{6}{5^{\frac{3}{2}}}\frac{1}{\sqrt{SC}}} \approx \frac{0.7326}{\sqrt{SC}} \tag{40}$$

$$a_c^T = \frac{1}{\sqrt{5}} \approx 0.4472 \tag{41}$$

In the section on nonlinear capillary-gravity oscillations, Table 1 accounts for numerical estimates at two different values of the gravitational acceleration, g and $10^{-4}\,g$.

Longitudinal Waves

To avoid some tedious and rather not so relevant complications, we first simplify the problem. We suppose that the deformability of the surface has a negligible influence on the longitudinal wave motion in agreement with Lucassens's finding.[8] Thus, we set the capillary number C to zero. Also, it is known that surfactant accumulation on the surface affects mainly high-frequency oscillatory convection, and the frequency of longitudinal waves normally is small; thus the surfactant accumulation number H can be neglected. With these assumptions Eqs. (13–17) reduce to

$$w = 0 \tag{42}$$

$$\frac{Ea^2}{S} c + \frac{\partial^2 w}{\partial z^2} = 0 \tag{43}$$

$$SH_z \frac{\partial c}{\partial t} = -\frac{\partial c}{\partial z} \tag{44}$$

The surfactant concentration on the surface is chosen as the relevant variable in the harmonic oscillator. Differentiating Eq. (12) with respect to z and using Eq. (44), we have

$$-SH_z \frac{\partial^2 c}{\partial t^2} = \frac{\partial w}{\partial z} + S^{-1}\nabla^2 \frac{\partial c}{\partial z} \tag{45}$$

Now we use the boundary conditions (42–44). From Eq. (42) it follows that

$$B = -A \qquad (46)$$

Note that here the rotational part is of the same order as the potential one near the surface. Thus, we can neither ignore rotational nor potential terms in the longitudinal wave case. Using Eq. (46) and the long-wavelength assumption $a^2 \ll \omega \ll 1$, we obtain

$$\frac{\partial w}{\partial z} \approx \frac{1}{m}\left(1 - \frac{a}{\sqrt{\lambda}}\right)\frac{\partial^2 w}{\partial z^2}$$

Replacing $\partial^2 w/\partial z^2$ by Eq. (43),

$$\frac{\partial w}{\partial z} \approx Ea^2 q\left(\frac{\partial c/\partial t}{S^{\frac{3}{2}}\omega^2} - \frac{ac}{Sqm\sqrt{\lambda}}\right) \qquad (47)$$

It also follows that

$$S^{-1}\frac{\partial}{\partial z}\nabla^2 c \approx q\frac{\partial c}{\partial t} \qquad (48)$$

Substitution of Eqs. (47) and (48) into (45) yields

$$-\frac{SH_z}{q}\frac{\partial^2 c}{\partial t^2} - \frac{Ea^3}{Sqm\sqrt{\lambda}}c = \left(1 + \frac{Ea^2}{S^{\frac{3}{2}}\omega^2}\right)\frac{\partial c}{\partial t} \qquad (49)$$

Here we have

$$-\frac{SH_z}{q}\frac{\partial^2 c}{\partial t^2} \approx \frac{H_z\sqrt{S}}{\sqrt{2\omega}}\left(\frac{\partial^2 c}{\partial t^2} + \omega\frac{\partial c}{\partial t}\right) \qquad (50)$$

$$-\frac{Ea^3}{Sqm\sqrt{\lambda}}c \approx \frac{Ea^3}{S\omega\sqrt{2S\omega}}\left(c + \frac{1}{\omega}\frac{\partial c}{\partial t}\right) \qquad (51)$$

Replacing Eqs. (50) and (51) into Eq. (49) and neglecting high-order small quantities, we obtain

$$\frac{1}{\sqrt{2S\omega}}\left(SH_z\frac{d^2 c}{dt^2} - \frac{Ea^3}{S\omega}c\right) = -\left(1 + \frac{Ea^2}{S^{\frac{3}{2}}\omega^2}\right)\frac{dc}{dt} \qquad (52)$$

This is the equation we expected. In contrast with transverse waves, the leading part in Eq. (52) corresponds to the terms with first-order derivative. Indeed,

$$\left|\frac{1}{\sqrt{2S\omega}}\left(SH_z\omega^2 + \frac{Ea^3}{S\omega}\right)\right| \ll \left|\omega\left(1 + \frac{Ea^2}{S^{\frac{3}{2}}\omega^2}\right)\right| \qquad (53)$$

This relation reveals the intrinsic character of a longitudinal wave: It is necessarily related to viscous convection. The potential part of Eq. (52), though small compared with the dissipative one, determines the oscillating feature of this convection. When energy dissipation and Marangoni work compensate each other,

$$1 + \frac{Ea^2}{S^{\frac{3}{2}}\omega^2} = 0 \tag{54}$$

The stability analysis for the oscillator gives

$$-E \begin{cases} > \dfrac{S^{\frac{3}{2}}\omega^2}{a^2} & \text{explosion} \\[3mm] < \dfrac{S^{\frac{3}{2}}\omega^2}{a^2} & \text{damped motion} \end{cases}$$

In the neutral case we also have

$$SH_z\omega^2 + \frac{Ea^3}{S\omega} = 0 \tag{55}$$

Thus, Eqs. (54) and (55) describe the characteristics of the longitudinal oscillation of the air-liquid interface. One expects that sustained longitudinal oscillations occur at the surface with dispersion relation

$$a_c = H_z\omega S^{\frac{1}{2}}$$

which, together with Eq. (54), generalizes earlier findings.[8,19]

Comparison of Eqs. (39) and (55) shows that both thresholds depend on the Schmidt (Prandtl) number. However, transverse waves are directly related to the interfacial deformation (measured by C) whereas longitudinal waves rather depend on Langmuir's adsorption (measured by H_z). Crossover from one to another mode of instability is given by the condition: when S is greater (smaller) than $C^3/(7.931\,H_z^2)^4$ we have longitudinal (transverse) waves first.

Transverse Waves not Obeying the Laplace Law
Numerical exploration of problem posed in the subsection of the disturbance equations shows that there exists a mode of transverse oscillation that does not follow from the Laplace equation, $f(Bo) = 0$. Let us explore this possibility using the harmonic oscillator approach developed for Laplace-Kelvin and Marangoni-Lucassen waves. From Eq. (26),

$$\frac{\partial^2\zeta}{\partial t^2} + \frac{Bo + a^2}{SC}a\zeta = -2a^2\frac{\partial z}{\partial t} + 2a\frac{\partial w}{\partial z} - \frac{Ea^2}{S}(c - \zeta) \tag{56}$$

It can be solved directly from Eqs. (13–21):

$$\frac{\partial w}{\partial z} = \frac{f(Bo)(a-m)-1}{2m-2a-\lambda/a}\,\lambda\,\frac{\partial \zeta}{\partial t} = \frac{a}{\lambda}[\lambda + 2ma + 2a^2 + f(Bo)(m+1)\lambda]\frac{\partial \zeta}{\partial t} \quad (57)$$

$$c - \zeta = -\frac{1}{q\lambda}\frac{\lambda f(Bo)\left(a+\dfrac{q-mS}{S-1}\right)-\lambda+2a\dfrac{q-m}{S-1}}{2m-2a-\lambda/a}\frac{\partial \zeta}{\partial t}$$

$$= \frac{a}{q\lambda^3}(m+a)\left[\lambda f(Bo)\left(a+\frac{q-mS}{S-1}\right)-\lambda+2a\frac{q-m}{S-1}\right]\frac{\partial \zeta}{\partial t} \quad (58)$$

Using these relations, Eq. (56) becomes

$$\frac{\partial^2 \zeta}{\partial t^2} + \frac{Bo+a^2}{SC}a\zeta = -4a^2\left[1+\frac{a(a+m)}{\lambda}\right]\frac{\partial \zeta}{\partial t}$$

$$- 2a^2 f(Bo)(m+a)\frac{\partial \zeta}{\partial t} - \frac{Ea^3}{Sq\lambda^3}(m+a)^2$$

$$\times \left[\lambda f(Bo)\left(a+\frac{q-mS}{S-1}\right)-\lambda+2a\frac{q-m}{S-1}\right]\frac{\partial \zeta}{\partial t} \quad (59)$$

Also assume here the high-frequency approximation $\omega \gg a^2 = 0(1)$. This permits a simplified exploration of the problem and is still reasonable as we search for transverse oscillatory modes that due to viscous damping penetrate little in the liquid layer. Then Eq. (59) becomes

$$\frac{\partial^2 \zeta}{\partial t^2} + \frac{Bo+a^2}{SC}a\zeta - \sqrt{2}\,\omega^{\frac{3}{2}}a^2 f(Bo)\zeta = -4a^2\frac{\partial \zeta}{\partial t} - \sqrt{2\omega}\,a^2 f(Bo)\frac{\partial z}{\partial t}$$

$$+ \frac{Ea^3}{Sq\lambda}\left[1-f(Bo)\frac{q-aS\sqrt{\lambda}}{S-1}\right]\frac{\partial \zeta}{\partial t} \quad (60)$$

As a particular case of this equation, one can easily obtain the harmonic equation for Laplace transverse waves by taking the Laplace approximation $f(Bo = 0)$. However, let us explore the case $f(Bo) \neq 0$. Consider the case of low Schmidt numbers $S \ll a^2/\omega$. Then we obtain

$$\frac{d^2 \zeta}{dt^2} + \left[\frac{Bo+a^2}{SC}a - \frac{Ea^2}{S}\right]\zeta = \left\{-4a^2 + \left(\frac{Ea}{2}-a^2\sqrt{2\omega}\right)f(Bo)\right\}\frac{d\zeta}{dt} \quad (61)$$

This equation describes once more a harmonic oscillator. As we did earlier, let us consider the neutral state. When the damping factor, i.e., the coefficient of the first derivative with respect to time, vanishes, we obtain

$$\frac{Bo+a^2}{SC}a - \omega^2 = \frac{Ea^2}{S} \quad (62)$$

$$\left[\frac{Ea}{2} - \sqrt{2\omega}\, a^2\right]\left[\frac{1}{a} - \frac{Bo + a^2}{SC\omega^2}\right] - 4a^2 = 0 \qquad (63)$$

It follows that

$$E^2 - 2a\sqrt{2\omega}\, E + 8S\omega^2 = 0 \qquad (64)$$

There are two roots (both of them are positive):

$$E = a\sqrt{2\omega}\,(1 \pm \sqrt{1 - 4S\omega/a^2}) \approx a\sqrt{2\omega}\left(1 \pm 1 \mp \frac{2S\omega}{a^2}\right) \qquad (65)$$

It appears that the branch $E \approx 2a(2\omega)^{\frac{1}{2}}$ has no minimum when the wave-length varies in the capillary length range. Therefore, it has no physical meaning and presumably is a spurious consequence of our simplifications. We consider the other root, $E = 2(2)^{\frac{1}{2}}S\omega^{\frac{3}{2}}/a$. Then the dispersion relations are

$$\frac{Bo + a^2}{SC}a - \omega^2 = \frac{Ea^2}{S} \qquad (66)$$

$$E = \frac{2\sqrt{2}S\omega^{\frac{3}{2}}}{a} \qquad (67)$$

These two equations determine the neutral state. By taking

$$\frac{dE[a,\omega(a)]}{da} = 0 \qquad (68)$$

we obtain the critical elasticity Marangoni number,

$$E_c = \frac{5a^2 - Bo}{2aC} \qquad (69)$$

that permits transverse waves to be sustained at the air-liquid surface without being controlled by the Laplace law, $f(Bo) = 0$. Interestingly enough, it has been shown[37] that the threshold for this mode of oscillation is lower than the threshold for standard Bénard convection (steady pat-terned convection), provided the capillary number is large enough and the Schmidt (or Prandtl) number is small enough, as in the case of liquid He-4 layer near the lambda line. Here we have a suggestion for an experimental test. If we really want to avoid that buoyancy masks the real phenomenon the experiment should be conducted under low/microgravity conditions.

Liquid-Liquid Interface

Capillary-Gravity Transverse Oscillations

Now let us consider the deformable interface separating two liquids of different densities, viscosities, and mass diffusivities and again assume the problem to be two dimensional, i.e., with horizontal and vertical coordinates x and z, respectively. If the liquid below is labeled "one," the evolution of infinitesimal disturbances upon the motionless state on either side of the liquid-liquid interface is given by a straightforward extension of the equations described in the preceding section. We have the following dimensionless equations:

$$\frac{\partial u_i}{\partial x} + \frac{\partial w_i}{\partial z} = 0 \qquad (i = 1, 2) \tag{70}$$

$$\frac{\partial w_1}{\partial t} = -\frac{\partial p_1}{dz} + \nabla^2 w_1 \tag{71a}$$

$$N_\rho \frac{\partial w_2}{\partial t} = -\frac{\partial p_2}{\partial z} + N_\eta \, \nabla^2 w_2 \tag{71b}$$

$$\frac{\partial C_1}{\partial t} - w_1 = \frac{1}{S} \, \nabla^2 C_1 \tag{72a}$$

$$\frac{\partial C_2}{\partial t} - w_2 = \frac{N_D}{S} \, \nabla^2 C_2 \tag{72b}$$

with $\nabla^2 = \partial^2/\partial x^2 + \partial^2/\partial z^2$ and where $C_i (i = 1, 2)$ denotes solutal or surfactant concentration in each volume; $w_i (i = 1, 2)$ is the vertical velocity along z; $u_i (i = 1, 2)$ is the horizontal velocity along x; p_i $(i = 1, 2)$ is the pressure; $S = v_1/D_1$ (Schmidt number), with v and D the viscosity and mass diffusivity, respectively; and $N_D = D_2/D_1$, $N\rho = \rho_2/\rho_1$ and $N_\eta = \eta_2/\eta_1$, with $\eta = \rho v$ the dynamic viscosity. Note that we have introduced only one of the two Schmidt numbers.

If from either side we allow for surfactant adsorption without considering the surfactant accumulation at the interface and call N_Z the ratio of the corresponding Langmuir adsorption slopes, the following bc must be satisfied at $z = 0$:

$$w_1 = w_2 \tag{73}$$

$$\frac{\partial w_1}{\partial z} = \frac{\partial w_2}{\partial z} \tag{74}$$

$$B\xi - \frac{\partial^2 \xi}{\partial x^2} + CS(p_2 - p_1) - 2CS(N_\eta - 1)\frac{\partial w}{\partial z} = 0 \tag{75}$$

$$\frac{E}{S}\frac{\partial^2(C_1 - \xi)}{\partial x^2} - \left(\frac{\partial^2}{\partial x^2} - \frac{\partial^2}{\partial z^2}\right)(N_\eta w_2 - w_1) = 0 \tag{76}$$

$$\frac{\partial C_1}{\partial z} = \frac{\partial C_2}{\partial z} \tag{77}$$

$$C_1 - \xi = N_z(C_2 - \xi) \tag{78}$$

where we also consider that the interfacial tension changes with the surfactant concentration:

$$E = - \frac{\partial \sigma}{\partial C_1} \frac{\beta_1 l^2}{\eta_1 D_1}$$

The variable E is again the surfactant or elasticity Marangoni number, and l is the given space scale in the problem. Once more this could be the capillary length but does not need to be. Here $B = gl^2(\rho_1 - \rho_2)/\sigma_0$ is the Bond number, with σ_0 a reference value for the interfacial tension and g the gravitational acceleration. $C = \eta_1 \rho_1 / \sigma_0 l$ is indeed the capillary number. Again, note that the capillary length is given by $l^2 = \sigma_0/g(\rho_1 - \rho_2)$, which corresponds to a Bond number equal to unity. β_i ($i = 1, 2$) is the surfactant gradient in phase i.

Solutions of the problem can be sought in the form indicated in the preceding section, but with twice the number of variables. We now have

$$w_1 = A_1 e^{az} + B_1 e^{m_1 z} \tag{79}$$

$$w_2 = A_2 e^{-az} + B_2 e^{-m_2 z} \tag{80}$$

$$p_1 = - \frac{\lambda A_1}{a} e^{az} \tag{81}$$

$$p_2 = - \frac{N_\rho \lambda A_2}{a} e^{-az} \tag{82}$$

$$C_1 = F_1 e^{q_1 z} + \frac{A_1}{\lambda} e^{az} + \frac{S B_1}{\lambda(S - 1)} e^{m_1 z} \tag{83}$$

$$C_2 = F_2 e^{-q_2 z} + \frac{A_2}{\lambda} e^{-az} + \frac{S N_D^{-1} B_2}{\lambda(S N_D^{-1} - N_\rho N_\eta^{-1})} e^{m_2 z} \tag{84}$$

with

$$m_1^2 = \lambda + a^2, \qquad m_2^2 = N_\rho N_\eta^{-1} \lambda + a^2$$

$$q_1^2 = S\lambda + a^2, \qquad q_2^2 = S N_D^{-1} \lambda + a^2$$

Note that to simplify the notation we have omitted in Eqs. (79–84) a common factor $\exp(iax + \lambda t)$, where a is the horizontal Fourier wave number, and λ is the time constant whose real part determines stability. For purely oscillatory motions, $\lambda = i\omega$, where ω is the dimensionless frequency.

Obviously, Eqs. (71a) and (71b) must be valid at the deformable interface $z = \zeta(x, t)$. Adding these two equations and denoting by w the liquid velocity at interface points, we obtain

$$(1 + N_\rho) \frac{\partial w}{\partial t} = - \frac{\partial(p_1 + p_2)}{\partial z} + \nabla^2 w_1 + N_\eta \nabla^2 w_2 \tag{85}$$

Now using Eqs. (75) and (76), Eq. (85) becomes

$$(1 + N_\rho) \frac{\partial^2 \xi}{\partial t^2} + \frac{B + a^2}{CS} a\xi$$

$$= 2a(N_\eta - 1) \frac{\partial w}{\partial z} - 2a^2 w + 2N_\eta \frac{\partial^2 w^2}{\partial z^2} - \frac{Ea^2}{S}(C_1 - \xi) \qquad (86)$$

where we have used the kinematic condition $w = \partial \xi / \partial t$.

For the evaluation of all terms in the right-hand side of Eq. (86), we must solve for A_i, B_i, and F_i using the equations and the bc given earlier. Then, using for simplicity the high-frequency approximation $\omega \gg a^2$ and the fact that for most practical purposes the Schmidt number is much larger than unity, after some elementary albeit lengthy calculus, Eq. (86) finally becomes

$$(1 + N_\rho) \frac{d^2 \xi}{dt^2} + \frac{B + a^2}{CS} a\xi = \frac{a \sqrt{2\omega}}{N_\rho^{-\frac{1}{2}} N_\eta^{-\frac{1}{2}} - 1}$$

$$\times \left[\frac{Ea^2(1 - N_D^{\frac{1}{2}} N_\eta^{-1})}{\omega^2 S^2 (1 + N_z^{-1} N_D^{-\frac{1}{2}})} - 2 \right] \frac{d\xi}{dt} \qquad (87)$$

Note that the interfacial disturbances in a viscous liquid penetrate about $(v/\omega)^{\frac{1}{2}}$; thus for $(v/\omega)^{\frac{1}{2}} \ll 1$ the $\omega \gg a^2$ assumption is justified.

Equation (87) is the harmonic oscillator equation for transverse motions, $\xi(x, t)$, of the deformable liquid-liquid interface due to the Marangoni effect. Its damping coefficient vanishes with a suitable value of E. Then the dispersion relation is given by

$$(1 + N_\rho)\omega^2 = \frac{B + a^2}{CS} a \qquad (88)$$

Now using the expression for N_ρ, B, C, S, $a = lk$, and $\omega = \Omega l^2 / v_1$, Eq. (88) yields

$$(\rho_1 + \rho_2)\Omega^2 = (\rho_1 + \rho_2)gk + \sigma_0 k^3 \qquad (89)$$

which is the standard dispersion relation for gravity-capillary waves at a liquid-liquid interface. For the air-liquid interface, $\rho_2 \approx 0$; thus $\rho_1 + \rho_2 \approx \rho_1 - \rho_2 \approx \rho_1 \approx \rho$. Note that Ω and k have the units of s^{-1} and cm^{-1}, respectively.

For the damping coefficient in Eq. (87) to vanish, we must have

$$Ea^2(1 - N_D^{\frac{1}{2}} / N_\eta) > 0 \qquad (90)$$

i.e.,

$$\text{sign}(E) = \text{sign}(1 - N_D^{\frac{1}{2}} / N_\eta) \qquad (91)$$

Generally the sign of the Marangoni number is given by the sign of the volume gradient of, say, the surfactant β_i. Then, if β_i is positive, i.e., the mass flux is from liquid one to liquid two (β_2 has always the same sign as β_1), Eq. (91) demands

$$\eta_2/\eta_1 - \sqrt{D_2/D_1} > 0 \qquad (\beta_1, \beta_2 > 0) \tag{92}$$

However, if the gradient is negative, we have

$$\eta_2/\eta_1 - \sqrt{D_2/D_1} < 0 \qquad (\beta_1, \beta_2 < 0) \tag{93}$$

Then denoting by subscripts f and t the transport direction from phase f to phase t, we see that Eqs. (92) and (93) are just the same condition,

$$D_f/D_t > (\eta_f/\eta_t)^2 \tag{94}$$

irrespective of the sign of the gradients. Condition (94) is the necessary condition for having sustained oscillations at the liquid-liquid interface. This is achieved when the Marangoni number reaches the critical value, i.e., the minimum value that produces a vanishing damping coefficient in Eq. (87). This value is

$$E_c = -\frac{4SN_\eta(1 + N_z^{-1}N_D^{-\frac{1}{2}})B^{\frac{1}{2}}}{C(1 + N_\rho)(N_D^{\frac{1}{2}} - N_\eta)} \tag{95}$$

for a frequency

$$\omega_c = \frac{\sqrt{2}\,B^{\frac{3}{4}}}{\sqrt{SC(1 + N_\rho)}} \tag{96}$$

and a wave number

$$a_c = \sqrt{B} \tag{97}$$

Note that, when $B = 1$, i.e., when we set the capillary length to 1, Eqs. (96) and (97) reduce to

$$\omega_c = \frac{\sqrt{2}}{\sqrt{SC(1 + N_\rho)}} \qquad \text{and} \qquad a_c = 1$$

respectively.

These are the dimensionless values of the parameters that correspond to the onset of overstability at deformable liquid-liquid interface. Again, as in the case of the open surface of a liquid, discussed in the subsection on transverse gravity-capillary waves, we have clear-cut predictions for an experimental test. As the Bond number, B, contains the gravitational acceleration, g, all results show quite a sensitive variation with changing g, thus indicating once more the relevance of our findings to space bound experiments.

Longitudinal Waves

Since the analysis follows quite the same pattern as that developed in the previous subsection, we shall limit ourselves to a discussion of the results found. The harmonic oscillator equation now refers to the quantity say, C_1. We have indeed the natural extension of the problem posed in the subsection on longitudinal waves at the open surface of a liquid, i.e.,

$$(1 + N_\rho^{\frac{1}{2}} N_\eta^{\frac{1}{2}}) \Gamma S^{\frac{1}{2}} \left(\frac{d^2 C_1}{dt^2} \right) + (2\omega)^{\frac{1}{2}}$$

$$\left[(1 + N_\rho^{\frac{1}{2}} N_\eta^{\frac{1}{2}}) \left(1 + \frac{1}{N_D^{\frac{1}{2}} N} \right) + \frac{Ea^2}{S^2 \omega^2} \Pi_1 \right] \frac{DC_1}{dt} +$$

$$+ a\omega \left[(1 + N_\eta) \left(1 + \frac{1}{N N_D^{\frac{1}{2}}} \right) + \frac{E\alpha^2}{S^2 \omega^2} \Pi_2 \right] C_1 = 0 \qquad (98)$$

with

$$\Pi_1 = \sqrt{\frac{N_\rho N_D}{N_\eta}} - 1 \qquad (99)$$

and

$$\Pi_2 = N_D^{\frac{1}{2}} - 1 \qquad (100)$$

Thus, increasing the value of the elasticity Marangoni number when the damping coefficient vanishes in Eq. (98), we have the possibility of sustained interfacial oscillations.

When the damping coefficient is set to zero, we have the following relationships:

$$(1 + N_\rho^{\frac{1}{2}} N_{r_1}^{\frac{1}{2}}) \left(1 + \frac{1}{N_D^{\frac{1}{2}} N} \right) + \frac{Ea^2}{S^2 \omega^2} \Pi_1 = 0 \qquad (101)$$

and

$$a \left[(1 + N_{r_1}) \left(1 + \frac{1}{N_D^{\frac{1}{2}} N} \right) + \frac{Ea^2}{S^2 \omega^2} \Pi_2 \right] = \omega (1 + N_\rho^{\frac{1}{2}} N_\eta^{\frac{1}{2}}) \Gamma S \qquad (102)$$

To be satisfied, Eq. (91) demands that

$$E \Pi_1 < 0 \qquad (103a)$$

e.g.,

$$\text{sgn}(E) = -\text{sgn} \left(\frac{N_\eta^{\frac{1}{2}} N_D^{\frac{1}{2}}}{N_\eta^{\frac{1}{2}} - 1} \right) \qquad (103b)$$

that together with $\text{sgn}(E) = \text{sgn}(\beta_1)$ yields the following consequence: To have oscillatory behavior, we must satisfy

$$(D_f / D_t) > (v_f / v_t) \qquad (104)$$

where again f and t stand for from and to, a way of indicating how the surfactant is being transported from and to the volume. Condition (104) is a condition for overstability in Marangoni convection obtained some time ago by Sanfeld and collaborators.[24] However, contrary to their findings and due here to the role of surfactant adsorption at the interface, this condition is a necessary, albeit not sufficient, condition for overstability. Further constraints must be satisfied and are specified later.

Lucassen[8] introduced a complex elasticity modulus ε, which is related to our elasticity Marangoni number by the following relationship:

$$E = -\frac{\varepsilon^2 l^2 (k^2 + i\Omega/v_1)^{\frac{1}{2}}}{\eta_1 D_1} = \frac{\varepsilon l^2 (i\Omega/v_1)^{\frac{1}{2}}}{\eta_1 D_1} \qquad (105)$$

with $a = kl$ and $\omega = \Omega l^2/v_1$. The quantity k can be assumed to be smaller than the inverse of the viscous penetration length. Then, using these new variables, Eq. (101) reduces to

$$\varepsilon mk^2 + i\eta_1\Omega m^2 (v_1/D_1) \frac{[1 + (\rho_2\eta_2/\rho_1\eta_1)^{\frac{1}{2}}](1 + N(D_1/D_2)^{\frac{1}{2}})}{(v_1 D_2/v_2 D_1)^{\frac{1}{2}} - 1} = 0 \qquad (106)$$

with $m^2 = k^2 + i\Omega/v_1 \approx i\Omega/v_1$. Equation (78) is a generalization of the particular case discussed by Lucassen[8] and Van den Tempel and Lucassen-Reynders.[10]

Then using both Eqs. (101) and (102) we get

$$E_c = -\left[\frac{(N_D^{\frac{1}{2}} + N_\eta)(N_\rho^{\frac{1}{2}} N_\eta^{\frac{1}{2}} - 1)}{S\Gamma\Pi_1\Pi_2}\right]\left[\frac{(1 + N^{-1}N_D^{-\frac{1}{2}})S^2}{\Pi_1}(1 + N_\rho^{\frac{1}{2}} N_\eta^{\frac{1}{2}})\right] \qquad (107)$$

and the dispersion relation

$$\omega_c = a_c \frac{(N_D^{\frac{1}{2}} + N_\eta)(N_\rho^{\frac{1}{2}} N_\eta^{-1/2} - 1)}{S^{\frac{1}{2}}\Gamma\Pi_1\Pi_2} \qquad (108)$$

The value E_c is the minimal value of the elasticity Marangoni number needed to sustain the longitudinal interfacial convective oscillations of frequency ω_c. Using the fact that both ω and a must be positive numbers, we get from Eq. (80) that the following relationship must be satisfied:

$$v_f/v_t < 1 \qquad (109)$$

in order to have oscillations.

Thus, putting together Eqs. (104) and (109), we have the sufficient conditions to sustain the longitudinal waves. Our results generalize Lucassen's earlier finding, since he only considered damped motions. Here we see that with strong enough dissipation, i.e., for Marangoni numbers larger than E_c, these longitudinal oscillations can be sustained along the interface even if it is not deformed. On the other hand, our results delineate a more restricted domain of (surfactant-induced) interfacial oscillations than the domains reported by earlier authors.[19]

Nonlinear Capillary-Gravity Oscillations

Nonlinear Disturbance Equations

If we now consider the nonlinear extension of the problem posed in the subsection on disturbance equations, thus limiting ourselves to transverse oscillations, we have in dimensionless form

$$\nabla^* \underline{v} = 0 \tag{110}$$

$$\frac{\partial w}{\partial t} + u \frac{\partial w}{\partial x} + w \frac{\partial w}{\partial z} = -\frac{\partial p}{\partial z} + \nabla^2 w \tag{111a}$$

$$\frac{\partial u}{\partial t} + u \frac{\partial u}{\partial x} + w \frac{\partial u}{\partial z} = -\frac{\partial p}{\partial x} + \nabla^2 u \tag{111b}$$

and

$$\frac{\partial \theta}{\partial t} + u \frac{\partial \theta}{\partial x} + w \frac{\partial \theta}{\partial z} = w + P^{-1} \nabla^2 \theta \tag{112}$$

where we have explicitly indicated the full nonlinear disturbance systems, but once more in a two-dimensional problem. Besides, in order to rely our analysis to the simplest experimental conditions, e.g., the standard Bénard problem, we take β as a temperature gradient rather than a surfactant gradient. β is positive when the layer is heated from below. Thus, Eq. (112) is Fourier's heat equation. P is the Prandtl number ($P = \nu/\kappa$, with κ the thermometric conductivity of the liquid). As already pointed out in the Introduction, the results can be transferred verbatim to the case of surfactants.

The bc (4–8) at the open surface needs to be extended to the nonlinear case. Thus, we now have

$$\frac{\partial \xi}{\partial t} = w - u \frac{\partial \xi}{\partial x} \tag{113}$$

$$p - \frac{Bo}{CP} \xi + \frac{1}{N^3} \left[\frac{1}{CP} - \frac{M}{P} (\theta - \xi) \right] \frac{\partial^2 \xi}{\partial x^2}$$
$$= \frac{2}{N^2} \left[\frac{\partial w}{\partial z} - \left(\frac{\partial u}{\partial z} + \frac{\partial w}{\partial x} \right) \frac{\partial \xi}{\partial x} + \frac{\partial u}{\partial x} \left(\frac{\partial \xi}{\partial x} \right)^2 \right] \tag{114}$$

$$-\frac{M}{P} \left[\frac{\partial (\theta - \xi)}{\partial x} + \frac{\partial \theta}{\partial z} \frac{\partial \xi}{\partial x} \right] = \frac{1}{N} \left\{ \left(\frac{\partial u}{\partial z} + \frac{\partial w}{\partial x} \right) \left[1 - \left(\frac{\partial \xi}{\partial x} \right)^2 \right] + 4 \frac{\partial w}{\partial z} \frac{\partial \xi}{\partial x} \right\} \tag{115}$$

and

$$\frac{\partial \theta}{dz} = 0 \qquad (116)$$

We see nonlinear contributions like the second term in the right-hand side of the kinematic bc (113). Equation (116) prescribes the heat flux at the open surface, $N = (1 + |\partial \xi / \partial x|^2)^{\frac{1}{2}}$. Again the air is assumed to be passive and weightless with respect to the liquid.

Nonlinear Oscillator Equation

The simplest approach to the nonlinear problem posed earlier is the single-mode analysis, which is expected to be a useful description in a small enough neighborhood of the onset of overstability. Moreover, the more we move into low gravity, the larger the capillary length becomes, thus providing greater relevance to the single-mode approximation. On the other hand as already emphasized, transverse interfacial disturbances are expected to penetrate little in the liquid; the penetration depth depends on the wavelength and frequency excited and on the viscosity of the liquid. The latter assumption gives once more relevance to the potential flow approximation to the time-dependent convection, or in other terms, to the limitation of the study to the high-frequency motions only. Thus, for an arbitrary disturbance $f(x,z,t)$, we set

$$f(x,z,t) \approx f(z,t) \exp(iax) \qquad \text{and} \qquad \omega_0^2 = (Bo + a^2)a/CP$$

The latter is Laplace's law (potential flow). Using them, Eq. (113) becomes

$$\frac{\partial \xi}{\partial t} = w + \xi \frac{\partial w}{\partial z} \qquad (117)$$

On the other hand, Eq. (110) at $z = \xi$ is

$$\frac{\partial w}{\partial t} = -\frac{B_0 + a^2/N^3}{PC} a\xi + \frac{M}{PN^3} (\theta - \xi)a^3\xi$$

$$- 2a\frac{1 + a^2\xi^2}{N^2} \frac{\partial w}{\partial z} - a^2\left(1 + \frac{2a\xi}{N^2}\right) w + \left(1 - \frac{2a\xi}{N^2}\right)\frac{\partial^2 w}{\partial z^2} \qquad (118)$$

Note that neglecting the nonlinear terms and using the high-frequency approximation, $\omega_0 \gg a^2$, Eq. (118) yields, as expected, the (Laplace) harmonic oscillator equation (for zero damping):

$$\frac{d^2\xi_0}{dt^2} + \frac{Bo + a^2}{CP} a\xi_0 = 0 \qquad (119)$$

The zeroth-order (linear) disturbances are ξ_0,

$$w_0 = \frac{\partial \xi_0}{\partial t} e^{az} \tag{120}$$

and

$$\theta_0 \approx \xi_0 e^{az} + \frac{a}{\sqrt{2P\omega_0}} \left(\frac{1}{\omega_0} \frac{\partial \xi_0}{\partial t} - \xi_0 \right) \exp(\sqrt{2P\omega_0}\, z) \tag{121}$$

where ω_0 denotes the harmonic frequency in Eq. (119). The subscript zero describes the linear solutions.

Consideration now of the nonlinear terms in Eqs. (117) and (118) up to cubic terms leads to

$$\frac{d\xi}{dt} = w(1 + a\xi) \tag{122}$$

and

$$\frac{dw}{dt} = -\frac{Bo + a^2}{CP} a\xi$$

$$+ \frac{Ma^3}{P\omega_0 \sqrt{2Pg\omega_0}} \left(\frac{d\xi}{dt} - \omega_0 \xi \right) \left(-1 + 3a\xi + \frac{5}{2} a^2 \xi^2 \right)$$

$$- 4a^2[1 + 2a\xi(1 - 2a\xi)]w + \frac{3a^5}{2CP} \xi^3 \tag{123}$$

which, after reduction to a single differential equation, become

$$\frac{d^2\xi}{dt^2} + \delta \frac{d\xi}{dt} + [\omega_0^2 - \omega_0(\delta - 4a^2)]\xi$$

$$= -\omega_0^2 a\xi^2 + a \left(\frac{d\xi}{dt} \right)^2$$

$$+ (\delta - 4a^2) \left(\frac{d\xi}{dt} - \omega_0 \xi \right) 2a\xi - 8a^3 \xi \frac{d\xi}{dt}$$

$$- a^2 \xi \left(\frac{d\xi}{dt} \right)^2 + \frac{11}{2} a^2 \xi^2 (\delta - 4a^2) \left(\frac{d\xi}{dt} - \omega_0 \xi \right) + 16a^4 \xi^2 \frac{d\xi}{dt} + \frac{3a^5}{2CP} \xi^3 \tag{124}$$

where

$$\delta = \frac{Ma^3}{\sqrt{2}\, (P\omega_0)^{\frac{3}{2}}} + 4a^2 \tag{125}$$

At $\delta = 0$ we have overstability from the linear analysis. Positive (respectively, negative) values of δ account for subcritical (respectively, supercritical) motions.

For universality in the presentation it is useful to rescale both space and time. Thus, with $\zeta = a\xi$ and $\tau = \omega_0 t$, and using the high-frequency limit (here $\omega_0 \gg a^2 \approx 1$) with $\omega_0^2 \approx 1/CP$, Eq. (124) reduces to

$$\frac{d^2\zeta}{d\tau^2} + \Delta \frac{d\zeta}{d\tau} + \zeta = -\zeta^2 + \alpha\zeta^3 + \left(\frac{d\zeta}{d\tau}\right)^2 - \beta\zeta\frac{d\zeta}{d\tau} - \zeta\left(\frac{d\zeta}{d\tau}\right)^2 - \gamma\zeta^2\frac{d\zeta}{d\tau} \quad (126)$$

where $\Delta = \delta/\omega_0$, $\alpha = 3a^3/2CP\omega_0^2$, $\beta = 16a^2/\omega_0$, and $\gamma = 6a^2/\omega_0$. Thus, Eq. (126) is the simplest nonlinear equation for a dissipative oscillator describing the limit cycle oscillations of the air-liquid interface. We have checked that, indeed, Eq. (124), as well as Eq. (126), possess a limit cycle solution. This has been done both with the computer and using the time-derivative expansion procedure.[38] Recently, the authors have extended Eq. (126) to the space-modulated case, thus providing a set of Boussinesq-like[5] equations for the nonlinear evolution of the open surface.[39]

Limit Cycle Oscillations and Predictions for Ground and Microgravity Conditions

The time-derivative expansion procedure permits one to obtain in a perturbation scheme both the amplitude and the period of the oscillation in terms of the initial condition and thus to assess the stability of the limit cycle.[38] On one hand, it can be shown that the limit cycle bifurcates supercritically for Δ (or δ) negative (i.e., we have a supercritical Hopf bifurcation), and, on the other hand, one obtains the amplitude,

$$\zeta_{max}^2 = -\frac{2\delta}{3a^2} \quad (127)$$

For an effective gravitational acceleration of, say, $10^{-4} g$, with g the standard value on Earth, the predicted period of oscillation at the onset is, according to Eq. (124), of the order 1–2 min for most liquids. According to Eq. (126), the temperature gradient at overstability is of the order of a few Kelvins per centimeter for mercury and other liquids, including water. Table 1 provides specific predictions that show genuine microgravity relevance.

As already mentioned in the section on the case of a liquid open to air, onboard a spacecraft, in a low/microgravity environment, the crucial test of our predictions can be obtained by making an experiment with a water-alcohol solution or some other liquids with a minimum of surface tension vs temperature.[44–47] The suggested experiment is indeed the standard experimental case in Bénard convection. Before the minimum is reached, one expects steady polygonal cells (Bénard cells),

Table 1 Interfacial oscillations: predictions for the case of an open surface of a liquid layer heated from the air side

	Water		Mercury		Tin		$N_a NO_3$	
β, K/cm	5.85	1.9×10^3	441	1.39	55	1.8	5.2×10^3	16.54
Period, s	143	0.14	0.12	119.50	0.17	146	0.14	137.8
Penetration depth, cm	0.48	1.5×10^{-2}	0.5×10^{-2}	0.15	6×10^{-3}	0.18	1.8×10^{-2}	0.56
Gravitational acceleration	$10^{-4} g$	g	g	$10^{-4} g$	g	$10^{-4} g$	g	$10^{-4} g$

β is negative for liquids with no minimum (or maximum) in their surface tension. When the liquid has a minimum in its temperature dependence of the surface tension the sign of β is reversed, corresponding to heating from the liquid side. Note the relevance to spacebound low microgravity experiments.

whereas past the minimum, i.e., in the region where the surface tension of the liquid increases with increasing temperature, oscillations are predicted. Since Δ (or δ) contains the Marangoni number, the suggested experiment could provide not only a clear-cut test of our threshold conditions but, moreover, a verification of the predicted values for the nonlinear regime.[38]

Solitons Excited by the Marangoni Effect

Introduction

In the preceding sections we discussed the onset and eventual nonlinear sustainment of interfacial oscillations. These oscillations were either transverse or longitudinal waves. The former may be capillary-gravity waves that in a closed container are expected to develop as standing transverse periodic motions at the open surface of a liquid or at the interface between two liquids when the Marangoni effect is operating. In all of the cases considered, with the small viscous penetration depth or the high frequency limit, the approximations used amount tacitly to assume the "infinite" depth of the layer. Now we turn to purely traveling motions in the form of solitary waves with restriction to shallow layers, although quite a similar viewpoint to that guiding us in the preceding sections. However, here the ideal building block is not a harmonic oscillator but rather the nonlinear and dispersive Korteweg-de Vries soliton equation,[5-7] to which we add dissipation (viscosity and, say, heat diffusivity) and the Marangoni effect. We shall see that the latter effect may very well again trigger and eventually sustain the interfacial soliton along the open surface of a liquid layer heated from above (or below, according to the liquid used). As in the section dealing with nonlinear limit cycle oscillations we here restrict consideration to the standard Bénard problem, i.e., to the purely thermal Marangoni effect.

Dissipative KdV Equation

We consider a shallow horizontal liquid layer of thickness h initially at rest and subjected to a transverse thermal gradient. Disturbances upon the quiescent state obey the continuity and Navier-Stokes equations, to which we add to a first approximation either Fourier's heat equation or Fick's mass diffusion equation. These equations are supplemented with the corresponding nonlinear boundary conditions at the bottom and at the open surface. As was done earlier, for simplicity we shall restrict consideration to a two-dimensional geometry with x and z denoting the horizontal and vertical coordinates, respectively. β is the thermal gradient that again corresponds either to heat or mass diffusion (it is always positive when the heating is from the liquid side). Again, θ denotes either temperature disturbance or surfactant concentration disturbance, so that D is heat or mass diffusivity. The variables η and v still account for dynamic and kinematic viscosity, respectively; $\eta = \rho v$, with ρ the liquid density. All other quantities have the meaning assigned in earlier sections. The variable ∇^2 represents the Laplacian, which with the subscript Σ restricts its action to the open surface. The "shallow layer approximation" means that $h \ll l$, together with the assumption that amplitudes remain much smaller than h. Thus, we disregard the buoyancy effects in the Navier-Stokes equations. Here l is the "wavelength" or maximum horizontal extent of the interfacial deformation. For self-consistency in this section we recall the equations to be used. They are

$$\nabla * v = 0 \tag{128}$$

$$\frac{\partial}{\partial t} v + v * \nabla v = -\frac{1}{\rho} \nabla p + v \, \nabla^2 v \tag{129}$$

$$\frac{\partial}{\partial t} \theta = \beta w + \kappa \, \nabla^2 \theta \tag{130}$$

with the bc at the rigid insulating bottom $z = 0$,

$$w = u = 0 \tag{131}$$

and

$$\frac{\partial \theta}{\partial t} = 0 \tag{132}$$

and the bc at the surface $z = h + \zeta$,

$$\frac{\partial}{\partial t} \zeta = w - u \frac{\partial}{\partial x} \zeta \tag{133}$$

$$p = \rho g \zeta + 2\mu \frac{\partial}{\partial z} w - \sigma_0 \nabla_\Sigma^2 \zeta \tag{134}$$

$$\left[\frac{\partial}{\partial T} \sigma \right] \nabla_\Sigma (\theta - \beta \zeta) - \eta \left[\frac{\partial}{\partial z} u + \frac{\partial}{\partial x} w \right] = 0 \tag{135}$$

and

$$\frac{\partial}{\partial z} \theta = 0 \tag{136}$$

Note that we have considered the heat equation as a linear disturbance upon the nonlinear Navier-Stokes problem. The velocity $\underline{v}$ can be decomposed into its potential (φ) and rotational (ψ) parts:

$$v = \nabla \varphi + \nabla x \underline{\psi} \tag{137}$$

For the two-dimensional problem,

$$\underline{\psi} = \psi \underline{i} \tag{138}$$

Thus,

$$u = \frac{\partial \varphi}{\partial x} - \frac{\partial \psi}{\partial z} \tag{139}$$

$$w = \frac{\partial \varphi}{\partial z} + \frac{\partial \psi}{\partial x} \tag{140}$$

Noting that $\nabla^2 \varphi = 0$, we set

$$\varphi = \sum_{j=0}^{\infty} \frac{(-)^j}{(2i)!} z^{2j} \varphi_0^{(2j)} \tag{141}$$

On the other hand,

$$\psi = \sum_{i=0}^{\infty} z^i \psi_i \tag{142}$$

The simplest choices to fit the bc are

$$\psi_0 = 0$$

$$\psi_1 = \varphi_{0,x}$$

and $\psi_2 = \psi_3 = \cdots = 0$, or

$$\psi = z\varphi_{0,x} \tag{143}$$

The subscripts x, z, and t denote a derivative with respect to x, z, and t, respectively. Thus,

$$u = -\frac{z^2}{2} \varphi_0^{(3)} + \frac{z^4}{24} \varphi_0^{(5)} - \cdots \tag{144}$$

$$w = \frac{z^3}{6} \varphi_0^{(4)} - \frac{z^5}{120} \varphi_0^{(6)} + \cdots \tag{145}$$

Assume now that

$$\theta = \theta_0 + z\theta_1 + z^2\theta_2 + \cdots \tag{146}$$

Then using Eq. (130) and the bc it appears that even powers disappear, and $\theta_1 = \theta_3 = 0$. Thus,

$$\theta = z^5\theta_5 + z^7\theta_7 + \cdots \tag{147}$$

or else

$$\theta = z^5\left(\frac{7}{5}h^2 - z\right)\theta_7 + \cdots \tag{148}$$

Differentiating now Eq. (130) with respect to z gives $k\theta_{zzz} = \beta u_x$ at $z = h$. Then Eq. (147) becomes

$$\theta = z^5\left(\frac{7}{5}h^2 - z^2\right)\frac{\beta}{252h^2\kappa}\varphi_0^{(4)} + \cdots \tag{149}$$

Using the nonlinear kinematic bc (133),

$$\zeta_t = \frac{1}{6}h^3\varphi_0^{(4)} + \frac{1}{2}h^2\zeta\varphi_0^{(4)} - \frac{1}{120}h^5\varphi_0^{(6)} + \frac{1}{2}h^2\varphi_0^{(3)}\zeta x \tag{150}$$

Now using Eqs. (144) and (145), together with Eq. (135) in Eq. (129) taken at the open surface and defining $f = \varphi_0^{(3)}$ we have

$$\zeta_t = \frac{1}{6}h^3 f_x + \frac{1}{2}h^2\zeta f_x - \frac{1}{120}h^5 f_{xxx} + \frac{1}{2}h^2 f\zeta_x \tag{151}$$

and

$$f_t - \frac{2g}{h^2}\zeta_x + 2\frac{\sigma}{\rho h^2}\zeta_{xxx} + \frac{2}{h}\zeta f_t - \frac{1}{12}h^2 f_{txx} - \frac{1}{6}h^2 ff_x = 2v(f_{xx} + f/h^2) \tag{152}$$

Using now the bc (135), Eq. (152) becomes

$$f_t - \frac{2g}{h^2}\zeta_x + 2\frac{\sigma}{\rho h^2}\zeta_{xxx} + \frac{2}{h}\zeta f_t - \frac{1}{12}h^2 f_{xx} - \frac{1}{6}h^2 f_{xx}$$

$$= \frac{2v}{h^2}\left[\frac{4}{3}h^2 f_{xx} - \left(\frac{\partial\sigma}{\partial T}\right)\frac{1}{\rho vh}(\theta_x - \beta\zeta_x)\right] \tag{153}$$

For convenience we rescale the unknowns. With a the maximum value of ζ, we set

$$\theta = \beta h\Theta, \qquad \zeta = a\xi, \qquad t = \tau/C_0, \qquad x = ly, \qquad \text{and} \qquad f = \frac{aC_0}{h^3}v$$

with

$$C_0^2 = 2gh$$

Then Eqs. (151) and (153) become

$$\xi_\tau = \frac{1}{6} v_y + \frac{1}{2} \varepsilon \xi v_y - \frac{1}{120} \delta^2 v_{yyy} + \frac{1}{2} \varepsilon v \xi_y \tag{154}$$

$$v_\tau - \frac{2gh}{C_0^2} \xi_y + \frac{2\sigma h}{\rho l^2 C_0^2} \xi_{yyy} + 2\varepsilon\varphi v_\tau - \frac{\delta^2}{12} v_{\tau yy} - \frac{\varepsilon}{6} v v_y$$

$$= \frac{2v}{C_0 l}\left[\frac{4}{3} v_{yy} - \left(\frac{\partial\sigma}{\partial T}\right)\frac{\beta l h}{a\rho v C_0}\theta'_y\right] \tag{155}$$

with

$$\varepsilon = a/h, \qquad \delta = h/1 \qquad \text{and} \qquad \theta'_y = \frac{aC_0 h}{3k1} v_{yy}$$

Thus, recalling the definitions of the Bond and Marangoni numbers given earlier, and denoting $Re = C_0 1/v$, we obtain

$$\xi_\tau - \frac{1}{6} v_y - \frac{1}{2} \varepsilon \xi v_y - \frac{1}{2} \varepsilon v \xi_y + \frac{1}{120} \delta^2 v_{yyy} = 0 \tag{156}$$

and

$$v_\tau - \xi_y + \frac{\delta^2}{Bo} \xi_{yyy} + 2\varepsilon\xi v_\tau - \frac{\varepsilon}{6} v v_y - \frac{\delta^2}{12} v_{\tau yy} = \frac{2}{630Re}[840 + M]v_{yy} \tag{157}$$

For later convenience we rescale again τ and v. We set $\tau = (6)^{1/2}t$ and $v = -(6)^{1/2}V$. Then Eqs. (156) and (157) become

$$\xi_t + V_y + 3\varepsilon\xi V_y + 3\varepsilon V\xi_y - \frac{1}{20}\delta^2 V_{yyy} = 0 \tag{158}$$

and

$$V_t + \xi_y + 2\varepsilon\xi V_t + \varepsilon V V_y - \frac{\delta^2}{12}V_{tyy} - \frac{\delta^2}{Bo}\xi_{yyy} = \frac{\sqrt{6}}{315Re}[840 + M]V_{yy} \tag{159}$$

At zero order we have

$$V^{(0)} = \xi(V_t + V_y = 0)$$

so that the general solution can be assumed in the form

$$V = V^{(0)} + \varepsilon V^{(1)} + \delta^2 V^{(2)} + \gamma V^{(3)} \tag{160}$$

with $\varepsilon = a/h$ and

$$\gamma = \frac{\sqrt{6}}{315 Re}[840 + M]$$

Introducing Eq. (160) into both Eqs. (158) and (159) and subtracting them, the result is

$$\xi_t + \xi_y + \frac{5}{2}\varepsilon\xi\xi_y - \frac{30 - Bo}{60 Bo}\delta^2\xi_{yyy} = \frac{\gamma}{2}\xi_{yy} \qquad (161)$$

where ξ can be removed by a Galilean transformation.

Thus, we see that by setting $\gamma = 0$, Eq. (161) provides the Korteweg-de Vries equation.[5-7] The expression $\gamma = 0$ demands either $\nu = 0$, i.e., no viscosity, which is an irrelevant case in our analysis, or $M = -4$. The latter result indicates that for such a negative value of the Marangoni number the open surface of the liquid layer heated from the air side is excitable in the form of a KdV soliton. Whether or not the soliton is stable can only be decided by studying the role of the nonlinear part in Eq. (130) omitted here. However, what we can safely say is that due to the Marangoni effect $M = -840$ defines the onset of a soliton excitation in a quiescent shallow liquid layer subjected to a transverse thermal gradient. However, if M is positive, we are in the case of Bénard convection, and we know[18,19] that at $M = 80$ there is the onset of steady Bénard convection with polygonal (mostly hexagonal) cell patterns. Thus, again an experiment with a liquid layer heated from below could provide a clear-cut test of our prediction. Once more it is sufficient to operate with a liquid having a minimum in the surface tension vs temperature curve. Before the minimum we expect steady patterned convection (Bénard cells), and past the minimum, as seen in our study, the prediction is the solitary traveling wave. Note that, since the KdV equation corresponds to a genuine nonlinear excitation, the experimenter must strongly excite the liquid with, say, a sudden jump in the temperature gradient or with strong evaporation, as was done many years ago by Linde.[30] Another possibility is to mechanically excite the interface, and then with heat or mass transfer the Marangoni effect is expected to help sustain this excitation.

Acknowledgments

This chapter is based upon research sponsored by CICYT (Spain) Grant PB 86-651 and by an EEC Grant. Both authors acknowledge fruitful discussions with Ph. Drazin, J. K. Koster, H. Linde, A. Sanfeld, R. Sani, and P. Weidman and M. Adler, M. Hennenberg and A. N. Garazo. The first author wishes to acknowledge the hospitality of the CNLS at Los Alamos Laboratory and the Center for Low-Gravity Fluid Mechanics, University of Colorado at Boulder. The second author wishes to acknowledge the hospitality at the Service de Chimie Physique, Université Libre de Bruxelles.

References

[1]Lamb, H., *Hydrodynamics*, Dover, New York, 1932.

[2]Landau, L. D., and Lifshitz, E. M., *Fluid Mechanics*, Pergamon, Oxford, UK, 1959.

[3]Levich, B. G., *Physicochemical Hydrodynamics*, Prentice-Hall, Englewood Cliffs, NJ, 1962.

[4]Miller, C. A., and Neogi, P., *Interfacial Phenomena*, Dekker, New York, 1985.

[5]Whitham, G. B., *Linear and Non-Linear Waves*, Wiley, New York, 1974.

[6]Lamb, G. L., *Elements of Soliton Theory*, Wiley, New York, 1980.

[7]Drazin, P. G., and Johnson, R. S., *Solitons: An Introduction*, Cambridge Univ. Press, Cambridge, UK, 1989.

[8]Lucassen, J., *Transactions of the Faraday Society*, Vol. 64, 1968, pp. 2221, 2231.

[9]Lucassen-Reynders, E. H., and Lucassen, J., *Advances in Colloid and Interface Science*, Vol. 2, 1969, pp. 347–395.

[10]Van den Tempel, M., and Lucassen-Reynders, E. H., *Advances in Colloid and Interface Science*, Vol. 18, 1983, pp. 281–290.

[11]García-Ybarra, P., and Velarde, M. G., *Physics of Fluids*, Vol. 30, 1987, pp. 1649–1655.

[12]Velarde, M. G., García-Ybarra, P., and Castillo, J. L., *Physicochemical Hydrodynamics*, Vol. 9, 1987, pp. 387–392.

[13]Velarde, M. G., and Chu, X.-L., *Phys. Scripta*, Vol. T25, 1989, pp. 231–237.

[14]Chu, X.-L., and Velarde, M. G., *Physicochemical Hydrodynamics*, Vol. 10, 1988, pp. 727–737.

[15]Chu, X.-L., and Velarde, M. G., *Journal of Colloid and Interface Science*, Vol. 131, 1989, pp. 471–484.

[16]Sternling, C. V., and Scriven, L. E., *AIChE Journal*, Vol. 5, 1959, pp. 514–523.

[17]Velarde, M. G., and Normand, C., *Scientific American*, Vol. 243, 1980, pp. 92–108.

[18]Velarde, M. G., and Castillo, J. L., *Transport and Reactive Phenomena Leading to Interfacial Instability in Convective Transport and Instability Phenomena*, edited by J. Zierep and H. Oertel, Jr., Braun-Verlag, Karlsruhe, FRG, 1982, pp. 235–246.

[19]Legros, J. C., Sanfeld, A., and Velarde, M. G., *Fluid Sciences and Materials Science in Space*, edited by H. U. Walter, Springer-Verlag, New York, 1987, pp. 83–140.

[20]Velarde, M. G. (ed), *Physicochemical Hydrodynamics. Interfacial Phenomena*, Plenum, New York, 1988.

[21]Chu, X.-L., and Verlarde, M. G., *Journal of Colloid and Interface Science*, Vol. 127, 1989, pp. 205–208.

[22]Hennenberg, M., Sørensen, T. S., and Sanfeld, A., *Journal of the Chemical Society, Faraday Transactions, 2*, Vol. 73, 1977, pp. 48–56.

[23]Hennenberg, M., Bisch, P. M., Vignes-Adler, M., and Sanfeld, A., *Dynamics and Instability of Fluid Interfaces*, edited by T. S. Sorensen, Springer-Verlag, Berlin, 1978, pp. 227–259.

[24]Sanfeld, A., Steinchen, A., Hennenberg, M., Bisch, P. M., Van Lamsweerde-Gallez, D., and Dale-Vedove, W., *Dynamics and Instability of Fluid Interfaces*, edited by T. S. Sorensen, Springer-Verlag, Berlin, 1978, pp. 168–204.

[25]Sørensen, T. S., Hansen, F. Y., Nielsen, J., and Hennenberg, M., *Journal of the Chemical Society, Faraday Transactions 2*, Vol. 73, 1977, pp. 1589–1601.

[26]Sørensen, T. S., Hennenberg, M., and Hansen, F. Y., *Journal of the Chemical Society, Faraday Transactions 2*, Vol. 74, 1978, pp. 1005–1025.

[27]Sørensen, T. S., *Dynamics and Instability of Fluids Interfaces*, edited by T. S. Sørensen, Springer-Verlag, Berlin, 1978, pp. 1–74.

[28]Linde, H., and Loeschcke, K., *Chem. Ing. Tech. (GDR)*, Vol. 39, 1966, pp. 65–72.

[29]Linde, H., and Kunkel, E., *Zeitschrift fuer Warme-Stoffubertr.*, Vol. 2, 1969, pp. 60–66.

[30]Linde, H., and Schwartz, P., *Chem. Tech (GDR)*, Vol. 26, 1974, pp. 455–461.

[31]Linde, H., *Dynamics and Instability of Fluid Interfaces*, edited by T. S. Sørensen, Springer-Verlag, Berlin, 1978, pp. 75–119.

[32]Reinchenbach, J., and Linde, H., *Journal of Colloid and Interface Science*, Vol. 84, 1981, pp. 433–443.

[33]Linde, H., *Convective Transport and Instability Phenomena*, edited by J. Zierep and H. Oertel, Braun-Verlag, Karlsruhe, FRG, 1982, pp. 256–296.

[34]Velarde, M. G., and Chu, X.-L., *Physics Letters A*, Vol. 131, 1988, pp. 430–432

[35]Velarde, M. G., and Chu, X.-L., *Il Nuovo Cimento D*, Vol. 11, 1989, pp. 707–716.

[36]Chu, X.-L., and Velarde, M. G., *Il Nuovo Cimento D*, Vol. 11, 1989, pp. 1615–1629.

[37]Chu, X.-L., and Velarde, M. G., *Il Nuovo Cimento D*, Vol. 11, 1989, pp. 1631–1643.

[38]Chu, X.-L., and Velarde, M. G., *Physics Letters A*, Vol. 136, 1989, pp. 126–130.

[39]Velarde, M. G., and Chu, X.-L., *Phase Transitions in Soft Condensed Matter*, edited by T. Riste and D. Sherrington, Plenum, New York, 1989, pp. 139–143.

[40]Chu, X.-L., and Velarde, M. G., *Physical Review A*, 1990, (submitted for publication).

[41]Velarde, M. G., Chu, X.-L., and Garazo, A. N., *Physica Scripta*, 1990 (submitted for publication).

[42]Velarde, M. G., and Chu, X.-L., *Entropie*, Vol. 26, 1990, pp. 82–88.

[43]Velarde, M. G., and Chu, X.-L., *Interfacial Instabilities*, World Scientific, London (to be published).

[44]Vochten, R., and Petré, G., *Journal of Colloid and Interface Science*, Vol. 42, 1973, pp. 320–327.

[45]Motomura, K., Iwanaga, S.-I, Hayami, Y., Uryu, S., and Matuura, K., *Journal of Colloid and Interface Science*, Vol. 80, 1981, pp. 32–38.

[46]Desré, P. J., and Joud, J. C., *Acta Astronautica*, Vol. 8, 1981, pp. 407–427.

[47]Legros, J. C., Limbourg-Fontaine, M. C., and Petré, G., *Acta Astronautica*, Vol. 11, 1984, pp. 143–150.

Chapter 4. Gravity Modulation Effects

Gravity Jitters: Effects on Typical Fluid Science Experiments

R. Monti*

Institute Umberto Nobile, University of Naples, Naples, Italy

Introduction

THROUGHOUT this paper we will denote by the symbol $\underline{g}$ the resultant acceleration vector ($\underline{a}$) divided by the reference acceleration of gravity ($g_0 = 9.81$ m/s^2) or, equivalently, the ratio of the resultant ($\underline{R}$) of the body forces acting at a distance (gravity, electrostatic, electromagnetic) on the body of mass m plus the d'Alembert forces (inertia forces) divided by the reference weight of the body at sea level (mg_0) (for a motionless body the g level on ground is 1, by definition):

$$\underline{g} = \underline{R}/mg_0 = \underline{a}/g_0 \tag{1}$$

The knowledge of the g field is very important on orbital and suborbital platforms dedicated to microgravity missions (that should be designed to limit the value and the occurrence of local $\underline{g}$). A force $\underline{F}$ on the platform applied at time t and at point X,Y,Z induces a local $\underline{g}(t,x,y,z)$ at the point where the experiment takes place. A dynamic analysis considering all sorts of links (elastic, damping) between the platform elements would allow $\underline{g}(t,x,y,z)$ to be evaluated from $\underline{F}(t,X,Y,Z)$. Many points exist at which disturbance forces ($\underline{F}$) can be applied, and many types of the vector forces exist; consequently many types of g disturbances can be produced (steady, periodic, pulse type, or a combination thereof). Each platform, orbiting at a given height, is characterized by some residual quasisteady acceleration due to external sources (aerodynamic drag, solar pressure, solar wind) and

*Professor, Department of Aerodynamics.

to the position of the facility on the platform, i.e., the gravity gradient (only facilities positioned in the platform center of mass would experience a complete balance between gravity and inertia forces). Pulsed actions may result from impacts with foreign bodies (dockings, meteorites). Station-keeping and stabilizing maneuvers (or waste dumpings) are responsible for transient accelerations (g jitters). Internal causes (machinery and/or crew activities) are typically of random character; these internally generated g jitters (dislocation of masses on the platforms) are characterized by zero time average values.

To summarize, the production sources can be classified in Refs. 1–4:

1) external actions (aerodynamic forces, solar pressure, solar wind, meteorite impacts, dockings);

2) mass ejection from the platform (rockets firings, mass dumpings); and

3) movements of masses inside the platform (crew motion, alternating and/or rotary motion of machineries).

In terms of their characteristics and of their possible effects on the microgravity experimentation, the g disturbances are more appropriately divided into the following:

1) steady (or quasisteady) g fields;

2) pulsed accelerations (single pulses or compensating pulses); and

3) periodic, zero time average g jitters.

The first type of accelerations can be defined as the residual g level, characteristic of each platform and its orbit and of the position of the experimental facility. The other two types of unsteady accelerations are somehow controllable and will be referred to as g jitters. Most of the time the experimentation in microgravity is motivated by the assumption that steady-state (or quasi-steady-state) conditions prevail. This assumption is unfortunately not often verified because reasons for unsteadiness exist, such as the following:

1) disturbance time compared with experiment characteristic times;

2) time needed from cause to the effect to take place;

3) time needed for the effects to decay into the "undisturbed" steady state; and

4) stability thresholds related to the onset and the decay of different regimes.

Steady g fields are amenable of rather simple analyses.[5–8] Steady-state g levels, coupled with finite-duration processes, may give rise to quasisteady conditions if the time needed to induce the distortions is shorter than the process characteristic time; however, in many cases the effect of a given steady g level is not constant during the process itself but depends on the process stage. When dealing with internally produced periodic g jitters, transient effects occur at the start of the disturbance and at its termination (decay time needed to go from the periodic to the undisturbed conditions). Things are more complicated when there are more than one pulse-type disturbance (say, with rectangular or triangular time profiles) because of the possible complex interactions between subsequent pulses. Because of the rather long transient times that may characterize the diffusion of

momentum, energy and, especially, species, the effect of a single perturbation tends to last much longer than the duration of the pulse. The
duration of these transient phenomena are also related to the undisturbed
(residual) microgravity conditions; the strong reduction of the residual
gravity also implies the suppression of restoring forces, resulting in longer
decay times. As a consequence of great relevance, it might well be that few
g pulse occurrences will prevent steady conditions ever to be reached.
Furthermore, subsequent pulses may induce enhancements of the disturbances and/or of the phenomena that take place because of these disturbances. Another point of major concern arises when space processings are
examined. It is difficult to assess and quantify the influence of a limited
time disturbance on a final product. There is no doubt that if few
disturbance events of a certain intensity take place the space "product" will
be affected to an extent that depends not only on the level of the
disturbance but also on the ratio of the total duration of the disturbances
over the process duration. These last may be assessed in terms of the
distortions in the velocity, temperature, and concentration fields; the duration of these distortions may well exceed the duration of the cause (g-jitter
duration).

In conclusion, in the g-jitter problem the time of occurrence of the pulse
(t_p), pulse duration (Δt), time interval between subsequent pulses (t_d), and
experiment duration (τ_e), together with the g field (g level and orientation),
all play important roles in the evaluation of the effect on each type of
process. Because of the complex picture and the great number of parameters, the evaluation of the effects of g jitters requires that some reference
scenario be assumed, in which the role of the possible g disturbances is
identified.

Reference Scenario for the Microgravity Environment
For each platform the reference state can be assumed to be the ideal
low-gravity environment corresponding to the residual g of the platform;
this represents the best possible microgravity environment (MGE) (non-
avoidable residual g due to external forces on the platform). The reference
scenario that is assumed throughout this paper consists of the following:
1) The experiment motivation is the strong suppression of the convective
flows in nonuniform temperature and/or concentration fields.
2) Gravity jitters represent a disturbance of the existing MGE.

In the following we limit our discussion to stable regimes that are
monotonically dependent on the g disturbances; within small ranges of
the causes (C) a linear dependence with the effects (E) can always be
written as

$$E = KC \qquad (2)$$

The effect (E) is no longer linearly dependent on the cause (C) near
stability limits; for example, transition from a purely diffusive regime to a
convective one through stability boundaries may indeed cause very large

278 R. MONTI

disturbances to the basic (diffusive) conditions. The *g* jitters are responsible, in nonuniform density fluid media, for buoyancy forces that induce velocity fields, displacements (e.g., sedimentations), and alteration of temperature and concentration fields. Each of these effects can be assumed to be a "distortion" of the basic reference conditions; the evaluation of these distortions is typically performed by comparing the induced velocities, displacements, temperatures, and concentrations with the corresponding parameters typical of the undisturbed microgravity process that takes place in absence of the *g* jitter.

Modelings play an essential role in the evaluation of the effects induced by residual *g* and by *g* jitters and help set up an appropriate scenario. With reference to Fig. 1, three types of interactive modelings are needed that all together are able to evaluate the effects on the process of a force applied at a certain location of the platform and a certain time. The reference space product may be taken to be either that corresponding to 0 *g* (ideal conditions) or that obtainable at the undisturbed platform conditions (optimum conditions in absence of any *g* jitter). The outputs of a model are used, together with other data, as inputs to the next model.

Fluid dynamic modelings (FDM) that compute velocity, temperature, and concentration fields are the only ones needed when fluid science experimentation is being considered[9]; they must be used in conjunction with process modelings, when dealing with MS, LS, or ES experimentations that lead to

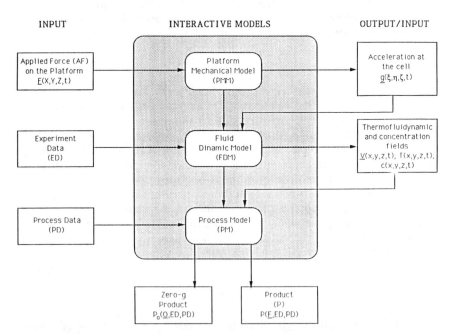

Fig. 1 Models necessary to evaluate the *g*-jitter effects in microgravity experimentation.

a product. If numerical FDM's are available (see Fig. 1), then it is possible
to analyze a number of problems:

1) Check that the expected MGE's offered by the different platforms are
suitable for the selected space processings.

2) Compare the effects of the different g disturbances on the final
products and identify what are the types of avoidable disturbances that
most likely unpair the space processing.

3) Try to establish "equivalence criteria" among different types of the
disturbances (steady g level, periodic, pulse, random) that help specify the
maximum allowable g jitter generated by space equipment operating on a
microgravity platform.

4) Establish the criteria for the additivity of the effects.

5) Identify the tolerability limits for different types of g jitters that are
applied simultaneous to the residual g level.

The modelings can be of different complexity; they range from a (rather)
simple order-of-magnitude analysis (OMA) type[5,6] to complex numerical
modeling (e.g., three-dimensional, unsteady).[9-11] To avoid time-consuming
computations for each combination of 1) the type of disturbance, 2) the
geometry and the fluid, and 3) the process, it would be suitable to perform
parametric computations for each category of disturbances and try to
evaluate the effect on the space processing from the results of these
computations. It would also be very helpful if some criteria could be
found, validated by numerical computations, that aid in finding guidelines in establishing tolerability limits for different classes of microgravity
experimentations.

Equivalence Criteria and Tolerability Limits

Restricting our attention to g jitters and assuming that additivity is
allowed, justified by the small values of the g disturbances and by the
linearization of the equations, then one would be able to analyze any
acceleration time profile $g(t)$ as a combination of different types of the
disturbances (steady, pulsed, periodic) and write that many simultaneous g
jitters (g_i) be equivalent to the summation of the single effects (E_i):

$$E\left(\sum \underline{g}_i\right) = \sum E_i(\underline{g}_i) \tag{3}$$

This assumption would allow a much easier use of the modelings and avoid
the need to run case-by-case combinations, as shown in Fig. 1. With
assumption (3) one could decompose any general acceleration time profile
$g(t)$ into three basic disturbances given earlier and then sum up the effects.
For instance, the velocity field, which results from a complex g disturbance,
could be evaluated by the summation of the fields induced by the component
disturbances (limited to small disturbances with no cross effects). Linear
analysis and additivity of the effects are unlikely to hold out all of the
possible conditions, especially for pulse-type disturbances. Distortions in the
TFD fields are induced during the application of the g-level distur-

bances and after termination of these disturbances. The presence on each microgravity platform of residual (unavoidable) g levels makes the tolerability to g jitters strongly dependent on the platform type and on the flight conditions. A case-by-case numerical analysis is probably necessary to evaluate the effects of transient g disturbances that act together with the steady g levels encountered in a space platform (say, $g = 10^{-6}$). In order to avoid specific computations of $E_i(g_i)$ for all possible types of disturbances, one would like to identify only a few types and resort, if possible, to equivalence criteria that state

$$E_j(\underline{g}_j) = E_i(\underline{g}_i) \tag{4}$$

Additivity [Eq. (3)] and equivalence criteria [Eq. (4)] (if at all possible) would substantially reduce the effort of the evaluation of the effects of g jitter on microgravity experimentation.

The identification of the maximum g levels that are tolerated by the different classes of experiments serves a number of purposes:

1) selection of the most appropriate platform on which the experiment should be embarked;

2) selection of the appropriate location (and/or of the orientation) of the experimental facility (e.g., distance from the center of mass) on each platform; and

3) identification of the specifications to the manufacturers of the maximum disturbances that any machinery on board may produce without impairing the microgravity experimentations.

In order to assess any tolerability limit,[4,8] as the maximum g jitters "bearable" by the experiment, a comparison should be made between the disturbed and the undisturbed products to identify the disturbance levels above which no benefits can derive from space processing. When dealing with periodic and/or pulse g disturbances, one must try to assess some criteria that compare the effects induced by these disturbances with the unavoidable (steady) g levels existing on the platform under consideration (undisturbed conditions). In fact, it is useless to eliminate internally produced disturbances if their effects are negligible compared with the steady acceleration caused, for instance, by aerodynamic drag or by gravity gradients. The same interacting modelings code shown in Fig. 1 could be also employed to evaluate the difference between the products obtained in real cases and those in the undisturbed conditions; in this case $\underline{F}(X,Y,Z,t)$ would only be the external steady (or quasisteady) forces applied to the platform.

Tolerability to steady g levels has been analyzed and criteria have been provided[6,8]; the problem is reduced at the identification of maximum effect (E) (e.g., described in terms of convective flowfields) that could be tolerated by a space processing. In particular, the maximum values of the residuals thermal Peclet and concentration Peclet numbers that would impair the process are identified. A similar analysis was made for simple sinusoidal g jitters: At each frequency a maximum g level was found as a tolerability limit, and a plot $g_{mx}(f)$ was identified for each category of experimentation; these plots (see, for instance, Fig. 17) are being given by the agencies as platform specifications. However, this information is not very helpful even if only two simultaneous sinusoidal g jitters are applied at different frequencies. For instance, one cannot be sure that the resulting flow distortion would be tolerable when the two periodic disturbances are both below the tolerability limits. The same uncertainty exists for periodic disturbances composed by different harmonics (as a result of a Fourier series expansion). One would be able to assess the tolerability of the experiment to two simultaneous disturbances if one could make use of an equivalence criterion stating that a disturbance level, at one frequency $[g_i(f_i)]$, is equivalent to another level at a different frequency $[g_{eqi}(f)]$. In that case one would write

$$\underline{g}_i(f_i) = \underline{g}_{eqi}(f); \qquad \underline{g}_j(f_j) = \underline{g}_{eqj}(f) \qquad (5)$$

and one could read the value $g_{mx}(f)$ and check that this is less than $(g_{eqi} + g_{eqj})$ [assuming that additivity of the effects hold; Eq. (3)].

For single-pulse g disturbances or for a number of subsequent pulses, the concept of tolerability can be borrowed from medicine: the "tolerable dose." Similar to what happen with X-rays, for which a threshold exists on the value of the product of the X power of the radiation and the time of exposition, beyond which a specific radiation becomes harmful for health, one may also think that every process (be it a material science or life science) could be characterized by a maximum tolerable dose (g level × duration time, for rectangular type pulses) beyond which the process is spoiled or is not worth being performed in space.

This criterion might sound very rough and could not be applicable in many instances especially when different "shapes" (or time profile) pulses are compared. However, during the time following a pulse disturbance, a near-field and a far-field distortion of the basic temperature, concentration, and fluid-dynamic fields (TCFD) can be identified, depending on the time elapsed from the application of the disturbing force. Typically the far field, which occurs after a sufficient time following the disturbance has elapsed (in practice at a time sufficiently larger than the characteristic momentum transfer time, L^2/ν), does not "remember" the details of the g disturbance. The concept of dose, as it will be shown by numerical results, applies fairly well to the far-field effects of the disturbance. The dose concept is also justified by the fact that the overall effect on a space product (or its distortion δP) is likely to depend not only on the distortion of the thermofluid-dynamic (and concentration) fields (induced by the g level) but also on the duration of the disturbing actions. Therefore, the "effective" distortion in the TCFD field (δF) (i.e., its time-integrated effect) might be related, in a first approximation, to the time integral of $\underline{g}(t)$ by

$$\delta F = f_F[g(t)] = U_D G \qquad (6)$$

where

$$G = \int_t^{t + \Delta t} |g(t)| \, dt$$

where Δt is the duration of the g disturbance that occurs at time t, the integral G is defined as the g dose, and U_D is the thermofluid-dynamic distortion caused by a unitary dose disturbance [applied in the same direction of $\underline{\underline{g}}(t)$]. The product distortion (δP) is assumed to depend on the thermofluid-dynamic effective distortion (δF). As in the medical practice the concept of dose implies also that there might be background values (residual g levels) that are not harmful because they are below a threshold level; at the same time, this background level may reduce the value of the tolerable δF. The preceding considerations gave the guidelines for the numerical analyses performed in Refs. 4 and 12 and were validated by the numerical results obtained for the study cases shown in the next section.

Study Cases

As mentioned earlier, the g-jitter effects on microgravity experiments cannot be generalized but should be referred to a specific class of experiments. The effects should be judged on the basis of the difference of the products obtainable in the presence and absence of g jitters; this implies that process modelings be available that link the TCFD distortions to the product features. If reference is made to specific processes, the results will be limited to that process only. It is therefore better to look at the effects of the g jitters on the TCFD fields for the experiments that are in greater demand of the lowest g fields. Previous works[6-8] have shown that the largest g-jitter disturbances are to be expected for the experiments that try to obtain purely diffusive conditions (thermal, concentration, or both) in otherwise quiescent fluid media.

Two study cases are considered as reference experiments: the first deals with thermal diffusion processes (as in some experiments of solution crystal growth), and the second deals with species diffusion processes (such as those occurring in experiments aimed at the measurement of the diffusion coefficient in liquids). An order-of-magnitude analysis (OMA) was made; in the OMA, first proposed by Napolitano,[5] the evaluation of the g-jitter effects consists of the comparison of the ideal (0 g) transport rates (of heat and masses) to the convective transport rates induced by g disturbance.[6] For energy and mass transport the ratio of the corresponding terms in the energy or species diffusion equation is the thermal Peclet number (VL/α) and the diffusion Peclet number (VL/D) that can be interpreted as the ratio between a reference velocity (V) and a diffusion (thermal or species) velocity. When convections are "created" by a steady gravitational field, the order of magnitude of the reference velocity becomes

$$V_g = g_0(g\Delta\rho L^2)/\mu \tag{7}$$

and the two Peclet numbers become the Rayleigh numbers for heat and mass diffusion:

$$Ra_T = g_0(gL^3\Delta\rho)/(\mu\alpha) \tag{8}$$

$$Ra_c = Ra_T/L_e = g_0(gL^3\Delta\rho)/(\mu D) \tag{9}$$

The first of these two numbers is relevant in heat diffusion (e.g., in measurements of thermal conductivity), and the second is relevant to mass diffusion (e.g., in the measurement of the diffusion coefficient); they are both relevant when energy and mass diffusion take place at the same time (e.g., in crystal growth from a solution or from a vapor). The ratio of these two numbers (i.e., the Lewis number D/α) gives an idea of the relative importance of the heat and mass diffusion processes in that fluid. The Lewis number for aqueous solutions is of the order 10^{-2} (liquid metals will have values of the order of 10^{-4}); typical values of the diffusion coefficient in liquids are $D = 10^{-5}$ cm^2/s.

The OMA shows that steady-state g level, which would cause a disturbance corresponding to a maximum tolerable value of the Rayleigh number $(Ra)_{mx}$ reads[8]

$$g_{mx}(Ra)_{mx}^{-1} = \mu\varphi/(g_0\Delta\rho L^3) \qquad (10)$$

where φ is the general diffusivity (α for heat and D for mass), and $(Ra)_{mx}$ is the corresponding maximum thermal or mass diffusion Rayleigh number (Ra_T or Ra_c). In liquids it is likely that the value of the maximum tolerable g level is rather low for mass diffusion processes, compared to that for heat diffusion processes. To avoid wall effects, one likes to utilize reactors (or pipes) of relatively large dimensions (L) that will imply smaller tolerability to g disturbances [Eq. (10)]. More specifically, the two cases studied in Refs. 4 and 12 deal with the following: a) crystal growth from a solution in the presence of temperature gradients responsible for buoyant forces (study case 1), and b) a diffusion coefficient measurement in isothermal conditions (study case 2).

The geometry of study case 1 is sketched in Fig. 2: A liquid specimen fills a cylindrical container (radius R, length L) whose boundaries are maintained at a constant temperature T_b. In a "cold finger" crystal growth experiment (e.g., from aqueous solutions), a crystal seed is located at the center of the cylinder; this is simulated by holding the central region of the cell at a constant temperature T_s ($T_s < T_b$). In the case of zero-gravity conditions ($g = 0$), a purely diffusive temperature (quasi-steady-state) conditions are established that correspond to a quiescent flow regime (almost spherical symmetry). In the presence of g levels (either residual gravity or g jitter), directed along the cell axis, the flow and temperature fields are distorted due to the induced convection caused by buoyancy forces. The time evolution of the disturbances, compared with the zero-gravity quiescent state, are analyzed together with the relaxation processes that take place after the g disturbance is over and that bring back the initial undisturbed, quiescent conditions.

Study case 2 is typical of an experiment envisioned for the measurement of the diffusion coefficient in liquids at isothermal conditions. The geometry cosidered is shown in Fig. 3. It consists of a two-dimensional cell: At the two sides of the diaphragm the liquid contains different concentrations of the diffusing species. At time $t = 0$ the diaphragm is eliminated and the diffusion process takes place. A concentration profile will evolve with time

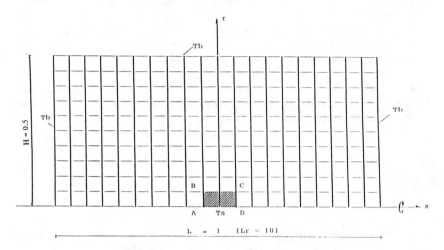

Fig. 2   Mesh and grid points for the simulation of crystal growth from solution.

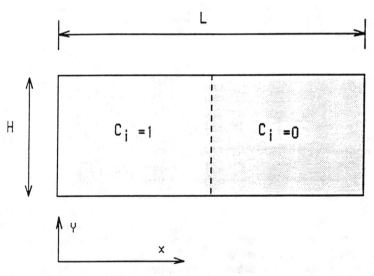

Fig. 3   Boundary and initial conditions for the measurement of the diffusion coefficient in liquids.

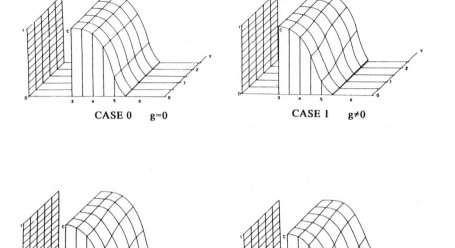

CASE 0   g=0   CASE 1   g≠0

CASE 2   g≠0   CASE 3   g≠0

Fig. 4   Concentration profiles at $0\,g$ and in the presence of g disturbances applied along y $(g = 0,\ 10^{-6},\ 10^{-5},\dots)$.

toward both sides of the cell; eventually, the concentration attains a constant value (along x and y). If the gravity is zero and there are no wall effects, the concentration distribution will be symmetrical with respect to the centerline, and an analytical solution will apply before the concentration "wave" reaches the end plates; the value of the diffusion coefficient can be correlated with the time profile of the concentration along the ampoule (or of the axial slope of the concentration in the midsection). If the residual gravity is not zero (e.g., when a g-jitter pulse is applied), the concentration field will be distorted (Fig. 4), and the accuracy of the measurement of the diffusion coefficient will be greatly affected (the concentration field is no longer correlatable with analytical solutions).

    Without going into the details of the full set of the Navier-Stokes equations, the numerical analysis, and the numerical codes used to solve the equations (for these points see Refs. 10 and 11), it is appropriate here to recall only the diffusion equation in two-dimensional, unsteady conditions (in the energy equation the viscous dissipation term is neglected). In nondimensional form this equation, which applies to both the study cases, reads

$$Pe\left(St_r\frac{\partial f}{\partial t} + u\frac{\partial f}{\partial x} + v\frac{\partial f}{\partial y}\right) - \left(\frac{\partial^2 f}{\partial x^2} + \frac{\partial^2 f}{\partial y^2}\right) = 0 \qquad (11)$$

where f is the variable associated with the diffusing quantity (either temperature T or concentration c), and $Pe = VL/\varphi$ is either the thermal Peclet number ($\varphi = \alpha$; $Pe = VL/\alpha$) or the concentration Peclet ($\varphi = D$; $Pe_c = VL/D$). The Stroual number $Str = L\omega/V$ is the parameter that measures the importance of the unsteady term with respect to the diffusive one (ω is the characteristic frequency of the disturbance). The values of the reference quantities V, L, D, and α are related to the experiment considered. The assumptions made in the course of the studies include the following: 1) Newtonian, nonmicropolar, double-component fluid; 2) constant transport properties; 3) Boussinesq approximation for the density; and 4) two-dimensional geometry and two-dimensional flow and concentration fields. The nondimensional general Navier-Stokes equations are expressed in terms of the vorticity and of the concentration c (or of the temperature T), integrated with a finite-difference scheme and solved by a semi-implicit technique.[4]

Results of the Fluid Dynamic Modelings

For both of the study cases extensive numerical experimentation is reported in Refs. 4 and 12 by considering the most usual g-jitter types encountered on microgravity platforms. In particular, five cases are examined:

1) constant residual $-g$ ($g = 0$, $g = 10^{-6}$, $g = 10^{-5}$) for the computation of the reference states;

2) single rectangular pulse for the simulation of thrusters firing (or of impact with external bodies) [this type of g jitter is characterized by the time of occurrence from the beginning of the process (tp), the time of duration (Δt), and the value of its amplitude (g)];

3) compensating double pulses for the simulation of astronaut steps[13] (or a mass dislocation) characterized by the shape of each pulse plus the time interval between the positive and negative pulse of the same dose, defined in Eq. (6);

4) periodic alternating rectangular pulses to simulate rotating machineries [the effect of this disturbance is characterized by the time of the start of the periodic disturbance (tp), by the duration of each single pulse (Δt), by the period (T), and by the value of the amplitude (g)]; and

5) sinusoidal pulses (from rotating machineries) characterized by the time of start of their occurrence (tp) by their period (T) (or, equivalently, by their frequency, $f = 1/T$) and by their amplitude.

The values of the g jitter assumed in the numerical experiments are taken from the typical values measured on Spacelab mission[1] and on the Salyut[3] for single pulse and for compensated double pulse. Their range is $10^{-5} \leqslant g \leqslant 10^{-3}$. The selection of the types of pulses was motivated by the need to assess the effects of g jitter on the experiment results, validate the dose concept, and propose some equivalence criteria, as mentioned earlier.

Study Case 1: Simulation of Solution Growth[4]

The liquid properties (water solution) around the crystal are $Pr = 6.9$, $v = 10^{-2}$ cm²/s, $\alpha = 1.45 \times 10^{-3}$ cm²/s, and $\beta r = 5.10^{-4}$ 1/K. A reference temperature difference is assumed to exist between the walls and the crystal surface: $\Delta(T)_r = 16$ K. A mesh with 11×21 grid points (11 along the radial y axis and 21 along the axial x axis) was used in the finite-difference scheme (Fig. 2). Most of the runs were made by considering a crystal growth from an aqueous solution that fills a cylinder with a height of 10 cm and a radius of 5 cm. It was assumed that concentration gradients did not create density gradients. For such cases the reference speed was $V_r = 10^{-3}$ cm/s and the reference time was $tr = 10^{-4}$ s.

The evaluation of the effects of the g jitters was made with reference to the distortions induced to the thermal and velocity fields.

During the application of the g disturbance, convective velocities are induced inside the cell; these values evolve in time and depend on the grid position. Two distortion parameters are suggested for an evaluation of the fluid dynamic disturbance with respect to the reference quiescent state (both of these parameters are zero at 0 g):

1) V^*_{max}, the maximum value of the nondimensional velocity induced in the field by the effect of the buoyancy; and

2) Ψ^*_{max}, the maximum value of the nondimensional stream function induced in the field by the effect of the buoyancy.

In particular, V^*_{max} indicates the maximum local disturbance effect, whereas Ψ^*_{max} is proportional to the global disturbance since it represents the amount of recirculation mass flow rate induced by buoyancy in the closed container.

The effect of the g disturbance on the thermal field is more critical because it depends on the space processing scopes; as a first attempt, similar to what was proposed for the flowfield, global and pointwise disturbance parameters are defined as

Maximum temperature distortion (D_m):

$$D_m = \max_i |(\Theta - \Theta_0)_i| \qquad (12)$$

where Θ is the local nondimensional temperature at a given time, and Θ_0 is the local nondimensional temperature at the reference quiescent field conditions $(g = 0)$. D_m is the maximum value (considering all of the grid points) of the nondimensional temperature difference between the temperature at a given time and the temperature that occurs at steady state (reference quiescent conditions, at $g = 0$, at the same field point).

Overall temperature distortion (D_T) (over the N grid points) is defined by

$$D_T = \sum_{i=1}^{N} |(\Theta - \Theta_0)_i| \qquad (13)$$

The identification of these disturbance parameters as the effects of the g disturbance is an essential step for the solution of the problem of the equivalence criteria [Eq. (4)]. Probably neither the value of D_T nor that of D_{max} is that which best represents the g-disturbance effect on the microgravity processing being examined. For instance, heat flux uniformity over a growing crystal probably has a more direct relevance over the quality of the crystal (or a uniform mass flux of a solute over a crystal being grown). In that case the disturbance parameter could be defined accordingly in terms of heat and species fluxes. However, careful evaluation of the heat flux variation and its distortion on the seeds surface is out of the scope of the present analysis because it is related to a specific process.

The values of the two Grashof numbers ($Gr = g_0 g \, \Delta\rho L^3/\mu\nu$) $Gr = 80$; $Gr = 800$, used in the computations, correspond to the two reference states $g = 10^{-6}$ and $g = 10^{-5}$. The corresponding thermal disturbance parameters are $D_m = 0.023$ and 0.202, $D_T = 0.49$ and 4.6. The computed disturbance parameters are almost proportional to the values of the level of the disturbances. This is true only for the assumed small (and steady) g levels. The evaluation of the disturbances, due to the residual gravity field (steady g levels), is important per se and for a comparison with the effects due to other transient disturbances. The value of the disturbance parameters, which correspond to the (steady) residual gravity levels, may, in fact, define a threshold limit for the acceptability in manned space station investigations for all of the g jitter generating the same values of the distortion parameters. From the tolerability point of view it is obvious that g-jitter disturbances, which generate distortion parameters having a value of the above order of magnitude, could (and should) be acceptable in the microgravity platform context for the examined experiment class.

Single-pulses g-jitter disturbances result from forces external to the S/C system (e.g., impacts) or from thrusters. The time profiles of the pulses were assumed to be rectangular waves of duration Δt (seconds) applied to otherwise 0-g conditions. The effects generated in the fluid (measured by the distortion parameters V_m^*, Ψ_m^*, D_t, and D_m) were evaluated during the forced phase (i.e., the pulse duration) and during the relaxation phase (i.e., when the g level returns to zero, and the system decays toward the quiescient equilibrium state). Figure 5a refers to the forced phase and shows the time behavior of the parameters V_m^* (Ψ_m^* has a similar profile) that correspond to three different g jitter profiles:

$$-Gr = 40,000, \qquad \Delta t = 4$$

$$-Gr = 80,000, \qquad \Delta t = 2$$

$$-Gr = 160,000, \qquad \Delta t = 1$$

where t is in seconds. The motivation of these experiments is self-evident: verification of the dependence of the disturbance parameters upon the pulse dose ($G = g \times \Delta t$). In Fig. 5b the values of D_m and D_T are shown in the near field. The results can be summarized as follows:

1) For small values of Δt the system behaves as a linear one with respect to G within the forced phase.

2) The fluid distortion parameters (V_m^* and Ψ_m^*) grow, within this phase, with a ramp angle proportional to Gr (and g) and reach a peak value proportional to G; after the end of the g pulse these values decay to zero.

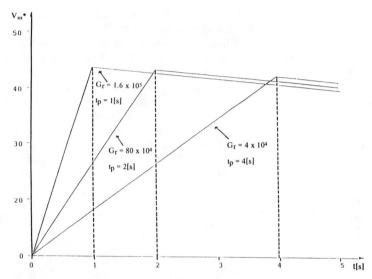

Fig. 5a **Velocity distortion parameters for single pulses of different duration but of equal dose** $(Gr \times \Delta t)$.

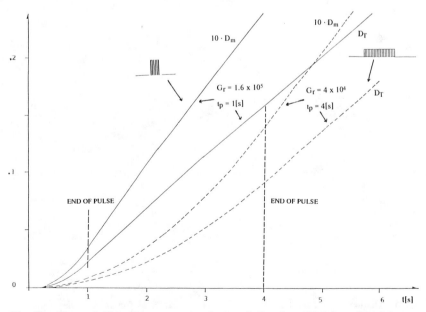

Fig. 5b **Thermal distortion parameter time evolution for equal dose disturbances.**

3) At the end of the pulse, the temperature distortion parameters (D_T and D_m) continue to grow.

The behavior of the disturbance parameters during the relaxation phase (far field) is shown in Fig. 6. The relaxation phase exhibits a clear exponential decay for the fluid disturbance parameters (V_m^* and Ψ_m^*) with a characteristic decay time of about 50 s (i.e., 1/20 of the reference time). The behavior of the thermal parameters (D_T and D_M) is quite different. Both parameters show an exponential growth, reaching a maximum at 70 s for the D_m, and at 150 s for the D_T. Only after these maxima, a thermal relaxation process takes place with a decay time of the order 22 s for D_T and of the order 300 s for D_m. The important aspect here is that the thermal disturbances keep on growing after the end of the g disturbance. The behavior of the three different pulses with the same dose is not distinguishable in the far field. It was shown that for liquids of different Pr numbers similar behaviors for D_m and D_T exist; the characteristic time for

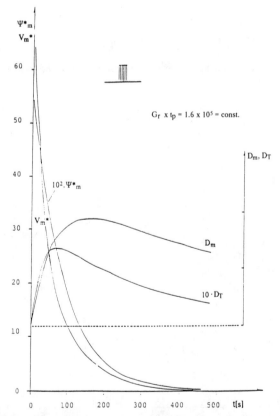

Fig. 6   Distortion parameter time profile in the far field for rectangular pulses of same dose $Gr \times Dt = 1.6 \times 10^5$.

the growth for the maxima and for the decay of the thermal distortion parameters depends on the Pr number. If one accepts as the threshold limit the disturbances generated by the residual gravitational field ($g = 10^{-6}$), one could use them as tolerable values ($V_m^* = 2$; $D_m = 0.02$), and we can then correlate the suggested limiting curves in the g-tp plane as follows:
Fluid distortion limits:

$$G_{mx} = g_{mx}\Delta t = (0.333\alpha/L_r\beta_T\Delta Tg_0) \qquad (14a)$$

Thermal distortion limits:

$$G_{mx} = g_{mx}\Delta t = (2.237v/L_r\beta_T\Delta Tg_0) \qquad (14b)$$

The curves for the maximum g level are plotted in Fig. 7 as g vs $1/t_p$ (a duration longer than 500 s gives rise to quasisteady conditions).

Compensated double-pulse g jitters typically result from mass dislocations internal to the platform: The total momentum of the system remains constant after the application of these disturbances. The shape of the g jitter is a positive rectangular profile followed by a corresponding negative rectangular profile, after a time delay (t_d). Figures 8 and 9 consider different values of t_d and show the disturbance parameters during the relaxation phase. During the forced period ($0 < t < t_d$) the behavior is exactly the same as the single-pulse case: When the negative pulse occurs, the velocity disturbances decay to zero almost immediately. The maximum of the overall disturbance in fluid velocity occurs at the end of the first

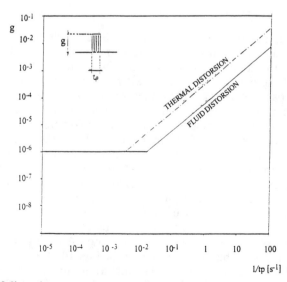

Fig. 7   Equal distortion parameter curves for single-pulse g jitter vs the inverse of the pulse duration.

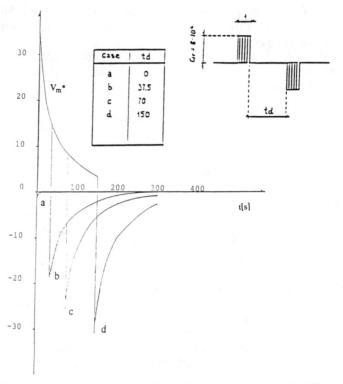

Fig. 8 **Time profile of the maximum velocity distortion parameter for different pulse intervals.**

positive pulse. Temperature disturbance parameters show a different trend due to their different growth after the first positive pulse. If the second negative pulse occurs after a short time, the overall disturbance is minimized, but if the second pulse occurs after the maximum of the growth of the thermal parameters ($t_d > 70$ s), then the maximum overall disturbance practically coincides with the ones predicted in the single-pulse case. Subsequent pulses that compensate each other (different shapes with the same dose) have also been considered; the results of the numerical experiments are very close to those related to pulses of identical time profiles. Saw-toothed profiles are used by NASA for describing the crew motion induced g-jitter contributions and for assessing their perturbation effects.[6] In particular, a crew step simulation has been considered: a positive $g = 10^{-2}$ triangular spike over 0.8 s, followed by a corresponding negative one after a delay of 1.6 s. Triangular profiles behave similarly to the equivalent square waves (having the same dose). A dose tolerability curve for compensating double pulses is shown in Fig. 10.

Repetitive single pulses may be very disturbing for microgravity experiments because their overall effects may be very large (even if their occurrence is few minutes apart) and because steady-state conditions are not

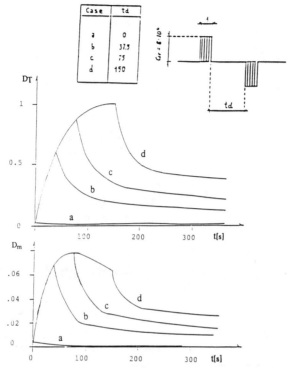

Fig. 9   Thermal distortion parameter time profile for different pulse intervals.

attained. The case of two subsequent pulses with the same sign, with a time interval t_d, was examined to simulate the effects of bits thrusters subsequent firings. The values of the temperature disturbance parameters may be greatly increased if the two pulses are applied at specific time intervals. This is the case shown in Fig. 11, there a delay of 150 s is assumed between the two pulses ($Gr = 80.000$; $\Delta t = 1$ s). The fluid distortion parameters have almost the same values of the single pulses; Fig. 12 shows, instead, how the temperature distortion parameters exhibit maxima that are almost two times those generated by the single pulses.

Rectangular periodic disturbances were also examined, and curves for the equal distortion parameters were found that look like the typical curves $gmx(f)$ of Fig. 17 (with some scattering). The periodic g jitters will be considered in more detail for the second study case and illustrated in the next section; however, the main results are similar for the two cases.

Study Case 2: Evaluation of the Diffusion Coefficient[12]
Microgravity experimentations aimed at the evaluation of the diffusion coefficient in a liquid medium have been simulated by numerical models. The initial conditions at the beginning of the diffusion process (in the liquid

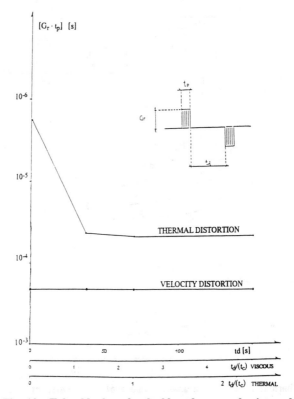

Fig. 10   Tolerable dose for double pulses vs pulse interval.

phase) are known, and the concentration field is assumed to be measured at different times during the diffusion process itself. It is difficult to perform the measurement of the concentration profiles along three directions. Probably average values along the light paths of the optical diagnostic equipment are more easily determined; hence, an effort was made to compute equivalent diffusion coefficients in terms of the measurable parameters. The concept of an equivalent diffusion coefficient also stems from the fact that in the described case one should compute diffusion coefficients from a concentration distribution (one or two dimension) that results from the time profile of the diffusion process plus an unknown amount of convection (induced by residual g and g jitters).

There are two possible ways of determining an equivalent diffusion coefficient: 1) from the differential balance equation, or 2) from the integrated diffusion equation. If the experimental microgravity concentration distribution $c(x,y,t)$ is available, without knowing $u(x,y,t)$ and $v(x,y,t)$, one could try to determine an equivalent diffusion coefficient (D_e) by observing the time variation of the local concentration value ($\partial c / \partial t$, for instance by a double-exposure interferometric method) and by measuring

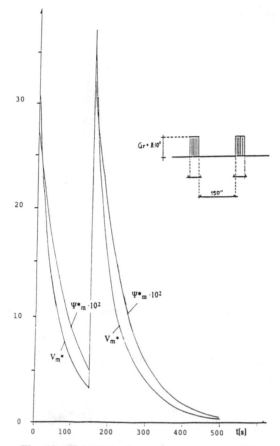

Fig. 11 Fluid distortion parameter time profile.

the concentration gradient (or the concentration derivative along the axis of the test cell, for instance, by a differential interferometer).

A possible definition of D_e can be made in terms of the concentration time rate of change with the help of the (purely diffusive) species balanced equation and to the real diffusion coefficient D by

$$\frac{\partial c}{\partial t} = D_e\left(\frac{\partial^2 c}{\partial x^2} + \frac{\partial^2 c}{\partial y^2}\right) = D\left(\frac{\partial^2 c}{\partial x^2} + \frac{\partial^2 c}{\partial y^2}\right) - \left(u\frac{\partial c}{\partial x} + v\frac{\partial c}{\partial y}\right) \qquad (15)$$

The difference between D_e and D is due to the convective motions:

$$D_e = D - \left(u\frac{\partial c}{\partial x} + v\frac{\partial c}{\partial y}\right)\bigg/\left(\frac{\partial^2 c}{\partial x^2} + \frac{\partial^2 c}{\partial y^2}\right) \qquad (16)$$

The second term represents the error committed when the convective

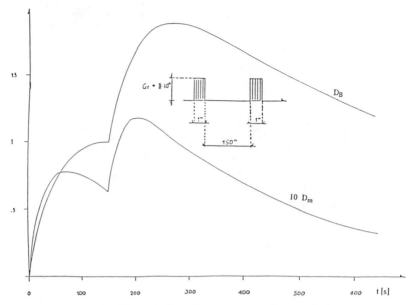

Fig. 12   **Thermal distortion parameters time evolution.**

velocities induced by the g disturbances are neglected. In a quasi-one-dimensional case, whenever

$$v\frac{\partial c}{\partial y} \ll u\frac{\partial c}{\partial x} \quad \text{and} \quad \frac{\partial^2 c}{\partial y^2} \ll \frac{\partial^2 c}{\partial x^2}$$

one can define an average cross-sectional concentration across the section of height H

$$C(x) = 1/L \int_0^H c(x,y)\, \mathrm{d}y \tag{17}$$

and write, in a finite-difference approximation,

$$D_e = \frac{\Delta C}{\Delta t} \bigg/ \frac{\partial^2 C}{\partial x^2} \tag{18}$$

The percentage error is

$$(D_e - D)/D = \Delta D/D = u/D[(\partial C/\partial x)/(\partial^2 C/\partial x^2)] \tag{19}$$

One would be able to compute the correct value of D if the velocity field, $\partial c/\partial x$ (or $\partial C/\partial x$) and $\partial^2 c/\partial x^2$ (or $\partial^2 C/\partial x^2$) are measured simultaneously. The value of $\Delta D/D$ defined in Eq. (19) plays the same role of the thermal disturbance parameters defined in the previous case and will be computed in the numerical cases. If the concentration field is available at different times in all of the specimens one could find the best "spots" and the best times at which the computation of D_e can be made by means of Eq. (16). For example, at points of relative minima or maxima of $C(\partial C/\partial x = 0)$, there would be no effect of the convective velocity along the direction of the extreme, and the computed value of D_e would coincide with D. Once the value of D_e is found from the knowledge of the concentration distribution $C(x,t)$ at two times, then a space- and time-averaged value could be found by

$$D_e = 1/(TL) \int_0^L \int_0^T De(t,x) \, dt \, dx \qquad (20)$$

Use of the results of the integrated equation is also possible at different times, as shown in Ref. 22. The error due to the convection is best evaluated by a least-squares method with the help of a numerical solution that simulates the presence of the convective motions. In a one-dimensional purely diffusion process, with initial boundary conditions (at $t = 0$),

$$c = c_0 \qquad x \leqslant 0$$

$$c = 0 \qquad x > 0$$

the concentration distribution and its axial gradient read

$$c(x,t) = c_0\left(1 - 2/\sqrt{\pi} \int_0^X \exp(\eta^2) \, d\eta \right) \qquad (21)$$

$$\partial c/\partial x = -(c_0/\sqrt{\pi Dt}) \exp[-x^2/(4Dt)] \qquad (22)$$

$$X = (x - x_0)/2\sqrt{Dt} \qquad (23)$$

An equivalent diffusion coefficient could also be defined as the value of D_e in Eq. (21) or (22) that best fits the experimental measured data taken at many points at a given time (t). It is obvious that the best fit of $(\partial c/\partial x)$ at t_f or of $C(x,t_f)$ will take into account the entire history of the convective events that take place during the time $0 \leqslant t \leqslant t_f$. Evaluation of the error $\Delta D/D$ was made by running numerical models in the presence of the g disturbances. Before the results of the numerical experimentations are discussed, two comments must be made:

1) The numerical experiments performed required a large amount of computation time for two reasons: 1) a diffusion experiment typically is quite long (its characteristic time is an order of hours); and 2) the fact that the interest was in determining the influence of gravity pulses having a duration (Δt) of an order of seconds forced us to select small integration time steps (typically $1/20\Delta t$).

2) The interest of a typical diffusion experiment is related mainly in the transient phase of the phenomenon. Thus, the analysis of the disturbances induced by a residual gravity or by a g jitter (pulsed or periodic) is greatly affected by the various characteristic times involved [i.e., the time of occurrence of the disturbance (tp), the characteristic time of the disturbance (duration, period, shape), the characteristic time of the diffusion process (L^2/D), and the characteristic time of the viscous/buoyant disturbance convective process (L^2/v)] and by the criterion of evaluating the effects caused by the gravity disturbance on the diffusion process.

Different criteria have been examined [e.g., concentration profiles (along x and y; Fig. 4), concentration gradients (x and y components) along x and y, velocity (x and y components) induced by buoyant effects, and distortion parameters $\Delta D/D$].

The numerical experiments performed are relative to diffusion problems characterized by the geometry and the initial conditions on the concentration fields shown in Fig. 3; the fluid considered is an aqueous solution with a diffusing component having a Schmidt number $Sc = 690$. The relevant reference parameters are

$$Lr = 5 \ (\text{cm}); \qquad v_r = 0.01 \ (\text{cm}^2/\text{s})$$

$$Vr = 2. \times 10^{-3} \ (\text{cm/s}); \qquad tr = 2500 \ (\text{s})$$

$$D = 1.5 \times 10^{-5} \ (\text{cm}^2/\text{s}); \qquad \beta_c = \partial \rho / \partial c = 10^{-2} \ (\text{cm}^{-3})$$

With the preceding values the following relation therefore exists between the Grashof number and the g level:

$$Gr = 1.22625 \times 10^7 g \qquad (24)$$

To simplify the evaluation of the results, the parameter that describes the effect of the disturbance is taken to be the axial gradient of the normalized concentration gradient at the midpoint of the cell:

$$F_c = \frac{\partial c}{\partial x_{(0.5,0.125)}} \qquad (25)$$

This value in the 0-g reference cases is related to the real and equivalent diffusion coefficients by

$$(F_c)_0 = C_0/\sqrt{\pi D t}; \qquad \Delta D/D = [(F_c)_0/F_c]^2 - 1 \qquad (26)$$

Space profiles and/or time profiles of various parameters are shown in Ref. 22 to give a better insight into the phenomenology considered.

The first numerical experiment was performed by assuming a 0-g environment to check the numerical code; it was run for a very long time (20,000 s), corresponding to a nondimensional time of 11.6. The outputs were checked against the theoretical one-dimensional analytical solution [Eqs. (21) and

(22)] and against the constancy of the total contents of the diffusing species inside the cell. Both checks were positive and confirmed the accuracy of the numerical computations.

A computation of the reference cases was made by assuming a residual gravity field of $g = 10^{-6}$ (free flier environment) and $g = 10^{-5}$ (to simulate a likely Columbus attached laboratory environment). The corresponding disturbance parameter values are utilized to determine the acceptance criteria for g disturbances. Figure 13 plots the transversal concentration profiles taken at the midplane ($x = 0.5$) for the three cases of constant g levels at time $t = 2500$ s. It can be noted the nonplanarity of the concentration front due to the gravity level, which induces a buoyant motion within the cell [the central point of the cell maintains the initial concentration value ($c = 0.5$)]. Figure 14 shows the transversal profile of the axial component of the concentration gradient, (F_c), taken at the midplane ($x = 0.5$) for the three cases of constant g level at time $t = 2500$ s. The difference of $F_c - (F_c)_{0.5}$ is quite low for the 0-g and the $g = 10^{-6}$ case. For the case of $g = 10^{-6}$ residual gravity an error of about 2% is introduced, and for the $g = 10^{-5}$ case an error of 54% is introduced in the measure of D. It is obvious that the error analysis is limited by the rather crude formula used for the evaluation of D.

Single-pulse g-jitter effects were evaluated; contrary to the case of a disturbance to an otherwise steady-state process, the time for occurrence of the pulse (tp) is a parameter of importance for the tolerability of the pulse. If a pulse is applied at the very beginning of the process, its effect is small; similarly, at the end of the growth, when the concentration is constant along the cell, the pulse has no effect in creating convective flows. There is a time at which the convection created by the pulse will be largest. Three rectangular single pulses of 4 s duration and of $g = 10^{-3}$ occurring at different (nondimensional) times tp (0.1, 0.5, and 0.75) are computed. In Fig. 15 their effects are shown. In the present case the concentration distortion is largest when the pulse is applied

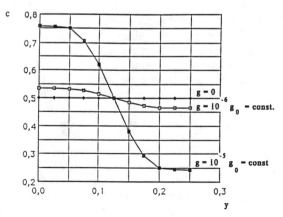

Fig. 13   Transversal concentration profiles at the middle of the cell ($t = 250$ s).

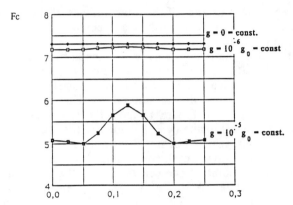

Fig. 14 Transverse profile of axial concentration gradient at the middle of the cell ($t = 2500$ s).

at about $t = 0.75$ (nondimensional time). However, the time-integrated effects of the pulse are larger for pulses that occur at the initial stages of the experiments; this can be seen in Fig. 15, which shows how the average concentration difference, with respect to the 0-g case, is largest at the observation time ($t_{obs} = 1$) for case 1, even though the $c(y)$ profile is flatter compared to cases 2 and 3. Therefore, one should expect a tolerability that is dependent on the time of the pulse occurrence and on the time of the observation (t_{obs}) of the g-pulse effect. This is obviously a great complication for these classes of experiments.

To check the dose concept, pulse durations of $\Delta t = 2$, 4, and 8 s have been considered in order to simulate propulsive actions on the platform (e.g., for attitude control). The same dose G [Eq. (8)] is being taken in each of the three cases ($G = 4 \times 10^{-4}$ s) that correspond to $g = 2 \times 10^{-4}$, 10^{-4}, and 0.5×10^{-4} (rather small disturbances). The effects are being evaluated at three times: 1) right after the pulse is applied ($t = tp + \Delta t = 2000$ s $+ \Delta t$, 2) at $t = 4000$ s, and 3) at $t = 8000$ s. A comparison with respect to the 0-g case is made. Figure 16 shows the values of F_c taken at $x/L = 0.5$ as a function of y/H at the indicated times. Within the range of accelerations and pulse duration considered, an equivalence among different pulses of the same dose (G) can be claimed, at least for the far field (i.e., sufficiently downstream in time, after the pulse is terminated). The reasons for this behavior may be found on the rather quick response of the fluid system to the g pulse in the velocity field and of the slow response in the concentration field, due to the low value of the diffusion coefficient and to the long period of time necessary to restore the diffusion-controlled process (existing before the pulse). A preliminary analysis made with an OMA method for simple periodic, sinusoidal disturbances[3] shows that typical tolerable g levels have a qualitative trend as function of the frequency as depicted in Fig. 17; this curve is being given to the aerospace industries as specifications for the maximum disturbances (or g jitters) to be created by the

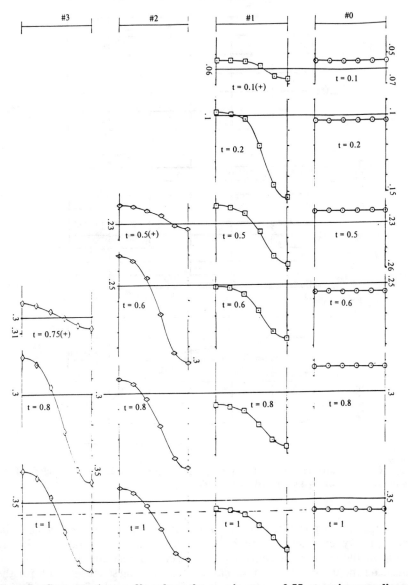

Fig. 15 Concentration profiles along the y axis at $x = 0.55$ at various nondimensional times for pulses of $g = 10^{-3}$ occurring at $t = 0.1$, 0.5, and 0.75. On the right is the reference case ($g = 0$).

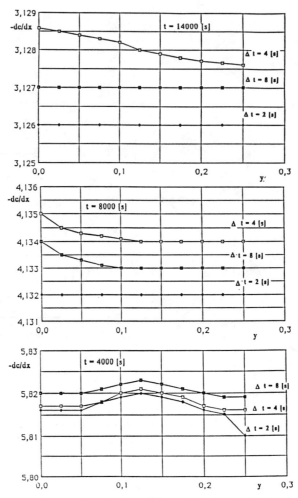

Fig. 16 Concentration gradient in the middle section ($x = 2.5$ cm) for pulses of the same dose $G = 4E - 4$ s applied at $t = 2000$ s.

payloads and transmitted to the microgravity platforms on which the microgravity experiments are hosted. All of the considerations already made in the section on equivalence criteria and tolerability limits, which illustrate the difficulty of a practical utilization of these curves both for the users and as specification to the manufacturers, apply for the diffusion processes examined here. A practical problem arises with microgravity environment specifications. The payload disturbance outputs will act as an input to the platform and subsequently to other facilities. It is difficult to guarantee the good quality of the MGE on the platform when it is in the presence of other payloads that are simultaneously active at other locations

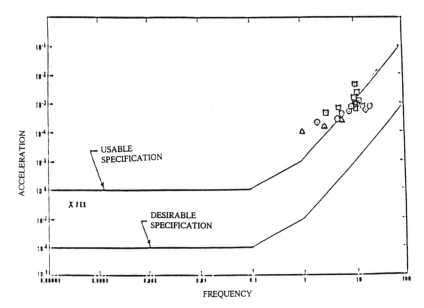

Fig. 17   Proposed acceleration limit specifications and measured values on Spacelab.

of the same platform. Decomposition along three axes, Fourier series expansion, and rms (or amplitude) measurements are all possible criteria that must be checked. Reference 12 examined the possibility of establishing first approximation criteria for the practical equivalence of nonsinusoidal (zero time average) g jitters to sinusoidal ones in an attempt to identify the equivalence parameters (frequency, amplitude, and rms or dose). For the study under consideration here, one can say that usual periodic disturbances characterized by a first harmonic period T and by a half-period dose,

$$G_H = \int_0^{T/2} g(t)\ \mathrm{d}t$$

are equivalent, in a very first approximation, to a sinusoidal disturbance of the same frequency ($f = 1/T$) and of the amplitude $g_f = \pi f G_H$, i.e.,

$$g(t) = \underline{g}_f \sin(2\pi f t)$$

Instead of decomposing an alternating periodic disturbance in Fourier series and measuring the corresponding amplitudes of each harmonic, the compliance with the sinusoidal $g_{mx}(f)$ curve could be found by two simple measurements on the g jitters: the period ($1/f$) and the half-period acceleration dose (G_H). These parameters will uniquely identify the amplitude of the equivalent sinusoidal disturbance. Numerical experiments were performed for the cases of alternating rectangular periodic disturbances and

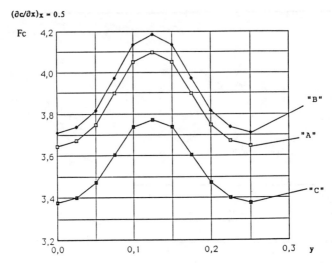

Fig. 18 Transverse profile of the axial components of concentration gradient at the middle cell and at $t = 2520$ s.

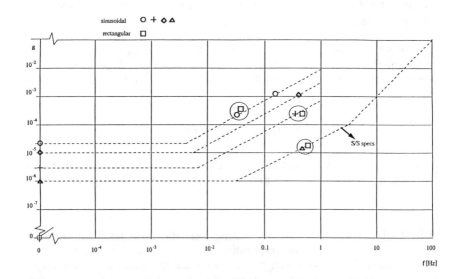

Fig. 19 Tolerable g level for periodic disturbances. Circles denote the three check points for the equivalence between sinusoidal and rectangular jitter with the same half-period dose.

for the case of sinusoidal disturbances. The aim is twofold: 1) to validate the equivalence of the dose against the rms concept for sinusoidal alternating disturbances, and 2) to correlate the disturbance effect, as measured in terms of the value of $\partial C / \partial x$ at the middle point, for the various periodic cases, and to try to ascertain the equivalent dose. For the alternating rectangular disturbances three cases were considered. Cases A and B have equivalent doses, cases A and C have equivalent rms, and for all cases the g jitters occur at $tp = 250$ s with a period of 20 s. The disturbance was observed during the period 2500–2520 s. Figure 18 plots the transverse profile of the axial component of the concentration gradient ($\partial c / \partial x$). Here we note larger differences between the three cases; from the analysis of all of these plots we can conclude that disturbances exhibiting the same half-period dose (G_H) for alternating, nonsinusoidal, periodic disturbances give a similar response; on the other hand the rms concept does not seem to give sufficiently similar responses. Figure 19 shows the equivalence of different sinusoidal and rectangular periodic pulses with steady g levels; as a very first approximation the plot could be taken as a check of the proposed equivalence criterion on the half-period dose. Extension of this criterion to other classes of experiments, to be shown by systematic (and time-consuming) numerical experimentations, would allow us to solve the intriguing problem of the utilization of the $g_{mx}(f)$ curve prepared for sinusoidal disturbances (which are more easily analyzed).

Conclusions

The studies of the effects of g jitters on microgravity experimentation presented here are related to the experimentations that look for diffusion processes (thermal or species) in an otherwise quiescent liquid medium and take into account points that have not been considered thus far. Very long characteristic times associated with energy diffusion (in high-Prandtl-number liquids) and with mass diffusion imply that short disturbances induce long-lasting effects due to the long decay phase following the disturbances. The long "memory" of the system to random g jitters creates difficulties in the attainment of steady processes. A first attempt is made here to establish equivalent criteria for different types of g jitters based on the consideration that space processings are affected by g disturbances, not only because of the distortions induced in the concentration, temperature, and velocity field, but also because of the duration of these disturbances. Disturbance parameters have been proposed for different experiments, and a first equivalence criterion for short g pulses was identified in terms of the pulse dose. The criterion, which was checked for two specific categories of experiment, seems to be valid, at least in the far field (i.e., some time after the disturbance occurred). This criterion might have an immediate utilization in prescribing thruster operations in terms of the maximum ejected propellant mass (for a given specific impulse), more than in terms of the maximum thrust. A further complexity arises in connection with unsteady processes (e.g., measurement of diffusion coefficients and crystal growth) for which the effect of a g jitter on the process may strongly depend on its time of occurrence from the beginning of the experiment. Two identical g pulses have different effects if they occur at different stages of the process. The usefulness of the maximum tolerable amplitudes of sinusoidal g jitters vs the frequency as provided by the agencies to payload manufacturers is questioned for very practical reasons: In the presence of more than one frequency (as for the case of a Fourier series decomposition), it is not possible to decide if the tolerability limits are trespassed. On the other hand, it is also difficult to check that a given payload is in compliance with this curve. The rms values of the periodic g jitters could be measured, or a Fourier series analysis could be made. Very preliminary numerical results seem to suggest that the half-period dose might be equivalence criterion between different shape periodic g jitters; this could represent an alternative to the Fourier analysis and to the rms experimental measurements.

In conclusion, the evaluation of the g jitter disturbances and temperature distributions in a liquid medium is a very difficult problem to solve, in general due to the large number of parameters involved, specifically, the characteristic times. The evaluation of the effect of these disturbances on the specific space products might be also very complex and is related to the fluid medium. Efforts should be made, through equivalence criteria, to reduce the number of the variables and to set guidelines for the experimenters, the Agencies, and the manufacturers of the payloads to be hosted on microgravity platforms.

Acknowledgments

The author wishes to thank Techno System for the support and C. Golia for the numerical configurations reported in Refs. 9 and 12.

References

[1]Hamacher, H., Jilg, R., and Merbold, U., "Analysis of Microgravity Measurements Performed During D1," *6th European Symposium on Material Sciences Under Microgravity Conditions*, 1986.

[2]Hamacher, H., Jilg, R., and Merbold, U., "Analysis of Microgravity Measurements Performed During the D-1 Mission," *Proceedings 6th European Symposium on Materials Sciences Under Microgravity Conditions*, European Space Agency Rept. SP 256, 1987, pp. 413–420.

[3]Sarychev, V. A. et al., "Determination of Microacceleration on the Salyut-6 and Salyut-7 Orbital Complexes," International Aeronautical Federation Paper IFA-87-233, 1987.

[4]Monti, R., "Tolerable g-Level for Fluid Science Experiments," European Space Agency R-66.525, TS Rept. TS-7-87, 1987.

[5]Napolitano, L. G., "Surface and Buoyancy Driven Free Convection," *Acta Astronautica*, Vol. 9, 1982, pp. 199–215.

[6]Monti, R., "g-Level Threshold Determination," European Space Agency Contract 5504/83/F/FS/SC, TS Rept. TS-7-84, 1984.

[7]Favier, J. J., and Camel, D., "Order of Magnitude Analysis of Solute Macrosegregation in Crystal Growth from the Melt," European Space Agency Rept. SP-191, 1983, pp. 295–299.

[8]Monti, R., Langbein, D., and Favier J. J., "Influence of Residual Acceleration on Fluid Physics and Materials Science Experiments," *Fluid Sciences and Materials Science in Space, An European Perspective*, edited by H. U. Walter, Springer-Verlag, Berlin, 1987.

[9]Monti, R., and Golia, C., "On the Effects of the Orbit Acceleration on Microgravity Experiments in Fluidynamics," *IX Congresso Nazionale AIDAA*, 1987.

[10]Crespo del Arco, E., Pulicant, J. P., Randriamampianina, A., Extrement, G. P., and Bontoux, P., "Complex Flow Regimes in the Melt of a Bridgman Model Multiple, Steady and Time Dependent, Solutions in Two and Three Dimensional, Low-Prandtl Number Convection" *VIIth European Symposium on Materials and Fluid Sciences in Microgravity*, 1989.

[11]Napolitano, L. G., Golia, C., and Vivani, A., "Effects of Variable Transport Properties on Thermal Marangoni Flows," *Acta Astronautica*, Vol. 13, 1986, pp. 661–667.

[12]Monti, R., "Study on Allowable g-Levels for Spacelab, Columbus and Eureca," European Space Agency Contract 7681/88/F/FL(SC), TS Rept. TS-9-89, 1989.

[13]Knabe, W., and Eilers, D., "Low-Gravity Environment in Space Lab," ERNO Rept. 1986.

Effect of Gravity Jitter on Natural Convection in a Vertical Cylinder

M. Wadih,* N. Zahibo,† and B. Roux‡

Institut de Mécanique des Fluides, Marseille, France

Introduction

MANY scientific experiments dealing with materials processing or fundamental fluid physics have been performed for several years by using the low level of gravitation onboard an orbiting spacecraft. Compared to 1 g on the ground, the residual microgravity (μg) level still affects physical processes; in addition, this μg vector field is characterized by a steady-state component and by a fluctuating contribution called g jitter. Both the steady-state and fluctuating components can affect the fluid phases subjected to thermal (or solutal) gradients. Recent studies by Hamacher and Merbold[1] and Hamacher et al.[2,3] have been devoted to measuring these components during the D1 Spacelab mission. These authors used two accelerometer systems. One operated in a peak detection mode (1 Hz) to reduce the total amount of data (peak values within intervals of 1 s were detected from an analog random response in the positive and negative directions of the coordinate considered), allowing a bandwidth of up to 100 Hz to be analyzed. The second operated in a high-rate sampling mode, suitable for performing a frequency analysis of up to about 5 Hz.

We are directly involved in collaborative work with groups doing experiments concerning materials science and fluid physics in space. Most of these experiments are performed in cylindrical furnaces, as in the D1 mission.[4-6] Several other experiments are now prepared. A complete

*Maître de Conférence.
†Edudiant.
‡Directeur de Recherche CNRS.

understanding of the role of the gravity modulation on the convective motions in fluids is needed to better analyze the available results and to better design future experiments.

According to previous workers,[2,3,7-9] two types of frequencies have to be considered. One type corresponds to attitude changes during the orbital motion (very small frequencies). The other, g jitter, which corresponds to spacecraft maneuvers and mechanical vibrations, gives rise to a random frequency band varying from 0.1 to 100 Hz.

Most of experiments to be considered are performed in long cylindrical containers of a small radius (typically one or a few centimeters) and in furnaces delivering axial temperature gradient. Different materials are used corresponding to fluids covering a wide range of Prandtl numbers (see Table 1).

The purpose of this study is to analyze the effect of gravity modulation on the onset of convection in differentially heated cylinders, considering a wide range of values of ω and Pr. As a first step, the study is limited to the case of vertical cylinders of infinite length, where a negative temperature gradient is maintained in the upward direction. After giving the governing equations for small perturbations in the section on governing equations and formulating the stability problem by using the Floquet theory and expansions of the variable in terms of Bessel functions in the section on problem formulation, we propose two techniques for solving these perturbation equations. The first one, which is analytical, is presented in the section on analytical approach (adiabatic case) for boundary conditions of adiabatic type. The second, called the matrix method, is presented for more general boundary conditions in the section on matrix methods. An asymptotic analysis has been considered for large frequencies; two methods are proposed, respectively in the subsection on the limiting case: $\omega \rightarrow \infty$ as an application of the analytical method for small-amplitude modulations, and in the section on asymptotic analysis for large values of w, through an original approach, valid for finite-amplitude modulations. Comparisons with the available results in the literature concerning the alteration of the stability threshold, resulting from different kinds of modulation (gravity or temperature surface) will be made, although most of these available results concern the classical Bénard problem (i.e., without confinement of side walls).

Table 1 Typical properties of physical fluids

	$v, \text{cm}^2 \cdot \text{s}^{-1}$	$\kappa, \text{cm}^2 \cdot \text{s}^{-1}$	Pr
Gases	10^{-1}	10^{-1}	1
Liquid metals	$10^{-3} - 10^{-1}$	$10^{-2} - 10^{0}$	$10^{-3} - 10^{-1}$
Organic liquids	$10^{-3} - 10^{-2}$	$10^{-4} - 10^{-3}$	$10^{-3} - 10^{-1}$
Molten glasses	10^{2}	10^{-2}	$10^{3} - 10^{4}$
Molten salts	$10^{-3} - 10^{-2}$	10^{-3}	$10^{0} - 10^{1}$
Silicone oils	$10^{-2} - 10^{1}$	10^{-4}	$10^{1} - 10^{4}$
Water	10^{-2}	10^{-3}	10^{1}

Presentation of the Problem

As reported by Ostrach,[8] it was found by previous authors[10-12] that vibrations can either substantially enhance or retard heat transfer and drastically affect convection by altering the transitions from quiescent to laminar flow (critical Rayleigh number) and from laminar to turbulent flow.

Many works have been devoted to the effect of unsteady constraints on the onset of convective motion. The mathematical difficulty lies in the fact that the equations describing the growth of initial disturbances are nonautonomous and the method of normal modes is not applicable. Several investigations have been made in the case of periodic constraints applied to different shear flows or buoyancy-driven motions.

Many authors have established the significant influence of a time-periodic excitation on the stability threshold. Hall[13] studied the linear and nonlinear stability of the modulated Couette flow by means of the small-parameter method. In addition, he considered the case of large frequencies and arbitrary amplitude. The modulated-Poiseuille flow has been studied by Grosch and Salwen,[14] who used a fourth-order Runge-Kutta method (as proposed by Ralston 1962) for the numerical integration of a system of (N first order) differential equations.

The problem of convective stability in the presence of a periodically varying parameter was first emphasized by Gershuni and Zhukhovitskii,[15] who established the alteration of the stability threshold by using the first-order approximation (one trial function) of Galerkin's method. This method, which reduces a first-order differential system to an ordinary differential equation with periodic coefficients, allows the problem to be handled with an arbitrary modulation amplitude. It has been used for the classical (unconfined) Bénard problem and for a flow confined in a vertical cylinder heated from the bottom (e.g., one considered herein) by Gershuni et al.[12] For large frequencies these authors established an analytical relation that will be discussed in the section on other nondimensionalizations.

Most of the other studies devoted to the modulation of the surface temperature or the gravity in buoyancy-driven flows only concern the (unconfined) Bénard problem with different kinds of boundary conditions. In the case of small-amplitude modulations a basic linear stability analysis has been done by Venezian,[16] who studied the effect of the modulation of the temperature gradient on two-dimensional small disturbances. Roppo et al.[17] also studied the effect of the same kind of modulation, but on three-dimensional disturbances; in addition, they considered the nonlinear stability analysis. For finite-amplitude modulations, in addition to Gershuni et al.,[12] who used only one trial function (that corresponds, in fact, to a separation of variables solution), several authors[18-21] used a Galerkin technique with a small set of trial functions. This small number appeared to be sufficient in the case of rigid and conducting horizontal walls considered by these authors, who all establish in different manners the alteration of the stability limit under the modulation of constraints.

The great difficulty lies in the choice of the stability criterion, as discussed by Homsy[22]; this paper does not deal with this problem or with the other fundamental question concerning the ability of the linear theory to predict instability. The use of the Floquet theory in this paper is based on the stability of the null solution in the sense of Liapunov and thus implies the prediction of the asymptotic stability. In the simpler case, when Mathieu's equation gives a good approximation of the equation of perturbation, one has a simple criterion for linear stability. However, in general, Mathieu's equation does not describe all of the properties of evolution systems of the Navier-Stokes type. In these cases the Floquet theory is the best approach to the problem. In other cases the energy method gives sufficient conditions of stability, as shown by Homsy,[22] for instance.

Governing Equations

We specifically consider the case of gravity modulation:

$$g = g_0 + \epsilon^* \cos\omega^* t^* \qquad (1)$$

where ϵ^*, ω^*, and t^* represent dimensional amplitude, frequency, and time, respectively, and g_0 represents the mean gravity (for space applications, g_0 could be $10^{-3} \, \mathrm{m \cdot s^{-2}}$ or less) applied to a column of a viscous incompressible fluid, in the case of a cylinder of infinite length, subjected to a constant temperature gradient,

$$dT/dz = -\gamma \qquad (2)$$

All fluid properties are constant, except that the density ρ varies linearly with the temperature in the buoyancy terms, according to the Boussinesq approximation:

$$\rho = \rho_0[1 - \alpha(T - T_0)]$$

where α is the volume expansion coefficient, and the subscript 0 represents the mean condition. Also, we neglect the viscous dissipation terms in the energy equation. Thus, following Bird et al., we write the governing equations as (see p. 338)

$$-\frac{D\underline{U}}{Dt} + v - {}^2U + [1 - \alpha(T - T_0)]g\mathbf{k} - \rho_0^{-1}\nabla P = 0 \qquad (3a)$$

$$\nabla \cdot \underline{U} = 0 \qquad (3b)$$

$$-DT/Dt + k \nabla^2 T = 0 \qquad (3c)$$

In Eqs. (3) $\underline{U}$, P, v, and k are the velocity, pressure, kinematic viscosity, and thermal diffusivity, respectively; k represents the unit vector upward, in the positive z direction (antiparallel to gravity); D represents the substantial derivative; and ∇^2 is the Laplacian of a vector field in Eq. (3a) and of a scalar one in Eq. (3c), the expressions of which are given in cylindrical coordinates by Bird et al. (1960) (see also Charlson and Sani[23]).

These equations admit an equilibrium solution in which $\underline{U} = 0$, $T = T(z,t)$ is a solution of

$$-\frac{\partial T}{\partial t} + k\, \nabla^2 T = 0$$

and the pressure $p(z,t)$ balances the buoyancy forces. Of course, the precise form of $T(z,t)$ depends on the boundary conditions.

In Eqs. (3), the following splitting of the variables can be used:

$$T = T(z,t) + \theta'(r,\phi,z,t)$$

and

$$P = p(z,t) + p'(r,\phi,z,t)$$

where θ' and p' represent (small) perturbations of the temperature and the pressure due to the convective motion. After linearization of Eqs. (3), we obtain the usual form of the small-perturbation equations (see, for example, Gershuni and Zhukhovitskii[15]). These equations for the velocity U of components (u,v,w) for θ and p, in nondimensional form, can be written in cylindrical coordinates (r,ϕ,z) as

$$-Pr^{-1}\frac{\partial U}{\partial t} + \nabla^2 U + R(t)\theta k - \nabla p = 0 \qquad (4a)$$

$$\nabla U = 0 \qquad (4b)$$

$$-\frac{\partial \theta}{\partial t} + \nabla^2 \theta + w = 0 \qquad (4c)$$

where the dependent variables have been nondimensionalized with r_0 (radius of the cylinder) for length, r_0^2/k for time, k/r_0 for velocity, $\rho v k/r_0^2$ for pressure, and γr_0 for temperature. Equations (4) contain two dimensionless parameters: the Prandtl number, defined as $Pr = v/k$, and the Rayleigh number, defined as $R(t) = \alpha g(t)\gamma r_0^4/(vk)$.

We consider the following boundary conditions on the side walls (at $r = 1$):

$$U = 0 \qquad (5)$$

$$\frac{\partial \theta}{\partial r} = -b\theta \qquad (6a)$$

where b is the Biot number. In Eqs. (6) we will mainly consider the two limit cases, corresponding to insulated or perfectly conducting side walls, respectively:

$$\frac{\partial \theta}{\partial r} = 0 \tag{6b}$$

or

$$\theta = 0 \tag{6c}$$

In addition, the solutions must be regular at $r = 0$.

Problem Formulation

We consider disturbances of the form

$$U(r,\phi,z,t) = \exp(iaz + in\phi)U(r,t) \tag{7a}$$

$$\theta(r,\phi,z,t) = \exp(iaz + in\phi)q(r,t) \tag{7b}$$

and

$$p(r,\phi,z,t) = \exp(iaz + in\phi)p(r,t) \tag{7c}$$

where a and n denote axial and azimuthal wave numbers. We also introduce two parameters, the mean Rayleigh number R_0 and the vibrational Rayleigh number ϵ:

$$R_0 = \alpha g_0 \gamma r_0^4/(vk) \tag{8a}$$

and

$$\epsilon = \alpha \epsilon^* \gamma r_0^4/(vk) = \epsilon_v R_0 \tag{8b}$$

where ϵ_v is the usual nondimensional modulation amplitude:

$$\epsilon_v = \epsilon^*/g_0 \tag{8c}$$

The aim is to calculate the critical values of R_0 as a function of ϵ and a for fixed Pr and ω. The minimum critical value of R_0, R_0^c, is such that

$$\partial R_0(a,\epsilon)/\partial a = 0 \tag{9}$$

We denote a_c the critical value of a at which condition (9) holds.

From a physical point of view, in the limit of small values of ϵ the effect of modulation is to alter the critical Rayleigh number such that

$$R_0^c(a_c,\epsilon) = R_{00}^c + \eta_c(a_c,\epsilon) \tag{10}$$

where R_{00}^c is the critical Rayleigh number in the unmodulated case ($\epsilon = 0$), and η_c is a function of a_c and ϵ such that η_c

$$(a_c, 0) = 0 \tag{11a}$$

Moreover, it can be shown that the change of ϵ into $-\epsilon$ corresponds to the translation of the time origin by a half-period; therefore, it does not change the physical problem. Thus, for small values of ϵ, η_c is of the form

$$\eta_c(a_c, \epsilon) = k\epsilon^2 + 0(\epsilon^4) \tag{11b}$$

By using the Taylor expansion of $\partial R_0 / \partial a$ and the power expansion of a_c,

$$a_c = a_0 + \epsilon a_1 + 0(\epsilon^2)$$

Venezian[16] demonstrated that $a_1 = 0$ and that a_0 is the critical wave number of the unmodulated case. In addition, for an infinite vertical cylinder Yih[22] proved that $a_0 = 0$. Then we have

$$a_c = 0(\epsilon^2) \tag{12}$$

Thus, the critical Rayleigh number [Eq. (10)] will be determined through Eq. (11b), in which the factor k will be computed for condition (12). However, the expression (11b) is not always the most convenient since ϵ depends on R_0^c. We can given an explicit formulation of Eq. (11b) in terms of ϵ_v; accounting for the definition (8b) of ϵ, we have

$$\eta^2 + \eta[2R_{00}^c - (k\epsilon_v^2)^{-1}] + R_{00}^{c2} = 0$$

This expression exhibits two solutions for any ϵ_v when $k < 0$. For $k > 0$ the solutions only exist in the domain $0 \leqslant \epsilon_v \leqslant \epsilon_v^0$, where $\epsilon_v^0 = 0.5 \, (R_{00}^c k)^{-\frac{1}{2}}$. The two solutions for small ϵ_v can be written as

$$\eta_1 = kR_{00}^{c2}\epsilon_v^2 \tag{13a}$$

$$\eta_2 = -2R_{00}^c + (k\epsilon_v^2)^{-1} - kR_{00}^{c2}\epsilon_v^2 \tag{13b}$$

In practice, since $\eta_1 < \eta_2$, only the first family [Eq. (13a)] is interesting for the determination of the first instability threshold.

We restrict this study to the case of an infinitely long cylinder for which u, v, and p are $\mathcal{O}(\epsilon)$.[15] From the definition (7c) the term $\partial p / \partial z$ is proportional to p and to a_c [equal to $\mathcal{O}(\epsilon^2)$ after Eq. (12)]. This term is $\mathcal{O}(\epsilon^3)$; it is canceled, compared to the terms in ϵ^2, in the Eq. (4a). Thus, defining a vector field X with components $w(r,t)$ and $\theta(r,t)$, we obtain from Eqs. (4)

$$\frac{dX}{dt} = M_0 X - \epsilon \cos\omega t N X \tag{14}$$

with

$$M_0 = \begin{bmatrix} PrP_n - PrR_0 \\ -1P_n \end{bmatrix} \tag{15a}$$

and

$$N = \begin{bmatrix} 0 & Pr \\ 0 & 0 \end{bmatrix} \tag{15b}$$

where P_n is the linear operator defined by

$$P_n = \frac{\partial^2}{\partial r^2} + r^{-1}\frac{\partial}{\partial r} - \frac{n^2}{r^2} \tag{16}$$

The use of the Floquet theory leads us to look for solutions of the following form:

$$X(r,t) = e^{\sigma t}x(r,t) \tag{17}$$

such that $x(r,t)$ is a period function in time. Here σ is the so-called Floquet exponent. If we define the small parameter η by

$$\eta = R_0 - R_{00} \tag{18}$$

where the double index 00 represents the unmodulated conditions, the substitution of Eqs. (17) and (18) into Eq. (14) gives the equation

$$\left(M_{00} - \frac{d}{dt}\right)x = \sigma x + \eta Nx + \epsilon \cos\omega t Nx \tag{19}$$

where M_{00} is defined by

$$M_{00} = \begin{bmatrix} PrP_n & -PrR_{00} \\ -1 & P_n \end{bmatrix} \tag{20}$$

In order to determine the critical stability conditions, we suppose that the effect of the modulation is weak when the amplitude ϵ is small. Thus, the parameter η remains small, and the solutions at $\mathcal{O}(\epsilon,\eta)$ are those of the unmodulated case:

$$x = x_{00} + \mathcal{O}(\epsilon,\eta) \tag{21a}$$

$$\sigma = \sigma_{00} + \mathcal{O}(\epsilon,\eta) \tag{21b}$$

where x_{00} is the solution of the eigenvalue problem:

$$\left(M_{00} - \frac{d}{dt}\right)x_{00} = \sigma_{00}x_{00} \tag{22}$$

with the appropriate boundary conditions.

According to the principle of exchange of stability, $\sigma_{00} = 0$ is a simple eigenvalue of Eq. (22). In order to investigate the effect of ϵ and h on the solution of Eq. (19), we assume

$$x = x_{00} + \sum_{p+q \geqslant 1} (\epsilon^p \eta^q x_{pq}) \tag{23a}$$

$$\sigma = \sum_{p+q \geqslant 1} (\epsilon^p \eta^q \sigma_{pq}) \tag{23b}$$

Here the critical conditions are obtained for a certain function $\eta = \eta_c(\epsilon)$ such that the real part of σ is zero:

$$\mathscr{R}\{\sigma[\epsilon, \eta_c(\epsilon)]\} = 0 \tag{24}$$

Substitution of Eqs. (23) into Eq. (19) and identification of coefficients of like power in ϵ, η, and ϵ^2 lead to the following recursive equations:

$$(M_{00} - d/dt)x_{00} = 0 \tag{25a}$$

$$(M_{00} - d/dt)x_{10} = \sigma_{10}x_{00} + \cos\omega t N x_{00} \tag{25b}$$

$$(M_{00} - d/dt)x_{01} = \sigma_{01}x_{00} + N x_{00} \tag{25c}$$

$$(M_{00} - d/dt)x_{20} = \sigma_{20}x_{00} + \sigma_{10}x_{10} + \cos\omega t N x_{10} \tag{25d}$$

Equation (25a) corresponds to the unmodulated case. In order for the other equations to have periodic solutions, the steady part of the right-hand side must be orthogonal to the null space of the adjoint operator of M_{00}. Denoting $\langle X|Y \rangle$ the scalar product, the solvability conditions of the Eqs. (25) are

$$\sigma_{10} = 0 \tag{26a}$$

$$\sigma_{01} = -\langle N x_{00}|x_{00}^* \rangle / \langle x_{00}|x_{00}^* \rangle \tag{26b}$$

$$\sigma_{20} = -\langle \overline{\cos\omega t N x_{10}|x_{00}^*} \rangle / \langle x_{00}|x_{00}^* \rangle \tag{26c}$$

where the bar denotes a time average. Thus, the expansion (23b) can be written as

$$\sigma = \eta\sigma_{01} + \epsilon^2\sigma_{20}$$

and the condition (24) becomes, since σ_{01} is real,

$$\eta_c = -\mathscr{R}(\sigma_{20})/\sigma_{01}\epsilon^2 + \mathscr{O}(\epsilon^4)$$

Thus, according to the notation (11b), we have

$$k = -\mathscr{R}(\sigma_{20})/\sigma_{01} \tag{27}$$

Representing x_{10} in the form

$$x_{10} = x^1_{10}\, e^{i\omega t} + \langle \text{conjugate} \rangle \tag{28}$$

we have to solve, from Eq. (25b), the following problem for $x^1 10$:

$$(M_{00} - i\omega)x^1_{10} = 0.5Nx_{00} \tag{29}$$

The system (25) will be solved in the next paragraphs with two different techniques based on the use of Bessel functions. An analytical approach is considered in the next paragraph for the adiabatic boundary conditions, whereas a more general method, called the matrix technique, is presented in the section on the matrix method.

Analytical Approach (Adiabatic Case)

In this section we consider an analytical approach for solving the problems of Eq. (25). In order to alleviate the presentation of the method, we will only consider the adiabatic conditions in Eq. (6b). For the unmodulated case [Eq. (25a)], we have

$$w_{00} = In(\xi r) - \mu Jn(\xi r) \tag{30}$$

$$\theta_{00} = [In(\xi r) + \mu Jn(\xi r)]\xi^{-2} \tag{31}$$

where

$$\xi = R_{00}^{\frac{1}{4}} \quad \text{and} \quad \mu = In(\xi)/Jn(\xi) \tag{32}$$

and Jn and In are, respectively, the Bessel function and the modified Bessel function of order n. The critical values of x are the ones satisfying the following characteristic relation:

$$In(\xi)J'n(\xi) + I'n(\xi)Jn(\xi) = 0 \tag{33}$$

where the prime denotes a first derivative. Then the critical value of Rayleigh number for the unmodulated case is given by

$$R_{00}^c = (\xi^c)^4 \tag{34}$$

For $n = 0$ and $n = 1$, respectively, we have the classical solutions

$$\xi^c = 4.611 \quad \text{and} \quad R_{00}^c = 452 \tag{35a}$$

$$\xi^c = 2.871 \quad \text{and} \quad R_{00}^c = 67.9 \tag{35b}$$

General Case (Finite w)

In order to solve system (25) we introduce the scalar product

$$\langle X | Y \rangle = \int X \cdot Y \, dr \tag{25}$$

In fact, it is convenient to multiply all the of Eq. (25) by r, since rP_n is self-adjoint. The adjoint operator of rM_{00} is

$$(rM_{00})^* = \begin{bmatrix} rPrP_n & -r \\ -rPrR_{00} & rP_n \end{bmatrix}$$

The null space of $(rM_{00})^*$ is generated by the vector x_{00}^* of components

$$w_{00}^* = w_{00} \quad \text{and} \quad \theta_{00}^* = PrP_n w_{00}$$

The solvability conditions (26) of Eq. (25), multiplied by r, are

$$\sigma_{01} = -\frac{\int rNx_{00} \cdot x_{00}^* \, dr}{\int rx_{00} \cdot x_{00}^* \, dr} \tag{36a}$$

$$\sigma_{20} = -\frac{\int r(\cos\omega t Nx_{10}) \cdot x_{00}^* \, dr}{\int x_{00} \cdot x_{00}^* \, dr} \tag{36b}$$

To $\mathcal{O}(\epsilon^4, \eta^2)$, due to the form of N, we only need to know the θ_{10}^1 component of x_{10}^1, which has to satisfy Eq. (29). We have to solve the following problem for θ_{10}^1:

$$(P_n^2 - \lambda P_n - \beta)\theta^1 10 = \theta_{00}/2 \tag{37}$$

with the two boundary conditions

$$\frac{\partial \theta_{10}^1}{\partial r} = 0 \quad \text{and} \quad P_n \theta^1 10 = 0 \quad \text{at} \quad r = 1 \tag{38}$$

where λ and β are defined by

$$\lambda = i\omega(1 + Pr^{-1}) \quad \text{and} \quad \beta = \omega^2 Pr^{-1} + \xi^4$$

Note that ξ represents, ξ^c.

If the discriminant of Eq. (38) is different from 0, i.e.,

$$\lambda^2 + 4\beta \neq 0 \quad \text{or} \quad \omega^2(1 - Pr^{-1})^2 - 4\xi^4 \neq 0 \tag{39}$$

the general solution of Eq. (38) is

$$\theta_{10}^1 = \alpha_1 In(\gamma_1 r) + \alpha_2 Jn(\gamma_2 r) + \beta_1 In(\xi r) + \beta_2 Jn(\xi r) \tag{40}$$

where β_1, β_2, γ_1, and γ_2 are defined by

$$\beta_1 = -[2\xi^2(\omega^2 Pr^{-1} + \lambda\xi^2)]^{-1}$$
$$\beta_2 = \mu[2\xi^2(-\omega^2 Pr^{-1} + \lambda\xi^2)]^{-1}$$
$$\gamma_1 = \{[\lambda + (\lambda^2 + 4\beta)^{1/2}]/2\}^{\frac{1}{2}}$$
$$\gamma_2 = \{[-\lambda + (\lambda^2 + 4\beta)^{\frac{1}{2}}]/2\}^{\frac{1}{2}} \tag{41}$$

Then the boundary conditions [Eq. (38)] give

$$\alpha_1 = (N1 - N2)/D1 \tag{42}$$

and

$$\alpha_2 = -[\xi\beta_1 I'n(\xi) + \xi\beta_2 J'n(\xi) + \gamma_1 I'n(\gamma_1)\alpha_1][\gamma_2 J'n(\gamma_2)]^{-1} \tag{43}$$

where

$$N1 = (i\omega - \xi^2)\beta_1 In(\xi) + (i\omega + \xi^2)\beta_2 Jn(\xi)$$
$$N2 = (\gamma_2^2 + i\omega)Jn(\gamma_2)\xi[\beta_1 I'n(\xi) + \beta_2 J'n(\xi)][\gamma_2 J'n(\gamma2)]^{-1}$$
$$D1 = (\gamma_1^2 - i\omega)In(\gamma_1) + (\gamma_2^2 + i\omega)Jn(\gamma_2)[\gamma_1 I'n(\gamma_1)][\gamma_2 J'n(\gamma_2)]^{-1}$$

Finally, the expression of k, from Eq. (27), is

$$k = -\xi^2 \mathcal{R}\{\alpha_1(K_1 - \mu K_1') + \alpha_2(K_2 - \mu K_2') + \beta_1(K_3 - \mu K_4)$$
$$+ \beta_2(K_4 - \mu K_4')\}/(K_3 - \mu^2 K_4') \tag{44}$$

where

$$K_1 = \int rIn(\gamma_1 r)In(\xi r) \, dr$$

$$K_1' = \int rIn(\gamma_1 r)Jn(\xi r) \, dr$$

$$K_2 = \int rJn(\gamma_2 r)In(\xi r) \, dr$$

$$K_2' = \int rJn(\gamma_2 r)Jn(\xi r) \, dr$$

$$K_3 = \int rIn^2(xr) \, dr$$

$$K_4 = \int rIn(xr)Jn(xr) \, dr$$

and

$$K'_4 = \int rJn^2(\xi r)\, dr \qquad (45)$$

These integrals [Eq. (45)] have been analytically solved. The computation of Eq. (44) requires an accurate evaluation of ξ when solving Eq. (33) and an accurate determination of the Bessel functions occurring in Eqs. (42), (43), and (45), mainly when $\omega \to 0$. The method has been presented for insulated walls [Eq. (6b)], but it could be extended to the case of perfectly conducting walls [Eq. (6c)]. The values of k have been calculated from Eq. (44) in the case $n = 1$ for a wide range of values of ω and for $10^{-3} \leqslant Pr \leqslant 10^3$. The results are presented in Figs. 1a–1d for $10^{-3} \leqslant Pr \leqslant 10^{-2}$, $10^{-2} \leqslant Pr \leqslant 2.10^{-1}$, $5.10^{-1} \leqslant Pr \leqslant 5$, and $5 \leqslant Pr \leqslant 100$,

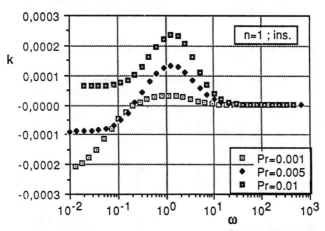

Fig. 1a Alteration coefficient vs ω for the insulated case; $10^{-3} \leqslant Pr \leqslant 10^{-2}$.

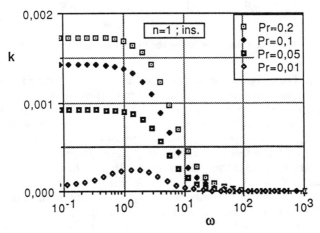

Fig. 1b The value k vs ω for the insulated case; $10^{-2} \leqslant Pr \leqslant 2.10^{-1}$.

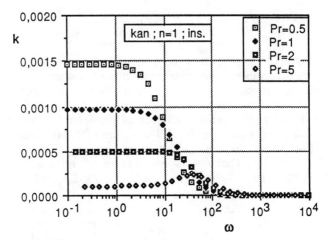

Fig. 1c   The value k vs ω for the insulated case; 0.5 ⩽ Pr ⩽ 5.

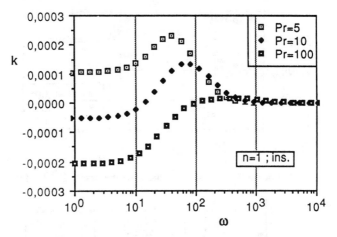

Fig. 1d   The value k vs ω for the insulated case; 5 ⩽ Pr ⩽ 100.

respectively. These figures show that, for the lowest and the highest values of Pr, k becomes negative for small ω, indicating that in this case the modulation of the gravity diminishes the critical value of the Rayleigh number for the onset of the convection compared to the unmodulated case, i.e., $R_0^c < R_{00}^c$. As k appears to tend asymptotically to a minimum when $\omega \to 0$, we considered this limit case, $k(0)$, to give precise domains of Pr for which k can take negative values (Fig. 2). In these domains (which correspond to $Pr < 8.10^{-3}$ and $Pr > 8$), k is negative from $\omega = 0$ up to a limit value of ω, which depends on Pr; typically, this limit is $\omega \cong 0.1$ for $Pr < 8.10^{-3}$ (Fig. 1a) and $\omega \cong Pr$ for $Pr > 8$ (Fig. 1d).

For high ω, k is positive for any Pr and goes to zero when $\omega \to \infty$. This behavior is better illustrated by the log-log plotting in Figs. 3a and 3b for

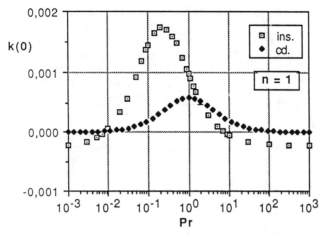

Fig. 2   The value $k(0)$ vs Pr for the insulated and conducting cases.

$0.01 \leqslant Pr \leqslant 0.2$ and $0.5 \leqslant Pr \leqslant 5$, respectively, which shows that the value of k behaves as ω^{-2} when $\omega \to \infty$. This property will be demonstrated in the next subsection for any Pr, in the specific case of the insulated wall, and under a more general hypothesis in the section on asymptotic analysis for large values of w.

Finally, we can note that the condition (40) is always satisfied for $Pr = 1$. But for $Pr \neq 1$, this condition is not satisfied for

$$\omega = \omega_d = 2\xi Pr/|Pr - 1| \qquad (46)$$

For such a frequency the two first terms of the right-hand side of Eq. (39) are no longer linearly independent. A specific study would have to be done in that case, but we observed that, in fact, expression (44) gives continuous results even for $\omega = \omega_d$. This continuity means that the value $\omega = \omega_d$ is not a singular point; this feature will be confirmed by the results of the matrix method developed in the next section.

Limiting Case: $\omega \to \infty$

If we use the asymptotic representation of Bessel functions for large values of the argument, i.e.,

$$In(x) = e^x[1 + 0(x^{-1})](2\pi x)^{-0.5} \qquad (47)$$

and

$$Jn(x) = [\cos(x - n\pi/2 - \pi/4) + \mathcal{O}(x^{-1})](x\pi/2)^{-0.5} \qquad (48)$$

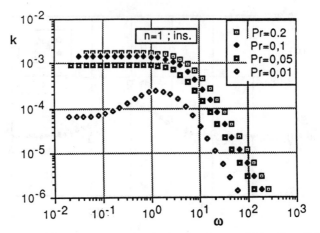

Fig. 3a   The value k vs ω for the insulated case; $0.01 \leqslant Pr \leqslant 0.2$.

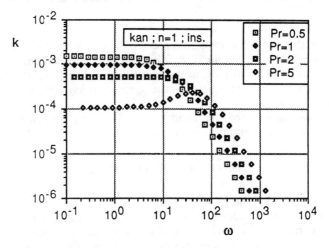

Fig. 3b   The value k vs ω for the insulated case; $0.5 \leqslant Pr \leqslant 5$.

we find that the most important term in the numerator of k in Eq. (44) is

$$\mathscr{R}\{\beta_1 K_3 - \mu\beta_2 K_4'\}$$

Thus,

$$k = \omega^2/[2Pr(\omega^4 Pr^{-2} - \lambda^2\xi^2)] + \mathcal{O}(\omega^{-4}) \tag{49a}$$

$$= 0.5Pr/\omega^2 + \mathcal{O}(\omega^{-4}) \tag{49b}$$

Relation (49b) confirms the behavior seen in Figs. 3a and 3b and shows, in addition, that k is proportional to Pr for large values of ω. Thus, the critical Rayleigh number for large ω is

$$R_0^c = R_{00}^c + 0.5 \, Pr[\epsilon/\omega]^2 \tag{50}$$

It is always greater than R_{00}^c, which is given by Eq. (34).

It can be noted that Eqs. (49) and (50) have been established under condition (39), which is discussed at the end of the previous subsection. For the large values of ω considered in this paragraph, condition (39) is always satisfied. In any case we will recover relation (50) in a quite general case by a direct asymptotic analysis in the section on asymptotic analysis for large values of w.

Matrix Method

Our aim in this section is to present a general method for solving system (14) by the use of trial functions. The advantage of this method is that it can be easily extended for general boundary conditions and geometries, especially for confined cylinders. Here, only the case of infinite cylinders is considered, with either conducting or insulated walls. We mainly focus on case $n = 1$, which corresponds to the basic mode of instability for long cylinders.[24,25] Then the system (14) can be written as

$$-Pr^{-1}\frac{\partial w}{\partial t} = P_1 w - R\theta \tag{51}$$

$$-\frac{\partial \theta}{\partial t} = -w + P_1 \theta \tag{52}$$

with the boundary conditions (5) and (6).

The linear operator P_1 in the Eqs. (51) and (52) is defined from Eq. (16) as

$$P_1 = \frac{\partial^2}{\partial r^2} + r^{-1}\frac{\partial}{\partial r} - r^{-2} \tag{53}$$

we use the expansion

$$w = \Sigma_i a_i(t) w_i(r) \tag{54}$$

$$\theta = \Sigma_i b_i(t) \theta_i(r) \tag{55}$$

where the trial functions w_i and θ_i satisfy the equations

$$P_1 w_i = -\xi_i^2 w_i \tag{56}$$

$$P_1 \theta_i = -\gamma_i^2 \theta_i \tag{57}$$

and the boundary conditions (5) and (6). The solutions are

$$w_i(r) = J_1(\xi_i r) \tag{58}$$

$$\theta_i(r) = J_1(\gamma_i r) \tag{59}$$

J_1 being the first-order Bessel functions and ξ_i and γ_i satisfying the relations

$$J_1(\xi_i) = 0 \tag{60}$$

and

$$J_1'(\gamma_i) = 0 \tag{61a}$$

for insulated walls, or

$$J_1(\gamma_i) = 0 \tag{61b}$$

for conducting walls, or

$$\gamma_i J_1'(\gamma_i) + b J_1(\gamma_i) = 0 \tag{61c}$$

for finite Biot number.

By substituting Eqs. (54) and (55) into Eq. (51) and (52), multiplying these equations by rw_j and $r\theta_j$, respectively, and then integrating on the interval $r \in [0, 1]$, we obtain the system

$$K\frac{\mathrm{d}X}{\mathrm{d}t} = M'X \tag{62}$$

with

$$K = \begin{bmatrix} K_1 & 0 \\ 0 & K_2 \end{bmatrix}$$

and

$$M' = \begin{bmatrix} A_1 & -RC \\ -{}^t C & A_2 \end{bmatrix} \tag{63}$$

where the matrices K_1, K_2, A_1, A_2, and C are defined by their general term

$$K_{1ij} = -\int_0^1 rw_i w_j \, \mathrm{d}r; \qquad K_{2ij} = \int_0^1 r\theta_i \theta_j \, \mathrm{d}r$$

$$C_{ij} = \int_0^1 r\theta_i w_j \, \mathrm{d}r$$

$$A_{1ij} = \int_0^1 rP_n w_i w_j \, \mathrm{d}r; \qquad A_{2ij} = \int_0^1 rP_n \theta_i \theta_j \, \mathrm{d}r$$

and where ${}^tC$ denotes the matrix transpose of C.

General Thermal Boundary Conditions

As in the section on problem formulation, we use the parameter η and the Floquet exponent σ. Thus, the perturbation equation [Eq. (19)] can be written as

$$\left(M'_{00} - \frac{d}{dt}\right)x = \sigma x + \eta N'x + \epsilon \cos\omega t N'x \qquad (64)$$

where M'_{00} and N' are defined by

$$M'_{00} = \begin{bmatrix} K_1^{-1}A_1 & -R_{00}K_1^{-1}C \\ K_2^{-1}\,{}^tC & K_2^{-1}A_2 \end{bmatrix}$$

$$N' = \begin{bmatrix} 0 & K_1^{-1}C \\ 0 & 0 \end{bmatrix} \qquad (65)$$

where K_1^{-1} and K_2^{-1} represent the inverse of K_1 and K_2.

We also look for solution of the form of Eqs. (23) and get recursive equations of the form of Eqs. (25). The solvability condition (26) for these equations can be written as

$$\sigma_{01} = \langle x_{00}^* | N'x_{00} \rangle / \langle x_{00}^* | x_{00} \rangle \qquad (66a)$$

$$\sigma_{20} = 0.5\langle x_{00}^* | \{N'(M'_{00} + i\omega I)^{-1}\}N'x_{00} \rangle / \langle x_{00}^* | x_{00} \rangle \qquad (66b)$$

In these expressions the bar denotes the standard scalar product, I is the identity matrix,

x_{00} is a solution of

$$M'_{00}x_{00} = 0$$

and x_{00}^* is a solution of

$$\,{}^tM'_{00}x_{00}^* = 0$$

Thus, from expressions (66a) and (66b), we can compute the values of k, which are defined by Eq. (27). These computations depend on R_{00}^c (the classical critical Rayleigh number for the unmodulated case), the values of which are 67.963 and 215.560, respectively, for insulated and conducting cases. The computations that have been done for a wide range of values of ω and Pr need a highly accurate algorithm to invert matrices, mainly for insulated walls. The influence of the number of trial functions, L, has been checked; most of the computations have been done with $L = 30$ for insulated walls, whereas $L = 5$ appeared sufficient for conducting walls.

For insulated walls the results well recover those of the analytical method of the section on problem formulation; they give exactly the same graphs as those in Figs. 1–3 and they are not repeated in this chapter.

328   M. WADIH ET AL.

For the conducting case the computations carried out in the range $0 < \omega < 10^4$ are plotted in Figs. 4a and 4b for $0.1 < Pr < 1$ and $1 < Pr < 10$, respectively; they show that k is always positive and reaches a maximum when $\omega \to 0$. The limit value of k when $\omega \to 0$, denoted $k(0)$, is also presented in Fig. 2; this graph, which is plotted as a function of Pr, shows a "symmetry" with respect to $Pr = 1$ (the results being identical for Pr and $1/Pr$), when $\omega \to 0$.

Particular Case ($n = 1$, Conducting Walls)
Because of the symmetry existing in the case $n = 1$, for conducting walls, the matrices K_1, K_2, A_1, A_2, and C are diagonal; thus, expressions (65) can be considerably simplified. Finally, expressions (66) can be analytically

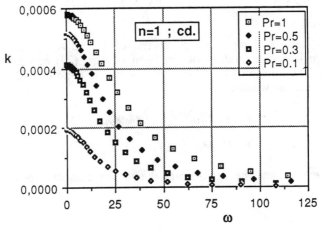

Fig. 4a   The value k vs ω for the conducting case; $0.1 \leqslant Pr \leqslant 1$.

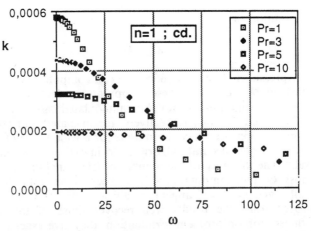

Fig. 4b   The value k vs ω for the conducting case; $1 \leqslant Pr \leqslant 10$.

written, and k can be written as

$$k = 0.5Pr[\omega^2 + R_{00}^c(Pr + 1)^2]^{-1} \qquad (67)$$

For a given Pr, k reaches a maximum when $\omega \to 0$. Then the limiting form of Eq. (67) is

$$k(0) = 0.5R_{00}^{c-1}(Pr^{-\frac{1}{2}} + Pr^{\frac{1}{2}})^{-2} \qquad (68)$$

It exhibits the "symmetry" with respect to $Pr = 1$, as mentioned earlier (see Fig. 2).

The relation (67) has been used to control the validity of the numerical code used in the preceding subsection to calculate k for the entire domain $0 < \omega < \infty$. The agreement is very good, with the three first decimal places being identical. This confirms the results plotted in Figs. 4a and 4b and proves that k is positive for any Pr and w. This is in contrast to the adiabatic case, where k takes negative values for small ω when Pr is small or large enough (see Fig. 2 and Figs. 1a–1d).

For $Pr = 1$, 10, and 0.1 and for $Pr = 0.001$ and 0.01 the graphs plotted in Figs. 6 and 7 are very similar to those given by Roppo et al.[17] for the Bénard problem with free and conducting horizontal surfaces subjected to a temperature modulation (see their Fig. 1). In addition, the results of Roppo et al. show that $R_0^c(Pr = 10)$ equals $R_0^c(Pr = 0.1)$ when $\omega \to 0$, in agreement with the "symmetry" property of Eq. (68).

Another very interesting similarity with the results available in the literature can be observed for the modulated Bénard problem with surface temperature modulation, studied by Venezian[16] and Rosenblat and Herbert.[19] By using Eqs. (8c), (10), and (11b) and putting $R = R_0^c/R_{00}^c$, we get the following quadratic relation for R:

$$kR_{00}^c\epsilon_v^2 R^2 - R + 1 = 0 \qquad (69)$$

As does Eqs. (13) for η, Eq. (69) admits two solutions (for $0 \leqslant \epsilon_v \leqslant \epsilon_v^0$, when $k > 0$) that are, for small ϵ_v,

$$R_1 = 1 + kR_{00}^c\epsilon_v^2 \qquad (70a)$$

$$R_2 = [kR_{00}^c\epsilon_v^2]^{-1} - 1 + kR_{00}^c\epsilon_v^2 \qquad (70b)$$

As $R_1 < R_2$, we only retain the first solution [Eq. (70a)]. Thus, for $\omega \to 0$ expression (67) for k gives

$$R = 1 + 0.5\epsilon_v^2 Pr(Pr + 1)^{-2} \qquad (71)$$

which is identical to the expressions obtained by the previous authors [relations (46) by Venezian[16] and (4.16) by Rosenblat and Herbert[19]] for modulations of surface temperature. Such an identity proves that for $\omega \to 0$ the stability alteration (with respect to the unmodulated case) is not only independent of the type of periodic modulation (surface temperature or gravity), as previously mentioned by Gershuni et al.,[12] but is also independent of the type of geometry (confined or not confined); this property was never mentioned before.

We have to point out that the quadratic relation (69) for R and the second family [Eq. (70b)] have also been obtained by Gershuni et al.[12] by the first-order approximation of a Galerkin method. All of the other authors only considered the first family [Eq. (70a)].

Asymptotic Analysis for Large Values of ω

In this section we present a general asymptotic analysis of equations of the following form:

$$\frac{dX}{dt} = M_0 X + \epsilon \cos\omega t N X \tag{72}$$

in which M_0 has distinct eigenvalues and N is a nilpotent matrix (i.e., $N^2 = 0$). Of course, the conditions for M_0 and N were already satisfied in Eq. (24). But they apply for more general situations (other kinds of periodic modulations or geometries); they are satisfied for most of the periodic modulations usually considered in the literature (gravity modulation or temperature modulation of surfaces). Note that, in addition, Eq. (72) is of the same type as Eq. (62) if K is invertible.

The present analysis is developed for large ω. But, in contrast to the previous paragraphs, where ϵ was assumed to be small, we now consider that ϵ may increase, but such that

$$\epsilon/\omega \ll 1 \tag{73}$$

Venezian[16] also developed an asymptotic analysis, but he assumed that ϵ remains small and explained the reasons for which the inequality (73) limits the validity of his own results. Another asymptotic analysis has been proposed by Gershuni et al.,[12] with finite-amplitude modulations, but in the frame of separation of variables solution; they derived an analytical relation that will be discussed further [through expression (87)].

We now introduce a new small parameter,

$$\delta = \epsilon/\omega \tag{74}$$

and look for solutions of the form

$$X = \exp[u_1(t)N + u_2(t)N_2]Y \tag{75}$$

with

$$u_1(t) = \delta \sin\omega t, \qquad u_2(t) = \delta^2 \sin2\omega t/4\omega$$

and

$$N_2 = NM_0 N \tag{76}$$

Then Eq. (72) can be written as

$$\frac{dY}{dt} = L_0 Y - (u_1 M_2 - u_2 M_3)Y - (2u_1 u_2 N_2 M_0 N + u_2^2 N_2 M_0 N_2)Y \tag{77}$$

with

$$L_0 = M_0 - 0.5\delta^2 N_2$$

$$M_2 = NM_0 - M_0 N$$

and

$$M_3 = N_2 M_0 - M_0 N_2$$

A $t0(\delta^3)$, Eq. (77) can be written as

$$\frac{dY}{dt} = L_0 Y - \delta \sin\omega t M_2 Y + \frac{\delta^2 \sin2\omega t}{4\omega} M_3 Y \tag{78}$$

Equation (78) can be solved as in the previous sections by using η and δ instead of ϵ:

$$L_0 = L_{00} - \eta N$$

$$Y = Y_{00} + \Sigma \delta^p \eta^q Y_{pq} \quad p + q \geqslant 1 \tag{79}$$

$$\sigma = \sigma_{00} + \Sigma \delta^p \eta^q \sigma_{pq} \quad p + q \geqslant 1 \tag{80}$$

where L_{00} is obtained from L_0 by taking R_{00} instead of R_0.

When the principle of exchange of stability is valid, critical conditions are obtained for

$$\eta_c = -\{0.5\langle N_2 Y_{00}|Y_{00}^*\rangle + \overline{\langle \sin\omega t M_2 Y_{10}|Y_{00}^*\rangle}\}\delta^2$$

$$\div \langle NY_{00}|Y_{00}^*\rangle + \mathcal{O}(\delta^4) \tag{81}$$

where the bar represents the time averaging.

If we note that the second term in Eq. (81) is of $\mathcal{O}(\omega^{-2})$, since

$$\overline{\sin\omega t M_2 Y_{10}} = 0.5 M_2 \mathcal{R}(L_{00} - i\omega)^{-1} M_2 Y_{00} \tag{82a}$$

$$= 0.5 \, M_2 L_{00}[(L_{00})^2 + \omega^2]^{-1} M_2 Y_{00} \tag{82b}$$

we have

$$\eta_c = -\{0.5 < N_2 Y_{00} | Y_{00}^* \rangle / \langle N Y_{00} | Y_{00}^* \rangle + \mathcal{O}(\omega^{-2})\}\delta^2 + \mathcal{O}(\delta^4) \tag{83}$$

This is quite a universal result in the sense that it applies for any kind of fluid layer geometries and periodic constraints, only under the assumptions that $N^2 = 0$ and that the principle of exchange of stability is valid. With appropriate operators this analysis can also be applied to bounded domains.

In the case of an infinite vertical cylinder, considered in the previous paragraphs, we have from Eq. (76),

$$N_2 = \begin{bmatrix} 0 & Pr \\ 0 & 0 \end{bmatrix}\begin{bmatrix} PrP_n & -PrR_0 \\ -1 & P_n \end{bmatrix}\begin{bmatrix} 0 & Pr \\ 0 & 0 \end{bmatrix} = \begin{bmatrix} 0 & -Pr^2 \\ 0 & 0 \end{bmatrix}$$

$$= \begin{bmatrix} 0 & -Pr^2 \\ 0 & 0 \end{bmatrix} = -PrN \tag{84}$$

Substitution of Eq. (84) into Eq. (83) gives

$$\eta_c = [0.5Pr + \mathcal{O}(\omega^{-2})]\delta^2 + \mathcal{O}(\delta^4)$$

or, in terms of ϵ,

$$\eta_c = \epsilon^2[0.5Pr\omega^{-2} + \mathcal{O}(\omega^{-4})] \tag{85}$$

Thus, using definition (11b) for k, we find

$$k = 0.5Pr\omega^{-2} \tag{86}$$

which is identical to Eq. (49b). Relation (85) can be compared to relation (4.4) of Gershuni et al.,[12] also for large ω, by again using $R = R_0^c/R_{00}^c$ and $\epsilon_v = \epsilon^*/g_0$. Thus, we have

$$R - 1 = 0.5R^2 k_v \epsilon_v^2 \qquad \text{with} \qquad k_v = R_{00}^c Pr\omega^{-2} \tag{87}$$

Expression (4.4) of Gershuni et al.[12] recovers Eq. (87), but with a different coefficient, $k_v = m^{-2}Pr\omega^{-2}$, where m is such that

$$R_{00}^c m^2 = 1 + (0.5\mu_1^2 - 4)/(b + 3)^2$$

where b is the Biot number as introduced in condition (5a), and μ_1 is the first solution of $J_1(\mu_1)$, i.e., $\mu_1 = 3.832$. Thus, the two expressions are identical for the limiting case $b\mathcal{Æ}_\infty$, where $R_{00}^c m^2 = 1$. This agreement is probably due to the fact that in this case (conducting walls) all of the matrices (K_1, K_2, A_1, A_2, and C) are diagonal, and the first-order approximation of Gershuni et al.[12] in their relation (1.17) is valid. But for finite b their expression (4.4) would involve an extra effect of the boundary conditions, whereas in Eq. (87) this effect of the boundary conditions only enters through R_{00}^c. Perhaps the first-order approximation of Gershuni et al.[12] is no longer valid for finite b.

We can note that expression (87), as well as relation (4.4) of Gershuni et al.,[12] admit two solutions: one valid for R_0^c close to R_{00}^c and the other for large R_0^c. The first solution, which has to be retained when $\epsilon_v \ll \omega$, can be re-expressed as

$$R = 1 + 0.5 k_v \epsilon_v^2 \tag{88}$$

or

$$\eta_c = 0.5 R_{00}^c k_v \epsilon_v^2 = 0.5 Pr \omega^{-2} R_{00}^{c2} \epsilon_v^2 \tag{89}$$

Other Nondimensionalizations

We can note that the asymptotic expression (86), for $\epsilon/\omega \ll 1$, can be expressed in terms of dimensional variables as

$$\eta_c = 0.5(\alpha\gamma)^2 (r_0^4/\nu\kappa)[\epsilon^*/\omega^*]^2 \tag{90}$$

or

$$R_0^c/R_{00}^c = 1 + 0.5\alpha\gamma g_0^{-1}[\epsilon^*/\omega^*]^2 \tag{91}$$

where $[\epsilon^*/\omega^*]$ represents a vibrational velocity, and $[g_0/\alpha\gamma]^{0.5}$ a reference velocity. Expression (91) is independent of r_0, ν, κ, and Pr, and expression (90) shows that $r_0^2/(\nu\kappa)^{0.5}$ should be taken as a reference time, as in the study of Gershuni et al.[12] Introducing the new nondimensionalized frequency $\Omega = \omega/Pr^{0.5}$, which corresponds to this reference time, the asymptotic expression (86) becomes independent of Pr, and we have

$$\eta_c = 0.5[\epsilon/\Omega]^2 \tag{92a}$$

or

$$\eta_c = 0.5[R_{00}^c \epsilon_v \Omega]^2 \tag{92b}$$

It would also be interesting, instead of k, to introduce the variable k'', defined by

$$k'' = k\Omega^2 \tag{93}$$

such that

$$\eta_c = k''[\epsilon/\Omega]^2 \qquad (94a)$$

or

$$\eta_c = k''[R_{00}^c \epsilon_v/\Omega]^2 \qquad (94b)$$

Then the limit, when $\Omega \to \infty$, is simply a constant, $k'' = 0.5$.

Another interesting feature is that for the conducting case expression (67) becomes

$$k''(\Omega,Pr) = 0.5[1 + R_{00}^c(Pr^{0.5} + Pr^{-0.5})^2\Omega^{-2}]^{-1} \qquad (95)$$

and exhibits a symmetry with respect to $Pr = 1$, i.e., $k''(\Omega,Pr) = k''(\Omega,1/Pr)$, and presents a maximum at $Pr = 1$. The graphs of k'' are plotted as a function of Ω in Fig. 5 for several values of Pr (i.e., $Pr = 1$, 10^{-1}, 10^{-2}, and 10^{-3}).

Another interesting nondimensionalization is to introduce the nondimensional frequency

$$\omega' = \omega/(R_{00}^c Pr)^{0.5} \qquad (96)$$

and the relative alteration factor

$$k' = kR_{00}^c \qquad (97)$$

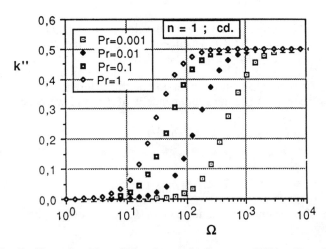

Fig. 5   The value k'' vs Ω for the conducting case; $0.001 \leqslant Pr \leqslant 1$.

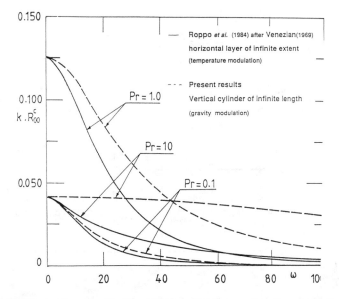

Fig. 6 Modified alteration factor kR_{00}^c vs frequency, for the conducting case; comparison with the results plotted by Roppo et al.[1] after Venezian[16] for the modulated Bénard problem, at $Pr = 0.1$, 1, and 10.

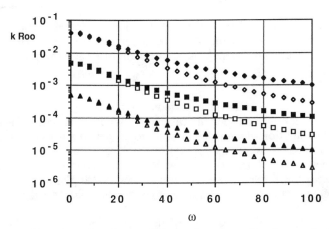

Fig. 7a Modified alteration factor kR_{00}^c vs frequency, for the conducting case; comparison with the results of Venezian[16] for the modulated Bénard problem, at $Pr = 0.001$, 0.01, and 0.1.

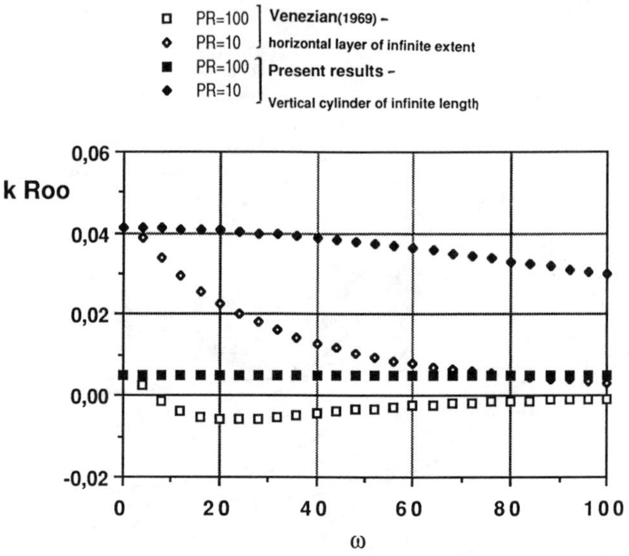

Fig. 7b Modified alteration factor kR^c_{00} vs frequency, for the conducting case; comparison with the results of Venezian[16] for the modulated Bénard problem, at $Pr = 10$ and 100.

The interest of this nondimensionalization, which corresponds to the reference time,

$$t_{ref} = 1/(\alpha g_0 \gamma)^{0.5} \qquad (98)$$

is to take account of the effect of the inertia forces. Also, for large frequencies these forces control the behavior of the system. This is shown in the Figs. 8a and 8b, where k' is independent of Pr and thermal conditions.

Case of Biperiodic Modulations

In this section we consider the case where the residual gravity is assumed to be a combination of sinusoidal oscillations around a nonzero mean value and of periodic fluctuations (peaks) of small amplitudes.

The method developed shows that the critical conditions can be simply determined for different wall conductances by applying the results of the previous sections for the bounded cylinder as well as for the unbounded one.

We consider a residual gravity of the form

$$g = \mu g_0 + \epsilon_0^* \cos\omega_0^* t^* + f(t^*) \qquad (99)$$

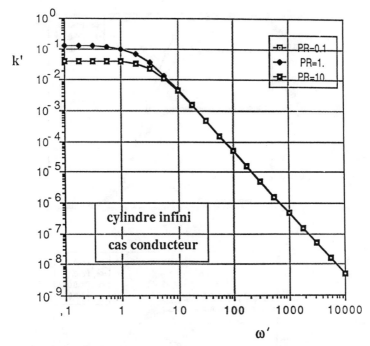

Fig. 8a Relative alteration factor K' according to ω' in the case of infinite cylinder for perfectly conducting walls.

where $f^*(t^*)$ is a continuous, differentiable, and periodic function characterizing infinitesimal fluctuations.

Also, we can write this function in the form

$$f(t^*) = \sum_{\substack{m=1 \\ 2L}}^{m=\infty} \epsilon_m^* \cos(\omega_m^* t^*) \qquad (100)$$

with

$$\epsilon_m^* = 1/L \int_0^L f(t^*) = \sum_{m=1} \epsilon_m^* \cos(\omega_m^* t^*)$$

with $2L$ being the period of $f(t^*)$: $2L = 2\Pi/\omega_1^*$ and $\omega_m^* = m\omega_1^*$; $m = 1, 2, \ldots$.

If we define a small parameter ϵ^* by

$$\epsilon_m^* = a_m \epsilon^*$$

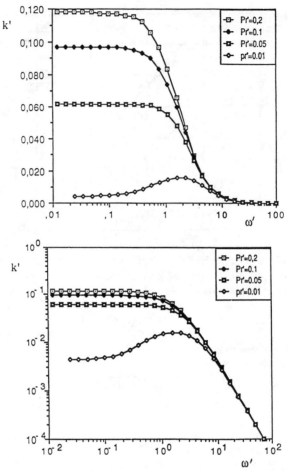

Fig. 8b   **Relative alteration factor K' according to ω' in the case of infinite cylinder for insulating walls.**

we can write the instantaneous Rayleigh number in the form

$$R(t) = R_0 + \epsilon_0 \cos\omega_0 t + \epsilon \sum_{m=1}^{m=\infty} a_m \cos\omega_m t \tag{101}$$

which introduces, as in the previous sections, the mean Rayleigh number

$$R_0 = \alpha g_0 \gamma r_0^4/\nu\kappa$$

and the vibrational Rayleigh numbers

$$\epsilon_0 = \alpha \epsilon_0^* \gamma r_0^4/\nu\kappa$$

for the basic oscillations and

$$\epsilon = \alpha \epsilon^* \gamma r_0^4 / \nu \kappa$$

for the fluctuations.

In order to solve system (4) we introduce the parameter ν and the Floquet exponent σ. Then the system becomes

$$\left(M_{00} \frac{d}{dt} \right) = \sigma X + \eta N X + \left(\epsilon_0 \cos\omega_0 t + \epsilon \sum_{m=1} a_m \cos\omega_m t \right) N X \qquad (102)$$

and the same routine used in the previous sections leads to the expression of η_c:

$$\eta_c = k(\omega_0)\epsilon_0^2 + \epsilon^2 \sum_{m=1} a_m^2 k(\omega_m) + 2a_{m0}k(\omega_0)\epsilon_0\epsilon \qquad (103)$$

where

$$k(\omega) = -R_e \langle Nf(\omega) | x_{00}^{0*} \rangle | \langle Nx_{00}^0 | x_{00}^{0*} \rangle \qquad (104)$$

with $f(\omega)$ being the solution of the system

$$(M_{00} - i\omega)f = \tfrac{1}{2} Nx_{00}^0 \qquad (105)$$

and x_{00}^0 and x_{00}^{0*} are, respectively, the solution and adjoint solution of the problem without modulations.

Expression (103) has been established only in the case where there is interference, $m0$ being defined, then, by $\omega_{m0} = \omega_0$.

Case of a Bounded Cylinder

Problem Formulation

We consider a cylinder of length L and radius $r0$, heated from below. We denote ΔT the difference of temperature $(T1 - T0)$, with temperature $T0$ on the upper boundary and $T1$ on the lower boundary, such as $T1 > T0$.

By introducing the nondimensional axial and radial variables, z' and r', respectively, by

$$z' = z/L \qquad \text{and} \qquad r' = r/L$$

and by defining the aspect ratio A by

$$A = r0/L$$

we obtain by applying the Boussinesq approximation[23] the linearized system in nondimensional variables (any confusion being possible, the prime is supressed for nondimensional variables),

$$\frac{\partial U}{\partial t} = -Pr\nabla_p + PrR(t)\theta e_z + Pr\nabla^2 U \tag{106a}$$

$$\nabla \cdot U = 0 \tag{106b}$$

$$\frac{\partial \theta}{\partial t} = Ue_z + \nabla^2\theta \tag{106c}$$

where e_z is the unit vector in the vertical upward direction. The coordinate origin takes on the middle of the cylinder $\{0 \leqslant r \leqslant 1 \text{ and } -\frac{1}{2} \leqslant z \leqslant \frac{1}{2}\}$. The instantaneous Rayleigh number is defined by

$$R(t) = \alpha g(t) \, \Delta t L^4/(vkr_0) \tag{107}$$

We consider the following boundary conditions:

$$U = 0, \qquad \theta = 0; \qquad 0 \leqslant r \leqslant 1 \tag{108}$$

for conducting walls on the top and bottom $(z \leqslant \frac{1}{2})$, and

$$U = 0, \qquad \theta = 0; \qquad -\frac{1}{2} \leqslant z \leqslant \frac{1}{2} \tag{109a}$$

for conducting side boundaries $(r = 1)$, or

$$U = 0; \qquad \frac{\partial \theta}{\partial r} = 0; \qquad -\frac{1}{2} \leqslant z \leqslant \frac{1}{2} \tag{109b}$$

for insulating side boundaries $(r = 1)$.

Charlson and Sani[23] have shown that, in the axisymmetric case $(n = 0)$, the solution separated into two classes according to parity in z, and it is the class giving and odd vertical velocity in z, which the first become instable. Thus, we directly use this result, which respects the following symmetries:

$$u_r(r,z) = -u_r(r,-z)$$

$$u_z(r,z) = u_z(r,-z)$$

$$\theta(r,z) = \theta(r,-z)$$

These considerations lead us to define a function $\phi(r,z,t)$ by

$$U = \mathrm{rot}(\phi e_\phi) \tag{110}$$

and the application of the Galerkin method leads us to set down

$$U = \sum_{ij} B_{ij}(t) U_{ij} \tag{111a}$$

$$\theta = \sum_{ij} C_{ij}(t) \theta_{ij} \tag{111b}$$

We multiply the Eq. (106a), respectively, by U_{k1} and T_{k1}, and we substitute Eqs. (106a) and (106b) into the equation obtained. The integration over the cylindrical domain

$$\Omega = \{0 \leqslant r \leqslant 1, -\tfrac{1}{2} \leqslant z \leqslant \tfrac{1}{2}\}$$

leads to the system

$$Pr^{-1} \sum_{ij} K1 \frac{dB_{ij}}{dt} = \sum_{ij} ZB_{ij} + R(t) \sum_{ij} M1B_{ij} \tag{112a}$$

$$\sum_{ij} K2 \frac{dC_{ij}}{dt} = S_{ij} M2B_{ij} + \sum_{ij} SB_{ij} \tag{112b}$$

with

$$K1 = \int U_{ij} \cdot U_{kl} \, d\Omega; \qquad Z = \int \nabla^2 U_{ij} \cdot U_{kl} \, d\Omega$$

$$M1 = \int T_{ij} e_z \cdot U_{kl} \, d\Omega; \qquad M2 = \int U_{ij} e_z \cdot T_{kl} \, d\Omega$$

$$K2 = \int T_{ij} T_{kl} \, d\Omega; \qquad S = \int \nabla^2 T_{ij} T_{kl} \, d\Omega \tag{112c}$$

As in the infinite cylinder, we introduce into the definition (107) of $R(t)$ the mean Rayleigh number and the vibrational Rayleigh number ϵ. In the case of a cylinder of finite length, these numbers have the following expressions:

$$R_0 = \alpha g_0 \, \Delta t L^4/(\nu k r_0) \qquad \text{and} \qquad \epsilon = \alpha \epsilon^* \, \Delta t L^4/(\nu k r_0) \tag{113}$$

where g_0 and ϵ^* are always the mean and the amplitude of modulations of the gravity, respectively. System (112) can finally be written in the form

$$K\frac{dX}{dt} = M_0 X + \epsilon \cos\omega t N X \qquad (114)$$

with

$$K = \begin{bmatrix} K_1 & 0 \\ 0 & K_2 \end{bmatrix} \quad M_0 = \begin{bmatrix} Z & R_0 M_1 \\ M_2 & S \end{bmatrix} \qquad (114a)$$

and

$$N = \begin{bmatrix} 0 & M_1 \\ 0 & 0 \end{bmatrix} \qquad (114b)$$

Resolution

Resolution in the Absence of Modulations
The determination of different matrices from general elements in i, j, k and l, and also the determination of trial functions, are easy. It is henceforth a classical case in the determination of threshold appearance of convective motions in a vertical cylinder under steady constraints.[23]
Indeed, the symmetrical considerations lead us to consider the function

$$\psi_{ij} = \phi_{ij}/r \qquad (115)$$

where ϕ_{ij} is defined in Eq. (110), and to lay down

$$\psi_{ij}(r,z) = \psi_i(r)F_j(z) \quad \text{and} \quad \theta_{ij}(r,z) = T_i(r)\theta_j(z) \qquad (116)$$

where F_j and θ_j are symmetrical in z. For $\psi_i(r)$, $F_j(z)$, and $\theta_j(z)$ regular functions, satisfying the boundary conditions and obeying the following equations:

$$A^2\psi_i = \lambda_i^4 \psi \qquad (117a)$$

with

$$\psi_i(1) = 0, \quad \frac{dT_i(r)}{dr(1)} = 0 \qquad (117b)$$

where A is an operator defined by

$$A = r\frac{d}{dr}\left(r^{-1}\frac{d}{dr}\right)$$

$$\frac{d^4F_j}{dz} = \delta_j^4 F_j \qquad (118a)$$

with

$$F_j(-\tfrac{1}{2}) = 0 \quad \text{and} \quad \frac{\mathrm{d}F_j}{\mathrm{d}z(\tfrac{1}{2})} = 0 \tag{118b}$$

$$PT_i = -\gamma_i^2 T \tag{119a}$$

with $T_i(1) = 0$ for the conducting case or $\mathrm{d}T_i/\mathrm{d}r(1) = 0$ for the insulating case. P is defined by

$$P = 1/r \frac{\mathrm{d}}{\mathrm{d}r}\left(r\frac{\mathrm{d}}{\mathrm{d}r}\right) \tag{119b}$$

$$\frac{\mathrm{d}^2\theta_j}{\mathrm{d}z^2} = -\alpha_j^2\theta_j \tag{120a}$$

with

$$\theta_j(-\tfrac{1}{2}) = 0 \quad \text{and} \quad \theta_j(\tfrac{1}{2}) = 0 \tag{120b}$$

The resolution of the last equations leads to the following functions:

$$\psi_i(r) = r[I_1(\lambda_i r) - \mu_i J_1(\lambda_i r)] \tag{121a}$$

with

$$\mu_i = I_1(\lambda_i)/J_1(\lambda_i) \tag{121b}$$

where λ_i, according to boundary conditions, check the equation

$$I_1(\lambda_i)/J_0(\lambda_i) - I_0(\lambda_i)/J_1(\lambda_i) = 0 \tag{121c}$$

and where J_0, J_1 and I_0, I_1 are the Bessel function and modified Bessel function of order 0 and 1, respectively.

$$T_i(r) = J_0(\gamma_i r) \tag{122a}$$

with γ_i checking

$$J_0(\gamma_i) = 0 \tag{122b}$$

for the conducting case, or

$$J_1(\gamma_i) = 0 \tag{122c}$$

for the insulating case.

$$F_j(z) = \text{ch}(\delta_j z)/\text{ch}(\delta_j/2) - \cos(\delta_j z)/\cos(\delta_j/2) \qquad (123a)$$

where δ_j check the equation

$$\text{sh}(\delta_j)/\text{ch}(\delta_j/2) + \sin(\delta_j)/\cos(\delta_j/2) = 0 \qquad (123b)$$

and

$$\theta_j(z) = \cos(2j - 1)\pi z \qquad (124)$$

Resolution in the Presence of Modulations

We apply directly the matrix method developed by Wadih and Roux (1988) for an infinite vertical cylinder. The introduction of the alternation function η is defined by

$$\eta = R_0 - R_{00}^c$$

where R_0 and R_{00}^c are, respectively, the mean Rayleigh number and the critical Rayleigh number in the unmodulated conditions, and the Floquet exponent σ leads to the system

$$K\frac{dX}{dt} = (M_{00} - \sigma 1)X + \eta NX + \epsilon \cos\omega t NX \qquad (125)$$

where M_{00} is obtained from M_0 by taking R_{00}^c instead of R_0. Similar to the case of an infinite cylinder dealt with by Wadih and Roux the resolution of the last system leads to the alteration factor k defined by

$$\eta_c = k\epsilon^2$$

where η_c corresponds to the critical conditions and thus gives

$$k = -\mathscr{R}(\sigma_{20}/\sigma_{01}) \qquad (126)$$

with

$$\sigma_{20} = 0.5 < x_{00}^* I\{N(M_{00} + i\omega 1)^{-1}\}Nx_{00}\rangle/\langle x_{00}^* Ix_{00}\rangle$$

$$\sigma_{01} = \langle x_{00}^* INx_{00}\rangle/\langle x_{00}^* Ix_{00}\rangle \qquad (127)$$

where x_{00} and x_{00}^* are, respectively, the solution of the steady system and the solution of the adjoint system. For the large values of the frequency, we have established an asymptotic general relation:

$$\eta_c = \{ - (\tfrac{1}{2}) < N_2 X_{00} | X_{00}^* > / < N X_{00} | X_{00}^* > + \mathcal{O}(\omega^{-2}) \} (\epsilon/\omega)^2 \tag{128}$$

with $N_2 = N M_{00} N$.

Results and Discussion

Case of an Infinite Cylinder

We have seen that a perfect agreement is found, in the case $n = 1$, between the values of k given by the analytical methods presented in and the subsection on the limiting case $(\omega \to \infty)$ and the section on asymptotic analysis for large values of ω, and by the section in which the matrix method is described when using 30 and 5 trial functions, respectively, for insulated and conducting walls.

The knowledge of k and R_{00}^c (which can be easily found) permits us to determine h_1 from Eq. (13a) or R_1 from Eq. (70a). In both cases, for $k > 0$ the solution only exists for $\epsilon_v \leqslant \epsilon_v^0$, where $\epsilon_v^0 = 0.5[kR_{00}^c]^{-\frac{1}{2}}$. Values of ϵ_v^0 can be simply derived in the following particular cases. For conducting walls we have seen in Eq. (67) that k is maximum at $\omega = 0$ for any Pr; thus, using Eq. (68) we have a lower limit of ϵ_v^0:

$$\epsilon_v^0(0) = 2^{-\frac{1}{2}}(Pr^{-\frac{1}{2}} + Pr^{\frac{1}{2}}) \tag{129}$$

For example, $\epsilon_v^0(0) = 2^{\frac{1}{2}}$ at $Pr = 1$. However, for large ω, in both conducting and adiabatic cases the use of Eq. (86) leads to

$$\epsilon_v^0 = \omega[2R_{00}^c Pr]^{-\frac{1}{2}} \tag{130}$$

For conducting walls the existence of a maximum enhancement for $\omega = 0$ and $Pr = 1$, as shown by Eq. (67), was also exhibited in previous studies concerning the Bénard problem with a surface temperature modulation. From a physical point of view it is well known that for very low frequencies surface temperature modulations affect the entire volume of the fluid exactly as gravity modulations, and then the same law must describe the alteration of the stability limit for both kinds of modulation. The new feature obtained herein is the absence of the effect of the geometry configuration (e.g., confined or not) on this alteration, whereas the boundary conditions (conducting or insulated walls) have a significant effect, as shown in Fig. 2.

For high frequencies it is physically difficult to make direct comparisons with the Bénard problem with surface temperature modulation, since under such a modulation only a thin layer near the walls is concerned, and then the equilibrium state tends to that of the unmodulated case. However, under gravity modulations high frequencies correspond to a renormalization of the static gravity field, and then, when frequency modulations are large enough in comparison with the characteristic time for the diffusion processes (temperature and vorticity), the buoyancy force takes a mean value that leads to the equilibrium state of the unmodulated case.

To our knowledge, there is no asymptotic relation for large ω in the previous works that allows, like here, a conclusion concerning the similarity of the behavior of systems under the modulations of surface temperature or gravity. Under the present assumptions (i.e., $\epsilon \ll \omega$), the alteration behaves as ω^{-2} for both kinds of modulation. For the gravity modulation the effect of the thermal boundary conditions appears clearly; the alteration is simply proportional to R_{00}^c. This result shows a difference with the results of Gershuni et al.,[12] which present an extra effect of these thermal conditions.

For insulated walls two domains of Pr have been found ($0 < Pr < 8.10^{-3}$ and $8 < Pr < \infty$), in which the effect of modulation is a destabilizing one for small ω. The existence of such a destablization, due to the modulation, has already been mentioned for the modulated Bénard problem in infinite horizontal layers, under special circumstances. Venezian[16] reported such an effect in the case of free and conducting horizontal surfaces, where the temperature of these surfaces is modulated in phase. Homsy,[21] also for free and conducting horizontal surfaces, but with a gravity modulation, showed that $R_0^c < R_{00}^c$ in the extreme cases where $Pr \to 0$ and $Pr \to \infty$. In addition, identical values of R_0^c were found by these authors in both of these extreme cases; a similar property can be seen in the present results for the adiabatic case for $\omega \to 0$ (Fig. 2).

We also proved in the section on asymptotic analysis for large values of ω that the asymptotic relation (50) established for small ϵ is still valid for finite ϵ, such that $\epsilon \ll \omega$ (or $\epsilon^* \ll \omega^*$). Relation (90) shows that the modulation always leads to a stabilization for large ω^*, even for finite ϵ^*. The inequality $\epsilon^* \ll \omega^*$, which limits the mathematical validity of the method presented herein, is not a severe physical limitation; in particular, it is easily satisfied for several classes of fluids (see Tables 1 and 2) in the case of the g jitter modulations.

Case of a Bounded Cylinder

We present first the calculation results for a particular aspect ratio $A = 2.665$ and for $Pr = 0.01$. In the presence of modulation we find the

Table 2   Typical microgravity conditions

g-Jitter frequency, Hz	$10^{-1} \leqslant \omega^* \leqslant 10^2$
g-Jitter amplitude, cm $\cdot$ s^{-2}	$10^{-2} \leqslant \epsilon^* \leqslant 10^{-1}$
Mean μg, cm $\cdot$ s^{-2}	$10^{-2} \leqslant g_0 \leqslant 10^{-1}$
Furnace radius, cm	$1 \leqslant r_0 \leqslant 10$
Temperature gradient, °C/cm	$1 \leqslant \gamma \leqslant 10$
Time reference	$t_{\text{ref}} = r_0^2/\kappa$
Nondimensional frequency	$\omega = \omega^*, \ t_{\text{ref}} = \omega^* r_0^2/\kappa$

critical Rayleigh numbers, $R_{00}^c = 1790.75$ and $R_{00}^c = 1802.96$, for insulating side walls and perfectly conducting walls, respectively. These values agree with those calculated by Charlson and Sani[23] and Buell and Catton. In the case of the gravity modulation the results obtained by the matrix method are very close to those given by the asymptotic relation for $\omega \to \infty$. Figures 9a and 9b show the variation of alteration factor values given, respectively, by the direct method used in this chapter and by an asymptotic analysis given by relation (83). A perfect agreement is observed for the axisymmetric case at which the study is restricted, as well in the case of perfectly conducting walls as in that of insulating walls, from a nondimensional frequency $\omega' \cong 40$.

On other hand, those figures confirm the analysis made about an infinite cylinder, concerning the independence of alteration according to thermal boundary conditions for the large frequencies, for the aspect ratio considered.

Figures 10a and 10b also show that the effect of modulation for insulating walls as well as for conducting walls is stabilizing for high

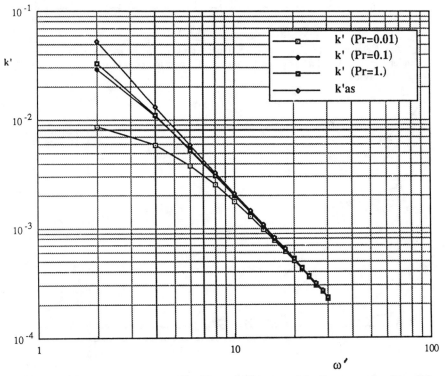

Fig. 9a   Relative alteration factor (K') according to ω' in the case of a bounded cylinder for $A = 4$ and k' as given by asymptotic relation (83) in the case of perfectly conducting walls.

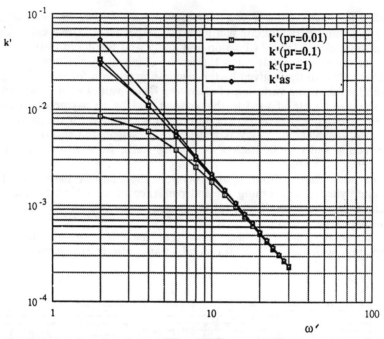

Fig. 9b Relative alteration factor (K') according to ω' in the case of a bounded cylinder for $A = 4$ and k' as given by asymptotic relation (83) in the case of perfectly insulating walls.

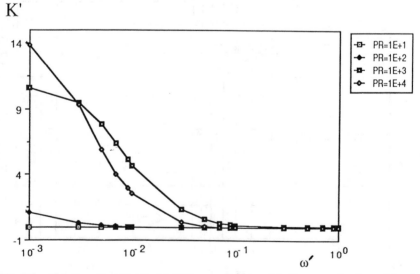

Fig. 10a Relative alteration factor (K') according to ω' in the case of a short-sighted cylinder for $n = 0$, $A = 6$, and $Pr = 10$, 100, 1000, and 10,000 in the case of insulating walls.

K'

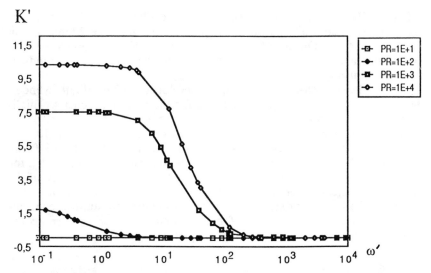

Fig. 10b   Relative alteration factor (K') according to ω' in the case of a short-sighted cylinder for $n = 0$, $A = 6$, and $Pr = 10, 100, 1000$, and $10,000$ in the case of perfectly conducting walls.

frequencies in the case of a short-sighted cylinder for the axisymmetric mode ($n = 0$) for the aspect ratio considered. This effect tends toward 0, when $\omega \to \infty$, as for the case of an infinite cylinder. This can be easily explained by the asymptotic relation (83). Indeed, as a result of the particular form of the operator N, the matrix product $N_2 = NM_0N$ does not depend on R_0 and thus does not depend on ω. The alteration function η_c is simply proportional to $(\epsilon/\omega)^2$. The behavior of the system in the case of a short-sighted cylinder therefore follows a law in $1/\omega^2$ for high frequencies, getting back the same physical interpretation as for an infinite cylinder.

Other calculations have been also made for an aspect ratio $A = 4$, corresponding to a flat cylinder. These calculations have been made for $Pr = 10, 100, 1000$, and $10,000$ in order to extract the first conclusion concerning the effect of Prandtl number in the case of flat cylinders on the stability of the systems under the modulation of gravity.

By analogy with the conclusion established in the case of an infinite vertical cylinder, we introduce the relative alteration coefficient K', which is defined by

$$K' = KR_{00}^c$$

where R_{00}^c is the value of the critical Rayleigh number for a given aspect ratio. The introduction of the nondimensional frequency ω' is defined likewise for the infinite cylinder by

$$\omega' = \omega/(Pr \cdot R_{00}^c)^{\frac{1}{2}}$$

Figures 10a and 10b show, respectively, the evolution of K' for conducting and insulated walls according to ω', for $10^{-3} \leqslant \omega' \leqslant 1$ and for an aspect ratio $A = 6$. On these figures we represented K' corresponding to the same four values of Prandtl number. We note that the effect of modulations is stabilizing.

In accordance with the same analysis developed for an infinite cylinder, the behavior of K' is independent of Pr for the large frequencies. This is confirmed in the presentation of K' according to ω' on Fig. 11, corresponding to insulating walls for the following values of Pr: 1, 0.1, 0.01, 10, and 100, and for an aspect ratio $A = 4$. In this figure, for each value of Pr we represented the curve K' according to ω' and the asymptotic curve K'_{as} for the same values of the frequency ω'. We note that K'_{as} is independent of the Prandtl number, and for the large frequencies K' tends towards K'_{as}, and thus toward a behavior that is independent of the Prandtl number, confirming the physical analysis drawn for the infinite cylinder concerning the supremacy of inertia forces for large frequencies.

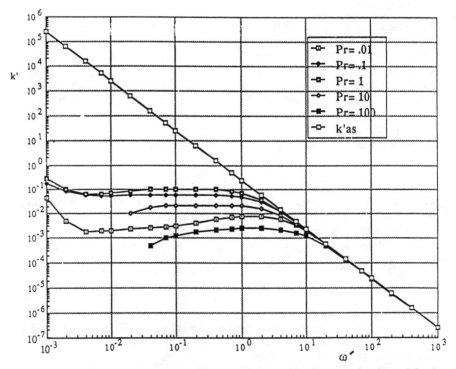

Fig. 11 Relative alteration factor (K') according to ω' in the case of a short-sighted cylinder for $n = 0$, $A = 4$, k' as given by asymptotic relation (83) in the case of insulating walls.

Conclusion

The case of gravity modulation in a long vertical cylinder has been examined. A method has been proposed for analyzing onset of convection in the case of periodic gravity modulation of small amplitude by using the Floquet theory. Since the stability threshold was connected to the one for the unmodulated case, R_{00}^c, by the relation $R_0^c = R_{00}^c + k\epsilon^2$, we only needed to determine the value of the alteration coefficient k. For certain cases an analytical approach is possible as presented in the section on the analytical approach (adiabatic case) for insulated walls. However, a quite general technique is described in the section on the matrix method for both conducting and insulated walls; this matrix technique has been used for the aximuthal wave number $n = 1$, which corresponds to the most unstable situation for unmodulated case in long cylinders.

The values of k given by both analytical and matrix approaches are identical; they have been obtained for a wide range of values of ω and Pr. Some situations have been found where the gravity modulation has a destabilizing effect; this occurs for insulated walls and small ω in the range of $0 < Pr < 8.10^{-3}$ and $8 < Pr < \infty$. In the other cases the gravity modulation is stabilizing.

In the conducting case a completely new and very simple expression of k, valid for any Pr and ω, has been derived. This expression [Eq. (67)] shows that k is always positive (stabilizing effect) and presents a maximum for $Pr = 1$ at $\omega \to 0$; this behavior agrees with previous results obtained by other authors for the modulated Bénard problem (without lateral confinement).

For large ω an asymptotic analysis has been done for small ϵ; it has been extended in the section on asymptotic analysis for large values of ω to the case of finite ϵ for quite general boundary conditions and periodic constraints. A universal law, $k = 0.5 Pr\omega^{-2}$, valid for adiabatic and conducting conditions (in fact, for any value of the Biot number) has been found that shows a stabilizing effect of the gravity modulation for any Pr.

References

[1]Hamacher, H., and Merbold, U., AIAA Paper 85-7026, 1985.

[2]Hamacher, H., Merbold, U., and Jilg, R., *37th International Astronautical Congress*, International Aeronautical Federation Paper IAF-86-268, 1986.

[3]Hamacher, H., Merbold, U., and Jilg, R., *15th International Symposium on Space Technology and Science*, Paper S-1-3, 1986.

[4]Billia, B., Jamgotchian, H., Favier, J. J., and Camel, D., European Space Agency Special Publ., Vol. 256, 1987, p. 377.

[5]Camel, D., Favier, J. J., Dupouy, M. D., and Le Maguet, R., European Space Agency Special Publ., Vol. 256, 1987, p. 317.

[6]Henry, D., and Roux, B., European Space Agency Special Publ., Vol. 256, 1987, p. 487.

[7]Malméjac, Y., Bewersdorff, A., Da Riva, I., and Napolitano, L. G., European Space Agency Rept., BR-05, 1981.

[8]Ostrach, S., European Space Agency Special Publ., Vol. 114, 1976, p. 41.

[9]Ostrach, S., *Annual Review of Fluid Mechanics*, Vol. 14, 1982, p. 313.

[10]Richardson, P. D., *Applied Mechanics Review*, Vol. 20, 1967, p. 201.
[11]Pak, H. Y., Winter, E. R. F., and Schoenals, R. J., *Journal of Fluid Mechanics*, Vol. 158, 1970.
[12]Gershuni, G. Z., Zhukhovitskii, E. M., and Iurkov, I. S., *Journal of Applied Mathematics and Mechanics (PMM)*, Vol. 34, 1970, p. 442.
[13]Hall, P., *Journal of Fluid Mechanics*, Vol. 67, 1975, p. 29.
[14]Grosch, C. E., and Sallwen, H., *Journal of Fluid Mechanics*, Vol. 34, 1968, p. 177.
[15]Gershuni, G. Z., and Zhukhovitskii, E. M., *Journal of Applied Mathematics and Mechanics (PMM)*, Vol. 27, 1963, p. 1197.
[16]Venezian, G. *Journal of Fluid Mechanics*, Vol. 35, 1969, p. 243.
[17]Roppo, M. N., Davis, S. H., and Rosenblat, S., *Physics of Fluids*, Vol. 27, 1984, p. 796.
[18]Gresho, P., and Sani, R. L., *Journal of Fluid Mechanics*, Vol. 40, 1970, p. 783.
[19]Rosenblat, S., and Herbert, D. M. *Journal of Fluid Mechanics*, Vol. 43, 1970, p. 385.
[20]Rosenblat, S., and Tanaka, G. A., *Physics Fluids*, Vol. 14, 1971, p. 1319.
[21]Yih, C. S., and Li, C. H., *Journal of Fluid Mechanics*, Vol. 54, 1972, p. 143.
[22]Homsy, M. G., *Journal of Fluid Mechanics*, Vol. 62, No. 2, 1974, p. 387.
[23]Charlson, G. S., and Sani, R. L., *International Journal of Heat and Mass Transfer*, Vol. 13, 1970, p. 1479.
[24]Yih, C. S., *Q. Appl. Math.*, 17, 1959, p. 25.
[25]Charlson, G. S., and Sani, R. L., *International Journal of Heat and Mass Transfer*, Vol. 14, 1971, p. 2157.

Chapter 5. Buoyancy, Capillary Effects, and Solidification

Double-Diffusive Convection and its Effect Under Reduced Gravity

C. F. Chen*

University of Arizona, Tucson, Arizona

Introduction

WHEN a fluid contains two diffusing components with different molecular diffusivities, convective motion may be generated when potential energy is released because of differential diffusion. Indeed, such motion may occur even when the overall density distribution is stable. An excellent presentation of the fundamentals of double-diffusive convection is given by Turner.[1] In solidification of binary alloys, temperature and concentration gradients occur naturally in the melt and in the mushy zone. Convection due to double-diffusive effects may be generated that affect the homogeneity of the resulting solid. Now that material processing in space is becoming a reality, there is a need to examine the effect of low gravity on the possible suppression of double-diffusive convection.

In this paper we review the recent developments in double-diffusive convection generated by diffusive instability due to sideways heating in the next section, by salt-finger instability in the section on behavior of salt fingers, and by diffusive instability due to bottom heating in the section on layer generation.

Layer Generation Due to Sideways Diffusive Instability

When stratified fluid stabilized in the vertical direction by a solute gradient is subjected to a lateral temperature gradient, vorticity is generated

*Professor, Department of Aerospace and Mechanical Engineering.

355

that leads to convective motion at supercritical conditions. The convection pattern consists of a series of horizontal convecting layers of constant temperature and solute concentration. This effect was first reported by Thorpe et al.[2] Their experiment was conducted in a narrow vertical tank filled with a salinity gradient. When a lateral temperature difference was imposed across the tank, a series of convecting cells appeared in the tank. They also performed an asympototic stability analysis in which the nondiffusive nature of the side walls and the boundary-layer flows along the walls were neglected. The critical condition was found to be

$$R_T = 2.76(R_S)^{\frac{5}{6}} \tag{1}$$

The thermal and solute Rayleigh numbers are defined as

$$R_T = \frac{g\alpha\Delta T D^3}{\kappa v} \tag{2}$$

$$R_S = -\frac{g\beta S_z D^4}{\kappa v} \tag{3}$$

in which g is the gravitational acceleration, $\alpha = -\rho^{-1}(\partial p/\partial T)$, ΔT is the temperature difference across the tank, D is the width of the tank, κ is the thermal diffusivity, v is the kinematic viscosity, $\beta = \rho^{-1}(\partial p/\partial S)$, and S_z is the vertical concentration gradient. Their experimental results, obtained at large Rayleigh numbers, agreed well with the results of stability analysis. Hart[3] improved the analysis by taking into account the proper boundary conditions. His results were applicable to all Rayleigh numbers. At the asymptotic state, the critical condition is similar to Eq. (1), except that the constant becomes 2.84 for a salt-heat system.

When the tank is tilted from the vertical, the situation becomes more complicated. This is because the isoconcentration lines must be perpendicular to a nondiffusive boundary. The bending of the isoconcentration lines causes a local buoyancy imbalance, thus generating a flow along the boundary. This effect was pointed out, independently, by Phillips[4] and Wunsch,[5] who showed that under steady conditions, when the buoyancy force is balanced by the viscous force, a steady buoyancy layer flow existed along the sloping walls. Chen[6] showed that motions of such buoyancy layers in a stably stratified fluid containing two solutes with different molecular diffusivities in an inclined slot are likely to generate instabilities, resulting in horizontal layer convection.

Paliwal and Chen[7] performed a series of experiments on the onset of cellular convection in an inclined slot filled with vertically stratified fluid. Because of the antisymmetric nature of the buoyancy layer flows along the top and bottom walls of the inclined slot, they found that the system is more stable when heating is applied from the lower wall and becomes less stable when heating is applied at the upper wall. This result is contrary to intuition. Their experimental results were confirmed by a linear stability analysis.[8]

When the stratified fluid fills a large region, or a wide tank, and is subjected to a temperature gradient from one of the side walls, neither the width of the tank nor the depth of the fluid is the suitable length scale. Chen et al.[9] have shown that the correct length scale is the height of rise h of a heated fluid parcel in the stratified background:

$$h = - \frac{\alpha \Delta T}{\beta \dfrac{\partial S}{\partial z}} \qquad (4)$$

By performing a number of experiments with different salinity gradients, they found that the critical Rayleigh number (based on h) for the onset of cellular convection is

$$R_T = \frac{g\alpha \Delta T h^3}{\kappa \nu} = (1.5 \pm 0.25) \times 10^4 \qquad (5)$$

The thickness of the convecting layers at onset, measured in six experiments up to 1.8 times the critical Rayleigh number, ranged from 0.69 to 0.85 h. These experiments were conducted in a tank 12.5 cm wide. It was found later by Huppert and Turner[10] and by Huppert et al.[11] that, at a supercritical Rayleigh number of $\sim 10^5$, the layer of thickness is reduced to approximately 0.6 h, and this result holds independent of Prandtl numbers and diffusivities of the fluid.

More recently, Jeevaraj and Imberger[12] performed a similar experiment in a wide tank, 170 cm in width. The experimental results show that the critical Rayleigh number ranges from 1.22 to 2.36 $\times$ 10^4, and the layer thickness ranged from 0.88 to 0.96 h in very good agreement with the results of Chen et al.[9] They also obtained the interesting result that the length of the convecting layers grew as $t^{\frac{1}{4}}$, in which time t is measured from the start of the experiment. Tanny and Tsinober[13] performed experiments in tanks 34 cm and 120 cm wide. The temperature of the side wall was raised in a programmed manner, either as an exponential with different time constants or as a straight line. The time constant varied from 200 to 800 s. It is to be noted that the time constant in the experiments of Chen et al.[9] was 180 s. The thickness of the convecting layer at onset agreed well with previously cited results; however, the critical Rayleigh number at the onset was approximately 50% larger than the values found by Chen et al.[9] and Jeevaraj and Imberger.[12] This may be due to the slow heating of the side wall. However, they found that, if they used the thermal diffusion scale $(\kappa t)^{\frac{1}{2}}$ as the width of a fictitious narrow tank, then the critical state was well predicted by the stability relation given by Thangam et al.[14] for a vertical slot.

When a constant heat flux is supplied through the side wall, the temperature of the side wall will increase indefinitely. Is it true, then, that the system will be unstable eventually? Narasawa and Suzukawa[15] have performed a series of experiments using different solutes and found that this is not true. They determined that the correct length scale is the thermal boundary thickness, and the critical parameter is the ratio of the thermal Rayleigh number to the solute Rayleigh number, both based on the characteristic length defined earlier. When this parameter exceeded the critical value, cellular convection appeared. It was found that the critical Rayleigh number ratio is an increasing function of $\tau(= \kappa_S/\kappa)$, the ratio of the solute diffusivity κ_S to the thermal diffusivity κ. The thickness of the convecting layers at the onset ranges from 0.6 to 0.9 of the characteristic length.

Behavior of Salt Fingers
When warm and solute-rich fluid is overlying cold and fresher fluid with an overall density distribution that is gravitationally stable, convection will be generated through the finger instability mechanism of the double-diffusive system. The critical condition is given by Stern[16] for a free-free layer:

$$\frac{R_s}{\tau} = R_T + \frac{27\pi^4}{4} \tag{6}$$

The Rayleigh numbers are based on the thickness of the fluid layer. In laboratory experiments and in practice, the convecting cells are long and narrow, aptly named fingers, with upward and downward flow in altering cells. Linear stability theory predicts that the convecting cells have an aspect ratio of 1 at the critical state. However, at supercritical states, which are normally found in practice, the wavelength of the fastest growing mode is small and $\propto (R_T)^{-\frac{1}{4}}$.

The transport of solute by the fingers in terms of buoyancy flux βF_S, where F_S is the flux of solute, is proportional to the initial density difference due to the solute[1]:

$$\beta F_s \propto (\beta \Delta S)^{\frac{3}{4}} \tag{7}$$

The ratio of the heat flux to the solute flux in buoyancy terms, $\alpha F_T/\beta F_S$, is a function of the density ratio R_ρ, which is the ratio of the density change due to temperature difference to that due to solute difference across the layer:

$$\frac{\alpha F_T}{\beta F_S} = f(R_\rho) \tag{8}$$

and

$$R_\rho = \frac{\alpha \Delta T}{\beta \Delta S} \tag{9}$$

It is found that generally the heat flux is less than the salt flux, and the ratio $\alpha F_T/\beta F_S$ remains mostly constant over a large range of density ratios $R_\rho > 2$. Turner[17] found the flux ratio to be 0.56 for a salt-heat system; Linden[18] found 0.12 for a sugar-heat system; and Schmitt[19] found a decreasing trend from 0.68 for $R_\rho < 2.5$ to 0.33 at $R_\rho > 6$. These results show a completely opposite effect from the molecular diffusion case in which the diffusivity of heat is nearly two orders of magnitude larger than the diffusivity of the solute. Furthermore, the heat transport is much larger than that in the diffusive case due to finger convection.

It is known that, when the flux of solute becomes excessive, the fingers become unstable collectively, and they break up into horizontal layer convection. This was demonstrated by experiments carried out by Stern and Turner.[20] They deposited a layer of sugar solution on top of a previously prepared salt solution gradient. The density of the sugar solution was made slightly smaller than that of the salt solution in the top layer. For suitable values of the salinity gradient, the finger field broke down into convecting layers. Stern[21] made a linear stability analysis of infinitely long two-dimensional fingers subjected to internal wave perturbations of long wavelength. By using some plausible though untested assumptions, he was able to arrive at the following stability criterion: When

$$\frac{\beta F_S - \alpha F_T}{\nu(\alpha T_z - \beta S_z)} > O(1) \tag{10}$$

collective instability of the fingers sets in. Holyer[22,23] made a rigorous mathematical analysis of the problem and arrived at a stability criterion similar to Eq. (10), with the constant on the right equal to $\frac{1}{3}$. More recently, she extended the analysis to three-dimensional salt fingers with square cross sections. The constant in the stability criterion, Eq. (10), became $\frac{2}{3}$ (Ref. 24). Experimental results in salt-heat and sugar-salt systems show that this ratio varies from 0.002 to 5 (see Refs. 18, 20, and 25–27). These differences may be attributed to the difference between the experimental conditions and those assumed for the theoretical analysis. But both types of investigations suggest that the important parameter governing the process is the ratio given in Eq. (10).

Recently, Howard and Veronis[28] considered the stability of salt fingers of finite extent spanning the region between two layers of fluid of different salt concentration and temperature. At the zeroth order, diffusion of salt is neglected (i.e., $\tau = 0$). Steady-state solutions were obtained for both two- and three-dimensional fingers. An interesting result is that the width of fingers corresponding to the maximum buoyancy flux is $1.7L$ for the two-dimensional case, where L is the buoyancy layer scale,

$$L = \left(\frac{4\nu\kappa}{g\alpha T_z}\right)^{\frac{1}{4}} = \frac{H}{R_T^{\frac{1}{4}}} \tag{11}$$

in which the subscript z denotes partial derivative with respect to z, and H is the thickness of the finger zone. The thermal Rayleigh number is based on H. This result agrees with Stern's[16] prediction that the wavelength of the fastest growing wave at large R_T is proportional to $R_T^{-\frac{1}{4}}$.

Howard and Veronis[28] also applied their theory to salt fingers in a Hele-Shaw flow or, equivalently, in a porous medium. The appropriate length scale is

$$\left(\frac{12\nu\kappa}{g\alpha T_z d^2}\right)^{\frac{1}{2}} \tag{12}$$

in which d is the gap width of the Hele-Shaw cell. In this case the size of the salt fingers is much more affected by the gravitational acceleration than in the viscous fluid case. Taylor and Veronis[29] carried out finger convection experiments in a Hele-Shaw cell using a sugar-salt system. In this case T_z in Eq. (12) refers to the vertical gradient of the salt concentration. When the apparatus was set at 10 deg from the horizontal, the two-dimensional fingers were approximately 0.8 cm wide. When the apparatus was raised to the vertical position, the size of the cells was readjusted through an instability mechanism to smaller scales. Veronis[30] presented a stability analysis of the observed process.

Imhof and Green[31] carried out salt finger experiments in a porous medium using a sugar-salt system. The porous medium consisted of glass beads 0.5–0.71 mm in diameter. The lower half of the tank was filled with salt solution, and the top half was filled with sugar solution that was mixed with a small amount of flourescent dye. As the sugar fingers extend downward, the salt gradient becomes smaller and smaller. As a consequence, the sugar fingers become wider near the bottom of the tank, in agreement with Eq. (12).

Layer Generation Due to Diffusive Instability by Heating from Below

When a layer of fluid whose density is stably stratified by solute concentration is heated from below and with free-free boundary conditions, the system becomes unstable in an oscillatory mode when the thermal Rayleigh number exceeds the critical value[32]:

$$R_T = \frac{Pr + \tau}{Pr + 1} R_S + (1 + \tau)\left(1 + \frac{\tau}{Pr}\right)\frac{27\pi^4}{4} \tag{13}$$

In contrast to the finger mode, the critical condition depends on the Prandtl number Pr and the ratio of diffusivities τ. Experimental verification of the result is difficult to carry out since the theory requires a constant solute gradient within the layer, even at the boundaries. Shirtcliffe[33,34] and Wright and Lochrke[35] did find experimentally that the onset of instability was in an oscillatory mode, but in both cases the solute boundary conditions were not exactly the same as those required by the stability analysis.

Turner[36] found that if heating continues beyond the critical point, a number of horizontally convecting layers were generated successively from the bottom heating wall. Turner[1] also developed a theory for the prediction of the growth of the first layer and the onset of the second layer. The theory assumes that the fluid is well mixed within the layer so that the temperature and solute concentration are uniform and that density jumps due to temperature and solute concentration just cancel each other at the interface. With these assumptions, he showed that the thickness of the first layer grows as $t^{\frac{1}{2}}$. He then argues that the first layer will stop growing when the thermal boundary layer ahead of the interface separating the layer from the rest of the fluid becomes unstable. From the experimental data it was determined that the critical Rayleigh number for the onset of instability in the thermal boundary layer was $\sim 2 \times 10^4$, one order of magnitude larger than the onset values for Bernard convection. Huppert and Linden,[37] using the same ideas and flux laws for the transport of heat and solute across diffusive interfaces,[38,39] constructed a model for the successive generation of layers in a deep solute-stratified fluid subjected to a constant heat flux from below. The results compared favorably with the experiments conducted in a deep tank.

Recently, Fernando[40] advanced a new theory for the determination of the height of the first mixed layer. He argued that, as the layer is growing, energetic eddies have large enough kinematic energy to overcome the potential energy of the stratified fluid, thus engulfing the latter into the mixed layer. When the layer is of sufficient thickness, the scale of the eddies becomes larger, and the potential energy increases to the same level as the kinetic energy of the eddies; then the growth stops and a second layer is started. His experimental results agreed well with his theoretical predictions.

Before the establishment of steady convection in horizontal layers, the diffusive system exhibits a rich variety of intermediate dynamic states. Huppert and Moore[41] examined the nonlinear motions by numerically integrating the two-dimensional equations governing the motion in a layer of stably stratified fluid heat from below. They found that, as R_T is increased for a given R_S, Pr, and τ, the quiescent state first gives way to periodic oscillatory motion. At a higher value of R_T, the motion becomes aperiodic. Finally, the motion becomes time independent when R_T is increased further. DaCosta et al.[42] investigated these transitions in more detail using a model equation which was obtained using an expansion scheme first suggested by Veronis.[32] The series expansion for the stream function was truncated at one term, and expansions for the temperature and salinity were truncated at two terms. The governing partial differential equations became a set of five simultaneous ordinary differential equations. Using these five-mode equations, they found that the transition process to reach the final time-independent state is more circuitous. As R_T increases, the system bifurcates from "symmetrical to asymmetrical oscillations, then followed by a number of period doubling bifurcations and ending in chaos." With further increases in R_T, the chaotic solution loses stability and the motion becomes stable. Recently, Knoblock et al.[43] numerically integrated the nonlinear governing equations on a supercomputer. Their results are quite similar to those found by DaCosta et al.[42] using the modal equations. The rich dynamic response of the system was carefully studied in this paper.

In an investigation into the stability of Langmuir circulation, Leibovich et al.[44] considered the diffusive instability problem with a constant heat flux bottom boundary. They found that the nonlinear dynamic response of the system is determined by the lateral boundary conditions. When the no-flux conditions are applied on the lateral boundaries, the flow exhibits complex temporal behavior, including chaos. When the periodic conditions are applied on the lateral boundaries, such complex flows are not realized.

Double-Diffusive Effects During Solidification

During solidification there is naturally a temperature gradient present. If there is more than one element present, whether due to alloying components or due to the presence of impurities, there will be a concentration gradient present in the melt. If the temperature gradient and the concentration gradient are aligned in the right manner, double-diffusive convection will ensue. In a recent review Chen[45] discussed the effect of double-diffusive convection on a number of solidification processes. In the following paragraphs several cases will be discussed to illustrate some of the double-diffusive phenomena presented earlier. In the geological context double-diffusive convection is likely to occur during a magma solidification. Many geological phenomena may be explained by such processes (see Ref. 46).

Chen and Turner[47] systematically investigated the effect of double-diffusive convection on the solidification process by a series of experiments. These experiments were performed with aqueous solutions of sodium carbonate, Na_2CO_3, which was chosen because the solution would solidify around room temperature. Cooling was effected in various configurations. When cooling was from the top, the crystals formed in horizontal layers, whether the initial fluid was homogeneous, was stratified with a constant gradient, or was stratified in several distinct layers. The dominant effect was the formation of a cold but light fluid layer against the cooled upper boundary as the denser crystals grew. The crystals were small and closely packed in the upper layer (which was vigorously convecting) and larger and more loosely packed in the lower, more quiescent fluid. In the case where there were preexisting layers, there was an abrupt increase in the growth rate of crystals as they crossed an interface. With side-wall cooling and a constant gradient, the lighter solution left behind after crystallization streamed upward in a thin boundary layer right to the top of the tank. The outer flow consisted of circulation in nearly horizontal convecting layers, each of which was depressed slowly as the light fluid collected at the top. When a constant gradient of Na_2CO_3 was cooled from below, crystallization produced finger instability, leading to the growth of a mixed layer in the fluid above.

Turner[48] subsequently showed that solidification of a solution of uniform concentration from a cooled side wall would result in a melt in which there was a concentration gradient because of the existence of the composition boundary layer. McBirney[49] used the results of a series of experiments based on the same principle to construct a convincing explanation of the unmixing of magma as evidenced by the abrupt change in composition of volcanic rocks. The countercurrent flow of a light compositional buoyancy layer flowing upward along the crystallization front and the adjacent thermal boundary layers flowing downward has been analyzed by Nilson.[50] Nonsimilar solutions were obtained using the method of asymptotic expansion under the assumption that $Pr \gg 1$ and $\tau \ll 1$. In a map of density ratio R_ρ and the diffusivity ratio τ, the regions of pure upflow, pure downflow, and counterflow are clearly delineated.

Layer generation due to a lateral temperature gradient has been observed in a computer simulation of a vertical Bridgeman crystal growth system as reported by Adornato and Brown.[51] The calculation was made for a silicon-germanium alloy. At the solidification front, germanium is rejected, thus making the melt heavier. Gradually, a density-stratified fluid is built up in the ampoule. The radial temperature gradient existing in the ampoule caused the occurrence of a number of horizontal convecting cells under the same mechanism as that discussed in the section on layer generation due to sideways diffusive instability. However, contrary to normal expectations, these convecting layers seemed to redistribute the silicon in the alloy more uniformly, approaching that of diffusion-controlled growth.

When a binary alloy is directionally solidified from below, double-diffusive instability may develop if the lighter element is being rejected at the solidification front. Coriell et al.[52] considered this problem together with the possibility of morphological instability[53] at a smooth interface. Results for a lead-tin alloy showed that double-diffusive convection is likely to happen at low growth velocities (10^{-5} to 10^{-3} cm/s), and morphological instability is likely to happen at higher growth velocities (10^{-2} to 10 cm/s). For growth velocities intermediate between the two ranges, the onset of instability is in the oscillatory mode. The critical wavelength for the double-diffusive case is approximately 10 times the concentration length scale; the critical wavelength for the morphological case is approximately equal to the concentration length scale.

During directional solidification of alloys with sufficiently high solutal concentrations, however, the planar freezing surface is morphologically unstable,[53] and solidification is dendritic. As a result, the melt is separated from the completely solid region consisting of solid dendrites and interdentritic liquid, generally referred to as the "mushy zone." The mushy zone may be regarded as a porous layer with a permeability that varies in the direction perpendicular to the freezing front, and its value is also directionally dependent. Under some circumstances, however, there are defects in the castings in the form of solute-rich, columnal regions extending longitudinally in the direction of solidification. These are known as "freckles," and their presence causes a deleterious effect on the strength of the castings. Although the general mechanism for the onset of freckles is known to arise from double-diffusive effects, the precise mechanism is not clearly understood (see reviews by Fisher[54] and Glicksman et al.[55]).

Recently, Chen and Chen[56] carried out a series of experiments on directional solidification to study the onset of plume convection, which is the precursor of freckles. These experiments were carried out using the analog casting system of NH_4Cl-H_2O solution by cooling it from below with a constant temperature surface ranging from $-31.5°C$ to $+11.9°C$. The NH_4Cl concentration was 26% in all solutions, with a liquidus temperature of 15°C. It was found that finger convection occurred in the fluid region right above the mushy layer in all experiments. Plume convection with associated chimneys in the mush occurred in experiments with bottom temperatures as high as $+11.0°C$. However, when the bottom temperature was raised to $+11.9°C$, no plume convection was observed, although finger convection was carrying on as usual. Based on these observations, a model of the flow is presented. The porosity of the mush was determined by computed tomography. Using the permeability value calculated by the Kozeny-Carman relationship, the critical solute Rayleigh number across the mush layer for onset of plume convection was estimated to be 170.

Double-Diffusive Effects Under Reduced Gravity

Convection in a double-diffusive system is driven by the release of potential energy in a gravitational field due to the difference in diffusivities. If such an experiment is carried out in a 0-g environment, no double-diffusive convection will be generated. Temperature and concentration gradients in the fluid will cause molecular diffusion down the gradient. In this case, the off-diagonal terms of the diffusivity tensor may become important, because the time scale involved in cross diffusion is of the same order as that for mass diffusion.[57]

However the gravitational environment provided by the Space Shuttle, is not at 0 g. At an orbital altitude of 200–400 km, the residual atmosphere offers a small amount of drag on the Space Shuttle, thus causing a constant deceleration. The Shuttle rotates about its center of mass, generating a contrifugal acceleration. Both of these are generally very small, in the range of 10^{-6} to $10^{-7}g_0$, where g_0 is the gravitational acceleration at sea level.[58] On the D-1 mission, the accelerometer system installed in the Shuttle had a low limit of $10^{-5}g_0$. As a result, no quasisteady acceleration was recorded. However, the record did show transient g loading up to $10^{-2}g_0$ due to crew motion, firings of vernier thrustors, and experiment operations.

Let us examine the likelihood of onset of the three types of double-diffusive instabilities discussed in the previous sections. To fix ideas, the property values of a lead-tin alloy as presented by Coriell et al.[52] will be used to evaluate the parameters:

$$v = 2.43 \times 10^{-3} \text{ cm}^2/\text{s}$$

$$\kappa = 0.108 \text{ cm}^2/\text{s}$$

$$\kappa_S = 3 \times 10^{-5} \text{ cm}^2/\text{s}$$

$$\tau = \kappa_S/\kappa = 2.78 \times 10^{-4}$$

$$\alpha = 1.15 \times 10^{-4} K^{-1}$$

$$\beta = 5.2 \times 10^{-3} \text{wt\%}^{-1}$$

$$Pr = v/\kappa = 2.25 \times 10^{-2}$$

For diffusive instabilities generated either by heating from the side or by heating from below, the critical conditions are given by Eqs. (5) and (13). Both require the thermal Rayleigh number to exceed a constant of the order 10^3 to 10^4. Since the thermal diffusivity of liquid metal is usually high, it is generally unlikely that the thermal Rayleigh number would exceed the critical value when g is reduced to 10^{-2} to $10^{-4}g_0$. However, the effect of a steady oscillating gravity of low level on diffusive instabilities is yet to be investigated.

However, the finger instability is likely to occur at reduced gravity because the critical condition for onset of such instability is that R_S/τ exceeds a level of order 10^3, as shown in Eq. (6). If we let $g = \epsilon g_0$, then we obtain

$$\epsilon \Delta S D^3 \geqslant \frac{10^3 \kappa_S \nu}{g_0 \beta} = 1.43 \times 10^{-5} \text{ wt\% cm}^3$$

If $D = 0.1$ cm, $\Delta S = 10\%$, the $\epsilon \cong 10^{-3}$ or finger convection may happen at $10^{-3}g_0$. If the layer thickness is larger, say, 1 cm, then such instability could occur at $10^{-6}g_0$. Based on the laboratory experiment with NH_4Cl-H_2O solution, Chen and Chen[56] reasoned that finger convection could occur for such a system at $10^{-3}g_0$ and plume convection could occur at $10^{-1}g_0$.

Since the Space Shuttle experiences acceleration levels at 10^{-2} to $10^{-4}g_0$ in an oscillating mode, the cumulative effect of g jitter on the onset of subsequent evolution of finger convection may be rather different from the phenomenon occurring at steady state. It is important that such effects be investigated. Furthermore, at reduced g levels, the finger convection occurs near the marginal state. This is a rather unusual situation, since, in most applications and experiments carried out at g_0, finger convection occurs at grossly supercritical states. Under these circumstances, the convection cells are long and narrow. However, at the marginal states, the cells will have an aspect ratio near 1. Furthermore, as shown by Straus,[59] the salt flux transported by finger convection near the marginal state is approximately one order of magnitude smaller than that obtained in laboratory experiments at supercritical states by Turner[17] and others. The effect of the reduced flux on the directional solidification process is not known at present, and further investigations are needed to determine whether it will adversely affect the solute distribution in the melt.

Conclusions

In this chapter we reviewed the recent results on the following three aspects of double-diffusive convection: diffusive instability caused by sideways heating, diffusive instability caused by bottom heating, and the salt-finger convection case. We highlighted some of these processes that are likely to occur during solidification. Using property values of lead-tin alloy, we concluded that diffusive instabilities are unlikely to occur at steady reduced gravity levels of 10^{-2} to $10^{-4}g_0$. However, salt finger convection is likely to be generated. In all these cases, the effect of an oscillatory gravitational acceleration at low level, the g jitter, should be investigated.

Acknowledgment
Financial support of this work through National Science Foundation Grant MSM-8702732 and NASA Grant NAG-3-723 is greatly appreciated.

References
[1]Turner, J. S., *Buoyancy Effects in Fluids*, Cambridge Univ. Press, Cambridge, UK, 1973.
[2]Thorpe, S. A., Hutt, P. K., and Soulsby, R., "The Effect of Horizontal Gradients on Thermohaline Convection," *Journal of Fluid Mechanics*, Vol. 38, 1969, pp. 375–400.
[3]Hart, J. E., "On Sideways Diffusive Instability," *Journal of Fluid Mechanics*, Vol. 49, 1971, pp. 279–288.
[4]Phillips, O. M., "On Flows Induced by Diffusion in a Stably Stratified Fluid," *Deep-Sea Research*, Vol. 17, 1970, pp. 435–443.
[5]Wunsch, C., "On Oceanic Boundary Mixing," *Deep-Sea Research*, Vol. 17, 1970, pp. 293–301.
[6]Chen, C. F., "Double-Diffusive Convection in an Inclined Slot," *Journal of Fluid Mechanics*, Vol. 72, 1974, pp. 721–729.
[7]Paliwal, R. C., and Chen, C. F., "Double-Diffusive Instability in an Inclined Fluid Layer. Part 1. Experimental Investigation," *Journal of Fluid Mechanics*, Vol. 98, 1980, pp. 755–768.
[8]Paliwal, R. C., and Chen, C. F., "Double-Diffusive Instability in an Inclined Fluid Layer. Part 2. Stability Analysis," *Journal of Fluid Mechanics*, Vol. 98, 1980, pp. 769–785.
[9]Chen, C. F., Briggs, D. G., and Wirtz, R. A., "Stability of Thermal Convection in a Salinity Gradient due to Lateral Heating," *International Journal of Heat and Mass Transfer*, Vol. 14, 1971, pp. 57–65.
[10]Huppert, H. E., and Turner, J. S., "Ice Blocks Melting into Salinity Gradient," *Journal of Fluid Mechanics*, Vol. 100, 1980, pp. 367–384.
[11]Huppert, H. E., Kerr, R. C., and Hallworth, M. A., "Heating or Cooling a Stable Compositional Gradient from the Side," *International Journal of Heat and Mass Transfer*, Vol. 27, 1984, pp. 1395–1401.
[12]Jeevaraj, C. G., and Imberger, J., "Experimental Study of Double-Diffusive Instability in Sidewall Heating," *Journal of Fluid Mechanics* (to be published).
[13]Tanny, J., and Tsinober, A. B., "The Dynamics and Structure of Double-Diffusive Layers in Sidewall-Heating Experiments," *Journal of Fluid Mechanics*, Vol. 196, 1988, pp. 135–156.
[14]Thangam, S., Zebib, A., and Chen, C. F., "Transition from Shear to Sideways Diffusive Instability in a Vertical Slot," *Journal of Fluid Mechanics*, Vol. 112, 1981, pp. 151–160.
[15]Narasawa, U., and Suzukawa, Y., "Experimental Study of Double-Diffusive Cellular Convection due to a Uniform Lateral Heat Flux," *Journal of Fluid Mechanics*, Vol. 113, 1981, pp. 387–405.
[16]Stern, M. E., "The "Salt-Fountain" and Thermohaline Convection," *Tellus*, Vol. 12, 1960, pp. 172–175.
[17]Turner, J. S., "Salt Fingers Across a Density Interface," *Deep-Sea Research*, Vol. 14, 1967, pp. 599–611.

[18] Linden, P. F., "On the Structures of Salt Fingers," *Deep-Sea Research*, Vol. 20, 1973, pp. 325–340.

[19] Schmitt, R. W., "Flux Measurements on Salt Fingers at an Interface," *Journal of Marine Research*, Vol. 37, 1979, pp. 419–435.

[20] Stern, M. E., and Turner, J. S., "Salt Fingers and Convecting Layers," *Deep-Sea Research*, Vol. 16, 1969, pp. 497–511.

[21] Stern, M. E., "Collective Instability of Salt Fingers," *Journal of Fluid Mechanics*, Vol. 35, 1969, pp. 209–218.

[22] Holyer, J. Y., "On the Collective Instability of Salt Fingers," *Journal of Fluid Mechanics*, Vol. 110, 1981, pp. 195–207.

[23] Holyer, J. Y., "The Stability of Long Steady, Two-Dimensional Salt Fingers," *Journal of Fluid Mechanics*, Vol. 147, 1984, pp. 169–185.

[24] Holyer, J. Y., "The Stability of Long, Steady, Three-Dimensional Salt Fingers to Long-Wavelength Perturbations," *Journal of Fluid Mechanics*, Vol. 156, 1985, pp. 495–503.

[25] Lambert, R. B., and Demenkow, J. W., "On the Vertical Transport due to Salt Fingers," *Journal of Fluid Mechanics*, Vol. 54, 1972, pp. 627–640.

[26] Griffith, R. W., and Ruddick, B. R., "Accurate Fluxes Across a Salt-Sugar Finger Interface Deduced from Direct Density Measurements," *Journal of Fluid Mechanics*, Vol. 99, 1980, pp. 85–95.

[27] McDougall, T. J., and Taylor, J. R., "Flux Measurements Across a Finger Interface at Low Values of the Stability Ratio," *Journal of Marine Research*, Vol. 42, 1984, pp. 1–14.

[28] Howard, L. N., and Veronis, G., "The Salt-Finger Zone," *Journal of Fluid Mechanics*, Vol. 183, 1987, pp. 1–23.

[29] Taylor, J., and Veronis, G., "Experiments in Salt Fingers in a Hele-Shaw Cell," *Science*, Vol. 231, 1985, pp. 39–41.

[30] Veronis, G., "The role of the Buoyancy Layer in Determining the Structure of Salt Fingers," *Journal of Fluid Mechanics*, Vol. 180, 1987, pp. 327–342.

[31] Imhof, P. T., and Green, T., "Experimental Investigation of Double-Diffusive Groundwater Fingers," *Journal of Fluid Mechanics*, Vol. 188, 1988, pp. 363–382.

[32] Veronis, G., "On Finite Amplitude Instability in Thermohaline Convection," *Journal of Marine Research*, Vol. 23, 1965, pp. 1–17.

[33] Shirtcliffe, T. G. L., "Thermosolutal Convection: Observation of an Overstable Mode," *Nature (London)*, Vol. 213, 1967, pp. 489–490.

[34] Shirtcliffe, T. G. L., "An Experimental Investigation of Thermosolutal Convection at Marginal Stability," *Journal of Fluid Mechanics*, Vol. 35, 1969, pp. 677–688.

[35] Wright, J. H., and Lochrke, R. I., "The Onset of Thermohaline Convection in a Linearly–Stratified Horizontal Layer," *Transactions of ASME C: Journal of Heat Transfer*, Vol. 98, 1976, pp. 558–563.

[36] Turner, J. S., "The Behavior of a Stable Salinity Gradient Heated from Below," *Journal of Fluid Mechanics*, Vol. 33, 1968, pp. 183–200.

[37] Huppert, H. E., and Linden, P. F., "On Heating a Stable Salinity Gradient from Below," *Journal of Fluid Mechanics*, Vol. 95, 1979, pp. 431–464.

[38] Turner, J. S., "The Coupled Turbulent Transports of Salt and Heat Across a Sharp Density Interface," *International Journal of Heat and Mass Transfer*, Vol. 8, 1965, pp. 759–767.

[39] Huppert, H. E., "On the Stability of a Series of Double-Diffusive Layers," *Deep-Sea Research*, Vol. 18, 1971, pp. 1005–1021.

[40] Fernando, H. J. S., "The Formation of a Layered Structure when a Stable Salinity Gradient is Heated from Below," *Journal of Fluid Mechanics*, Vol. 182, 1987, pp. 525–541.

[41] Huppert, H. E., and Moore, D. R., "Nonlinear Double–Diffusive Convection," *Journal of Fluid Mechanics*, Vol. 78, 1976, pp. 821–854.

[42] DaCosta, L. N., Knobloch, E., and Weiss, N. O., "Oscillations in Double-Diffusive Convection," *Journal of Fluid Mechanics*, Vol. 109, 1981, pp. 25–43.

[43] Knobloch, E., Moore, D. R., Toomre, J., and Weiss, N. O., "Transitions to Chaos in Two-Dimensional Double–Diffusive Convection," *Journal of Fluid Mechanics*, Vol. 166, 1986, pp. 409–448.

[44] Leibovich, S., Lele, S. K., and Moroz, I. M., "Nonlinear Dynamics in Langmuir Circulations and in Thermosolutal Convection," *Journal of Fluid Mechanics*, Vol. 198, 1989, pp. 471–511.

[45] Chen, C. F., "Double-Diffusive Effects During Solidification," *Interdisciplinary Issues in Materials Processing and Manufacturing*, Vol. 2, edited by S. K. Samata, R. Komanduri, R. McMeeking, M. M. Chen, and A. Tseng, American Society of Mechanical Engineers, New York, 1987, pp. 527–540.

[46] Huppert, H. E., "The Intrusion of Fluid Mechanics into Geology," *Journal of Fluid Mechanics*, Vol. 173, 1986, pp. 557–594.

[47] Chen, C. F., and Turner, J. S., "Crystallization in a Double-Diffusive System," *Journal of Geophysical Research*, Vol. 85, 1980, pp. 2573–2593.

[48] Turner, J. S., "A Fluid Dynamical Model of Differentiation and Layering in Magma Chambers," *Nature (London)*, Vol. 285, 1980, pp. 213–215.

[49] McBirney, A. R., "Mixing and Unmixing of Magmas," *J. Volcanology and Geothermal Research*, Vol. 7, 1980, pp. 357–371.

[50] Nilson, R. H., "Countercurrent Convection in a Double-Diffusive Boundary Layer," *Journal of Fluid Mechanics*, Vol. 160, 1985, pp. 181–210.

[51] Adornato, P. M., and Brown, R. A., "Convection and Segregation in Directional Solidification of Dilute and Non-Dilute Binary Alloys: Effect of Ampoule and Furnace Design," *Journal of Crystal Growth*, Vol. 80, 1987, pp. 155–190.

[52] Coriell, S. R., Cordes, M. R., and Boettinger, W. J., "Convective and Interfacial Instabilities During Undirectional Solidification of a Binary Alloy," *Journal of Crystal Growth*, Vol. 49, 1980, pp. 13–28.

[53] Mullins, W. W., and Sekerka, R. F., "Stability of Planar Interfaces During Solidification of a Dilute Binary Alloy," *Journal of Applied Physics*, Vol. 35, 1964, pp. 444–451.

[54] Fisher, K. M., "The Effects of Fluid Flow on the Solidification of Industrial Castings and Ingots," *PhysicoChemical Hydrodynamics* Vol. 2, 1981, pp. 311–326.

[55] Glicksman, M. E., Coriell, S. R., and McFadden, G. B., "Interaction of Flows with Crystal-Melt Interface," *Annual Review of Fluid Mechanics*, Vol. 18, 1986, pp. 307–335.

[56] Chen, C. F., and Chen, F., "Experimental Study of Directional Solidification of Aqueous Ammonium Chloride Solution," *Journal of Fluid Mechanics* (to be published).

[57] Turner, J. S., "Multicomponent Convection," *Annual Review of Fluid Mechanics*, Vol. 17, 1985, pp. 11–44.

[58] Hamacher, H., Fitton, B., and Kingdon, J., "The Environment of Earth-Orbiting Systems," *Fluid Sciences and Materials Science in Space*, edited by H. W. Walter, Springer-Verlag, Berlin, 1987, pp. 1–50.

[59] Straus, J. M., "Finite Amplitude Doubly Diffusive Convection." *Journal of Fluid Mechanics*, Vol. 56, 1972, pp. 353–374.

Instability During Directional Solidification: Gravitational Effects

S. R. Coriell* and G. B. McFadden†
National Institute of Standards and Technology, Gaithersburg, Maryland

Introduction

MANY materials are made by freezing the liquid phase to form a solid, which may be single phase or multiphase. Examples include the preparation of silicon by the Czochralski process, the directional solidification of metallic superalloys for turbine blades, and the freezing of water to form ice. In this article we will discuss recent research on the solidification (or crystal growth) of liquids containing two components to form a single-phase crystal, for example, water containing a small concentration of sugar to form ice.

The phase diagram of a binary alloy gives the freezing point of the alloy as a function of the solute concentration. It also relates the solute concentration in the liquid to the solute concentration in the crystal; in general, these concentrations are not equal, and as the crystal grows solute is either rejected or preferentially incorporated. This gives rise to a gradient of solute near the crystal-melt interface. The release of latent heat during the transformation from melt to crystal likewise gives rise to temperature gradients. In a gravitational field density variations due to gradients of temperature and solute can cause fluid flow in the melt, which in turn alters the thermal and concentration fields. Since solute diffusion coefficients are orders of magnitude smaller than kinematic viscosities, even very small fluid flow velocities can significantly modify the solute distribution.

Copyright © 1990 by the American Institute of Aeronautics and Astronautics, Inc. All rights reserved.
*Research Chemist, Metallurgy Division.
†Mathematician, Applied and Computational Mathematics Division.

For many materials solute diffusion coefficients in the crystal are many orders of magnitude smaller than those in the liquid. Thus, as the crystal grows, the solute distribution is frozen in place; i.e., the solute distribution in the solidified material is determined by the solute distribution in the melt at the crystal-melt interface. The properties of the material depend strongly on the spatial distribution of solute, and one of the goals of materials processing research is to be able to predict and control this distribution. Although the distribution can be modified subsequent to solidification by high-temperature annealing and mechanical working, such processes add to the time and cost of preparing the material.

The prediction of the solute distribution in the solidified material requires the calculation of the fluid flow, temperature, and concentration fields in the melt and the temperature and concentration fields in the crystal. The usual difficulties encountered in natural convection computations are further amplified since the crystal-melt interface is a free boundary with shape and position that must also be calculated. In addition, the crystal-melt interface is subject to morphological instabilities even in the absence of fluid flow. These morphological instabilities typically are on the scale of micrometers, whereas the fluid flow occurs over a length scale associated with the container, typically centimeters. Another factor that makes this an interesting but difficult field is the large number of dimensionless parameters that arise. Finally, for many technologically important materials, especially those with high melting points, the thermophysical properties are poorly known.

The advantage of experiments in space for the solidification processing of materials and for the study of fundamental solidification phenomena is the significant reduction in gravitational force and hence in buoyancy-driven convection. Since the role of convection on solidification phenomena is not fully understood, it is not possible to definitively predict the differences between Earth-based and space-based experiments at present. In space, there is a reduction of roughly six orders of magnitude in the static component of the gravitational acceleration. However, in space there can be larger variations in the direction and magnitude of the gravitational acceleration with time, for example, g jitter. Other sources of fluid flow, such as surface-tension-driven flows, may become dominant in space. Many of the important theories of solidification and crystal growth are zero-gravity theories; that is, they assume that the fluid is stagnant. Under certain circumstances they are apparently good approximations on Earth and should be even better in the microgravity environment of space. However, comparison between experiments on Earth and experiments in space provide invaluable insight concerning the role of convection in solidification processes and important guidance for the improvement of current theories.

In this chapter, we give an elementary introduction to solidification, discuss recent research that attempts to understand the interaction of fluid flow with the crystal-melt interface through linear stability analysis, and briefly discuss recent numerical calculations of fluid flow and interface morphology during alloy solidification.

Directional Solidification of Binary Alloys

In directional solidification a crystal is grown by being pulled at a steady velocity V through a furnace that maintains a controlled temperature profile in the melt and crystal. In the hotter portion of the furnace the solute in the melt is assumed to be well-mixed, with a concentration c_∞. The density of the melt depends on the temperature and the solute concentration. It is well known that any horizontal density variation gives rise to fluid flow.[1] Hence, to minimize flow, growth should be in the vertical direction. For growth in the vertical direction in an idealized furnace that is free from imperfections, there is a steady-state solution of the relevant differential equations in the reference frame of the furnace, in which there is no convection in the melt and the temperature and solute concentration vary only with distance from the crystal-melt interface. The interface is then planar, with a uniform temperature and concentration. The stability of this solution depends on the parameters V, c_∞, and G_L, which define the growth conditions for the crystal, where G_L is the temperature gradient in the liquid at the crystal-melt interface.

An alloy of lead and tin provides an interesting example for the discussion of convective instabilities. The lead-tin phase diagram has a eutectic so that for nearly pure lead, tin is rejected during freezing, whereas for nearly pure tin, lead is rejected during freezing. The density of liquid lead is larger than the density of pure tin, and the density as a function of composition is approximately linear.[2] Thus, during the freezing of tin containing lead, the heavier lead is rejected, giving rise to a heavy layer of fluid near the interface. Since the liquid density decreases with temperature, during freezing vertically upward the density of the tin containing lead alloy will decrease with height, and we would not expect a convective instability since both the temperature and solute fields promote a liquid density that decreases with height.

For an alloy of lead containing tin the situation is more complicated since the temperature and solute fields act in opposing ways; the lighter tin is rejected, and there is the possibility of a convective instability.[3] Before we discuss these convective instabilities we discuss morphological instabilities that arise even in the absence of flow and thus are possible for both alloys of lead containing tin and tin containing lead.

Interfacial Instabilities

In 1953 Rutter and Chalmers[4] and Tiller et al.[5] interpreted their experimental observations of solute inhomogeneities during directional solidification of a tin alloy in terms of the instability of the crystal-melt interface and proposed a heuristic explanation for the instability that they called "constitutional supercooling."

Many materials can be cooled in the liquid phase to temperatures that are more than 20% below their freezing point. The supercooled liquid is metastable, in the sense that, if it is placed in contact with a crystal seed, the crystal spontaneously grows into the liquid with a speed that is determined by the rate at which the latent heat that is released as the liquid solidifies can diffuse into the liquid ahead of the interface. A physical explanation for interface instability in the case of growth into a supercooled melt can be based on the conservation of energy at the interface.[6] The magnitude of the local temperature gradient is increased where a perturbation extends into the liquid, which results in an increase of the local interface velocity at the bump, so that the perturbation grows. (A competing force for stability is provided by the interfacial surface energy, as will be described later.)

In contrast, in the directional solidification of a binary alloy, the temperature gradient in the liquid at the interface is positive; thus, the liquid temperature increases with distance from the interface. However, the liquid ahead of the interface may still be supercooled because of the variation of the freezing point with concentration. A prediction of interface instability is obtained from the thermodynamic condition that the liquid ahead of the interface is *constitutionally* supercooled; that is, the temperature lies below the freezing temperature as adjusted by the presence of solute. In more quantitative terms the unperturbed solute profile in the liquid is given by

$$c(z) = c_\infty - \frac{D}{V} G_c \exp\left(-\frac{Vz}{D}\right) \tag{1}$$

where D is the solute diffusivity in the liquid, and G_c is the solute gradient at the interface. The solute field exhibits a length scale D/V over which the solute field approaches its far-field value. The temperature gradient G_L in the melt is approximately constant, and the constitutional supercooling criterion for instability is[5]

$$\frac{m(k-1)c_\infty V}{kD} > G_L \tag{2}$$

where m is change in melting point with solute concentration (slope of liquidus on the phase diagram), and k is the distribution coefficient, that is, the ratio of the solute concentration in the crystal to that in the liquid at the crystal-melt interface. With some modifications the constitutional supercooling criterion is found to give good agreement with experiments for low growth speeds. For higher growth speeds, as the length scale D/V becomes smaller, surface energy is found to play an important role in delaying the onset of instability, as was shown by Mullins and Sekerka in 1964.[7] The role of surface energy in their analysis is to cause a further modification to the equilibrium temperature of the interface, T_I, as described by the Gibbs-Thomson equation

$$T_I = T_M + mc_L - T_M \Gamma \mathcal{K} \tag{3}$$

where T_M is the melting point of the pure material, c_L is the liquid solute concentration at the interface, Γ is the capillary length on the order of angstroms, and $\mathcal{K}$ is the mean curvature of the interface.

Mullins and Sekerka[7] performed a linear stability analysis of the planar interface during directional solidification, including the effects of surface energy, based on solutions to the diffusion equations for temperature and solute. The resulting dispersion relation for the *morphological stability* of the interface reduces at low velocities to a *modified constitutional supercooling* criterion, in which the temperature gradient in the liquid, G_L, that appears in the stability criterion 2 of Tiller et al.[5] is replaced by the average of the liquid and solid temperature gradients, weighted by the thermal conductivities of each phase. At higher velocities, the destabilizing effect of the solute gradient is balanced by the effects of surface energy, and the system is more stable than would be predicted by the constitutional supercooling criterion. Their stability analysis also provides the wavelength of the fastest growing instability; the wavelength is influenced by both D/V and Γ. In Fig. 1 we plot the critical concentration and wavelength at the onset of morphological instability for an alloy of tin containing a small amount of lead as a function of the solidification velocity. The interface is morphologically unstable for concentrations above those indicated by the solid curve. The wavelength at the onset of instability decreases with increasing velocities and is much smaller than the typical container even for very low velocities. The low-velocity regime is well described by the

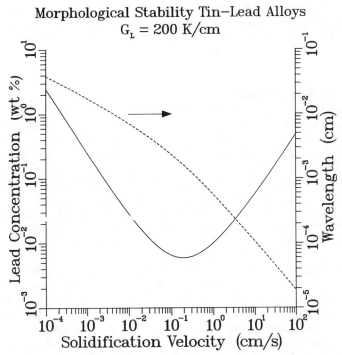

Fig. 1   The critical concentration of lead in tin (solid curve) above which interface instability occurs, as a function of solidification velocity for directional solidification with a temperature gradient in the liquid of 200 K/cm. The wavelength at the onset of instability is shown by the dashed curve.

modified constitutional stability criterion. Velocities above about 0.1 cm/s cannot be obtained by conventional directional solidification, but are possible by electron beam or laser annealing. The thermophysical properties used in the calculations are given in Table 1 of Coriell and McFadden[8]; for simplicity the slope of the liquidus curve and the distribution coefficient were assumed constant in the calculations. For lead-tin alloys these properties depend on composition.[2] Although it is not difficult to treat this composition dependence, all of the calculations reported here will neglect it.

For growth conditions for which the planar crystal-melt interface is morphologically unstable, cellular growth or dendritic growth occurs. Steady-state cellular growth gives rise to solute segregation transverse to the growth direction, and dendritic growth gives rise to both transverse and longitudinal segregation. Research on dendritic growth is extremely active,[9] but will not be discussed here. Near the onset of instability of the planar interface, cellular growth can be studied by nonlinear stability analyses,[10,11] that is, expansions in the amplitude of the deviation of the interface from planarity. Numerical computations have followed the evolution of two-dimensional cells to large degrees of instability,[12] and we have calculated the interface shape and solute concentration for shallow three-dimensional cells.[13] Such numerical calculations are restricted to a few wavelengths even on supercomputers. Solutions are found for a range of wavelengths, and at present the mechanism of wavelength selection (or even whether there *is* wavelength selection at all) is unclear.

Double-Diffusive Convection

To minimize buoyant convection in the melt, directional solidification is normally performed with growth along the vertical axis in order to avoid as much as possible the horizontal density gradients that necessarily produce natural convection. The density of the liquid is a function of both the temperature and concentration, and for one-dimensional planar growth both the temperature gradient and the solute gradient are perpendicular to the interface and aligned with the gravity vector. In the absence of horizontal density gradients, a quiescent base state is then possible; however, with diffusion of both temperature and solute occurring in the melt, the conditions for the stability of this base state are more complex than is the case for a single-component system.[3]

Even if the net density in the liquid decreases with height, convective instability may occur due to the different rates of diffusion of heat and solute. If we consider the case of solidification vertically upward with rejection at the interface of a solute that is lighter than the solution, then the temperature field tends to stabilize the system, since hotter (lighter) fluid overlies colder (heavier) fluid and the solute field tends to destabilize the system (solute-rich fluid underlying solute-poor fluid). For metallic alloys and semiconductors the rate at which heat diffuses through the liquid is orders of magnitude faster than that of the solute. A physical mechanism for instability in such a configuration can be given in terms of the forces that act on a parcel of fluid that is displaced from its equilibrium position.[3] For example, a vertical displacement of such a parcel carries it to a region of hotter fluid with lower solute concentrations. The temperature of the parcel equilibrates rapidly with the surroundings, but the parcel remains rich in solute, so that the displaced parcel is lighter than its surroundings. If the resulting buoyancy forces exceed the viscous forces, the vertical displacement will be reinforced, and convective instability results. This instability was first considered in oceanographic studies, where the radiant heating of salt water can lead to large-scale convective motions.

For directional solidification linear stability with respect to thermosolutal convection has been considered by a number of authors.[14-21] In Figs. 2–4 we show the results of calculations for lead containing tin for various gravitational accelerations, g, in units of the Earth's gravitational acceleration, $g_e = 980 \text{ cm/s}^2$. The thermophysical properties used in the calculations are given in Table 1 of Coriell et al.[15] In Fig. 2 the critical tin concentration above which convective instability occurs is shown for a temperature gradient in the liquid G_L of 200 K/cm and solidification velocities

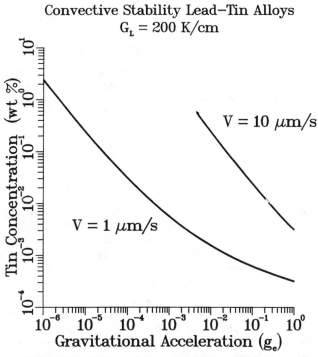

Fig. 2   The critical concentration of tin in lead above which convective instability occurs as a function of gravitational acceleration (in units of the Earth's gravitational acceleration) for directional solidification at 10 and 1 μm/s for a temperature gradient in the liquid of 200 K/cm.

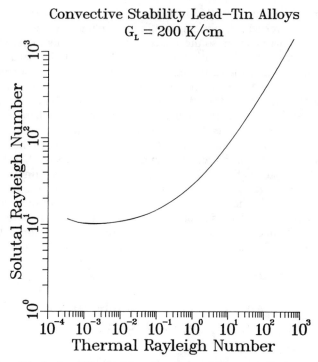

Fig. 3   The critical solutal Rayleigh number above which convective instability occurs as a function of the thermal Rayleigh number for direction solidification of lead-tin alloys at 10 and 1 μm/s for a temperature gradient in the liquid of 200 K/cm. The curves for the two different velocities are indistinguishable.

V of 1 and 10 μm/s. In general, the convective and morphological modes are actually coupled, but in the parameter range shown in Figs. 2–4 the coupling is weak, and the results shown correspond to convective instability. In an alloy of lead containing tin, for sufficiently large tin concentrations there is certainly instability since increasing the tin concentration promotes both the convective and morphological instability. The interesting question is which instability occurs first, i.e., at the lowest tin concentration. The classic Mullins-Sekerka morphological instability is independent of gravitational acceleration and for $G_L = 200$ K/cm occurs at 7.8 and 0.80 wt % for $V = 1$ and 10 μm/s, respectively. Thus, for the lower velocity the convective mode occurs at lower solute concentrations than the morphological mode for all of the gravitational accelerations shown in Fig. 2. The curve for the higher solidification velocity has been terminated at solute concentration below the concentration for morphological instability. When the critical concentrations for convective and morphological instabilities are similar, there is coupling between the modes, and the behavior can be fairly complicated.[15,22,23] For experiments on Earth it is clear from Fig. 2 that only very small concentrations of tin in lead are required to cause

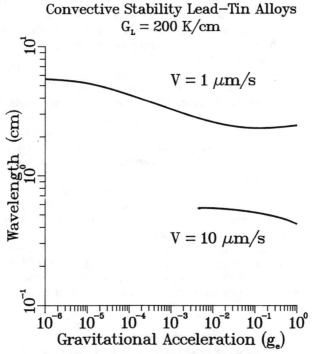

Fig. 4 The wavelength at the onset of convective instability as a function of gravitational acceleration (in units of the Earth's gravitational acceleration) for directional solidification at 10 and 1 μm/s for a temperature gradient in the liquid of 200 K/cm.

convective instability. The results in Fig. 2 are more easily understood by introducing the dimensionless thermal and solutal Rayleigh numbers, namely,

$$Ra = \frac{g\alpha G_L (D/V)^4}{\nu\kappa} \tag{4}$$

$$Rs = \frac{g\beta c_\infty (1-k)(D/V)^3}{\nu D k} \tag{5}$$

which measure the ratio of the thermal and solutal buoyancy forces to the viscous forces. In these equations α is the relative density change with temperature, ν is the kinematic viscosity, κ is the thermal diffusivity, and β is the relative density change with composition. The destabilizing influence of the solute field is balanced by the thermal field and viscous forces. In Fig. 3 we use the same results as those in Fig. 2 and plot the solutal Rayleigh number as a function of the thermal Rayleigh number. The results for the two velocities are shown, but the curves are indistinguish-able. For small values of the thermal Rayleigh number the convective instability occurs when the solutal Rayleigh number exceeds about 10. The wavelengths at the onset of convective instability are shown in Fig. 4. For a given velocity the wavelengths are fairly weak functions of the gravita-tional acceleration; the wavelengths are inversely proportional to velocity.

These linear stability calculations assume a laterally infinite system; for the lower velocity, especially at small gravitational accelerations, the wave-length at the onset is almost 6 cm. This is large compared to the size of a typical container, and the container walls may stabilize the system with respect to convective instability. Guerin et al.[24] have recently treated the effect of confining walls in two dimensions for the case of vanishing thermal Rayleigh number. For side walls of width L, the system is stabilized with respect to convection when the dimensionless Peclet num-ber, $Pe = VL/D$, is less than about 10. The stabilization increases rapidly with decreasing Pe; for example, for $L = 1.0$ mm and $V = 1.0\ \mu m/s$ the solutal Rayleigh number at the onset of convection is about 40,000. The strong lateral confinement results in flow up one wall and down the other wall at the onset of convection. The stabilizing thermal field may further enhance the stability of the system, but these calculations have not yet been performed. An interesting example of destabilization of a double diffusive

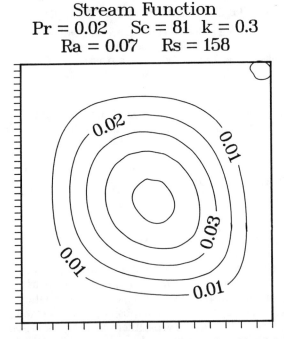

Stream Function
Pr = 0.02   Sc = 81   k = 0.3
Ra = 0.07   Rs = 158

Fig. 5   **The steady-state stream function as a function of position for the directional solidification of lead containing tin.**

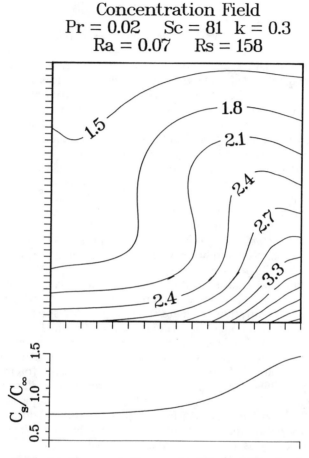

Fig. 6   The steady-state tin concentration as a function of position for the directional solidification of lead containing tin. The lower plot shows the lateral variation in the solid concentration.

system by side walls that are perfectly conducting to heat has been studied.[25]

In Figs. 5 and 6 we show the result of a two-dimensional calculation of the stream function and solute distribution in the melt for conditions of convective instability for lead containing tin. In this calculation[26] the crystal-melt interface is assumed planar, and the upper boundary remains a fixed distance (1.0 cm) from the interface. The growth conditions correspond to $V = 1.0\,\mu\text{m/s}$, $G_L = 200\,\text{K/cm}$, $L = 1.0\,\text{cm}$, $g = 10^{-4}g_e$, and $c_\infty = 0.36$ wt % or, in terms of the dimensionless numbers, $Ra = 0.07$, $Rs = 158$, and $Pe = 3.3$. The maximum flow velocity is only about $4\,\mu\text{m/s}$, but this is sufficient to cause a twofold variation in the solute distribution at the crystal-melt interface and hence in the crystal. Flow transitions and time-periodic behavior as the degree of instability increases have been calculated.[27-29]

For lead containing tin, it is clear that, except for extremely small concentrations of tin, or for small containers, it is impossible to avoid solute inhomogeneities on Earth. For growth vertically upward, with a temperature gradient of 200 K/cm, the maximum concentration is about 0.2 wt % tin at a solidification velocity of about 40.0 μm/s. This is the velocity at which morphological and convective instabilities occur at the same concentration.[15] One could try growth vertically downward; for lead containing tin the solute field would be stabilizing and the thermal field destabilizing with respect to convection. This case has not been extensively studied theoretically; since the destabilizing thermal field extends over the length of the sample, the onset of convection will depend on the sample length. The stabilizing solute field near the interface might alleviate some of the effects of convection away from the interface. This case corresponds to the "diffusive" regime of double diffusive convection,[3] and it is likely that the onset of instability will be oscillatory in time. Growth vertically downward in 3-mm-diam tubes has been shown to eliminate macrosegregation in eutectic alloys.[30]

For fundamental studies of morphological instability (and the resulting cellular growth) in Earth-based experiments, alloys for which both the thermal and concentration fields are stabilizing with respect to convection, for example, tin containing lead, would seem to offer the best possibility of avoiding convective influences. However, even for these alloys the avoidance of flow due to horizontal temperature gradients in actual furnaces is extremely difficult.[31,32] In fact, a number of experiments for this type of alloy in the cellular and dendritic growth regime have shown a severe macroscopic interface curvature.[33-36] In fact, this macroscopic curvature or "steepling" occurs when growth is such that the solute field is stabilizing and does not occur when the solute field is destabilizing.[35] The effect has been eliminated by using a ternary alloy such that the net density along the liquidus is nearly independent of temperature and composition.[34]

Even in the absence of horizontal gradients, there can be convective effects on morphological stability for growth with stabilizing temperature and concentration fields. The perturbed crystal-melt interface causes lateral gradients in temperature and solute, which induce fluid flow that alters the temperature and solute profiles and hence the morphological stability of the system. In Fig. 7 we show the results of a linear stability calculation for tin containing lead for three gravitational accelerations. Although there is only a small change in the critical concentration at the minima of the curves as the gravitational acceleration changes from zero to the Earth's gravitational acceleration, there is an order of magnitude change in the wavelength $\lambda = 2\pi/\omega$ at the onset of convection. The minima of the curves corresponding to $g = 10^{-2}g_e$ and $g = 0$ are essentially identical, although convection again slightly destabilizes a small wave-number perturbations. For an alloy of lead containing tin, the effect of convection on Earth is less important as the solidification velocity increases from 1.0 μm/s.[8] The largest effect occurs in a hypothetical system with the properties of tin

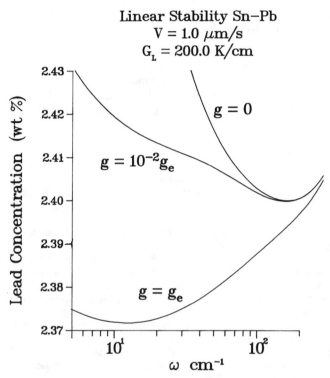

Fig. 7   The concentration of lead in tin above which interface instability occurs as a function of the wave number of a sinusoidal perturbation for a solidification velocity of 1.0 μm/s and a temperature gradient in the liquid of 200 K/cm for three values of the gravitational acceleration.

containing lead except that the melt density change with composition vanishes. For this hypothetical system there is a long wavelength instability with a critical composition that is several orders of magnitude smaller for the Earth's gravitational acceleration than for the zero-gravity case. The wavelength at the onset of this convectively modified morphological instability is comparable to the container size.

It is clear that it is extremely difficult and for some alloys impossible to avoid convection during directional solidification experiments on the Earth. A central problem is then to understand the effect of fully developed convection on the morphology of the crystal-melt interface. The disparate length scales between the convective flows and the morphological features makes a completely numerical simulation impossible at present. There have been a number of linear stability analyses in which a simple (usually analytic) flowfield has interacted with the crystal-melt interface.[22,23,37-40] In certain cases a strong interaction between convection and morphological

382 S. R. CORIELL AND G. B. McFADDEN

stability has been observed. In other cases the interaction is relatively weak. The number of dimensionless groups and the complexity of flowfields make general statements impossible.

Acknowledgment

This work was conducted with the support of the Microgravity Science and Applications Division of the National Aeronautics and Space Administration.

References

[1] Joseph, D. D., *Stability of Fluid Motions II*, Springer-Verlag, Berlin, 1976, p. 18.

[2] Poirier, D. R., "Densities of Pb-Sn Alloys During Solidification," *Metallurgical Transactions*, Vol. 19A, 1988, pp. 2349–2354.

[3] Turner, J. S., *Buoyancy Effects in Fluids*, Cambridge University Press, Cambridge, UK, 1973, pp. 251–287.

[4] Rutter, J. W., and Chalmers, B., "A Prismatic Substructure Formed During Solidification of Metals," *Canadian Journal of Physics*, Vol. 31, 1953, pp. 15–39.

[5] Tiller, W. A., Jackson, K. A., Rutter, J. W., and Chalmers, B., "The Redistribution of Solute Atoms During the Solidification of Metals," *Acta Metallurgica*, Vol. 1, 1953, pp. 428–437.

[6] Langer, J. S., "Instabilities and Pattern Formation in Crystal Growth," *Reviews of Modern Physics*, Vol. 52, 1980, pp. 1–28.

[7] Mullins, W. W., and Sekerka, R. F., "Stability of a Planar Interface During Solidification of a Dilute Binary Alloy," *Journal of Applied Physics*, Vol. 35, 1964, pp. 444–451.

[8] Coriell, S. R., and McFadden, G. B., "Buoyancy Effects on Morphological Instability During Directional Solidification," *Journal of Crystal Growth*, Vol. 94, 1989, pp. 513–521.

[9] Glicksman, M. E., Winsa, E., Hahn, R. C., Lograsso, T. A., Tirmizi, S. H., and Selleck, M. E., "Isothermal Dendritic Growth—A Proposed Microgravity Experiment," *Metallurgical Transactions*, Vol. 19A, 1988, pp. 1945–1953.

[10] Coriell, S. R., McFadden, G. B., and Sekerka, R. F., "Cellular Growth During Directional Solidification," *Annual Review of Materials Science*, Vol. 15, 1985, pp. 119–145.

[11] Sriranganathan, R., Wollkind, D. J., and Oulton, D. B., "A Theoretical Investigation of the Development of Interfacial Cells During the Solidification of a Dilute Binary Alloy: Comparison with the Experiments of Morris and Winegard," *Journal of Crystal Growth*, Vol. 62, 1983, pp. 265–283.

[12] Ramprasad, N., Bennett, M. J., and Brown, R. A., "Wavelength Dependence of Cells of Finite Depth in Directional Solidification," *Physical Review B*, Vol. 38, 1988, pp. 583–592.

[13] McFadden, G. B., Boisvert, R. F., and Coriell, S. R., "Nonplanar Interface Morphologies During Unidirectional Solidification of a Binary Alloy. II. Three-Dimensional Computations," *Journal of Crystal Growth*, Vol. 84, 1987, pp. 371–388.

[14] Caroli, B., Caroli, C., Misbah, C., and Roulet, B., "Solutal Convection and Morphological Instability in Directional Solidification of Binary Alloys," *Journal de Physique (Paris)*, Vol. 46, 1985, pp. 401–413.

[15] Coriell, S. R., Cordes, M. R., Boettinger, W. J., and Sekerka, R. F., "Convective and Interfacial Instabilities During Unidirectional Solidification of a Binary Alloy," *Journal of Crystal Growth*, Vol. 49, 1980, pp. 13–28.

[16] Coriell, S. R., and Sekerka, R. F., "Effect of Convective Flow on Morphological Stability," *PhysicoChemical Hydrodynamics*, Vol. 2, 1981, pp. 281–293.

[17] Glicksman, M. E., Coriell, S. R., and McFadden, G. B., "Interaction of Flows with the Crystal-Melt Interface," *Annual Review of Fluid Mechanics*, Vol. 18, 1986, pp. 307–335.

[18] Hennenberg, M., Rouzaud, A., Favier, J. J., and Camel, D., "Morphological and Thermosolutal Instabilities Inside a Deformable Solute Boundary Layer During Directional Solidification. I. Theoretical Methods," *Journal de Physique (Paris)*, Vol. 48, 1987, pp. 173–183.

[19] Hurle, D. T. J., Jakeman, E., and Wheeler, A. A., "Effect of Solutal Convection on the Morphological Stability of a Binary Alloy," *Journal of Crystal Growth*, Vol. 58, 1982, pp. 163–179.

[20] Schaefer, R. J., and Coriell, S. R., "Convective and Interfacial Instabilities During Solidification of Succinonitrile Containing Ethanol," *Materials Processing in the Reduced Gravity Environment of Space*, edited by G. E. Rindone, Elsevier, Amsterdam, 1982, pp. 479–489.

[21] Young, G. W., and Davis, S. H., "Directional Solidification with Buoyancy in Systems with Small Segregation Coefficient," *Physical Review B*, Vol. 34, 1986, pp. 3388–3396.

[22] Coriell, S. R., McFadden, G. B., Boisvert, R. F., and Sekerka, R. F., "Effect of a Forced Couette Flow on Coupled Convective and Morphological Instabilities During Unidirectional Solidification," *Journal of Crystal Growth*, Vol. 69, 1984, pp. 15–22.

[23] McFadden, G. B., Coriell, S. R., and Alexander, J. I. D., "Hydrodynamic and Free Boundary Instabilities During Crystal Growth: The Effect of a Plane Stagnation Flow," *Communications on Pure and Applied Mathematics*, Vol. 41, 1988, pp. 683–706.

[24] Guerin, R. Z., Billia, B., Haldenwang, P., and Roux, B., "Solutal Convection During Directional Solidification of a Binary Alloy: Influence of Side Walls," *Physics of Fluids*, Vol. 31, 1988, pp. 2086–2092.

[25] McFadden, G. B., Coriell, S. R., and Boisvert, R. F., "Double-Diffusive Convection with Sidewalls," *Physics of Fluids*, Vol. 28, 1985, pp. 2716–2722.

[26] McFadden, G. B., Rehm, R. G., Coriell, S. R., Chuck, W., and Morrish, K. A., "Thermosolutal Convection During Directional Solidification," *Metallurgical Transactions A*, Vol. 15, 1984, pp. 2125–2137.

[27] Heinrich, J. C., "Numerical Simulations of Thermosolutal Instability During Directional Solidification of a Binary Alloy," *Computer Methods in Applied Mechanics and Engineering*, Vol. 69, 1988, pp. 65–88.

[28] McFadden, G. B., and Coriell, S. R., "Thermosolutal Convection During Directional Solidification. II. Flow Transitions," *Physics of Fluids*, Vol. 30, 1986, pp. 659–671.

[29] McFadden, G. B., and Coriell, S. R., "Solutal Convection During Directional Solidification," AIAA Paper 88-3635-CP, 1988, pp. 1572–1578.

[30] Boettinger, W. J., Biancaniello, F. S., and Coriell, S. R., "Solutal Convection Induced Macrosegregation and the Dendrite to Composite Transition in Off-Eutectic Alloys," *Metallurgical Transactions A*, Vol. 12, 1981, pp. 321–327.

[31] Brattkus, K., and Davis, S. H., "Directional Solidification with Heat Losses," *Journal of Crystal Growth*, Vol. 91, 1988, pp. 538–556.

S. R. CORIELL AND G. B. McFADDEN

Brown, R. A., "Theory of Transport Processes in Single Crystal Growth from the Melt," *AIChE Journal*, Vol. 34, 1988, pp. 881–911.

Burden, M. H., Hebditch, D. J., and Hunt, J. D., "Macroscopic Stability of a Planar, Cellular or Dendritic Interface During Directional Freezing," *Journal of Crystal Growth*, Vol. 20, 1973, pp. 121–124.

McCartney, D. G., and Hunt, J. D., "Measurements of Cell and Primary Dendrite Arm Spacings in Directionally Solidified Aluminum Alloys," *Acta Metallurgica*, Vol. 29, 1981, pp. 1851–1863.

Verhoeven, J. D., Mason, J. T., and Trivedi, R., "The Effect of Convection on the Dendrite to Eutectic Transition," *Metallurgical Transactions A*, Vol. 17, 1984, pp. 991–1000.

Tunca, N., and Smith, R. W., "Variation of Dendrite Arm Spacing in Al-rich Zn-Al Off-Eutectic Alloys," *Journal of Materials Science*, Vol. 23, 1988, pp. 111–120.

Brattkus, K., and Davis, S. H., "Flow Induced Morphological Instability: Stagnation Point Flows," *Journal of Crystal Growth*, Vol. 89, 1988, pp. 423–427.

Delves, R. T., "Theory of Interface Stability," *Crystal Growth*, edited by B. R. Pamplin, Pergamon, Oxford, UK, 1974, pp. 40–103.

Fang, Q. T., Glicksman, M. E., Coriell, S. R., McFadden, G. B., and Boisvert, R. F., "Convective Influence on the Stability of a Cylindrical Solid-Liquid Interface," *Journal of Fluid Mechanics*, Vol. 151, 1985, pp. 121–140.

McFadden, G. B., Coriell, S. R., Boisvert, R. F., Glicksman, M. E., and Fang, Q. T., "Morphological Stability in the Presence of Fluid Flow in the Melt," *Metallurgical Transactions A*, Vol. 15, 1984, pp. 2117–2124.

Segregation and Convection in Dendritic Alloys

D. R. Poirier*

University of Arizona, Tucson, Arizona

Introduction

The susceptibility of cast alloys to macrosegregation is dealt with on a day-to-day basis by foundrymen, ingot-makers, and continuous casting engineers alike. For the most part, the approach to solving macrosegregation problems is to evaluate a number of full-scale production runs to determine acceptable processing parameters for each new alloy. Although segregation is sometimes reduced by these empirical methods, it usually persists and often is found to reach unacceptable levels leading to expensive scrap. In many processes macrosegregation in castings, ingots, or billets is the overriding factor in limiting the size and production rate of the cast product. Therefore, there is strong incentive to model macrosegregation in dendritic alloys, and such work has been underway since the late 1960s.

It is now abundantly clear that, in order to model macrosegregation in solidified alloys, it is necessary to analyze convection, as well as the partitioning of solutes, that occur within the solid plus liquid region (i.e., "mushy zone") of a solidifying alloy. However, in understanding macrosegregation phenomena in dendritic alloys, it is instructive to first consider the partitioning of the solutes between the solid and the liquid, while isolating this aspect of the overall process from the convective effects. Solute partitioning takes place within the dendritic spacings, which are typically 10–400 μm, and the segregation that results is "microsegregation." On the other hand, macrosegregation occurs on a length scale comparable to a macroscopic dimension of the cast product. In some cases, the macrosegregation is localized, but even so it is (or can be made) visually apparent. Even though the length scales of microsegregation and macrosegregation

*Professor, Department of Materials Science and Engineering.

are quite different, the two are interwoven. Therefore, before macrosegregation/convective phenomena are described, microsegregation in dendritic alloys is discussed.

Microsegregation

In crystal growing processes the desired interface between the solid and liquid is planar, which is a necessary criterion for avoiding lateral segregation of solute and minimizing crystal defects. However, most solidification processes (e.g., shaped castings, ingots, weldments, continuous castings, etc.) involve alloys, relatively rich in solutes, that solidify at solidification rates and thermal gradients that cause the interface not to be planar. There is instead a complex dendritic interface that occupies the mushy zone.

Solidification with No Thermal Gradient

In castings that solidify in molds of relatively insulating materials (i.e., so-called sand molds), thermal gradients are quite shallow, and, indeed, the temperature within a significant portion of the casting is approximately uniform (but not constant, of course). For this scenario, a solute balance is

$$(C_L - C_s^*) \, df_s = (1 - f_s) \, dC_L \tag{1}$$

where

C_L = concentration of solute in the liquid, wt%
C_s^* = concentration of solute in the solid at the interface, wt%
f_s = weight fraction of solid

The solute balance, Eq. (1), is also shown in Fig. 1.

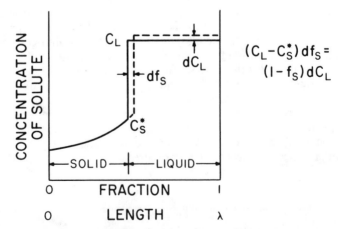

Fig. 1 Solute partitioning during solidification.

The left-hand side of Eq. (1) is the solute rejected by the solid, as solidification proceeds by the increment df_s. Notice that the concentration of solute in the solid behind the interface does not change, because inherent in this simple model no diffusion in the solid is assumed. This is a realistic assumption for solutes in many alloys because the diffusion coefficient is typically 10^{-8} to $10^{-10}\,cm^2 \cdot s^{-1}$. However, there are instances where diffusion in the solid cannot be ignored, but this does not introduce a qualitative change in this presentation.

The right-hand side of Eq. (1) is the increase of solute in the liquid that must result because solute is rejected at the interface. In this example C_s^* is less than C_L. In Fig. 1, as solidification proceeds by df_s, C_L increases to $C_L + dC_L$, but it does so while maintaining spatial uniformity. The uniformity of C_L is really an approximation, because, in general, there must be an enrichment in C_L near the solid-liquid interface as solidification proceeds. However, the diffusivity in the liquid is typically $10^{-5}\,cm^2 \cdot s^{-1}$ (3–5 orders of magnitude greater than the diffusivity in the solid), and the length scale of the interdendritic liquid is only 5–200 μm. Thus, the difference in C_s^* (the concentration of the solute at the interface) and the average concentration of solute in the liquid is typically quite small; thus C_L is approximated to be uniform.

In the absence of convection, the overall concentration of solute, in both phases taken together, remains at C_0 at any location and any time within the casting as it solidifies. Then, with the equilibrium partition ratio defined as

$$k = C_s^* / C_L \tag{2}$$

Eq. (1) is integrated, with $C_L = C_0$ at $f_s = 0$, to give

$$C_L = C_0(1 - f_s)^{k-1} \tag{3}$$

or, alternatively,

$$C_s^* = kC_0(1 - f_s)^{k-1} \tag{4}$$

A plot of Eq. (4) is shown as Fig. 2. Such a plot describes the microsegregation in the as-cast structure, which persists because there is no diffusion in the solid.

Equation (4) has been verified many times, at least qualitatively, by means of microanalysis. Specifically, the dendritic structure can be revealed by suitable metallographic preparation, so that the solid that formed first can be readily identified. For solutes with $k < 1$, this solid invariably has a solute concentration less than C_0. In the interdendritic regions, where solidification occurred last, the solute concentration is typically much greater than C_0.

There is another manifestation of microsegregation that is shown in Fig. 2 but is not directly indicated by either Eq. (3) or Eq. (4). That is, in many alloys, an eutectic mixture of two solids forms from the liquid at the end of

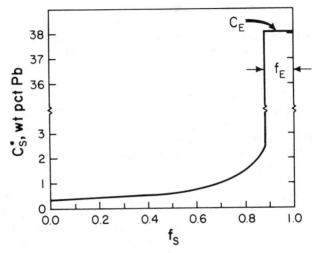

Fig. 2   Microsegregation in Sn-10 wt% Pb alloy, according to Eqs. (4) and (5).

solidification. Again, when $k < 1$, C_L increases as solidification progresses. Eventually, $C_L = C_E$ and the remaining liquid $(1 - f_s = f_E)$ solidifies. For a binary, the solidification of the eutectic liquid is isothermal. Thus, Eq. (3) can be written as

$$f_E = (C_E/C_0)^{1/(k-1)} \tag{5}$$

Equation (5) gives the fraction of eutectic constituent in the completely solidified alloy. For the simple case considered, with no convection anywhere in the solidifying alloy, the amount of eutectic in any metallographic field of view would be the same.

Solidification with a Thermal Gradient

When the thermal gradient is shallow, as in the scenario of the previous section, the resulting as-cast structure compromises dendritic equiaxial grains. Often dendritic solidification is intentionally effected in a thermal gradient in order to increase production rate or to bring about an aligned dendritic structure. The as-cast structure comprises dendritic columnar grains, or, indeed, dendritic single crystals can be produced if special techniques are employed.

Figure 3 shows a mushy zone in a thermal gradient. In the absence of convection, only a negligible amount of solute is rejected by the solidifying dendrite tips into the all-liquid zone. The partitioning of solute is all behind the dendrite tips and into the interdendritic liquid. Therefore, for analyzing microsegregation the volume element shown in Fig. 3 applies. With a laboratory coordinate system, Eqs. (1), (3), and (4) apply, and microsegregation in the completely solidified alloy is again as shown in Fig. 2.

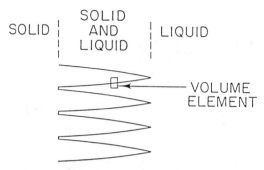

Fig. 3 Mushy zone in a thermal gradient and the volume element selected for the conservation equations.

In addition to thermal gradient across the mushy zone, in Fig. 3, there is also a concentration gradient, so that the solute balance of Eq. (1) should be modified to include diffusion in the liquid in the direction of the gradients. The relative importance of this diffusion to the solute balance can be examined by solving for the concentration of solute in the liquid at the tip of the dendrites, C_t; it is[1]

$$C_t = (1 - a)C_0 \qquad (6)$$

where

$a = DG/m_L R C_0$
D = diffusivity of the solute in the liquid
G = thermal gradient in the mushy zone
m_L = slope of the liquidus on the equilibrium phase diagram, dT/dC_L
R = solidification rate

For most dendritic solidification processes, $0.98 \leqslant C_t/C_0 < 1$, so that the integration, with $C_L = C_0$ for $f_s = 0$, that leads to Eqs. (3), (4), and (5) is valid.

Convection of the Interdendritic Liquid

The admission of convection of the interdendritic liquid is the starting point in understanding many macrosegregation phenomena. The continuity equation for the unit element depicted in Fig. 3 is

$$\frac{\partial}{\partial t}(\rho_s g_s + \rho_L g_L) = -\nabla \cdot \rho_L g_L v \qquad (7)$$

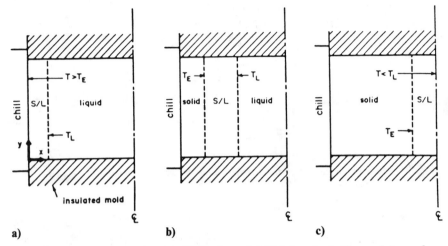

Fig. 4   Horizontal directional solidification: a) initial transient, b) complete mushy zone, c) final transient. (From Maples and Poirier.[14])

where

ρ_s = mass density of the dendritic solid
ρ_L = mass density of the interdendritic liquid
v = velocity of the interdendritic liquid
g_L = volume fraction of the interdendritic liquid
g_s = volume fraction of the dendritic solid

In writing Eq. (7) there is no motion of the solid.
The solute conservation equation is now

$$\frac{\partial C_L}{\partial t} + \left[\frac{(1-k)C_L}{g_L}\right]\left(\frac{\rho_s}{\rho_L}\right)\frac{\partial g_L}{\partial t} + \frac{u}{g_L}\nabla C_L = 0 \qquad (8)$$

In Eq. (8) the superficial velocity ($u = g_L v$) is used. With $\rho_s = \rho_L$ the volume fractions are the same as weight fractions, and if at the same time $u = 0$, then Eq. (8) reduces to Eq. (1). Equation (8) is the so-called "local solute redistribution" equation, first presented by Flemings and Nereo.[2] In deriving Eq. (8) it is assumed that ρ_s is constant, and there is no diffusion in the liquid in the direction of the thermal gradient.

In the earliest works[2-5] on modeling macrosegregation, convection in the all-liquid zone was ignored, and convection of the interdendritic liquid was assumed to be only that which satisfied continuity [i.e., Eq. (7)]. The first model, which included gravity-driven convection of interdendritic liquid, was done by Mehrabian et al. in 1970.[6] Their approach, with physical and numerical improvements, was subsequently used to model and study macrosegregation in cylindrical remelted ingots,[7-9,11-13,15] under thermal conditions in which the width of the mushy zone varies[10,14] and for the multicomponent alloys.[10,12]

Examples of these types of calculations are shown in Figs. 4–7. Figure 4 shows a binary alloy undergoing horizontal solidification from a chill surface in an otherwise insulated mold. Initially the mold is liquid with a uniform composition of C_0. When heat is extracted from the alloy, it solidifies away from the chill, and during solidification there is zone containing solid and liquid (i.e., the mushy zone). The momentum equation for the mushy zone is simply given by Darcy's law; thus,

$$v = -\frac{K}{\mu g_L} (\nabla p + \rho_L g) \qquad (9)$$

where

μ = viscosity of the interdendritic liquid
p = pressure
g = gravitational acceleration
K = permeability

The solution was obtained by solving for the pressure field. The equation for pressure was derived by combining Eqs. (7–9), resulting in a second-order differential equation.[14]

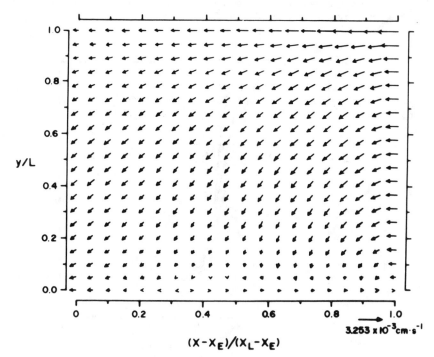

Fig. 5   Flow vectors (v) of the interdendritic liquid in Al-4.5 wt% Cu alloy during horizontal solidification. (From Maples and Poirier.[14])

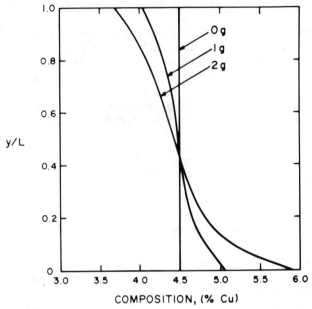

Fig. 6   Effect of gravity on macrosegregation in Al-4.5 wt% Cu alloy after horizontal solidification. (From Maples and Poirier.[14])

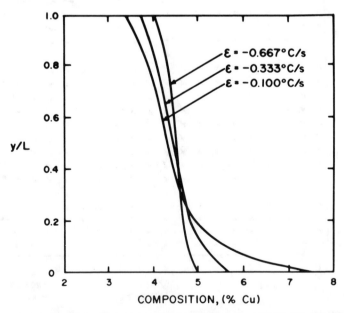

Fig. 7   Effect of cooling rate on macrosegregation in Al-4.5 wt% Cu alloy after horizontal solidification. (From Maples and Poirier.[14])

The top and bottom of the mushy zone are in contact with the mold; thus,

$$v_y = 0 \quad \text{at} \quad y = 0 \quad \text{and} \quad y = L$$

$$\text{for} \quad x_E \leqslant x \geqslant x_L \tag{10}$$

where L is the height of the mold, and x_L is the position of the dendritic tips.

At the eutectic isotherm, there is a finite amount of liquid of eutectic composition that solidifies at a constant temperature. The densities of the eutectic-liquid and the eutectic-solid are not equal, so that there is flow to compensate for solidification shrinkage or expansion of the eutectic. This requirement is

$$v_x = -\frac{\rho_{SE} - \rho_{LE}}{\rho_{LE}} U_E \quad \text{at} \quad x = x_E$$

$$\text{for} \quad 0 \leqslant y \leqslant L \tag{11}$$

where U_E is the speed of the eutectic isotherm. Conditions (10) and (11) can be rewritten to provide boundary conditions for pressure by substituting the velocity components into Eq. (9).

Because it is assumed that there is no convection in the bulk liquid, then

$$p = p_0 + \rho_{L0} g (L - y) \quad \text{at} \quad x = x_L$$

$$\text{for} \quad 0 \leqslant y \leqslant L \tag{12}$$

where p_0 is the ambient pressure, and ρ_{L0} is the density in the all-liquid zone.

After solidification at a point (x,y) is complete, the local average composition is given by

$$\bar{C}_s(x,y) = \frac{\rho_s \int_{t_L}^{t_E} C_s^*(x,y,t) \, dg_s(x,y,t) + \rho_{SE} g_E C_E}{\rho_s(1 - g_E) + \rho_{SE} g_E} \tag{13}$$

where the liquidus isotherm has passed the point at time t_L, and the eutectic isotherm has passed at time t_E.

With zero gravity, flow of the interdendritic liquid is due solely to solidification shrinkage, and flow is exactly normal to the isotherms. Also, after solidification the composition from top to bottom is uniform and equal to C_0.

Figures 5 and 6 show the effects of the gravity force in the mushy zone of an Al-4.5 wt% Cu alloy. In this alloy, the density of the interdendritic liquid increases with decreasing temperatures; hence, the effect of gravity is to cause downward flow within much of the mushy zone, with some reversal of flow in the lower right areas of Fig. 5. Hence, the solute-rich liquid flows away from the upper portion toward the lower portion of the ingot resulting in negative segregation at the top and positive segregation at the bottom, as shown in Fig. 6.

For a given gravity force the most important process variable, which influences macrosegregation, is cooling rate. This is shown in Fig. 7 for three different cooling rates. One of the curves is that given in Fig. 6 for a cooling rate of $-0.333°C \cdot s^{-1}$. When the cooling rate is doubled ($-0.667°C \cdot s^{-1}$), the segregation is substantially reduced. However, when the cooling rate is reduced to $-0.100°C \cdot s^{-1}$, segregation is increased significantly.

The effect of cooling rate on segregation can be explained by comparing the magnitudes of the velocities of interdendritic liquid to the magnitudes of the isotherm velocities. With the slower cooling rate the magnitude of the velocity of the interdendritic liquid is less than that at a greater cooling rate, but relative to the velocity of the isotherms, there is substantially more flow at the slower cooling rate. Consequently, macrosegregation increases with decreasing cooling rate as shown in Fig. 7.

Conservation Equations

In the previous section the energy equation was not used in the analysis of convection and macrosegregation within the mushy zone. Instead, the thermal field was known a priori, either by a decoupled calculation or by measurement. Thus, C_L was known, because C_L and T in the mushy zone are linked by the equilibrium phase diagram. Furthermore, convection in the all-liquid zone was neglected. Such a simplification is in order, when macrosegregation is in the form of a "surface to center" or a "top to bottom" variation in composition and thermosolutal convection is not important. However, by admission of convection in the all-liquid zone, it follows that there is macrosegregation of solute in the all-liquid zone as solidification proceeds. Therefore, the composition of the liquid at the dendrite tips is not restricted to C_0, and, more likely, the plane defining the position of the dendrite tips is not planar. The scenario depicted by Figs. 4–7 is oversimplified, and a more comprehensive model should be used.

Szekley and Jassal[16] developed a model for dendritic solidification in which the mushy zone was analyzed as a porous medium; they represented the flow of interdendritic liquid with a modified Navier-Stokes equation, with the viscous term replaced by the Darcy term. More recently, Bennon and Incropera[17] developed a continuum model to represent the transport processes in the mushy zone during solidification. Their model is based on mixture theory, in which the mushy zone is viewed as an overlapping continuum. That is, any location in the mushy zone can be occupied by solid and liquid simultaneously. On this basic premise, they derived the conservation equations for the mushy zone.

According to Hassanizadeh and Gray,[18] mixture theory is "an intuitive extension of classical single-phase relations." Hence, it is better to formulate such problems with volume averaging. Beckermann[19] and Beckermann and Viskanta[20] stated that they used volume averaging to derive the momentum equation for dendritic solidification, but important details and some assumptions were not given in their derivations.

To typify these type of calculations, with the full momentum equation written for the mushy zone, the results of Bennon and Incropera[17,21] are briefly reviewed. Continuum equations governing the conservations of mass, momentum, energy, and solute were written in two-dimensional Cartesian coordinates.[17] The conservation equations were applicable to any of three zones in a solidifying system (i.e., the all-solid, mushy, and all-liquid zones). Thermosolutal effects in the momentum equation were included by the usual Boussinesq approximation, with the gravity force term approximated with linear terms for the density variations with respect to temperature and solutal concentration.

Bennon and Incropera[17] chose to express the energy equation in terms of enthalpies. In their calculations, however, they did not include the compositional effect on the enthalpy of either phase. Prior to their work, solute conservation equations were typically written in terms of C_L; Bennon and Incropera worked with the local average concentration of solute in both phases. By this maneuver, they were able to compute the liquid mass fraction at any location within the mushy zone, based on the average composition of both phases taken together. However, they assumed that the solid as well as the liquid existing at a given temperature were uniform in concentrations.

Bennon and Incropera[21] did calculations for the horizontal solidification of 30% NH_4Cl in water. Some of their calculated results are shown in Figs. 8a and 8b. In Fig. 8a velocity vectors are shown. The tips of the dendrites are not aligned along a vertical plane, as depicted previously in Figs. 4 and 5 for the oversimplified model. Particularly noteworthy are the discrete flow channels within the mushy zone. These channels are highlighted by dashed lines in Fig. 8a. These channels contain solute-rich liquid, so that there is localized melting of dendrites. With less solid, the permeability increases; hence, there is yet more flow of the solute-rich liquid to the channels. The macrosegregation is more complex (Fig. 8b) than was previously depicted (Fig. 7).

Conservation of Momentum

The works of Bennon and Incropera,[17,21] as well as those of Beckermann and Viskanta,[20] Szekley and Jassal,[16] and Voller and Prakash,[22] have clearly indicated the need for analyzing dendritic solidification with a comprehensive set of conservation equations. But in reviewing those works it was found that there were as many different representations of the momentum equation as there were individual works.[23] This indicated the need for a systematic approach of deriving a rigorous representation of the conservation of momentum in the mushy zone.

Ganesan and Poirier[23] applied volume averaging[18,24-27] to derive the conservation of momentum for the flow of liquid in a dendritically

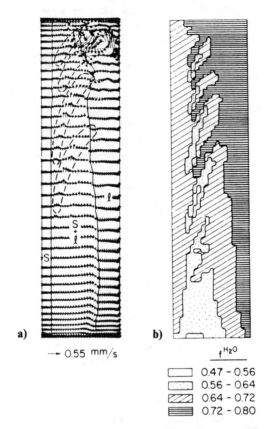

a)

→ 0.55 mm/s

b)

f^{H_2O}

	0.47 - 0.56
	0.56 - 0.64
	0.64 - 0.72
	0.72 - 0.80

Fig. 8 a) Flow vectors during solidification; b) macrosegregation after complete solidification in NH₄Cl-70 wt% H₂O ingot. (From Bennon and Incropera.[21])

solidifying alloy with the result

$$\rho_L g_L \left(\frac{\partial v}{\partial t} + v\nabla v\right) = -g_L \nabla p + \rho_L g_L g + \mu \nabla^2(g_L v) + \mu(-g_L^2 K^{-1}v + C: vv)$$

(14)

The permeability K is a second-order tensor. It can be written with three components for the directional solidification of a single crystal, when the principal directions are parallel and perpendicular to the primary dendrite arms. For a columnar dendritic structure that is polycrystalline, there are two components: a permeability component for flow parallel to the primary dendrite arms and a permeability component for flow perpendicular to the primary dendrite arms.[28] Finally, in equiaxial structures (small grains), the permeability is isotropic.

In Eq. (14) C is a resistance coefficient (third-order symmetric tensor). In isotropic equiaxial structures this coefficient is zero. It has not been measured for columnar structures, but the ratio of the resistance, given by the Darcy term, to the resistance, associated with the second term in the brackets on the right-hand side of Eq. (14), is estimated to be 5–50 near the dendrite tips ($g_s = 0.1$) and 10^3 to 10^5 in the interior of the mushy zone ($g_s = 0.5$). Hence, it appears that taking $C = 0$ is a good approximation, but a more careful evaluation of this point is probably in order.

Conservation of Energy

The energy equation, as applied to dendritic solidification with flow of interdendritic liquid, is given by Yeum and Poirier.[29] With respect to a stationary coordinate system, it is

$$\bar{\rho}\frac{\partial \bar{H}_s}{\partial t} + \rho_L g_L \frac{\partial L}{\partial t} - L\frac{\partial L}{\partial t}(\rho_s g_s) = \nabla \cdot (\kappa \nabla T) - \rho_L g_L v(\nabla \bar{H}_s + \nabla L) \quad (15)$$

where

$\bar{\rho}$ = average density of the solid plus liquid mixture
κ = thermal conductivity of the solid plus liquid mixture
$\bar{H}_s$ = enthalpy of the dendritic solid
$L = H_L - \bar{H}_s$ (an effective latent heat of fusion)
H_L = enthalpy of the liquid

There are several features of Eq. (15) that are not generally recognized. First, because of microsegregation of solute, there is microsegregation of enthalpy within the local solid; hence, the overbar in $\bar{H}_s$ indicates the average enthalpy of the solid. Second, L is not the true thermodynamic latent heat of fusion; more importantly, it varies throughout solidification because both $\bar{H}_s$ and H_L are functions of temperature *and* concentrations. Poirier and Nandapurkar[30] give a method of calculating the enthalpies $\bar{H}_s$ and H_L during dendritic solidification of a binary alloy. The major point is that the heat of mixing can be significant, so that enthalpy should be treated as a state function. Thus, its differential should be written as

$$dH = (\partial H/\partial T)_{x_1} \, dT + (\partial H/\partial X_1)_T \, dX_1 \quad (16)$$

for a binary, with the atom fraction of component 1, X_1. In formulations of the energy equation as applied to solidification, the second term of Eq. (16) generally has not been included.

Conservation of Solute

It is instructive to review all of the assumptions used by Flemings and Nereo[2] in the derivation of Eq. (8):
1) There is no diffusion in the solid.
2) There is local equilibrium at the solid-liquid interface.

3) The concentration of solute in the local interdendritic liquid (i.e., at a given isotherm) is uniform.

4) There is no diffusion of solute in the liquid in the direction of the thermal gradient.

5) No pores form, so that $g_s + g_L = 1$.

6) The density of the solid is constant.

If we keep assumptions 2–6 and replace assumption 1 with the assumption that there is complete diffusion in the local solid (i.e., the composition of the dendritic solid at a given temperature is uniform), then the conservation of solute is

$$\left[1 + k\left(\frac{\rho_s g_s}{\rho_L g_L}\right)\right]\frac{\partial C_L}{\partial t} + \left[\frac{(1-k)C_L}{g_L}\right]\left(\frac{\rho_s}{\rho_L}\right)\frac{\partial g_L}{\partial t} + \frac{\boldsymbol{u}}{g_L}\nabla C_L = 0 \qquad (17)$$

Equation (17) should be used for those solutes that diffuse rapidly in the solid (i.e., interstitial alloy elements). A typical example would be the solidification of steel, in which carbon is the solute of interest.

In the previous discussion pertaining to Eq. (6), it was mentioned that the diffusion of solute in the interdendritic liquid in the direction of the thermal gradient can usually be ignored. However, there are circumstances when this diffusion should be included. Therefore, to make the solute conservation equation more comprehensive, assumption 4 is relaxed. It can be assumed, instead, that this diffusion is Fickian; thus the flux is

$$\boldsymbol{j} = -\rho_L D_e \nabla C_L \qquad (18)$$

where the effective diffusivity D_e is related to D by

$$D_e = g_L D / \tau \qquad (19)$$

and τ is a tortuosity factor that accounts for the irregular geometry of the interdendritic diffusion paths.

For some alloys it is true that ρ_s is approximately constant during solidification, but this is not generally valid. Therefore, assumption 6 should also be relaxed for the sake of making the solute conservation equation more general. Finally, there are circumstances wherein the extent of diffusion in the solid is neither negligible nor complete.

When all the added factors are incorporated into the solute conservation equation, the final result is

$$\left[1 + 2\alpha^* k\left(\frac{\rho_s g_s}{\rho_L g_L}\right)\right]\frac{\partial C_L}{\partial t} - (C_L - \bar{C}_s)\left(\frac{g_s}{\rho_L g_L}\right)\frac{\partial \rho_s}{\partial t}$$

$$-\left(\frac{\rho_s C_L}{\rho_L g_L}\right)(1-k)\frac{\partial g_s}{\partial t} + \left(\frac{\boldsymbol{u}}{g_L}\right)\nabla C_L = \left(\frac{1}{\rho_L g_L}\right)\nabla(\rho_L D_e \nabla C_L) \qquad (20)$$

where $\bar{C}_s$ is the average concentration of solute in the solid, and α^* is the instantaneous diffusion parameter defined by Ganesan and Poirier.[31]

When $\alpha^* = 0$, there is no diffusion in the solid, and the coefficient of $\partial C_L/\partial t$ reduces to unity as in Eq. (8). When $\alpha^* = 1/2$, there is complete diffusion of solute in the local solid, and the coefficient of $\partial C_L/\partial t$ reduces to its counterpart in Eq. (17). Thus, $0 \leqslant \alpha^* \leqslant 0.5$. More generally, Ganesan and Poirier[31] showed that

$$\alpha^* = f(\alpha,k,f_s) \qquad (21)$$

where α is the Fourier diffusion number. The Fourier diffusion number is

$$\alpha = D_s t_f/\lambda^2 \qquad (22)$$

where D_s is the diffusivity in the solid, t_f is the local solidification time, and λ is one-half of the characteristic dendrite arm spacing. Because D_s varies exponentially with the reciprocal of the absolute temperature, the value of α^* can vary significantly (but always $1/2 \leqslant \alpha^* \leqslant 0$) during solidification. For some simplicity in applying Eq. (20), it is recommended that the extent of diffusion in the solid be decoupled from Eq. (20). The numerical scheme given by Yeum et al.[32] can be used for estimating the extent of the diffusion in the solid.

Directional Solidification and Thermosolutal Convection

Sometimes an important manifestation of flow of interdendritic liquid is the formation of localized segregates, which are called "channel segregates" or "freckles." Freckles were shown to form when the interdendritic liquid causes a local remelting within the mushy zone.[7,8,10,33-35] However, freckles as they form in vertical directionally solidified (DS) castings have not been modeled. Currently, it is believed that it is necessary to consider thermosolutal (or double-diffusive) convection during solidification in order to model freckles, and this is the subject of research in progress. Here a brief description of the initial modeling of the thermosolutal convection in a directional solidification process of a dendritic alloy is described.[36,37]

Consider the upward vertical solidification of a dendritic alloy, with a complete mushy zone underlying an all-liquid zone. The casting has a uniform cross section with insulated sides. The mushy zone extends from the eutectic isotherm ($z = 0$) to the isotherm along the dendrite tips ($z = z_t$). With the solidification rate R less than approximately $0.1 \text{ cm} \cdot \text{s}^{-1}$, which is certainly appropriate in most solidification processes of dendritic alloys, we can ignore the effects of curvature at the dendrite tips on both the solutal field and the thermal field in the vicinity of $z = z_t$. Along the isotherm at the tip of the dendrites, the composition C_t must be slightly greater than the far-field composition C_∞ in the all-liquid zone, because there is diffusion in the liquid.

We make the following assumptions: 1) the mushy zone moves with a constant solidification speed R (in the z direction); 2) there is no mass diffusion in the solid; 3) the solid and liquid phases have constant and equal densities ρ, except in the Boussinesq approximation for the buoyancy force; 4) gradients of g_L in the momentum equation are negligible; 5) the permeability is isotropic; and 6) C in Eq. (14) is zero. With these assumptions, Eq. (14) reduces to the momentum equation given by Beckermann and Viskanta,[20] with the effective viscosity taken to be the reference dynamic viscosity μ_0. Then the continuity and the momentum equations in a coordinate system moving with velocity R in the z direction are

$$\frac{\partial u}{\partial x} + \frac{\partial w}{\partial z} = 0 \tag{23}$$

x Momentum:

$$\frac{\partial u}{\partial t} - R\frac{\partial u}{\partial z} + \frac{1}{\phi}\left(u\frac{\partial u}{\partial x} + w\frac{\partial u}{\partial z}\right) = -\frac{\phi}{\rho_0}\frac{\partial p}{\partial x} + v_0\nabla^2 u - \frac{v_0}{K}u \tag{24}$$

z Momentum:

$$\frac{\partial w}{\partial t} - R\frac{\partial w}{\partial z} + \frac{1}{\phi}\left(u\frac{\partial w}{\partial x} + w\frac{\partial w}{\partial z}\right) = -\frac{\phi}{\rho_0}\frac{\partial p}{\partial z} + v_0\nabla^2 w - \frac{v_0\phi}{K}w - \phi\frac{\rho}{\rho_0}g \tag{25}$$

The components of u (i.e., the superficial velocity) are u and w, and ϕ is the weight fraction of liquid; $\phi = f_L = g_L$ because of assumption 3.

Equations (24) and (25) subsume both the Darcian and non-Darcian (Brinkman) terms to account for the porous medium behavior over the entire range of volume fraction of liquid within the mushy zone. It is important to note that, in the vicinity of the dendrite tips, use of the Darcian term alone would introduce serious error, whereas by including the Brinkman term, we permit viscous shear. Note that K is a function of ϕ,[28] and K approaches infinity as ϕ approaches 1. The reference state is the liquid at its liquidus temperature T_0 and solute concentration C_0. The Boussinesq approximation for the buoyancy term is included by the variable density ρ, which depends on solute concentration and temperature. Finally, v_0 is the kinematic viscosity of the liquid at its reference state (C_0, T_0).

For the energy equation, additional assumptions include following: 1) the local temperature of the solid and liquid are equal $(T_s = T_L = T)$ 2) the thermal conductivity within the mushy zone can be represented as an effective conductivity κ; and 3) the latent heat of fusion L throughout the solidification temperature range is constant. It follows from assumption 3 that $dH_s = dH_L = \hat{c}\, dT$, where $\bar{H}_s$ and H_L are the intensive enthalpies of the dendritic solid and interdendritic liquid, respectively, and $\hat{c}$ is the heat capacity of the mushy zone.

With all of these assumptions, the energy equation is

$$\frac{\partial T}{\partial t} + u\frac{\partial T}{\partial X} + (w - R)\frac{\partial T}{\partial z} = \bar{\alpha}\nabla^2 T - \frac{L}{\hat{c}}\frac{\partial \phi}{\partial t} + \frac{RL}{\hat{c}}\frac{\partial \phi}{\partial z} \tag{26}$$

where $\bar{\alpha} = \kappa/\rho_0\hat{c}$.

For the solute conservation equation, Eq. (8) is used but with no solidification shrinkage and with constant density. Added to the terms in Eq. (8), we include Fickian diffusion of solute. The final form is

$$\phi \frac{\partial C}{\partial t} + u \frac{\partial C}{\partial x} + (w - \phi R) \frac{\partial C}{\partial z} = D \left[\phi \nabla^2 C + \frac{\partial C}{\partial z} \frac{\partial \phi}{\partial x} + \frac{\partial C}{\partial x} \frac{\partial \phi}{\partial x} \right]$$

$$- (1 - k) \frac{\partial \phi}{\partial t} C + (1 - k) R \frac{\partial \phi}{\partial z} C \qquad (27)$$

where D is the solute diffusivity in the liquid, and C is C_L (to simplify the notation).

Each of Eqs. (23–27) reduces to the usual form for the all-liquid zone as $\phi \to 1$. Thus, we are assured of continuity between the variables in the mushy zone and the variables in the all-liquid zone. The boundary conditions are

At $z = 0$:

$$T = T_E; \qquad C = C_E; \qquad u = w = 0 \qquad (28)$$

At $z = \infty$ (far-field conditions):

$$C = C_\infty; \qquad T = T_\infty; \qquad u = w = 0 \qquad (29)$$

where C_E and T_E are concentration of solute and temperature at the eutectic isotherm, respectively.

Although many simplifying assumptions have been introduced in the writing of Eqs. (23–27) they preserve the basic terms for a linear stability analysis for the onset of convection. For the linear stability analysis, the basic state (nonconvecting state) was perturbed and the evolution of the perturbations was studied. The subscript s is for the dependent variables in their basic state.

By definition of our basic state, $u_s = w_s = 0$, and Eqs. (24) and (25) reduce to

$$\frac{\partial p_s}{\partial x} = 0 \qquad (30)$$

and

$$\frac{\partial p_s}{\partial x} = -\rho g \qquad (31)$$

The solute equation for the basic state comes from Eq. (27):

$$\frac{d^2 C_s}{dz^2} + \frac{R}{D} \frac{dC_s}{dz} + \frac{1}{\phi_s D} \frac{d\phi_s}{dz} \frac{dC_s}{dz} + \frac{(1 - k)R}{\phi_s D} \frac{d\phi_s}{dz} C_s = 0 \qquad (32)$$

The energy equation for the basic state is derived from Eq. (26). For the mushy zone it is convenient to write it in terms of concentration by making use of $m_L = dT/dC$, which is the slope of the liquidus on the equilibrium phase diagram. The result $(0 \leqslant z \leqslant z_t)$ is

$$\frac{d^2 C_s}{dz^2} + \frac{R}{\bar{\alpha}} \frac{dC_s}{dz} + \frac{RL}{m_L \hat{c} \bar{\alpha}} \frac{d\phi_s}{dz} = 0 \tag{33}$$

C_s and ϕ_s also satisfy

$$C_s(0) = C_E \tag{34}$$

$$C_s(z_t) = C_t \tag{35}$$

and

$$\phi_s(z_t) = 1 \tag{36}$$

Equations (32–36) must be solved numerically to obtain $C_s(z)$ and $\phi_s(z)$ within the mushy zone. With $C_s(z)$ known, $T_s(z)$ follows from the liquidus of the equilibrium phase diagram. C_t is determined as part of the solution in the all-liquid zone.

For the all-liquid zone, $z_t < z < \infty$, C_s is given by

$$\frac{C_s - C_\infty}{C_t - C_\infty} = \exp\left[-\frac{R}{D}(z - z_t) \right] \tag{37}$$

where

$$C_t = C_\infty + \frac{DG_L}{mR} \tag{38}$$

and G_L is the thermal gradient at the tip of the dendrites. The temperature in the all-liquid zone is

$$T_s = T_t + G_L \frac{\bar{\alpha}}{R} \exp\left[-\frac{R}{\bar{\alpha}}(z - z_t) \right] \tag{39}$$

To obtain differential equations for the perturbed quantities in the linear stability analysis, the dependent variables, expressed as the sum of the steady-state values and the perturbations, are substituted into the differential equations for continuity, x and z momentum, energy, and solute conservation. For a linear stability analysis, the inertial terms in Eqs. (24) and (25) and products of perturbed quantities are neglected. In the usual manner, the basic state equations are subtracted from these equations. By this operation, and subsequent operations to eliminate the perturbed pressure, there results the final set of equations for the perturbed quantities (for details, see Ref. 36).

The values of the parameters and the physical properties, used in the computations, correspond to the reference state $C_0 = C_\infty = 20$ wt% Sn and $T_0 = 280°C$, the liquidus temperature for the alloy; they can be found in Nandapurkar et al.[36]

For the permeability the following expression was used[28]:

$$K = \frac{4.70 \times 10^{-4} d_1^2 \phi^3}{(1 - \phi)} \tag{40}$$

where d_1 is the primary dendrite arm spacing, given by[38]:

$$d_1 = 556 G_L^{-0.311} \tag{41}$$

for small growth rates, with d_0 in microns and G_L in $K \cdot cm^{-1}$. The calculations were performed for temperature gradients ranging between 0.5 and $100 \, K \cdot cm^{-1}$.

Marginal stability curves for various fractions of the gravitational constant are shown in Figs. 9–11. The regions identified with S and U represent regions in which the basic nonconvecting states are stable and unstable, respectively.

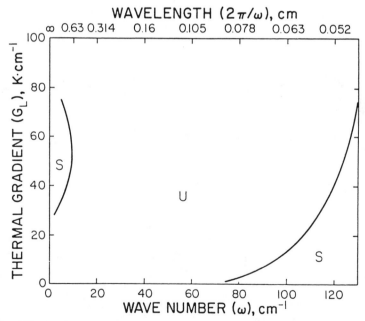

Fig. 9   Marginal stability curves for the dendritic solidification of Pb-20 wt% Sn for a solidification rate of 2×10^{-3} cm $\cdot$ s^{-1} with a gravitational constant of $1 g_0$. Stationary convection is predicted in the region U. The system is stable for the wavelengths in the regions marked S. (From Nandapurkar et al.[36])

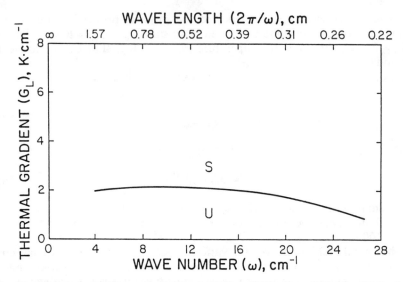

Fig. 10 Marginal stability curve for the dendritic solidification of Pb-20 wt% Sn for a solidification rate of 2×10^{-3} cm · s^{-1} with a gravitational constant of $0.1\,g_0$. (From Nandapurkar et al.[36])

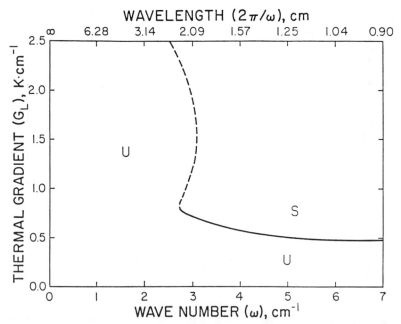

Fig. 11 Marginal stability curve for the dendritic solidification of Pb-20 wt% Sn for a solidification rate of 2×10^{-3} cm · s^{-1} with a gravitational constant of $0.01\,g_0$. The region *U* under the solid curve is for stationary convection, and the region *U* to the left of the broken curve is for oscillatory convection. (From Nandapurkar et al.[36])

For a gravitational constant of $1\,g_0$ ($g_0 = 9.8\,\mathrm{m \cdot s^{-2}}$; see Fig. 9), there are two branches that define the marginal stability of the system. These curves do not exhibit a region where the system is unconditionally stable for all wave numbers; thus, unless the horizontal width of the container holding the solidifying alloy is quite narrow, convection is expected. However, it would be possible to suppress convection by effecting solidification in narrow molds. For example, with $G_L = 10\,\mathrm{K \cdot cm^{-1}}$, the system is stable with respect to perturbations with wavelengths less than 0.66 cm. Thus, in molds of lesser widths, there would be no convection during solidification. The marginal stability curves for a gravitational constant of $0.5\,g_0$ are qualitatively similar to those at $1\,g_0$; quantitatively the region of instability is compressed. Calculations indicate that the system would be convectively stable in molds of widths less than 0.95 mm for a temperature gradient of $10\,\mathrm{K \cdot cm^{-1}}$.

The marginal stability curves change dramatically with a further reduction in the gravitational constant. With the gravitational constant equal to $0.1\,g_0$ (Fig. 10), solidification without convection is predicted with a gradient of $2\,\mathrm{K \cdot cm^{-1}}$. Also, for gradients greater than about $2.5\,\mathrm{K \cdot cm^{-1}}$, the system is always convectively stable.

When the gravitational constant is $10^{-2}\,g_0$, the system exhibits two branches (Fig. 11). The marginal stability branch, drawn with the solid curve, is for stationary convection; the region to the left of the broken curve is unstable with respect to oscillatory convection. At this reduced level of the gravitational constant, the system is stable provided $G_L > 0.75\,\mathrm{K \cdot cm^{-1}}$ in molds less than 2.0 cm in width. Finally, the linear stability calculations for a gravitational constant of $10^{-4}\,g_0$ showed that the system is stable for all wave numbers, at least for the range of thermal gradients analyzed (viz., $G_L \geqslant 2.5\,\mathrm{K \cdot cm^{-1}}$).

Calculations of the nonlinear convection during solidification in molds of finite widths were also done. The nonlinear calculations agree with the linear stability results, in that when the stability analysis predicts an unstable case, convection develops in the nonlinear system, and when a stable system is predicted, the perturbations die out in the nonlinear model.[37]

An example of supercritical convection is shown in Fig. 12. In this case, there is convection in the upper part of the mushy zone, and convection produces a strong disturbance in the concentration field at the tip of the dendrites. However, the temperature field remains almost unaffected by the convection. An organized cell system can be observed in the all-liquid zone, and the effect of convection in the upper part of the mushy zone is evident. In all of the nonlinear calculations[37] it was observed that convection in the mushy zone is only significant in the upper 20% of the mushy zone, if there is any convection at all. Furthermore, convection in the mushy zone is almost entirely driven by convection in the all-liquid region.

The model presented here is only valid at the onset of convection. After some time, constitutional equilibrium is not satisfied, because it is assumed that the distribution of the volume fraction liquid remains constant with

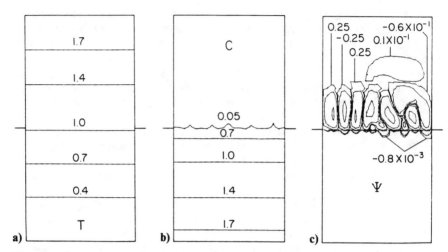

Fig. 12   Convection in Pb-10 wt% Sn: a) isotherms $(T - T_E)/(T_t - T_E)$; b) solutal isoconcentrates; c) streamline contours. (From Heinrich et al.[37])

time. For this reason, calculations have been performed for a short time after the onset of convection that describe the main qualitative features of the dynamics of the flow in the vicinity of the dendrite tips. Extension of the present model by relaxing the conditions that the fraction of interdendritic liquid in the mushy zone is constant with time is currently being pursued with the expectation that it will show the formation of the severe localized segregates known as channel segregates or freckles.

Acknowledgments
The author is grateful for sponsorship from the Microgravity Sciences and Applications Division of NASA, Grant NAG 3-723, and from the National Science Foundation Grant MSM-8702732. His collaborators are J. C. Heinrich, P. Nandapurkar, K. Yeum, and S. Felicelli of the University of Arizona. The author also acknowledges, with appreciation and fondness, his earlier macrosegregation research at the Massachusetts Institute of Technology with M. C. Flemings, S. Kou (currently at the University of Wisconsin-Madison), and T. Fujii of Kawasaki Steel Corporation. Discussions and/or collaborations with V. Laxmanan, at General Motors Research Laboratory; A. Hellawell and J. R. Sarazin, at Michigan Technological University; W. D. Bennon, at Alcoa Technical Center, S. Ganesan of the University of Arizona; and A. L. Maples (formerly of General Electric Co.), on the general topic of segregation phenomena, are greatly appreciated.

References

[1] Flemings, M. C., *Solidification Processing*, McGraw-Hill, New York, 1974, pp. 77–83.

[2] Flemings, M. C., and Nereo, G. E., "Macrosegregation, Part I," *Transactions of The Metallurgical Society-American Institute of Mining, Metallurgical and Petroleum Engineering*, Vol. 239, No. 9, Sept. 1967, pp. 1449–1461.

[3] Kirkaldy, J. S., and Youdelis, W. V., "Contribution to Theory of Inverse Segregation," *Transactions of the Metallurgical Society of American Institute of Mining, Metallurgical and Petroleum Engineering*, Vol. 212, No. 6, Dec. 1958, pp. 833–840.

[4] Flemings, M. C., Mehrabian, R., and Nereo, G. E., "Macrosegregation, Part II," *Transactions of The Metallurgical Society of American Institute of Mining, Metallurgical and Petroleum Engineering*, Vol. 242, No. 1, Jan. 1968, pp. 41–49.

[5] Flemings, M. C., and Nereo, G. E., "Macrosegration, Part III," *Transactions of the Metallurgical Society of American Institute of Mining, Metallurgical and Petroleum Engineering*, Vol. 242, No. 1, Jan, 1968, pp. 50–55.

[6] Mehrabian, R., Keane, M., and Flemings, M. C., "Interdendritic Fluid Flow and Macrosegregation; Influence of Gravity," *Metallurgical Transactions*, Vol. 1, No. 5, May 1970, pp. 1209–1220.

[7] Kou, S., Poirier, D. R., and Flemings, M. C., *Electric Furnace Proceedings, Iron and Steel Society of American Institute of Mining, Metallurgical and Petroleum Engineering*, New York, 1977, Vol. 35, pp. 221–228.

[8] Kou, S., Poirier, D. R., and Flemings, M. C., "Macrosegregation in Rotated Remelted Ingots," *Metallurgical Transactions B*, Vol. 98, No. 4, Dec. 1978, pp. 711–719.

[9] Ridder, S. D., Reyes, F. C., Chakravorty, S., Mehrabian, R., Nauman, J. D., Chen, J. H., and Klein, H. J., "Steady State Segregation and Heat Flow in ESR," *Metallurgical Transactions B*, Vol. 9B, No. 3, Sept. 1978, pp. 415–425.

[10] Fujii, T., Poirier, D. R., and Flemings, M. C., "Macrosegregation in a Multicomponent Low Alloy Steel," *Metallurgical Transactions B*, Vol. 10B, No. 3, Sept. 1979, pp. 331–339.

[11] Petrakis, D., Flemings, M. C., and Poirier, D. R., *Modeling of Casting and Welding Processes*, edited by H. D. Brody and D. Apelian, The Metallurgical Society-American Institute of Mining, Metallurgical and Petroleum Engineering, Warrendale, PA, 1981, pp. 285–312.

[12] Jeanfills, C., Chen, J. H., and Klein, H. J., *Modeling of Casting and Welding Processes*, edited by H. D. Brody and D. Apelian, The Metallurgical Society of American Institute of Mining, Metallurgical and Petroleum Engineering, Warrendale, PA, 1981, pp. 313–332.

[13] Ridder, S. D., Kou, S., and Mehrabian, R., "Effect of Fluid Flow on Macrosegregation in Axi-Symmetric Ingots," *Metallurgical Transactions B*, Vol. 12B, No. 3, Sept. 1981, pp. 435–447.

[14] Maples, A. L., and Poirier, D. R., "Convection in the Two-Phase Zone of Solidifying Alloys," *Metallurgical Transactions B*, Vol. 15B, No. 1, March 1984, pp. 163–72.

[15] Poirier, D. R., Kou, S., Fujii, T., and Flemings, M. C., *Electroslag Remelting*, Army Materials and Mechanics Research Center, Watertown, MA, Rept. AMMRC TR 78–28, June 1978.

[16] Szekley, J., and Jassal, A. S., "Experimental and Analytical Study of the Solidification of a Binary Dendritic System," *Metallurgical Transactions B*, Vol. 9B, No. 3, Sept. 1978, pp. 389–398.

[17]Bennon, W. D., and Incropera, F. P., "Continuum Model for Momentum, Heat and Species Transport in Binary Solid-Liquid Phase Change Systems—I. Model Formulation," *International Journal of Heat and Mass Transfer*, Vol. 30, Oct. 1987, pp. 2161–2170.

[18]Hassanizadeh, M., and Gray, W. G., "General Conservation Equations for Multiphase Systems: 1. Averaging Procedure," *Advances in Water Resources*, Vol. 2, No. 3, Sept. 1979, pp. 131–141.

[19]Beckermann C., "Melting and Solidification of Binary Systems with Double-diffusive Convection in the Melt," Ph. D. Thesis, Purdue Univ., West Lafayette, IN, 1987.

[20]Beckermann, C., and Viskanta, R., "Double-Diffusive Convection during Dendritic Solidification of a Binary Mixture," *Physicochemical Hydrodynamics*, Vol. 10, No. 2, 1988, pp. 195–213.

[21]Bennon, W. D., and Incropera, F. P., "The Evolution of Macrosegregation in Statically Cast Binary Ingots," *Metallurgical Transactions B*, Vol. 18B, No. 3, 1987, pp. 611–616.

[22]Voller, V. R., and Prakash, C., "Fixed Grid Numerical Modelling Methodology for Convection-Diffusion Mushy Region Phase-Change Problems," *International Journal of Heat and Mass Transfer*, Vol. 30, No. 8, Aug. 1987, pp. 1709–1719.

[23]Ganesan, S., and Poirier, D. R., "Conservation of Mass and Momentum for the Flow of Interdendritic Liquid During Solidification," *Metallurgical Transactions B*, Vol. 21B, Feb. 1990 (to be published).

[24]Whitaker, S., "Advances in Theory of Fluid Motion in Porous Media," *Industrial and Engineering Chemistry*, Vol. 61, No. 12, Dec. 1969, pp. 14–28.

[25]Gray, W. G., and Lee, P. C. Y., "On the Theorems for Local Volume Averaging of Multiphase Systems," *International Journal of Multiphase Flow*, Vol. 3, No. 4, June 1977, pp. 333–340.

[26]Slattery, J. C. "Single Phase Flow Through Porous Media," *American Institute of Chemical Engineers-Journal*, Vol. 15, No. 6, Nov. 1969, pp. 866–872.

[27]Whitaker, S., "Diffusion and Dispersion in Porous Media," *American Institute of Chemical Engineers-Journal*, Vol. 13, No. 3, May 1967, pp. 420–427.

[28]Poirier, D. R., "Permeability for Flow of Interdendritic Liquid in Columnar-Dendritic Alloys," *Metallurgical Transactions B*, Vol. 18B, March 1987, pp. 247–255.

[29]Yeum, K. S., and Poirier, D. R., "Modelling Directional Solidification of a Dendritic Alloy," *Cast Metals*, Vol. 1, No. 3, 1988, pp. 161–170.

[30]Poirier, D. R., and Nandapurkar, P., "Enthalpies of a Binary Alloy during Solidification," *Metallurgical Transactions A*, Vol. 19A, Dec. 1988, pp. 3057–3061.

[31]Ganesan, S., and Poirier, D. R., "Solute Redistribution in Dendritic Solidifications with Diffusion in the Solid," *Journal of Crystal Growth*, Vol. 97, No. 3/4, Oct. 1989, pp. 851–859.

[32]Yeum, K. S., Laxmanan, V., and Poirier, D. R., "Efficient Estimation of Diffusion during Dendritic Solidification," *Metallurgical Transactions A*, Vol. 20A, No. 12, Dec. 1989, pp. 2847–2856.

[33]Giamei, A. F., and Kear, B. H., "On the Nature of Freckles in Nickel Base Superlloys," *Metallurgical Transactions*, Vol. 1, No. 8, Aug. 1970, pp. 2185–2192.

[34]Sample, A. K., and Hellawell, A., "Mechanisms of Formation and Prevention of Channel Segregation During Alloy Solidification," *Metallurgical Transactions A*, Vol. 15A, No. 12, Dec. 1984, pp. 2163–2173.

[35]Sarazin, J. R., and Hellawell, A., "Channel Formation in Pb-Sn-Sb Alloy Ingots and Comparison with the System NH_4Cl-H_2O," *Metallurgical Transactions A*, Vol. 19A, No. 7, July 1988, pp. 1861–1871.

[36]Nandapurkar, P., Poirier, D. R., Heinrich, J. C., and Felicelli, S., "Thermosolutal Convection during Dendritic Solidification of Alloys: Part I. Linear Stability Analysis," *Metallurgical Transactions B*, Vol. 20B, Oct. 1989, pp. 711–721.
[37]Heinrich J. C., Felicelli, S., Nandapurkar, P., and Poirier, D. R., "Thermosolutal Convection during Dendritic Solidification of Alloys: Part II. Nonlinear Convection," *Metallurgical Transactions B*, Vol. 20B, Dec. 1989, pp. 833–891.
[38]Mason, J. T., Verhoeven, J. D., and Trivedi R., "Primary Dendrite Spacing. I. Experimental Studies," *Journal of Crystal Growth*, Vol. 59, No. 3, 1982, pp. 516–524.

Fluid Flow and Microstructure Development

Jonathan A. Dantzig* and Long-Sun Chao*
University of Illinois, Urbana, Illinois

Nomenclature

c, c_0, c_{ref} = concentration, base concentration, and reference composition, respectively

c_s, c_l = concentration in solid and liquid, respectively

c_{ps}, c_{pl} = specific heat of solid and liquid, respectively

D_s, D_l = solid and liquid diffusion coefficient, respectively

D = rate of strain tensor

$f(r)$ = function defined by Eq. (32)

F = body force vector

g = gravity vector

G, G_s, G_l = temperature gradient, solid temperature gradient, and liquid temperature gradient, respectively

h = enthalpy

k, k_s, k_l = thermal conductivity, solid thermal conductivity, and liquid thermal conductivity, respectively

k_0 = distribution coefficient

K = mean curvature of surface

L_f = latent heat of fusion

m_l, m_s = slope of liquidus and solidus curves, respectively

n = normal vector to a surface

p = pressure

Pe_h, Pe_m = peclet number for heat and mass transfer, respectively

Q = body heating

*Associate Professor, Department of Mechanical and Industrial Engineering.

411

R = ratio of temperature gradient in solid to gradient in liquid
Re = Reynolds number
$S(t)$ = interface position
t = time
T,T_s,T_l = temperature, solid temperature, and liquid temperature, respectively
T_{liq} = liquidus temperature
T_m,T_0 = freezing point of pure material and scale reference temperature, respectively
u = velocity vector
U_{tr} = translation velocity for cells into a shear flow
V = interface velocity
x = coordinate axis
X = length of domain in example problem
$z = x - Vt$
α = thermal diffusivity
α' = generalized diffusivity
β_T = thermal expansion coefficient
β_S = solutal expansion coefficient
γ_{sl} = solid-liquid surface energy
$\dot{\gamma}$ = shear rate
ζ = dimensionless coordinate
λ = cell spacing
μ = viscosity
ν = kinematic viscosity
ρ = density
ρ_l = density of liquid
ρ_s = density of solid
σ = dummy variable for transcendental equations
Φ = viscous dissipation
ψ = cell inclination angle

Introduction

METAL and semiconductor manufacturing represents an important and highly contested fraction of the U.S. economy. The typical production route for these materials begins with melting and alloying raw materials to a predetermined composition, followed by controlled solidification. The solidification process produces a microstructure that provides the materials with the properties they will have in service. Once all of the material is solidified, it is very difficult to change any of the gross features of the microstructure, because the transport processes that change the structure occur much more rapidly when one of the phases is liquid. Thus, it is desirable to control the microstructure during freezing, which necessitates that one have a good understanding of the process by which the structure forms.

In this article we will examine the interaction of macroscopic processing conditions with microstructure evolution, paying particular attention to the role of fluid flow during solidification. Most theories describing microstructure development during solidification are based on diffusive transport of heat and mass, and the presence of fluid flow acts to change the solute field, rather than fundamentally altering the phase change process. This makes the microgravity environment attractive for studying these phenomena because it represents an opportunity to study phase change processes with much less convective transport.

Solidification begins with nucleation of the parent crystal followed by continued growth. Nucleation is affected most by the purity of the solidifying material and is not strongly affected by fluid flow. However, the microgravity environment can be very important for studying nucleation phenomena, because it affords the possibility of observing nucleation in experiments done without a container, which is a possible source of contamination. These phenomena are discussed in other chapters in this volume; hence, this chapter will focus on growth processes, emphasizing the use of analysis to study and control structure development.

Metals and semiconductors can be considered to be incompressible. For that reason we may write the following equations representing the balance of mass, momentum, energy, and solute in the solidifying material:
For the fluid,

$$\nabla \cdot \boldsymbol{u} = 0 \tag{1}$$

$$\rho_l\left(\frac{\partial \boldsymbol{u}}{\partial t} + \boldsymbol{u} \cdot \nabla \boldsymbol{u}\right) = -\nabla p + \nabla \cdot (\mu D) + \rho_l \boldsymbol{F}$$
$$+ \rho_l \boldsymbol{g}\{1 - \beta_T(T - T_{\text{ref}}) - \beta_S(c - c_{\text{ref}})\} \tag{2}$$

$$\rho_l c_{pl}\left(\frac{\partial T}{\partial t} + \boldsymbol{u} \cdot \nabla T\right) = \nabla \cdot (k_l \nabla T) + Q \tag{3}$$

$$\frac{\partial c_l}{\partial t} + \boldsymbol{u} \cdot \nabla c_l = \nabla \cdot (D_l \nabla c_l) \tag{4}$$

and for the solid,

$$\rho_s c_{ps} \frac{\partial T}{\partial t} = \nabla \cdot (k_s \nabla T) + Q \tag{5}$$

$$\frac{\partial c_s}{\partial t} = \nabla \cdot (D_s \nabla c_s) \tag{6}$$

where it has been implicitly assumed that the solid cannot move ($\boldsymbol{u} = 0$). The momentum balance equation for the fluid includes terms on the right-hand side representing the variation in density of the fluid as a function of its temperature and composition. These terms represent driving forces for convection due to these buoyant effects. Note that each of these terms is proportional to the magnitude of the gravity vector; hence, the microgravity environment provides a unique means to suppress convection.

It is not sufficient for most purposes to simply take an experiment into low Earth orbit and assume that convection will be suppressed. After the effects of fluid flow on microstructure are examined, conditions under which buoyant forces produce significant convection will be studied. To provide a frame of reference, some of the basic analytical problems in solidification will be introduced first.

Consider first the solidification of a binary alloy where gravity is zero. In this "thought experiment" there will be no buoyant forces, and we can examine the role of the transport porperties independently of convection. Figure 1 shows a typical phase diagram for a dilute binary alloy. A vertical line in this diagram represents a single composition, and the various regions in the diagram indicate which phases are present at various temperatures. The equilibrium thermodynamics implied by the phase diagram usually holds true at least at the liquid-solid interface, so that the following boundary conditions apply at the interface: Thermodynamic equilibrium yields a relation between temperature, composition, and the local interface curvature

$$T_s = T_l = T_m + m_l c_l - \frac{\gamma_{sl}}{L_f} K \tag{7}$$

and the compositions of the two phases are related by the phase diagram

$$c_s = k_0 c_l \tag{8}$$

The mechanical and thermal boundary conditions that apply at the interface specify that mass is conserved, and there is no slip of the fluid on the solid:

$$\rho_l \boldsymbol{u} \cdot \boldsymbol{n} = (\rho_l - \rho_s) \boldsymbol{V} \cdot \boldsymbol{n} \tag{9}$$

where V is the velocity of the liquid-solid interface, and u is the fluid

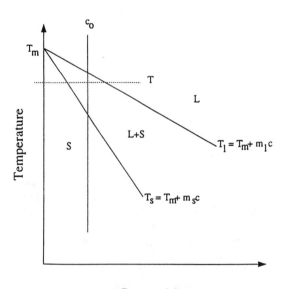

Fig. 1   Portion of a typical phase diagram for a binary system.

velocity. Heat is also conserved at the interface:

$$k_s \nabla T_s \cdot \boldsymbol{n} - k_l \nabla T_l \cdot \boldsymbol{n} = L_f V \cdot \boldsymbol{n} \tag{10}$$

Taken together, Eqs. (1–10) completely specify the solidfication field problem. Solution of the full system of equations is intractable in general; thus, the equations must be reduced to a smaller and simpler set to understand the important phenomena. The next several paragraphs will develop several of the basic problems governing heat and mass flow with phase change, and these problems will serve as a basis for examining the more complex phenomena encountered in materials processing.

Solution of the simplest heat flow problem in solidification, the solidification of a pure material in one dimension (see Fig. 2) was first presented by Stefan.[1] The experiment begins with a semi-infinite region at the equilibrium freezing temperature, T_m. At time zero, one end of the region is suddenly cooled to a lower temperature, T_0, corresponding to the solid. At some point after the beginning of the experiment, there will be two regions, one that is solid next to the boundary and another that is liquid and is separated by an interface. Considering all of the thermophysical properties to be independent of temperature, the governing equations for this problem reduce to

$$\rho c_p \frac{\partial T_s}{\partial t} = k \frac{\partial^2 T_s}{\partial x^2}; \qquad 0 \leqslant x \leqslant S(t) \tag{11a}$$

$$T(x = 0, t) = T_0 \tag{11b}$$

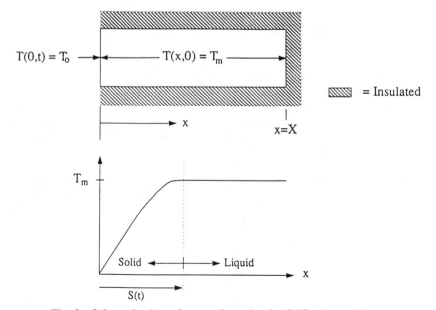

Fig. 2   Schematic view of a one-dimensional solidification problem.

$$T(x,t = 0) = T_m \tag{11c}$$

$$\frac{\partial T}{\partial x}(x = X,t) = 0 \tag{11d}$$

The temperature in the liquid ahead of the interface is constant at T_m so that no field problem needs to be solved there.

At the liquid-solid interface, $S(t)$, Eqs. (7–10) reduce to

$$T = T_m \tag{12}$$

and

$$k_s \frac{\partial T_s}{\partial x}[x = S(t),t] = L_f \frac{dS}{dt} \tag{13}$$

The position of the solidification boundary is thus a solution to a differential equation and is coupled to the solution of the temperature field in the liquid and solid. To complete the specification of the problem, an initial condition for the position of the interface is needed:

$$S(t = 0) = 0 \tag{14}$$

When the initial temperature of the liquid is equal to the equilibrium freezing temperature; this problem has an exact solution given by the following equations:

$$T(x,t) = T_0 + \frac{T_m - T_0}{\text{erf}\ (\sigma)} \text{erf}\left(\frac{x}{2\sqrt{\alpha t}}\right) \tag{15a}$$

$$\left[\sigma e^{\sigma^2\sigma^2} \text{erf}(\sigma)\right] = \frac{\rho_s c_{ps}(T_m - T_0)}{L_f\sqrt{\pi}} \tag{15b}$$

$$S(t) = 2\sigma\sqrt{\alpha t}; \quad \frac{dS}{dt} = \frac{\sigma\sqrt{\alpha}}{\sqrt{t}} \tag{15c}$$

Important observations about this solution that have applicability outside this simplified problem are that the thickness of the solidified layer is proportional to the square root of time measured from the beginning of the experiment, and the corresponding interface velocity is then inversely proportional to the square root of time. Considering, as an example, the solidification of pure aluminum ($T_m = 660°C$) against a perfectly conducting wall at 25°C, and replacing the material properties in this equation by their numerical values, we see that the solidification rate is given by the following equations:

$$\alpha = 0.981 \text{ cm}^2/\text{s} \tag{16a}$$

$$\sigma = 0.693 \tag{16b}$$

$$\frac{dS}{dl} = \frac{0.687 \text{ cm}^2/\text{s}^{\frac{1}{2}}}{\sqrt{t}} \tag{16c}$$

Note the (artificial) singularity in the velocity of the interface as $t \to 0$ caused by the instantaneous change in temperature at the boundary $x = 0$.

Much of the work in this field has correlated the microstructure with the interface velocity and temperature gradient during freezing. However, the apparently simple problem just presented is already too complex to use for such experiments, because the interface speed and temperature gradient change continuously during the experiment. An apparatus was developed by Bridgman to solidify samples at controlled interface velocities and interface velocities, as illustrated in Fig. 3.

The use of such an apparatus suggests a second basic problem in solidification: steady-state freezing (i.e., at constant velocity V) of a dilute binary alloy. In this case, solute transport at the interface must be considered in addition to the heat transfer. The assumption that the solidfication proceeds in steady state makes it more convenient to translate the coordinate system with the interface. A new variable, z, is defined as

$$z = x - Vt \tag{17}$$

Once again assuming temperature and composition independent properties, and one-dimensional solidification, the governing equations for energy and solute transport reduce to

$$-V \frac{\partial T_l}{\partial z} = k_l \frac{\partial^2 T_l}{\partial z^2} \tag{18}$$

$$-V \rho_l c_{pl} \frac{\partial c_l}{\partial z} = D_l \frac{\partial^2 c_l}{\partial z^2} \tag{19}$$

$$-V \rho_s c_{ps} \frac{\partial T_s}{\partial z} = k_s \frac{\partial^2 T_s}{\partial z^2} \tag{20}$$

$$-V \frac{\partial c_s}{\partial z} = D_s \frac{\partial^2 c_s}{\partial z^2} \tag{21}$$

The problem can be simplified still further after the governing equations are scaled. Choosing a length scale L and appropriate dimensionless scales for temperature and concentration yields a scaled version for each equation that looks like the prototype:

$$-\frac{L|V|}{\alpha'} \frac{\partial \phi}{\partial \zeta} = \frac{\partial^2 \phi}{\partial \zeta^2} \tag{22}$$

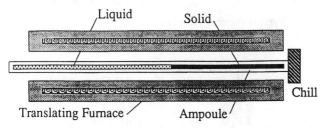

Fig. 3 Schematic view of a Bridgman-type directional solidification furnace.

where α' is the appropriate thermal or chemical diffusivity, L is a length scale, ϕ represents a concentration or temperature, and ζ is the scaled coordinate. Choosing typical values for the various quantities [e.g., for an aluminum-copper alloy (see Table 1)] shows us that the temperature field must be nearly linear in both phases at the interface and that diffusion in the solid state can be neglected with respect to diffusion in the liquid. This yields temperature and concentration fields shown schematically in Fig. 4. If the solidification is to proceed in the steady state, then the composition of the solid at the interface must be equal to c_0, otherwise, there would be continuous change of composition with solidification. Equation (7) implies that the temperature at the interface must be T_s, as shown.

The insertion of typical values for the diffusion coefficient for the liquid and the solid (see Table 1) reveals that the diffusive process is approximately 10,000 times slower in the solid state than it is in the liquid state. Accordingly, the diffusion in the solid state can be ignored in comparison to that in the liquid. If thermodynamic equilibrium is to be satisfied at the liquid-solid interface, then the composition in the liquid must be equal to c_0/k_0 and the composition as $z \to \infty$ is c_0. Solving this equation to obtain the concentration distribution in the liquid,

$$c_l = c_0\left(1 + \frac{1 - k_0}{k_0} e^{-Vz/D}\right) \qquad (23)$$

If we examine the concentration distribution ahead of the interface and ask the question "What is the equilibrium freezing point of the material in the

Table 1   Properties and typical values for a dilute
alloy of Cu in Al

Properties for Al-2% Cu Alloys[2]	
k_0	0.14
D_l	$3 \times 10^{-9} \text{ m}^2/\text{s}$
D_s	$3 \times 10^{-13} \text{ m}^2/\text{s}$
k_l	95 W/mK
k_s	210 W/mK
ρ_l	$2.39 \times 10^3 \text{ Kg/m}^3$
ρ_s	$2.70 \times 10^3 \text{ Kg/m}^3$
c_{pl}	$1.08 \times 10^3 \text{ J/m}^3\text{K}$
c_{ps}	$1.17 \times 10^3 \text{ J/KgK}$
Typical process variables	
L	10^{-2} m
V	10^{-3} m/s
Transport coefficients	
LV/α_l	0.27
LV/α_s	0.15
LV/D_l	3.3×10^3
LV/D_s	3.3×10^7

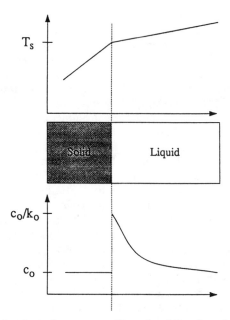

Fig. 4   Schematic of steady-state one-dimensional freezing of a binary alloy.

liquid ahead of the interface?" we can obtain that temperature by substituting the equation for the liquidus curve on the phase diagram into the left-hand side, yielding

$$\frac{T_{liq} - T_0}{m_l} = c_0\left(1 + \frac{1 - k_0}{k_0}e^{-Vz/D}\right) \tag{24}$$

Solving for T_{liq} and evaluating the gradient yields

$$T_{liq} = T_0 + m_l c_0\left(1 + \frac{1 - k_0}{k_0}e^{-Vz/D_l}\right) \tag{25a}$$

$$\frac{\partial T_{liq}}{\partial z} = -\frac{m_l c_0(1 - k_0)V}{k_0 D_l} \qquad (z = 0) \tag{25b}$$

Two possibilities exist ahead of the interface. Either the true temperature exceeds the equilibrium liquidus temperature, in which case any perturbation that appears on the interface will remelt, or the true temperature is below the equilibrium freezing temperature, in which case any perturbation that appears on the interface will grow. The limiting case is one where

$$\frac{\partial T}{\partial z} = G_l = \frac{\partial T_{liq}}{\partial z} \tag{26}$$

This argument was advanced by Tiller et al.[3] in the early 1950s and is known as constitutional supercooling. This work was formalized and extended in the early 1960s by Mullins and Sekerka[4,5] who performed a linear stability analysis for the interface under steady-state plane front growth. They also considered mechanical equilibrium of the interface including the energy associated with curvature of the interface in Eq. (7). They found two limits of stability: 1) a long-wavelength limit, corresponding to the constitutional supercooling argument given earlier, and 2) a short-wavelength limit where all of the undercooling at the interface supplies the energy for curvature. The latter was called absolute stability and has been encountered at high solidification rates. Above the marginal stability limit (i.e., at larger interface speeds or lower gradients), cells and eventually dendrites appear. Trivedi and Somboonsuk[6] have published some exceptionally illuminating photomicrographs of these transitions in succinonitrile-acetone transparent alloys. The geometry of the fully developed interfaces is sufficiently complex that analytical treatments similar to the work of Mullins and Sekerka have not been achieved.

The process of setting the microstructure is thus essentially one of the interaction of the thermal and solute fields. Using typical values for the dilute aluminum-copper system, we see that a typical boundary-layer thickness for the thermal field $[(\alpha_{sl})/V]$ is quite large, whereas for the solute $[(D_{sl})/V]$ it is a few microns. For this reason solidification problems can be thought of as ones where the thermal field is set by the processing conditions and the solute field follows through diffusion.

When fluid flow is present, the same concepts apply. Flow may affect the morphological stability and scaling of the microstructure, and it can alter the existing thermal and solute fields to alter the microstructure. This chapter will consider several problems where the altered thermal and solute field cause microstructure changes. The flow can have effects on both the local (microscopic) and the global (macroscopic) scale. In the next section we will examine these effects, using the three simple problems just described as a basis. However, one important exception to this viewpoint is the area of coupled instabilities for solidification and fluid flow. A recent review by Glicksman et al.[7] provides an excellent summary of recent advances in the coupling of fluid flow and morphological stability.

Role of Fluid Flow

Microscopic Scale

The effect of fluid flow on the morphological stability of the liquid-solid interface has been considered by several authors. In each case, a linear stability analysis was performed similar to that of Mullins and Sekerka, but a basic state including some fluid flow ahead of the interface was also considered. Delves[8] considered a simple shear flow parallel to the advancing interface and found that the processing conditions, i.e., the interface velocity and temperature gradient in the liquid that produced morphological instability, were essentially the same as those when no flow was present.

However, the interface instability in this case came as a set of traveling waves across the interface. Solutions admitted waves that traveled parallel or antiparallel to the fluid flow.

Coriell et al.[9] considered the stability of a binary alloy solidified antiparallel to the gravity vector for the case where the rejected solute was less dense than the parent alloy, admitting the possibility of double diffusive instabilities. They found extended regions of instability, but almost no coupling between the morphological and double-diffusive modes. Davis et al.[10] examined the coupling of the natural convection with a solidifying interface by considering an ice layer frozen from above, i.e., parallel to the gravity vector. The liquid-solid interface served as a record for the Rayleigh-Bénard cells present in the fluid.

Brattkus and Davis[11] considered a stagnation flow approaching the interface. They found traveling waves similar to those found by Delves, but they also found stationary waves that were not quite sinusoidal. The interface was distorted from an initially sinusoidal disturbance to one that became more like a sawtooth. The mechanism for this distortion was that the advancing flow compressed the solute boundary layer on the leading edge of the interface disturbance and expanded it behind the disturbance. This led to a higher concentration gradient on the front side of the interface disturbance than on the back, making the bump asymmetric.

There has been relatively less attention paid to the interaction of fluid flow with fully developed cellular and/or dendritic interfaces. Dantzig and Chao[12] considered the case of a shear flow perpendicular to an advancing cellular array. The physical model for this work is shown in Fig. 5. Several dilute alloys of lead in tin were considered, under several different processing conditions. Assuming the process to be steady and using the scales

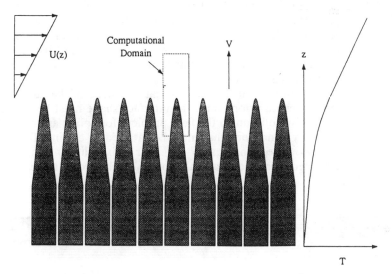

Fig. 5   Schematic view of physical model of shear flow near the interface.

length: λ, velocity: $\dot{\gamma}\lambda$, concentration: c_0, and temperature: T_0 reduces the governing equations to

$$\left(\frac{\dot{\gamma}\lambda}{\nu}\right)u^* \cdot \nabla^* u^* = -\nabla^* p^* + \nabla^{*2} u^* \tag{27a}$$

$$\left(\frac{\dot{\gamma}\lambda}{D}\right)u^* \cdot \nabla^* c = \nabla^{*2} c \tag{27b}$$

$$\left(\frac{\dot{\gamma}\lambda}{\alpha}\right)u^* \cdot \nabla^* \theta = \nabla^{*2}\theta \tag{27c}$$

Note that the buoyancy terms were neglected because the experimental configuration is stably stratified with respect to both thermal and solutal disturbances. For typical values of the shear rate, $\dot{\gamma} = 10\,\text{s}^{-1}$ and $\lambda = 30\,\mu\text{m}$, the dimensionless premultipliers of the advective terms are

$$\left(\frac{\dot{\gamma}\lambda}{\nu}\right) = Re = 0.045 \tag{28a}$$

$$\left(\frac{\dot{\gamma}\lambda}{D}\right) = Pe_m = 15 \tag{28b}$$

$$\left(\frac{\dot{\gamma}\lambda}{\alpha}\right) = Pe_h = 0.0025 \tag{28c}$$

This indicates that the effect of the flow on the thermal field can be neglected, but the flow has an important effect on the solute field.

The velocity and concentration fields were solved in the domain within $\pm 3D/V$ of the cell tip using the finite-element method, implemented in FIDAP.[13] As a result of the symmetry of the interface, a single cell was modeled, using a finite-element mesh such as that shown in Fig. 6. The mesh consisted of $9 \times 21 \times 25$ nodes, joined into eight-noded trilinear brick elements. See FIDAP manuals[13] or standard textbooks for details of the theory of the numerical methods.

The biggest difficulty in this problem is that the shape of the liquid-solid interface is not known a priori; thus, the interface shape must be found as a consequence of the solution to the field problem for solute and temperature. The boundary conditions that must be satisfied at the interface were given in the previous section, Eqs. (7–10). For a fixed linear temperature profile, as suggested from the small value of Pe_h, only two of these remain, Eqs. (7) and (8).

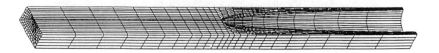

Fig. 6 FEM mesh for the cell.

There are several strategies available in the literature for constructing solutions to free boundary problems. Most of these involve first selecting a shape for the interface and constructing a solution using one of these boundary conditions, then updating the shape based on the error in satisfying the other boundary condition. In this work an approximate technique was used to select a shape for the cell. Bower et al.[14] considered directional solidification and noted that, deep within the cell, the remaining liquid is in such a small area that it could be assumed to be completely mixed. For thermodynamic equilibrium to hold at the interface, the liquid composition must be G/m_l deep within the cells. By assuming that this condition held all the way up to the cell tips they were able to obtain an equation for the fraction solid as a function of alloy concentration and temperature.

For a linear temperature profile this reduces to

$$f_s = 1 - \left[\frac{(k_0 - 1)(T_{\text{base}} + Gz - T_m) - ac_0}{c_0(k_0 - 1 - ak_0)} \right]^{1/(k_0 - 1)}$$

$$a = \frac{DG}{m_l V c_0}$$

(29)

Considering a round cell inside a square array, the radius is related to the fraction solid by

$$r(z) = \lambda \sqrt{\frac{f_s(z)}{\pi}}$$

(30)

where λ is the cell spacing. Combining Eqs. (29) and (30) yields a shape for the interface. This shape was then corrected to include the undercooling associated with curvature. Computing the mean curvature K as

$$K = \frac{1}{r(z)[1 + r'(z)^2]^{\frac{1}{2}}} + \frac{r''(z)}{[1 + r'(z)^2]^{\frac{3}{2}}}$$

(31)

then combining the four preceding equations yields a nonlinear ordinary differential equation for the cell shape:

$$\frac{\mathrm{d}^2 r}{\mathrm{d}z^2} = \frac{2}{G}[f(r) - Gz]\left[1 + \left(\frac{\mathrm{d}r}{\mathrm{d}z} \right)^2 \right]^{\frac{3}{2}}$$

$$f(r) = m_l c_0 \left[\frac{a}{k_0 - 1} + \frac{1 - ak_0}{k_0 - 1} \left(1 - \frac{\pi r^2}{\lambda^2} \right)^{k_0 - 1} \right] - \frac{G}{2r} + T_m - T_{\text{base}}$$

(32)

Integration of this equation by a Runge-Kutta scheme was performed using the unmodified cell shape far from the cell tip as a boundary condition. An example cell shape computed in this way is shown in Fig. 7, compared to an unmodified cell shape.

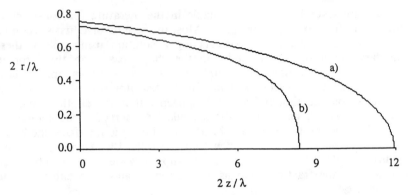

Fig. 7   **Cell shapes near the cell tip; a) solution of Bower et al.[14]; b) corrected for curvature by Eq. (32).**

The boundary conditions on the computational domain were as follows:

Vertical boundaries:

$$\frac{\partial c}{\partial n} = 0$$

Bottom:

$$\frac{\partial c}{\partial z} = \frac{G}{m_l}$$

Top:

$$-D_l \frac{\partial c}{\partial z} = V(c - c_0)$$

The boundary condition explicitly satisfied on the interface was that of local equilibrium, and the solute balance equation was used to update the cell shape and microstructure as described later.

In the absence of any shear flow, the concentration field will be symmetric with respect to the centerline of the cell, and the lateral sides will be symmetry planes. When the shear flow exists, solute is convected from the leading edge, around the cell, and appears on the trailing edge. This leads to asymmetry in the concentration field, and in particular makes the concentration gradient on the leading edge greater than that on the trailing edge. The solute balance equation for the interface tells us, then, that the normal velocity on the front face of the cell must be greater than on the trailing face; i.e., the cell translates into the flow. The resulting microstructure will consist of cells tilted toward the flow direction, with an inclination angle give by

$$\psi = \tan^{-1} \frac{U_{tr}}{V} \qquad (33)$$

The translation velocity U_{tr} was applied as a boundary condition on the cell. A series of processing conditions was examined using dilute alloys of 0.5 and 1.0 without Pb in Sn. Detailed results are included for just one case to demonstrate the phenomena, and the rest are summarized in Fig. 8.

Note that in steady growth the lateral sides must be both periodic and no-flux boundaries; hence, the no-flux boundary condition was used to reduce the computational load. The procedure used to determine ψ was to use trial values until a concentration field was found that was again symmetric with respect to the cell centerline. This was done by integrating the solute flux over the entire leading edge and over the entire trailing edge and comparing them. If the flux on the leading edge exceeded that on the trailing edge, the angle was increased, whereas if the flux on the leading edge was smaller than on the trailing edge, the angle was decreased.

The results are summarized in Fig. 8. The trends in the data can be interpreted as showing that the primary indicator for the angle is the "openness" of the structure. For fixed concentration, increasing the temperature gradient or the interface velocity results in a less porous structure, and thus smaller inclination angles. Similarly, increasing the concentration of the alloy produces less solidification per unit volume for the same processing conditions; hence, these alloys show larger inclinations. This interpretation was integrated into the representation in Fig. 8 by correlating the inclination angle with the product $\lambda c_0^{0.25}$, which is a measure of the openness of the structure. There is no special significance to the exponent 0.25.

The broadest conclusion from this study is that fluid flow can affect the solidification microstructure, but it does so mainly by altering the solute fields near the interface. For this reason it is important that the causes of

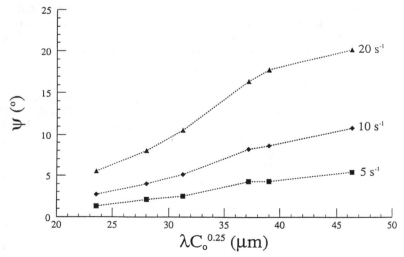

Fig. 8 **Computed inclination angle of advancing cells for various processing conditions.**

fluid flow be well understood and controlled. Most of these flows originate on the macroscopic scale, and the next sub-section will consider these problems.

Macroscopic Scale

Having shown that the microstructure can be strongly affected by flow in the melt, we will now consider means for determining and controlling buoyant flows during solidification. Note that the reduced gravity to be expected under space processing conditions is not sufficient to completely eliminate this convection. As in the previous sections, we will consider Bridgman growth experiments, this time for GaAs crystals.

To properly analyze samples produced in directional solidification, the investigator must know the existing temperature, solute, and velocity fields throughout the sample during the entire experiment. What is usually measured, however, is a relatively small number of temperature histories using thermocouples, most of which reside in the furnace or on the outside of the ampoule. There is really no way to avoid this, because embedding large numbers of thermocouples within the samples would undoubtedly contaminate the sample and overly influence the experiment, obscuring the effects that the experiment was intended to observe. Thus, the investigator must try to infer from the limited amount of recorded data and the sample just what actually occurred in the experiment. This arrangement usually provides far less information than is needed to properly interpret the resulting sample. Perhaps more important is the fact that, because there is no sufficiently accurate means for predicting what the experimental conditions will be before performing the experiment, it is difficult to properly design the experiment so that it will be performed under the conditions that the experimenter wishes to investigate.

In our early simulations of the crystal growth process, we found that the liquid-solid interface was highly curved. The curvature of the interface induces natural convection in the melt. Mixing of the fluid by the convection causes segregation. Since the interface is often an isoconcentrate, this situation causes radial segregation in the final crystal. The following analysis demonstrates that the primary cause of nonplanarity of the interface is the difference in thermal conductivity of the solid and liquid GaAs, and this fact also suggests means for improving the planarity of the interface by altering the applied furnace temperature distribution. Under steady-state conditions, the energy balance at the liquid-solid interface given in Eq. (10) becomes

$$k_s \nabla T_s \cdot \boldsymbol{n} - k_l \nabla T_l \cdot \boldsymbol{n} = 0 \tag{34}$$

When the interface is flat and perpendicular to the gravity vector in our models,

$$\boldsymbol{n} = \hat{z} \tag{35}$$

where $\hat{z}$ is the unit vector along the crystal growth axis. Consider first the

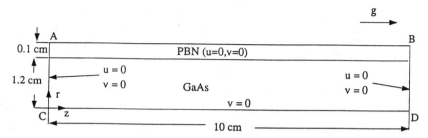

Fig. 9 Simplified model of growth ampoule with temperature distribution imposed on the boundary.

simplified model shown in Fig. 9, with insulated ends and a temperature profile specified on the outer diameter. For a constant temperature gradient imposed on the boundary, the temperature distribution near the interface can be expressed as

$$T_s = T_l = T_{\text{ref}} + Gz \tag{36}$$

where G is the temperature gradient on the boundary, then combining Eqs. (35–37) yields

$$k_s G - k_l G = 0 \tag{37}$$

which implies that

$$k_s = k_l \tag{38}$$

However, in GaAs, k_l is not equal to k_s; thus, Eqs. (35–37) cannot be satisfied simultaneously. If Eq. (36) is imposed as a processing condition, and the material ensures that Eq. (34) is satisfied, then the radial component of the unit normal vector of the interface cannot be zero and the interface is not flat. If we wish to have a flat interface when k_l is not equal to k_s, by combining Eqs. (34) and (36), the energy balance at the interface becomes

$$k_s G_s = k_l G_l \tag{39}$$

with G_s not equal to G_l. This suggests that the interface could be flattened by applying a bilinear temperature profile with gradients satisfying Eq. (40) and joined at the melting temperature on the boundary of the crystal.

Because of the presence of the PBN container, there is a small difference between the temperature distribution near the interface and that applied on the boundary. Trial and error was used to obtain the temperature profiles on the boundary leading to a maximum out-of-plane error in the interface less than 1.0×10^{-3} cm. In a typical application, either G_s or G_l was fixed, while the other one was changed to achieve a flat interface.

In all of the calculations that follow, G_l was set to be 5 K/cm and G_s was varied to achieve a flat interface. The dimensionless parameter R is defined as the ratio of the two gradients. From Eq. (40) an estimate for R that would yield a flat interface is

$$R \approx \frac{G_s}{G_l} = \frac{k_l}{k_s} = 2.06 \qquad (40)$$

The results are summarized in Fig. 10, where three trial temperature gradient ratios were applied on the boundary producing these steady temperature solutions. These results were obtained considering conduction only. The figure shows that for $R = 1$ (constant gradient in both phases), the interface has strong concave curvature. When $R = k_l/k_s$, it can be seen that the interface shape has been overcorrected and is significantly convex. By using trial-and-error methods to vary R, a flat interface was obtained when $R = 1.91$.

Figure 11 shows the results from an analysis of a similar problem, but now also including natural convection ($g = 0.0981$ m/s^2). When $R = 1$, there are two circulations, but the circulation near the interface disappears for the flat interface. Note that convection is eliminated only near the interface, whereas the end effect of the insulated boundary still induces some buoyant flow.

Figure 12 shows a schematic of the real apparatus used in the experiments to grow GaAs in space. A radiative heat-transfer boundary condition, instead of an imposed temperature, applies on the ampoule boundary facing the furnace wall. This was implemented in FIDAP by modifying the radiation heat-transfer element so that, for any time and for any location on the boundary, the reference temperature could be calculated from measured temperature data. All the other boundaries were treated as though they were insulated. The temperature contours and streamlines of the steady solutions for $R = 1$ and $R = 2.67$ are shown in the Fig. 13, in

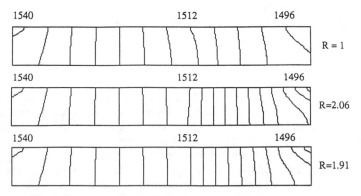

Fig. 10 **Steady-state temperature distributions computed for various imposed temperature distributions** ($G_l = 5$ K/cm).

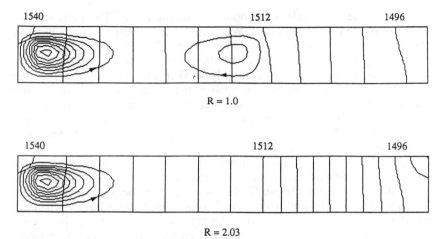

R = 1.0

R = 2.03

Fig. 11   Computed results for idealized geometry when convection is included in the calculation ($G_l = 5$ k/cm, $g = 0.0981$ cm/s).

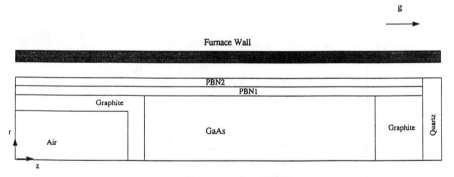

Fig. 12   Schematic view of the model for the experiment used to grow GaAs by GTE.

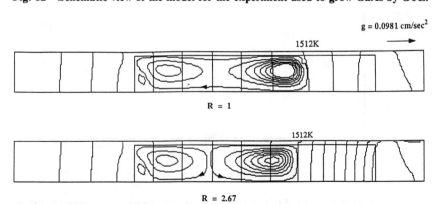

R = 1

R = 2.67

Fig. 13   Computed isotherms and streamlines for GaAs in the GTE space experiment apparatus. Note that the maximum absolute value of the stream function was 1.7×10^{-7}.

which the inner rectangle is the boundary of the GaAs. The same stream-lines are plotted in both figures. When $R = 1$, the interface is not flat and there is a strong clockwise circulation. When $R = 2.67$, somewhat above the value predicted on the basis of the thermal conductivity ratios, the interface is flat and there is a slightly weaker circulation. Note that the primary reason for the difference between the value of R to obtain the flat interface for the earlier cases and that for the present one is not the different boundary conditions for temperature; rather, it is the different surrounding materials and geometry, which cause different boundary effects, especially at the ends.

Figure 14 shows the value of R needed to produce a flat interface under steady-state conditions at different locations in the crystal. The range of the z-coordinate corresponding to GaAs is 4.4–13.2 cm. Because of end effects, it was not possible to find a value of R to produce a flat interface when the interface was near either end. From the figure it can be seen that increasing g necessitates a larger value for R to flatten the interface. This is because larger gravitational accelerations induce stronger convective heat transfer in the liquid, which in turn induce larger values for G_l in Eq. (41); therefore, a larger value of R is needed to compensate.

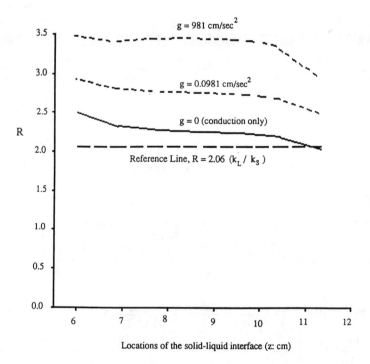

Fig. 14   Values of R required to produce a flat interface under steady-state conditions at various locations in the crystal and under several different gravitational accelerations.

Consider next the case where the crystal is being solidified over time. The initial condition for the simulation was taken to be the steady-state solution, with the radiation reference temperatures given by those at time equal to zero, corresponding to approximately half of the GaAs being molten. The temperature profile on the furnace was then slid up the crystal to simulate transient growth.

Solutions for this model using a single temperature gradient, i.e., $R = 1$, are illustrated in Fig. 15, which show the computed streamlines and successive interface positions under reduced gravity ($g = 9.8 \times 10^{-4}$ m/s^2). It is clear that the interface is far from flat, and this situation will lead to radial segregation on slices cut perpendicular to the growth axis. To extend the earlier steady-state analysis to transient problems, the heat balance at the interface is that given in Eq. (10). If the interface is flat, using Eq. (35) yields

$$k_s G_s = k_l G_l + \rho L_f |V| \qquad (41)$$

In a manner similar to that described in the preceding paragraphs, we may use this equation to adjust the temperature profile on the boundary to obtain a flat interface. The estimated value for R corresponding to the flat

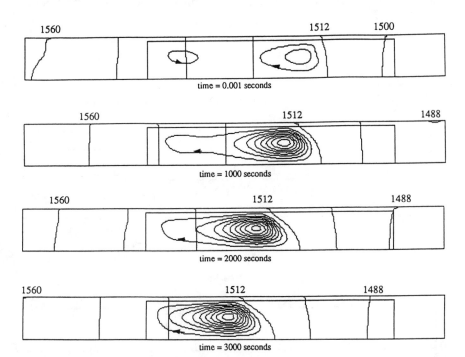

Fig. 15   Streamlines and successive interface positions for GaAs crystal growth under $10^{-4} \times$ the Earth's gravity.

interface for a given G_l is

$$R \approx \frac{k_l}{k_s} + \frac{\rho|V|L_f}{k_s G_l} \tag{42}$$

The first term on the right-hand side of this equation corresponds to steady state, and the second term represents the heat liberated by the moving interface. Simple trial-and-error methods for calculating R to produce a flat interface in transient problems proved to be difficult. The best results were obtained using

$$R = R_{\text{steady}} + \frac{\rho|V|L_f}{k_s G_l} \tag{43}$$

Note that, if $|V|$ is $10\,\mu\text{m/s}$ and G_l is $5\,\text{K/cm}$, the second term on the right-hand side is 11.66, compared to R_{steady}, which was 2.67. Figure 16 shows that this method significantly improves the shape of the moving interface. However, the interface near the initial point (between A and B in Fig. 16b) was not very flat. The initial condition used was the steady solution with the heat source at the initial location, $R = 2.67$. As soon as the heat source leaves the initial point, R jumps up to 14.33. This inconsistency caused some curvature in the interface near the starting point (see Fig. 17b.)

Time (seconds): A - 0.01　B - 500　C - 1000　D - 1500　E - 2000　F - 2500　G - 3000　H - 3500

$g = 0.0981$ cm/sec^2

$R = 1$

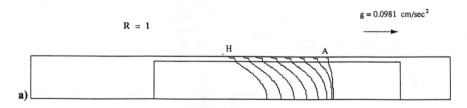

a)

$R = R_{\text{steady-state}} + \rho L\, v_{\text{hs}}\, /\, k_S\, (G_R)\, L$

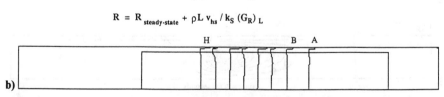

b)

Fig. 16　Computed locations of liquid-solid interface and streamlines during transient simulation.

This situation was improved by ramping the speed of the heat source from 0 to 10 μm/s (with constant acceleration) in the beginning of processing. During this period the value of R was changed according to Eq. (43). The corresponding result (Fig. 17c) is much improved over the result with a sudden change in speed (Fig. 17b). The final results are illustrated in Fig. 18. The streamlines in this figure and in Fig. 15 correspond to the same values of the stream function; hence, it can be seen that application of this method not only flattens the interface considerably but also decreases the intensity of the natural convection near the interface.

This model could also be applied to processing on Earth. Figure 19 shows the result for $g = 981$ cm/s^2. The resulting interface could be made as flat as in $10^{-4} g_e$, but there remains significantly more convection. Please

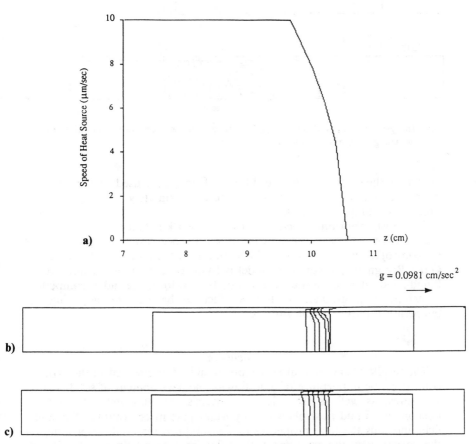

Fig. 17   Computed locations of liquid-solid interface and streamlines in the early stages of transient simulation using a ramp-up in speed: a) speed profile; b) interface positions with no ramp-up; c) interface positions with ramp-up.

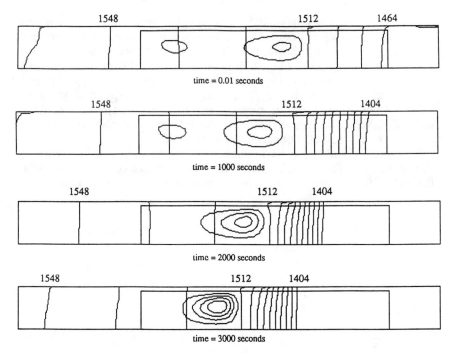

Fig. 18   Streamlines and successive interface positions for GaAs crystal grown at $10^{-4} g_e$ using controlled temperature profiles.

note that these streamlines are different from those used in the previous figure. At $981 \, \text{cm/s}^2$, the velocities were approximately 1000 times larger than those computed for $10^{-4} g_e$.

The most important conclusion from this work is that a model considering fluid flow and heat transfer can be used to specify processing conditions for controlling fluid flow and solidification near the liquid-solid interface in GaAs Bridgman growth. The model is based on the energy balance at the liquid-solid interface and uses this condition to force the radial component of temperature gradient at the interface to be zero by adjusting the temperature profile of the furnace.

Discussion

The results of the preceding section should be interpreted in the broader context of the importance of fluid flow in the development of solidification structures, as well as the role that analysis can play in studying these phenomena. Fluid flow plays an important part in determining the solute field and thus the final properties of the solidified material. Depending on the objectives of the experiment, fluid flow may be a help or a hindrance.

The importance of the analytical methods is that they allow experimenters to separate the phenomena into diffusive and convective effects.

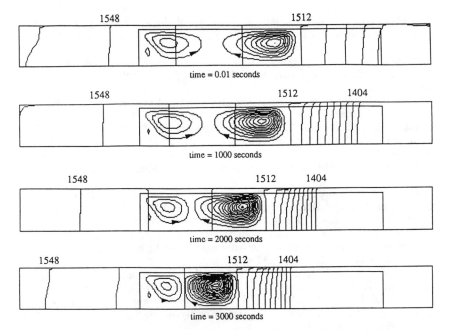

Fig. 19   Streamlines and successive interface positions for GaAs crystals grown at 1 g_e using controlled temperature profiles.

They also allow one to better design and interpret experiments under a variety of processing conditions. It was readily apparent in the second example that significant convection can be observed even at $10^{-4} g_e$ if care is not taken to carefully control the experimental conditions. At the same time, the analysis showed that, if the goal of the crystal grower were simply to achieve a flat growth interface, this could be done on Earth.

Acknowledgment

The authors would like to thank the National Science Foundation, which supported the work on cellular microstructures under Grant DMR 83-03917; NASA, which supported the work on furnace modeling under Grants NAGW-1683 and NAG 3-947; and the National Center for Supercomputing Applications at the University of Illinois for grants of computer time for both problems.

References

[1]Stefan, J., *Annals of Physics and Chemistry*, Vol. 42, 1891, pp. 269–286.

[2]Kurz, W. F., and Fisher, D. J., *Fundamentals of Solidification*, Trans-Tech, Aedermannsdorf, Switzerland, 1984.

[3]Tiller, W. A., Jackson, K. A., Rutter, J. W., and Chalmers, B., "The Redistribution of Solute Atoms During the Solidification of Metals," *Acta Metallurgical*, Vol. 1, 1953, pp. 419–437.

[4]Mullins, W. W., and Sekerka, R. F., "Morphological Stability of a Particle Growing by Diffusion or Heat Flow," *Journal of Applied Physics*, Vol. 34, 1963, pp. 323–329.

[5]Mullins, W. W., and Sekerka, R. F., "Stability of a Planar Interface during Solidification of a Dilute Binary Alloy," *Journal of Applied Physics*, Vol. 35, 1964, pp. 444–451.

[6]Trivedi, R., and Somboonsuk, "Constrained Dendritic Growth and Spacing," *Material Science and Engineering*, Vol. 65, 1984, pp. 65–74.

[7]Glicksman, M. E., Coriell, S. R., and McFadden, G. B., "Interactions of Flows with the Crystal-Melt Interface," *Annual Review of Fluid Mechanics*, Vol. 18, 1987, pp. 307–335.

[8]Delves, R. T., *Crystal Growth*, edited by B. R. Pamplin, Pergamon, New York, 1974, pp. 40–103.

[9]Coriell, S. R., Cordes, M. R., Boettinger, W. J., and Sekerka, R. F., "Solutal Convection Induced Macrosegregation and the Dendrite to Composite Transition in OFF-Eutectic Alloys," *Journal of Crystal Growth*, Vol. 66, 1980, pp. 512–524.

[10]Davis, S. H., Muller, U., and Dietsche, C., "Pattern Selection in Single-Component Systems Coupling Benard Convection and Solidification," *Journal of Fluid Mechanics*, Vol. 144, 1984, pp. 133–151.

[11]Brattkus, K., and Davis, S. H., "Induced Morphological Instabilities: Stagnation Point Flows," *Journal of Crystal Growth*, Vol. 89, 1988, pp. 423.

[12]Dantzig, J. A., and Chao, L. S., *Proceedings of the Tenth U.S. National Congress of Applied Mechanics*, American Society of Mechanical Engineers, New York, 1986, pp. 249–255.

[13]Engelman, M. S., *FIDAP Theoretical Manual*, Fluid Dynamics International, Evanston, IL, 1987.

[14]Bower, T. F., Brody, H. D., and Flemings, M. C., "Measurements of Solute Redistribution in Dendritic Solidification," *Transactions of AIME*, Vol. 236, 1966, pp. 624.

Analysis of Convective Situations with the Soret Effect

D. Henry*

Ecole Centrale de Lyon, Ecully Cédex, France

Introduction

THE thermal diffusion in a mixture (often called the Soret effect) corresponds to the diffusion of the constituents of the mixture under the action of a temperature gradient, leading to the separation of these constituents. This effect is counterbalanced by ordinary diffusion (caused by the concentration gradient in creation) called backdiffusion, which limits this separation. The global effect, which leads to a stationary state corresponding to a separation called "Soret separation," is characterized by the Soret coefficient $St = D'/D$, where D is the ordinary diffusion coefficient and D' the thermal diffusion coefficient.

This Soret effect has been studied for two main reasons: 1) as a possible separation method for the components of a mixture, isotopes, or polymers, particularly with the development of thermogravitational columns, where the efficiency of the separation is increased by a slow laminar flow in the perpendicular direction; and 2) as a fundamental physical process that it is interesting to explain in terms of theoretical models of the liquid state. Another more recent motivation for its study is that it could be the cause of perturbations in processes using a temperature gradient as the crystal growth of semiconductors.

Several papers have been devoted to the experimental determination of the Soret coefficient of a binary mixture, and various techniques have been proposed for it. In most of these experiments the mixture is placed in a container subjected to a temperature gradient that is generally vertical. The separation between the species that results from the Soret effect is measured by different kinds of techniques (direct concentration measurement, optical or electrical measurements, etc.). The results reported in the literature show

---

*C.N.R.S. Researcher, Laboratoire de Mécanique des Fluides et d'Acoustique.

437

large discrepancies[1] and even opposite signs of the Soret coefficient. Thomaes[2] mentions that, in contrast to the predictions of kinetic, thermodynamic, and statistical theories, only a few systems have been found to have negative St. Most of these discrepancies are attributed to convection that is generated by the action of the gravity field on the density gradients present in the cell,[3-5] and that it is difficult to suppress, particularly when a large temperature gradient is imposed.

In view of these difficulties for experiments at the ground level, several experiments have been planned in a space environment in order to reduce the natural convection.[6-15] In fact, in a space environment the gravity is only reduced by a factor 10^3–10^4, and the convection is not always negligible. More precisely, the microgravity acceleration is composed by quickly transient accelerations characterized by their compensated and random nature (g jitter caused by external forces as vibrations or crew activities), superposed on a smaller steady acceleration level caused by external forces as atmospheric drag. The intensity and the orientation of the gravity vector with respect to the container is then not controlled.

In all cases (ground or space situations), it was then interesting to study the convection arising in such systems and the way it affects the Soret separation.

Since the ground experiments were performed in vertical situations, many papers have been first devoted to the stability analysis of the steady state of a mixture subjected to a vertical temperature gradient. Most of them concern the linear stability in horizontal (nonconfined) layers (e.g., Refs. 16 and 17) and more recently in confined vertical cylinders (e.g., Refs. 18 and 19). At the same time experiments have been performed to analyze the onset of convection and the flow structure of a mixture subjected to the Soret effect (e.g., Refs. 20 and 21).

Our contribution in this domain is a linear stability study,[22] which is focused on long vertical cylinders and concerns both stationary and oscillatory thresholds. We also wanted to show the difficulties inherent in the measurement of the Soret coefficient of a binary mixture, since even layers with a stable stratification in a cylinder heated from above proved to be experimentally perturbated by convection. In fact, because the experiments require strong temperature differences in order to achieve significant species separation, the smallest geometric defect produces a significant nonvertical temperature gradient and generates a buoyancy-driven flow that induces a remixing of the species and limits the Soret separation. To study such a defect, we made the three-dimensional numerical simulation of some cases with a small cell inclination γ (quasivertical cylinders), in which we analyzed the flow structure and the remixing rate of a binary mixture.[23,24]

Concerning the experiments performed in space, it has been shown that the transient accelerations of the g jitter have no perturbating effect on the separation of constituents.[25] A direct numerical simulation was then required to study the steady flow corresponding to different orientations of the cell. No results were available on Soret processes in such situations. Our work has been devoted to the characterization of the flow and to the remixing process in horizontal and inclined cylinders for low-g conditions, i.e., for small Grashof numbers ($0 \leqslant Gr \leqslant 10$), and for extended values of the characteristic parameters.[26-28]

Our results could help experimenters estimate for given parameters the effect of convection on the expected separation and thus better analyze their results obtained either on ground or in space, or design an experiment where the flow will have little effect on the separation. A secondary goal of this global study on Soret situations is to have a general insight into double-diffusive convection and separation processes over a large domain of different situations. Moreover, the combination of a stability analysis and a numerical simulation is a good tool for analyzing the transition from stability to instability, the presence of imperfect bifurcations, and the relation between inclined and vertical situations. This global study corresponds to a large part to the thesis of Henry.[25]

Hypothesis and Mathematical Description of the Problem

We consider a binary Newtonian fluid mixture confined in a cylinder with an aspect ratio $A = L/R$ (L length, R radius). The orientation of the cylinder is given by the inclination γ with respect to the gravity vector g (see Fig. 1). We use the simplified Boussinesq approximation; the physical properties are assumed to be constant, except the density $\bar{\rho}$, which is assumed to be a linear function of the temperature $\bar{T}$ and of the mass fraction $\bar{X}$ (of the denser component) in the buoyancy term:

$$\bar{\rho} = \bar{\rho}_0[1 - \alpha(\bar{T} - \bar{T}_0) + \beta(\bar{X} - \bar{X}_0)] \tag{1}$$

where α and β are the thermal and solutal expansion coefficients, respectively, and an overbar denotes a dimensional quantity. We use the phenomenological equations relating the heat and mass fluxes ($\bar{J}_Q$ and $\bar{J}_X$) to the thermal and solutal gradients:

$$\bar{J}_Q = -\lambda\nabla\bar{T} - \bar{\rho}D_f\nabla\bar{X} \tag{2}$$

$$\bar{J}_X = -\bar{\rho}D'\bar{X}(1 - \bar{X})\nabla\bar{T} - \bar{\rho}D\nabla\bar{X} \tag{3}$$

The contribution of the temperature to the flux of matter (Soret effect) is the leading phenomenon in our study since it creates the separation of the constituents. The normal Soret effect ($D' > 0$) occurs when the denser component migrates to the colder region. The contribution of the mass fraction to the flux of heat (Dufour effect corresponding to D_f), which is small particularly for liquids, will be neglected.

We consider realistic boundary conditions: The walls are assumed to be rigid and impervious to mass transfer; both ends of the cylinder are kept at a fixed temperature ($\bar{T}_c$ and $\bar{T}_h$), and the lateral walls are either conducting or adiabatic.

The equations that govern the motion of the fluid in the thermal diffusion cell describe the conservation of the global mass, momentum,

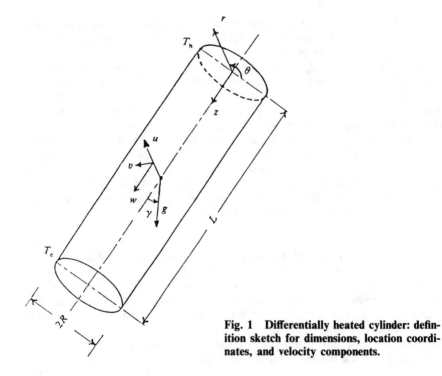

Fig. 1 Differentially heated cylinder: definition sketch for dimensions, location coordinates, and velocity components.

energy, and mass of the denser component. The introduction of the density variation [Eq. (1)] and of the phenomenological equations [Eqs. (2) and (3)] leads to the system to be solved with the boundary conditions.

Stability Analysis of Vertical Situations

In a vertical cylinder submitted to an axial temperature difference, a mixture evolves to a stationary state that is a state of rest characterized by vertical and constant temperature and mass fraction gradients (Soret equilibrium corresponding to zero diffusion flux). In fact, such an equilibrium state can be convectively stable or unstable depending on the values of the characteristic parameters of the problem.[22,25]

Perturbation Equations and Stability Method

To study the stability of the system (linear study), we consider the evolution (decrease or increase) of infinitesimal perturbations applied on the equilibrium state. The amplitude of the perturbations of the different variables is given by the following set of linear dimensionless equations [T nondimensionalized by $(\bar{T}_h - \bar{T}_c)/A$, X by $(\bar{X}_h - \bar{X}_c)_{\text{Sor}}/A = [D'\bar{X}_0(1 - \bar{X}_0)(\bar{T}_h - \bar{T}_c)]/(DA)$, r by R, z by L, V by v/R, and t by R^2/v]:

$$\frac{\partial V}{\partial t} = -\nabla p + Gr(T - SX)e_z + \nabla^2 V \tag{4}$$

$$Pr\frac{\partial T}{\partial t} = PrV_z + \nabla^2 T \tag{5}$$

$$Sc\frac{\partial X}{\partial t} = -ScV_z + \nabla^2 T + \nabla^2 X \tag{6}$$

$$\nabla \cdot V = 0 \tag{7}$$

with the Grashof number $Gr = (-e_g \cdot e_z)\alpha g\,\Delta\bar{T}R^3/(Av^2)$ (In the instability study as we deal with vertical situations, the gravity term is given as a function of the unit vector e_z rather than e_g, but as a consequence the Grashof number has a sign given by $(-e_g \cdot e_z)$ indicating the orientation of the cell with regard to the gravity, $Gr \geqslant 0$ when the heating is from the bottom, and $Gr \leqslant 0$ when the heating is from the top. In the rest of the study we will keep e_g, and Gr will be by definition positive), the Prandtl number $Pr = v/\kappa$, the Schmidt number $Sc = v/D$, and the separation or Soret parameter $S = D'\bar{X}_0(1 - \bar{X}_0)\beta/(D\alpha)$. The associated boundary conditions are $T = 0$ at $z = 0$ and $z = 1$; $T = 0$ at $r = 1$ (conducting sides) or $\partial T/\partial r = 0$ at $r = 1$ (adiabatic sides); $V = 0$ and $\partial T/\partial n + \partial X/\partial n = 0$ on all of the boundaries.

To satisfy more easily the zero flux condition on the boundaries, the mass fraction X is replaced by a new dimensionless variable Xc such that $X = -T + Xc$. In fact, at steady state without a convective effect, it can be shown from Eq. (6) that $X = -T$ on the whole domain, i.e., that the deformations of the temperature lead to identical deformations of the mass fraction. Thus, this change of variables expresses that the perturbation X can be decomposed at steady state in a part proportional to the perturbation of the temperature and in another part Xc that can be considered as the "self"-deformation of the mass fraction due to the convective transport term.

The perturbations T and Xc depend on Pr and Sc, respectively. We can define new perturbations T' and Xc' that are of the same order of magnitude and verify identical stationary equations by the following change: $T = PrT'$ and $Xc = ScXc'$. We then have:

$$\frac{\partial V}{\partial t} = -\nabla p + Gr[Pr(1 + S)T' - ScSXc']e_z + \nabla^2 V \tag{8}$$

$$Pr\frac{\partial T'}{\partial t} = V_z + \nabla^2 T' \tag{9}$$

$$-Pr\frac{\partial T'}{\partial t} + Sc\frac{\partial Xc'}{\partial t} = -V_z + \nabla^2 Xc' \tag{10}$$

Equation (8) indicates, then, that the buoyancy forces that generate the motion are proportional to the perturbations of the temperature T' with the factor $GrPr$ $(1 + S)$ and to the perturbations of the mass fraction Xc' with the factor $GrScS$. It must be pointed out that the term containing the perturbations of the temperature, T', corresponds to the perturbations of the temperature but also to its repercussion on the mass fraction field by Soret effect, whereas the term containing the perturbations of the mass fraction, Xc', corresponds only to the self-perturbations of the mass fraction due to the convective transport term. Now that T' and Xc' are perturbations of the same order of magnitude, the relative importance of the two terms will directly depend on the factors $GrPr(1 + S)$ and $GrScS$ and will be correctly represented by the parameter $\psi = ScS/[Pr(1 + S)]$, which has been first proposed by Schechter et al.[16] and Gutkowicz-Krusin et al.[17]

If $\psi \to 0$, the main influence comes from the perturbation of the temperature, whereas if $\psi \to \infty$, the main influence comes from the self-perturbation of the mass fraction. Equation (8) can then be written as a function of this parameter ψ and of the Rayleigh number $\tilde{R} = R(1 + S) = GrPr(1 + S)$:

$$\frac{\partial V}{\partial t} = -\nabla p + \tilde{R}(T' - \psi Xc')e_z + \nabla^2 V \tag{11}$$

The comprehension of the role of ψ will help in the interpretation of the stability curves. This formulation is mainly interesting for the stationary instability thresholds $\tilde{R}_{cr}^{st}$, because they will only depend on ψ, whereas the oscillatory instability thresholds $\tilde{R}_{cr}^{osc}$ will also depend on Pr and Sc.

The final system is solved with a Galerkin method where the trial functions depending on the cylindrical coordinates verify the boundary conditions. The use of circular harmonics in θ [$\exp(in\theta)$] reduces the three-dimensional problem to a two-dimensional one as the different modes separate. We only considered the modes corresponding to the onset of the first instability, i.e., the axisymmetric mode ($n = 0$) for flat cells and the one-roll asymmetric mode ($n = 1$) for long cells. We mainly gave interest to long cells where the confinement is strong, since this situation corresponded to our experimental situations and had not been considered in detail.

Results of the Stability Analysis

The results can be given by "universal" $\tilde{R}$ curves as a function of ψ, only depending on the aspect ratio A and on the thermal boundary conditions. They represent the relation between the critical global density gradient (or the critical global force due to T) and the ratio of the two forces acting in the buoyancy term. These two forces can be stabilizing or destabilizing. The global force due to T is destabilizing for $\tilde{R} > 0$ and stabilizing for $\tilde{R} < 0$. The self-force due to X acts in the same way as the global force due to T if $\psi > 0$ and in the opposite way if $\psi < 0$.

Steady Stability Curves for A = 6 (Figs. 2 and 3)

On the quadrant $\tilde{R} < 0$, $\psi > 0$, the two forces are stabilizing, and no instability occurs.

On the quadrant $\tilde{R} > 0$, $\psi > 0$, the two forces are destabilizing, but, depending on the value of ψ, each of them will have a different importance. For $\psi = 0$ the global force due to T is preponderant, whereas for $\psi \to \infty$, the self-force due to X is so. Between these two values of ψ the two forces will contribute to the destabilization. The value of $\tilde{R}_{\mathrm{cr}}$ (representative of the global force due to T) will decrease regularly (because of the increasing importance of the self-force due to X) from the value R_0 corresponding to the purely thermal case for $\psi = 0$ ($R_0 = R_{\mathrm{co}}$ for conducting lateral thermal boundaries and $R_0 = R_{\mathrm{ad}}$ for adiabatic ones) until the value zero for $\psi \to \infty$, indicating that, because the global force due to T does not play any role, the instability will occur independently of its value.

The stability curves of the previous domain continue for $\psi < 0$ (in the vicinity of $\psi = 0$), but the thresholds increase strongly ($\tilde{R}_{\mathrm{cr}} > R_0$) because of the stabilizing contribution of the self-force due to X and quickly become quite meaningless because of the presence of more unstable oscillatory behaviors.

In the quadrant $\tilde{R} < 0$, $\psi < 0$, for $|\psi| \to \infty$ the destabilizing self-force due to X is preponderant, and we again obtain $\tilde{R}_{\mathrm{cr}} = 0$. When $|\psi|$ decreases, the

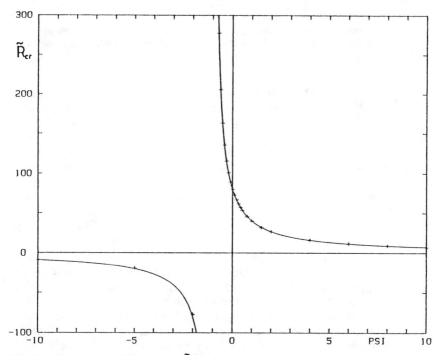

Fig. 2 Stationary stability curves $\tilde{R}_{\mathrm{cr}}^{\mathrm{st}}$ as a function of ψ; $A = 6$: +, results obtained in the adiabatic case; solid line, curve $\tilde{R} = R_{\mathrm{ad}}/(1 + \psi)$.

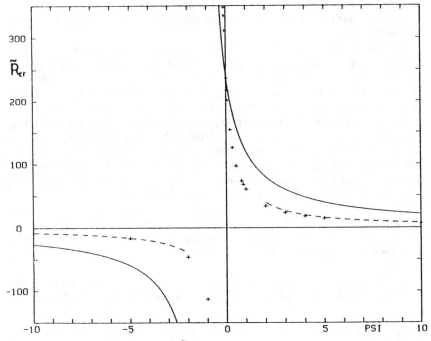

Fig. 3 Stationary stability curves $\tilde{R}_{cr}^{st}$ as a function of ψ; $A = 6$: +, results obtained in the conducting case; solid line, curve $\tilde{R} = R_{co}/(1 + \psi)$; dotted line, asymptotic curve ($|\psi| \to \infty$) $\tilde{R} = R_{ad}/\psi$.

influence of the stabilizing global force due to T increases more and more, but as long as $|\psi| > 1$, the destabilizing self-force due to X remains preponderant. The instability still occurs, but the critical thresholds $|\tilde{R}|$ increase regularly. For ψ in the vicinity of -1, the two forces are opposite and have a same strength but an instability can occur since the boundary conditions on Xc' of a Neumann type can lead to stronger perturbations Xc' compared to T', especially for conducting boundaries. The instabilities are then present on a certain domain for $|\psi| < 1$, mainly for conducting boundaries.

Particular Behaviors

Asymptotic behavior. The preponderance of the destabilizing self-force due to X observed in two quadrants for $|\psi| \to \infty$ corresponds to an asymptotic behavior $\tilde{R}_{cr}\psi = cst$. In fact, by consideration of the governing equations [Eqs. (9–11)] we have shown that in the case of long cylinders this constant is close to the value R_{ad} (Figs. 2 and 3). We observe from our results that this relation is valid as soon as $|\psi| > 10$. In fact, considering a (R,S) formulation, the extent of validity of this relation will depend on Sc/Pr. For $Sc/Pr = 1$ (gases), this asymptotic domain is degenerate and only corresponds to the point $S = -1$. But for $Sc/Pr = 100$ (liquids), it corresponds to a large domain of S values: $R > 0$, $S > 0.1$ and $R < 0$, $S < -0.1$. In this case the relation that can be written $R_{cr}S = R_{ad}/100$ indicates that on the asymptotic domain two situations with reverse heating and opposite values of S will present an instability for the same critical Rayleigh number $|R_{cr}|$. This phenomenon, which is then valid for $|S| > 0.1$, corresponds to the fact that on the asymptotic domain only X, which is in the two cases identically stratified, participates to the creation of the motion.

Analytical Expression for Adiabatic Boundaries. For adiabatic boundaries, in the case of long cylinders, the variation of $\tilde{R}_{cr}$ as a function of ψ is correctly represented by an analytic relation $\tilde{R}_{cr} = R_{ad}/(1 + \psi)$, valid on almost the whole domain of ψ except near $\psi = -1$ (Fig. 2). Such a relation is not valid for conducting boundaries (Fig. 3).

Influence of Thermal Boundary Conditions. In situations with a Soret effect we must take into account that, at stationary state, the influence of the deformation of T, because of its action on X by Soret effect, is a global influence that will be represented by $R(1 + S) = \tilde{R}$ and not by R. Then, for $\tilde{R} > 0$ this influence is destabilizing and a deformation of T will increase the motion, whereas for $\tilde{R} < 0$ the influence is stabilizing, and a deformation of T will decrease the motion. Thus, the presence of conducting boundary conditions, which limits the deformations of T, will be stabilizing for $\tilde{R} > 0$ and destabilizing for $\tilde{R} < 0$ (compared to adiabatic conditions). The non-influence of the thermal boundary conditions at the transition between these two domains ($\tilde{R} = 0$) corresponding theoretically to $S = -1$ is in fact valid on the whole asymptotic domain since there only the self-deformation of X participates to the creation of the motion.

Oscillatory Stability Curves

The quadrant $\tilde{R} > 0$, $\psi < 0$, where the global force due to T is destabilizing and the self-force due to X is stabilizing, corresponds in its larger part to a domain of oscillatory instabilities. For moderate values of ψ, the thresholds of stationary and oscillatory instability are given in Fig. 4 for different Sc and Pr values, conducting boundaries, and an aspect ratio of 6. For almost all of the negative values of ψ, the oscillatory instability sets in before the stationary instability. The transition between the two modes occurs for small values of $|\psi|$, which get closer to zero and correspond to smaller thresholds when Sc/Pr increases. For large values of Sc/Pr as in most liquids, $\tilde{R}_{cr}$ does not depend much on ψ and, on Sc and Pr, and corresponds to a Rayleigh number close to the pure fluid value.

Some explanations can be given on these phenomena. Since the motion is generated by the deformations of T that are destabilizing, it can be initiated only beyond the critical threshold corresponding to the purely thermal case, so that the buoyancy forces due to T overtake the viscous forces. But near this threshold an even small extra negative force (due to a deformation of X that is stabilizing) can prevent the motion from getting to its stationary state. The motion will then decrease, but if Sc/Pr is large, the deformations of X will already be important; then, when the system will be almost convectively stable, a deformation of X will still be present and

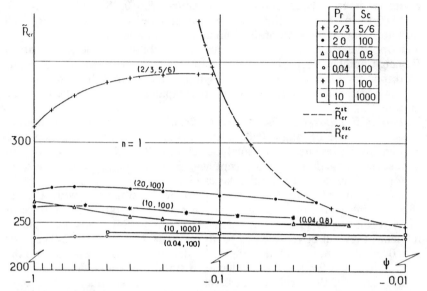

Fig. 4 Stationary and oscillatory stability curves in the quadrant $\tilde{R} > 0$ and $\psi < 0$; $A = 6$, conducting case.

drive the motion in the reverse direction. For large Sc/Pr we can then understand that the oscillatory motion could settle near the purely thermal threshold and for small contributions of X. For small Sc/Pr values, as the deformations of X are smaller, the contribution of X must be larger and the threshold values higher. We can also deduce some indications on the frequencies. An increase of Sc/Pr corresponding to larger deformations of X will increase the period of the oscillations leading to smaller frequencies. An increase of $|S|$ giving more influence to X will allow the initial motion created by T to be counterbalanced more quickly, which will then lead to larger frequencies.

Numerical Simulation of Convection in a Thermal Diffusion Cell

When the cell is not vertical or for supercritical vertical situations, a motion is created in the cell. This convection will be studied by a three-dimensional numerical simulation for different situations corresponding either to space experiments or to ground experiments.

Equations, Numerical Method, and Main Notations

The equations defined in the section on hypothesis and mathematical description of the problem are expressed in terms of the vorticity vector $Z(Z_r, Z_\theta, Z_z)$, the temperature T, the mass fraction X, and the velocity vector $V(u,v,w)$:

$$\frac{\partial Z}{\partial t} = Gr\nabla \otimes (V \otimes Z) - \nabla \otimes (Te_g) + S\nabla \otimes (Xe_g) - \nabla \otimes (\nabla \otimes Z) \quad (12)$$

$$Pr\frac{\partial T}{\partial t} = -GrPr\nabla \cdot (TV) + \nabla^2 T \quad (13)$$

$$Sc\frac{\partial X}{\partial t} = -GrSc\nabla \cdot (XV) + \nabla^2 T + \nabla^2 X \quad (14)$$

$$\nabla^2 V = -\nabla \otimes Z \quad (15)$$

The associated boundary conditions are $u = v = w = 0$ and $(J_X)_n = \partial T/\partial n + \partial X/\partial n = 0$ on the boundaries; $Z_1 = -\partial v/\partial z$, $Z_2 = \partial u/\partial z$, $Z_3 = 0$ at $z = 0$ and $z = A$; $Z_1 = 0$, $Z_2 = -\partial w/\partial r$, $Z_3 = \partial v/\partial r$ at $r = 1$; $T(z = 0) = A/2$, $T(z = A) = -A/2$, and $T = 0$ (perfectly conducting) or $\partial T/\partial r = 0$ (adiabatic) at $r = 1$.

The scaling factors used in this formulation are R, $Grv/$R, and R^2/v for length, velocity, and time. The reference quantities for temperature and mass fraction are $\bar{T}_{ref} = (\bar{T}_h - \bar{T}_c)/A$ and $\bar{X}_{ref} = (D'\bar{X}_0(1 - \bar{X}_0)(\bar{T}_h - \bar{T}_c))/(DA)$. The dimensionless temperature and mass fraction are taken as $T = (\bar{T} - \bar{T}_0)/\bar{T}_{ref}$ and $X = (\bar{X} - \bar{X}_0)/\bar{X}_{ref}$, where $\bar{X}_0$ is the initial mass fraction at the mean temperature $\bar{T}_0 = (\bar{T}_h + \bar{T}_c)/2$. The mass fraction is nondimensionalized with a value corresponding to the rate of separation per unit length in the "perfect" Soret case (without motion), giving the values $+A/2$ and $-A/2$ at the cold and warm ends, respectively. The rate of separation, $\partial X/\partial z$ or X_z, is then equal to 1. The dimensionless parameters have been already defined in the preceding section. In this section Gr is by definition positive.

To solve the governing system (12–15), the equations are developed in cylindrical coordinates in a finite-difference form, and an iterative false transient method for vorticity, temperature, and mass fraction equations is used with an alternating-direction implicit (ADI) scheme. The velocity is then evaluated from the vorticity by a Fourier series method. The method initially developed by Leong and de Vahl Davis[29] has been adapted for an implementation on a Cray computer.[25,26,30]

The presentation and analysis of the numerical results will be done mainly through velocity, isotherm, and iso-mass fraction fields in the vertical symmetry plane (denoted as the V plane). For the velocities, together with V, we will use $\bar{V}$ or $\tilde{V} = \bar{V}/(v/R)$; e.g., we will consider the maximum of the velocity in the V plane, $\tilde{V}_{max}$. For a quantitative interpretation of isotherm and iso-mass fraction fields, we need the position of the zero contour (the one passing through the center) and the isocontours spacing, DX or DT, that will be given for each graph. We need to express the mass fraction gradients in the middle of the cavity, $(X_z)_{mid}$ and $(X_r)_{mid}$, or in the whole central part, $(X_z)_{cp}$, and to consider the curves giving the absolute value of the mean mass fraction in a z plane as a function of z,

$Xm(z)$. [In the figures, Xm will be given as a function of k, position of the mesh points in the z direction, and only for half of the cylinder, since the middle of the cavity is a center of symmetry: from $k = 1$ corresponding to the hot end of the cylinder ($z = 0$), to the center ($z = A/2$).] The global separation is then characterized by $Xm(z = 0)$, or by Xbot, the ratio between this actual separation and the perfect separation $A/2$ without motion: Xbot $= Xm(z = 0)/(A/2)$.

For a better understanding of the physical mechanisms, it is useful to consider the expressions for the two contributions (thermal and solutal) of the buoyancy forces, F_T and F_X, which are responsible for the motion in the V plane. These expressions can be written as:

$$F_T = -T_z \sin(\gamma) + T_r \cos(\gamma) \qquad (16)$$

$$F_X = S[X_z \sin(\gamma) - X_r \cos(\gamma)] \qquad (17)$$

It is also interesting to consider the angle α_X of the zero mass fraction contour with the vertical (or the deformation β_X so that $\alpha_X = \beta_X + 90 \deg - \gamma$). Similarly, for the isotherm $T = 0$, we can consider the angles α_T and β_T. In that case we have

$$F_T = -T_n \cos(\alpha_T) = T_n \sin(\beta_T - \gamma) \qquad (18)$$

$$F_X = SX_n \cos(\alpha_X) = -SX_n \sin(\beta_X - \gamma) \qquad (19)$$

where the subscript n denotes a derivative in the direction normal to the zero-X or zero-T contour. Thus, the sign of F_i changes when $\alpha_i \geqslant 90 \deg$ (or $\beta_i \geqslant \gamma$), i.e., when zero-X or zero-T contours overpass the horizontal.

Parametric Study of Convective Situations Related to Space Experiments

We consider numerous situations relevant to space experiments: 1) rather small values of Gr, $0.01 \leqslant Gr \leqslant 10$; 2) all possible cell inclinations, $0 \deg \leqslant \gamma \leqslant 180 \deg$; and 3) long cells, $A \geqslant 6$. The values of Pr and Sc correspond to an experiment with AgI-KI, i.e., $Pr = 0.6$ and $Sc = 60$. The values of the separation parameter S are chosen from -0.75 to 1, corresponding to many real fluids.[26-28]

Preliminary Comments on the Results

Some general features useful for the discussion can be obtained from the results:

1) The fluid presents a one-roll motion in the V plane.

2) The isotherms contours are nearly the same in all cases, parallel to the end walls, with the convection having no important effect on them since Pr is moderate (conducting regime).

3) The convection influence on the mass fraction repartition can be strong (as a result of the high Sc values). The evolution of the iso-mass fraction contours in the V plane is given in the horizontal case ($\gamma = 90 \deg$)

as a function of Gr (Fig. 5) and as a function of S (Fig. 6) and for some Gr and S values as a function of γ (Fig. 7). In the last figure DX is 0.2 except for a few cases in Fig. 7b (i.e., $S = 0$ at $\gamma = 90$ deg and $S = 0.5$ for 67.5 deg $\leqslant \gamma \leqslant 135$ deg), where $DX = 0.18$. It can be seen that an increase of Gr increases the deformations of X. The effect of S is not so evident. An increase of S increases the deformations in the horizontal case for the Gr considered (Fig. 6). In fact, for small Gr (i.e., $Gr = 1$), this behavior is observed for any γ and the inclination of the zero iso-mass fraction

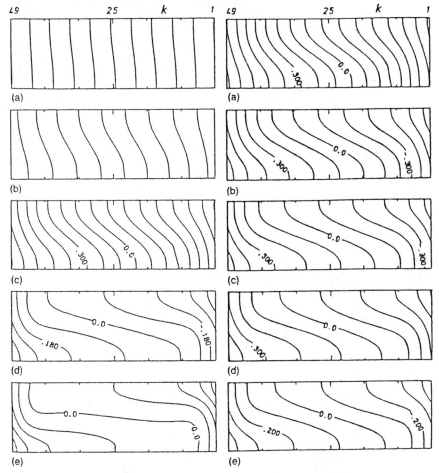

Fig. 5 Iso-mass fraction contours in the V plane for different Gr; $A = 6$, $\gamma = 90$ deg, $S = 0$: a) $Gr = 0.1$, $DX = 0.2$, $Xbot = 0.999$; b) $Gr = 0.5$, $DX = 0.2$, $Xbot = 0.915$; c) $Gr = 1$, $DX = 0.1$, $Xbot = 0.733$; d) $Gr = 3$, $DX = 0.09$, $Xbot = 0.308$; e) $Gr = 5$, $DX = 0.07$, $Xbot = 0.196$.

Fig. 6 Iso-mass fraction contours in the V plane for different S; $A = 6$, $\gamma = 90$ deg, $Gr = 2$, $DX = 0.1$: a) $S = -0.75$, $Xbot = 0.761$; b) $S = -0.5$, $Xbot = 0.571$; c) $S = 0$, $Xbot = 0.447$; d) $S = 0.5$, $Xbot = 0.389$; e) $S = 1$, $Xbot = 0.352$.

450   D. HENRY

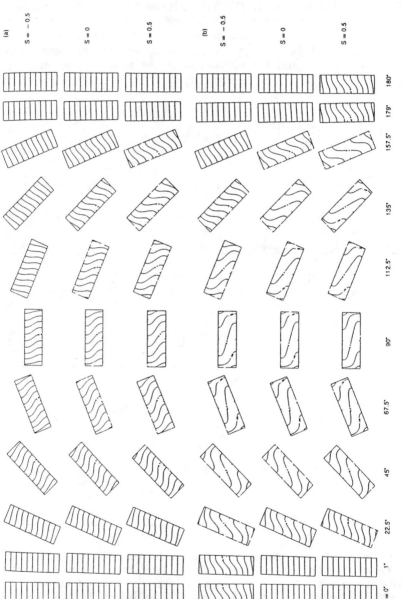

Fig. 7 Iso-mass fraction in the V plane, for different inclinations ($0 \deg \leqslant \gamma \leqslant 180 \deg$) and different S: a) $Gr = 1$; b) $Gr = 3$.

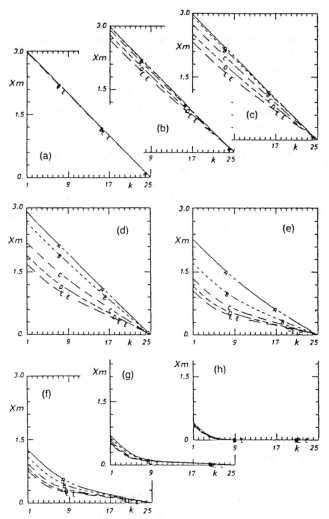

Fig. 8 *Xm* vs *k*; *A* = 6; (A) *S* = −0.75; (B) *S* = −0.5; (C) *S* = 0; (D) *S* = 0.5; (E) *S* = 1; a) *Gr* = 0.1; b) *Gr* = 0.3; c) *Gr* = 0.5; d) *Gr* = 1; e) *Gr* = 2; f) *Gr* = 3; g) *Gr* = 5; h) *Gr* = 10.

$\alpha_X < 90$ deg (Fig. 7a). But for larger *Gr* (i.e., *Gr* = 3; Fig. 7b), this behavior is no longer valid for small inclinations up to a certain limit inclination (denoted γ_{ov}) where we have $\alpha_X > 90$ deg (slope of the zero iso-mass fraction overshooting the horizontal).

4) A central part can be defined where the velocity is independent of *z* and where the separation rate $|\partial Xm/\partial z|$ is constant. (See Fig. 8.)

Characteristics of the Flow
  General Results and Analytical Expressions. In all cases the motion of the fluid corresponds to a single roll, going up at the hot wall and down at the

cold wall, and flowing in the core of the cavity parallel to the lateral boundaries. The velocity profile is quite independent of z along the cavity, except near the two ends where the recirculation of the flow occurs on a length close to one diameter. The velocity profile in the central part can be characterized by its maximum velocity V_{max} or $\tilde{V}_{max}$, which is also quite independent of z.

Such a situation is a characteristic of long cavities. It corresponds to a core-driven regime with little influence of the end parts and of the inertia terms. In our case these inertia terms are negligible, since the results obtained in solving Navier-Stokes equations or associated Stokes equations (neglecting inertia terms) differ only by less than 0.1% on the value of V_{max}. Moreover, with the small value of Pr ($Pr = 0.6$), the convective term in the energy equation is also negligible, leading to a purely conductive profile for T (denoted T_0) with $T_z = -1$ and $T_r = 0$ almost everywhere. The equation of vorticity [Eq. (12)] is then reduced to

$$\frac{\partial Z}{\partial t} = -\nabla \otimes (T_0 e_g) + S\nabla \otimes (Xe_g) - \nabla \otimes (\nabla \otimes Z) \qquad (20)$$

If we denote e_z the unit vector in the z direction and e_x the unit vector perpendicular to e_z in the plane of g, we have

$$e_g = \cos(\gamma)e_z + \sin(\gamma)e_x$$

Since T_0 is only a function of z, we have $\nabla \otimes (T_0 e_z) = 0$. Then Eq. (20) can be written as

$$\frac{\partial Z}{\partial t} = -\sin(\gamma)\nabla \otimes (T_0 e_x) + S\nabla \otimes (Xe_g) - \nabla \otimes (\nabla \otimes Z) \qquad (21)$$

1) For $S = 0$, if we denote V_γ the velocity nondimensionalized by $Gr \sin(\gamma)\nu/R$, we have

$$\frac{\partial Z_\gamma}{\partial t} = -\nabla \otimes (T_0 e_x) - \nabla \otimes (\nabla \otimes Z_\gamma) \qquad (22)$$

This equation has a unique solution at steady state, which we can denote by $V_\gamma = KV_{\gamma 0}$, where $V_{\gamma 0}$ is a velocity field such that $(V_{\gamma 0})_{max} = 1$. We can then write $\tilde{V}$,

$$\tilde{V} = KGr \sin(\gamma)V_{\gamma 0} = \tilde{V}_{max} V_{\gamma 0} \quad \text{with} \quad \tilde{V}_{max} = KGr \sin(\gamma) \qquad (23)$$

Thus, all of these cases with $S = 0$, which correspond to a uniform buoyancy throughout the cavity, have a unique structure. They only differ by the global intensity of the motion.

In the asymptotic situation $A \rightarrow \infty$ and for the horizontal case[31] an analytical expression of the profile $V(r)$ has been obtained for $S = 0$. The maximum of this profile is $V_{max} = 1/[12\sqrt{(3)}] = 0.0481$. The results obtained in our case with $A = 6$ for $S = 0$ give $V_{max} = 0.0468$ in the horizontal case, a value not too far from the asymptotic one. For all of the different inclinations, the relation

$$\tilde{V}_{\max} = 0.0468 Gr \sin(\gamma) \tag{24}$$

is also correctly verified. For $S \neq 0$ a similar structure of flow can only be obtained if the mass fraction field is undisturbed by the convection. In that case $X = X_0 = -T_0$, and by the same considerations as those for $S = 0$, we can obtain

$$\tilde{V}_{\max} = KGr(1 + S) \sin(\gamma)$$

2) In the other cases the buoyancy is no longer uniform in the cavity; thus, we will not have a definite structure of flow. But since we are still in core-driven regimes, we can try to relate in the V plane the velocity to the buoyancy forces corresponding to the θ vorticity, Z_θ, which determines the motion in the V plane, the plane of main circulation. For the buoyancy force the relations (16) and (17) applied with the specific properties of T give

$$F = F_T + F_X = \sin(\gamma) + S[X_z \sin(\gamma) - X_r \cos(\gamma)]$$

Since X_z is almost constant in the central part and equal to $(X_z)_{\mathrm{mid}}$, F will be constant there if $\gamma = 90$ deg, i.e., the horizontal situation. Following Eq. (24), we can propose

$$\tilde{V}_{\max} = 0.0468 Gr[1 + S(X_z)_{\mathrm{mid}}] \tag{25}$$

This relation is quite well verified since the maximum difference with the computed values is less than 1%, except for $S = -0.75$ in the range $1.5 \leqslant Gr \leqslant 3$, where $(X_z)_{\mathrm{mid}}$ varies strongly with Gr and where this difference reaches 3%.

If $\gamma \neq 90$ deg, the buoyancy force in the V plane is not constant in the central part and vary between the two following values:

$$F_{\mathrm{mid}} = \sin(\gamma) + S[(X_z)_{\mathrm{mid}} \sin(\gamma) - (X_r)_{\mathrm{mid}} \cos(\gamma)] \tag{26}$$

$$= \sin(\gamma) + S(X_n)_{\mathrm{mid}} \cos(\alpha_X) \tag{27a}$$

$$= \sin(\gamma) + S(X_n)_{\mathrm{mid}} \sin(\gamma - \beta_X) \tag{27b}$$

and

$$F_{\mathrm{wall}} = \sin(\gamma)[1 + S(X_z)_{\mathrm{mid}}] \tag{28}$$

We do not expect an exact expression for $\tilde{V}_{\max}$ in these cases, but only to situate it between $KGrF_{\mathrm{mid}}$ and $KGrF_{\mathrm{wall}}$. Figures 9a and 9b, corresponding to $S = -0.5$ and 0.5 at $Gr = 3$, show that in all cases $\tilde{V}_{\max}$ is well located in this interval, being closer to the expression $KGrF_{\mathrm{mid}}$. Indeed, when $\gamma = 90$ deg the two expressions are identical and correspond to the expression obtained for the horizontal cylinder.

3) Finally, the expression

$$\tilde{V}_{\max} = KGr\{\sin(\gamma) + S[(X_z)_{\mathrm{mid}} \sin(\gamma) - (X_r)_{\mathrm{mid}} \cos(\gamma)]\} \tag{29}$$

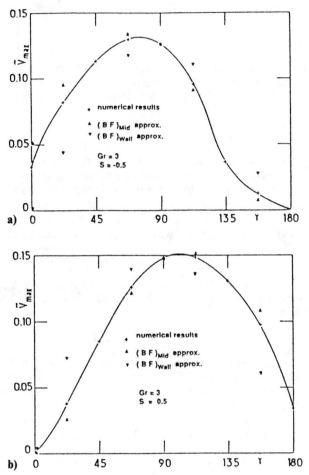

Fig. 9 $\tilde{V}_{max}$ vs γ at $Gr = 3$: a) $S = -0.5$; b) $S = 0.5$. Comparison between numerical results; F_{mid} approximation, Δ [Eq. (26)]; F_{wall} approximation, ∇ [Eq. (28)].

(with $K = 0.0468$ for $A = 6$ and $K \to 0.0481$ for $A \to \infty$) is quite a good approximation in all cases. But in cases where the temperature and mass fraction are unperturbed by convection, the expression is exact and $\tilde{V}_{max}$ is characteristic of the intensity of the flow which has a well-defined structure. In the other cases the expression is approached and the structure of the flow is modified in each case depending on γ and S. We will observe in the V plane that in the horizontal case the length of the central part is modified. In other cases a change can be also observed in the velocity profile $V(r)$.

Influence of S and γ on $\tilde{V}_{max}$. The evolution of $\tilde{V}_{max}$ vs γ for different S is given in Figs. 10a and 10b for $Gr = 1$ and $Gr = 3$, respectively. For $Gr = 3$ these curves intersect at a certain γ, which is close to γ_{ov}.

For $Gr = 1$ and for $Gr = 3$ in the range $\gamma_{ov} < 180$ deg, where $\alpha_X < 90$ deg, the solutal contribution of F given in Eq. (27a) would be "regular"; i.e., positive S would increase the flow, and negative S would decrease it. This behavior is confirmed in Fig. 10 by the regular influence of S on $\tilde{V}_{max}$ in these cases and corresponds to what is observed for horizontal cells in our study. For $Gr = 3$ the behavior observed in Fig. 10b in the range $0 < \gamma < \gamma_{ov}$ (where $\alpha_X > 90$ deg) is the opposite, as predicted by Eq. (27a), and corresponds to what was observed on iso-mass fraction contours. Thus, the S influence is always regular for small values of Gr, whereas for higher Gr a critical inclination γ_{ov} exists that delimits two domains of opposite S influence.

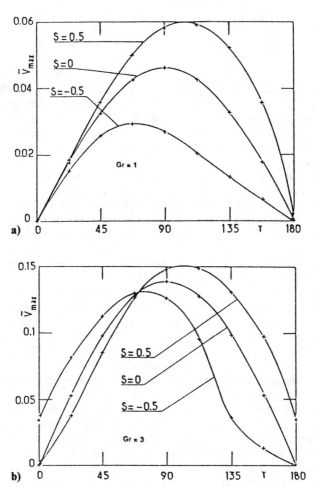

Fig. 10 $\tilde{V}_{max}$ vs γ for different S at $A = 6$, $Pr = 0.6$, and $Sc = 60$: a) $Gr = 1$; b) $Gr = 3$.

For $S = 0$ expression (24) expresses a symmetry with respect to $\gamma = 90$ deg and indicates that the horizontal cell orientation leads to the largest convective effects. This symmetry is quite well verified: the small differences can be explained by the slight deformation of the isotherms neglected in Eq. (24).

The maxima $\tilde{V}_{max}$ differences between the cases $S \neq 0$ and $S = 0$ observed for $\gamma > 90$ deg correspond to the angle γ for which the zero iso-mass fraction at the middle point is vertical ($\alpha_X = 0$) and then has the largest influence (Fig. 10). The position of these maxima is close to $\gamma = 90$ deg if the deformation β_X is small for $\gamma = 90$ deg (Fig. 10a for $Gr = 1$) and farther away if β_X is larger for $\gamma = 90$ deg (Fig. 10b for $Gr = 3$).

Another interesting point to notice in Fig. 10 is the position of the maximum of $\tilde{V}_{max}$, which is located at $\gamma = 90$ deg for $S = 0$ and shifted toward larger γ when $S > 0$ and smaller γ when $S < 0$. When $S \neq 0$, the buoyancy force [Eq. (27b)] has two contributions: one from the temperature proportional to $\sin(\gamma)$ and the other from the mass fraction, which is proportional to $SX_n \sin(\gamma - \beta_X)$. In the vicinity of $\gamma = 90$ deg, $\sin(\gamma)$ remains close to 1, whereas $\sin(\gamma - \beta_X)$ increases if $\gamma > 90$ deg and decreases if $\gamma < 90$ deg. Thus, the contribution of T is almost constant, whereas the contribution of X increases with γ, giving an additional extra contribution if $S > 0$ and an extra negative contribution if $S < 0$ (we assume that X_n is slowly varying in this domain). As a consequence, the maximum of $\tilde{V}_{max}$ occurs at $\gamma > 90$ deg for $S > 0$ and at $\gamma < 90$ deg for $S < 0$.

Influence of S on the Length of the Central Part of the Convective Roll. For $S \neq 0$ the buoyancy force depends on the deformation of the iso-mass fractions, which although it is quite constant in the central part, decreases in the end parts. This smaller deformation in the end parts leads to a larger or smaller buoyancy force depending on the original α_X in the central part and on the sign of S. Compared to the case with $S = 0$, where the buoyancy force is constant everywhere, this increase or decrease of the buoyancy force for $S \neq 0$ will tend to increase or decrease the length of the parallel flow domain obtained for $S = 0$.

This behavior is shown in Figs. 11 and 12 through velocity vector fields and particle tracks in the V plane. The flows obtained with two opposite values of S (-0.5 and 0.5) can be compared at $Gr = 3$ for two interesting examples, $\gamma = 22.5$ deg (Fig. 11) and $\gamma = 112.5$ deg (Fig. 12). In the first example (Fig. 11), we have $\gamma < \gamma_{ov}$, indicating that $\alpha_X > 90$ deg and that the S effect is opposite. The fact that outside the central part the iso-mass fractions are less deformed indicates that they are moved in the horizontal direction, leading to a diminution of their effect. Since their primary effect in the central part was the opposite one, the diminution of this effect outside the central part will correspond to a regular influence; positive S will then contribute to the increase of the flow outside the central part and thus increase the length of this central part (negative S has an opposite effect). In the second example (Fig. 12), $\gamma > \gamma_{ov}$, the effect of S is the regular one. The smaller deformation of iso-mass fraction outside the central part moves them toward the vertical (without overshooting it),

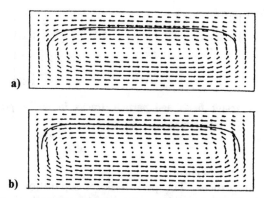

Fig. 11 Velocity vector fields and particle tracks in the V plane at $Gr = 3$ and $\gamma = 22.5$ deg: a) $S = -0.5$; b) $S = 0.5$.

increasing their effect. This increase of the primary effect, which was a regular one, also leads to a regular influence; positive S contributes to the increase in the length of the central part.

Even in the case of instability for $\gamma = 0$ or 180 deg (or for γ close to these values), we find a similar behavior. Since the isotherms cannot be deformed by the (small) convection and are then almost horizontal, the buoyancy force and thus the motion can only be induced by iso-mass fraction deformations. Because the main deformations are located in the central part of the cell, the buoyancy forces will be dominant in this part, leading to a flow characterized by a rather short convective roll (Fig. 13).

Similar considerations can also explain changes in the velocity profile $V(r)$ for $S \neq 0$, e.g., the shift of the maximum velocity toward the axis or toward the lateral walls.

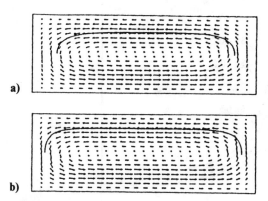

Fig. 12 Velocity vector fields and particle tracks in the V plane at $Gr = 3$ and $\gamma = 112.5$ deg: a) $S = -0.5$; b) $S = 0.5$.

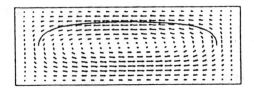

Fig. 13 Velocity vector fields and particle tracks in the V plane at $Gr = 2.8$, $\gamma = 0$ deg, and $S = -0.5$.

Characteristics of the Separation

General Considerations. The perfect Soret separation exists only for a fluid at rest, where the variation of X is linear in z (with $X_z = 1$ everywhere) corresponding to the constant temperature gradient T_z equal to -1. But for all values of S and γ, one can find a Gr small enough that the motion does not significantly disturb the mass fraction field and the separation. This will be called the "separation regime." For the values of S chosen herein ($-0.75 \leqslant S \leqslant 1$), values of Gr less than 0.01 at $Sc = 60$ correspond well to this separation regime (see Figs. 14 and 15 giving, respectively, Xbot as a function of Gr for $\gamma = 90$ deg and Xbot as a function of γ). When the Grashof number is larger, the motion created in the cell begins to be strong enough to deform the iso-mass fractions and reduce the separation.

To have a qualitative understanding of what is happening in this case, it is useful to consider the flow behavior mentioned in the previous paragraph, that is, the simple definition of three zones in the cylinder: a central zone, 1 diam long for $A = 6$, with a constant velocity profile; and two identical zones of recirculation, 1 diam long, near the two ends.

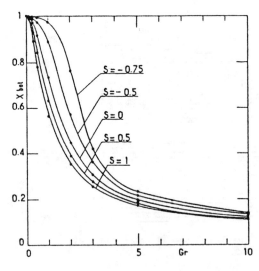

Fig. 14 Xbot vs Gr for different S; $A = 6$; $\gamma = 90$ deg.

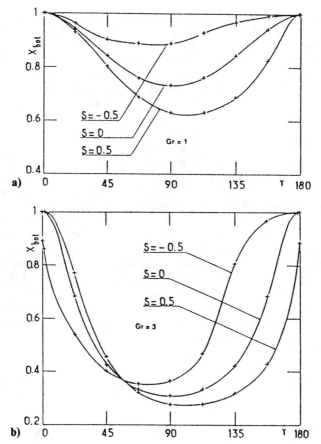

Fig. 15 Xbot vs γ for different S at $A = 6$, $Pr = 0.6$, and $Sc = 60$: a) $Gr = 1$; b) $Gr = 3$.

In the central part the two main variables responsible for the mass fraction separation [see Eq. (14)], the temperature gradient ($T_z = -1$), and the velocity profile V are identical for all z. This indicates that the way the separation will set in will be the same all along this central part, leading to a constant separation rate and then to variations of X linear in z. We verified from the numerical results that X_z is almost constant inside the whole central part [the value noted $(X_z)_{cp}$], even for high Gr, where the mass fraction profiles were deformed by the motion. This constant is equal to 1 for small Gr and diminishes to zero for increasing Gr. It corresponds to the absolute value of the slope of the linear part of the Xm curves, $|\partial Xm/\partial z|_{lin}$ (Fig. 8). More generally, we observed that $(X_z)_{cp} = (X_z)_{mid} = |\partial Xm/\partial z|_{lin}$.

For all cases considered herein, the simple definition of the central part and recirculation zones (1 diam long) does not correspond exactly to the extent of the linear and nonlinear parts of $Xm(z)$ (Fig. 8). But this choice is convenient because it ensures us that in all cases the variation of X is perfectly linear in the whole central part.

Since the temperature gradient is constant everywhere ($T_z = -1$), we can expect that $(X_z)_{cp}$ or $(X_z)_{mid}$ will depend essentially on the velocity intensity and particularly on $\tilde{V}_{max}$. Figure 16, which summarizes some $(X_z)_{mid}$ results vs $\tilde{V}_{max}$, obtained for different Gr, S, and γ, seems to confirm this behavior. We see also that, as soon as the convective motion has a certain influence on the separation, the deformation of the iso-mass fractions and the decrease of $(X_z)_{mid}$ with $\tilde{V}_{max}$ are strong.

There is no such direct dependency for the global separation Xbot, since it results from the separation outside the central part (not connected to $\tilde{V}_{max}$) and inside the central part (directly connected to $\tilde{V}_{max}$ but proportional to the length of this central part that strongly depends on S). Since this central part has the smallest separation rate (Fig. 8), the longer this part is, the smaller the global separation will be (for a given $\tilde{V}_{max}$). Nevertheless, in Fig. 15 we find again for Xbot the two particular points mentioned for $\tilde{V}_{max}$, i.e., the shifting of the extrema (here minima) from 90 deg toward higher γ for $S > 0$ and smaller γ for $S < 0$ and the presence of a crossing point between the curves for different S if Gr is large enough to generate important deformations at $\gamma = 90$ deg. Compared to the $\tilde{V}_{max}$ curves in Fig. 10, extrema positions are almost the same, but crossing point positions are different, situated in the range 67.5 deg $< \gamma < 90$ deg for $\tilde{V}_{max}$ and 45 deg $< \gamma < 67.5$ deg for Xbot.

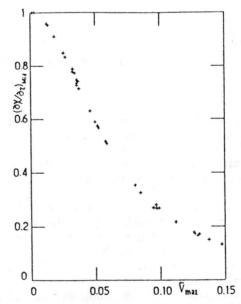

Fig. 16 $(X_z)_{mid}$ vs $\tilde{V}_{max}$ for various Gr, γ, and S.

Close to the end walls, the velocity slows down, and the separation rate $|\partial Xm/\partial z|$ reaches its maximum, equal to one, for any S and Gr (Fig. 8). This means that, except perhaps for very high Gr (much higher than the ones considered herein), a separation always exists near the end walls. This is particularly evident for the highest Gr (3, 5, and 10) in Fig. 8. It is also illustrated by the fact that the Xbot curves (Fig. 14) tend almost to an asymptotic value, close to 0.1, when Gr is increased up to 10, for any S between -0.75 and 1.

The influence of S on Xbot follows that on $\tilde{V}_{max}$ with only small differences. For small values of Gr, an increase of S will decrease the separation, whereas for higher Gr, a value close to γ_{ov} determines two domains, one with the same influence of S for γ larger than this value and the other with opposite influence.

Determination of the Separation for $S = 0$. Since a simple relation giving $\tilde{V}_{max}$ as a function of the parameters of the problem exists for $S = 0$, it would be particularly interesting to find a correspondence between $\tilde{V}_{max}$ and Xbot in order to easily estimate Xbot. In fact, for $S = 0$ the mass fraction equation [Eq. (14)] can be used independently of the others. With the specific flow structure and temperature field, we can write

$$Sc\,\frac{\partial X}{\partial t} = -Sc\,\tilde{V}_{max}\nabla\cdot(XV_{\gamma 0}) + \nabla^2 T_0 + \nabla^2 X \tag{30}$$

Then, at steady state for a given Sc, the mass fraction field and as a consequence the separation Xbot are only dependent on $\tilde{V}_{max}$. When $\tilde{V}_{max}$ increases, the convective transport term increases and the separation decreases.

A curve giving Xbot as a function of $\tilde{V}_{max}$ has been obtained for $S = 0$ in the horizontal case ($\gamma = 90$ deg) with different Gr (Fig. 17). The results

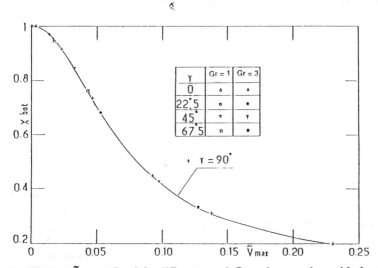

Fig. 17 Xbot vs $\tilde{V}_{max}$ at $S = 0$ for different γ and Gr, and comparison with the curve corresponding to $\gamma = 90$ deg.

obtained for other values of γ are located close to this curve, indicating the validity of our hypothesis. For $S = 0$, which corresponds to situations where X has only little influence on the motion, the calculation of $\tilde{V}_{max}$ [Eq. (24)] and the use of the curve in Fig. 17 will give a good estimate of the global separation for inclined cells. (By drawing the curve of Fig. 17 as a function of a new velocity integrating Sc in its dimensionalization, it would have been possible to include the effect of Sc.)

Influence of the Thermal Properties of Lateral Boundaries

All of the results mentioned in this study have been obtained with perfectly conducting lateral walls. To see the influence of the thermal boundary conditions, we studied some cases with adiabatic lateral boundaries. This influence was expected to be small, since all cases correspond to very small deformations of isotherms. Effectively, in all cases the mass fraction profiles are very similar and the separations of the same order of magnitude. The difference reaches a maximum of 5% for $Gr = 10$ and decreases to 1.5% for $Gr = 3$ and to less than 1% for $Gr = 0.3$. The results given in the case of perfectly conducting lateral walls are a good approximation for all possible thermal boundary conditions, especially if we are interested in the domain of strong separation (small Gr) allowing a good measurement of Soret coefficients.

Influence of Pr and Sc

At steady state Pr is only present in the energy transport equation [Eq. (13)], as a factor in the convective term. Because the isotherms are not deformed for $Pr = 0.6$, corresponding to a negligible convective term, we can expect the same behavior for smaller values of Pr. In this case it will be possible to use the previous results without any correction.

Considering the Sc influence, it is useful to write the equations in simple form, taking into account the specific properties mentioned in the previous paragraphs:

$$\frac{\partial Z}{\partial t} = -\nabla \otimes (T_0 e_g) + S\nabla \otimes (X e_g) - \nabla \otimes (\nabla \otimes Z) \tag{31}$$

$$Sc \frac{\partial X}{\partial t} = -GrSc\nabla \cdot (XV) + \nabla^2 T_0 + \nabla^2 X \tag{32}$$

$$\nabla^2 V = -\nabla \otimes Z \tag{33}$$

For a given A and γ the solution (V,T,X) at steady state only depends on S and on the product $GrSc$. Thus, two cases with the same S and the same product $GrSc$ will give the same dimensionless results. It is then possible to extend the previous result at $Sc = 60$ to different Sc; we only have to use a transformed Gr: $Gr_{tr} = GrSc/60$ (e.g., in Fig. 14). There is no special limitation to the validity of this procedure, except that the temperature gradient and the inertia term must keep the same particular properties. But this will be the case if the Pr is smaller or not very different from $Pr = 0.6$ and if the actual Gr is not too far from our domain of study.

Influence of the Aspect Ratio and Extension of the Results to Larger Aspect Ratios

Comparison of Separation Processes for Different A. We consider situations with larger aspect ratios ($A \geqslant 6$), since they correspond to the space experimental situations. In that case the properties of long cavities already mentioned for $A = 6$ are still valid. We can distinguish in all cases a central zone (which becomes longer when A is increased) with a constant velocity profile and two recirculating zones 1 diam long near the ends (a length equal to 2 in dimensionless form). Moreover, in the case of thermal convection ($S = 0$), the intensity of the flow, $\tilde{V}_{max}$, in this central part for a given Gr has been shown, in the subsection on characteristics of flow, to be quite the same for different aspect ratios as soon as $A \geqslant 6$. Thus, we compared the separation processes in cylinders with different aspect ratios but for the same Gr.

We made some additional calculations for cylinders with aspect ratio $A = 9$ and 12 for different values of Gr and S. For $S = 0$ we found for a given Gr the same velocity profile in the central part (which increases with A) as for the case $A = 6$. Moreover, even for $S \neq 0$, the $(X_z)_{mid}$ value and the maximum velocity $\tilde{V}_{max}$ are almost identical, the differences being less than 1%. Thus, let us compare for a given Gr and a given S the behavior of the mass fraction for the three different aspect ratios:

1) In the central parts the temperature gradient is constant (-1 in all cases) and the $\tilde{V}$ profiles all along z are identical for the three aspect ratios, indicating that the mechanism of separation will be the same throughout these central parts. This behavior is confirmed by comparing the Xm curves plotted for A equal to 6, 9, and 12 in Fig. 18. This figure illustrates well, near the center, the linear portion of the curves with the same slope for the same Gr, but varying in length.

2) In the end parts, for the three A values the flow seems to be identical; it corresponds to a recirculation over a common length of 2 of an identical flow for the same Gr with an intensity $\tilde{V}_{max}$. We can expect that the separation in these end parts will be quite similar in the three cases. By comparing the Xm curves, we see a slightly smaller separation for large A. In fact, this is due to a loss of computational accuracy for large A. Some more accurate calculations confirm that in the end parts the separation evolves similarly for the different A.

Extension of the Results to Larger A. From the mechanisms presented earlier, it is possible to find a rather good procedure to obtain the separation in long cavities directly from the results obtained with $A = 6$. Because of the centrosymmetry, this procedure has to be developed only for a half-part of the cylinder.

For $A = 6$ we consider $Xm(z = 0)$ (which can be deduced from Xbot) and the separation at the limit of the central part which corresponds to $Xm(z = 2)$, but also nearly to $(X_z)_{mid}$ plotted as a function of Gr for

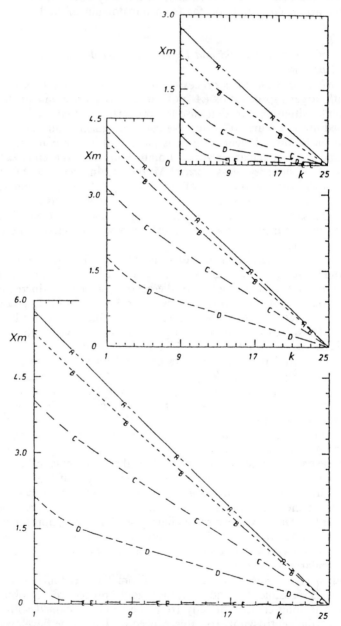

Fig. 18   *Xm* vs *k*; *S* = 0; a) *A* = 6 [(A) *Gr* = 0.5, (B) *Gr* = 1, (C) *Gr* = 2, (D) *Gr* = 3, (E) *Gr* = 5]; b) *A* = 9 [(A) *Gr* = 0.3, (B) *Gr* = 0.5, (C) *Gr* = 1, (D) *Gr* = 2]; c) *A* = 12 [(A) *Gr* = 0.3, (B) *Gr* = 0.5, (C) *Gr* = 1, (D) *Gr* = 2, (E) *Gr* = 10].

different S in Fig. 19. We can deduce the separation in each end part, which is $Xm(z = 0) - Xm(z = 2)$, and the separation rate in the central part is given by $|\partial Xm/\partial z|_{\text{lin}}$ or $(X_z)_{\text{mid}}$. If we consider a cell with an aspect ratio A, the length of half the central part is $[(A/2) - 2]$, leading to a separation of $(X_z)_{\text{mid}}[(A/2) - 2]$, whereas the end part will induce approximately the same separation as for $A = 6$, that is, $Xm(z = 0) - Xm(z = 2)$. We then obtain a global separation that is

$$Xm_A(z = 0) = (X_z)_{\text{mid}}[(A/2) - 2] + Xm(z = 0) - Xm(z = 2)$$

or

$$Xm_A(z = 0) = Xm(z = 0) + (X_z)_{\text{mid}}[(A/2) - 3] \tag{34}$$

where the second term in the right-hand side represents the additional separation that results from the longer central part. If we compare for $A = 9$ and 12 the results deduced by this procedure to the results obtained directly by the simulation with a similar accuracy, a good agreement is found, in all cases better than 1%.

Influence of A on the Separation. These procedures can give roughly the influence of the aspect ratio on the global separation and more particularly on Xbot, in different convective situations. The formula (34) written with Xbot, compares a case with an aspect ratio A with the initial case $A_0 = 6$:

$$X\text{bot}_A = Xm_A(z = 0)/(A/2) = [X\text{bot} + (X_z)_{\text{mid}}(A/A_0 - 1)]/(A/A_0) \tag{35}$$

From that formula we found that the aspect ratio has no influence on Xbot for slightly convective situations where the separation is perfect. When Gr

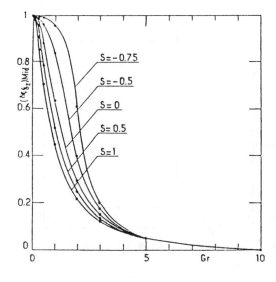

Fig. 19 $(X_z)_{\text{mid}}$ vs Gr for different S; $A = 6$, $\gamma = 90$ deg.

increases, corresponding to a higher deformation of the iso-mass fractions and a decrease of the separation, Xbot depends on the aspect ratio, being slightly smaller for larger aspect ratios. For larger Gr, corresponding to strongly convective situations, Xbot is inversely proportional to the aspect ratio. At last, $(X_z)_{\text{mid}}$ is found as asymptotic value for A approaching infinity, which agrees with the fact that in this case the central part plays a dominant role, whereas the end parts have a negligible effect.

Study of the Convective Perturbation of Ground Experiments: A Quasivertical Cylinder

The Soret ground experiments, even performed in perfectly stable situations (vertical cell with top heating for positive S), proved to be difficult to realize. Small technical defects can generate a buoyancy-driven flow that perturbates the separation. In this study we consider the perturbation created by a small cell inclination γ (quasivertical cylinder), more precisely $\gamma = 1$ deg. The values of S are taken in the range $0 \leqslant S \leqslant 0.673$ and those of Gr in the range $482 \leqslant Gr \leqslant 4820$. The aspect ratio of the cylinder A is equal to 3. The study is limited to the case of conducting walls because we expect that the isotherms will not be too much distorted by the flow since γ is small.[23,24]

Preliminary Comments on the Results

The temperature fields drawn in the V plane are given in Fig. 20a and 20b for two extreme convective situations: $Gr = 482$ for $S = 0.673$ and $Gr = 4820$ for $S = 0$. In the first case, which is characterized by a higher value of S, Fig. 20a shows that the isotherms remain practically parallel to the end walls ($T_z = -1$, $T_r = 0$). In the second example (Fig. 20b), the

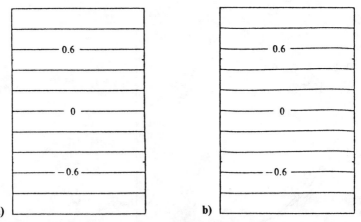

Fig. 20   Isotherm patterns in the V plane: a) $Gr = 482$ and $S = 0.673$; b) $Gr = 4820$ and $S = 0$.

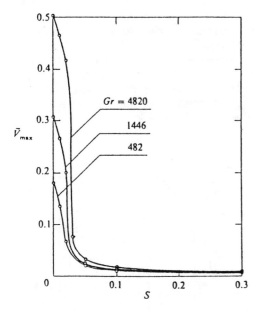

Fig. 21 Maximum of the velocity $\tilde{V}_{max}$ vs Soret parameter S for various Gr.

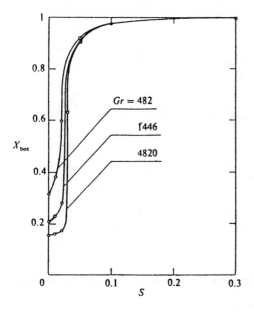

Fig. 22 Global species separation Xbot vs Soret parameter S for various Gr.

468   D. HENRY

isotherms appear to be slightly distorted by the flow, but this is enough to have a significant effect on the convective flow. In Figs. 21 and 22, respectively, we plot $\tilde{V}_{\max}$ and Xbot vs S for three values of Gr: 482, 1446, and 4820. These figures show two regimes: one corresponding to small values of S ($0 \leqslant S \leqslant 0.03$) for which $\tilde{V}_{\max}$ reaches high values (increasing with Gr), and the separation is perturbated, and the other corresponding to higher S ($S \geqslant 0.05$) for which $\tilde{V}_{\max}$ remains small and the separation is good (Xbot close to 1) for any value of Gr. We will separately analyze these two regimes in detail.

Small Soret Parameter ($0 \leqslant S \leqslant 0.03$)

The velocity fields presented in Fig. 23 show that the fluid motion in the V plane corresponds to a single roll flowing parallel to the lateral wall and

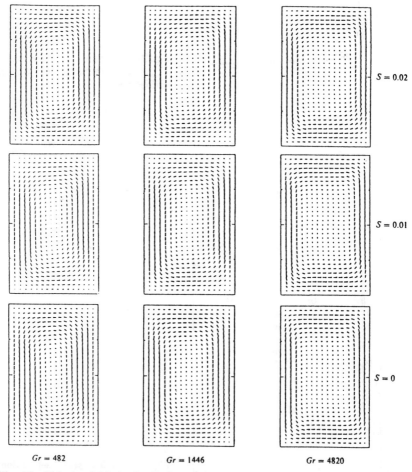

$S = 0.02$

$S = 0.01$

$S = 0$

$Gr = 482$ $Gr = 1446$ $Gr = 4820$

Fig. 23   Velocity fields in the V plane for small Soret parameters ($0 \leqslant S \leqslant 0.02$) at various Grashof numbers.

recirculating at the end walls. When Gr increases, the flow seems to concentrate near the lateral wall, leaving a large area of dead fluid in the center. The flow intensity given by $\tilde{V}_{max}$ is not proportional to Gr (even for $S = 0$), in contrast to the case of low Gr considered in the previous paragraphs. The zero velocity in the center is not an inertial effect, but is due to the vanishing of the buoyancy forces. For $S = 0$, where only the thermal contribution plays a role, we can verify that the isotherm deformation in the center, β_T, is close to 1 deg giving $F_T \sim 0$ [Eq. (18)]. This deformation ($\beta_T = 1$ deg in the center) exists over a longer r distance when Gr increases. Near the lateral walls, the conducting boundary conditions involve isotherm deformations in the reverse sense and favor the motion. In fact, the main motion is generated there and cannot correspond to usual core-driven flow.

Although the cylinder is not long ($A = 3$), there is an area in the central region where the flow is uniform (z independent) and the separation has a constant rate. Iso-mass fraction contours are strongly distorted by the motion, as shown in Fig. 24 for the V plane. The deformations increase with Gr, being already important at $Gr = 482$ for $S = 0$. In all cases they are associated with a very poor separation rate, as can be observed in Fig. 22.

For $S = 0$ and $482 \leqslant Gr \leqslant 4820$, we observe that $\beta_X > 90$ deg in the center. For small but finite S, iso-mass fraction contours are still strongly distorted and generate a force opposite to F_T (as $\beta_X = 90$ deg and $\gamma = 1$ deg, i.e., $\beta_X > \gamma$), but since S and (X_n) are small, F_X is less than F_T and slightly slows down the motion, which always presents the same structure (with a central region at rest). This leads to a smaller deformation of the X contours and a better separation rate, but this influence is only significant for very small values of Gr (see Fig. 24).

Moderate Soret Parameter ($0.05 \leqslant S \leqslant 0.673$)

General Comments. The velocity fields in the V plane (see Fig. 25) present some differences for higher S ($0.05 \leqslant S \leqslant 0.3$), compared to those presented in Fig. 23 for $0 \leqslant S \leqslant 0.02$, but they all show a central zone at rest. Some particle tracks in the V plane, given in Fig. 26 for $S = 0.673$, again exhibit more clearly such a central zone, mainly for high Gr ($Gr = 4820$). For high Gr, Figs. 25 and 26 also show a recirculating zone (contrarotative rolls) near the two end walls.

The iso-mass fraction contours (in Fig. 27) exhibit a uniform inclination throughout the cell (except near the lateral walls); this structure is quite different from the one (of remixing type) observed in Fig. 24 for low S. In addition, we can see that the value of β_X that corresponds to this inclination decreases when S increases.

In any part of the cylinder the velocity appears to be small compared with that of the case $S = 0$ (purely thermal buoyancy), meaning that F_T [Eq. (16)] given in this case by

$$F_T = \sin(1 \text{ deg}) \tag{36}$$

470 D. HENRY

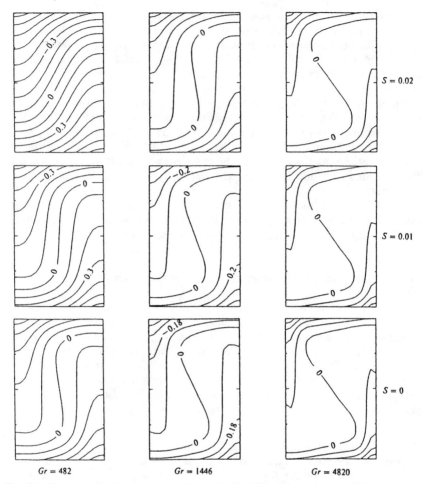

Fig. 24 Iso-mass fraction contours for small S: $DX = 0.1$, except at $GR = 1446$ for $S = 0$ where $DX = 0.09$ and at $Gr = 4820$ where $DX = 0.09$ for $S = 0.02$, $DX = 0.08$ for $S = 0.01$ and $DX = 0.07$ for $S = 0$.

is completely neutralized by F_X, whose sign is opposite in any case ($\beta_X > \gamma$). Using Eqs. (36) and (19), the mechanical equilibrium condition can be written as

$$F_T + F_X = \sin(1 \deg) - SX_n \sin(\beta_X - 1 \deg) = 0 \qquad (37)$$

When $X_n \sim 1$, which is true for moderate S, we obtain

$$\beta_X - 1 \deg = \arcsin(\sin(1 \deg)/S) \sim 1 \deg/S \qquad (38)$$

Thus, $(\beta_X - 1 \deg)$, which represents the inclination of the iso-mass fraction

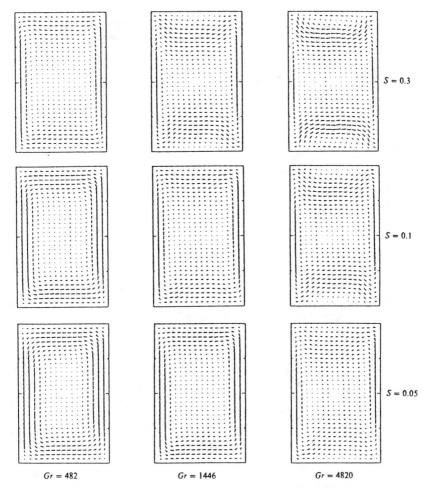

Fig. 25   Velocity fields in the V plane for moderate Soret parameter ($0.05 \leqslant S \leqslant 0.3$) at various Grashof numbers.

contours with respect to the horizontal, on the axis, is inversely proportional to S. To confirm this interpretation, we compare the numerical value of this inclination ($\beta_X - 1$ deg) with Eq. (38) in Table 1. The computed value of ($\beta_X - 1$ deg) for $S = 0.03$ is also mentioned in Table 1, but, in fact, Eq. (38) does not apply correctly to such a small S. Nevertheless, Table 1 shows that β_X increases when S diminishes and leads to a worse separation rate (X_z or X_n), which in turn requires a higher β_X to reach the equilibrium given by Eq. (37). To be more rigorous, we would have to take the isotherm deformations, β_T, into account and thus use Eq. (18) instead of Eq. (36); the equilibrium [Eq. (37)] would be more easily satisfied, but qualitatively the result would be the same.

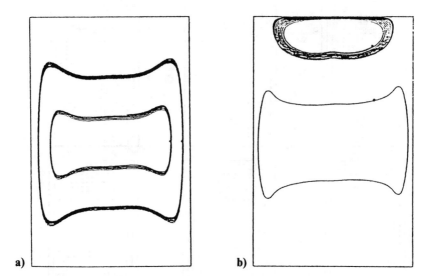

Fig. 26 Particle tracks in the V plane at $S = 0.673$. The particles are released at different points (r,z): a) for $Gr = 482$, at $(r = 0.75, z = 1.5)$ and $(r = 0.9, z = 1.5)$; b) for $Gr = 4820$, at $(r = 0.5, z = 1)$ and $(r = 0.5, z = 0.35)$.

This mechanical equilibrium corresponding to weak flows and inclined mass fraction field allows good separation rates. The perturbation of the global separation Xbot remains small (Xbot close to 1) as long as the X contour inclination is not too great. We can see in Fig. 22 that this separation is excellent for $S \geqslant 0.3$. It is still good for $0.1 \leqslant S \leqslant 0.3$ and acceptable for $S = 0.05$. Thus, at $\gamma = 1$ deg, the limit for performing an accurate measurement of the Soret coefficient would correspond to $S = 0.1$, which leads to a limit value $\beta_X \sim 10$ deg for the X contour inclination.

Influence of the Walls. In the main part of the cylindrical cell the velocity is small, but, as already mentioned, a significant flow exists near the walls. On the lateral walls, the no-mass-flux condition leads to $T_r + X_r = 0$.

Table 1   Values of $(\beta_X - 1)$ in degrees obtained at different Gr and from Eq. (38)

	\multicolumn{3}{c}{Gr, deg}		
S	482	1446	4820	Eq. (38), deg
0.673	1.5	1.5	1.5	1.5
0.3	3.3	3.4	3.4	3.3
0.1	10.2	10.1	10.3	10
0.05	21.4	20.9	21.7	20
0.03	——	——	49.4	33.3

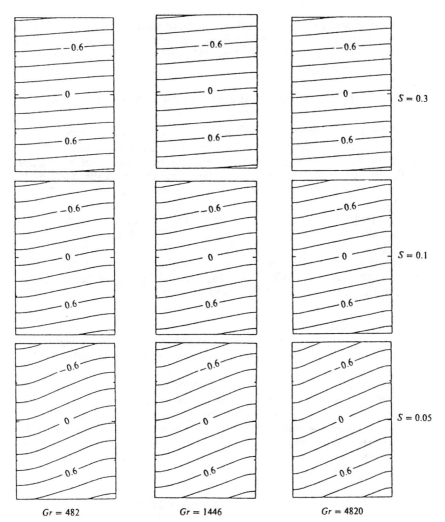

Fig. 27   Iso-mass fraction contours for moderate Soret parameters; $DX = 0.2$.

Therefore, since the isotherms are not deviated by the flow ($T_r \sim 0$), this condition gives $X_r \sim 0$, which means that the iso-mass fraction contours, perpendicular to the lateral walls, are not deformed by the flow. In particular, in the V plane we have $\beta_X = 0$ deg ($\beta_X < \gamma$), and then Eq. (19) shows that the solutal buoyancy, F_X, is smaller than F_T but acts in the same way. Thus, a flow is generated in a small layer adjacent to the lateral wall; the thickness of this layer diminishes when S increases (Fig. 25), because the inclination of the X contours in the central region needed for the equilibrium [Eq. (37)] is weak and the transition to the "unperturbed" zone near the lateral wall is rapid. This thickness is also smaller for higher Gr (Fig. 25) (equilibrium reached on a larger part of the cell).

On the end walls the no-mass-flux condition leads to $T_z + X_z = 0$. But, since the isotherms are not deviated by the flow ($T_z \sim 1$), this condition gives $X_z \sim 1$. In that case, using Eqs. (17) and (36), the total buoyancy force can be written as

$$F = F_T + F_X = \sin(1 \text{ deg}) + S[\sin(1 \text{ deg}) - X_r \cos(1 \text{ deg})]$$

In our study the solutal and thermal buoyancy forces are often opposite (since $\beta_X > \gamma$), but the thermal contribution generally remains higher ($F > 0$). In fact, near the end walls, we can have the reverse situation ($F < 0$), giving a (small) contrarotating flow near the end walls. This can be seen for $Gr = 4820$ in Fig. 25 with $S = 0.3$ and more clearly in Fig. 26 with $S = 0.673$. This contrarotating flow occurs when, near the end walls, $|F_X|$ is markedly larger than F_T. When $|F_X| \sim F_T$, the flow only slows down near the end walls, as can be observed in Fig. 25 for $Gr = 4820$ with $S = 0.1$, for $Gr = 1446$ with $S = 0.1$ and $S = 0.3$, and for $Gr = 482$ with $S = 0.3$.

For moderate S the flow in a quasivertical cylinder is mainly driven near the walls; the same behavior was observed in the steady basic state of double-diffusive regimes by Paliwal and Chen.[32]

Transition Domain

It is interesting to look at the transition between the two domains described in the previous sections since they exhibit quite different flow behaviors.

Figures 21 and 22 show that the domain of moderate S (for which an equilibrium has been seen to exist in the central region) can also be characterized by $\tilde{V}_{\max}$ and Xbot being independent of Gr. We can give a theoretical extreme limit of this domain by considering that F_X cannot exceed the value corresponding to vertical X contours, i.e., ($\beta_X - 1 \text{ deg}$) = 90 deg. Taking $X_n = 1$, the equilibrium expressed by Eq. (37) would then be satisfied when

$$S \geq \sin(1 \text{ deg})/\sin(90 \text{ deg}) = 0.01745$$

In fact X_n diminishes with the inclination and $X_n \ll 1$ for ($\beta_X - 1 \text{ deg}$) = 90 deg; thus, this limiting value of S is not very accurate but appears to be qualitatively acceptable.

Another means to estimate the transition domain is to calculate F_X in the center and compare it to F_T, which is assumed constant and equal to $\sin(1 \text{ deg})$, as in Eq. (36). The comparison indicates that an obvious equilibrium is obtained down to $S = 0.05$. Below this value the transition clearly occurs for high Gr ($Gr = 4820$ in the interval $0.02 < S < 0.03$), whereas it appears smoother for smaller Gr.

This transition can also be seen from a phenomenological point of view. For moderate S the theoretical inclination β_X given by Eq. (38) to reach the equilibrium is weak, X_n is close to 1, and thus Eq. (37) is well satisfied. When S diminishes, β_X given by Eq. (38) increases, inducing a reduction of

X_n. The value needed to reach the equilibrium, which is given in that case by Eq. (37), is still higher. In addition, as β_X increases, the thickness of the layer adjacent to the lateral walls increases and the central zone of constant X-contour inclination reduces. As we have seen earlier, when S is too small ($S < 0.03$), the equilibrium cannot be maintained, except in a very small domain near the center, because of a slight deformation of the isotherms. This transition will be seen in the subsection on vertical situations to correspond to an imperfect bifurcation.

Influence of the Main Parameters
  Grashof Effect. The influence of Gr in the range $482 \leqslant Gr \leqslant 4820$ considered herein is not important, since the same main characteristics are found in all cases. For smaller Gr the differences observed between the two domains will be less important, since the deformations of the iso-X are smaller for $S = 0$; until small values of Gr for which no transition occurs for $S \geqslant 0$, the separation Xbot being always close to 1.
  Aspect Ratio Effect. Some numerical simulations that have been done for higher A (namely, $A = 6$) show that increasing A does not have a large influence on the general behaviors discussed in the previous sections. Some changes are therefore found for small S, where a certain longitudinal motion exists, since the confinement is different.
  Influence of the Cell Inclination. The interpretation of the results for $\gamma = 1$ deg can be extended to any (small) value of γ. Expressions (37) and (38) for the equilibrium are easily generalized by

$$F_T + F_X = \sin(\gamma) - SX_n \sin(\beta_X - \gamma) = 0 \tag{39}$$

and

$$\beta_X - \gamma = \arcsin[\sin(\gamma)/S] \sim \gamma/S \tag{40}$$

Consequently, at high Gr and for any small γ, we will have the two domains of S mentioned earlier, but the localization of the transition will depend on γ. For a given γ we can calculate the lower limit value, S_{lim}, of this equilibrium state domain. We have seen that this limit would correspond to $\beta_X \sim 10$ deg and $X_n \sim 1$. According to Eq (39), we have

$$S_{\text{lim}} = \sin(\gamma)/\sin(10 \text{ deg} - \gamma) \tag{41}$$

For small γ ($\gamma \leqslant 1$ deg), Eq. (41) gives

$$S_{\text{lim}} \sim \gamma/10 \text{ deg} \tag{42}$$

A good separation is then only possible for $S > \gamma/10$ deg (except for cases where F_T is not strong enough to generate a perturbating flow and where there is no limit).

Similarly, at high *Gr* and for a given *S*, we can introduce a limit inclination, $\gamma_{\lim}$ (in degrees),

$$\gamma_{\lim} \sim 10S/(1 + S) \qquad (43)$$

such that the equilibrium can be reached in a certain domain of γ defined by $0 \leqslant \gamma \leqslant \gamma_{\lim}$.

Study of Vertical Situations or Close to the Vertical Situations

An interesting particularity of this work was the possibility of analyzing vertical and inclined situations at the same time.

Vertical Situations

In the paragraph concerning space situations ($Pr = 0.6$, $Sc = 60$), we considered different inclinations (from 0 deg to 180 deg) for three values of S ($S = -0.5$, 0, and 0.5) and two values of Gr ($Gr = 1$ and 3). The results of the stability analysis (see the section on stability analysis of vertical situations) indicate that for $\gamma = 0$ deg (heating from the top) and $S \geqslant 0$ any perturbation of the equilibrium (initial motion) will be damped. For $\gamma = 180$ deg (heating from the bottom) and $S \leqslant 0$ the domain of low *Gr* will now allow permanent convective situations (high oscillatory or stationary thresholds). It will only be possible to find situations in which the perturbations are not damped for $\gamma = 0$ deg when $S < 0$ and for $\gamma = 180$ deg when $S > 0$. We will see later that for such situations the perturbations definitively generate a convective flow. The stability analysis gives the same critical value of *Gr* ($Gr_c = 2.66$) for $\gamma = 0$ deg at $S = -0.5$ and $\gamma = 180$ deg at $S = 0.5$, respectively. This common value of Gr_c for opposite S corresponds to the fact that the two cases belong to the asymptotic domain ($\psi \to \infty$) defined in the subsection on particular behaviors. It is also clear if we write the buoyancy force, which, since T is not deformed by the convection motion, is reduced to

$$F = -SX_r \cos(\gamma)$$

The iso-mass fraction is then responsible for the motion and acts proportionally to the product $S \cos(\gamma)$, which is identical for opposite S and opposite cell inclinations (0 and 180 deg).

The case $Gr = 1$ corresponds to stable vertical situations at $\gamma = 0$ and $\gamma = 180$ deg for all S in the range $-0.5 \leqslant S \leqslant 0.5$. On the contrary, the case $Gr = 3$ must be unstable for $S = -0.5$ at $\gamma = 0$ deg and for $S = 0.5$ at $\gamma = 180$ deg. For $Gr = 1$ the simulation confirms the stability of all the vertical situations for the three values of S, since perturbations applied on the system cannot induce a definite permanent flow. For $Gr = 3$ the same phenomenon is observed, except in the "unstable" situations ($\gamma = 0$ deg for $S = -0.5$ and $\gamma = 180$ deg for $S = 0.5$), where flow settles with quite identical characteristics in both cases (Xbot $= 0.891$ and 0.882, respectively), in good agreement with the fact that these two situations present the same Gr_c.

To give an indication of the evolution of the flow behavior and of the separation when increasing the "distance" from the critical situation, we carried out calculations for $S = 0.5$ with $Gr = 5$ (nearly twice the Gr_c) and for $S = 1$ with $Gr = 3$ (the corresponding Gr_c being close to 1.33). In both cases the flow increases and corresponds to more perturbed situations with a poorer separation (Xbot $= 0.610$ and 0.536, respectively).

Transition from Inclined to Vertical Situations

The transition from inclined to vertical situations is regular. For $Gr < Gr_c$ the flow speed regularly decreases to zero, the iso-mass fraction deformation decreases, and the separation Xbot reaches 1; the tangent of the different Xbot curves vs γ at $\gamma = 0$ or $\gamma = 180$ deg is horizontal (Fig. 15). In contrast, for $Gr > Gr_c$ a flow already exists for a vertical cell (despite the fact that the isotherms are horizontal) and generates a deformation of the iso-mass fractions. A slight inclination in γ will then increase this flow, leading to an extra deformation of these iso-mass fractions and a diminution of the separation Xbot. For $S = -0.5$ a direct simulation at $\gamma = 1$ deg confirms that there is no horizontal tangent at $\gamma = 0$ deg.

Correspondence Between Instability and Crossing of $\tilde{V}_{max}$ and Xbot Curves

As we saw earlier (Figs. 7 and 10), the crossing of $\tilde{V}_{max}$ (and Xbot) curves (as a function of γ) for different S roughly corresponds to the fact that the iso-mass fractions slope in the central part overshoots the horizontal (the occurrence of such an overshooting for $S = 0$ is a sufficient condition for that crossing). For $Gr = 3$ such a crossing is necessary at $S = -0.5$, since we have an instability for this value of S for $\gamma = 0$ deg giving rise to $\tilde{V}_{max} > 0$ (and Xbot < 1). Since it corresponds to a general change in the influence of S, it will occur for all S, independently of whether the corresponding vertical situation is stable or not. For example, at $S = -0.4$, which corresponds to a stable vertical situation ($Gr_c = 3.33$), we obtain Xbot $= 0.962$ for $\gamma = 1$ deg, indicating that the curves corresponding to $S = -0.4$ and 0 have to cross.

More generally, the crossing depends on Gr, because the convection needs to be strong enough to change the iso-mass fraction slope up to reach the horizontal. For a given Gr it must occur as soon as a negative S gives an unstable situation for $\gamma = 0$ deg. In fact, the limiting negative value of S to consider is -1, since below this value the mass fraction would play a dominant role and impose the direction of the flow (generally opposite), leading to a completely different behavior with γ. For this value $S = -1$ the stability analysis of the section on stability analysis of vertical situations gives $Gr_c = 1.33$. Consequently, for $Gr < 1.33$ we would have no crossing of the $\tilde{V}_{max}$ curves for different S as well as no instability for $\gamma = 0$ deg, and the influence of S will be the usual one for all γ. These conclusions agree with the results presented for $Gr = 1$ (Fig. 10a). For any $Gr > 1.33$ the curves for different S will cross for certain $\gamma = \gamma_{ov}$ (close to or far from 90 deg, depending on the value of Gr), which delimits two domains with opposite influence of S, and we will have unstable situations at $\gamma = 0$ deg below a certain negative S [in agreement with the results presented for $Gr = 3$ (Fig. 10b)].

Transition from Stability to Instability for (or close to) Vertical Situations
We wanted to get more information about the unstable vertical situations. Since the two unstable situations studied for $Gr = 3$ correspond to the same process (the same X stratification generating instability), we focused our attention on the one at $\gamma = 0$ deg for $S = -0.5$. We tried to obtain a numerical confirmation on the value of Gr_c given by the stability analysis ($Gr_c = 2.66$). For $Gr = 2.8$ a convective situation is exhibited, with Xbot = 0.933. For $Gr = 2.5$ the initially imposed flow slowly decreases with the iterations and already gives quite a weak flow, with Xbot = 0.999, even stopping the calculations for a weak computational accuracy. We can conclude that the equilibrium is lost for Gr between 2.5 and 2.8, in good agreement with the value of 2.66 given by the stability analysis.

From the results obtained at $S = -0.5$, for $\gamma = 0$ and 1 deg, it is possible to gain some understanding of the transition between stable and unstable situations, e.g., when Gr is increased (Fig. 28). For Gr corresponding to very stable situations (as $Gr = 1$), the differences between the values of $\tilde{V}_{max}$ (or between the values of Xbot) for $\gamma = 1$ and $\gamma = 0$ deg are small. When Gr increases up to $Gr_c = 2.66$, the differences increase quite strongly as the values at $\gamma = 0$ deg are unchanged, whereas those at $\gamma = 1$ deg present a rather quick variation. On the contrary, when $Gr > Gr_c$ ($Gr = 3$

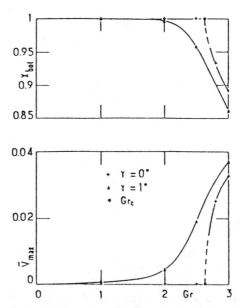

Fig. 28 Transition from stable to unstable situations for increasing Gr, observed on Xbot and $\tilde{V}_{max}$, for vertical ($\gamma = 0$ deg) and nearly vertical ($\gamma = 1$ deg) cells.

or 5), the values of $\tilde{V}_{max}$ and Xbot at $\gamma = 0$ deg change suddenly, reducing the difference with the values at $\gamma = 1$ deg. For $\gamma = 0$ deg the transition then occurs abruptly at $Gr = 2.66$, whereas for $\gamma = 1$ deg the transition develops smoothly through an imperfect bifurcation, with rapid increase in the flow and the mixing (corresponding to a decrease of separation) for $2 < Gr < Gr_c$.

This bifurcation, which corresponds to the crossing of the lower curve (heating from the top and $S < 0$) of the instability diagram (Gr,S) (which could be obtained from Fig. 3) when $|Gr|$ is increased at constant S, could have been also considered when S is varied (from 0 to negative values) at $Gr = 3$, for example. Such a consideration will give a new insight into the study of quasivertical cylinders in ground experiments, shown in the subsection on the study of the convective perturbation of ground experiments. In the last paragraph we have shown that for $Gr < 1.33$ no instability occurs at $\gamma = 0$ deg (in the domain $S > -1$), and for any $\gamma \neq 0$ deg a positive S increases the flow. For larger Gr there is a negative critical S ($-1 < S_c < 0$) beyond which an instability occurs at $\gamma = 0$ deg and, at least in the vicinity of $\gamma = 0$ deg, a positive S decreases the flow. In that case the bifurcation obtained at $\gamma = 0$ deg for S_c (sudden increase of $\tilde{V}_{max}$ and decrease of Xbot) will give with $\gamma = 1$ deg a smoother imperfect bifurcation beginning for values of S slightly higher than S_c. For small Gr, as $Gr = 3$ ($S_c = -0.45$), this imperfect bifurcation begins for negative values of S and then does not affect the values of Xbot for $S > 0$. But for larger Gr (corresponding to small $|S_c|$), as $Gr \geqslant 482$, this imperfect bifurcation begins or even occurs mainly in the domain of positive S. The sudden variations observed in the results of the subsection on the study of the convective perturbation of ground experiments (Figs. 21 and 22) can then be understood as the signature of an imperfect bifurcation.

Soret Experiments

Presentation of the Experiments

Experiments on Molten Salts

An experiment on an ionic molten salt mixture, AgI-KI, has been realized by Dupuy and Bert during the D1 flight in November 1985.[6-10] The experiment was performed in confined cylindrical cells ($A = 12$ or 20) with a strong thermal gradient (70°C/cm). The characteristics of the mixture (72% in moles of AgI; see Ref. 33) lead to the following dimensionless parameters: $Pr = 0.6$; $Sc = 60$; $Gr = 1.69\,g$ for $A = 12$; $Gr = 0.219\,g$ for $A = 20$, with g expressed in centimeters squared per second.

The measurement of the separation was realized first by in situ electromotive force measurements during the experiment, then by direct chemical analysis of the samples after the experiment. The results obtained by the two techniques indicate a positive value of S corresponding to an enrichment in Ag at the cold end. But the actual value of S cannot be obtained because the evolution with time of the electromotive force shows that the experiment has been stopped before the Soret steady state has been obtained. Some experiments realized on ground seem to have been perturbated, since they present a smaller global separation and an irregular separation profile along the cell.[10]

Experiments on Metal Alloys
   Some experiments on metal alloys have been realized by Malmejac and Praizey[12] during the F.S.L.P. mission in December 1983 and by Praizey[14] during the D1 flight in November 1985. The experiments concerned different alloys based on Sn (Sn-Co, Sn-Ag, Sn-Bi, Sn-Au) and were performed in large-aspect-ratio cylinders ($A = 18$) under large temperature gradients. The general characteristics of the mixtures correspond to those of Sn and give the following values for the dimensionless parameters: $Pr = 0.01$, $Sc = 34$, and $Gr = 0.55\,g$.
   The separation has been measured by neutronical activation on the samples that have been divided in six parts (in the liquid state) by a mechanical device at the end of the experiment. The results are given in Malmejac and Praizey[12] and Praizey.[13,14] They are compared with the results obtained during ground experiments. In all cases the separation measured in space is larger than the one measured on ground.

Application of the Numerical Results to the Analysis of the Perturbation of the Experiments

Space Experiments
   The different results of the subsection on parametric study of convective situations related to space experiments can be used to analyze the perturbation of the space experiments. They are valid for molten salts as well as for metal alloys, since in both cases $Pr \leqslant 0.6$. In the second case we must use a transformed Grashof, since the value of Sc is different from 60. Practically, for $S = 0$ the relation (24), the curves in Figs. 16 and 17, and the relation (35) easily give the separation Xbot for any γ, Gr, and A. For $S \neq 0$ the separation Xbot can be obtained by using the curves in Figs. 14 and 19 and the relation (35) for any Gr and A in the horizontal case (see Ref. 15 for an example of application).
   In fact, for a quick information, in Fig. 29 we have drawn on the domain $(GrSc, S)$ the curves corresponding to given percentages of perturbation on the separation Xbot, namely, 1, 5 and 10%. These curves are given in each case for $A = 6$ and $A \to \infty$ and correspond to the horizontal case, where the perturbations are nearly the largest. If we consider that in space the steady acceleration level is less than $10^{-4}\,g_0$ (where g_0 is gravity on the ground), we have for the first experiments $GrSc < 10$ and for the second ones $GrSc < 1.9$. In both cases, for the experimental values of S that are in general not very large, the perturbation is less than 1%, indicating the reliability of the results.

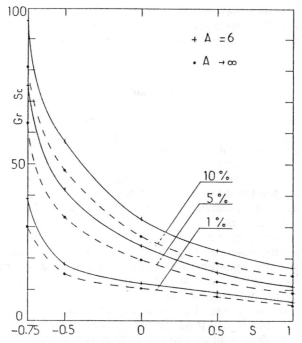

Fig. 29   Curves corresponding to different percentages of perturbation on the separa-tion Xbot (1, 5, and 10%) in the $(GrSc,S)$ representation; $A = 6$ and $A \to \infty$.

Ground Experiments

The ground experiments that are realized in vertical situations with top heating can be theoretically unstable only for mixtures with negative values of S and in supercritical conditions. Only the experiment of Praizey on Sn-Co is near such a state (experimental Gr near Gr_c); it is the most perturbated of all.[14]

In fact, most ground experiments are realized in theoretically stable conditions, and their perturbation can only be explained by some technical defects such as a small inclination of the cell or radial thermal gradients. No estimation of verticality defects is available. An estimation of the radial temperature differences ΔT_r has been done by Praizey for his experiments. An indication of the perturbation induced can be obtained from the study on quasivertical cylinders (see the subsection on the study of the convective perturbation of ground experiments). In that study relation (43) gives (as a function of S) the limit inclination γ_{lim} to have a good separation. In fact, the inclination of the cell γ can be related easily to a radial temperature difference if we know the axial temperature difference ΔT_{axial} and the characteristics of the cell:

$$\Delta T_r = \Delta T_{\text{axial}} R \, \sin(\gamma)/L \tag{44}$$

With Eqs. (43) and (44) we can then define for each experiment a limit ΔT_r, to which the experimental ΔT_r can be compared to estimate the perturbation. Some practical applications can be found in Praizey.[15]

Conclusions

The objective of this chapter was to get a better knowledge of convective phenomena that could appear during thermal diffusion experiments and perturbate the separation in order to permit experimenters to better analyze their results. We were mainly interested in the experiments planned in a space environment, which are characterized by confined cylindrical cells. The problem has been approached in two ways: a linear stability study for the determination of the stability thresholds in vertical situations and a three-dimensional numerical study for the simulation of the convective situations.

Concerning the space experiments, we found on the domain of low Gr a separation regime, convenient for good measurements since the flow does not perturbate the separation. Outside this domain, the experiments will be perturbated, but we are able to give, for extended values of the parameters of the problem, the perturbation induced on the separation by the convection. Concerning the ground experiments, they can be theoretically unstable when heated from the top, only for negative Soret parameter S. But some defects difficult to avoid, such as a misverticality or radial thermal gradients can strongly perturbate the system, mainly for small S values. These results have been applied to the experiments realized recently in space by French laboratories.

More generally, this work has given new insights into double-diffusive phenomena in low-convective and confined situations. We analyzed the intensity and the structure of the flow as a function of the parameters of the problem and the way it affects the separation of the constituents. Moreover, the double approach mentioned earlier allowed the consideration of the vertical situations in continuity with inclined situations. We could then analyze the transition from stability to instability and the presence of imperfect bifurcations: one of the most striking behaviors is the sudden transition, observed in quasivertical cylinders between two very different domains, which is in fact the signature of such an imperfect bifurcation.

Acknowledgments

The author thanks the Centre National d'Etudes Spatiales (Division Microgravité Fondamentale et Appliquée) for giving financial support and the Centre de Calcul Vectoriel pour la Recherche for providing him with computing time on the Cray-1S computer. He is indebted to B. Roux, J. Dupuy, J. Bert, R. Sani, and J. P. Praizey for numerous and fruitful discussions.

References

[1]Velarde, M. G., and Schechter, R. S., "Thermal Diffusion and Convective Stability: A Critical Survey of Soret Coefficient Measurements," *Chemical Physics Letters*, Vol. 12. 1971, pp. 312–315.

[2]Thomaes, G., "The Bénard Instability in Liquid Mixtures," *Advances in Chemical Physics*, Vol. 32, 1975, pp. 269–279.

[3]De Groot, S. R., *L'effet Soret*, North-Holland, Amsterdam, 1947.

[4]Agar, J. N., and Turner, J. C. R., "Thermal Diffusion in Solutions of Electrolytes," *Proceedings of the Royal Society of London, Series A*, Vol. A255, 1960, pp. 307–330.

[5]Dulieu, B., Chanu, J., and Walch, J. P., "A propos de l'influence de la convection sur la mesure de l'effet Soret: le cas d'un défaut d'horizontalité," *Journal of Chemical Physics*, Vol. 78, 1981, pp. 193–201.

[6]Bert, J., Henry, D., Layani, P., Chuzeville, G., Dupuy J., and Roux, B., "Space Experiment on Thermal Diffusion: Preparation and Theoretical Analysis," *Proceedings of the Fifth European Symposium on Material Sciences Under Microgravity*, European Space Agency, Noordwijk, The Netherlands, ESA SP-222, 1984, pp. 347–351.

[7]Bert, J., Moussa, I., Henry, D., and Dupuy, J., "Space Thermal Diffusion Experiment in a Molten AgI-KI Mixture," *Proceedings of the Nordeney Symposium on Scientific Results of the German Spacelab Mission D1*, 1986, pp. 152–160.

[8]Bert, J., Moussa, I., and Dupuy, J., "Space Thermal Diffusion Experiment in a Molten AgI-KI Mixture," *Proceedings of the Sixth European Symposium on Material Sciences Under Microgravity*, European Space Agency, Noordwijk, The Netherlands, ESA SP-256, 1987, pp. 471–475.

[9]Bert, J., Henry, D., Mellon, H., and Dupuy, J., "Space Thermal Diffusion Experiment in a Molten AgI-KI Mixture—Theoretical Convection Approach and Relation with in Situ Measurement Results," *Proceedings of the Joint International Symposium on Molten Salts*, Electrochemical Society, Vol. 87–7, 1987, pp. 340–352.

[10]Moussa, I., "Analyse des résultats d'une expérience de thermodiffusion réalisée en microgravité (Spacelab D1) dans le mélange eutectique fondu $AgI_{0.72}KI_{0.28}$," PhD Dissertation, Université Claude Bernard, Lyon, France, 1989.

[11]Praizey, J. P., "Technical Implications of a Flight Design," *Proceedings of the Fourth European Symposium on Material Sciences Under Microgravity*, European Space Agency, Noordwijk, The Netherlands, Rept. ESA SP-191, 1983, pp. 127–132.

[12]Malmejac, Y. and Praizey, J. P., "Thermomigration of Cobalt in Liquid Tin," *Proceedings of the Fifth European Symposium on Material Sciences Under Microgravity*, European Space Agency, Noordwijk, The Netherlands, Rept. ESA SP-222, 1984, pp. 147–152.

[13]Praizey, J. P., "Thermomigration in Liquid Metallic Alloys. XXVIth COSPAR," *Advances in Space Research*, Vol. 6, 1986, pp. 51–60.

[14]Praizey, J. P., "Thermomigration dans les alliages métalliques liquides," *Proceedings of the Sixth European Symposium on Material Sciences Under Microgravity*, European Space Agency, Noordwijk, The Netherlands, Rept. ESA SP-256, 1987, pp. 501–508.

[15]Praizey, J. P., "Intérêt de la microgravité pour la mesure des coefficients de thermodiffusion dans les alliages métalliques liquides," CEA Rept. CEA-R-5449, 1988.

[16]Schechter, R. S., Prigogine, I., and Hamm, J. R., "Thermal Diffusion and Convective Stability," *Physics of Fluids*, Vol. 15, 1972, pp. 379–386.

484   D. HENRY

[17]Gutkowicz-Krusin, D., Collins, M. A., and Ross, J., "Rayleigh-Bénard Instability in Nonreactive Binary Fluids. I. Theory. II. Results," *Physics of Fluids*, Vol. 22, 1979, pp. 1443–1460.
[18]Crespo, E., and Velarde, M. G., "Two-Component Bénard Convection in Cylinders," *International Journal of Heat and Mass Transfer*, Vol. 25, 1982, pp. 1451–1456.
[19]Hardin, G. R., Sani, R. L., Henry, D., and Roux, B., "Buoyancy Driven Instability in a Vertical Cylinder: Binary Fluid with Soret Effect. Part 1. General Theory and Stationary Stability Results," *International Journal Num. Math. Fluids*, Vol. 10, 1990, pp. 79–117.
[20]Hurle, D. T. J., and Jakeman, E., "Soret-Driven Thermosolutal Convection," *Journal of Fluid Mechanics*, Vol 47, 1971, pp. 667–688.
[21]Abernathey, J. R., and Rosenberger, R., "Soret Diffusion and Convective Stability in a Closed Vertical Cylinder," *Physics of Fluids*, Vol. 24, 1981, pp. 377–381.
[22]Henry, D., and Roux, B., "Stationary and Oscillatory Instabilities for Mixture Subjected to Soret Effect in Vertical Cylinder with Axial Temperature Gradient," *Proceedings of the Fourth European Symposium on Material Sciences Under Microgravity*, European Space Agency, Noordwijk, The Netherlands, Rept. ESA SP-191, 1983, pp. 145–152.
[23]Henry, D., and Roux, B., "Numerical Study of the Perturbation of Soret Experiments by Three-Dimensional Buoyancy Driven Flows," *Proceedings of the Sixth European Symposium on Material Sciences Under Microgravity*, European Space Agency, Noordwijk, The Netherlands, Rept. ESA SP-256, 1987, pp. 487–491.
[24]Henry, D., and Roux, B., "Soret Separation in a Quasi-Vertical Cylinder," *Journal of Fluid Mechanics*, Vol. 195, 1988, pp. 175–200.
[25]Henry, D., "Simulation numérique 3D des mouvements de convection thermosolutale d'un mélange binaire—étude paramétrique de l'influence de la convection sur la séparation des espèces du mélange, par effet Soret, dans un cylindre incliné," PhD Dissertation, Université Claude Bernard, Lyon, France, 1986.
[26]Henry, D., and Roux, B., "Three-Dimensional Numerical Study of Convection in a Cylindrical Thermal Diffusion Cell: Its Influence on the Separation of Constituents," *Physics of Fluids*, Vol. 29, 1986, pp. 3562–3572.
[27]Henry, D., and Roux, B., "Numerical Simulation of 3D Convective Motion Disturbing the Soret Separation of the Two Components of a Binary Fluid Mixture. XXVIth COSPAR," *Advances in Space Research*, Vol. 6, 1986, pp. 141–146.
[28]Henry, D., and Roux, B., "Three-Dimensional Numerical Study of Convection in a Cylindrical Thermal Diffusion Cell: Inclination Effect," *Physics of Fluids*, Vol. 30, 1987, pp. 1656–1666.
[29]Leong, S. S., and de Vahl Davis, G., "Natural Convection in a Horizontal Cylinder," *Proceedings of the First International Conference on Numerical Methods in Thermal Problems*, Pineridge, Swansea, UK, 1979, pp. 287–296.
[30]Smutek, C., Roux, B., Bontoux, P., and de Vahl Davis, G., "3D Finite Difference for Natural Convection in Cylinders," Proceedings of the Fifth Gesellschaft für Angewandte Mathematik und Mechanik Conference, *Notes on Numerical Fluid Mechanics*, Vol. 7, Vieweg, 1984, pp. 338–345.
[31]Bejan, A., and Tien, C. L., "Fully Developed Natural Counterflow in a Long Horizontal Pipe with Different End Temperatures," *International Journal of Heat and Mass Transfer*, Vol. 21, 1978, pp. 701–708.

[32]Paliwal, R. C., and Chen, C. F., "Double-Diffusive Instability in an Inclined Fluid Layer. Part 1. Experimental Investigation. Part 2. Stability Analysis," *Journal of Fluid Mechanics*, Vol. 98, 1980, pp. 769–785.

[33]Mellon, H., "Diffusion thermique dans un mélange d'électrolytes fondus sous gravité réduite: recherche des conditions optimales de l'expérience," PhD Dissertation, Université Claude Bernard, Lyon, France, 1978.

Complex (and Time-Dependent) Natural Convection in Low-Prandtl-Number Melts

P. Bontoux,* A. Roux,† J. P. Fontaine,‡ A. Randriamampianina,*
and G. P. Extremet*
Institut de Mécanique des Fluides, Marseille, France
E. Crespo del Arco§
Universidad Nacional de Educacion a Distancia, Madrid, Spain
and
J. P. Pulicani‡
University of Nice, Nice, France and
University of Alabama in Huntsville, Huntsville, Alabama

Introduction

THE hydrodynamic aspects of crystal growth from melts have been the topic of a number of papers for the last ten years.[1-3] The striations in the grown crystals were shown to result from the temperature fluctuations in the liquid phase.[4,5] The difficulties of the computation of complex systems involving thermosolutal convection coupled with interface morphology were evoked by Ettouney and Brown,[6] Crochet et al.,[7] and Adornato and Brown.[8] The stability of the low-Prandtl-number fluid layers has been analyzed by Hart,[9,10] Gill,[11] and Laure and Roux[12] in the case of a single-cell flow thermally driven by an imposed horizontal temperature gradient.

Numerous theoretical and experimental studies have been devoted to the understanding of the behavior of low-Prandtl-number convection, partly for their applications to space processing. Hurle et al.[13] reported observations of oscillations in a differentially heated boat of liquid gallium with a

*Director & Chargés de Recherche, C.N.R.S.
†Graduate Student.
‡Research Associate.
§Professor titular.

free surface. The stability of the steady flow of low-Pr liquids to oscillatory perturbations was studied by Winters[14] in a rectangular cavity, and the results were assessed by direct numerical simulations.[15] The difficulties arising from the computation of the time-dependent convection and the discrepancies observed between the various results already published in the literature have been the topic of a workshop.[16] For this purpose experiments were performed by Hart and Pratte,[17] Hung and Andereck,[18] and Wang et al.,[19] and they mainly concern the behavior of liquid mercury in a variety of closed rigid containers. The accuracy of the solutions was widely discussed in former works.[15,20-23] Here we are particularly concerned in the validity of the physical models used for the computations with respect to the experimental results.

The melt in the floating-zone technique consists in a liquid bridge maintained between two rigid supports of the same material. A number of liquid bridge configurations were investigated experimentally.[24,25] The temperature gradient imposed to the liquid bridge generates a surface tension mechanism that drives the convection in the melt, particularly in a microgravity environment (see Ref. 26). Time-dependent behaviors have been observed due to surface tension (see Refs. 27-29). The thermocapillary convection can be controlled using different techniques as, for example, the rotation of the rods[30-32] or a magnetic field.[33] The influence and the possible deformation of the interface have been investigated by Fu and Ostrach,[34] Zebib et al.,[35] and Cuvelier and Driessen.[36]

The first purpose of this chapter is to report on the ability to compute the time-dependent instabilities. The study is made in the case of Bridgman models with simple, rigid, and/or planar stress-free boundaries. The second goal refers to preliminary studies devoted to the computation of steady but complex phenomena: the shape of a free boundary submitted to thermocapillary convection and the combined effects on the melt flow of multiple driving mechanisms. Some situations corresponding to the surface-tension-driven convection and to buoyancy- and rotation-dominated convection are presented in the case of a floating-zone model.

Physical Model

The main physical model (related to the Bridgman technique) is a cavity or a box: The bottom and side walls are rigid, and the upper wall is either rigid (R-R case) or stress-free and flat (R-F case). The dimensionless sizes are 1 : 4 for the cavity (two-dimensional model) in the x (vertical) and y (longitudinal) directions (length-to-depth aspect ratio $A = L/H = 4$) and 1 : 4 : 1 for the box (three-dimensional model) in the x, y, and z (transversal) directions (width-to-depth aspect ratio $Az = 1$). The vertical walls (at $y = 0$ and 4) are isothermal, hot, and cold, respectively (with temperature difference δT). The side walls can be either conducting (C) or insulating (A). The governing equations for a Boussinesq fluid are the Navier-Stokes equations coupled to the energy equation. The surface tension effect was neglected; thus, the physical parameters are the Grashof number, $Gr = g\alpha H^4(\delta T/L)/v^2$, and the Prandtl number, $Pr = v/\kappa$. In this chapter we consider $Pr = 0$ and 0.015, and the characteristic scale is H^2/v for the time variable.

The second physical model refers to the floating-zone technique. The liquid bridge is confined between two rigid walls and bounded by a free surface (Fig. 14a). The melt-crystal interfaces are assumed to be fixed. The (height-to-radius) dimensionless size is $A = H/R$. The axisymmetry is assumed. The lower support is cold, and the free surface is submitted to either a monotonic temperature profile or a profile involving a plateau at midheight (see Fig. 14c). The gravity (when considered) acts along the axial direction, and the Grashof number is defined by $Gr = g\beta\Delta TR^3/v^2$. Two kinds of situations are investigated: 1) the effect of surface tension on the shape of the free surface is considered for a $Pr = 0.1$ melt under zero-gravity conditions; and 2) the effects of buoyancy and rotation (of supports) are considered for a $Pr = 0.015$ melt assuming a stress-free and "planar" liquid-gas interface. The additional parameters are the Marangoni number, $Ma = (d\sigma/dT)R\delta T/\alpha\mu A$ (equivalent to a Peclet number), the Reynolds-Marangoni number, $Re_M = Ma/Pr$, and the capillary number, $Ca = (d\sigma/dT)\delta T/\sigma$, where σ is the surface tension coefficient. When the rotation is considered, the associate Reynolds numbers are $Re_\Omega = \Omega R^2/v$ and $Re_{s\Omega} = s\Omega R^2/v$, where s is the rotation ratio of the two supports (at rotation velocities $s\Omega$ and Ω, see Ref. 31).

Numerical Solution

The computations were preformed using spectral (for the two-dimensional problem) and finite-element methods (for both two- and three-dimensional ones). The spectral Tau-Chebyshev methods are developed for the time-dependent streamfunction (ψ) and vorticity (ω) formulation (cf. Refs. 15, 22, and 37). The methods are based on the matrix diagonalization technique and on multistep schemes for the time-integration. The finite-element method[38] uses the primitive variable formulation with a penalty approach for the pressure. Four-noded quadrilateral elements and eight-noded brick elements have been considered. The integration is based on a combination of successive substitution scheme and of quasi-Newton algorithm.[23]

The validity of the two-dimensional numerical solutions was assessed elsewhere by comparing the results of the two different spectral methods obtained with 17×31 to 81×101 resolutions and those of the finite-element method obtained with uniform and nonuniform 31×81 grids. For the three-dimensional ($1:4:1$) box, $15 \times 21 \times 10$ and $15 \times 41 \times 10$ meshes (in x, y, and z) were employed for the full and half-cavity, respectively, assuming symmetry with respect to the vertical midplane.

With a Cray 2 computer about $10-20$ min of CPU time is necessary for the computation of a quasiperiodic (QP) solution during 70 periods of time ($1/f$), starting from rest and using 17×31 Chebyshev polynomials. About 45 min is needed to compute 150 periods for $Gr = 26,000$ starting from a QP state and with the same resolution. For the three-dimensional problem the CPU times are about 12 min for the computation of steady solutions with the $15 \times 21 \times 10$ mesh and from 15 to 35 min with the $15 \times 41 \times 10$ mesh.

For the floating-zone problem the finite-element code was also used. (Refer to Engelman[38] for the details on the modeling and on the numerical method.) For the situations corresponding to $Pr = 0.1$, the interface shape is computed using contact angles of 90 deg at the liquid-gas-solid corners. The computations were performed using 21×31 and 21×51 meshes (Fig. 14b).

Time-Dependent Buoyancy-Driven Convection in a Bridgman Model

In the R-F cavity the two kinds of steady solutions correspond to one- and two-roll patterns and are denoted by S1 and S11, respectively (Fig. 1). In the R-R cavity three center-symmetric steady solutions were found for $Gr \leqslant 40{,}000$. Hereafter we denote them by S1, S12, and S2, corresponding to one, three, and two corotating rolls, respectively (Fig. 2). Two quasi-periodic (asymmetric) solutions have been found, both with three corotating rolls. The spectrum of the solutions denoted by QP shows two frequencies, f and f' (see Fig. 11), and the spectrum of QP2 shows f, f', and $f/2$ (this $f/2$ is not visible in the center of the cavity). Finally, we denote by Pn the solutions (in both R-R and R-F cases) that have a spectrum with a fundamental frequency f and subharmonics f/n (see Fig. 10).

Results

Subharmonic Instability in the Case of an Upper Stress-Free Surface (R-F)

The accurate determination of the onset of the oscillatory regimes (P1) was reported in Pulicani et al.[15] The computations were carried out using

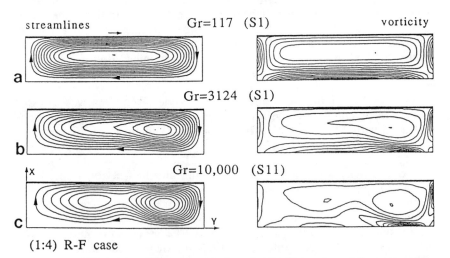

(1:4) R-F case

Fig. 1   Two-dimensional steady flow solutions (streamline and vorticity patterns) in the R-F case for $Pr = 0$: a) $Gr = 117$; b) $Gr = 3124$; and c) $Gr = 10{,}000$.

streamlines   Gr=25,000   (S12)   vorticity

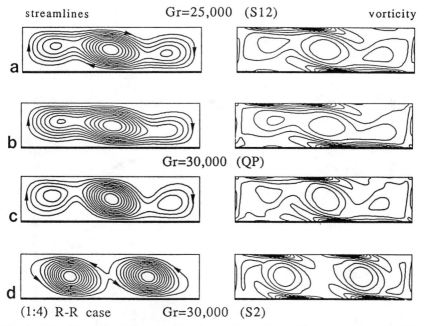

(1:4) R-R case   Gr=30,000   (S2)

Fig. 2   **Two-dimensional flow solutions (streamline and vorticity patterns) in the R-R case for $Pr = 0$ and for a) $Gr = 25,000$ (steady S12); b,c) $Gr = 30,000$ (quasiperiodic QP at two different steps of time); and d) $Gr = 30,000$ (steady S2).**

a spectral method based on 17×33 to 33×65 Chebyshev polynomial expansions. Details on the computed solutions are summarized in Table 1, where the fundamental frequencies f are given and the time dependence is identified from phase plane projections. We observe that the fundamental f increases regularly with Gr until a $f/2$ behavior (P2) is reached preceding a chaotic behavior (NP).

An illustration of spurious chaos due to insufficient spatial resolution is shown at $Gr = 60,000$ with two solutions obtained with different 17×33 and 21×49 spatial resolutions (see also Curry et al.[39]). In Fig. 3 we plotted (for $M \times N = 17 \times 33$) the time histories of the streamfunction at the centerpoint of the cavity, $\psi(t)$, and the corresponding power spectrum density, together with the evolution of the system in a phase plane constructed with $\psi(t)$ and a delayed series $\psi(t + \Delta t)$. For $Gr = 60,000$ the results indicate a period doubling bifurcation: The power spectrum density of time series $\psi(t)$ emphasizes two peaks associated with frequencies $f = 27.26$ and $f/2$ and their harmonics. The representation $\psi(t + \Delta t)$ vs $\psi(t)$ (with a time difference $\Delta t = 2.10^{-2}$) confirms the occurrence of $f/2$ frequency behavior (P2). At $Gr = 70,000$, the time series, phase plane evolution, and power spectrum of the streamfunction solution at the midpoint obtained are then characteristic of a chaotic behavior. The solution obtained at $Gr = 60,000$, with a 21×49 resolution, is monoperiodic

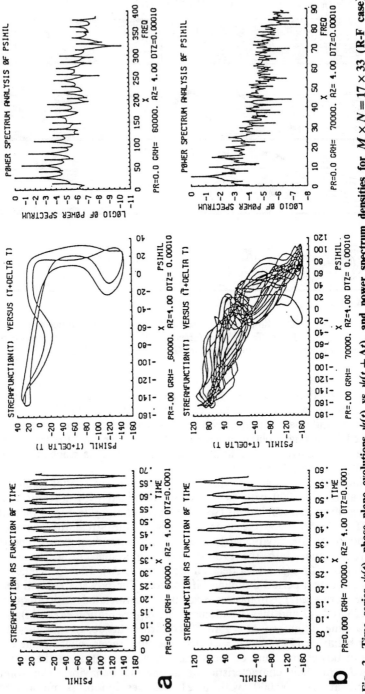

Fig. 3 Time series $\psi(t)$, phase plane evolutions $\psi(t)$ vs $\psi(t + \Delta t)$, and power spectrum densities for $M \times N = 17 \times 33$ (R-F case): a) $Gr = 60,000$ (P2); and b) $Gr = 70,000$ (NP).

Table 1   Fundamental frequencies at various *Gr* for different spectral $M \times N$ resolutions and definition of regimes

Gr	17 × 33		21 × 49		25 × 65	
13,100	——	S11	——	——	——	——
13,500	12.32	P1	——	——	——	——
15,000	13.23	P1	13.23	P1	——	——
20,000	15.93	P1	16.15	P1	——	——
30,000	21.10	P1	21.20	P1	——	——
40,000	26.10	P1	25.75	P1	——	——
60,000	27.26	P2	33.82	P1	——	——
70,000	24.40	NP	37.43	P1	——	——
80,000	——	——	40.65	P1	——	——
90,000	——	——	44.19	P1	——	——
125,000	——	——	52.08	P1	——	——
130,000	——	——	54.27	P1	——	——
132,000	——	——	53.40	P1	——	——
133,000	——	——	52.77	P1	——	——
135,000	——	——	53.55	P2	——	——
140,000	——	——	53.94	P2	53.76	P1
160,000	——	——	57.54	NP	59.6	P1
180,000	——	——	59.64	NP	63.91	P1
200,000	——	——	56.60	NP	63.33	P1
210,000	——	——	——	——	68.24	P1
Gr	25 × 65		31 × 65		33 × 65	
215,000	——	——	——	——	69.0	P3
220,000	70.58	NP	70.89	NP	70.06	NP

Periodicity is observed in phase plane projections: P1, periodic; P2, double periodic; P3...; NP, chaotic.

with $f = 33.82$. This shows that the period doubling obtained at $Gr = 60,000$ is spurious and results from an insufficient resolution. A range of flow motion was investigated at higher *Gr* with the 21 × 49 resolution (Table 1). The analysis using phase plane projections has shown periodic motion up to $Gr = 133,000$. The period-doubling bifurcation for this spatial resolution is observed at *Gr* slightly lower than $Gr = 135,000$. The periodic motion (P1) obtained for $Gr = 132,000$ is detailed by time series $\psi(t)$ and phase projection (Fig. 4). Also, we illustrate in this figure the double periodic motion (P2) at $Gr = 140,000$ and the chaotic motion (NP) at $Gr = 160,000$.

The solution is still periodic at $Gr = 210,000$ with $M \times N = 25 \times 65$ and becomes chaotic at $Gr = 220,000$ (Fig. 5). The spatial resolution was further refined, and simulations made at $Gr = 220,000$ with finer 31 × 65 and 33 × 65 resolutions exhibit a time behavior that suggests chaotic motion. The flow at $Gr = 220,000$ corresponds to the competition between two-cell and five-cell patterns (see Fig. 6). The detailed results have revealed that the chaotic motion already exists at

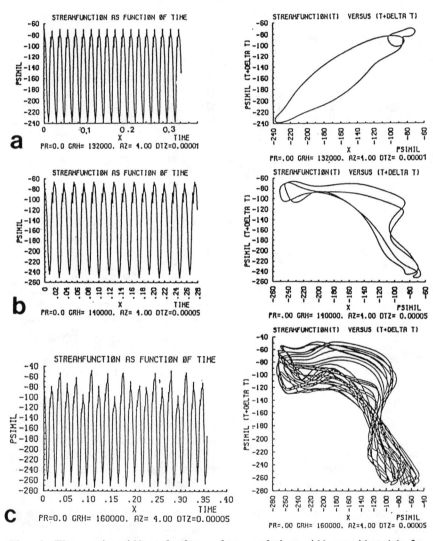

Fig. 4 Time series $\psi(t)$ and phase plane evolutions $\psi(t)$ vs $\psi(t + \Delta t)$ for $M \times N = 21 \times 49$ (R-F case): a) $Gr = 132,000$ (P1); b) $Gr = 140,000$ (P2); and c) $Gr = 160,000$ (NP).

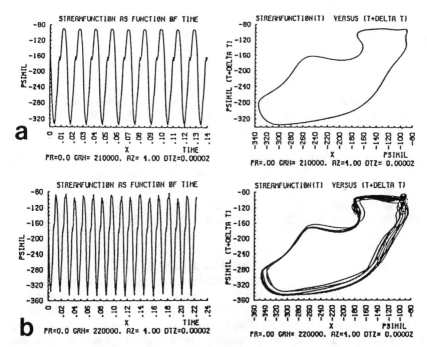

Fig. 5   Time series $\psi(t)$ and phase plane evolutions $\psi(t)$ vs $\psi(t + \Delta t)$ for $M \times N = 25 \times 65$ (R-F case): a) $Gr = 210{,}000$ (P1); and b) $Gr = 220{,}000$ (NP).

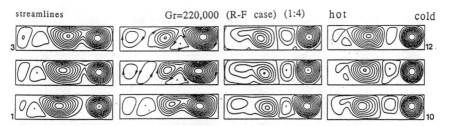

Fig. 6   Iso-ψ contours at different steps of time corresponding to the nonperiodic flow for $Pr = 0$ and for $Gr = 220{,}000$ (R-F case).

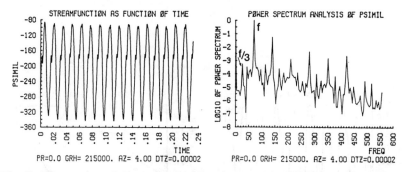

Fig. 7   Time series of the streamfunction ψ_{mid} at the midpoint of the cavity and corresponding power spectrum for $Pr = 0$ and for $Gr = 215,000$ (R-F case).

$Gr = 218,000$, and the transition via a subharmonics cascade was illustrated at $Gr = 215,000$ by a $f/3$ behavior (Fig. 7). A synthesis of numerical simulations is presented in Fig. 8, where the Grashof number at the onset of chaotic motion, Gr_{NP}, is displayed in terms of the $M \times N$ resolution. The results tend to indicate that saturation was obtained for Gr_{NP} when $M \times N > 1500$.

Multiple Flow Solution in a Rigid Wall Model (R-R)

A partial hysteresis cycle was originally determined in a former paper,[15] and we investigate more deeply here some aspects of the hysteresis diagram. The main results are summarized in Fig. 9, which shows the evolution of ψ_{max} (steady solution) and $\psi_{max,min} = \min_t(\psi_{max})$ (time-dependent solution) for the various regimes with respect to Gr. Several flow structures are found for the same Gr numbers, suggesting the existence of

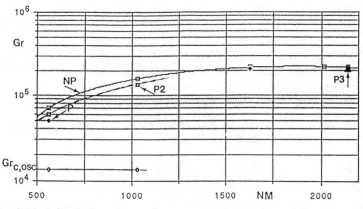

Fig. 8   Critical Grashof number at the onset of subharmonics and chaotic motions vs the number of degrees of freedom, $M \times N$ (R-F case, $Gr_{c,osc} = 13,500$; cf. Ref. 21).

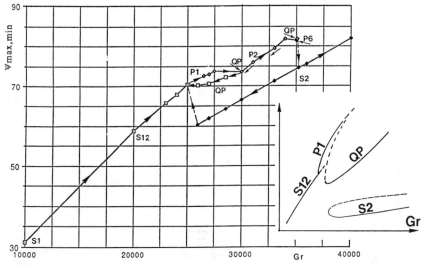

Fig. 9   Hysteresis cycle $\psi_{\text{max,min}}$ vs Gr with a sketch of the possible bifurcations explaining the coexistence of regimes (R-R case).

different branches or of inverse bifurcations. Two oscillators are located in the center of the cavity and near the side walls. Characteristic displays for ψ_{mid} and ψ_{quart} are given in Figs. 10 and 11.

The different flows are identified in Fig. 9 when $10,000 \leqslant Gr \leqslant 40,000$. As soon as $Gr \neq 0$, the basic solution corresponds to the Hadley circulation,[9] and the first steady instability superimposes to it with wavelength $\lambda_x = 2.34$ at $Gr \cong 8000$ for an infinite layer.[16] Its size decreases with increasing Gr, and this allows the rise of two secondary circulations near the side walls for $A = 4$. This S12 solution shown in Fig. 2a succeeds to the former S1-type solution. The steady solutions S1 and S12 are obtained for $Gr \leqslant 25,000$ when using as an initial condition either the rest or the asymptotic solution or the steady solution at lower Gr. The first Hopf bifurcation occurs for $Gr \cong 26,000$ to an oscillatory solution (P1), which keeps the same (center-symmetric) S12-type spatial structure. The oscillatory solution P1 is obtained for $26,000 \leqslant Gr \leqslant 30,000$ when starting from the same kind of initial conditions. The succeeding transitions to time-dependent solutions correspond to a break of symmetry. At Gr close to 30,000, the time-dependent solution is governed by two incommensurate frequencies (quasiperiodic QP solution). The QP solution (Figs. 2b and 2c) is obtained in particular when starting from rest: After a rapid transient (about one fundamental period) the solution is first monoperiodic (P1) for about 30 periods; then a QP solution establishes and is set after about 50 periods from the beginning. The time elapsed to get the QP solution increases with larger resolution.

A second periodic behavior different from P1 is obtained from the QP solution when $31,000 \leqslant Gr \leqslant 34,000$ (P2 solution): The solution oscillates with a frequency $f/2$ between a two- (S2) and three-cell (S12) structures.

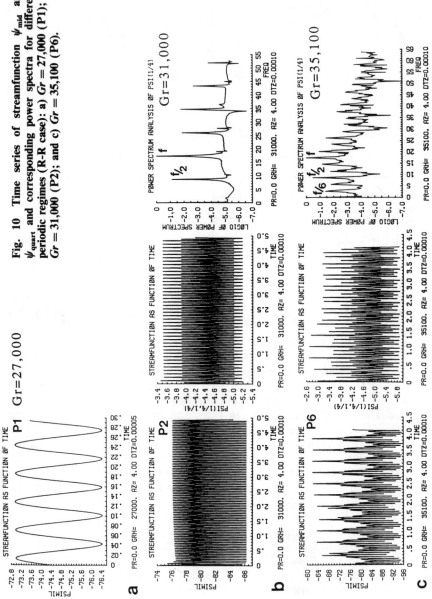

Fig. 10 Time series of streamfunction ψ_{mid} and ψ_{quart} and corresponding power spectra for different periodic regimes (R-R case): a) $Gr = 27,000$ (P1); b) $Gr = 31,000$ (P2); and c) $Gr = 35,100$ (P6).

This competition between different modes of convection suggests the occurrence of pairing behavior[15] as encountered in many situations when there is a breakdown of (odd or even) symmetry. When starting with the P2 solution at $Gr = 31,000$, a QP2 behavior is observed for $35,000 \leqslant Gr \leqslant 35,050$ as the time history of ψ_{mid} shows again a time modulation in addition to the fundamental f and $f/2$ oscillations (Fig. 11). For $Gr = 35,000$ and $35,050$, the secondary frequency f' observed in the power spectrum of ψ_{mid} is nearly $f/6.55$ and $f/6.29$, respectively. When Gr is increased to $35,100$, the solution then exhibits a P6 behavior corresponding to a phase locking of the frequencies, $f' \cong f/6$. When Gr is further increased from $35,100$ to $35,200$, the solution tends first to oscillate during about 30 fundamental periods, then changes suddenly into a different steady mode of convection (S2), which involves two distinct cells in the horizontal direction (Fig. 2d). This kind of solution is found to be stable up to at least $Gr = 50,000$.

When Gr is decreased, this S2 solution persists down to $Gr = 25,000$, for which a S12 solution still exists. The S12 solution was obtained after a long transient from an initial S2 condition at $Gr = 28,500$, $27,000$, and $26,000$. The S2 solution remains in a first stage, then changes suddenly to a S12 structure through an oscillatory damped process. The results obtained for $25,000 \leqslant Gr \leqslant 35,200$ prove the existence of an hysteresis effect. Different solutions can be obtained simultaneously in this range: on one side S2, on the other side either S12 or P1 or P2 or QP. The lines with arrows in Fig. 9 indicate the transition existing between the two kinds of solutions. Successive states QP → P6 → S2 are observed for $35,000 \leqslant Gr \leqslant 35,200$.

When Gr is decreased from $31,000$ (P2) to $30,000$, we obtain the same QP solution as that obtained when Gr is increased. When Gr is decreased from $30,000$ down to $27,000$, the QP solution persists suggesting a second hysteresis cycle. At $Gr = 26,000$, a complex QP-type behavior is obtained (Fig. 11). In the solution the major frequency is not the same at different regions of the cavity. When Gr is decreased from $26,000$ to $25,000$, the solution tends to become steady (S12) after a periodic transient.

Coming back to the QP regime obtained for $Gr = 30,000$, the power spectrum of ψ_{mid} exhibits a second frequency f' with an amplitude of one order of magnitude smaller than the fundamental one, f. This f' frequency is also observed in the power spectrum of ψ_{quart}, but with an amplitude of three orders of magnitude smaller than f. Moreover, it reveals the presence of two frequencies f_1 and f_2, which are one order of magnitude smaller than f. The relations found between these frequencies are $f_1 - f_2 = f'$ and $f_1 + f_2 = f$ (cf. Ref. 15) We note that f_1 and f_2 are close to $f/2$, and the transition from a QP to a P2 solution suggests the occurrence of a strong resonance phenomenon. Then, as Gr is increasing, $f' \to 0$; thus, f_1 and f_2 lock to $f/2$, and the resulting flow is periodic P2.

Recently, with nonlinear stability analysis, Winters and Jack[40] have found the disconnected S2 branch that arises at a threshold very close to the Hopf bifurcation S12 → P1 and that confirms the coexistence of the S2

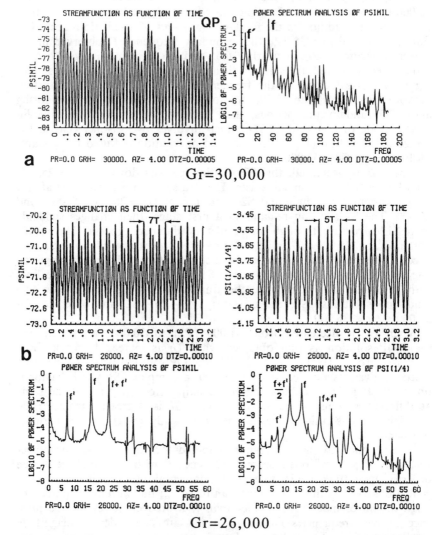

Fig. 11 Time series of streamfunction ψ_{mid} and ψ_{quart} and corresponding power spectra for different quasiperiodic regimes (R-R case): a) $Gr = 30,000$ (QP); b) $Gr = 26,000$ (QP); and c) $Gr = 35,000$ (QP2).

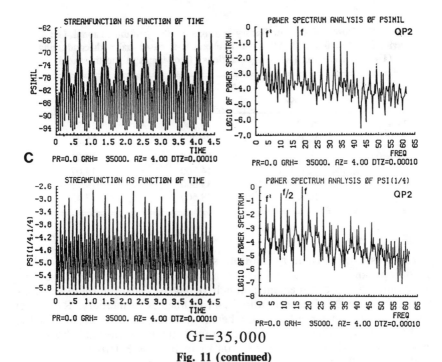

Gr=35,000

Fig. 11 (continued)

and P1 regimes. In the sketch of Fig. 9 we represent these results and also suggest an inverse bifurcation from the centersymmetric periodic flow P1 to an asymmetric quasiperiodic regime QP in order to explain the second hysteresis.

Analysis and Discussion

The analysis and the discussion are developed following the three points: 1) the steady states and the transition to an oscillatory motion, 2) the succeeding transitions derived from an oscillatory state when Gr is increased, and 3) the particular behavior when Gr is decreased.

Flow Regimes at Low Values of Gr

The two-dimensional numerical solutions have been obtained for several (dynamic and thermal) boundary conditions and for $Pr = 0$ and 0.015. The steady flow patterns are shown in Fig. 1 for $Pr = 0$ and $Gr = 117$, 3124, and 10,000 in the R-F case. At low Gr, there is one roll (Fig. 1a), and when Gr increases, two rolls are observed (Fig. 1c). In the R-R case the structure of the flow shows one roll at very low Gr and three rolls at higher values (see $Gr = 25,000$ in Fig. 2a). The solutions obtained for $Pr = 0.015$ in the R-R and R-F cases have shown very similar structures.

Table 2   Onset of the oscillatory motions in the two-dimensional (1 : 4) cavity for $Pr = 0$ and 0.015 in the R-F and R-R cases and with conducting (C) and insulating (A) wall conditions (frequencies at the threshold)

	Case	Onset of oscillatory motion	Frequency
$Pr = 0$	R-F	13,500	12.32
	R-R	26,000	16.31
$Pr = 0.015$	R-F-A	19,000	14.91
	R-F-C	14,700	13.07
	R-R-A	33,500	19.72
	R-R-C	28,500	17.45

The onset of oscillatory motion slightly depends on the boundary conditions and on Pr as it can be seen in Table 2. The effect of the three-dimensional confinement is much stronger. Thus, the onset of oscillations in a cavity of aspect ratio 1 : 4 : 1 (filled with mercury, R-R case) has been found at $Gr = 199,000$, whereas it occurs at $Gr = 42,300$ for the 1 : 4 : 2 cavity.[17] Details of the computed three-dimensional flows are given in Figs. 12 and 13 for $Pr = 0.015$ in both R-F ($Gr = 60,000$) and R-R ($Gr = 150,000$) cases (see Ref. 23). Pathlines and velocity fields are projected in different planes for these values of Gr. The closeness of the lateral walls ($Az = 1$) has a strong stabilizing effect on the flow, which does not show any major instability such as is observed in the corresponding two-dimensional solutions.

Second Transition
   As Gr increases, the frequency of the oscillations increases until a second transition occurs. In the R-F case the oscillatory flow develops into a subharmonics cascade.[21] The spectrum of the streamfunction, ψ_{mid} (at the

Table 3   Second transition of the time-dependent motions in the two-dimensional (1 : 4) cavity for $Pr = 0$ and 0.015 in the R-F and R-R cases and with conducting (C) wall conditions (types of the solutions and frequencies at the transition)

	Case	Gr	Type	Frequency
$Pr = 0$	R-F	215,000	P3	69.0
	R-R	29,500	QP	17.45
$Pr = 0.015$	R-R-C	35,000	QP	19.85

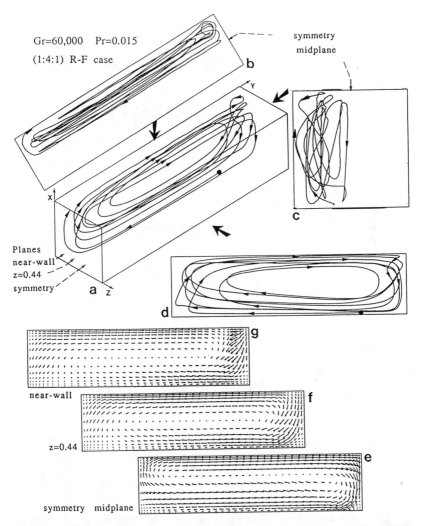

Fig. 12   Three-dimensional flow patterns in the cavity (1 : 4 : 1) for _Gr_ = 60,000, _Pr_ = 0.015, and the R-F condition: a) particle path; b) projections in horizontal plane; c) projections in vertical transversal plane; and d) projections in longitudinal plane. Velocity fields in three vertical planes: e) the symmetry plane; f) the intermediate (_z_ = 0.44) plane; and g) the near-wall lateral plane. The maximum velocities in the planes are e) 0.932, f) 0.708 and g) 0.657.

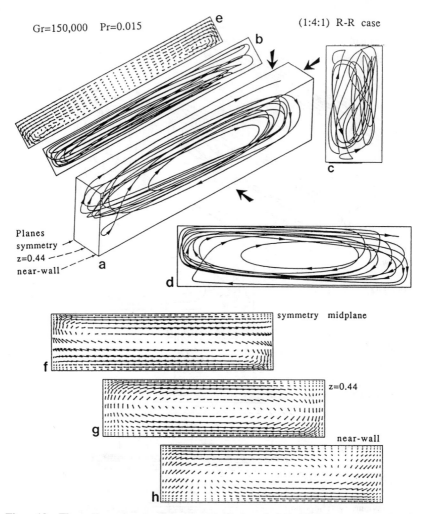

Fig. 13 Three-dimensional flow patterns in the half-cavity (1 : 4 : 1) for $Gr = 150{,}000$, $Pr = 0.015$, and the R-R condition: a) particle path; b) projections in horizontal plane; c) projections in vertical transversal plane; d) projections in longitudinal plane. Velocity fields in one horizontal plane and three vertical planes: e) the horizontal midplane; f) symmetry plane; g) intermediate ($z = 0.44$) plane; and h) near-wall lateral plane. The maximum velocities in the planes are e) 0.363, f) 0.719, g) 0.698, and h) 0.515.

Table 4 Characteristics of the solutions in the R-R case for
$Pr = 0$ when the Gr number is increased
(frequencies and type of solution)

$Gr\uparrow$	f	Subharmonics	f'	Type
25,000	Steady	——	——	S12
26,500	17.12	——	——	P1
27,000	17.34	——	——	P1
27,500	17.6	——	——	P1
30.000*	17.46	——	2.54	QP
31,000	17.33	$f/2$	——	P2
33,000	17.48	$f/2$	——	P2
33,500	17.56	$f/2$	——	P2
34,000*	17.63	$f/2$	——	P2
35,000*	17.35	$f/2$	2.65	QP2
35,050	17.36	$f/2$	2.76	QP2
35,000*	17.47	$f/2$ and $f/6$	——	P6
35,200	Steady	——	——	S2

For Gr^* see time histories and power spectra in Figs. 7 and 8.

midpoint of the cavity), for $Gr = 215,000$ shows peaks for f, $f/3$, and $2f/3$ (see Fig. 7). The motion is chaotic for $Gr \geqslant 218,000$ and the iso-ψ contours for $Gr = 220,000$ are displayed in Fig. 6 at different steps of time. No evidence of hysteresis has been found in this case.

In the R-R case a quasiperiodic motion (denoted QP in Table 3) is found for $Gr \geqslant 30,000$ (Figs. 2b and 2c). When Gr increases (see Table 4) the second frequency goes to zero $f' \rightarrow 0$, and there is a locking phase of the frequencies: $(f + f')/2$ and $(f - f')/2 \rightarrow f/2$ at $Gr = 31,000$. This $f/2$ peak is visible in the power spectrum of the streamfunction ψ_{quart} (near a corner of the cavity at $x = 1/4$ and $y = A/4$) for $Gr = 34,000$ and up to $Gr = 35,100$ during the P2 $\rightarrow$ QP2 $\rightarrow$ P6 transition. At $Gr = 35,200$, a steady regime involving two corotating rolls (denoted S2) is obtained and persists up to $Gr = 50,000$ (at least).

Table 5 Characteristics of the solutions in the R-R case for
$Pr = 0$ when the Gr is decreased (frequencies and type of solution)

$Gr\downarrow$	f	Subharmonics	f'	Type
30,000*	17.46	——	2.54	QP
28,500	16.98	——	3.92	QP
27,000	16.50	——	6.42	QP
26,000*	15.99	——	6.93	QP
25,000	Steady	——	——	S12

For Gr^* see time histories, power spectra, and phase diagrams in Figs. 7 and 9.

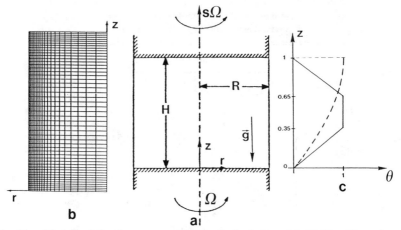

Fig. 14   Modelized floating-zone problems: a) physical model; b) 21 × 51 mesh; and
c) monotonic and symmetric temperature profiles.

Hysteresis in the R-R Case
   Decreasing Gr from $Gr = 35,200$, the two-roll steady S2 solution (Fig.
2d) is stable down to $Gr = 25,000$ (see Ref. 15). The QP solution at
$Gr = 30,000$ was also used as initial condition to compute the solution for
$Gr = 28,500, 27,000$, and $26,000$. In all of these cases the QP state persisted.[41] The details of this second hysteresis are summarized in Table 5.
Concerning the complex QP solution at $Gr = 26,000$, the ratio of the major
frequencies in the flow, f (ψ_{mid}) and $(f + f')/2$ (ψ_{quart}), is very close to a
rational number, then $f/f' \cong 7/3$. In order to study the QP solution for
$Gr = 26,000$, two different initial QP conditions were considered at
$Gr = 30,000$ and $27,000$. In both cases a QP solution was obtained. This
QP solution shows different frequencies in two points of the cavity.

Surface-Tension-, Buoyancy-, and Rotation-Driven Convection in a Floating-Zone Model

Surface-Tension-Driven Convection
   Under a zero-gravity condition and with the monotonic temperature
profile, the flow is driven by the surface tension, and the thermocapillary
convection develops from the hot point toward the cold one along the
interface (Fig. 14). The solutions have been computed for $Pr = 0.1$ and for
two values of the aspect ratio, $A = 0.5$ and $A = 2$ (Fig. 15). For $A = 0.5$ the
shape of the free boundary is very little modified (Figs. 15a and 15b):
When Re_M is increased from 10 to 1000, the distortion effect is limited from
1 to 3% (with respect to the depth of the layer). When the interface
deformation is not allowed, the convection cell similarly moves toward the

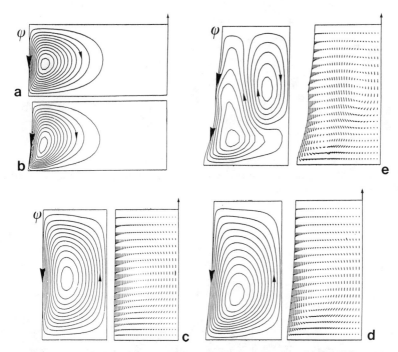

Fig. 15   Shaped liquid-gas interface submitted to surface-tension-driven convection in the floating-zone model for $Pr = 0.1$ and under zero-gravity condition. Streamline patterns (ψ) and velocity fields. Aspect ratios, Reynolds-Marangoni and capillary numbers: $A = 0.5$—a) $Re_M = 10$, b) $Re_M = 1000$; $A = 2.0$—c) $Re_M = 10$, $Ca = 0.02$, d), $Re_M = 160$, $Ca = 0.085$, and e) $Re_M = 640$, $Ca = 0.17$.

cold corner for this range of Re_M, but with strong discrepancies on the corresponding velocity magnitudes (see Ref. 42). For $A = 2$ the shape of the free boundary depends more strongly on Re_M and Ca. The distortion rate increases from 10% at $Re_M = 160$ to 20% at $Re_M = 640$ (Figs. 15c–15e). The magnitude of the velocity is increased by 30% between $Re_M = 160$ and 640, where two counter-rotating vortices appear. Near the cold wall the free boundary rises, whereas it drops near the hot upper corner and a slight thickening of the layer is exhibited at middistance between the two walls.

Buoyancy- and Rotation-Driven Convection

In this subsection the planar stress-free model is considered at the interface liquid-gas for $Pr = 0.015$ and with the symmetric temperature gradient (Fig. 14c). We focus the results on some effects of Gr and Re_Ω. In a terrestrial environment the convection induced by buoyancy shows small distortions of the streamlines up to $Gr = 10^6$ (Fig. 16). The changes are more important in the isotherm and pressure patterns. In particular, the

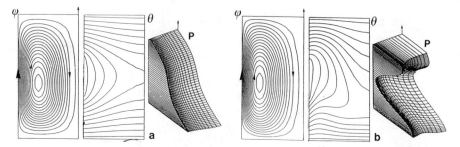

Fig. 16   Effect of Grashof number under terrestrial conditions for $Pr = 0.015$ and $A = 2$. Streamline (ψ) and isotherm (θ) patterns and pressure profiles at a) $Gr = 10^4$ and b) $Gr = 10^6$.

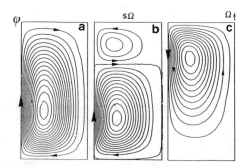

Fig. 17   Combined effects of buoyancy and rotation. Streamlines at $Gr = 100$, $Pr = 0.015$, and $Re_\Omega = 10$ for three values of the rotation rate: a) $s = 0$; b) $s = -1$; and c) $s = -5$.

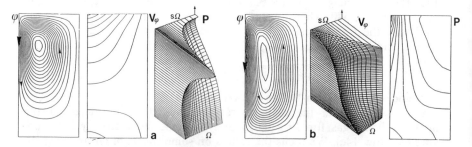

Fig. 18   Combined effects of buoyancy and rotation for $Pr = 0.015$ and $s = -5$. Streamline patterns (ψ), patterns and profiles of the azimuthal velocity (v_ϕ), and of the pressure (p) at a) $Gr = 100$ and $Re_\Omega = 10$; and b) $Gr = 1600$ and $Re_\Omega = 40$.

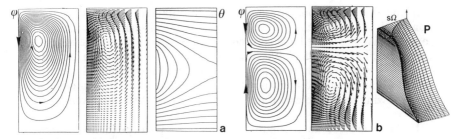

Fig. 19   Rotation-driven convection under zero-gravity and gravity conditions. Streamline patterns (ψ) and velocity fields, isotherm pattern (θ), and pressure profile (p) for $Re_{s\Omega} = 10$ ($Re_\Omega = 0$): a) $Gr = 0$ and b) $Gr = 100$.

pressure profile at $Gr = 10^6$ exhibits two overpressures near the rigid walls and an underpressure at midheight, suggesting possible break of the bridge under real conditions.

The influence on the streamline pattern of the acceleration due to the differential rotation of the rods is shown in Fig. 17 for $Gr = 100$, $Re_\Omega = 10$ and when the rotation rate is varied from $s = 0$ to -5. The solutions show that the basic buoyancy-driven cell is reduced by the rotation of the upper wall, which generates a counter-rotating one. The streamline patterns are shown in Fig. 18, together with the azimuthal velocity and pressure patterns and profiles for $s = -5$ and for $Gr = 100$ and $Re_\Omega = 10$, and for $Gr = 1600$ and $Re_\Omega = 40$. When Re_Ω and Gr are increased (as $Re_\Omega = Gr^{\frac{1}{2}}$), the cell is stretched along the free surface while the azimuthal velocity field remains dominated by the diffusion and the pressure profile then exhibits an overpressure near the top corner under the effect of the (centrifugal) flow driven by the rotation of the upper rod. The possible inversion of the cell flow is also illustrated under zero-gravity and gravity conditions when only the upper rod is rotating at $Re_\Omega = 10$ (Fig. 19). The pressure profile in Fig. 19b also shows that the rotation of the rods tends to reduce the pressure variation along the free surface (compare to the pressure profile at zero rotation in Fig. 16a). These results exhibit behaviors similar to the ones predicted theoretically by Harriot[32] under zero-gravity conditions.

Conclusions

The study of convective motion in the fluid phase during crystal growth of liquid metals is quite a difficult task. From the experimental point of view only temperature measurements have been reported. The nontransparency of the liquid phase preserves a direct observation of the motion inside the fluid. On the other hand, the numerical computation in such systems requires the use of simplifying assumptions. Considerable physics insight can be obtained from studying simple models and from the progressive inclusion of more realistic phenomena.

In the Bridgman configurations we have obtained time-dependent solutions with a two-dimensional model. The detailed characterization of the route to chaos requires large computing times especially because of the high spatial resolution needed. We have not made any attempt to provide an exhaustive study of the dynamic behavior of our systems. Some of our results obtained with direct numerical computation can be compared with the following available experimental and theoretical results given by other authors:

1) The onset of an oscillatory motion and its dependence on several parameters (such as the lateral confinement) was reported by Hurle et al.[13] In the particular R-R case for aspect ratio $1:4:2$ and for $Pr = 0.015$, Hung and Andereck[18] measured the frequency. At $Gr = 39,000$, the measured frequency was $f = 27.06$, which was relatively close to our computed value $f = 22.42$ obtained in the two-dimensional case for $Gr = 40,000$.

2) Some dynamic behaviors that we have found in our simulations have also been observed experimentally. Thus, Hung and Andereck[18] claimed that the transition to chaos via subharmonic instabilities seems to be characteristic for liquid metals (see also Refs. 5 and 43).

3) Specifically, for a R-R cavity, Hart and Pratte[17] found some quasiperiodic states showing period doubling in the center of the cavity, which was not observed at other locations (compare with our Fig. 10).

4) Another dynamic feature of the R-R case is the hysteresis cycle. For instance at $Gr = 27,000$ there are three solutions depending on the initial conditions: steady S2, oscillatory P1, or quasiperiodic QP. No evidence of hysteretic effects has been reported in experiments. However, nonlinear stability analysis has recently confirmed the coexistence of two branches, steady and periodic ones, both of them close to the Hopf bifurcation.[40]

The results obtained in the floating-zone model are limited to steady regimes but include other phenomena such as surface-tension-driven convection, rotation, and deformation of the interface. The preliminary results show the potential capability of the numerical methods to enlighten the study of coupled mechanisms in crystal growth systems.

Acknowledgments
Works were performed jointly with R. Peyret (University of Nice), R. L. Sani (University of Colorado), and A. Chaouche (I.M.F.M.). A number of results were obtained in the workshop on numerical modelings for low-Pr convection organized by B. Roux (I.M.F.M.). The authors acknowledge Dr. M. A. Rubio (U.N.E.D.-Madrid) for fruitful discussions. Collaborations with the Center for Low-Gravity Fluid Mechanics (University of Colorado-Boulder). The computations were carried out on Cray 2 computer with the support of the C.C.V.R. Support was also provided by C.N.R.S., M.R.E.S., D.R.E.T., C.N.E.S., and Conseil Régional P.A.C.A., in France, and from D.G.I.C.Y.T. in Spain.

References

[1] Pimputkar, M., and Ostrach, S., "Convective Effects in Crystals Grown from Melts," *Journal of Crystal Growth*, Vol. 55, 1981, pp. 614–646.

[2] Langlois, W. E., "Buoyancy Driven Flows in Crystal-Growth Melts," *Annual Review of Fluid Mechanics*, Vol. 17, 1985, pp. 191–215.

[3] Ben Hadid, H., Roux, B., Randriamampianina, A., Crespo del Arco, E., and Bontoux, P., "Onset of Oscillatory Convection in Horizontal Layers of Low-Prandtl-Number Melts," *Physicochemical Hydrodynamics Interfacial Phenomena*, edited by M. G. Velarde, Vol. 174, NATO-ASI Series B, Plenum, New York, 1988, pp. 997–1028.

[4] Azouni, M. A., "Time-Dependent Natural Convection in Crystal Growth Systems," *Phys. Chem. Hydrodyn.*, Vol. 2, 1981, pp. 295–309.

[5] Favier, J. J., Rouzaud, A., and Comera, J., "Influence of Various Hydrodynamics Regimes in a Melt on a Solidification Interface," *Rev. Phys. Appl.*, Vol. 22, Aug. 1987, pp. 713–718.

[6] Ettouney, H. M., and Brown, R. A., "Finite-Element Methods for Steady Solidification Problems," *Journal of Computational Physics*, Vol. 49, Jan. 1983, pp. 118–150.

[7] Crochet, M. J., Geyling, F. T., and Van Schaftingen, J. J., "Numerical Solution of the Horizontal Bridgman Growth of a Gallium Arsenide Crystal," *Journal of Crystal Growth*, Vol. 65, 1983, pp. 166–172.

[8] Adornato, P. M., and Brown, R. A., "Petrov-Galerkin Methods for Natural Convection in Directional Solidification of Binary Alloys," *International Journal for Numerical Methods in Fluids*, Vol. 7, 1987, pp. 761–791.

[9] Hart, J. E., "Stability of Thin Non-Rotating Hadley Circulations," *Journal of the Atmospheric Sciences*, Vol. 29, 1972, pp. 687–697.

[10] Hart, J. E., "A Note on the Stability of Low Prandtl Number Hadley Circulation," *Journal of Fluid Mechanics*, Vol. 132, July 1983, pp. 271–281.

[11] Gill, A. E., "A Theory of Thermal Oscillations in Liquid Metals," *Journal of Fluid Mechanics*, Vol. 64, July 1974, pp. 577–588.

[12] Laure, P., and Roux, B., "Synthèse des Résultats Obtenus par l'Etude de Stabilité des Mouvements de Convection dans une Cavité Horizontale de Grande Extension," *Comptes Rendus de l'Académie des Sciences*, Paris, Vol. 305, Series II, 1987, pp. 1137–1143.

[13] Hurle, D. T., Jakeman, E., and Johnson, C. P., "Convective Temperature Oscillations in Molten Gallium," *Journal of Fluid Mechanics*, Vol. 64, July 1974, pp. 565–576.

[14] Winters, K. H., "Oscillatory Convection in Liquid Metals in a Horizontal Temperature Gradient," *International Journal for Numerical Methods in Fluids*, Vol. 25, 1988, pp. 401–414.

[15] Pulicani, J. P., Crespo del Arco, E., Randriamampianina, A., Bontoux, P., and Peyrets, R., "Spectral Simulations of Oscillatory Convection at Low Prandtl Number," *International Journal for Numerical Methods in Fluids*, Vol. 10, April 1989, pp. 481–517.

[16] Roux, B., (ed), "Numerical Simulation of Oscillatory Convection in Low Pr Fluids," *Notes Num. Fluid Mechanics*, Vol. 27, Vieweg, Braunschweig, 1989.

[17] Hart, J.E., and Pratte, J. M., "A Laboratory Study of Oscillations in Differentially Heated Layers of Mercury," 1989, pp. 329–337.

[18] Hung, M. C., and Andereck, C. D., "Subharmonic Transitions in Convection in a Moderately Shallow Cavity," 1989, pp. 338–343.

[19] Wang, T. M., Korpela, S. A., Hung, M. C., and Andereck, C. D., "Convection in a Shallow Cavity," *Notes Num. Fluid Mechanics*, Vol. 27, Vieweg, Braunschweig, 1989, pp. 344–353.

[20] Crespo del Arco, E., "Contribucion al Estudio de Inestabilidades Termohidro-dinamicas en Fluidos Newtonianos y Viscoelasticos en Capas Horizontales y Cilindros, Thesis, UNED, Madrid, Spain, July 1987.

[21] Crespo del Arco, E., Randriamampianina, A., and Bontoux, P., "Two-Dimensional Simulation of Time-Dependent Convective Flow of a $Pr \to 0$ Fluid. A Period Doubling Transition to Chaos," *Synergetics, Order & Chaos*, edited by M. G. Velarde, World Scientific, London, 1988, pp. 244–257.

[22] Randriamampianina, A., Crespo del Arco, E., Fontaine, J. P., and Bontoux, P., "Spectral Methods for Two-Dimensional Time–Dependent $Pr \to 0$ Convection," *Notes Num. Fluid Mechanics*, Vol. 27, Vieweg, Braunschweig, 1989, pp. 244–255.

[23] Extremet, G. P., Fontaine, J. P., Chaouche, A., and Sani, R. L., "Two- and Three-Dimensional Finite Element Simulations for Buoyancy-Driven Convection Inside a Confined $Pr = 0.015$ Liquid Layer."

[24] Napolitano, L. G., Monti, R., and Russo, G., "Some Results of the Marangoni Free Convection Experiments," *5th European Symposium on Material Sciences Under Microgravity*, Schloss Elmau, ESA-SP 222, 1984.

[25] Martinez, L., "Liquid Bridge Modeling of Floating Zone Processing," *Physicochemical Hydrodynamics Interfacial Phenomena*, Vol. 174, edited by M. G. Velarde, NATO ASI Series B, Plenum, New York, 1988.

[26] Schwabe, D., "Surface-Tension-Driven Flow in Crystal Growth Melts," *Crystals* 11, Springer, Heidelberg, 1988.

[27] Schwabe, D., Scharmann, A., and Preisser, F., "Experiments on Surface Tension Driven Flow in Floating Zone Melting," *Journal of Crystal Growth*, Vol. 43, 1978, pp. 305–312.

[28] Schwabe, D., "Marangoni Effects in Crystal Growth Melts," *Physico-Chemical Hydrodynamics*, Vol. 2, 1981, pp. 263–280.

[29] Schwabe, D., and Scharmann, A., "Measurements of the Critical Marangoni Number of the Laminar Oscillatory Transition of the Thermocapillary Convection in Floating Zones." *5th European Symposium on Material Sciences under Microgravity*, Schloss Elmau, ESA-SP 222, 1984.

[30] Chun, C. H., and Schwabe, D., "Marangoni Convection in Floating Zones," *Convective Transport and Instability Phenomenon*, edited by J. Zierep and H. Oertel, Braun, Karlshruhe, 1982, pp. 297–317.

[31] Saghir, M. Z., "A Study of the Marangoni Convection on the Germanium Floating Zone," *Low Gravity Sciences*, Vol. 67, edited by J. N. Koster, Sciences and Technology Series, Advances in the Astronautical Sciences, American Astronautical Society, 1986, pp. 77–100.

[32] Harriot, G. M., "Laminar Mixing in a Small Floating Zone," *Low Gravity Sciences*, edited by J. N. Koster, Sciences and Technology Series, Adv. Astr. Sciences, Vol. 67, American Astronautical Society, 1986, pp. 77–100.

[33] Ochiai, J., et al., "Experimental Study on Marangoni Convection," *5th European Symposium on Material Sciences under Microgravity*, Schloss Elmau, ESA-SP 222, 1984, pp. 291–295.

[34] Fu, B. I., and Ostrach, S., "Numerical Solutions of Thermocapillary Flows in Floating Zones," *Transport Phenomena in Materials Processings*, American Society of Mechanical Engineers, PED-Vol. 10, HTD-Vol. 29, 1983.

[35] Zebib, A., Homsy, G. M., and Meiburg, E., "High Marangoni Number Convection in a Square Cavity," *Physics of Fluids*, Vol. 28, 1985, p. 3467.

[36] Cuvelier, C., and Driessen, J. M., "Thermocapillary Free Boundaries in Crystal Growth," *Journal of Fluid Mechanics*, Vol. 169, Aug. 1986, pp. 1–26.

[37] Crespo del Arco, E., Peyret, R., Pulicani, J. P., and Randriamampianina, A., "Spectral Calculations of Oscillatory Convective Flows," *Finite Element Analysis*,

edited by T. J. Chung and G. R. Karr, UAH Press, Huntsville, AL, 1989, pp. 1464–1472.

[38]Engelman, M. S., *FIDAP Users Manual*, Fluid Dynamics, Evanston, IL, 1986.

[39]Curry, J. H., Herring, J. R., Loncaric, J., and Orszag, S. A., "Order and Disorder in Two- and Three-Dimensional Bénard Convection," *Journal of Fluid Mechanics*, Vol. 147, Oct. 1984, pp. 1–38.

[40]Winters, K. H., and Jack, R. O., "Anomalous Convection at Low Prandtl Number," *Comm. Applied Num. Meth.* (to be published).

[41]Crespo del Arco, E., Pulicani, J. P., and Randriamampianina, A., "Complex Multiple Solutions and Hysteresis Cycles near the Onset of Oscillatory Convection in a $Pr = 0$ Liquid Submitted to a Horizontal Temperature Gradient," *Comptes Rendus de l'Académie des Sciences*, Paris, Serie II, Vol. 309, 1989, pp. 1869–1876.

[42]Roux, A., "Etude Numérique des Mouvements Thermoconvectifs dans des Systèmes présentant une Surface Libre par une Méthode aux Eléments Finis," *Rapport de D.E.A.*, I.M.F., Marseille, France, 1989.

[43]Gollub, J. P., and Benson, S. V., "Many Routes to Turbulent Convection," *Journal of Fluid Mechanics*, Vol. 100, Oct. 1980, pp. 449–470.

Fluid Dynamics and Solidification
of Levitated Drops and Shells

E. H. Trinh*

California Institute of Technology, Pasadena, California

Introduction

THE reduction in magnitude of the gravitational acceleration allows the use of benign sample levitation methods to study the capillarity-dominated phenomena affecting the dynamic and static behavior of three-dimensional liquid-gas and liquid-liquid interfaces. In addition, the virtual elimination of buoyancy-driven convection unmasks phenomena controlled by thermocapillary effects, which have been shown to dominate internal fluid flows under microgravity conditions. Finally, the avoidance of physical contact between the observed material sample and container walls facilitates control of contamination and reduces the probability for heterogeneous nucleation in an undercooled melt. Such a reduction usually implies an enhanced capability for attaining deep undercooling. Scientific interest in the deeply undercooled liquid state includes, but is not restricted to, the determination of thermophysical properties, the partial control of cooling rate during solidification, the selection of possible metastable solid structures upon crystallization of alloys, and the study of the subtle effects influencing nucleation of the solid phase (such as small-amplitude mechanical perturbation and flows).

In this chapter the fluid dynamic investigation of simple free liquid drops will first be described through ground-based and low-gravity experimental results. The behavior of compound drops and liquid shells, as described in recent theoretical and experimental studies, will then be discussed. The experimental investigations using both levitation devices and drop tubes will be considered in the case of 1-g laboratory investigations in a discussion highlighting the advantages and drawbacks of both techniques.

---

*Project Scientist, Jet Propulsion Laboratory.

Recent results obtained in quasistatic undercooling and physical properties measurement experiments will be discussed for both ground-based and low-gravity investigations.

Promising areas of investigation using microgravity flight instrumentation will be mentioned in the areas of fluid dynamics and materials science where single droplets and shells are the systems of interest. Suggestions regarding the important technical capabilities required for carrying out rigorous experiments and the needed diagnostic techniques and instrumentation will be provided in the conclusion.

Fluid Dynamics of Drops and Liquid Shells

Levitation Studies of Drops and Shells

Oscillatory Dynamics of Simple Drops

The fascination with the perfection and apparent simplicity of a liquid drop probably arose even before the pioneering efforts of Rayleigh[1] and Lamb[2] who sought to ellucidate scientifically the details of the workings of surface forces. The liquid drop may serve as a first-order model for natural systems as disparate as planets and atomic nuclei, but it also bears the scrutiny of specialists in meteorology, chemical engineering, and physics. A resurgence in interest has been recently induced by the ability to perform containerless experiments in low gravity using small forces for both position control and for probing of the free drop. Acoustic as well as electrostatic and electromagnetic techniques can be used to levitate and manipulate liquid samples held together by surface tension. Acoustic and ultrasonic methods have been developed the most extensively and have demonstrated some versatility in positioning, oscillating, and rotating liquid samples, particularly at room temperature. Ground-based experimental investigations of levitated droplets in a gas medium can easily be carried out when droplets range from 0.5 to 8 mm diam and when ultrasonic as well as electrostatic levitators are used.[3,4] In general, power levels required for 1-g levitation of samples are such that varying degrees of drop distortion always exist, and the effects of the levitation stresses are not negligible. The minimization of this interference is generally accomplished by reducing the sample size and investigating high-surface-tension liquids. The influence of a nonuniform distribution of surface charges on a droplet behavior has not yet been consistently studied, and the effects of electrostatic levitation on the sample dynamics have yet to be assessed.

Although the equilibrium shape of a free liquid sample under the sole action of surface tension is spherical, interesting phenomena are more often observed when secondary stresses act on the drop due to perturbations such as those induced by natural sources (e.g., gas flows, interparticle collision, and electrical charges) or nonnatural effects (e.g., forces used for levitation and positioning). Static distortion of nonrotating drops due to surface charges and distributed electric fields have been investigated both theoretically and experimentally.[5-7] New interest on the effects of acoustic radiation stresses on droplets has recently generated theoretical investigations of the shape of liquid samples when submitted to a variety of acoustic fields.[8,9] Preliminary experimental studies have confirmed that a good understanding of the linear range (small deformation from the spherical shape) was available, but that larger deviations from sphericity could not yet be reliably calculated.[10] This is illustrated by the results shown in Fig. 1, where experimental measurements have been obtained with a quasi-one-dimensional acoustic field and are compared with a one-dimensional analytical treatment.[8] The approximate agreement in the small deformation region has suggested the use of a noncontact method for the measurement of the surface tension of levitated drops[11] not involving any dynamic analysis of sample shape. Results for much larger deformation are also shown in Fig. 2.

The study of the stability of a droplet shape within a distributed stress field such as that found in an acoustic standing wave has only just begun and has been motivated by experimental observations revealing the onset of macroscopic shape oscillations for low-viscosity liquids when a threshold for the intensity level of the stress field is exceeded. The various modes of shape oscillations are excited in turn with the higher-lobed modes appearing at increasingly higher amplitude for the sound field. No rigorous

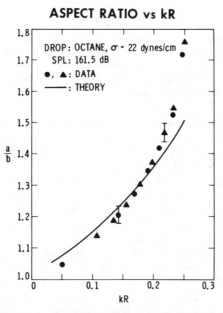

Fig. 1 Experimental data from the measurement of the deformation of levitated droplets in 1 g. The drops are oblate with a horizontal-to-vertical axial ratio denoted by a/b. The drop size is given in terms of the kR parameter, which is the circumference of a spherical drop of the same volume divided by the acoustic wavelength. In this particular case the wavelength was 1.6 cm. The levitation is achieved by a single-axis ultrasonic levitator.

Drop Shape vs. Size

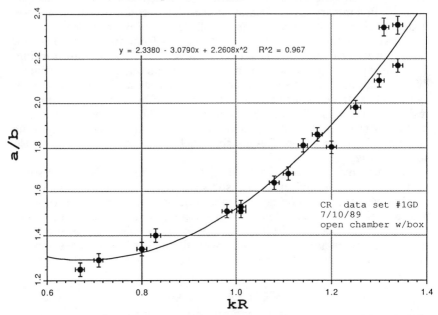

y = 2.3380 - 3.0790x + 2.2608x^2   R^2 = 0.967

CR data set #1GD
7/10/89
open chamber w/box

Fig. 2 Much more severe deformation is shown in this case for a higher acoustic pressure and larger droplets. No analytical predictions are available for such significant distortions in this particular approximate one-dimensional acoustic field. The acoustic wavelength is 1.2 in this case. The solid curve plotted through the data is obtained by a least-squares fit procedure.

understanding of this droplet-acoustic field interaction has yet been obtained. The theoretical treatment of this problem is complicated by the time-varying shape of the drop, which affects the boundary conditions as well as the acoustic stress distribution.

A more controlled study of the oscillatory dynamics of free drops is provided by time modulating the acoustic (or other appropriate) forces used for positioning to reveal the spectrum of resonant modes. According to the existing linear theories,[1,2,12,13] the value of the liquid surface tension can be directly calculated from the frequency of the normal modes if the volume of the spherical drop is known. However, both theoretical and experimental evidence show that this simple relationship breaks down as the amplitude of the shape oscillations increases.[14-17] Present evidence obtained in ground-based laboratories with drops immersed in an immiscible liquid host as well as liquids levitated in air indicate a marked soft nonlinearity where the resonance frequency of the fundamental (oblate-prolate) mode decreases with increasing oscillation amplitude. These results hold for both forced as well as free oscillations. However, these ground-based data must not be interpreted as final, since they refer to systems including two immiscible liquids or to significantly distorted droplets.

Measurement of the resonant frequencies of shape oscillation and of the free decay rate of such oscillations provides information on the value of both the surface tension and viscosity of a pure liquid. Application of this technique to melts at very high temperature has now become one of the goals of future microgravity flight experiments, although some preliminary results obtained in 1 g already exist in the case of electromagnetically levitated melts of ferrous alloys.[18] However, care must be exercised because of the nonlinear effects mentioned in the preceding paragraph and because of possible sample rotation. This latter phenomenon has been thoroughly investigated both analytically[19,20] and experimentally.[21] A significant increase in the value of the frequency of resonant modes of shape oscillations has been predicted and measured for increasingly higher rotation rates. In addition to this increase, a decoupling of the degenerate axisymmetric modes is induced by the drop rotation, thus yielding different modes with frequencies varying with the rotation rate. The interpretation of a Fourier transform spectrum of the drop oscillations will be much more difficult, and the extraction of surface tension values will require quantitative information on both oscillation amplitude as well as drop rotation velocity.

Another potential application of the dynamics of drop shape oscillations is the rheological study of interfacial phenomena including chemical surfactants. Recent experimental results[22] have demonstrated that the time dependence of the fundamental resonance frequency can be used to monitor the evolution of the concentration of contaminants at the liquid drop-host medium interface. The present study again involves two immiscible liquid systems having a decidedly intractable theoretical description. A simplified system involving a liquid drop immersed in a gas host will greatly reduce the complexity level of the theoretical treatment. The possibility thus that some much needed information can be gained on interfacial phenomena involving complex physical parameters such as the interfacial viscosity and elasticity. Experimental investigation of the interfacial characteristics of contacting drops with an intervening surfactant layer and of the mechanism of coalescence is realizable with current drop positioning and manipulation capabilities. Figure 3 displays photographs of the interface between two contacting drops held together by ultrasonic radiation pressure stresses prior to coalescence.

Rotational Dynamics of Levitated Charged and Uncharged Drops

The discovery by T. Wang of a controllable steady-state acoustic torque[22,23] has made it possible to drive a freely suspended drop into a solid-body rotation. Thus, Wang has proposed to experimentally investigate the long-standing problem of equilibrium shape of rotating liquid masses. An initial attempt had been carried out by Plateau at the turn of the century[24] and involved a droplet suspended in an immiscible liquid host by a flat plate mounted on a thin shaft. Various figures of equilibrium were obtained by Plateau, including a toroidal configuration. This experiment has been revisited in the 1970s, providing additional figures of equilibrium.[25] However, this approach introduced perturbations such as the

Fig. 3 Photograph of two ultrasonically levitated droplets of silicone oil immersed in distilled water containing a surfactant. The acoustic radiation pressure drives the two particles together, creating an axisymmetric flow in the interfacial film that can be resolved in the photographs. The coalescence induction mechanism has not been identified in this case.

outer liquid mass, the solid drop suspension system, and probably differential rotation. A more rigorous experiment can be carried out by acoustically and electrostatically levitating droplets in air and making use of the acoustic torque. Such a ground-based investigation will be described later, and the microgravity flight experimental results will be discussed in a later section.

The thrust of the experimental effort has been to provide data for comparison with the available theoretical treatments of Chandrasekhar[26] and Brown and Scriven.[27] Ground-based experimentation can provide both qualitative and quantitative information under conditions slightly different from those postulated by the theoretical analysis. Electrostatic levitation used in 1 g requires the presence of a net charge on the sample surface (although its magnitude can be a small fraction of the Rayleigh limit), and ultrasonic levitation induces a significant static distortion of ordinary liquid drops having a diameter greater than 1 mm. Furthermore, the interpretation of the experimental results obtained through acoustically induced rotation must be analyzed with an understanding of the effects of both the levitating as well as the torque-inducing acoustic and electrostatic stress fields. A rigorous experimental investigation in microgravity allowing the minimization of these secondary factors still appears as the ultimate test of the available theories.

Rhim and co-workers[28] have succeeded in measuring the critical bifurcation angular velocity at which the equilibrium shape symmetric with respect to the rotation axis becomes unstable, and transition to a nonaxisymmetric two-lobed shape takes place. A comparison with the analysis of Brown and Scriven has revealed a good agreement under the assumption of constant angular momentum despite the presence of electrical charges. A detailed examination of the drop shapes reveals minor effects due to gravity, but the sequence of evolution of the equilibrium shape of the rotating drop is qualitatively similar to that observed for uncharged drops in low gravity (results that will be described later) for a charge level that is small with respect to the Rayleigh limit. However, a qualitative disagreement is observed at the very last stage of fission, where a decrease in the drop deformation is predicted just before separation into two distinct drops. Only the two-lobed configuration has been observed thus far, in agreement with the low-gravity experimental results for uncharged drops. These results have been obtained by electrostatically levitating the charged drop and by rotating it through the action of acoustic torque operating at audio frequency.

Trinh and Leung[29] have recently initiated the ground-based experimental study of ultrasonically levitated uncharged drops having diameters in the millimeter range and have confirmed some of the observations on charged drops in 1 g and on uncharged drops in microgravity. The full effect of the Earth's gravitational field and the subsequent high ultrasonic pressure level required to levitate a millimeter-size droplet cause the droplet's equilibrium static shape to approach that of an oblate spheroid with an axial ratio of at least 1.1. Application of an acoustic torque has allowed the measurement of the rotation-induced droplet axisymmetric distortion as well as the determination of the onset of bifurcation into the nonaxisymmetric two-lobed configuration. Preliminary results confirm the predictions of Brown and Scriven[27] for the critical bifurcation angular velocity within the present experimental accuracy. The principal experimental uncertainty arises from the difficulty in obtaining accurate measurements of the rotation rate of a transparent spinning levitated droplet and of the surface tension of the droplet material. Additional experimental data obtained include the measurement of the change of the fundamental mode of shape oscillations with rotation rate in the axisymmetric drop shape region and the change in the critical bifurcation angular velocity with the static equilibrium shape of an initially nonrotating droplet. The comparison between these ground-based results with data obtained in microgravity will allow the rigorous separation of the true gravitational effects from those associated with the containerless positioning and manipulation techniques. Figure 4 reproduces some

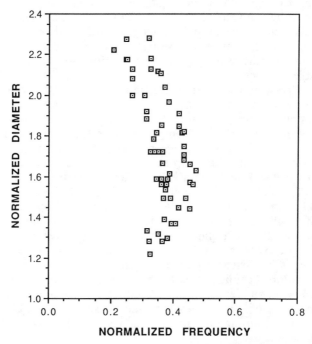

Fig. 4 Results of the measurement of the deformation of a rotating levitated droplet of glycerin as a function of the reduced rotation velocity. Along the vertical axis are plotted measured values for the ratio of the largest drop dimension to the resting circular diameter for nonrotating drops as seen along the rotation axis. Along the horizontal axis are values of the rotation rate divided by the theoretical fundamental frequency of shape oscillation. These data have been obtained with highly distorted drops in the oblate shape. Results of runs for several different drops with varying amount of oblate deformation have been plotted. Theoretical predictions for bifurcation reduced velocity specify 0.57 for the bifurcation velocity. The bifurcation velocity is obviously lowered by static drop distortion.

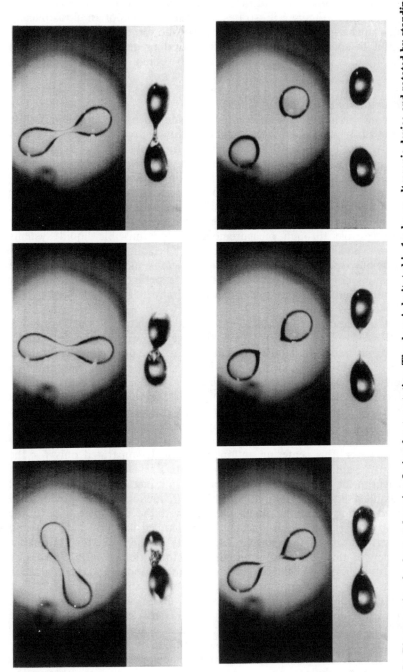

Fig. 5 Photographs of a drop undergoing fission due to rotation. The drop is levitated in 1 g by an ultrasonic device and rotated by standing waves at audio frequencies.

of the results obtained with ultrasonically levitated drops with acoustically induced rotation. Figure 5 illustrates the fission process of an ultrasonically levitated uncharged drop rotated by an acoustic torque. The behavior near the onset of splitting confirms the theoretical predictions of Brown and Scriven[27] and is in agreement with Spacelab results.

Compound Drops and Liquid Shells

A second immiscible liquid completely enclosed by the first one, or the presence of an undissolved gas bubble within a drop, presents a second fluid-fluid interface and further complicates the analysis of the dynamics of these fluid particles. A basic treatment of the compound drop problem has been authored by Saffren and co-workers,[30] who used the coupled oscillators approach to calculate the various modes of shape oscillations for different volume ratios of the two fluids. The theoretical predictions have been experimentally confirmed in the restricted region of the parameter space accessible through measurements with immiscible liquid systems using water shells immersed in silicone oil.[30] The splitting of the fundamental mode of shape oscillation into in-phase (bubble mode) and out-of-phase (sloshing mode) resonant modes has been experimentally confirmed for both charged and uncharged near-neutrally buoyant liquid shells immersed in a host liquid.

Levitation studies of liquid shells in air have been carried out both in 1 g as well as in the reduced-gravity environment during airplane parabolic flights[31,32] as well as on ballistic rocket flights.[33] The detailed observation of forced oscillations of thick and thin liquid shells (soap bubbles) through modulated acoustic radiation pressure in 1 g is hampered by the noncentered configuration due to gravity, but has revealed the detailed flow mechanisms within such oscillating systems. Observations carried out in low gravity have demonstrated the centering effects of shape oscillations for thick as well as thin shells and the decentering action of residual acceleration. However, this centering action has been demonstrated only for the case of large-amplitude oscillations. A definitive experimental evaluation of this effect in the absence of near-contacting interfaces is yet to be obtained for gas-filled liquid shells. The short-lived reduction of the gravity level achieved during parabolic flights has allowed the demonstration of the capability for experimental studies of the detailed interface behavior and the determination of the relative insensitivity of the oscillatory resonance frequency of thin shells to static shape deformation. Figure 6 provides snapshots of liquid shells undergoing shape oscillation in 1 g, centered thick and thin shells in low gravity, and rotating liquid shells in 1 g.

The controlled positioning of a simple liquid shell has allowed the experimental observation of other phenomena relevant to free surface behavior. Such an instance is the nonlinear characteristics of capillary waves on a thin liquid layer (such as the surface of a thin shell). Evidence of transition from linear small amplitude behavior to larger displacement, nonlinear and turbulent regimes has been gathered through the observation of ultrasonically induced capillary waves. A violent and chaotic motion of

Fig. 6 Photographs of liquid shells levitated in 1 g.

the whole liquid surface is observed due to the growth in amplitude of the capillary waves and eventually leads to a redistribution of the liquid mass in the shell and to a subsequent sharp transition to a quiescent state of the shell surface. This cycle periodically repeats as time elapses. Figure 7 shows photographs displaying a shell in a quiescent state as well as during excitation of high-amplitude capillary waves. The effect of gravity on such large-scale chaotic phenomena has not yet been clearly analyzed.

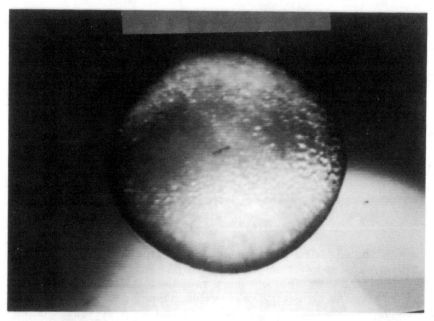

Fig. 7 Photographs of a thin liquid shell in a quiescent state and under the effect of large-amplitude capillary waves.

Microgravity Flight Experiments

Precursor microgravity experiments onboard rockets in ballistic trajectories have been carried out using audio frequency acoustic positioning systems operating at room temperature and ultrasonic systems operating at elevated temperatures.[34-36] These investigations have been designed more for flight hardware technology development efforts rather than for scientific experiments; consequently, the results have been at best qualitative when the apparatus did function nominally. The average duration of these flights has been on the order of 4 to 5 min, sufficient time for the deployment and the investigation of a single sample. The data thus obtained have allowed the refinement of the capabilities of Shuttle-based experiments that followed a short time later.

Both man-operated and fully automated experiments have been flown in the Space Shuttle to carry out drop physics and materials processing investigations. Room temperature drop dynamics results have been gathered during the Spacelab 3 flight of the Drop Dynamics Module[37,38] and the STS 61-C mission of the Triple Axis Acoustic Levitator. The former investigation has provided the first experimental results on the equilibrium shapes of acoustically rotated simple drops, which have indicated a slightly lower value for the observed critical angular velocity at bifurcation than theoretically predicted. A reflight of this investigation will provide the opportunity to confirm this result. The latter flight experiment (3AAL) has been designed to investigate the dynamics of oscillating and rotating liquid shells and the thermocapillary flows within a drop having been spot heated on the surface. Subnominal performance of the gas injector system has not allowed quantitative studies of the liquid shells dynamics, and failure of the camera system did not allow the performance of the second scientific objective.

New flight experiments have been proposed, and some have been accepted for a future Spacelab flight (USML-1). A new facility, the Drop Physics Module, will be available to perform both near-ambient temperature drop dynamics and statics investigations, as well as higher temperature material science experiments (up to 1000°C). A group of investigators will take advantage of two available acoustic positioning chambers to perform controlled studies of the vibrational and rotational dynamics of free drops and liquid shells, of the properties of surfactant-laden interfaces, and of the time-dependent surface properties of chemically processed soil samples precursors for ceramic materials.

Higher-temperature experiments have already been attempted during past Shuttle missions using fully automated devices and electromagnetic as well as acoustic and ultrasonic positioners. The results of these experiments will be briefly described in a later section dealing with the undercooling of levitated samples.

Free-Fall Experiments in Drop Tubes

Dynamics of Liquid Shells

Short-lived periods of reduced gravity can be obtained during free fall in drop tubes installed in the laboratory. Periods of 1–4 s of a 0.00001-*g* acceleration level have been obtained in existing facilities and have provided scientists with the opportunity to study the behavior of single and multiple falling drops in liquid shells.

The initial major technological impetus for the study of the dynamics of liquid shells was provided by the Laser Fusion Target Fabrication Program, which required thin, spherical, and concentric solid shells of submillimeter size to be used as fuel carriers. The technological program included the development of a reliable method for producing concentric liquid shells and for solidifying them into the required fuel carriers. The generation of submillimeter shells has been accomplished by ultrasonic excitation of a hollow liquid jet[39] and that of millimeter-size shells by the hollow jet instability,[40] a mechanism similar to the Rayleigh instability. The removal of the effects of gravity has been thought to allow both the experimental studies of the bubble centering mechanism as well as the manufacture of spherical concentric shells.

The primary effort has been directed toward the understanding of the bubble-centering mechanism based on shape oscillations.[41,42] Experimental confirmation of this effect has been provided by investigations using levitated drops in 1 g and in low gravity and for the case of large-amplitude shape oscillations where the liquid interfaces separation is greatly reduced during the oscillations. Existing evidence appears to indicate a consistent concentricity for relatively thin liquid shells falling in a *drag free* drop tube[40] and the initial oscillations induced by the formation of the shells due to the hollow jet instability could provide the centering action. The absence of any significant residual acceleration will preserve the concentric configuration. The aspect ratio of the shells studied in these drop tube investigations is on the order of 10 : 1 or less (the thickness is about 10% of the radius).

Solidification Studies

Mainly experimental investigations have been carried out on the solidification of shells during free-fall in drop tubes for the purpose of manufacturing laser fusion targets from a variety of materials such glasses, metals, polymers, and even solid hydrogen.[43] The rigorous requirements for sphericity (1%), concentricity (1%), and surface smoothness (100–300Å) have been satisfied in some cases for a shell diameter of a few hundred microns to a few millimeters. Mostly empirical methods have been used to determine the optimal and practical experimental parameters required to approach the desired goals. The specific processes involved in the melting, shell forming, and solidification of the particles are different and involve different regions of the nondimensional fluid and transport parameters space. Pure metal shell (tin and aluminum) crystallizing during free-fall has shown a varying degree of concentricity and sphericity (1% sphericity can be obtained in certain cases), but always displays an uneven surface finish due to cellular dendritic crystal structures and perhaps to the rheological characteristics of the melt during the shell formation and solidification phases. Protuberances remain as relics of the initial liquid jet pinch-off during the shell formation stage, but can be eliminated in a tin shell by the use of surface active agents. The only parameter controlled has been the cooling rate within a limited range.

The manufacture of hollow shells made of metallic glass has been of special interest because of the high-density metals involved, the desired smooth amorphous surface finish, and the hardness of the metal alloy. Shells of Au-Pb-Sb alloys on the order of 1 mm diam have been classified, and selected samples have been shown to be characterized by an aspect ratio of 16 : 1, variations in sphericity and concentricity of no more than 1%, and a surface finish with less than ± 250 Å variations.[44]

Available analytical treatments of containerless solidification has been restricted to the problem of a simple drop of generally spherical geometry. The case of a liquid shell in free-fall in a geseous environment and at a moderately high temperature has not yet been analyzed. The available experimental data are also sketchy and restricted to special cases. Additional experiments with detailed temperature-time history, fluid dynamic description of the surface flows, and characterization of the surface chemistry are needed for further progress in this area. Such investigations may be carried out with levitation techniques in 1 g, in microgravity with longer observation times and tighter control on the environmental conditions, and more accurate diagnostics. A detailed study of the interaction between the solidification mechanism and the residual fluid flow within the molten shell will be needed in order to provide information on the factors controlling the manufacture of spherical and concentric solid shells.

Studies of Undercooling and Solidification

Accessing the heart of the metastable (undercooled and superheated) liquid state is not a trivial experimental endeavor because of the pervasive influence of heterogeneous nucleation. Reduction of the material sample size and its isolation from any contact with catalytic material or disturbance have been the obvious strategy adopted by those interested in the properties of the metastable liquid state. Thus, single or multiple isolated droplets of highly purified materials have often been in the form of the liquid samples to be undercooled or superheated. In general, the scientific objectives have been to either probe the limits of undercooling or superheating, to measure the properties of the liquids in the metastable state, or to solidify the undercooled melt under controlled conditions in order to investigate the resulting solid microstructures. Existing containerless experimentation techniques allow the isolation of the sample from contact with possible catalytic substances and also permit the modest reduction of the sample size down to the millimeter scale, but they have introduced secondary perturbations associated with the levitation or positioning mechanisms. For example, ground-based electromagnetic levitation on conducting materials is notorious for inducing significant convective flows inside the suspended melt; ultrasonic levitation devices are characterized by high ac as well as dc host gas velocities; and electrostatic methods introduce surface charges on the sample. However, there is no definitive evidence that any of these factors significantly influences the onset of nucleation in substantially undercooled melts, but the current ultrasonic levitation results on limits of undercooling of ordinary liquids, such as water and low melting pure metals, have been much more modest than those obtained with emulsion techniques.[17] On the other hand, a rigorous comparison cannot be made bacause the levitation experiments were carried out with sample volumes one million times greater than those used with the emulsion method, suggesting that the difference might be explained by heterogeneous nucleation due to physical impurities in the melt.

Investigations conducted in microgravity might provide precious information on this problem by allowing the tremendous reduction in the magnitude of these secondary perturbations. Measurements made in microgravity with the same samples used in ground-based experiments and in identical experimental environments will provide rigorous data required for the evaluation of the viability of these "dynamic nucleation" effects. In support of an eventual microgravity experiment, precursor ground-based studies must deal with the accurate description of the behavior of the melt that is being manipulated by the various mechanisms of levitation. For example, residual internal flow caused by radio frequency (RF)-induced eddy currents in conducting melts as a function of the RF field magnitude, or internal flow or surface molecular motion in an acoustically positioned liquid sample must be accurately measured. The initial physical state of the melt droplets is of primary interest, and all of the relevant information, even on model materials usable at room temperature, must be available for a sensible assessment of the results of microgravity experiments.

Similarly, the problem of a solidifying droplet in the absence of a container, but in a specific gaseous or vacuum environment, must be resolved in order to gain a reasonable understanding of and control over the solidification process. This knowledge and control capability is essential to the understanding of the formation of resulting solid microstructures.

The development of noninvasive diagnostic techniques for samples processed without a container is also based on an exact understanding of the behavior of the liquid droplets (or shells) being studied. Optical, thermal, and surface properties of drops within the processing environment and under the influence of various applied stresses must be known or quantitatively measurable under different environmental conditions.

Ground-Based Levitation Experiments

Ground-based levitation of molten metals at high temperatures (higher than 1200°C) using RF techniques has been achievable for at least three decades.[45] The present interest in microgravity applications has initiated a technological development effort in the redesign of these electromagnetic levitators. A major improvement has been the adoption of quasi-independent heating and levitation RF coils working at different frequencies.[46] Operation in low gravity would, in principle, allow the decoupling of the heating and positioning mechanism, as well as the reduction of the accompanying convection induced in the melt. However, ground-based applications are still limited to high temperatures because of the high power levels required for sample levitation.

Recent investigations have been concerned with undercooling and sample properties measurement in the levitated state and have consequently concentrated on the dynamic behavior of the levitated melt droplet. Measurement of surface tension of iron-nickel alloys,[47] pure copper, nickel, and nickel alloys,[48] of the emissivity and temperature of copper, sliver, gold, nickel, palladium, platinum, and zirconium,[49] and of the solidification velocity in heterogeneously nucleated iron-nickel alloys[50] have recently been obtained using electromagnetic levitators. In most cases significant undercooling has been observed even in the experiments in 1 g, where the sample continuously undergoes translational and large-amplitude shape oscillations at the same time as a nonaxisymmetric rotational motion (tumbling motion). The results of the surface tension measurements have not been consistently correlated to existing data. A possible explanation could be provided by the droplet constant motion, which would decrease the measurement accuracy or could violate the theoretical assumptions required for the accurate determination of the surface tension through the small-amplitude shape oscillations theory.

The rigorous correlation between specific coil design and the melt oscillatory or rotational motion has not been made available. The origin of the sample instabilities cannot yet be determined and cannot yet be empirically eliminated in ground-based devices. However, one might anticipate that the low-gravity environment will improve the sample stability, and some evidence obtained during tests of a German-built device conducted on board the KC-135 airplane flying parabolic trajectories has shown a definite reduction in the rotational instability. In any case, a quantitative description of the dynamic behavior of an electromagnetically positioned sample will be required in order to fully analyze the results obtained in any of the investigations in nucleation or solidification.

Ground-based ultrasonic levitation methods have been used to study undercooling and solidification mechanisms at a much lower temperature and under slightly more controlled conditions in terms of sample stability.[17,51] Water, succinonitrile, O-terphenyl, gallium, indium, and tin samples of 2–3 mm diam have been undercooled between 10 and 21% of their absolute melting point. Quasistatic thermal conditions were used with very low cooling rate, and the sample was distorted into an oblate spheroid of axial ratio of at least 1.2 when levitated in a gaseous environment. The distortion was negligible when the sample was levitated in an immiscible liquid medium (supercooled water in decahydronapthalene). However, data on the variation of the nucleation temperature with drop distortion have revealed no noticeable effect with a drop axial ratio of 1.2–2.0. Substantial convective gas flow around the levitated sample was also induced by acoustic streaming and caused very slow liquid motion on the droplet surface. Experimental measurement of these flows for ordinary water drops have yielded velocity magnitude on the order of 0.1 mm/s for ground-based levitation. Experimental evidence also indicated that phase transition could be triggered in a significantly undercooled melt (greater than 10% undercooling) by inducing shape oscillations. Except for this last observation, there exists no compelling evidence of catalytic action caused by the levitation forces. Under these circumstances and for these specific experimental conditions, the onset of solidification was probably controlled by heterogeneous nucleation due to physical impurities included in the melt. Another possible catalytic effect might be caused by gaseous cavitation of existing gas bubbles suspended in the melt; these bubbles might arise because of trapped gas that would be released during containerless melting of the sample. Acoustic waves are highly attenuated inside the suspended droplet, and the sample is generally placed at an acoustic pressure nodal position. These two considerations would imply a weak acoustic pressure within the melt droplet and rule out any transient cavitation. The transmitted acoustic pressure would still be enough to induce stable cavitation of existing gas or vapor bubbles. A transition from a stable to a transient cavitation regime could conceivably occur and cause a pressure increase high enough to induce nucleation. No experimental evidence has yet been provided to support the existence of such a mechanism. Figure 8 displays a series of photographs showing the various stages of solidification of O-terphenyl into a polycrystalline solid.

Microgravity Flight Experiments

The opportunity to carry out preliminary materials science investigation using freely positioned droplets in low gravity onboard the Space Shuttle. The early version of an ultrasonic positioner to be used for glass samples processing at up to 1500°C and in a totally automated mode without human intervention has been flown twice, with partial success on the second attempt. The demonstration of the high-temperature capabilities of acoustic positioning has been obtained through the deployment, melting, and solidification of a simple glass droplet. The observation of the motion of a gas bubble included in the glass and of its eventual removal from the melt has been tentatively attributed to flows driven by thermal gradients arising from the uneven heating of the furnace.

Two other automated flight experiments using audio frequency acoustic and electromagnetic positioning have been marred by equipment failure causing the curtailment of the scientific investigations. A silica-glass-coated sample of Fe-Ni alloy was melted and solidified in the early version of the electromagnetic positioner, but the failure of the power system forced the early cool down and loss of position control of the sample which solidified upon contacting the restraining cage. Substantial undercooling has been obtained, and the retrieval of the processed sample has allowed its microstructural analysis. Partial dewetting of the glass coating was observed during the processing of the sample.

Several rocket-borne flight experiments have also been carried out both in the United States and in Europe. Technology development and preliminary scientific tests were the goals of these short-duration, automated investigations. Although the results have thus far been of mixed quality, the continuing use of lower-cost rocket and airplane-based tests will surely improve the quality and reliability of the experimental apparatuses and procedures.

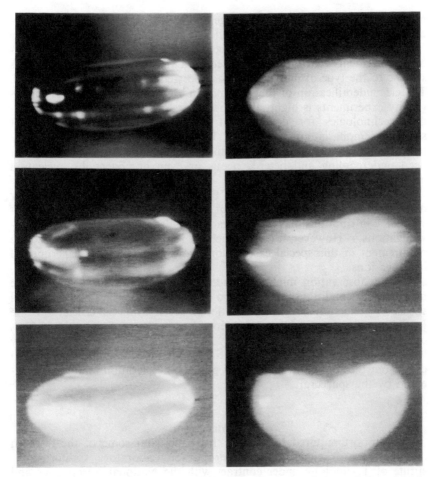

Fig. 8   Various stages during the solidification of a levitated drop of *O*-terphenyl.

Concluding Remarks

The   full exploitation of the advantages of low-gravity conditions in space in the area of the fluid dynamics and materials science involving single drops or shells will depend on availability of the appropriate flight facilities, the relevant diagnostic instrumentation, and the capability of altering experimental procedures on-orbit. For near-ambient temperature investigations the existing containerless positioning technology appears adequate if rigorous control is also available of environmental parameters such as temperature, humidity, gas composition, contamination control, and lighting, to name a few. The principal scientific objectives at this time appear to be the observation of surface-tension-dominated phenomena in the absence of any extraneous stimulus. Other crucial experimental factors include the ability to accurately image and record the dynamic phenomena of interest for analysis during as well as after the flight.

High-temperature materials science investigations have more demanding requirements that must be satisfied by a variety of different techniques specific to the type of materials to be processed in a containerless manner. The clear identification of the scientific and technological goals for microgravity experiments is highly desirable prior to the selection of the candidate technologies for development into spaceflight facilities. Present scientific emphasis appears to be based on the study of very-high-temperature behavior of materials and on the possibility of carrying out deep undercooling and heterogeneous or even homogeneous nucleation investigations.

Theoretical analytical modeling of the three-dimensional often nonlinear dynamics with free boundaries is not yet always available, although substantial contributions have been made with respect to microgravity investigations.[52] However, the guidance of reliable models is an absolute requirement in this special case of long lag time and costly experimental investigations.

A final observation might be made regarding the requirement for the microgravity conditions characteristic of experiments using free drops and shells. A reduction of the gravitational acceleration is crucial for the feasibility of these investigations. However, because of the containerless conditions valid scientific results may be obtained even if only milligravity levels are achieved. Active compensating techniques can be implemented to cancel out the residual transient and periodic acceleration impulses experienced by the spacecraft.

Acknowledgment

The support required to prepare this paper and some of the research described was provided at the Jet Propulsion Laboratory, California Institute of Technology under contract with the National Aeronautics and Space Administration.

References

[1] Rayleigh, L., "Equilibrium of Revolving Liquid Under Capillary Forces," *Philosophical Magazine*, Vol. 28, 1914, pp. 161–180.

[2] Lamb, H., *Hydrodynamics*, Cambridge Univ. Press, Cambridge, UK, 1932.

[3] Trinh, E. H., "Compact Acoustic Levitation Device for Studies in Fluid Dynamics and Materials Science in the Laboratory and in Microgravity," *Review of Scientific Instruments*, Vol. 56, 1985, pp. 2059–2065.

[4] Rhim, W. K., Chung, S. K., Trinh, E. H., and Elleman, D. D., *Mat. Res. Soc. Symp. Proc.*, Vol. 87, 1987, pp. 329–339.

[5] Tsamopoulos, J. A., and Brown, R. A., "Resonant Oscillations of Inviscid Charger Drops," *Journal of Fluid Mechanics*, Vol. 147, 1984, pp. 373–395.

[6] Brazier-Smith, P. R., Brook, M., Latham, J., Saunders, C. P. R., and Smith, M. H. "The Vibration of Electrified Drops," *Proceedings of the Royal Society of London*, Vol. 322, 1971, pp. 523–534.

[7]Pruppacher, H., Rasmussen, R., and Walcek, C., "Proceedings of the Second International Colloquium on Drops and Bubbles," Jet Propulsion Lab.-NASA Publ. 82-7, 1982, pp. 239–243.

[8]Marston, P. L., LoPorto, A., and Pullen, G., *Journal of the Acoustical Society of America*, Vol. 69, 1981, pp. 1499–1503.

[9]Jackson, W., and Barmatz, M., "Equilibrium Shape and Location of a Liquid Drop Acoustically Positioned in a Resonant Rectangular Chamber," *Journal of the Acoustical Society of America*, Vol. 84, 1988, pp. 1845–1862.

[10]Trinh, E. H., and Hsu, C. J., "Equilibrium Shape of Acoustically Levitated Drops," *Journal of the Acoustical Society of America*, Vol. 79, 1985, pp. 1335–1338.

[11]Trinh, E. H., and Hsu, C. J., NASA Technical Briefs, NPO-16746, 1985.

[12]Miller, C., and Scriven, L., *Journal of Fluid Mechanics*, Vol. 32, 1968, pp. 417–433.

[13]Prosperetti, A., "Free Oscillations of Drops and Bubbles: The Initial-Value Problem," *Journal of Fluid Mechanics*, Vol. 100, 1980, pp. 333–347.

[14]Tsamopoulos, J., and Brown, R., *Journal of Fluid Mechanics*, Vol. 147, 1984, pp. 373–395.

[15]Foote, G. B., PhD Dissertation, Univ. of Arizona, Tucson, 1971.

[16]Trinh, E., and Wang, T. G., "An Experimental Study of Large Amplitude Drop Shape Oscillations," *Journal of Fluid Mechanics*, Vol. 122, 1982, pp. 315–338.

[17]Trinh, E. H., Gaspar, M., Robey, J., and Arce, A., "Experimental Studies in Fluid Mechanics and Materials Science Using Acoustic Levitation," *Mat. Res. Soc. Proc.*, Vol. 87, 1987, pp. 57–66.

[18]Schade, J., McLean, A., and Miller, W. A., *115th TMS Annual Meeting Proceedings*, Vol. 81, 1986.

[19]Chandrasekhar, S., "The Stability of a Rotating Liquid Drop," *Proceedings of the Royal Society of London*, Vol. 286, 1965, pp. 1–26.

[20]Busse, F., *Journal of Fluid Mechanics*, Vol. 142, 1984, pp. 1–12.

[21]Annamalai, P., Trinh, E., and Wang, T. G., "Experimental Study of the Oscillations of a Rotating Drop," *Journal of Fluid Mechanics*, Vol. 158, 1985, pp. 317–327.

[22]Lu, H., "Study of Interfacial Dynamics of Drops in the Presence of Surfactants and Contaminants," PhD Dissertation, Yale Univ., New Haven, CT, 1988.

[23]Wang, T. G., Kanber, H., and Rudnick, I., "First Order Acoustic Torque," *Physical Review Letters*, Vol. 38, 1977, pp. 3–8.

[24]Plateau, J., "The Figures of Equilibrium of a Liquid Mass," *Smithsonian Report*, Vol. 270, 1863.

[25]Wang, T. G., Tagg, R., Cammack, L., and Croonquist, A., "Second International Colloquium on Drops and Bubbles Proceedings," Jet Propulsion Lab.-NASA Publ. 82-7, 1982, pp. 203.

[26]Chandrasekhar, S., *Ellipsoidal Figures of Equilibrium*, Yale Univ. Press, New Haven, CT, 1969.

[27]Brown, R. A., and Scriven, L., "Liquid Drops," *Proceedings of the Royal Society of London*, Vol. 371, 1980, pp. 331–357.

[28]Rhim, W. K., Chung, S. K., Trinh, E. H., and Elleman, D. D., "Charged Drop-dynamics Experiments Using an Electrostatic Acoustic Hybrid System," *Mat. Res. Soc. Symp. Proc.*, Vol. 87. 1987, pp. 329–337.

[29]Trinh, E. H., and Leung, E., "Ground-Based Studies of the Vibrational and Rotational Dynamics of Acoustically Levitated Drops and Shells," AIAA Paper 90-0315, 1990.

[30]Saffren, M., Elleman, D. D., and Rhim, W. K., "Second International Colloquium on Drops and Bubbles Proceedings," Jet Propulsion Lab.-NASA Publ. 82-7, 1982, pp. 7–14.

[31] Trinh, E., "Fluid Dynamical Study of Free Liquids," *IEEE Proceedings on Sonics and Ultrasonics*, Vol. 2, IEEE, New York, 1983, pp. 1143–1150.

[32] Lee, M. C., Feng, I. A., Elleman, D. D., Wang, T. G., and Young, A. T., "Second International Colloquium on Drops and Bubbles," Jet Propulsion Lab.-NASA Publ. 82-7, 1982, pp. 107–111.

[33] Croonquist, A., Rhim, W. K., Elleman, D. D., and Wang, T. G., "SPAR VIII Experiment 77-18 Flight 2 Report," Jet Propulsion Lab.-NASA Rept. 1983.

[34] Day, D. E., and Ray, C. S., "Final Report for MEA/A-1 Experiment 81F01," NASA Contract NAS-8-34758 (1986).

[35] Jacobi, N., Croonquist, A., Elleman, D. D., and Wang, T. G., "Second International Colloquium on Drops and Bubbles," Jet Propulsion Lab.-NASA Publ. 82-7, 1982, pp. 31–38.

[36] Piller, J., Knauf, R., Preu, P., Lohofer, G., and Herlach, D., "Proceedings of the 6th European Symposium on Materials Science Under Microgravity Conditions," European Space Agency Rept. SP-256, 1987, pp. 437–443.

[37] Wang, T. G., Trinh, E. H., Croonquist, A., and Elleman, D. D., "Shapes of Rotating Free Drops: Spacelab Experimental Results," *Physical Review Letters*, Vol. 56, 1986, pp. 452–455.

[38] Trinh, E. H., Wang, T. G., Croonquist, A., and Elleman, D. D., "Spacelab 3 Drop Dynamics Experiments," AIAA Paper 86-0196, 1986.

[39] Calliger, R., Turnbull, R., and Hendrick, C. D., "Hollow Drop Production by Injection of Gas Bubbles into a Liquid Jet," *Review of Scientific Instruments*, Vol. 48, 1977, pp. 846–850.

[40] Kendall, J., "Proceedings of the 2nd International Colloquium on Drops and Bubbles," Jet Propulsion Lab.-NASA Publ. 82-7, 1982, pp. 79–87.

[41] Saffren, M., Elleman, D. D., and Rhim, W. K., *2nd International Colloquium on Drops and Bubbles*, Vol. 7, JPL-NASA Publ. 82-7, 1982, pp. 7–16.

[42] Lee, C. P., and Wang, T. G., "The Centering Dynamics of Thin Liquid Shells in Capillary Oscillations," *Journal of Fluid Mechanics*, Vol. 188, 1988, pp. 411–435.

[43] Hendricks, C., "Proceedings of the 2nd International Colloquium on Drops and Bubbles," Jet Propulsion Lab.-NASA Publ. 82-7, 1982, pp. 88, 124.

[44] Lee, M. C., "Metal Shell Technology," *SAMPE Journal*, Vol. 19-6, 1983, pp. 7–11.

[45] Okress, E. C., Wroughton, D., Comenetz, D., Brace, G., and Kelley, J. C. R., "Electromagnetic Levitation of Solid and Molten Metals," *Journal of Applied Physics*, Vol. 23, 1952, pp. 545–552.

[46] Lenski, H., and Willnecker, R., Rept. on the Workshop on Containerless Experimentation in Microgravity, JPL Publ. D-7142, 1990.

[47] Schade, J., McLean, A., and Miller, W. A., *115th TMS Annual Meeting Proceedings*, Vol. 81, 1986.

[48] Hansen, G. P., Krishnan, S., Hauge, R., and Margrave, *J. Met. Transactions*, Vol. 19A, 1988, pp. 1939–1951.

[49] Krishnan, S., Hauge, R., and Margrave, J., "Spectral Emissivity and Optical Constants of Electromagnetically Levitated Liquid Metals as a Function of Temperature and Wavelength," "Proceedings of the 2nd Non-Contact Temperature Measurement Workshop," Jet Propulsion Lab.-NASA Publ. 89-16, 1989, pp. 110–140.

[50] Willnecker, R., Herlach, D., and Feurbacher, B., "Containerless Undercooling of Bulk Fe-Ni Melts," *Applied Physics Letters*, Vol. 49, 1986, pp. 1339–1341.

[52] Bauer, "Natural Frequencies and Stability of Immiscible Spherical Liquid System," *Applied Microgravity Technology*, Vol. 1, 1988, pp. 90–102.

Chapter 6. Separation Phenomena

Separation Physics

Paul Todd*

National Institute of Standards and Technology, Boulder, Colorado

Nomenclature

A = cross-sectional area of electrophoresis or extraction chamber
A_M = surface area of separand molecule of particle
a = particle radius
B = constant of proportionality for partition coefficient
b = half thickness of FFE chamber
b_{ij} = monomer interaction coefficients
$C(x)$ = concentration of density solute in density gradient
C_i = concentration of the ith solute
C_{ij} = osmotic virial coefficient
c = concentration of solute
D = diffusion coefficient
E = electric field strength
e = charge on the electron
g = acceleration of Earth's gravity
Hj = Hjertén number
I = electric current
J_i = flow of ith substance
K = dissociation constant
k = electrical conductivity
k, l = coordinates of the sample zone center in horizontal rotation
k_B = Boltzmann constant
L = number of species
M = molecular weight
m = dimensionless molality
N = number of monomers per polymer molecule

This work, performed for the National Institute of Standards and Technology, an agency of the United States Government, is not subject to copyright.

*Physicist, Center for Chemical Technology.

n = particles or molecules/volume (number density)
R = droplet radius, column radius, universal gas constant
R_i = production rate of solute species i
Ra = Rayleigh number
r = radial distance variable
S = surface stress vector
T = temperature, K
t = time
v = velocity
x = distance in the direction of separand migration
y = distance in the direction of FFE chamber thickness
z = distance in the direction of flow in FFE
z_i = valence of ith solute
α = thermal diffusivity
β = thermal expansion coefficient
γ = time constant for settling of a rotating zone
γ_{pb} = interfacial free energy between separand and bottom phase
γ_{pt} = interfacial free energy between separand and top phase
$\Delta\gamma$ = interfacial free energy
ϵ = dielectric constant
ζ = zeta potential (at surface of slipping plane)
η = viscosity
κ = Debye-Hückel constant (inverse of Debye length)
λ = Brønsted partitioning coefficient
μ = electrophoretic mobility (velocity/field)
μ_j = potential of jth type
v = kinematic viscosity
v_i = rigidity exponent of polymer i
ρ = density
τ = residence time in a separation system
ω = angular frequency, angular velocity

Subscripts
av = average
eo = electroendosmosis
D = droplet
i = solute
j = field
m = maximum distance in rectangular chamber
n = number average
s = sample stream
w = weight average
z = in the z direction
zm = maximum value in the z direction
0 = medium or fluid
1 = water in biphasic system
2 = one polymer in biphasic system
3 = other polymer in biphasic system

Introduction

SEPARATION science is the characterization and development of processes such as chromatography, magnetic filtration, sedimentation, filtration, electrophoresis, and extraction used in the purification of chemical products. Most separation science research in low gravity has been restricted to biochemical applications of electrophoresis, extraction, and, to a lesser extent, filtration. This chapter reviews the basic physics and chemistry of preparative electrophoresis and isoelectric focusing, certain aspects of extraction and filtration, with emphasis on gravity-related aspects of these processes, and low-gravity studies of these processes to date. Certain aspects of this subject were reviewed by Snyder in 1986; although very few experiments have been performed in orbital spaceflight since that time,[1] a number of significant papers have appeared, and low-gravity research on aircraft and sounding rockets has intensified.

Electrokinetic Separations

The scale up of electrokinetic separation methods, electrophoresis, isotachophoresis, and isoelectric focusing has encountered obstacles for decades.[2-4] These obstacles can be considered in two categories: 1) those that are caused by inertial acceleration (gravity) and 2) those that are not. Gravity-dependent processes include 1) convection; 2) zone, or droplet, sedimentation; and 3) in some cases, particle sedimentation.[5,6]

Gravity-independent obstacles include electroendosmosis,[7] electrohydrodynamics,[8,9] and Joule heat[4,7] (most biological separands cannot withstand high temperatures, and temperature gradients cause poor separations by causing viscosity and conductivity gradients).

Electrophoresis

Electrophoresis is the motion of particles (molecules, small particles, and whole biological cells) in an electric field and is one of several electrokinetic transport processes. The velocity of a particle per unit applied electric field is its electrophoretic mobility μ; this is a characteristic of individual particles and can be used as a basis of separation and purification. This separation method is a rate (or transport) process.

The four principal electrokinetic processes of interest are electrophoresis (motion of a particle in an electric field), streaming potential (the creation of a potential by fluid flow), sedimentation potential (the creation of a potential by particle motion), and electroendosmosis (the induction of flow at a charged surface by an electric field, also called electro-osmosis). These phenomena always occur, and their relative magnitudes determine the practicality of an electrophoretic separation or an electrophoretic measurement. It is generally desirable, for example, to minimize motion due to electroendosmosis in practical applications. A brief discussion of the general electrokinetic relationships follows.

The surface charge density of suspended particles prevents their coagulation and leads to stability of lyophobic colloids. This stability determines the successes of paints and coatings, pulp and paper, sewage and fermentation, and numerous other materials and processes. The surface charge also leads to motion when such particles are suspended in an electric field. The particle surface has an electrokinetic ("zeta") potential ζ, proportional to σ_e, its surface charge density—a few millivolts at the hydrodynamic surface of stable, nonconducting particles, including biological cells, in aqueous suspension. If the solution has electrical permittivity ϵ (7×10^{-9} F/m in water at 25°C), the electrophoretic velocity is

$$v = \frac{2\zeta\epsilon}{3\eta}E \qquad (1)$$

for small particles, such as molecules, whose radius of curvature is similar to that of a dissolved ion (Debye-Hückel particles), and

$$v = \frac{\zeta\epsilon}{\eta}E \qquad (2)$$

for large (von Smoluchowski) particles, such as cells and organelles. The variable η is the viscosity of the bulk medium. These relationships are expressed in rationalized MKS(SI) units. At typical ionic strengths ($0.01 - 0.2$ equiv/L), particles in the nanometer size range usually have mobilities (v/E) that are different from those specified by Eqs. (1) and (2) (Ref. 10). A summary of equations is given in Table 1.

Sedimentation Potential

If one charged particle moves with respect to another, a potential will be created, and this potential will impart motion to other charges in the environment, including dissolved ions. Although the ζ potential of a stationary particle is only "felt" by charges up to 7 Å or so away (the "Debye length"), an electric field is swept through a greater distance when the particle moves. If a particle is caused to move by the acceleration of gravity (upward or downward), the strength (V/cm) of the electric field generated is

$$E = \frac{4\zeta\epsilon(\rho - \rho_0)g}{3\eta\kappa} \qquad (3)$$

where κ is measured in units of inverse length (cm^{-1}) and is directly proportional to the ionic strength of the surrounding medium and inversely related to the distance over which a charge can exert a force in a solution. This potential is also known as "counterstreaming potential" and as the "Dorn effect." This potential could be as great as 20 mV, which is comparable to the negative charge at the surface of a typical cell, organelle, or colloidal particle. A streaming potential can also be developed by passing fluid over a charged surface.

Table 1   List of equations presented in the text and their meaning

Description	Equation	Text no.
Electrophoretic velocity of a solute molecule	$v = \dfrac{2\zeta\epsilon}{3\eta} E$	(1)
Electrophoretic velocity of a suspended particle	$v = \dfrac{\zeta\epsilon}{\eta} E$	(2)
Sedimentation potential	$E = \dfrac{4\zeta\epsilon(\rho - \rho_0)g}{3\eta}$	(3)
Density gradient electrophoresis	$dx/dt = \mu(x)E(x) - 2a^2 g[\rho - \rho_0(x)]/9\eta(x)$	(8)
Electric field strength	$E(x) = I/[Ak(x)]$	(9)
Standard electrophoretic mobility	$\mu(x) = \mu_s \eta_s / \eta(x)$	(10)
Time to travel distance x in density gradient electrophoresis	$t(x) = \displaystyle\int_0^x \dfrac{Q(x)\exp(\alpha x)}{L(x)} \, dx$	(11)
Droplet density	$\rho_D = \rho_0 + \dfrac{a^3}{R^3} n(\rho - \rho_0)$	(12)
Coordinates of center of sample zone in rotating electrophoresis	$(x - k)^2 + (y - 1)^2 = r^2 \exp(2\gamma t)$	(14)
Inverse time constant for centrifugal motion	$\gamma = \dfrac{V_D(\rho_D - \rho_0)}{6\pi\eta R/\omega^2 + 32\pi R^5 \rho\rho_0/27\eta}$	(15)
Coordinates of sample zone during centrifugal motion	$\sqrt{k^2 + l^2} = \dfrac{V_D(\rho_D - \rho_0)g}{\sqrt{[(4\pi R^3/3)^2\omega^2\rho_0^2 + (6\pi\eta R)^2]}}$	(16)
Electro-osmotic velocity	$v = \dfrac{\zeta\epsilon}{\eta}$	(18)
Radial electro-osmotic velocity distribution	$v = \dfrac{\zeta\epsilon}{\eta} E(r^2/R^2 - \tfrac{1}{2})$	(19)
Electro-osmotic velocity distribution, rectangular chamber	$v_{eo}(y) = \tfrac{1}{2}v_{eo}(0)[3(y - b)^2/b^2 - 1]$	(21)
Flow velocity distribution in rectangular system	$v_z(y) = v_{zm} - (v_{zm}/b^2)(y - b)^2$	(22)
Maximum flow velocity relative to average flow velocity	$v_{zm} = 3v_{zav}/2$	

(Table 1 continued on next page)

Table 1 (continued)

Description	Equation	Text no.
Residence-time distribution in FFE	$\tau(y) = z_m / v_z(y)$	
	$= z_m / [(3v_{zav}/2) - (3v_{zav}/2b^2)(y-b)^2]$	(23)
Migration distance distribution in the absence of EEO in FFE	$x(y) = \mu E \tau(y)$	
	$= \mu E z_m / [(3v_{zav}/2) - (3v_{zav}/2b^2)(y-b)^2]$	(24)
Surface stress due to electro-hydrodynamics	$S = \dfrac{E^2 \epsilon [\epsilon/\epsilon_s \{(k_s/k)^2 + (k_s/k) + 1\} - 3]}{4\pi\{(k_s/k)^2 + 1\}^2} \cos 2\theta$	(25)
Migration distance in FFE vs y	$x(y) = \dfrac{2E z_m(\mu - \mu_{eo}/2)}{3v_{zav}[1 - (y-b)^2/b^2]}$	(26)
Critical Rayleigh number for FFE	$Ra_c = \dfrac{g\beta \, |\nabla T| \, x_m^4}{\alpha} < \pi^4$	(27)
Chemical potentials in two-phase polymer solutions	$\Delta\mu_2 = -RT[\ell n m_2 + 2C_{22}m_2 + 2C_{23}m_3]$	(34a)
	$\Delta\mu_3 = -RT[\ell n m_3 + 2C_{23}m_2 + 2C_{33}m_3]$	(34b)
	$\Delta\mu_1 = -RT[m_2 + m_3 + C_{22}m_2 + 2C_{23}m_2 m_3 + C_{33}m_3^2]$	(34c)
Osmotic virial coefficients of polymers in two-phase systems	$C_{22} = b_{22}N_2^{3v2}[1 + (2/9)\,\ell n(M_{w2}/M_{n2})]$	(35a)
	$C_{33} = b_{33}N_3^{3v3}[1 + (2/9)\,\ell n(M_{w3}/M_{n3})]$	(35b)
	$C_{23} = b_{23}\{N_2^{3v2}[1 + (2/9)\,\ell n(M_{w2}/M_{n2})]$	
	$\times [N_3^{3v3}[1 + (2/9)\,\ell n(M_{w3}/M_{n3})]\}^{\frac{1}{2}}$	(35c)
Partition coefficient (Brønsted)	$K = \exp(\lambda M/k_B T)$	(36)
Partition coefficient	$K = B\exp\{-\gamma_{tb}A[1 - (\Delta\gamma + \sigma_e\Delta\psi/\gamma_{tb}]^2/4k_B T\}$	(38)

Combined Fields

Most objects are acted upon by a combination of forces. To deal with this fact, all types of flow (mass, charge, magnetic flux, etc.) are assumed to be interdependent, and transport relationships are described by a flow-and-field matrix. All flows J_i are caused by a potential, generalized as $\Delta\mu_j$, in proportion to a coefficient L_{ij} that relates them:

$$J_i = L_{ij}\Delta\mu_j \qquad (4)$$

Thus, in the powerful Onsager relationships flow can be generalized on the basis of what is flowing, J_i, for the ith component and the potentials causing the flow, μ_j, for the jth type of potential. More than one type of field can cause more than one type of flow; thus, in general, as suggested by

L. Onsager, one has a matrix of motions or flows:

$$J_1 = L_{11}\Delta\mu_1 + L_{12}\Delta\mu_2 + L_{13}\Delta\mu_3 + \cdots$$

$$J_2 = L_{21}\Delta\mu_1 + L_{22}\Delta\mu_2 + L_{23}\Delta\mu_3 + \cdots$$

$$J_3 = L_{31}\Delta\mu_1 + L_{32}\Delta\mu_2 + L_{33}\Delta\mu_3 + \cdots$$

or

$$J_i = \sum_j L_{ij}\Delta\mu_j \qquad (5)$$

This means, for example, that electric potentials can move charged masses and inertial potentials can move charges associated with mass. In this example (a falling charged particle) one can determine the downward mass flux J_m and the electric current $I = J_e$:

$$J_m = L_{11}\Delta\mu_g + L_{12}\Delta\mu_e \qquad (6a)$$

$$J_e = L_{21}\Delta\mu_g + L_{22}\Delta\mu_e \qquad (6b)$$

In most cases, L_{21} and L_{12}, the cross-term coefficients (the effect of gravity on a current and the effect of the electric field on sedimentation, respectively) are considered small compared to L_{11} and L_{22}. However, most physicists will point out that at subcellular dimensions $\Delta\mu_e \gg \Delta\mu_g$; thus, it may not be possible to ignore cross terms in small-particle transport. The solution of Eq. (6) at steady state leads to[11]

$$J_m = \frac{32\pi^2 a^5(\rho - \rho_0)^2 cg}{27\eta} + \frac{4\pi\zeta\epsilon a^3(\rho - \rho_0)cE}{3\eta} \qquad (7a)$$

$$J_e = \frac{4\pi\zeta\epsilon a^3(\rho - \rho_0)gc}{3\eta} + kE \qquad (7b)$$

where k is conductivity, and c is concentration of particles. Each of these terms is recognizable, from the top, left to right, as Stokes' sedimentation of a particle of radius a and density ρ in a medium of density ρ_0, electrophoresis [Eq. (1)], streaming potential [Eq. (3)], and Ohm's law.

Preparative Electrophoresis

Analytical electrophoresis in a gel matrix is very popular, because convection is suppressed; however, high sample loads cannot be used because of the limited volume of gel that can be cooled sufficiently to provide a uniform electric field. Capillary zone electrophoresis, a powerful, high-resolution analytical tool,[12] depends on processes at the micrometer scale and is not applicable to preparative electrophoresis. Therefore, preparative electrophoresis must be performed in free fluid. The two most frequently used free-fluid methods are zone electrophoresis in a density gradient and free-flow (or continuous-flow) electrophoresis (FFE or CFE). Other methods, which are less popular and are not studied in low-gravity experiments, will not be discussed in detail. These are described in partial reviews by Ivory[4] and by Mosher et al.[13]

Physical Constraints Imposed by Living Cells and Biomolecules

Buffers

General compositions and characteristics of typical categories of buffers used with each of the methods are listed in Table 2, where it can be seen that a wide range of ionic strengths has been used, partly to avoid the application of high currents in preparative separations, which require low-conductivity buffers to avoid convective mixing due to Joule heating. At very low ionic strength, phosphate buffers are found to be harmful to cells, and this is the principal reason that triethanolamine was introduced by Zeiller and Hannig[14] as a general carrier buffer for free-flow electrophoresis. When cells are present, neutral solutes such as sucrose or mannitol must be added for osmotic balance. These buffers have undergone testing in analytical and preparative cell electrophoresis.

Temperature

Temperature control is generally imposed on all free electrophoresis systems as a means of suppressing thermal convection and avoiding viscosity gradients (see below), and biological separands impose additional constraints on thermoregulation of electrophoretic separators. With a few notable exceptions, living cells do not tolerate temperatures above 40°C. Additionally, mammalian cells are damaged by being allowed to metabolize in non-nutrient medium; hence, it is necessary to reduce the metabolic rate by reducing temperature, typically to 4–10°C. Proteins are more thermotolerant, but at 37°C the enzymes of homeothermic organisms and many bacteria (including *E. coli*) are active, and unwanted hydrolysis reactions occur in samples containing mixtures of proteins.

Electric Field

It is possible to move membrane proteins in cells immobilized in an electric field[15] and in sufficiently strong fields (>3 kV/cm) to produce dielectric breakdown of the plasma membrane in a few microseconds, resulting in a potential for cell-cell fusion[16] or pores through which macromolecules can readily pass.[17] The strength of such fields, 1–15 kV/cm, considerably exceeds the range of field strength used for electrophoresis.

Role of Cell Size and Density

Over a wide range of radii the mobility of a given cell type is independent of cell size, as predicted by electrokinetic theory [Eq. (2)]. Isolated electrophoretic fractions collected after density gradient electrophoresis have been analyzed individually with a particle size distribution analyzer, and no

Table 2 Examples of buffers used in free electrophoresis

Buffer name	Buffer ion	Concn, mM	Other	Concn, mM	k, mS/cm	Ref.
Standard saline	HCO_3^-	0.3	NaCl	145	12.5	1
A-1	$H_2PO_4^-$	0.367	NaCl	6.42	0.96	2
	HPO_4^-	1.76	EDTA	0.336		
			Glucose		222	
			Glycerol		540	
D-1	$H_2PO_4^-$	0.367	NaCl	6.42	0.9	3
	HPO_4^-	1.76	EDTA	0.336		
			Glucose		222	
			DMSO	630		
C-1	$H_2PO_4^-$	1.47	KCl	2.65	1.5	4
	HPO_4^-	8.10	$MgCl_2$	0.48		
			Glucose		55.5	
			Sucrose	199		
1 mM	$H_2PO_4^-$	0.15	KCl	2.65	0.5	5
	HPO_4^-	0.81	$MgCl_2$	0.48		
			Glucose		55.5	
			Sucrose	199		
STS-6	Barbital	2.0			0.16	6
Hannig-1	Triethanolamine	9.0	$MgCl_2$	0.0025	0.9	7
			$CaCl_2$	0.005		
	K-acetate	9.0	Glucose	5.0		
			Sucrose	285		
STS-7	NaPropionate	2.25			0.14	6
Tris-borate	Tris(hydroxy methylamino) methane	0.3			0.0126	8
	Boric acid	0.3				
STS-8	Triethanolamine	0.65	Glycine	30	0.06	9
			$MgCl_2$	0.30		
			$CaCl_2$	0.027		
	K-acetate	0.20	Glycerol	220		
			Sucrose	44.0		
R-1	K Phosphate	2.25	pH 7.2		0.90	10

1, Seaman 1965. 2, Allen et al. 1977. 3, Snyder et al. 1983. 4, Plank et al. 1982. 5, Plank unpublished personal communication. 6, Snyder and Rhodes 1989. 7, Heidrich and Dew 1983. 8, Hannig 1969. 9, Hymer et al. 1987. 10, Todd et al. 1989.

consistent relationship between size and electrophoretic fraction was found.[18] This finding implies that, despite the gravity and electrokinetic vectors being antiparallel (or parallel) in density gradient electrophoresis, size plays very little role in determining the electrophoretic migration of cultured cells. As Eq. (8) implies, density plays an important role in cell sedimentation, and, when $\rho - \rho_0$ is excessive (>0.03), free electrophoresis in a density gradient is dominated by sedimentation, whereas separation by free-flow electrophoresis is less so.[19,20]

Density Gradient Electrophoresis

Density gradient electrophoresis has been used for protein and virus separation[21] and has been shown to separate cells on the basis of electrophoretic mobility as measured by microscopic electrophoresis.[22,23] Methods and applications of this technique have been reviewed.[24] Downward electrophoresis in a commercially available device utilizing a Ficoll gradient was first used to separate immunological cell types.[25–27] Similar gradient-buffer conditions are employed in the Boltz-Todd apparatus[28,29] in which electrophoresis of cells and other separands is usually performed in the upward direction (Fig. 1). A limited number of separation problems can be solved by combining cell sedimentation and electrophoretic "levitation" at low (<10 V/cm) field without a density gradient.[30,31]

The use of a density gradient counteracts three effects of gravity: 1) particle sedimentation, 2) convection, and 3) zone, or droplet, sedimentation. However, the migration rates of cells in a density gradient are quantitatively different from those observed under typical analytical electrophoresis conditions,[32] in which physical properties are constant. These differences are expected for at least the following reasons: 1) the sedimentation component of the velocity vector changes as the cells migrate upward and the density of the suspending fluid decreases; 2) the viscosity of the suspending Ficoll solution decreases rapidly as the cells migrate upward[28]; 3) the conductivity of the suspending electrolyte increases as the cells migrate upward, consistent with the viscosity reduction; 4) carbohydrate polymers, including Ficoll, typically used as solutes to form density gradients increase the zeta potential of cells[33,34]; and 5) cells applied to the density gradient in high concentrations undergo droplet sedimentation, [29,35] thereby increasing the magnitude of the sedimentation component of the migration velocity [see Eq. (12)].

All of these variables have been studied.[19,22,28,34,36,37] In addition, cell population heterogeneity with respect to both density and size (as in cells of the anterior pituitary) contributes to the distribution of migration rates.[19]

The instantaneous velocity of a particle (assumed negatively charged) undergoing upward electrophoresis in a vertical density gradient may be described by

$$dx/dt = \mu(x)E(x) - 2a^2g[\rho - \rho_0(x)]/9\eta(x) \qquad (8)$$

where, in typical situations of interest, inertial and diffusion terms are

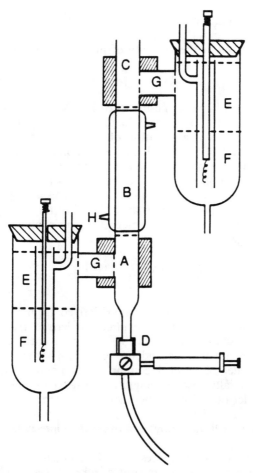

Fig. 1   Schematic diagram of a density gradient electrophoresis apparatus. A: Dense "floor" solution that supports a sample zone and density gradient B, which is surrounded by a water jacket; C: low-density "ceiling" solution that provides electrical continuity to upper side-arm electrode vessel; D: inlet system for gradient, sample and floor; E: electrophoresis buffer (low conductivity) overlay to prevent ions from saturated salt solution F from entering column through gel plugs G. Bright platinum wire electrodes are submerged within a chimney that facilitates the evolution of electrolysis gases (adapted from Ref. 29).

neglected. In Eq. (8) x is the vertical distance from the starting zone, μ is the anodic electrophoretic mobility, and E is usually determined from

$$E(x) = I/[Ak(x)] \qquad (9)$$

in which I is the constant applied current, k is the conductivity of the gradient solution, and A is the cross-sectional area of the gradient column. The second term of Eq. (8) represents the Navier-Stokes sedimentation velocity for spheres of density ρ and radius a under the influence of the gravitational acceleration g. The x-dependent variables are explicitly indicated in Eqs. (8) and (9). Standard electrophoretic mobilities, μ_s, are commonly expressed at the viscosity of pure water at 25°C, η_s. Thus,

$$\mu(x) = \mu_s \eta_s / \eta(x) \qquad (10)$$

Explicit expressions have been established for the x-dependent variables to allow integration of Eq. (8) to yield the distance of migration up the column as a function of time of application of the electric field. These empirical relationships consist of treating density and mobility as linear functions of $C(x)$, the polymer (or density gradient solute) concentration, conductivity as a quadratic function of $C(x)$, and viscosity as an exponential function.

Substitution of these functions into Eq. (8) gives an explicit relationship between velocity and migration distance. The time $t(x)$ to migrate a distance x may be found by numerical integration or analytically in terms of exponential integrals[38]:

$$t(x) = \int \frac{Q(x)\,\exp(\alpha x)}{L(x)}\,\mathrm{d}x \qquad (11)$$

The term $Q(x)$ is a quadratic function derived from the dependence of measured conductivity on $C(x)$, $\exp(\alpha x)$ is derived from the measured dependence of viscosity on $C(x)$, and $L(x)$ is a linear function of $C(x)$ derived from the measured dependence of electrophoretic mobility on $C(x)$. This simple model of vertical column electrophoresis can satisfactorily predict experimentally observed migration distance vs time plots at least for times of the order of 2–3 h without using fitted parameters.[19,34] In cases of a narrow cell density distribution the gradient may be tailored so that the cells are close to neutral buoyancy over an appreciable distance of the column.

Cell separation by electrophoresis in a Ficoll density gradient is typically accomplished in a 2.2 × 7.0-cm water-jacketed glass column using an isotonic gradient from 1.7 to 6.2% (wt/vol) Ficoll. It is found theoretically and experimentally that sedimentation, in addition to electrophoretic migration, plays a significant role in cell separation. Typically, fractions are collected by pumping "floor" (high-density solution) into the bottom of the column, thereby displacing the gradient and cells through the outlet at the top. Separands are typically collected in 0.5-ml fractions, so that, in a typical configuration (20- to 100-ml gradient), this is a low-volume batch process. Broadening of the bands of separands is expected to occur during the collection procedure. A calculation of Reynolds number for typical flow rates in the outlet tube leads to values less than 20; thus, laminar flow is expected, leading to some dispersion of the band. Because no fluid flows are present during the separation process, sample removal is the only source of significant sample band dispersion. When sample loads are too high, additional band broadening is caused by droplet sedimentation,[29] a gravity-dependent phenomenon that adversely affects all free electrophoresis processes in which the sample is applied as a zone.[35]

Droplet Sedimentation

The diffusion coefficients of small molecules are in the range of 10^{-6}–10^{-5} cm^2/s, of macromolecules 10^{-7}–10^{-6}, and the diffusion coefficients of whole cells and particles are 10^{-12}–10^{-9}. If a small zone, or droplet, of radius R contains n particles of radius a inside, whose diffusivity is much less than that of solutes outside, then rapid diffusion of solutes in and slow diffusion of particles out of the droplet (with conservation of mass) leads to a locally increased density of the droplet (Fig. 2):

$$\rho_D = \rho_0 + \frac{a^3}{R^3} n(\rho - \rho_0) \qquad (12)$$

If $\rho_D > \rho_0$, then the droplet falls down; if $\rho_D < \rho_0$, it is buoyed upward.[35] Droplet sedimentation (or buoyancy) is a special case of convection, and its occurrence is governed by the same Rayleigh-Taylor stability rules as convection, in which convection sets in when

$$d\rho/dx > 67.94vD/gR^4 \qquad (13)$$

where v = kinematic viscosity. In column electrophoresis droplet sedimentation cannot be overcome,[29] and the only way to scale column electrophoresis is to increase column cross-sectional area.[24] Under conditions of droplet sedimentation particles still behave individually unless the ionic environment also permits aggregation.[39,40] In the case of erythrocytes there is sufficient electrostatic repulsion among cells to permit the maintenance of stable dispersions up to at least 3×10^8 cell/ml.[41,42] Droplet sedimentation is an avoidable initial condition in zone processes, but it can be created during processing when separands become highly concentrated as in isoelectric focusing or interfacial extraction (see below). The absence of droplet

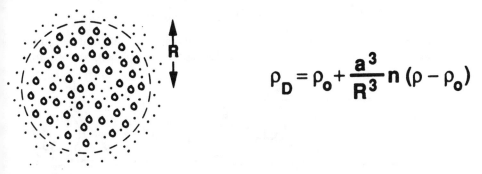

Fig. 2   Representation of droplet sedimentation. Zone density is increased as small molecules diffuse into the zone rapidly while large molecules or particles diffuse out slowly.[35]

sedimentation in low gravity is one of the most significant attractions of low-gravity separation science.

Low-Gravity Column Electrophoresis

The earliest electrophoresis experiments in low gravity were performed in a static cylindrical column on Apollo 14 and Apollo 16 lunar missions.[43-45] Horizontal column electrophoresis had been previously performed in rotating tubes of 1-3 mm i.d.[7] The small diameter minimized convection distance and facilitated heat rejection, whereas rotation counteracted all three gravity-dependent processes (i.e., convection, zone sedimentation, and, where applicable, particle sedimentation). This concept is illustrated in Fig. 3, which shows that the total vertical velocity vector oscillates as a sample zone moves in the electric field, so that spiral motion results with a radius vector that can be derived from equations of motion in which the sample zone is treated as a solid particle. Actually, the same zone is more like a sedimenting droplet, which can be treated as a particle with density ρ_D (see the section on droplet sedimentation). As a result of gravity and centrifugal acceleration, the center of the spiral (coordinates k, l) is not the center of the tube, and the vertical circle in x and y described by the sample zone is

$$(x - k)^2 + (y - l)^2 = r^2 \exp(2\gamma t) \tag{14}$$

in which a zone is stabilized in suspension when $k^2 + l^2$ and γ, the inverse time-constant for centrifugal motion are minimized. Hjertén solved the equations of motion for these values and found that neither has a minimum value as a function of ω or of any other controllable variable; hence, reasonable values must be determined from the solutions

$$\gamma = \frac{V_D(\rho_D - \rho_0)}{6\pi\eta R/\omega^2 + 32\pi R^5 \rho\rho_0/27\eta} \tag{15}$$

$$\sqrt{k^2 + l^2} = \frac{V_D(\rho_D - \rho_0)g}{\omega\sqrt{[(4\pi R^3/3)^2\omega^2\rho_0^2 + (6\pi\eta R)^2]}} \tag{16}$$

All vertical motion is assumed to be orthogonal to horizontal, electrokinetic motion; thus, it is possible to define conditions of zone density ρ_D

Fig. 3   Trajectory followed by a separand particle in free zone electrophoresis in a rotating tube. In the presence of gravity the center of the spiral is below the center of rotation.[7,40]

and radius R, solution density ρ_0, viscosity, and angular velocity ω such that the zone will be maintained in suspension. For example, if the Hjertén number

$$Hj \equiv \gamma\tau \qquad (17)$$

(defined here for the first time) is <1.0, then acceptable conditions for stability of the sample zone exist. In a typical separation τ, the residence time is between 1000 and 10,000 s, and γ is between 10^{-5} and 10^{-3} s^{-1}. On the other hand, if g is set to zero in Eq. (16), the sample zone is not displaced vertically from its original position except by centrifugal acceleration; hence, in the absence of g the tube must not rotate, and $\omega = 0$ in Eq. (15), giving $\gamma = 0$ and no centrifugal motion. Thus, low gravity represents an improvement over the rotating tube in terms of vertical sample zone stability (this analysis is also useful in low-gravity simulations on clinostats such as those used in gravitational biology research.[46]

In an open system, in which fluid flow in the direction of the applied electric field is possible, the net negative charge on the tube wall results in plug-type electroendosmotic flow (EEO, see above) that enhances total fluid transport according to the Helmholtz relationship[47] integrated over the double layer at the chamber wall

$$v = \frac{\zeta\epsilon}{\eta} E \qquad (18)$$

relationship; in a closed system, such as that used in almost all practical cases, flow at the walls must be balanced by return flow in the opposite direction along the center of the cylindrical tube, so that the flow profile is a parabola[48-50]:

$$v = \frac{\zeta\epsilon}{\eta} E(r^2/R^2 - \tfrac{1}{2}) \qquad (19)$$

Thus, if the electro-osmotic mobility μ_{eo} [coefficient of E in Eq. (18)] is high (of the order 10^{-4} cm^2/V-s), then sample zones will be distorted into parabolas as shown in the simulation of Fig. 4. It has also been shown that an additive parabolic pattern is superimposed if there is a radial temperature gradient that causes a viscosity gradient. A high temperature in the center of the chamber results in low viscosity and higher velocity due to both EEO return flow and faster migration of separands. Increased temperature also increases conductivity. Excessive power input under typical operating conditions can result in 25% higher separand velocity in the center of the tube even in the absence of EEO.[7] This undesirable, strictly thermal effect is independent of gravity and can be expected in low-gravity operations.

In such small-bore tubes with closed ends, scrupulous care must be taken to suppress EEO by using a nearly neutral, high-viscosity coating,[7] and rapid heat rejection and/or low current density is necessary to minimize

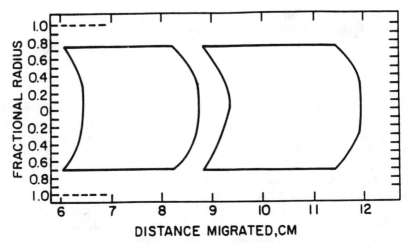

Fig. 4 Contour lines of sample bands predicted in simulation of human (right) and rabbit (left) erythrocyte migration, from left to right, in a cylindrical column in low gravity, using simulation method of Vanderhoff and Micale.[52] Assumptions were $\tau = 1.0$ h, $\mu_{eo} = -0.06$, μ (human) $= -2.05$, μ (rabbit) $= -1.05 \times 10^{-4}$ cm^2/V-s, $E = 18.6$ V/cm, sample width $= 0.75 \times$ chamber diameter. Courtesy of Dr. F. J. Micale.

radial thermal gradients. Such care was not taken in the Apollo electrophoresis experiments, and, although all gravitational effects were proved to be absent, EEO was comparable to the velocity of the test-particle separands.[44,45] Photographs taken during in-flight experiments were consistent with extreme examples of the parabolic distortions of the types modeled by Micale et al.,[51] Vanderhoff and Micale,[52] and Vanderhoff and Van Oss[6] and shown in Fig. 4.

A semiautomated static column electrophoresis system was designed for flight on the Apollo-Soyuz mission[53,54] and later used on Space Shuttle flight STS-3.[41,55,56] A sketch of its general features is given in Fig. 5. The system design included a procedure for minimizing EEO by silylating the heavily cleaned glass chamber walls and coating them with methylcellulose.[57] A low-conductivity buffer A-1 (Allen et al., 1974) was developed to minimize current density per unit field and hence minimize Joule heating. No protein separations were done with this system, but several cell types were separated on the Apollo-Soyuz flight in an experiment designated MA-011. To minimize sample processing on orbit, cells were frozen in a delrin disk that could be inserted into the electrophoresis columns (up to 8 disks and 8 columns). Columns were 6 mm in diameter and 15 cm long, to accommodate particles with $\mu = -1.0$ to -3.0×10^{-4} cm^2/V-s during a 1-h separation. Temperature control was provided by a thermoelectric cooler operated in two modes: thermostating at 25°C during electrophoresis and freezing the entire column after completion of each run. Each column could then be stored frozen for return to Earth with separated cells

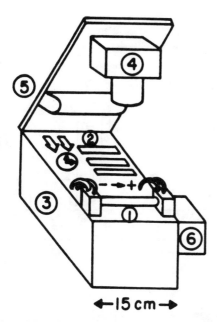

←15 cm→

Fig. 5 Sketch of free zone electrophoresis apparatus used on Apollo-Soyuz test project (experiment MA-011) and Space Shuttle flight STS-3 (experiment EEVT). The hinged cover held an illuminator (5) and a camera (4) that viewed the clock, voltmeter, ammeter (2) and, when uncovered, the electrophoresis column (1). Columns were interchangeable and were served by connectors to electrode compartments connected to a buffer recirculating system inside the housing (3). An accelerometer (6) was attached during STS-3 mission.

at different positions in the column to be sliced (like a loaf of bread) while still frozen for the collection of separated cell subpopulations. All of the electrophoresis fluids contained glycerol for the cryopreservation of living and fixed cells. Aldehyde-fixed erythrocytes[58] were used as test particles. Despite severely attenuated viability after flight, it was possible to obtain electrophoretic mobility distributions and unique subpopulations of cultured human kidney cells, for example, as Fig. 6 indicates.[23,59]

The same apparatus was used in Space Shuttle flight STS-3 in an experiment designated EEVT and designed to confirm the kidney cell separations achieved on Apollo-Soyuz with six more separations and to test the hypothesis that very high concentrations of test particles (fixed erythrocytes) can be separated in low gravity because of the lack of particle and zone sedimentation. Gravity-independent effects such as cell-cell interaction, reduced conductivity, and electrostatic repulsion could be sought in low-gravity experiments at a high cell concentration.[60] Thanks to in-flight photography during separations, the two erythrocyte samples could be tracked through the course of a 1-h separation[41] as shown in Fig. 7. It was generally found that particles (fixed erythrocytes) separated at high concentration on orbit separated in the same fashion as low-concentration cells in

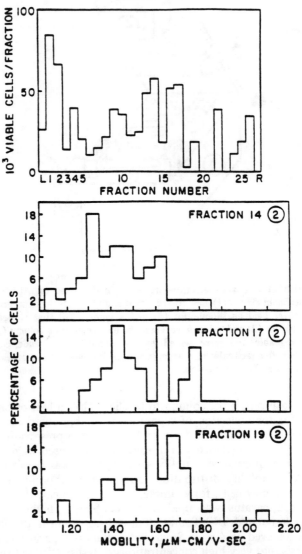

Fig. 6   Electrophoretic mobility distributions of human embryonic kidney cells cultured from three fractions separated by free zone electrophoresis in low gravity on Apollo-Soyuz test project (lower three panels). Top: Viable cell content of each of the 26 fractions collected from electrophoresis immediately after return from orbit. Migration was from left to right (redrawn from Refs. 23 and 59). Courtesy of NASA and CRC Press. Cells from different fractions had different mobility distributions and different functions.[53,54]

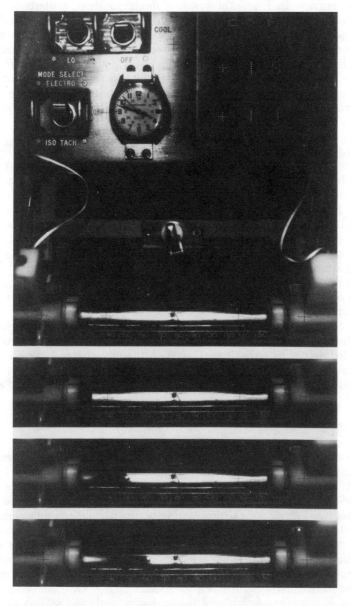

Fig. 7 Photographs showing top panel of free zone electrophoresis unit during STS-3 mission during 51 min of separation of a test mixture of concentrated human and rabbit erythrocytes. The column cooling cover was removed and photographed at 11, 22, 41, and 51 min. A broad band of cells is visible and moving from left to right (Ref. 41).

normal gravity. In Fig. 7, however, a single parabolic pattern of human and rabbit red cells is seen migrating from left to right, when two distinct bands were expected. The leading and trailing edges corresponded to the μ of human and rabbit red cells, respectively.[41,56] Simulation experiments in a rotating tube indicated that the observed pattern could occur if the human : rabbit cell ratio was 2 : 1 and significant (but undetermined) EEO was present.[39] Because of the loss of cryogenic liquid after sample return, there was no opportunity to slice separated cell subpopulations from the column.[55] Thus, the preservation of the position of the zones of cells by freezing proved to be detrimental to data collection on the Apollo-Soyuz test project (ASTP) and fatal on STS-3. It was subsequently established that similar separation experiments would be feasible in a colloidal sol that could be regelled after separation and remain solid without freezing or refrigeration.[61,62]

The Soviet space electrophoresis unit Tavriya did not depend upon immobilization of separands after separation, but it was designed with lateral outlets along one side of the column and with sample and buffer injection units at opposite ends, so that fraction collection could proceed during spaceflight onboard Salyut and Mir orbiting spacecraft.[63-65] The resolution of this system was established on the basis of the number of fixed lateral outlets. It was tested on Salyut-7 using samples of rat bone marrow cells, which separated into five visible bands prior to harvesting.

The Soviet space electrophoresis unit Genom was apparently designed to accommodate a very diluted gel in which DNA molecules could be separated according to molecular mass. Although separands were immobilized in the gel, a series of needles was used to harvest fractions while still on orbit in the Salyut-7 Space Station, possibly because a very weak gel was used.[64,66]

Free-Flow Electrophoresis

Free-flow electrophoresis has not found favor as an industrial-scale purification process. The process cannot be scaled up in size because convection occurs as a consequence of inadequate heat removal from current-carrying electrolyte solutions. The process cannot be scaled up in mass because sample zone sedimentation occurs when the sample concentration is too high. Various approaches to this problem have been utilized. In one system the buffer curtain flows upward in the annular zone between a vertical stationary and rotating cylinder. This system and conventional FFE systems have been compared by Mosher et al.[13] An equally heroic method uses the microgravity environment of spaceflight.[67,68] Of these two methods the first sacrifices resolution, and the second is costly and seldom available. Conventional free-flow electrophoretic separators have used downward flow and brine cooling to counteract those problems. In the case of the Continuous Flow Electrophoresis System (CFES) designed for operation at 1 g, upward flow over a long distance (>100 cm) with low field strength (<20 V/cm) and use of the electrode buffer as a coolant counteract these two problems. A generic free flow separator is sketched in Fig. 8.[69-71]

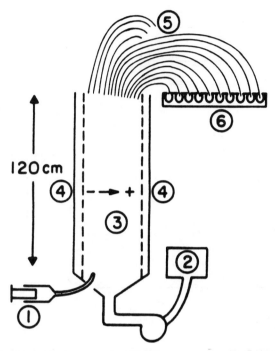

120 cm

Fig. 8 Sketch of generic free-flow electrophoresis system. Sample stream (1) enters chamber (3) near the bottom (in this case) of the upward-flowing, pumped (2) buffer stream, and its component separands migrate from left to right in the electric field between electrodes (4) according to their electrophoretic mobility and exit at the top of the chamber through individual outlets, 20–200 in number (5), depending on chamber design, for collection into fraction tubes (6).

A continuous-flow electrophoresis system is sketched diagrammatically in Fig. 9. Several investigators have constructed mathematical models of FFE.[72–82] The main ingredients of such a model include the following variables[76]: electrophoretic migration of separands, Poiseuille flow of pumped carrier buffer, electroendosmosis at the front and back chamber walls, horizontal and vertical thermal gradients, conductivity gradients, diffusion, sample input configuration, and fraction collection system. Figure 9 indicates the path followed by three categories of separands (low, medium, and high mobility) and indicates the effects of Poiseuille flow and electroendosmosis when the sample is injected at the bottom, electrophoresis is to the right, and fractions are collected at the top. A coordinate system is shown in which the z axis is the direction of pumped fluid (buffer) flow, the y axis is the chamber thickness, and the x axis is the chamber width and direction of migration of separands in the applied, horizontal electric field. Typical dimensions of available FFE chambers are $x_m = 6$–15 cm, $y_m = 0.03$–0.15 cm, (this dimension can be increased to 0.30 cm in low gravity), and $z_m = 22$–110 cm, where the subscript m designates the

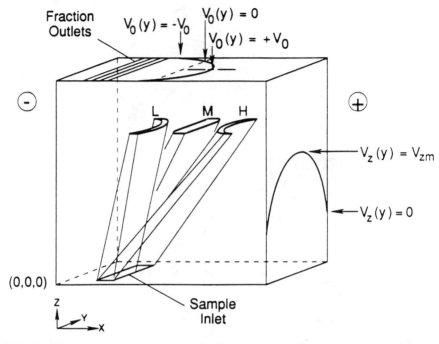

Fig. 9 Schematic diagram (not to scale) of an upward-flowing free-flow electrophoresis chamber. A coordinate system with (0,0,0) at the lower front left corner is shown. Sample stream is shown as a slit (circular sample streams are commonly used). Anodal migration of low, medium, and high (L,M,H) mobility separands results in crescent-shaped bands due to Poiseuille retardation of particles near the walls [where $v_z(y)$ is low] in the case of high-mobility and electro-osmotic flow near the walls [where $v_0(y)$ is high] in the case of low mobility. These two parabolic distortions balance in the case of medium-mobility separands.

distance to the opposite wall from the origin. The motion of a particle through such a chamber is complex. The migration distance x is the sum of electrophoretic migration and electro-osmotic flow:

$$x = \mu E\tau(y) + v_{eo}(y)\tau(y) \qquad (20)$$

But $\tau(y)$ and $v_{eo}(y)$ are explicit functions of y.[83] Under some conditions (nonuniform conductance) E is also a function of x and y.[8,84] None of these dependencies is gravity dependent; thus, their effects on resolution and capacity are the same in unit gravity and in low gravity. The specific dependencies are the parabolic distribution of electroendosmotic backflow, given by von Smoluchowski's equation for a rectangular chamber having thickness much smaller than its width or height:

$$v_{eo}(y) = \tfrac{1}{2}v_{eo}(0)[3(y-b)^2/b^2 - 1] \qquad (21)$$

where b is half the chamber thickness and the parabolic distribution of residence times, governed by the y dependence of the velocity along the z axis, assuming no diffusive or electrokinetic (see below) motion in the y direction and using the notation of Fig. 9:

$$v_z(y) = v_{zm} - (v_{zm}/b^2)(y - b)^2 \qquad (22)$$

into which can be substituted $v_{zm} = 3v_{zav}/2$, where v_{zav} is the average fluid flow velocity, which can be measured. This results in a residence time distribution given by

$$\tau(y) = z_m/v_z(y) = z_m/[(3v_{zav}/2) - (3v_{zav}/2b^2)(y - b)^2] \qquad (23)$$

and a migration distance distribution (in the absence of EEO) given by

$$x(y) = \mu E \tau(y) = \mu E z_m/[(3v_{zav}/2) - (3v_{zav}/2b^2)(y - b)^2] \qquad (24)$$

which is a statement that separands will arrive at the outlet end of the chamber distributed in nested crescents in which separand particles nearest the front and back walls migrate the farthest, as indicated in Fig. 9. However, the y dependencies can be minimized by using a thick chamber and limiting the sample stream to a central zone near the midpoint of the chamber in the y direction. A thicker chamber is allowed in low gravity where thermal distortions of flow presumably do not occur (ignoring Marangoni flow due to a small thermal gradient in the z direction and the viscosity gradient mentioned earlier). Figure 10 indicates how sharper bands can be achieved in a thicker chamber. However, such bands can be unsharpened by electrohydrodynamic phenomena when the conductivity of the sample zone is not properly matched to that of the carrier buffer, resulting in $E(x, y)$ not being constant, because $k(x,y)$ is not constant.[8,85-87] This phenomenon is also independent of gravity, although any effect of gravity on heat transfer will also affect $k(x,y)$, since k depends linearly on temperature. The general relationship for field distortion is derived from Taylor's "electrohydrodynamic" concept applied to a spherical drop[88]; it assumes a sample zone of circular cross section that is small compared to the y dimension of the chamber and having different conductivity k_s, viscosity η_s, and permittivity ϵ_s from that of the carrier buffer. The scalar magnitude of the surface stress S is given by

$$S = \frac{E^2 \epsilon \{\epsilon/\epsilon_s[(k_s/k)^2 + (k_s/k) + 1] - 3\}}{4\pi[(k_s/k)^2 + 1]^2} \cos 2\theta \qquad (25)$$

where θ is the angle subtended by S and the applied electric field lines. Thus, at any angle with respect to the direction of migration, the sign of the quantity in braces and that of the cosine function combine to determine whether the sample stream is pulled outward or compressed inward. This function predicts that under conditions of $k_s > k$ the sample stream will be pulled outward at $\theta = 90$ deg (toward the chamber walls) and compressed

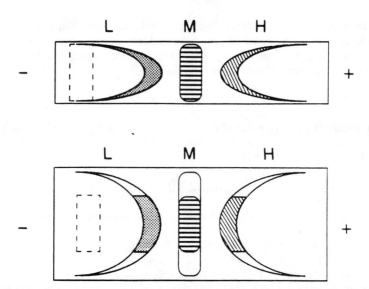

Fig. 10 A thicker FFE chamber allows the separation of separands in narrower bands, resulting in higher output purity. Preventing the sample stream from approaching the front and back chamber walls results in less band distortion due to Poiseuille and electro-osmotic flow. Dashed rectangle represents cross section of original sample stream at the inlet.

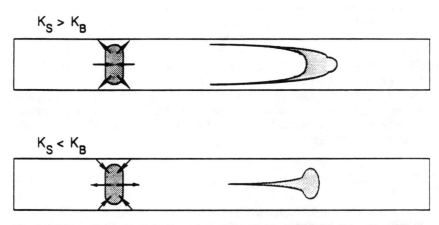

Fig. 11 Electrohydrodynamic distortion of sample bands in FFE. From Eq. (25) the stress vector at the "surface" of a high-conductivity sample stream is outward perpendicular to the electric field and inward parallel to the field lines (top diagram) and vice versa for a sample stream with lower conductivity than the carrier buffer (lower diagram).

in the perpendicular direction, as in the upper example of Fig. 11 and vice versa as in the lower example. By applying an ac field [the field appears as E^2 in Eq. (25)] to a thick chamber and using end-on photography of sample streams of latex particles, Snyder and Rhodes[8] were able to confirm the predictions of Eq. (25) and their consequent effects on FFE. They also pointed to an exacerbation of this phenomenon when one of the current-carrying ions in the sample stream has much higher mobility than the other, and the conductivity within the sample stream becomes nonuniform.

By ignoring Eq. (25) and assuming laminar flow, approximate relationships can be derived for the overall motion of a separand particle in FFE by substituting Eqs. (21–24) into Eq. (20) to give

$$x(y) = \frac{2Ez_m(\mu - \mu_{eo}/2)}{3v_{zav}[1 - (y - b)^2/b^2]} \tag{26}$$

which states that the direction of the parabola depends on the relative magnitudes of the separand mobility and the electro-osmotic mobility of the chamber wall.[78] In the example illustrated in Fig. 9, high-mobility separands (H) migrate farthest to the right near the chamber walls, whereas low-(L) or zero-mobility separands move farthest to the left near the chamber walls.

By applying Newton's laws of motion to Eq. (25) to determine the rate of formation of the structures depicted in Fig. 11 and combining with Eq. (26), the gravity-independent motions of separands in the total process could be modeled. The effects of temperature gradients on conductivity and viscosity can be added.[78,89]

The preceding relationships deal only with the gravity-independent components of separand motion. Gravity-dependent variables are treated more qualitatively, and the Rayleigh number,

$$Ra = \frac{g\beta|\nabla T|x_m^4}{\alpha\nu} \tag{27}$$

must generally not exceed ~8 to prevent convective instability due either to solutal zone sedimentation [Eq. (12)] or thermal convection [Eq. (27)].[4,73,90,91] Dimensionless analyses by Ostrach[90] and Saville[73] predicted asymmetric stationary instabilities (and corresponding temperature profiles) at lower Ra when carrier buffer and coolant are both flowing downward so that the chamber is warmer at the bottom than at the top. Deiber and Saville[74] pointed out that the cathodic side of the chamber will be more effectively cooled by the action of electroendosmosis, which sweeps the carrier buffer inward from the walls at the cathode side and outward from the center at the anode side producing approximately linear ∇T in the x direction, the condition to which Eq. (27) applies. This coupling of convection to electroendosmosis results in predicted asymmetric flow distortions. As Eq. (27) states, thermal distortions of flow are completely dependent on gravity and very sensitive to chamber width (x_m^4). At 1 g, practical chamber thicknesses are limited to about 0.5 mm. Ivory[4] has calculated that, by allowing the chamber to heat to 40°C when $g = 0$, greater chamber thickness (y_m) is feasible, and a 440-fold increase in sample capacity could be realized in FFE. This is similar to the experimentally demonstrated enhancements in protein separations in low gravity.[68,92,93]

In laboratory experiments Rhodes and Snyder[94] were able to induce and measure axial (z direction) and lateral (x direction) temperature gradients and to carefully control the Rayleigh number in the axial direction. It is generally assumed that the flow velocity $v_z(x) =$ const over most of x, except very close to $x = 0$ and $x = x_m$, the width of the chamber, where no-slip conditions presumably exist. Incidentally, the edge flow can also be controlled by using microporous electrode membranes and regulating electrode buffer pressure (A. Strickler, unpublished observations). Rhodes and Snyder visualized flow by injecting fluorescent dye uniformly across the chamber width and photographing the downstream fluorescence front. Sharp lateral gradients, with the center 0.4°C cooler than the edges, caused a two-fold increased velocity at the edges relative to the center, and, if $Ra > +10$, backflow occurred in the center. Similarly, an axial gradient giving $Ra > 5.1$ (0.9°C from top to bottom of the column) at the edges caused backflow at the edges. The results are fully consistent with the predictions of Ostrach and Saville that backflow will occur in descending FFE around $Ra > 8$. Some stability can be gained by concurrent flow of carrier buffer and coolant upward, so that a small temperature rise (3°C over a 100-cm column length) reduces the density at the top of the column.[68] Construction features of the CFES are sketched in Fig. 12.[8]

Free-Flow Electrophoresis in Low Gravity

Free-flow electrophoresis in low gravity, along with electrophoresis in low gravity in general, was probably born as a concept in Wyeth Laboratories in 1969.[76] This separation method was first implemented in spaceflight in an experiment named MA-014, July 16, 1975, on the ASTP. The stated purpose of this experiment was "to verify the theoretically expected better and higher capacity separations in the free-flow electrophoresis system under zero-g conditions."[67] The specific objectives included the characterization of temperature and velocity effects in a wide gap, the separation of preparative quantities of living cells, and an engineering study for potential future Spacelab equipment. To fit the envelope of the crowded spacecraft, the fully automated system served a separation chamber that was 3.8 mm thick (not considered feasible at 1 g) × 28 mm wide × 180 mm long (electrode length). No fractions were collected, but the system was monitored optically using a halogen lamp and a scanning detection system to measure light absorbance with a 128-sector photodiode array. All test samples were cell suspensions [rat bone marrow, mixed human and rabbit erythrocytes (RBC), rat spleen, and rat lymph node with human RBC markers]. Starting samples approaching 10^8 cells/ml were used in all cases, and separations were compared with 1-g counterparts at lower cell concentrations (0.1 those used in spaceflight).[95] With one exception (bone marrow), distributions were essentially the same in low gravity and at 1 g. Effects of

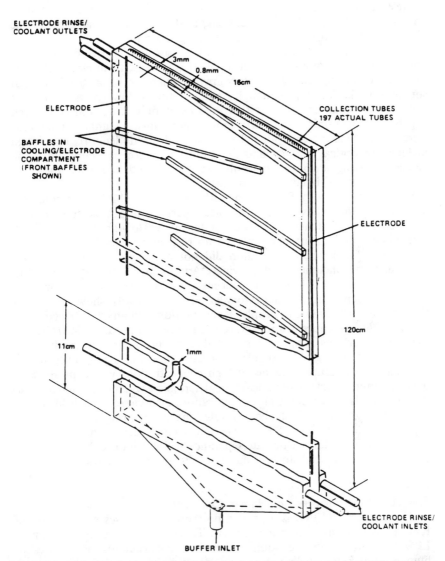

ELECTRODE RINSE/
COOLANT OUTLETS

3mm

0.8mm

16cm

ELECTRODE

COLLECTION TUBES
197 ACTUAL TUBES

BAFFLES IN
COOLING/ELECTRODE
COMPARTMENT
(FRONT BAFFLES
SHOWN)

ELECTRODE

120cm

11cm

1mm

ELECTRODE RINSE/
COOLANT INLETS

BUFFER INLET

Fig. 12 Diagram of construction features of McDonnell Douglas CFES chamber, showing upward buffer flow, sample inlet, sample outlets (top), ribbon electrodes, combined electrode compartments and cooling jacket, with alternating sloped baffles to suppress thermoconvective transport of electrode buffer by creating a serpentine pathway for its flow. The chamber width is 16 cm in low gravity and 6 cm on the ground. Electrode gases are exchanged off-line using membrane cartridges.[8] Courtesy of R. S. Snyder.

low gravity on resolution and throughput were not quantitated. One interesting feature of this research was the initial confounding of data analysis by unexpected light intensities, evidently due to the effect of the absence of thermal convection on the halogen lamp.

McCreight et al.[76] designed and built a low-gravity free-flow electrophoresis unit for sounding-rocket flight. It was to perform electrophoretic separations during the 7-min low-gravity missions of a Black Brandt sounding rocket (1973). Although only 5–7 min of low gravity is a short time, a small chamber, such as that used by Hannig et al., was considered capable of functioning and performing separations. Chamber dimensions were 5 mm thick × 5 cm wide × 10 cm long (electrode length). The design includes an optical scanner and an arrangement for collecting up to 50 fractions, giving a collection resolution of 2% and hence an ability to measure widths of bands of separands.

Snyder et al.[9] and Snyder and Rhodes[8] were able to demonstrate qualitatively the behavior of particulate separands in sample streams with conductivity unmatched to carrier buffer under low-gravity conditions, where convective disturbances played no role, and under laboratory conditions (ac field) where separation did not occur but photography was feasible. They concluded the following: "Additional effort is also needed to determine the role of particle concentration, if any, in free-flow electrophoretic separations. Laboratory experiments clearly show limitations that are not entirely due to droplet sedimentation." In any case, discontinuities in *buffer* ion concentrations have been shown to produce electrohydrodynamic distortions of sample bands that seriously compromise separations, as illustrated in Fig. 11. Soluble separands that add significantly to the conductivity of sample streams at high concentrations would presumably have the same effect. It remains to be determined whether or not particulate separands would, at high concentration, also cause such field distortions. Because very highly concentrated fixed erythrocytes migrated at expected velocities in the STS-3 experiment cited earlier, there exists one piece of evidence that particulate separands will not induce, by themselves, electrohydrodynamic distortion during electrophoretic separation.

Free-Flow Electrophoresis of Human Cell Proteins

Erythropoietin (EPO) is a protein hormone that causes proliferation of RBC precursors and their maturation. The kidney is the main organ of EPO biosynthesis in the adult[96]; therefore, supernatants of cultures of human adult kidney cells are a source from which to purify this drug. A severe "sample stream problem" (a rare product in a very dilute solution with over 100 other solutes) existed, and a major separation project was designed around this and other high-value products through a Joint Endeavor Agreement (JEA) with NASA, in which publicly supported scientists would use the CFES in orbit in exchange for launch services provided to the private sector. The Space Shuttle flights that were utilized for this purpose and the corresponding objectives are summarized in Table 3.

Table 3 Low-gravity experiments in free-flow electrophoresis on the U.S. Space Shuttle (STS)

STS no.	Date	Objectives	Results
4	6/82	Systems test at 0 g. Separation of albumins at high concentration.	Satisfactory function. 400-Fold increase in system capacity compared to ground.
6	4/83	Test purification of cell culture product.	700-Fold increased capacity. 4-Fold increased resolution.
		Test electrophoresis of hemoglobin and polysaccharide.	Retrograde migration of concentrated protein.
7	6/83	Test purification of cell culture product.	
		Test separation of latex microspheres.	Effect of conductance discontinuities.
8	8/83	Test separations of living cells from rat pituitary, dog pancreas, human kidney.	Functional subfractions of cells returned for study.
41-D	8/84	Purification of EPO for preclinical study.	Adequate amount of material purified but high endotoxin level.
51-D	4/85	Purification of EPO for preclinical study. Sterility tests.	EPO separated with minimum endotoxin.
61-B	8/85	Purification of EPO for preclinical study.	Adequate amount of material purified.

Early experiments demonstrated a 400-fold increase in capacity in a two-protein separation (rat serum albumin and ovalbumin). This is indicated in Fig. 13. Similarly, a 718-fold increase in capacity was obtained in the separation of proteins from a partially purified, highly concentrated culture supernatant solution.[92] An electrophoretic profile of this mixture, separated in the CFES in orbit, is shown in Fig. 14 (Ref. 93). The migration distribution of kidney cell culture supernatant medium subjected to electrophoresis at 1 g is shown at the top of the figure.[76] Eventually this material's sample stream problem was solved in other laboratories by engineering cultured cells that produce EPO at high concentrations after the EPO gene was isolated.[97]

As a provision of the agreement, it was possible for NASA-supported scientists to perform basic separation experiments and cell separation experiments using the low-gravity CFES. By using a mixture of polystyrene latex particles with three different mobilities (distinguishable by size and color), Snyder et al.[9,98] and Snyder and Rhodes[8] were able to discover the applicability of Eq. (25) to the distortion of sample bands having different

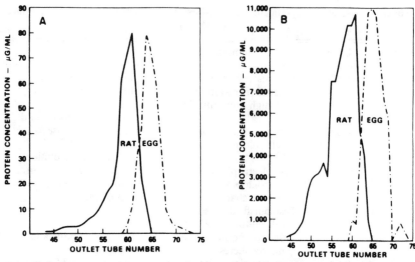

Fig. 13   Protein content of collected fractions after continuous flow electrophoresis in the laboratory (A) and in spaceflight (B) of a mixture of rat serum albumin (solid line) and ovalbumin (dashed line) at 0.1% (wt/vol) of each protein (A) or 12.5% of each protein (B). The product of sample concentration and flow rate in B is 463 times that in A (Hymer et al.[68]; courtesy of Humana Press).

conductivities from those of the carrier buffer and, subsequently, to evaluate the influence of high protein concentrations on sample band spreading. In particular, the retrograde migration of hemoglobin (toward the cathode) at high concentration, shown in Fig. 15, was explainable on the basis of a combination of the differential mobilities of the cation (Na^+) and the anion (hemoglobin$^-$), leading to the establishment of conditions to which Eq. (25) would apply.[8]

Free-Flow Electrophoresis of Living Cells

FFE has also been applied to kidney cell electrophoresis for a number of years.[53,54,59,68,99-106] This method will be used when large numbers of cells separated at high resolution are required.[107] Although FFE has lower resolution than density-gradient electrophoresis, the net combination of resolution and yield is advantageous. Early-passage cultures of human embryonic kidney cells contain a small fraction of cells that produce plasminogen activators, and those that produce urokinase consistently appear in a high electrophoretic mobility fraction.[18] Such cell populations were separated at high concentrations during electrophoresis in low gravity onboard Shuttle flight STS-8, and their capacities to produce plasminogen activators was evaluated.[108] During the same series of experiments on orbit, rat anterior pituitary cells, which are known to include a high-mobility fraction that is rich in growth hormone production,[19] were separated according to electrophoretic mobility, and separate fractions rich in growth

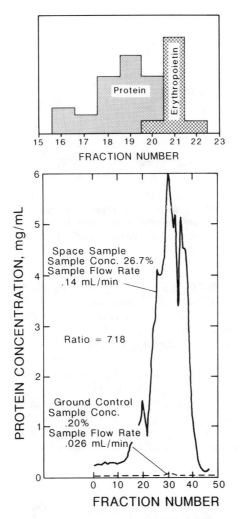

Fig. 14   Separation by free electrophoresis of kidney cell culture fluid. Top graph: laboratory separation (McCreight et al.[76]); bottom graph: CFES separation in space.[93]

hormone and prolactin production were characterized.[20,68,109,110] Also, at the same time, dog pancreatic islet cells were subjected to electrophoretic separation, and fractions rich in insulin and glucagon production were characterized (Hymer et al., 1989). The principal results of these three living-cell separations are summarized in Fig. 16.

Current and Future Low-Gravity Electrophoresis Research

One approach to simultaneously increasing capacity and resolution is the elimination of all flows that result in nonuniform migration velocity of separands vs y and the use of a thick chamber with large y_m. The former

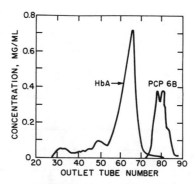

Fig. 15 Electrophoresis of hemoglobin and *Pneumococcus* capsular polysaccharide in CFES during Space Shuttle mission STS-6. Fraction 36 corresponds to zero migration distance. The occurrence of hemoglobin at lower fraction numbers is considered indicative of electrohydrodynamic band distortion as in Fig. 11 (see Ref. 8). Courtesy of R. S. Snyder and CRC Press.

can be achieved by setting $v_z(0) = v_{zm}$ by causing the wall to move at the same velocity as buffer flow[8,78,111]; this causes $\tau(y)$ to be constant in Eq. (23). If, at the same time, the zeta potential of the wall is set close to zero, v_{eo} becomes negligible, and there is no electroendosmotic distortion of any bands of separands. Increasing chamber thickness still requires elimination of convection, and it has been recommended that this be accomplished in low gravity as a component of the U.S. effort in low-gravity science. A diagram of this device is shown in Fig. 17.

A new free-flow electrophoresis apparatus was developed by German firms to be used in low-gravity experiments on the Space Shuttle.[95] The chamber is 20 cm long, 7 cm wide, and 0.5 mm thick on the ground and 4 mm thick in low gravity, based on calculations by Boese.[80] The unit can

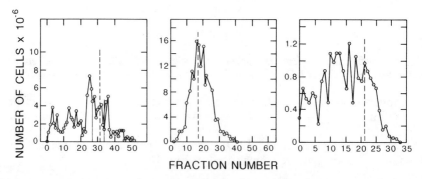

Fig. 16 Separation of three populations of lining cells into fractions. Left: adult rat pituitary cells, center: adult canine pancreas cells, right: human embryonic kidney cells. In each case the vertical dashed line indicates the fraction of cells having the highest production of growth hormone, insulin and plasminogen activator, respectively (adapted from Hymer et al.[68]).

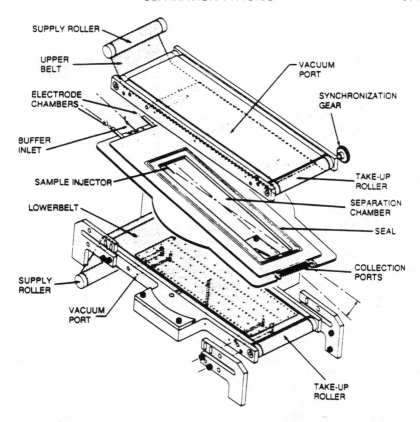

SUPPLY ROLLER

UPPER
BELT

VACUUM
PORT

ELECTRODE
CHAMBERS

SYNCHRONIZATION
GEAR

BUFFER
INLET

SAMPLE INJECTOR

TAKE-UP
ROLLER

LOWERBELT

SEPARATION
CHAMBER

SEAL

COLLECTION
PORTS

SUPPLY
ROLLER

VACUUM
PORT

TAKE-UP
ROLLER

Fig. 17   Schematic drawing of an example of a moving-wall electrophoresis system.[111] Courtesy of R. S. Snyder.

operate at 120 V/cm on the ground, and separands are collected in seven fractions. It is part of the German space shuttle research program Biotex. This system was operated on a TEXUS sounding rocket in May 1988. Optical monitoring during ascent and coast phases demonstrated separations of test particles, namely, rat, rabbit, and guinea pig erythrocytes.[95]

With the addition of the Kvant module to the Soviet Mir Space Station in April 1987, the Svetlana automated electrophoresis apparatus became the first permanently orbiting separation research facility. Products separated using this system have been turned over to centers in Moscow and Leningrad for use as standards in vaccine manufacture. A gel electrophoresis apparatus Svetoblok has also been operated on Mir.[112]

The Centre National d'Études Spatiales (CNES, Paris), along with European partners, proposed a free-flow electrophoresis facility Recherches Appliquées sur les Methodes de Séparation Electrophorétique Spatiales (RAMSES) for installation on the Spacelab International Microgravity Laboratory (IML-2) for a 1993 flight and subsequent flights. A system of interchangeable separation chambers is planned, and monitoring facilities are to include cross-sectional illumination, which is found to be very useful in ground-based studies.[8] Proposed free-flow chamber dimensions are 30 cm long × 6 cm wide × 3.0 mm thick (low gravity) or 1.5 mm thick (1 g).[113,114] Processes to be studied include free-flow zone electrophoresis, isoelectric focusing, and electrohydrodynamic effects. Eventually, a capacity to study membrane processes and chromatography is to be included.[115] Experiments planned by two U.S. laboratories are expected to be accommodated on Spacelab flights involving this apparatus.

A miniaturized free-flow electrophoretic separator with an optical monitoring system has been constructed for use in Japanese-supported Spacelab experiments.[116]

Isoelectric Focusing and Isotachophoresis

Isoelectric focusing (IEF) is the movement of a separand through a pH gradient to a pH at which it has zero net charge, at which point it ceases to move through the separating medium. Isoelectric focusing is considered an *equilibrium* process. The steady-state condition that occurs when a separand has reached equilibrium can be characterized quantitatively on the basis of the dissociation constants of water, ampholytes (amphoteric electrolytes used to establish the pH gradient), and separands; this has been done in some detail by Palusinski et al.[117] Equally interesting is the *rate* process whereby water, ampholytes and separands reach equilibrium by electrophoresis, and this has been modeled as well.[118] The overall result is a generalized mathematical description of electrokinetic separation processes.[119] Briefly summarized, this model makes it possible to calculate the geometrical distribution of any ion at any time in a uniform electric field applied to a stable solution in a column.[120] As developed to date the model is not applicable to flowing separation systems or circumstances in which gravitationally or electrically driven flows occur. The key relationships are the dissociation constants of the ampholytes A_j to the + and − forms,

$$K_{j1} = [A_j^0][H^+]/[A_j^+] \qquad (28a)$$

$$K_{j2} = [A_j^-][H^+]/[A_j^0] \qquad (28b)$$

the Einstein relationship between the diffusion coefficient and mobility of species i

$$D_i = RT\mu_i/e \qquad (29)$$

the flux of the ith species due to electrophoresis and diffusion,

$$-J_i = \mu_i z_i C_i(x)E(x) + (RT\mu_i/e)(dC_i/dx) \qquad (30)$$

the mass balance on the ith species (assuming no bulk flow),

$$dC_i/dt = -dJ_i/dx + R_i \qquad (31)$$

the conservation of charge among L ampholytes, assuming no net production of charges within the column,

$$e \sum_{1}^{3L+2} z_i R_i = 0 \qquad (32)$$

and the preservation of electroneutrality,

$$e \sum_{1}^{3L+2} z_i C_i = 0 \qquad (33)$$

where all variables are as previously defined, and R_i is the rate of production of the ith species, with the net production of the three species defined in Eq. (28) being zero. These relationships are combined in a computer program that performs numerical simulations using as inputs the physical dimensions of the column, the applied current, electrophoretic mobilities, and dissociation constants of the i components. The output includes the concentration profiles of all charge and uncharged species of all of the components, especially including pH, the profile of conductivity and of $E(x)$ [see Eq. (9)].

Three patented free-fluid devices have been designed to counteract the effects of the resulting concentration discontinuities and their subsequent convective phenomena. The recycling isoelectric focusing method (RIEF) uses a small focusing chamber and a large-capacity gang of reservoirs in a heat exchanger to minimize the volume within which the electric current is applied. Further stabilization is achieved by the separation of the focusing chamber into 10–20 compartments using nylon mesh screens.[121] This system has been tested industrially for the purification of interferons.[122] A schematic representation is shown in Fig. 18. A similar system, recycling free-flow focusing (RF3), utilizes stabilizing shear flow rather than screens in the focusing chamber.[3] As a consequence of low-gravity simulation studies, a successful effort was made to scale up the rotating free zone system of Hjertén[7] described earlier for IEF. The resulting apparatus (Rotofor) consists of a cylindrical chamber divided into 20 subcompartments by a parallel array of nylon meshes. It has a total capacity of 40 ml, and its central axis contains a cooling finger.[3,123] This is now a commercially available system, originally developed for the simulation of low gravity studies but now the most common instrument for free-fluid electrophoretic purification of proteins in the world.

The ampholytes normally used to establish natural pH gradients are expensive and often harmful to separands, especially living cells. Nonampholyteric, harmless buffer systems have been developed to create artificial pH gradients.[124,125] This separation method is of limited use in the purification of biological cells, most of which are isoelectric at a lower pH than they can tolerate.[126–128]

Isotachophoresis can be modeled, and in some cases implemented, in the same way as IEF.[119] Isotachophoresis is the separation of solutes on the basis of electrophoretic mobility when they migrate in the same direction at the same velocity at steady state. The boundaries between species of

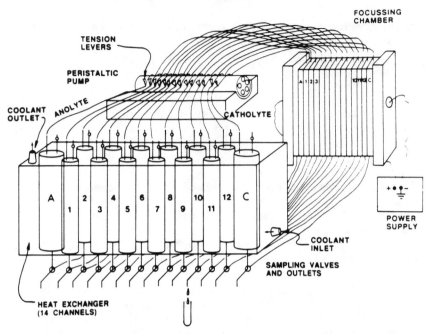

Fig. 18 Schematic representation of recycling isoelectric focusing system, showing heat exchanger with fraction reservoirs (front), small focusing cell with compartments separated by screens (rear), ganged peristaltic pump (top), and catholyte and anolyte compartments (C and A).

different mobility are sharply defined and stabilized by electrical forces,[129] even to some extent against convection.[130] A discontinuous electrolyte system is used at the site of sample injection. Typically, a system consists of a sample and "spacer ions" that will migrate between separands and enhance the distance between them at steady state, with a high-mobility "leading ion" reaching the anode first and a low-mobility "terminating ion" reaching the anode last (or vice versa when the separands are cations). It is possible, for example, to separate solutes on this principle using fast-flow stabilized recycling methods similar to RIEF.[131] Because this is a quasiequilibrium process operating at a final steady state, it shares many of the characteristics of isoelectric focusing, including concentration and density discontinuities.

Low-Gravity Experiments in Isoelectric Focusing and Isotachophoresis

The principal objectives to date of low-gravity studies in isoelectric focusing have been to explore means of counteracting "slumping" or sedimentation of concentrated bands of separands and to characterize EEO in the absence of gravity-related confounding factors such as thermal and solutal convection.[132]

The model relationships (28–32) do not account for the existence of net bulk fluid flow, as in EEO. Since two of the devices mentioned earlier (RIEF and Rotofor) depend on a system of screens to stabilize electrofocused separands and pH gradients near equilibrium, and during transient conditions it is important to know the influence of the screens, and their shape and composition, on EEO in rotationally or flow-stabilized IEF. Although net transport through the screens can be characterized on the ground, this flow is a sum of electroendosmotic and convective flows. The latter is expected to dominate near equilibrium when concentrated separands increase the local density; therefore, it was considered necessary to measure EEO in the absence of convection in low gravity.

The low-gravity experimental design consisted of an array of eight IEF chambers in which ampholytes would be brought to equilibrium in the presence or absence of a screen of different materials and configuration within each chamber. Chambers were "horizontal" cylinders of glass (coated or uncoated) 4.5 cm long and 0.625 cm i.d. At constant voltage (75 V dc from a battery) to each chamber, the current was monitored as a function of time, since current normally drops to less than 10% of its initial value as steady state is approached. Photography of the progress of hemoglobin (red) and albumin (dyed blue) also monitored progress toward the steady state. A diagram of the apparatus is shown in Fig. 19.[133]

Focusing experiments were designed to last up to 90 min on Shuttle flight STS-11. After approximately 20 min, disturbances occurred, and these resulted in internal flows that continued in all chambers for the rest of the experiment. It is speculated that sharp gradients of conductivity and of dielectric constant, as they developed in the absence of gravity (and hydrostatic pressure), may have lead to electrohydrodynamic body forces[134] only previously observed in immiscible liquid systems.[123]

The Tavriya electrophoresis unit was operated in IEF mode on Salyut-7 in 1982 (Ref. 64); however, an artificial pH gradient was formed using the borate-polyol method of Troitsky and Azhitsky[124] in the place of a natural pH gradient usually formed by the focusing of ampholytes (see above). High-resolution separation of five variant forms of human serum albumin was achieved. This appears not to have been possible at 1 g.[63]

Isotachophoresis experiments were performed on the Apollo-Soyuz flight.[54] The test separands were fixed or fresh human and rabbit erythrocytes. The columns were similar to those used for electrophoresis (15.24 cm long × 0.64 cm diam) and were operated in the MA-011 apparatus with a palladium cathode and a silver anode. Ground control experiments were performed using a rotating device (see above). It was possible to note that cells stayed in their bands in low gravity and that the bands had a sharper front than at 1 g, and these were predictable results, but, because of intermittent power to the column the separation did not reach steady state, and the quality of bands and their migration could not be evaluated.

The electrophoresis column Tavriya was used in isotachophoresis mode on the flight of Salyut-7, and a high-resolution separation of five variant forms of human serum albumin was achieved. This appears not to have been possible at 1 g.[63,64] Usually, sample bands in a cylindrical column are

operational characteristics:

Sample size:	5.08 L x 0.64 cm Dia
Sample volume:	0.82 cm^3
Column assembly (8):	2.54 L x 2.54 W x 9.91 cm H
Column volume:	63.9 cm^3
Assembly size:	53.24 L x 48.26 W x 22.86 cm H
Assembly weight:	28.7 kg

carrier: Orbiter middeck - 1 stowage locker

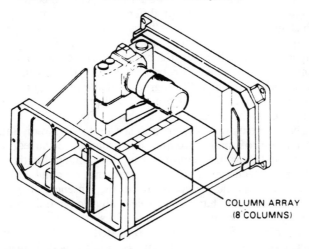

COLUMN ARRAY
(8 COLUMNS)

Fig. 19 Isoelectric focusing apparatus for low-gravity experiments onboard the Space Shuttle in middeck locker space. Eight compartments are viewed by the camera; each can be used to test specific hypotheses concerning stability of fluids in isoelectric focusing. Dimensions and other characteristics are given in the diagram.[133]

cylindrical zones with flat ends; in this case it was noted that after 3 h of clean separation the individual sample bands became spherical.

Biphasic Aqueous Extraction

Biphasic extraction is one of the most popular purification methods used in the chemical industry today. It has found limited popularity in bioprocessing because of the damaging effects of organic solvents on biomolecules and cells, whereas aqueous two-phase systems, due to their high water content, are biocompatible.[135,136] Moreover, these systems are reported to have provided stability to biologically active substances, such as enzymes.[137] Because of their similar physical properties, immiscible aqueous phases do not separate rapidly in large volumes, as in production-scale purifications. Despite some 800 papers on this subject,[138] large-scale commercial applications are not widespread. The high cost of lower-phase polymers is another deterrent to its widespread use. Cost containment has been recently affected by the introduction of low-cost polymer aqueous phase systems.[139]

In practice, multistage extractions are performed to achieve high-resolution separations[140]; however, low-gravity research to date has been restricted to making physical measurements on single-stage extraction systems. The physical problems associated with biphasic extraction research can be divided into two major categories, although they bear certain thermodynamic similarities: *phase separation*, the formation of two phases from a dispersion; and *partitioning*, the preferential transfer of a separand into one phase. The investigation of these two processes in low gravity is the subject of the next chapter on phase partitioning, in which a full introduction to this subject may be found.

Phase Separation

When two polymers A and B are dissolved in an aqueous solution at concentrations that cause phase separation, an upper phase forms that is rich in A and poor in B, and a lower phase forms that is rich in B and poor in A. Typically, A is polyethylene glycol (PEG), which is considered a relatively hydrophobic solute, and B is dextran or a similar polysaccharide. Polymer B can also be a salt at high concentration. The phase separation process is

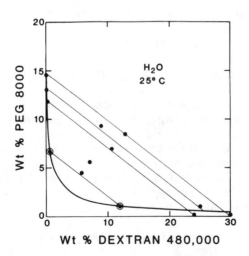

Fig. 20 Phase diagram of the dextran-water-polyethylene glycol system showing tie lines and binodial curve. Vertical limb of the binodial curve gives compositions of upper phases, and horizontal limb gives compositions of lower phases at equilibrium. Tie line length increases with polymer concentration and with polymer molecular weight, consistent with Eqs. (34) and (35). Courtesy of D. C. Szlag.

described by a two-dimensional phase diagram, such as in Fig. 20, in which high concentrations of A and B cause the formation of top-phase solutions with compositions given by points in the upper left and bottom-phase solutions with compositions given by points in the lower right. Each combination of A and B falls on a "tie line" connecting the resulting top and bottom phase compositions at equilibrium. Higher concentrations of polymers result in longer tie lines on the phase diagram. Any initial or total composition that lies on a tie line will result in the same equilibrium compositions, and the ratio of the volumes of the two phases is related to the position of the initial composition on the tie line. The curve that forms the envelope connecting the ends of the tie lines, the "binodial," shown in Fig. 20, separates the one- and two-phase regions on the diagram. As an example, one two-phase extraction system is described by the locations of the encircled points on the phase diagram of the PEG/dextran/water system at 25°C shown in Fig. 20. The top and bottom phase densities of this system are 1.0164 and 1.1059 g/cm^3, respectively, and the corresponding viscosities are 0.0569 and 4.60 P.

Such phase diagrams are strictly experimental; however, they represent thermodynamic equilibria, and they should be predictable on the basis of thermodynamic principles. Cabezas et al.[141] chose to apply statistical mechanics via the solution theory of Hill.[142] At equilibrium, the chemical potential of each substituent (typically dextran, PEG, and water) is the same in the top and bottom phase, and the chemical potential of each can be determined from their fractional molalities m_i and their osmotic virial coefficients C_{ij} by

$$\Delta\mu_2 = -RT[\ln m_2 + 2C_{22}m_2 + 2C_{23}m_3] \tag{34a}$$

$$\Delta\mu_3 = -RT[\ln m_3 + 2C_{23}m_2 + 2C_{33}m_3] \tag{34b}$$

$$\Delta\mu_1 = -RT[m_2 + m_3 + C_{22}m_2^2 + 2C_{23}m_2 m_3 + C_{33}m_3^2] \tag{34c}$$

where the subscripts 1, 2, and 3 correspond to water, PEG, and dextran, respectively. The same relationships apply to any pair of polymers, but not to polymer-salt combinations that form two phases, since electrostatics must be added to account for the chemical potentials of salts.

The osmotic virial coefficients C_{ij} are for polymers and constitute additional unknowns in Eqs. (34). These can be derived from group renormalization theory as applied to polymer solutions[143] by using monomer-monomer interaction coefficients b_{ij}:

$$C_{22} = b_{22}N_2^{3\nu_2}[1 + (2/9)\ln(M_{w2}/M_{n2})] \tag{35a}$$

$$C_{33} = b_{33}N_3^{3\nu_3}[1 + (2/9)\ln(M_{w3}/M_{n3})] \tag{35b}$$

$$C_{23} = b_{23}\{N_2^{3\nu_2}[1 + (2/9)\ln(M_{w2}/M_{n2})][N_3^{3\nu_3}[1 + (2/9)\ln(M_{w3}/M_{n3})]\}^{\frac{1}{2}} \tag{35c}$$

where C_{23} is based on an empirical application of the geometric mean rule. At equilibrium, the chemical potential of each constituent in the top phase is equal to its chemical potential in the bottom phase. These equalities result in a system of equations with the same number of equations as unknowns that can be solved for m_2 and m_3 to obtain equilibrium concentrations in both phases. These concentrations have been found to successfully predict phase diagrams, including the one shown in Fig. 20 (Ref. 141).

In addition to these equilibrium phenomena, rates of phase separation ("demixing") are of technical interest.[145] When two polymers are dissolved in aqueous solution at concentrations that cause phase separation, certain dissolved ions such as phosphate are unequally partitioned between the phases,[146] leading to a Donnan potential across the interface[147] and an electrokinetic (zeta) potential at the interface.[148-150] As a consequence of the latter, droplets of one phase move in the continuous phase in the presence of an externally applied electric field, not necessarily in a direction related to the Donnan electrical potential or in obedience to the rules of colloidal particle electrokinetics. These findings were the result of a search for a method for driving phase separation in low-gravity experiments.[151] It should be possible to control dimixing rates and possibly enhance demixing of phases with small density differences by the application of an electric field to two-phase polymer emulsions in practical applications.[152] To test this hypothesis, a vertical electrophoresis column of the type shown in Fig. 1 was used to measure the rate of demixing of PEG-rich top phases from dextran-rich or maltodextrin-rich lower phases. It was found that, at typical density differences, the rate of demixing of dispersions is increased severalfold in an electric field applied to 100-ml phase systems. When fields of reverse polarity were applied, in which the effect of field strength opposes the direction of buoyancy of PEG droplets, even greater demixing rates were observed.[153] An example is given in Fig. 21.

Partitioning

Partitioning of molecules between phases during demixing is also considered a *thermodynamic* process and not a rate process. This means that scale up under relatively nonhostile conditions should be feasible, and this is the main promise of biphasic aqueous extraction as a purification method. Partitioning was originally modeled by Brønsted,[154] who noted that, at the very least, the partition coefficient K should depend on the molecular weight of the separand:

$$K = \exp(\lambda M / k_B T) \tag{36}$$

but there are at least four properties that determine a molecule's partition coefficient: molecular weight, hydrophobicity, charge density, and binding affinity. Brooks and others[135,155,156,157] consider the surface area A_M (which depends on molecular weight) of a partitioning molecule and its interfacial free energy per unit area Δ_γ as the significant measureable variables in determining the partition coefficient:

$$K = \exp(\Delta\gamma A_M / k_B T) \tag{37}$$

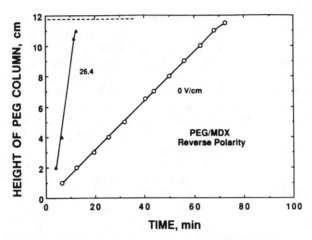

Fig. 21 Migration plots indicating the height of the upper (PEG-rich) phase at the top of the column as a function of time of demixing in the presence (left curve) and absence (right curve) of an electric field of 26.4 V/cm. The polymer mixture was PEG and maltodextrin with 0.1 M phosphate ions.

Although such a relationship seems valid for solute molecules, it does not satisfactorily describe the partitioning of particles, such as cells. In one view, Eq. (31) would only be satisfactory if T could be set to 10^4-10^5 K. Such a high $k_B T$ implies randomizing effects due to some nonthermal force, such as shear forces caused by gravity-driven motion of phase drops.[155–157] Furthermore, it is necessary to account for electrokinetic transport as a means of reaching equilibrium when an electrical potential, $\Delta\Psi$, exists between the two phases.[135] A particle can interact with both phases simultaneously; thus, the difference between its interfacial free energies with respect to the top phase γ_{pt} and with respect to the bottom phase γ_{pb}, which is $\Delta\gamma$, is a determinant of partition coefficient. The interfacial free energy between the top and bottom phase, γ_{tb}, plays a role in excluding particles from the interface. The suggested overall thermodynamic relationship for partitioning out of the interface is

$$K = B \exp\{-\gamma_{tb} A[1 - (\Delta\gamma + \sigma_e \Delta\Psi/\gamma_{tb}]^2/4k_B T\} \qquad (38)$$

When salts, especially salts of sulfate and phosphate, are dissolved in PEG-dextran solutions, an electrochemical potential is developed between the phases.[146,158] Arguments derived from equilibrium thermodynamics suggest that this could be a Donnan potential developed by the unequal partitioning of ions between the two phases.[147,148] It can be argued that ion adsorption at the interface will also lead to an electrical dipolar interfacial surface, thereby producing a zeta potential analogous to that found on colloidal particles.[150] Although the former hypothesis is consistent with the measured potential between phases at equilibrium,[152,159] and the latter hypothesis is consistent with the rapid migration of phase droplets in an electric field,[148,149] neither has been proven to explain the partitioning of cells between the phases on the apparent basis of electrophoretic mobility.[160-162]

The biphasic aqueous partitioning principle can also be applied to chromatographic columns. By immobilizing one phase in a particulate matrix, Müller[163] has succeeded in effecting chromatographic separations of proteins, and Skuse et al.[164] have reported the separation of cells. Chromatography has not been subjected to gravitational studies, since it is considered to consist (to a close approximation) of a combination of equilibrium processes. The extent to which this is the case for particulate separands may merit investigation.

Phase Separation in Biphasic Aqueous Systems in Low Gravity

Four broad categories of low-gravity phase separation experiments have been performed: 1) isopycnic experiments, in which a third polymer is added to equalize phase densities on the ground, 2) short-term demixing experiments during 20-s low-gravity periods in aircraft flight, 3) basic transport experiments during 7-min low-gravity periods on sounding rockets, and 4) orbital experiments to characterize full-length demixing events and phase equilibria. Experiment goals were generally aimed at characterizing the forces and processes that cause phase separation in low gravity and in general.

It was demonstrated on Space Shuttle flight STS-51D that polyethylene glycol and dextran will demix and undergo phase separation in the absence of gravity.[149,156,165,166] In hydrophilic vessels dextran forms a "shell" around the PEG phase, and in hydrophobic vessels PEG forms a shell around the dextran phase.[151] Thus, surface forces, presumably interfacial tension and gradients thereof and wall wetting, can drive phase separation in the absence of buoyancy. This supposition was confirmed in laboratory experiments in which added polymers were used to make the densities of the two phases as nearly equal as possible[151] and in which vessel surfaces were deliberately coated with hydrophilic polymers.[167]

Modifying the relative volumes of the two phases affected the rate of phase demixing in low gravity. If the volume of the wetting phase was low, less time was required to complete demixing. If the volume of the nonwetting phase was low, more time was required for the wetting phase to fully coalesce into a shell around the nonwetting phase, and systems with very low nonwetting volumes took hours to demix completely.[168]

Further details are given in the next chapter "Phase Partitioning" by Bamberger and Van Alstine, in which the relative roles of the processes of ripening, coalescence, and spinodal decomposition are also discussed.

Interfacial tension gradients can be generated by thermal gradients,[169] solute (especially surfactant) gradients,[170] and electrical gradients.[171] From this fact one might conclude that any thermodynamic variable that can generate an interfacial tension gradient could drive phase separation and the partitioning of particles, even in low gravity.

Physics of Phase Separation in Low Gravity

In low gravity a highly concentrated solute that changes the wetting properties of a solution will, if it exists as a gradient, create an interfacial tension gradient that may drive flow along the wettable walls of a vessel when not opposed by other forces, such as hydrostatic pressure. In mixtures of molten metals (e.g., in the zone refining process)[172] and in mixtures of molten salts (e.g., in directional solidification of electronic materials), solute-dependent convective motions occur near the solid-liquid interface. These flow phenomena are considered to be exaggerated in low gravity where buoyancy-driven convection and hydrostatic pressure are absent. Most studies of thermocapillary convection (Marangoni flow) are carried out in liquid systems in which at least one surface is open to a vapor phase,[173] so that a surface-tension-controlled surface exists, and surface tension gradients can be established by solute or temperature gradients.[174,175] It is normally assumed that surface tension gradients at a liquid-solid or liquid-liquid interface will not produce flow,[176] since the no-slip condition of Newtonian flow "clamps" the interface in the absence of preexisting momentum. However, the assumed no-slip condition need not apply when the driving force originates at the interface, as in electroendosmosis, described earlier, and liquid-solid interfacial flow is observed tangential to solidifying surfaces.[172,177]

In view of these notions the following hypothesis was framed: Two-phase aqueous systems should change configuration from two end-associated cylinders to two concentric phases when inertial acceleration is absent and surface tension gradients dominate. An experiment was designed to test this hypothesis in low-gravity experiments in which an interface was formed between previously separated top and bottom phases without salts. It was presumed that PEG solution would flow to the plastic walls of the cylindrical container while dextran solution stayed in the middle. The experiment was performed in the absence of demixing and coalescence to determine whether or not the interfacial tension gradient force was adequate to form concentric phases from apposed phases in a time period of the order of minutes, the duration of a sounding-rocket flight. This experiment also serves as an essential control for other interfacial transport experiments, in which immiscible phases with different interfacial tensions are apposed, and in which it is essential to know to what extent solutes are carried along by interfacial flow, if any, which could be misinterpreted as some other form of solute transport.

Rocket-borne experiments were performed using an automated minilab, the Materials Dispersion Apparatus (MDA) (Fig. 22), which consists of two blocks of inert plastic with multiple wells drilled at precise intervals in them. It functions by bringing wells in opposite blocks into contact with each other in response to an electronic signal. The version used in the sounding-rocket experiments had a layout of wells that permitted performance of experiments involving diffusive and electrokinetic transport, crystallization, and thin membrane formation (see below). In all cases the opposing wells had three specific relationships to one another: launch, coasting, and re-entry positions. It is estimated that the Reynolds number

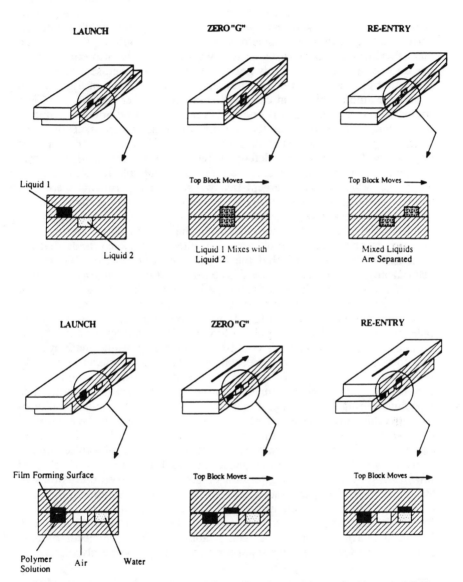

Fig. 22 System for contacting two fluids during low-gravity excursions on aircraft, sounding rockets, and orbital vehicles. Top: pairs of wells for bringing preformed top and bottom phases into contact. Bottom: system for forming a thin film for membrane casting. Cavities in one block are filled from the open side, whereas cavities in the opposite block are filled through channels (not shown).[179]

is less than 10 for the fluid motion as wells come into contact. In the execution of this experiment one well is filled with a previously separated dextran-rich solution and test dye, and an opposing well is filled with the corresponding equilibrium polyethylene glycol-rich solution. This arrangement produces an interface when the two wells are slid into contact with one another at the onset of low gravity. The wells are then separated from each other during re-entry of the sounding rocket, as shown in Fig. 22, step 3, and the amount of dye in each well is measured after landing.[178,179] For comparison (another control) one well is loaded with an aqueous solution of a detergent that increases the wetting (decreases the contact angle) of water on plastic combined with a dye at very dilute concentration. The opposite well is loaded with detergent-free colorless solution. If flow were to occur, upon opening the chambers the detergent solution should creep along the walls of the detergent-free wells. Both experiments were performed on the ground and during a 6.6-min sounding-rocket flight, and no differences in total dye transport between the wells was observed.[180] The immiscible phases did not reorient during the 6.6-min flight, and a detergent-rich layer did not "creep" across an interface between fluids that are miscible. It was concluded that the rate of fluid transport by this mechanism, if any, was too slow to be detected in a 6.6-min low-gravity period.

In the Space Shuttle STS-51D experiment (see the next chapter), it would have been of some interest to analyze the separated phases, which were photographed,[151] but no facility for chemical measurement was available on orbit, and phase re-equilibration at 1 g occurred upon return to the ground. Composition analysis could be facilitated by using phase systems that form gels. Research was undertaken to seek gelling media and conditions that would be suitable for the preservation of the spatial configuration of cell suspensions and macromolecular solutions after separation in free fluid during low gravity. Preliminary studies on the immiscibility of various gelling and nongelling polymers were carried out in the context of the broader goals of this field in space research.[161,165] A table of immiscibilities and relative phase volumes was constructed, and significant combinations in which the top phase gelled included mixtures of Ficoll and low-melting agarose. In an extension of the 19th century observations of Beijerinck[181,182] that agar and gelatin form immiscible solutions, mixtures of agarose and gelatin were also found to form two phases, both of which form gels. Both phases can be liquified at temperatures consistent with bioproduct activity, including cell viability. Certain molecular mass maltodextrins form gels after separation from polyethylene glycol.[61]

Partitioning of Separands Between Phases in Low Gravity

In examining the thermodynamics of the partitioning of particles between two liquid phases, as formalized by Albertsson and co-workers,[135] Brooks et al.[149] found that the factor $k_b T$ is more than 10 times its normal value at ambient temperature. This discrepancy is thought to be explained by the influence of gravity on the passage of particles between phases during coalescence during demixing. One way to test this hypothesis is to allow demixing to occur in the absence of gravity[152,156] so that sedimentation and convection are absent and to measure the equilibrium concentrations of separands. However, in such an experiment two processes are occurring: partitioning and demixing. In the case of particulate separands equilibrium concentrations may be related to transport rates.

There are four processes responsible for separand transport across the interfaces under study: diffusion, electrophoresis, buoyancy (including sedimentation), and interfacial tension. In low-gravity experiments it is possible to adjust each of these transport processes independently. It was found that diffusion of dissolved dyes across the interface between PEG-rich and dextran-rich immiscible aqueous solutions is the same on Earth and under microgravity when Earth-based experiments utilized phases with adequate density differences to prevent convective transport.[180]

Low gravity offers an opportunity to study transport processes individually and in the absence of demixing. Forming an interface between previously separated phases in low gravity eliminates the phase separation process and allows the examination of the partitioning process in isolation. For example, the total chemical potential required for partitioning can be supplied by the electric field generated by the Donnan potential between the two liquid phases plus the concentration gradient of the separand.[183] Transport across the corresponding interface without an electrochemical potential (no phosphate ions) is diffusive. Such a preformed interface, when formed in low gravity, does not require density stabilization (lower phase denser than upper phase); thus, a role of convective transport can be completely eliminated.

The following pair of mutually exclusive hypotheses was framed and tested: Charged molecules should be driven across an inteface by the electrochemical gradient or be repelled by the zeta potential at the surface of the interface. As Fig. 23 reiterates, in the presence of phosphate ions the PEG phase is positive with respect to the dextran phase, and PEG drops

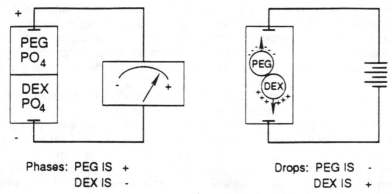

Phases: PEG IS +   Drops: PEG IS -
        DEX IS -   DEX IS +

Fig. 23 Illustration of electric field polarities in the PEG-dextran-water system with phosphate ions. PEG-rich (upper) phases are usually more positive, whereas PEG-rich drops behave in an electric field like high-mobility negative particles.

are negatively charged at their surface of hydrodynamic shear.[147,152,153] At
a preformed interface between the phases, a negative dye should experience
enhanced transport into the PEG-rich phase due to the positive Donnan
potential or decreased transport into the PEG-rich phase if repelled by the
negative electrokinetic potential.

Several controls are required for such an experiment in low gravity: a
measurement of fluid transport due to turbulent mixing during motion of
the two wells with respect to one another, transport due to diffusion in the
absence of an electrokinetic potential, and transport due to phase reorientation, if any.

In experiments utilizing the MDA (Fig. 22), it was found that adding
0.1 molar phosphate ions enhances the transport of anionic dye (Trypan
Blue) both in the laboratory and in 6.6 min of microgravity.[180] This result
indicates that there must be an electrochemical potential across the interface and that the observed electrophoresis of dye ions across that interface
is gravity independent.

Polymer and Ceramic Membrane Casting

Membrane filtration is one of the most heavily used separation processes.
Microporous membranes are used for removing particulates (0.1–1000 μm)
from solvents, dialysis membranes for separating macromolecules
(mol wt = 10^3–10^6), and reverse osmosis membranes and certain ceramic
membranes for selecting small molecules (mol wt = 10–1000), including
water and gases. This last category is currently of high interest because of
the low energy cost of membrane filtration compared to distillation, for
example, and the processes involved in making such membranes are the
subject of intense research.

Semipermeable polymer films are normally formed by spreading a thin
layer of polymer solution onto a smooth (typically glass) surface, allowing
the solvent to evaporate for a specified number of seconds or minutes (a
convection-dependent process under most conditions) and flooding the
freshly formed surface with water or an aqueous solution, into which the
solvent is extracted by convection and diffusion, resulting in gelatin of the
film. This procedure produces a dense skin at the upper surface of the film.
Beneath this skin the polymer film contains macrovoids. After the gelation
of a porous polymer film, a controlled temperature/pressure profile shrinks
the pores to their final morphology.

In experiments at 1 g in which the specific gravity of the solvent and of
the quenching (swelling) agent were modified, it was found to be possible
to modify the structure of cellulose acetate, polysulfone, and other polymeric membranes,[184] and Cabasso[185] has shown, in experiments monitored
by schlieren optics, that convective transport is the principal means of
removing solvent into the quenching fluid. Vera[186,187] proposed an experiment for orbital spaceflight in which solvent is transported into the vapor
phase and into the quenchant without convection.

The most popular technique for forming ceramic thin-film membranes is
the so-called sol-gel method. An example of a sol is a peptized solution of

boehmite (γ-AlOOH). Supported membranes are prepared by dipping one side of a support into the sol. As the sol is drawn into the support as a result of capillary forces, the boehmite particles enter at a slower velocity than the water, and the concentration of boehmite at the entrance of the pores increases. At a certain concentration, the sol converts into a gel. This gel film is then calcined to form a stable film with a selected pore size.

The gel formation process in both of these membrane-forming methods depends on capillary forces in order to change the local concentrations and cause phase separations to occur. In a reduced-gravity environment the ultimate capillary rise and its rate will be increased. Additionally, the multiple phases can significantly differ in density, viscosity, surface tension, and surface potential. In a normal-gravity environment, the density-driven convection processes will most likely overwhelm the forces due to interfacial and electrostatic phenomena. Low-gravity experiments permit a wider variety of test conditions and can potentially be used to obtain different structures. Polymer films, including binary membranes, formed in low gravity; hence, the absence of convective transport of solvent should be more uniform and lack irregularities such as macrovoids and aggregates. Inorganic filter membranes, which form by capillary forces, should have a different structure when capillary flow is not opposed by gravitational acceleration.

Recent research has focused on the development of very thin (10-μm) sulfonyl fluorocarbon membranes for pH-dependent high-flux transport of amino acids. These membranes are very asymmetric, and an understanding of the dependence of their transport properties on asymmetry is of considerable interest. Their amino acid permeability and selectivity can be modified by simple changes of pH. It has been further found that binary membranes composed of this material and chitosan (a natural polysaccharide) of various molecular masses (polymer chain length) have interesting transport properties. The mechanism of layering in these binary membranes is not well understood, and the relative roles of convection and phase separation need to be characterized. This is efficiently achieved in low gravity, where density-driven convection is absent, and these variables can be separated from one another.

Polymer Film Formation in Low Gravity

The question being asked in low-gravity film-forming experiments consists of the following: Do semipermeable membranes cast in the absence of convective removal of polymer solvent and in the absence of hydrostatic resistance to capillary flow have different structural and filtration properties from their counterparts formed in the presence of gravity? Two categories of experiments have been designed to address this question: one in which small numbers of large (15 × 40 cm) membranes are cast on orbit[186,187] and one in which several small (0.64-cm-diam) membranes can be cast using low-cost, low-volume hardware[179] on orbit, in low-gravity aircraft, or on sounding rockets.

In the latter case the MDA automatically functions to lay a thin film on a surface, expose the film to subsequent conditions during low gravity, and preserve the film for study upon return from flight. A sketch of this arrangement is shown in the lower diagram of Fig. 22, where it is seen that polymer or ceramic solution is contained in a cavity that contacts a glass or ceramic surface before or during the achievement of low gravity during flight; a pair of blocks moves the surface into contact with a final solution (or air) during rocket descent or continuation of orbital flight.

Boehmite ceramic membranes are cast on the surface of microporous ceramic disks after the achievement of low gravity. Capillary forces act to form the membrane for a fraction of a minute in low gravity. The freshly formed film is then moved to an empty cavity where solvent is allowed to evaporate during the remainder of the low gravity period. The microporous ceramic disks can be removed with the boehmite film, calcined and tested for structure by scanning microscopy and for function in permeation experiments.

In the case of polymer films the polymer solution is in contact with a glass disk at launch time. Upon achievement of microgravity the disk is slid to an open cavity and solvent evaporates into air at standard temperature and pressure for a fraction of a minute. The disk is then slid to the quenching solvent so that the polymer solvent is extracted in the absence of convection during the remaining low-gravity period.

Preliminary experiments of this type have been conducted during a sounding-rocket flight. It is thought that basic information obtained in low gravity film-forming experiments can significantly enhance our understanding of the role of gravity in membrane-forming process.

Other Bioprocessing Applications

In addition to separation physics, other bioprocessing applications of the low-gravity environment are being pursued. Examples are the crystallization of proteins and cell technology.[103,104,188,189] A brief summary of these activities follows.

Protein Crystal Growth

Protein crystal growth is affected by the same gravity-dependent processes as separation processes: 1) convection, 2) zone sedimentation, and 3) particle sedimentation. Crystals more dense than the mother liquor sediment away from the zone of crystallization, whereas those less dense float away from this zone. Sedimentation against a vessel wall modifies the structure of the crystal, typically due to lattice dislocations. Rapid nucleation on a dialysis membrane or vessel wall sometimes leads to large numbers of small crystals. The removal of solute from solution at the crystal surface can lead to convective disturbance at the lattice-forming surface.[190] Ideally, motionless, contactless crystal growth is desired, and the microgravity environment of spaceflight comes very close to providing these conditions.

Protein crystal growth is a popular field in microgravity research. Early studies on triglycine sulfate crystallization were very encouraging,[191] and Littke and John,[192] of the University of Freiburg, using the weightless environment of Spacelab I, obtained large, high-quality crystals of lysozyme, a standard in the protein crystal growth field, and β-galactosidase, which had not been satisfactorily crystallized before. The U.S. program has centered on the development of optimized equipment and methods, and Bugg and co-workers have tested the vapor diffusion method and the dialysis method as well as procedures for maintaining crystal integrity throughout spaceflight.[193,194]

Additional equipment and experiments have been developed by American, Japanese, and European workers.[178,179,195,196] The relative simplicity of spaceflight "hardware" for protein crystal growth will make it possible for large numbers (hundreds or thousands) of crystallization experiments to be performed on a single flight. It is also possible that the rigorous quantitative studies that are accompanying spaceflight crystallization research will lead to improvements in ground-based methods that may render microgravity crystallization unnecessary. Either way, pharmaceutical companies will have unprecedented access to these precious research products as a consequence of microgravity research.

Cell Biology and Technology

Space experiments have been performed in the area of applied cell biology,[188,197] and such experiments have been considered in terms of physical fundamentals.[11,198,199] In a Soviet-Hungarian program increased interferon secretion by human cells and increased growth rate of ciliated protozoa have been reported, respectively, by Talas et al.[200] and Tixador et al.[201] Early work in the U.S. Space program indicated little or no effect of microgravity on the growth of human fibroblasts in vitro.[202]

There is interest in maintaining cells in space for microgravity bioprocessing purposes.[188,203,204] With this purpose in mind, Tschopp et al.[205] found that cultured human kidney cells attach normally to microcarrier beads in culture in microgravity, whereas Hymer and Grindeland found anomalies in growth hormone production in preliminary experiments.[20,109] One of the most notable effects of gravity at the cell level was reported by Cogoli et al.,[206,207] who observed a 95% reduction in the incorporation of radioactive thymidine into DNA of stimulated human lymphocytes. This effect did not occur on a 1-g centrifuge in orbital spaceflight, and it has been observed in three spaceflight experiments. The notion of bioreactors in space has been considered for some time,[208] and these bioreactors might serve microgravity purification facilities by providing raw material, or an on-site culture facility for purified living cells.

Achnowledgments

It is a pleasure to acknowledge the participation of L. D. Plank, W. C. Hymer, S. Hjertén, D. R. Morrison, J. L. Sloyer, Jr., J. J. Pellegrino;

R. A. Gaines, B. E. Sarnoff, M. E. Kunze and R. M. Stewart in the performance of several aspects of the research described in this chapter. Advice and guidance over several years from Drs. M. Bier, R. S. Snyder, D. E. Brooks, F. J. Micale, and numerous others is greatly appreciated. S. R. Rudge, R. A. Mosher, J. M. Van Alstine, J. W. Lanham, and S. Bamberger provided valuable guidance during the preparation of the manuscript.

References

[1]Van Brunt, J., "Microgravity Workers Adjust to Disaster," *Bio/Technology*, Vol. 4, May 1986, p. 382.

[2]Bier, M., "Bioprocessing: Prospects for Space Electrophoresis," *Bioprocessing in Space*, edited by D. R. Morrison, NASA TM X-58191, Jan. 1977, pp. 117–124.

[3]Bier M., "Effective Principles for Scaleup of Electrophoresis," *Frontiers in Bioprocessing*, edited by S. K. Sikdar, M. Bier, and P. Todd, CRC, Boca Raton, FL, 1989, pp. 235–243.

[4]Ivory, C. F., "The Prospects for Large-Scale Electrophoresis," *Separation Science and Technology*, Vol. 23, 1988, pp. 875–912.

[5]Seaman, G. V. F., "Applications of Space Flight in Materials Science and Technology the Future—Biological Materials Applications," *Applications of Space Flight in Materials Science and Technology*, edited by S. Silverman and E. Passaglia, NBS Special Publ. 520, U.S. National Bureau of Standards, Gaithersburg, MD, 1977, pp. 83–88.

[6]Vanderhoff, J. W., and van Oss, C. J., "Electrophoretic separation of biological cells in microgravity," *Electrokinetic Separation Methods*, edited by P. G. Righetti, C. J. van Oss, and J. W. Vanderhoff, Elsevier/North-Holland, Amsterdam, 1979, pp. 257–274.

[7]Hjertén, S., *Free Zone Electrophoresis*, Almqvist and Wiksells Uppsala, Sweden, 1962.

[8]Snyder, R. S., and Rhodes, P. H., "Electrophoresis Experiments in Space," *Frontiers in Bioprocessing*, edited by S. K. Sikdar, M. Bier, and P. Todd, CRC, Boca Raton, FL, 1989, pp. 245–258.

[9]Snyder, R. S., Rhodes, P. H., Miller, T. Y., Micale, F. J., Mann, R. V., and Seaman, G. V. F., "Polystyrene Latex Separations by Continuous Flow Electrophoresis on the Space Shuttle," *Separation Science Technology*, Vol. 21, 1986, pp. 157–185.

[10]O'Brien, R. W., and White, L. R., "Electrophoretic Mobility of a Spherical Colloidal Particle," *Journal of the Chemical Society, Faraday Transactions 1*, Vol. 74, 1978, pp. 1607–1626.

[11]Tobias, C. A., Risius, J., and Yang, C.-H., "Biophysical Considerations Concerning Gravity Receptors and Effectors Including Experimental Studies on *Phycomyces blakesleeanus*," *Life Sciences and Space Research*, Vol. 11, 1973, pp. 127–140.

[12]Jorgenson, J. W., and Lukacs, K. D., "Capillary Zone Electrophoresis," *Science*, Vol. 222, 1983, pp. 266–272.

[13]Mosher, R., Thormann, W., Egen, N. B., Couasnon, P., and Sammons, D. W., "Recent Advances in Preparative Electrophoresis," *New Directions in Electrophoretic Methods*, edited by J. W. Jorgenson and M. Phillips, American Chemical Society, Washington, DC, 1987, pp. 247–262.

[14]Zeiller, K., and Hannig, K., "Free-Flow Electrophoretic Separation of

Lymphocytes. Evidence for Specific Organ Distributions of Lymphoid Cells," *Hoppe-Seylers Zeitschrift fuer Physiologische Chemie*, Vol. 152, 1971, pp. 1162–1167.

[15]McLaughlin S., and Poo, M. M., "The role of Electro-Osmosis in the Electric-Field-Induced Movement of Charged Macromolecules on the Surfaces of Cells," *Biophysical Journal*, Vol. 34, 1981, pp. 85–93.

[16]Zimmerman, U., "Electrical Breakdown, Electropermeabilization and Electro-fusion," *Rev. Physiol. Biochem. Pharmacol.*, Vol. 105, 1986, pp. 175–252.

[17]Sowers, A. E., *Biophysical Journal*, Vol 47, 1985, p. 171a.

[18]Todd, P., Plank, L. D., Kunze, M. E., Lewis, M. L., Morrison, D. R., Barlow, G. H., Lanham, J. W., and Cleveland, C., "Electrophoretic Separation and Analysis of Living Cells from Solid Tissues by Several Methods. Human Embryonic Kidney Cell Cultures as a Model," *Journal of Chromatography*, Vol. 364, 1986, pp. 11–24.

[19]Plank, L. D., Hymer, W. C., Kunze, M. E., and Todd, P., "Studies on Preparative Cell Electrophoresis as a Means of Purifying Growth-Hormone Producing Cells of Rat Pituitary," *Journal of Biochemical and Biophysical Methods*, Vol. 8, 1983, pp. 273–289.

[20]Hymer, W. C., Grindeland, R., Hayes, C., Lanham, J. W., Cleveland, C., Todd, P., and Morrison, D., "Heterogeneity in the Growth Hormone Pituitary Gland 'System' of Rats and Humans: Implications to Microgravity Based Research," *Microgravity Science and Applications Flight Programs*, NASA TM-4069, 1988, Vol. 1, pp. 47–88.

[21]Poulson, A., and Cramer, R., "Zone Electrophoresis of Type 1 Poliomyelitis Virus," *Biochemica et Biophysica Acta*, Vol. 29, 1958, pp. 187–192.

[22]Boltz, R. C., Jr., Todd, P., Gaines, R. A., Milito, R. P., Docherty, J. J., Thompson, C. J., Notter, M. F. D., Richardson, L. S., and Mortel, R., "Cell Electrophoresis Research Directed Toward Clinical Cytodiagnosis," *Journal of Histochemistry and Cytochemistry*, Vol. 24, 1976, pp. 16–23.

[23]Todd, P., Kurdyla, J., Sarnoff, B. E., and Elsasser, W., "Analytical Cell Electrophoresis as a Tool in Preparative Cell Electrophoresis," *Frontiers in Bio-processing*, edited by S. K. Sikdar, M. Bier, and P. Todd, CRC, Boca Raton, FL, 1989, pp. 223–234.

[24]Tulp, A., "Density Gradient Electrophoresis of Mammalian Cells," *Methods of Biochemical Analysis*, Vol. 30, 1984, pp. 141–198.

[25]Griffith, A. L., Catsimpoolas, N., and Wortis, H. H., *Life Sciences*, Vol. 16, 1975, pp. 1693–1702.

[26]Platsoucas, C. D., Good, R. A., and Gupta, S., *Proceedings of the National Academy of Sciences of the United States of America*, Vol. 76, 1979, pp. 1972–1976.

[27]Platsoucas, C. D., Beck, J. D., Kapoor, N., Good, R. A., and Gupta, S., *Cellular Immunology*, Vol. 59, 1981, pp. 345–354.

[28]Boltz, R. C., Jr., Todd, P., Streibel, M. J., and Louie, M. K., "Preparative Electrophoresis of Living Mammalian Cells in a Ficoll Gradient," *Preparative Biochemistry*, Vol. 3, 1973, pp. 383–401.

[29]Boltz, R. C., and Todd, P., "Density Gradient Electrophoresis of Cells in a Vertical Column," *Electrokinetic Separation Methods*, edited by P. G. Righetti, C. J. van Oss, and J. Vanderhoff, Elsevier/North-Holland, Amsterdam, 1979, pp. 229–250.

[30]Gillman, C. F., Bigazzi, P. E., Bronson, P. M., and van Oss, C. J., "Preparative Electrophoresis of Human Lymphocytes. I. Purification of Nonimmunoglobulin-Bearing Lymphocytes by Electrophoretic Levitation," *Preparative Biochemistry*, Vol. 4, 1974, pp. 457–472.

[31]Van Oss, C. J., and Bronson, P. M., "Vertical Ascending Cell Electrophoresis," *Electrokinetic Separation Methods*, edited by P. G. Righetti, C. J. van Oss, and J. W. Vanderhoff, Elsevier/North-Holland, Amsterdam, 1979, pp. 251–256.

[32]Seaman, G. V. F., *The Red Blood Cell*, edited by D. M. Surgenor, Academic, New York, 1975, pp. 1135–1229.

[33]Brooks, D. E., and Seaman, G. V. F., "The Effect of Neutral Polymers on the Electrokinetic Potential of Cells and Other Charged Particles," *Journal of Colloid and Interface Science*, Vol. 43, 1973, pp. 670–686.

[34]Todd, P., Hymer, W. C., Plank, L. D., Marks, G. M., Hershey, M., Giranda, V., Kunze, M. E., and Mehrishi, J. N., "Separation of Functioning Mammalian Cells by Density Gradient Electrophoresis," *Electrophoresis '81*, edited by R. C. Allen and P. Arnaud, de Gruyter, Berlin, 1981, pp. 871–882.

[35]Mason, D. W., "A Diffusion-Driven Instability in Systems that Separate Particles by Velocity Sedimentation," *Biophysical Journal*, Vol. 16, 1976, pp. 407–416.

[36]Gaines, R. A., "A Physical Evaluation of Density Gradient Cell Electrophoresis," PhD Dissertation, Pennsylvania State University, University Park, PA, 1981.

[37]Plank, L. D., Kunze, M. E., and Todd, P., "Electrophoretic Migration of Animal Cells in a Vertical Ficoll Gradient. Theory and Experiment," personal communication, 1989.

[38]Plank, L. D., Todd, P., Kunze, M. E., and Gaines, R. A., "Electrophoretic Mobility of Cells in a Vertical Ficoll Gradient," *Electrophoresis '81, Book of Abstracts*, 1981, p. 125.

[39]Todd, P., "Microgravity Cell Electrophoresis Experiments on the Space Shuttle: A 1984 Overview," *Cell Electrophoresis*, edited by W. Schütt and H. Klinkmann, de Gruyter, Berlin, 1985, pp. 3–19.

[40]Todd, P., and Hjertén, S., "Free Zone Electrophoresis of Animal Cells. I. Experiments on Cell-Cell Interactions," *Cell Electrophoresis*, edited by W. Schütt and H. Klinkmann, de Gruyter, Berlin, 1985, pp. 23–31.

[41]Snyder, R. S., Rhodes, P. H., Herren, B. J., Miller, T. Y., Seaman, G. V. F., Todd, P., Kunze, M. E., and Sarnoff, B. E., "Analysis of Free Zone Electrophoresis of Fixed Erythrocytes Performed in Microgravity," *Electrophoresis*, Vol. 6, 1985, pp. 3–9.

[42]Omenyi, S. N., Snyder, R. S., Absolom, D. T., Neumann, A. W., and van Oss, C. J., "Effects of Zero van der Waals and Zero Electrostatic Forces on Droplet Sedimentation," *Journal of Colloid and Interface Science*, Vol. 81, 1981, pp. 402–409.

[43]Snyder, R. S., "Electrophoresis Demonstration on Apollo 16," NASA TM X-64724, 1972.

[44]Snyder, R. S., Bier, M., Griffin, R. N., Johnson, A. J., Leidheiser, H., Micale, F. J., Ross, S., and van Oss, C. J., "Free Fluid Particle Electrophoresis on Apollo 16," *Separation and Purification Methods*, Vol. 2, 1973, pp. 258–282.

[45]McKannan, E. C., Krupnick, A. C., Griffin, R. N., and McCreight, L. R., "Electrophoretic Separation in Space—Apollo 14," NASA TM X-64611, 1971.

[46]Gordon, S. A., and Shen-Miller, J., "Simulated Weightlessness Studies by Compensation," *Gravity and the Organism*, edited by S. A. Gordon and M. J. Cohen, University of Chicago Press, Chicago, 1971, pp. 415–426.

[47]Abramson, H. A., Moyer, L. S., and Gorin, M. H. *Electrophoresis of Proteins*, Reinhold, New York, 1942.

[48]Bangham, A. D., Flemans, R., Heard, D. H., and Seaman, G. V. F., *Nature (London)*, Vol. 182, 1958, p. 642.

[49]Brinton, C. C., Jr., and Laufer, M. A., "The Electrophoresis of Viruses, Bacteria, and Cells and the Microscope Method of Electrophoresis," *Electrophoresis*, edited by M. Bier, Academic, New York, 1959, pp. 427–492.

[50]Seaman, G. V. F., "Electrophoresis Using a Cylindrical Chamber," *Cell Electrophoresis*, edited by E. J. Ambrose, Little, Brown, Boston, 1965, pp. 4–21.

[51]Micale, F. J., Vanderhoff, J. W., and Snyder, R. S., *Separation and Purification Methods*, Vol. 5, 1976, pp. 361–383.

[52]Vanderhoff, J. W., and Micale, F. J., "Influence of Electroosmosis," *Electrokinetic Separation Methods*, edited by P. G. Righetti, C. J. van Oss, and J. W. Vanderhoff, Elsevier/North-Holland, Amsterdam, 1979, pp. 81–93.

[53]Allen, R. E., Barlow, G. H., Bier, M., Bigazzi, P. E., Knox, R. J., Micale, F. J., Seaman, G. V. F., Vanderhoff, J. W., van Oss, C. J., Patterson, W. J., Scott, F. E., Rhodes, P. H., Nerren, B. H., and Harwell, R. J., "Electrophoresis Technology," Apollo-Soyuz Test Project Summary Progress Rept. Vol. 1, NASA SP-412, 1977, pp. 307–334.

[54]Allen, R. E., Rhodes, P. H., Snyder, R. S., Barlow, G. H., Bier, M., Bigazzi, P. E., van Oss, C. J., Knox, R. J., Seaman, G. V. F., Micale, F. J., and Vanderhoff, J. W., "Column Electrophoresis on the Apollo-Soyuz Test Project," *Separation and Purification Methods*, Vol. 6, 1977, pp. 1–59.

[55]Morrison, D. R., and Lewis, M. L., "Electrophoresis Tests on STS-3 and Ground Control Experiments: A Basis for Future Biological Sample Selections," *International Astronautical Federation*, Paper 82-152, 1983.

[56]Sarnoff, B. E., Kunze, M. E., and Todd, P., "Electrophoretic Purification of Cells in Space: Evaluation of Results from STS-3," *Advances in Astronautical Sciences*, Vol. 53, 1983, pp. 139–148.

[57]Patterson, W. J., "Development of Polymeric Coatings for Control of Electro-Osmotic Flow in ASTP MA-011 Electrophoresis Technology Experiment," NASA TM X-73311, 1976.

[58]Heard, D. H., and Seaman, G. V. F., "The Action of Lower Aldehydes on the Human Erythrocyte," *Biochemica et Biophysica Acta*, Vol. 53, 1961, pp. 366–372.

[59]Barlow, G. H., Lazer, S. L., Rueter, A., and Allen, R., "Electrophoretic Separation of Human Kidney Cells at Zero Gravity," *Bioprocessing in Space*, edited by D. R. Morrison, NASA TM X-58191, Jan. 1977, pp. 125–132.

[60]McGuire, J. K., and Snyder, R. S., "Operational Parameters for Continuous Flow Electrophoresis of Cells," *Electrophoresis '81*, edited by C. Allen and P. Arnaud, de Gruyter, Berlin, 1981, pp. 947–960.

[61]Todd, P., Szlag, D. C., Plank, L. D., Delcourt, S. G., Kunze, M. E., Kirkpatrick, F. H., and Pike, R. G., "An Investigation of Gel Forming Media for Use in Low Gravity Bioseparations Research," *Advances in Space Research*, Vol. 9, 1989, pp. 97–103.

[62]Plank, L. D., Kunze, M. E., Gaines, R. A., and Todd, P., "Density Gradient Electrophoresis of Cells in a Reversible Gel," *Electrophoresis*, Vol. 9, 1988, pp. 647–649.

[63]Mitichkin, O. V., Vavirovsky, L. A., Azhitsky, G. Y., Serebrov, A. A., Savitskaya, S. E., and Fokin, V. E., "Biotechnological Experiment Tavriya Carried Out Aboard Salyut-7," *Gagarin Scientific Readings on Cosmonautics and Aviation*, Nauka, Moscow, 1984.

[64]Avduyevsky, V. S., *Manufacturing in Space: Processing Problems and Advances*, Mir, Moscow, 1985.

[65]Babsky, V. G., Zhukov, M. Y., and Yudovich, V. I., "Electrophoresis of Biopolymers Under Low Gravity," *Fluid Mechanics and Heat and Mass Transfer in Low Gravity*, Nauka, Moscow, 1982.

[66]Egorov, B. B., and Peredkov, V. A., "Prospects for the Development of Purification Processes of Biologicals Under Low Gravity," *Gagarin Scientific Readings on Cosmonautics and Aviation*, Nauka, Moscow, 1984.

[67]Hannig, K., Wirth, H., and Schoen, E., "Electrophoresis Experiment MA-014," *Apollo-Soyuz Test Project Summary Science Report*, Vol. 1, NASA SP-412, 1977, pp. 335–352.

[68]Hymer, W. C., Barlow, G. H., Cleveland, C., Farrington, M., Grindeland, R., Hatfield, J. M., Lanham, J. W., Lewis, M. L., Morrison, D. R., Rhodes, P. H., Richman, D., Rose, J., Snyder, R. S., Todd, P., and Wilfinger, W., "Continuous Flow Electrophoretic Separation of Proteins and Cells from Mammalian Tissues," *Cell Biophysics*, Vol. 10, 1987, pp. 61–85.

[69]Hannig, K., "The Application of Free-Flow Electrophoresis to the Separation of Macromolecules and Particles of Biological Importance," *Modern Separation Methods of Macromolecules and Particles*, edited by T. Gerritsen, Wiley Interscience, New York, 1969, pp. 45–69.

[70]Hannig, K., "New Aspects in Preparative and Analytical Continuous Free-Flow Cell Electrophoresis," *Electrophoresis*, Vol. 3, 1982, pp. 235–243.

[71]Strickler, A., "Continuous Particle Electrophoresis: A New Analytical and Preparative Capability," *Separation Science*, Vol. 2, 1967, p. 335.

[72]Hannig, H., Wirth, H., Neyer, B., and Zeiller, K., "Theoretical and Experimental Investigations of the Influence of Mechanical and Electrokinetic Variables on the Efficiency of the Method," *Hoppe-Seyler's Zeitschrift fuer Physiologische Chemie*, Vol. 356, 1975, pp. 1209–1223.

[73]Saville, D. A., "The Fluid Mechanics of Continuous Flow Electrophoresis," *Physicochemical Hydrodynamics*, edited by D. B. Spalding, Advanced Publications, UK, Vol. 2, 1978, pp. 893–912.

[74]Deiber, J. A., and Saville, D. A., "Flow Structure in Continuous Flow Electrophoresis Chambers," *Materials Processing in the Reduced Gravity Environment of Space*, edited by G. E. Rindone, North-Holland, New York, 1982, pp. 217–224.

[75]McDonnell Douglas Astronautics Company, "Feasibility of Space Manufacturing—Production of Pharmaceuticals. Volume II: Technical Analysis," NASA CR-161325, 1978.

[76]McCreight, L. R., "Electrophoresis for Biological Production," *Bioprocessing in Space*, edited by D. R. Morrison, NASA TM X-58191, Jan. 1977, pp. 143–158.

[77]Giannovario, J. A., Griffin, R., and Gray, E. L., "A Mathematical Model of Free Flow Electrophoresis," *Journal of Chromatography*, Vol. 153, 1978, pp. 329–352.

[78]Rhodes, P. H., "High Resolution Continuous Flow Electrophoresis in the Reduced Gravity Environment," *Electrophoresis '81*, edited by R. C. Allen and P. Arnaud, de Gruyter, Berlin, 1981, pp. 919–932.

[79]Babsky, V. G., Zhukov, M. Y., Yudovich, V. I., *Mathematical Theory of Electrophoresis*, Naukova Dumka, Kiev, USSR, 1983.

[80]Boese, F. G., "Contributions to a Mathematical Theory of Free Flow Electrophoresis," *Journal of Chromatography*, Vol. 438, 1988, pp. 145–170.

[81]Biscans, B., Alinat, P., Bertrand, J., and Sanchez, V., "Influence of Flow and Diffusion on Protein Separation in a Continuous Flow Electrophoresis Cell: Computation Procedure," *Electrophoresis*, Vol. 9, 1988, pp. 84–89.

[82]Clifton, M. J., and Marsal O., "Heat Transfer Design of an Electrophoresis Experiment," *Acta Astronautica*, 1989 (to be published).

[83]Strickler, A., and Sacks, T., "Continuous Free-Film Electrophoresis. The Crescent Phenomenon," *Preparative Biochemistry*, Vol. 3, 1973, pp. 269–277.

[84]Rhodes, P. H., and Snyder, R. S., "Sample Band Spreading Phenomena in Ground and Space-Based Electrophoretic Separators," *Electrophoresis*, Vol. 7, 1986, pp. 113–120.

[85]Miller, T. Y., Williams, G. P., and Snyder, R. S., "Effect of conductivity and Concentration on the Sample Stream in the Transverse Axis of a Continuous Flow Electrophoresis Chamber," *Electrophoresis*, Vol. 6, 1985, pp. 377–381.

[86]Snyder, R. S., "Separation Techniques," *Materials Sciences in Space — A Contribution to the Scientific Basis of Space Processing*, edited by B. Feuerbacher, H. Hamacher, and R. J. Naumann, Springer-Verlag, Berlin, 1986, p. 465.

[87]Rhodes, P. H., Snyder, R. S., and Roberts, G. O., "Electrohydrodynamic Distortion of Sample Streams in Continuous Flow Electrophoresis," *Journal of Colloid and Interface Science*, Vol. 129, 1989, p. 78.

[88]Taylor, G. I., "Studies in Electrohydrodynamics. I. The Circulation Produced in a Drop by an Electric Field," *Proceedings of the Royal Society of London, Series A*, Vol. A291, 1966, pp. 159–167.

[89]Vanderhoff, J. W., Micale, F. J., and Krumrine, P. H., "Continuous Flow Electrophoresis," *Electrokinetic Separation Methods*, edited by P. G. Righetti, C. J. van Oss, and J. W. Vanderhoff, Elsevier/North-Holland, Amsterdam, 1979, pp. 121–141.

[90]Ostrach, S., "Convection in Continuous Flow Electrophoresis," *Journal of Chromatography*, Vol. 140, 1977, pp. 187–197.

[91]Saville, D. A., and Ostrach, S., "Fluid Mechanics," National Aeronautics and Space Administration Final Rept., Contract NAS8-31349, Code 361, 1978.

[92]Walker, C., "Toward Pharmaceutical Processing in Orbit," *Pharm Tech Conference '85 Proceedings*, Aster, Springfield, OR, 1985, p. 27.

[93]Clifford, D. W., "Commercial Prospects for Bioprocessing in Space," *Symposium on Commercial Opportunities in Space; Roles of Developing Countries*, McDonnell Douglas Astronautics, St. Louis, MO, 1987.

[94]Rhodes, P. H., and Snyder, R. S., "The Effect of Small Temperature Gradients on Flow in a Continuous Flow Electrophoresis Chamber," *Materials Processing in the Reduced Gravity Environment of Space*, edited by G. E. Rindone, North-Holland, New York, 1982, pp. 217–224.

[95]Hannig, K., and Bauer, J., "Free Flow Electrophoresis in Space Shuttle Program (Biotex)," *Advances in Space Research*, Vol. 9, No. 11, 1989, pp. 91–96.

[96]Jacobsen, L. O., Goldwasser, E., Fried, W., and Plazak, L., "Role of the Kidney in Erythropoietin," *Nature (London)*, Vol. 179, 1957, pp. 633–634.

[97]Jacobs, K., Shoemaker, C., Rudersdorf, R., Neill, S. D., Kaufman, R. J., Mufson, A., Seehra, J., Jones, S. S., Hewick, R., Fritsch, E. F., Kawakita, M., Shimizu, T., and Miyake, T., "Isolation and Characterisation of Genomic and cDNA Clones of Human Erythropoietin," *Nature, (London)*, Vol. 313, 1985, pp. 806–810.

[98]Snyder, R. S., Rhodes, P. H., and Miller, T. Y., "Continuous Flow Electrophoresis System Experiments on Shuttle Flights STS-6 and STS-7," *Microgravity Science and Applications Flight Programs, January–March 1987, Selected Papers*, Vol. 1, NASA TM-4069, 1988, pp. 27–46; also NASA TP-2778, 1987.

[99]Heidrich, H.-G., and Dew, M. E., "Homogeneous Cell Populations from Rabbit Kidney Cortex," *Journal of Cell Biology*, Vol. 74, 1983, pp. 780–788.

[100]Kreisberg, J. I., Sachs, G., Pretlow II, T. G., and McGuire, R. A., "Separation of Proximal Tubule Cells from Suspensions of Rat Kidney Cells by Free-Flow Electrophoresis," *Journal of Cell Physiology*, Vol. 93, 1977, pp. 169–172.

[101]Morrison, D. R., Barlow, G. H., Cleveland, C., Grindeland, R., Hymer, W. C., Kunze, M. E., Lanham, J. W., Lewis, M. L., Sarnoff, B. E., Todd, P., and Wilfinger, W., "Electrophoretic Separation of Kidney and Pituitary Cells on STS-8," *Advances in Space Research*, Vol. 4, No. 5, 1984, pp. 67–76.

[102]Morrison, D. R., Lewis, M. L., Cleveland, C., Kunze, M. E., Lanham, J. W., Sarnoff, B. E., and Todd, P., "Properties of Electrophoretic Fractions of Human Embryonic Kidney Cells Separated on Space Shuttle Flight STS-8," *Advances in Space Research*, Vol. 4, No. 5, 1984, pp. 77–79.

[103]Todd, P., "Pharmaceutical Technology as a Component of the U.S. Microgravity Science and Applications Program," *Pharm Tech Conference '85 Proceedings*, Aster, Springfield, OR, 1985, pp. 5–12.

[104]Todd, P., "Space Bioprocessing," *Bio/Technology*, Vol. 3, 1985, pp. 736–790.

[105]Todd, P., Hymer, W. C., Morrison, D. R., Goolsby, C. L., Hatfield, J. M., Kunze, M. E., and Motter, K., "Cell Bioprocessing in Space: Applications of Analytical Cytology," *The Physiologist*, Vol. 31, No. 1, 1988, pp. S52–S55 (Supplement).

[106]Todd, P., Morrison, D. R., Barlow, G. H., Lewis, M. L., Lanham, J. W., Cleveland, C., Williams, K., Kunze, M. E., and Goolsby, C. L., "Kidney Cell Electrophoresis in Space Flight: Rationale, Methods, Results and Flow Cytometry Applications," *Microgravity Science and Applications Flight Programs, January–March 1987, Selected Papers*, Vol. 1, NASA TM-4069, 1988, pp. 89–118.

[107]Pretlow II, T. G., and Pretlow, T. P., "Cell Electrophoresis," *International Review of Cytology*, Vol. 61, 1979, pp. 85–128.

[108]Barlow, G. H., Lewis, M. L., and Morrison, D. R., "Biochemical Assays on Plasminogen Activators and Hormones from Kidney Sources," *Microgravity Science and Applications Flight Programs, January–March 1987, Selected Papers*, NASA TM-4069. 1988, pp. 175–193.

[109]Hymer, W. C., Grindeland, R., Lanham, J. W., and Morrison, D., "Continuous Flow Electrophoresis (CFE): Applications to Growth Hormone Research and Development," *Pharm Tech Conference '85 Proceedings*, Aster, Springfield, OR, 1985, pp. 13–18.

[110]Hymer, W. C., Grindeland, R., and Lanham, J. W., "Continuous Flow Electrophoresis (CFE) at Unit and Microgravity: Applications to Growth Hormone (GH) Research and Development," *Microgravity Science and Applications*, National Academy Press, Washington, 1986, pp. 197–211.

[111]Rhodes, P. H., and Snyder, R. S., "Preparative Electrophoresis for Space," *Microgravity Science and Applications Flight Programs, January–March 1987, Selected Papers*, Vol. 1, NASA TM-4069, 1988, pp. 27–46.

[112]Johnson, N. L., *The Soviet Year in Space 1987*, Teledyne Brown Engineering, Colorado Springs, CO, pp. 85–95.

[113]*Electrokinetics in Microgravity: RAMSES a Facility for IML 2*, Centre National d'Études Spatiales, Paris, 1987.

[114]Escale, F., "Étude du fonctionnement d'une cellule d'electrophorese du zone a ecoulement continu, au sol, en vue d'experimentations en microgravité," PhD Dissertation, Université de Paul Sabatier, Toulouse, France, 1987.

[115]Bozouklian, H., Sanchez, V., Clifton, M., Marsal, O., and Esterle, "Electrokinetic Bioprocessing Under Microgravity in France as Illustrated by Space Bioseparation: a Programme Initiated in France and in Cooperation with Belgium and Spain," *Advances in Space Research*, Vol. 9, No. 11, 1989, pp. 105–109.

[116]*Experiment Facilities for Science and Applications under Microgravity*, National Space Development Agency (Japan), Tokyo, 1985.

[117]Palusinski, O. A., Allgyer, T. T., Mosher, R. A., Bier, M., and Saville, D. A., "Mathematical Modeling and Computer Simulation of Isoelectric Focusing of Simple Ampholytes," *Biophysical Chemistry*, Vol. 14, 1981, pp. 389–397.

[118]Palusinski, O. A., Bier, M., and Saville, D. A., "Mathematical Model for Transient Isoelectric Focusing of Simple Ampholytes," *Biophysical Chemistry*, Vol. 14, 1981, pp. 389–397.

[119]Bier, M., Palusinski, O. A., Mosher, R. A., and Saville, D. A., "Electrophoresis: Mathematical Modeling and Computer Simulation," *Science (Washington DC)*, Vol. 219, 1983, pp. 1281–1287.

[120]Mosher, R. A., Palusinski, O. A., and Bier, M., "Theoretical Studies in Isoelectric Focusing," *Materials Processing in the Reduced Gravity Environment of Space*, edited by G. E. Rindone, Elsevier, Amsterdam, 1982, pp. 255–260.

[121]Bier, M., and Egen, N., "Large Scale Recycling Isoelectric Focusing," *Electrofocusing*, edited by H. Haglund, J. G. Westerfeld, and J. T. Ball, Elsevier/Holland, New York, 1979, pp. 35–48.

[122]Nagabhushan, T. L., Sharma, B., and Trotta, P. P., "Application of Recycling Isoelectric Focusing for Purification of Recombinant Human Leukocyte Interferons," *Electrophoresis*, Vol. 7, 1986, p. 552.

[123]Bier, M., "Hormone Purification by Isoelectric Focusing in Space," *Microgravity Science and Applications Flight Programs, January–March 1987, Selected Papers*, NASA TM-4069, 1988, pp. 133–145.

[124]Troitsky, G. V., and Azhitsky, G. Y., *Isoelectric Focusing of Proteins in Natural and Artificial pH Gradients*, Kiev Naukova Dumka, Kiev, USSR, 1984.

[125]Boltz, R. C., Jr., Miller, T. Y., Todd, P., and Kukulinsky, N. E., "A Citrate Buffer System for Isoelectric Focusing and Electrophoresis of Living Mammalian Cells," *Electrophoresis '78*, edited by N. Catsimpoolas, Elsevier/North-Holland, Amsterdam, 1978, pp. 345–355.

[126]Boltz, R. C., Jr., Todd, P., Hammerstedt, R. H., Hymer, W. C., Thompson, C. J., and Docherty, J. J., "Initial Studies on the Separation of Cells by Density Gradient Isoelectric Focusing," *Cell Separation Methods*, edited by H. Bloemendal, Elsevier/North-Holland, Amsterdam, 1977, pp. 145–155.

[127]Sherbet, G. V., *The Biophysical Characterisation of the Cell Surface*, Academic, London, 1978.

[128]McGuire, J. K., Miller, T. Y., Tipps, R. W., Snyder, R. S., and Righetti, P. G., "New Experimental Approaches to Isoelectric Fractionation of Cells," *Journal of Chromatography*, Vol. 194, 1980, pp. 323–333.

[129]Jovin, T. M., "Multiphasic Zone Electrophoresis. I. Steady-State Moving Boundary Systems Formed by Different Electrolyte Combinations," *Biochemistry*, Vol. 12, 1972, p. 871.

[130]Bier, M., Hinckley, J. O. N., Smolka, A. J. K., "Potential Use of Isotachophoresis in Space," *Protides of the Biological Fluids: 22nd Colloquium*, edited by H. Peeters, Pergamon, New York, 1975, pp. 673–678.

[131]Sloan, J. E., Thormann, W., Bier, M., Twitty, G. E., and Mosher, R., "Recycling Isotachophoresis: a Novel Approach to Preparative Protein Fractionation," *Electrophoresis '86*, edited by M. J. Dunn, VCH, Deerfield Beach, FL, 1986, pp. 696–701.

[132]Bier, M., Egen, N. B., Mosher, R. A., and Twitty, G. E., "Isoelectric Focusing in Space," *Materials Processing in the Reduced Gravity of Space*, edited by G. E. Rindone, North-Holland, New York, 1982, pp. 261–268.

[133]*Accessing Space. A Catologue of Process, Equipment and Resources for Commercial Users 1988*, NASA NP-118, 1988, pp. 4–32.

[134]Melcher, J. R., and Taylor, G. I., *Annual Review of Fluid Mechanics*, Vol. 1, 1969, p. 111.

[135]Albertsson, P.-Å., *Partition of Cell Particles and Macromolecules*, 2nd ed., Wiley-Interscience, New York, 1971.

[136]Albertsson, P.-Å., *Partition of Cell Particles and Macromolecules*, 3rd ed., Wiley-Interscience, New York, 1986.

[137]Shanbhag, V. P., "Diffusion of Proteins Across a Liquid-Liquid Interface," *Biochimica et Biophysica Acta*, Vol. 320, 1973, pp. 517–527.

[138]Walter, H., Brooks, D. E., and Fisher, D., (eds.), *Partitioning in Aqueous Two-Phase Systems*, Academic, New York, 1985.

[139]Szlag, D. C., and Giuliano, K. A., "A Low-Cost Aqueous Two Phase System for Enzyme Extraction," *Biotechnology Techniques*, Vol. 2, 1988, pp. 277–282.

[140]Van Alstine, J. M., Snyder, R. S., Karr, L. J., Harris, J. M., "Cell Separation with Counter-Current Chromatography and Thin-Layer Countercurrent Distribution in Aqueous Two-Phase System," *Journal of Liquid Chromatography*, Vol. 8, 1985, pp. 2293–2313.

[141]Cabezas, H., Jr., Evans, J., and Szlag, D. C., "Statistical Thermodynamics of Aqueous Two-Phase Systems," *Downstream Processing and Bioseparation. Recovery and Purification of Biological Products*, edited by J.-F. P. Hamel, J. B. Hunter, and S. K. Sikdar, American Chemical Society, Washington, DC, 1990, ACS Symposium Series 419, pp. 38–52.

[142]Hill, T. L., "Theory of Solutions I," *Journal of the American Chemical Society*, Vol. 79, 1957, pp. 4885–4890.

[143]Freed, K. F., *Renormalization Group Theory of Macromolecules*, Wiley, New York, 1987.

[144]Cabezas, H. Jr., Evans, J. D., and Szlag, D. C., "A Statistical Mechanical Model of Aqueous Two-Phase System," *Fluid Phase Equilibria*, Vol. 53, 1989, pp. 453–462.

[145]Baird, J. K., "Application of the Theory of Ostwald Ripening to Microgravity Experiments," *Proceedings of the 5th European Symposium on Materials Science Under Microgravity*, European Space Agency SP-222, 1984, pp. 319–324.

[146]Johansson, G., "Partition of Salts and Their Effects on Partition of Proteins in a Dextran-Poly(ethylene glycol)-Water Two-Phase System," *Biochimica et Biophysica Acta*, Vol. 221, 1970, pp. 387–390.

[147]Bamberger, S., Seaman, G. V. F., Brown, J. A., and Brooks, D. E., "The Partition of Sodium Phosphate and Sodium Chloride in Aqueous Dextran Poly(ethylene glycol) Two Phase Systems," *Journal of Colloid and Interface Science*, Vol. 99, 1984, pp. 187–193.

[148]Brooks, D. E., Sharp, K. A., Bamberger, S., Tamblyn, C. H., Seaman, G. V. F., and Walter, H., "Electrostatic and Electrokinetic Potentials in Two Polymer Aqueous Phase Systems," *Journal of Colloid and Interface Science*, Vol. 102, 1984, pp. 1–13.

[149]Brooks, D. E., Bamberger, S. B., Harris, J. M., and Van Alstine, J., "Rationale for Two Phase Polymer System Microgravity Separation Experiments," *Proceedings of the 5th European Symposium on Material Sciences Under Microgravity*, European Space Agency SP-222, 1984, pp. 315–318.

[150]Levine, S., "A Theory of Electrophoresis of Emulsion Drops in Aqueous Two-Phase Polymer Systems," *Materials Processing in the Reduced Gravity of Space*, edited by G. E. Rindone, North-Holland, New York, 1982, pp. 241–248.

[151]Van Alstine, J. M., Karr, L. J., Harris, J. M., Snyder, R. S., Bamberger, S. B., Matsos, H. C., Curreri, P. A., Boyce, J., and Brooks, D. E., "Phase Partitioning in Space and on Earth," *Immunobiology of Proteins and Peptides IV*, edited by M. Z., Atassi, Plenum, New York, 1987, pp. 305–326.

[152]Brooks, D. E., and Bamberger, S., "Studies on Aqueous Two Phase Polymer Systems Useful for Partitioning of Biological Materials," *Materials Processing in the Reduced Gravity of Space*, edited by G. E. Rindone, North-Holland, New York, 1982, pp. 233–240.

[153]Raghava Rao, K. S. M. S., Stewart, R. M., and Todd, P., "Electrokinetic Demixing of Two-Phase Aqueous Polymer Systems. I. Separation Rates of Polyethylene Glycol-Dextran Mixtures," *Separation Science and Technology*, 1990 (to be published).

[154]Brønsted, J. N., *Z. Phys. Chem. A*, Bodenstein Festband 193, p. 257.

[155]Brooks, D. E., Sharp, K. A., and Fisher D., "Theoretical Aspects of Partitioning," *Partitioning in Aqueous Two-Phase Systems*, edited by H. Walter, D. E. Brooks, and D. Fisher, Academic, 1985, pp. 11–84.

[156]Brooks, D. E. Van Alstine, J., Snyder, R. S., Bamberger, S., and Harris, J. M., "Isolation of Biologicals by Partition in Two-Phase Polymer Systems," *Pharm Tech Conference '85 Proceedings*, Aster, Springfield, OR, 1985, pp. 20–21.

[157]Sharp, K. A., "Theoretical and Experimental Studies on Erythrocyte Partition in Aqueous Polymer Two Phase Systems," Ph.D Thesis, Univ. of British Columbia, Vancouver, 1985.

[158]Bamberger, S., Brooks, D. E., Sharp, K. A., Van Alstine, J. M., and Webber, T. J., "Preparation of Phase Systems and Measurement of Their Physicochemical Properties," *Partitioning in Aqueous Two-Phase Systems*, edited by H. Walter, D. E. Brooks, and D. Fisher, Academic, New York, 1985, pp. 85–130.

[159]Reitherman, R., Flanagan, S. D., and Barondes, S. H., "Electromotive Phenomena in Partition of Erythrocytes in Aqueous Two Phase Systems," *Biochimica et Biophysica Acta*, Vol. 297, 1973, pp. 193–202.

[160]Brooks, D. E., Seaman, G. V. F., and Walter, H., "Detection of Differences in Surface Charge-Associated Properties of Cells by Partition in Two-Polymer Aqueous Phase Systems," *Nature New Biol.*, Vol. 234, 1971, pp. 61–62.

[161]Brooks, D. E., "Cell Partitioning in Two Polymer Phase Systems: Towards Higher Resolution Separations," *Frontiers in Bioprocessing*, edited by S. K. Sikdar, M. Bier, and P. Todd, CRC, Boca Raton, FL, pp. 260–270.

[162]Walter, H., and Coyle, R. P., "Effect of Membrane Modification of Human Erythrocytes by Enzyme Treatment on Their Partition in Aqueous Dextran-Polyethylene Glycol Two-Phase Systems," *Biochimica et Biophysica Acta*, Vol. 165, 1968, pp. 540–543.

[163]Müller, W., "New Phase Supports for Liquid-Liquid Partition Chromatography of Biopolymers in Aqueous Poly(ethylene glycol)-Dextran Systems," *European Journal of Biochemistry*, Vol. 155, 1986, pp. 213–222.

[164]Skuse, D., Müller, W., and Brooks, D. E., "Column Chromatographic Separation of Cells Using Aqueous Polymeric Two-Phase Systems," *Analytical Biochemistry*, Vol. 174, 1988, p. 628.

[165]Brooks, D. E., Boyce, J., Bamberger, S. B., Harris, J. M., and Van Alstine, J. M., "Separation of Biological Materials in Microgravity," *Workshop on SPACE: Biomedicine and Biotechnology*, National Research Council of Canada, Ottowa, 1986.

[166]Bamberger, S., Van Alstine, J. M., Harris, J. M., Baird, J. M., Snyder, R. S., Boyce, J., and Brooks, D. E., "Demixing of Aqueous Polymer Two-Phase Systems in the Absence of Gravity," *Separation Science Technology*, Vol. 23, 1987, pp. 17–34.

[167]Harris, J. M., Brooks, D. E., Boyce, J. F., Snyder, R. S., and Van Alstine, J. M., "Hydrophilic Polymer Coatings for Control of Electroosmosis and Wetting," *Dynamic Aspects of Polymer Surfaces: Proceedings of the 5th Rocky Mountain American Chemical Society Meeting*, edited by J. D. Andrade, American Chemical Society, Washington, 1987.

[168]Brooks, D. E., Bamberger, S. B., Harris, J. M., Van Alstine, J., and Snyder, R. S., "Demixing Kinetics of Phase Separated Polymer Solutions in Microgravity," *Microgravity Science and Applications Flight Programs, January–March 1987, Selected Papers*, NASA TM-4069, 1988, pp.119–131.

[169]Barton, K., and Subramanian, R. S., "The Migration of Liquid Drops in a Vertical Temperature Gradient," *Journal of Colloid and Interface Science*, Vol. 133, 1989, p. 211.

[170]Chang, L. S., and Berg, J. C., "Electroconvective Enhancement of Mass or Heat Exchange Between a Drop or Bubble and Surroundings in the Presence of an Interfacial Tension Gradient," *AIChE Journal*, Vol. 31, 1985, pp. 149–151

[171]Chang, L. S., and Berg, J. C., "The Effect of Interfacial Tension Gradients on the Flow Structure of Single Drops or Bubbles Translating in an Electric Field," *AIChE Journal*, Vol. 31, 1985, pp. 551–557.

[172]Carlson, F. M., Chin, L.-Y., Fripp, A. L., and Crouch, R. K., "Finite Element Analysis of the Effect of a Non-Planar Solid-Liquid Interface on the Lateral Solute Segregation During Unidirectional Solidification," *Materials Processing in the Reduced Gravity of Space*, edited by G. E. Rindone, Elsevier/North-Holland, New York, 1982, pp. 629–638.

[173]Napolitano, L. T., "Marangoni Convection in Space Microgravity Environments," *Science (Washington DC)*, Vol. 225, 1984, pp. 197, 198.

[174]Kamotani, Y., Ostrach, S., and Lowry, S., "An Experimental Study of Heat Induced Surface-Tension Driven Flow," *Materials Processing in the Reduced Gravity of Space*, edited by G. E. Rindone, Elsevier/North-Holland, New York, 1982, pp. 161–172.

[175]Sen, A. K., Smith, M. K., and Davis, S. H., "Steady Thermocapillary Flows and Their Stability," *Materials Processing in the Reduced Gravity of Space*, edited by G. E. Rindone, Elsevier/North-Holland, New York, 1982, pp. 173–176.

[176]Adamson, A. W., *Physical Chemistry of Surfaces*, 4th ed., Wiley, New York, 1982.

[177]Schaeffer, R. J., and Coriell, S., "Convective and Interfacial Instabilities During Solidification of Succinonitrile Containing Ethanol," *Materials Processing in the Reduced Gravity of Space*, edited by G. E. Rindone, Elsevier/North-Holland, New York, 1982, pp. 479–489.

[178]Cassanto, J. M., Ziserman, H. I., Chapman, D. K., Korszun, Z. R., and Todd, P., "Simulation of Launch and Re-Entry Acceleration Profiles for Testing of Shuttle and Unmanned Microgravity Research Payloads," *Advances in Space Research*, Vol. 8, No.12, 1988, pp. 141–146.

[179]Cassanto, J. M., Holemans, W., Moller, T., Todd, P., Stewart, R. M., and Korszun, Z. R., "A Low-Cost, Low-Volume Carrier (Minilab) for Biotechnology and Fluids Experiments in Sounding Rockets and Orbital Space Flight," *AIAA Journal*, 1990 (to be published).

[180]Todd, P., Stewart, R. M., Cassanto, J. M., and Holemans, W., "Solute Transport Across Liquid-Liquid Interfaces in Low Gravity," *Advances in Space Research*, 1991 (to be published).

[181]Beijerinck, M. W., "Ueber eine Eigentumlichkeit der Löslischen Stärke," Centralblatt fur Bakteriologie, *Parasitenkünde u. Infektionskrankheiten*, Vol. 2, 1896, pp. 696–699.

[182]Beijerinck, M. W., "Ueber Emulsionsbildung bei der Vermischung Wasseriger Lösungen Gewisser Gelatinienden Kolloide," *Zeitz. für Chemie u. Industrie der Koll.* Vol. 7, 1910, pp. 16–20.

[183]Rudge, S. R., and Todd, P., "Applied Electric Fields for Downstream Processing," *Fundamentals and Practice of Large-Scale Protein Purification*, edited by M. R. Ladisch, R. C. Willson, C-d. C. Painton, and S. E. Builder, American Chemical Society, Washington, 1990, ACS (Symposium Series 427), pp. 244–270.

[184]Aptel, P., and Cabasso, I., "Novel Polymer Alloy Membranes Composed of Poly(4-vinylpyridine) and Cellulose Acetate. I. Asymmetric Membranes," *Journal of Applied Polymer Science*, 1980, pp. 1969–1989.

[185]Cabasso, I., "Ultrastructure of Asymmetric and Composite Membranes," Amer. Chem. Soc. Symp. Ser. 153: *Synthetic Membranes: Desalination*, Vol. 1, edited by A. F. Tubark, American Chemical Society, New York, 1981, pp. 267–291.

[186]Vera, I., "The Casting and Mechanism of Formation of Semipermeable

Membranes in a Microgravity Environment," *Advances in Space Research*, Vol. 6, No. 5, 1986, pp. 65–68.

[187]Vera, I., Cassanto, J. M., Todd, P., Korszun, Z. R., and Rock, L. R., *26th Plenary Meeting of the Committee on Space Research COSPAR 86*, Paris, 1986 (Abstract).

[188]Morrison, D. R., (ed.), *Bioprocessing in Space*, NASA TM X-58191, Jan. 1977.

[189]Morrison, D. R., and Todd, P., "Microgravity Sciences and Applications: an Overview of NASA's Program and Experiments in Microgravity Biotechnology," *Microgravity Science and Applications*, National Academy Press, Washington, DC, 1986, pp. 168–196.

[190]Naumann, R. J., and Herring, H. W., "Materials Processing in Space: Early Experiments," NASA SP-433, 1980.

[191]Owen, R. B., Kroes, R. L., and Witherow, W. K., "Results and Further Experiments Using Spacelab Holography," *Optics Letters*, Vol. 11, 1986, pp. 407–409.

[192]Littke, W., and John, C., "Protein Single Crystal Growth Under Microgravity," *Science (Washington DC)*, Vol. 225, 1984, pp. 203, 204.

[193]Bugg, C. E., "The Future of Protein Crystal Growth," *Journal of Crystal Growth*, Vol. 76, 1986, pp. 535–544.

[194]DeLucas, L. J., Suddath, F. L., Snyder, R., Naumann, R., Broom, M. B., Pusey, M., Yost, V., Herren, B., Carter, D., Nelson, B., Meehan, E. J., McPherson, A., and Bugg, C. E., "Preliminary Investigations of Protein Crystal Growth Using the Space Shuttle," *Journal of Crystal Growth*, Vol. 76. 1986. pp. 681–693.

[195]Plaas-Link, A., and Cornier, J., "Concepts of Crystallization of Organic Materials Under Microgravity," *Applied Microgravity Techniques*, Vol. 1, 1988, pp. 123–132.

[196]Todd, P., Sikdar, S. K., Walker, C., and Korszun, Z. R., "Application of Osmotic Dewatering to the Controlled Crystallization of Biological Macromolecules and Organic Compounds," *Journal of Crystal Growth*, 1990 (to be published).

[197]Taylor, G. R., "Cell Biology Experiments Conducted in Space," *BioScience*, Vol. 27, 1977, pp. 102–108.

[198]Todd, P., "Gravity and the Cell: Intracellular Structures and Stokes Sedimentation," *Bioprocessing in Space*, edited by D. Morrison, NASA TM X-58191, 1977.

[199]Todd, P., "Essay: Gravity-Dependent Phenomena at the Scale of the Single Cell," *Amer. Soc. Grav. Space Biol. Bull.*, Vol. 2, 1989, pp. 95–113.

[200]Talas, M., Batkai, L., Stoger, I., Nagy, K., Hiros, L., Konstantinova, I., Rykova, M., Mozgovaya, I., Giseva, O., and Kozarinov, V., "Results of Space Experiments Program Interferon I," *Acta Microbiologica Hungarica*, Vol. 30, 1983, pp. 53–61.

[201]Tixador, R. G., Richoilley, G., Grechko, G., Nefedov, Y., and Planel, H., "Multiplication de *Paramecium aurelia* au bord du vaisseau spatial Saliout-6." *Comptes Rendus Acad. Sci. Paris*, Vol. 287, 1978, pp. 829–832.

[202]Montgomery, P. O'B., Jr., Cook, J. E., Reynolds, R. C., Paul, J. S., Hayflick, L., Stock, D., Shulz, W. W., Kimzey, S., Thirolf, R. G., Rogers, T., Campbell, D., and Murrell, J., "The Response of Single Human Cells to Zero-Gravity," *Biomedical Results from Skylab*, edited by R. S. Johnston and L. F. Dietlein, NASA, 1977, pp. 221–234.

[203]Cogoli, A., and Tschopp, A., "Biotechnology in Space Laboratories," *Advances in Biochemical Engineering, Space and Terrestrial Biotechnology*, Vol. 22, edited by A. Riechter, Springer-Verlag, New York, 1982, pp. 1–49.

[204]Morrison, D. R., "Cellular Effects of Microgravity," *Microgravity Science and*

Applications Flight Programs, January–March 1987, Selected Papers, NASA TM-4069, 1988, pp. 217–226.

[205]Tschopp, A., Cogoli, A., Lewis, M. L., and Morrison, D. R., "Bioprocessing in Space: Human Cells Attach to Beads in Microgravity," *Journal of Biotechnology*, Vol. 1, 1984, pp. 281–293.

[206]Cogoli, A., Tschopp, A., and Fuch-Bislin, P., "Cell Sensitivity to Gravity," *Science (Washington DC)*, Vol. 225, 1984, pp. 228–230.

[207]Cogoli, A., Bechler, B., Muller, O., and Hunzinger, E., "Effect of Microgravity on Lymphocyte Activation," *Biorack on Spacelab D1*, European Space Agency Rept. SP-1091, 1988, pp. 89–100.

[208]Nyiri, L. K., "Some Questions of Space Bioengineering," *Bioprocessing in Space*, edited by D. R. Morrison, NASA TM X-58191, 1977, pp. 159–180.

Bibliography

Brooks, D. E., Bamberger, S. B., Harris, J. M., Van Alstine, J., and Snyder, R. S., "Demixing Kinetics of Phase Separated Polymer Solutions in Microgravity," *Proceedings of the 6th Symposium on Material Sciences Under Microgravity*, European Space Agency Rept. SP-256, 1985, p. 131.

Bugg, C. E., "Protein Crystal Growth in a Microgravity Environment," *Microgravity Science and Applications Flight Programs, January–March 1987, Selected Papers*, NASA TM-4069, 1988, pp. 249–266.

Egen, N. D., Twitty, G. E., Thormann, W., and Bier, M., "Fluid Stabilization During Isoelectric Focusing in Cylindrical and Annular Columns," *Separation Science and Technology*, Vol. 22, 1987, pp. 1383–1403.

Haglund, H., "Isotachophoresis—a Principle for Analytical and Preparative Separation of Substances such as Proteins, Peptides, Nucleotides, Weak Acids, Metals," *Science Tools*, Vol. 17, No. 1, 1970, pp. 2–13.

Hannig, K., and Wirth, H., "Free Flow Electrophoresis in Space," *Progress in Astronautics and Aeronautics*, Vol. 52, AIAA, NY, 1977, pp. 411–422.

Mehrishi, J. N., "Molecular Aspects of the Mammalian Cell Surface," *Progr. Biophys. Molec. Biol.*, Vol. 25, 1972, pp. 1–20.

Sourirajan, S., and Matsuura, T., *Reverse Osmosis and Ultrafiltration*, Division of Chemistry, National Research Council of Canada, Ottowa, 1984.

Voss, E. W., Jr., "Prolonged Weightlessness and Humoral Immunity," *Science (Washington DC)*, Vol. 225, 1984, pp. 214–215.

Walter, H., Krob, E. J., and Brooks, D. E., "Membrane Surface Properties Other Than Charge Involved in Cell Separation by Partition in Polymer, Aqueous Two-Phase Systems," *Biochemistry*, Vol. 15, 1976, pp. 2959–2964.

Phase Partitioning in Reduced Gravity

Stephan Bamberger*
NASA Marshall Space Science Laboratory, Huntsville, Alabama
James M. Van Alstine†
University of Alabama in Huntsville, Huntsville, Alabama
Donald E. Brooks‡ and John Boyce‡
University of British Columbia, Vancouver, Canada
and
J. Milton Harris¶
University of Alabama in Huntsville, Huntsville, Alabama

Introduction

THERE is a recognized need for improved separations technology.[1] This is especially true in biotechnology,[2] where, as illustrated in Fig. 1, much of the cost associated with a product is related to its isolation and purification.[3] Most biologicals are maintained in aqueous environments during their purification, and many biological separation techniques are affected by gravity-induced phenomena such as sedimentation and convection.[4] Low-gravity environments afford an opportunity to investigate the influence of gravity on separation techniques. They also allow study of interactions, such as surface wetting, that are masked or otherwise difficult to characterize in unit gravity. Low-gravity research provides an opportunity to improve bioprocessing techniques and carry out unique separations.

*National Research Council Research, Associateship Program.
†Corresponding author, Center for Materials Development in Space and Department of Biological Sciences.
‡Departments of Chemistry and Pathology.
¶Deparment of Chemistry.

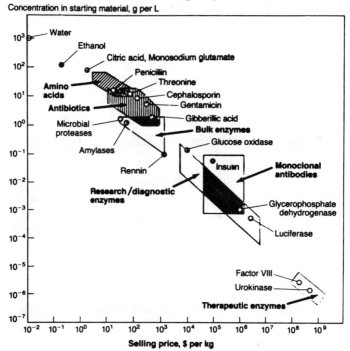

Separation contributes to biological product costs

Concentration in starting material, g per L

The selling price of products is a strong function of product concentration and consequently the cost of separation and/or purification. The reactor cost is typically less than 25% of total production cost. Three distinct categories are evident.

Fig. 1 Impact of separation/purification science on biotechnology. Relationship between product selling price and concentration in starting material.[3] Reproduced from *Chemical and Engineering News*, July 11, 1988, p. 30.

When pairs of polymers are dissolved in water at approximately 10% (wt/wt), two immiscible liquid phases often form. Such systems possess low liquid-liquid (L-L) interfacial tensions (Table 1) and readily emulsify upon gentle agitation. On Earth these emulsions rapidly demix due to phase density difference.[5,6] The glucose polymer dextran (D) and poly(ethylene glycol) (PEG) yield an aqueous two-phase polymer system (ATPPS) that demixes to a PEG-enriched phase floating on top of a dextran-enriched phase. ATPP systems can be buffered and rendered isotonic via the addition of salts. Such systems are highly compatible with biological molecules and particles such as cells. As a system demixes, added biological molecules tend to partition (i.e., distribute) asymmetrically between the phases. Because of their surface area bioparticles typically partition between one or both phases and the L-L interface. This differential partitioning forms the basis for a powerful L-L separation technique that can be carried out in single step or multistep regimes such as countercurrent distribution (CCD).[5,6] The

Table 1 Aqueous polymer two-phase systems studied in low gravity

System, %wt/wt (D:PEG:F)	Buffer[a]	PEG type[a] kDa	Dextran type[a] kDa	Viscosity (cP)		Interfacial tension, μN/m	Density, g/ml		Tie line length, %wt/wt	STS flight[b]
				PEG/F phase	D phase		PEG/F phase	D phase		
A (4:3.2:0)	I	20	500	4.9	33.3	5.7	1.0618	1.0466	9.57	26
				3.7[c]	24.6[c]					
B (5:3.5:0)	I	8	500	3.9	24.3	7.4	1.0239	1.0517	9.41	26, 51
				2.9[c]	19.8[c]					
C (5:4:0)	I	8	500	4.0	32.4	9.7	1.0195	1.0529	10.77	26, 51
D (6:4:0)	I	8	500	3.8	44.3	19.2	1.0286	1.0545	11.59	51
E (4.6:5:0)	I	20	40	7.1	10.7	11.5	1.0247	1.0583	12.69	26
				5.4[c]	7.9[c]					
F (7:5:0)	I	8	40	5.7	14.6	14.3	1.0289	1.0642	—	26
G (7:5:0)	II	8	40	5.7	14.2	6.2	1.0284	1.0624	11.69	51
H (8:4:0)	I	8	500	4.4	69.1	34.7	1.0239	1.0697	14.11	51
I (5.5:0.7:9.5)	III	8	500	39.6	19.4	3.9	1.069	1.069	—	26
J (7:0.3:12)	III	8	500			16.0	1.085	1.085	—	51

Systems at 22°C.
[a]Buffer salt concentrations: I = 109 mM Na$_2$HPO$_4$, 35 mM NaH$_2$PO$_4$, pH 7.2; II = 150 mM NaCl, 7.3 mM Na$_2$HPO$_4$, 2.3 mM NaH$_2$PO$_4$, pH 7.2; III = 84.6 mM Na$_2$HPO$_4$, 25.4 mM NaH$_2$PO$_4$, pH 7.5. PEG and D mol wt in kDa; mol wt of F = 400 kDa.
[b]Values for STS-26 and STS-51D.
[c]Systems at 30°C.

systems are so benevolent that cells are often cultured in the same phase system ultilized to separate them from their (macro)molecular products.[5-7]

Phase partitioning is emerging as the technique of choice for a number of biotechnical applications. Partition research groups exist at a number of biotechnical companies; national research institutes; and major universities in Sweden, Germany, Canada, Great Britain, Spain, the USSR, Japan, and the United States. Widespread interest is due to the cost effectiveness of this relatively simple, safe, and versatile technique.[7,8]

Although phase partitioning exhibits much promise for the purification of biological macromolecules, its greatest potential may lie in the isolation of particle subpopulations. Members of the same functional subpopulation often possess similar surface properties but differ in size and shape. To isolate such subpopulations, separation techniques sensitive to cell surface properties are required. Phase partitioning is one of the few techniques that separate cells predominantly on the basis of surface properties and can be scaled up for biotechnological requirements. Cell partition behavior is sensitive to surface properties since these properties determine particle interaction with the phase solutions and L-L interface structures. Such interactions can be altered by varying phase system physical properties that are isothermally controlled via system composition (Table 1).

Dissolving (asymmetrically partitioning) salts such as phosphates or citrates in a system often results in a Donnan potential between the bulk phases. In such systems cell partition can be related to cell surface charge measured electrophoretically. Such potentials are not present in systems where NaCl is the predominant salt.[9]

Affinity ligands can be prepared by covalently coupling PEG or dextran to substances with specific affinities for target cells or macromolecules. Addition of small amounts of polymer-derivatized ligands to phase systems selectively alters partition of the target population into the PEG-rich phase. Figure 2 illustrates how the addition of micromolar quantities of a PEG-derivatized fatty acyl compound (PEG-linoleate) completely shifts red blood cell partition into the upper phase. The basis for this shift appears to be increased (hydrophobic) interaction between the derivatized PEG and the cell surface. The resulting PEG-coated cells exhibit reduced interfacial free energy in the PEG-enriched phase. "Hydrophobic affinity partition" has been used to differentiate subpopulations of cancer cells and bacteria samples in a manner related to their ability to cause disease.[10,11] Recently, PEG-derivatized antibodies were developed for the immunospecific partitioning of cells and macromolecules. Immunospecific separation techniques such as fluorescent-activated cell sorting (FACS) are sensitive to particle surface structure but are often expensive, complicated, deleterious to cells, and technically difficult to scale up. Electrophoretic separation techniques also suffer such drawbacks. "Immunoaffinity partition" promises to efficiently isolate viable cell subpopulations in a simple test tube. The technique can easily process 10^8 cells within 10 min in a single step and is readily adaptable to large- or small-scale CCD.[12-14]

Solutes typically partition between the two phases with interfacial localization of only a few percent.[15] Partition is therefore expressed as the

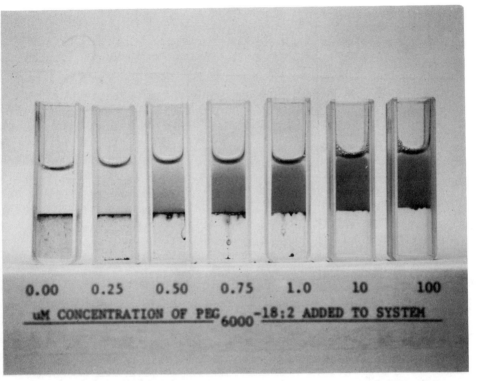

Fig. 2 Hydrophobic affinity partition of human red blood cells in a NaCl-enriched (5 : 4 : 0) system containing various micromolar amounts of PEG 6000–18 : 2 linoleate fatty acid ester.

coefficient K representing the ratio of solute of interest in the two phases. Solute partition follows the Boltzmann equation with

$$K = \exp[-\Delta E/kT] \qquad (1)$$

where ΔE is the energy necessary for solute to transfer from one liquid phase compartment to the other, k is the Boltzmann constant, and T is the absolute temperature. As expected, $\log K$ has been shown to vary with system salt composition and solute properties such as isoelectric pH.[5,6,16]

In contrast to solutes, particulates such as biological cells usually partition between one phase (typically the upper phase; Fig. 2) and the phase interface. Partition is expressed in terms of a coefficient, K, defined as the ratio of the fraction of cells added that partition into one of the bulk phases to the fraction not in that phase. $\log K$ has been shown to vary directly with cell surface properties such as charge, or system properties such as affinity ligand concentration, bulk phase potential, or interfacial tension (Table 1). Figure 3 illustrates how small increases in interfacial

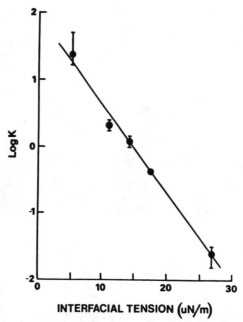

Fig. 3 Logarithm of the partition coefficient of *A. laidlawii* B bacterial cells vs interfacial tension for a series of NaCl-enriched two-phase systems.[18]

tension result in large increases in the adsorption of cells at the L-L interface with concomitant decreases in log$K$.[17]

If the partition of particles was determined solely by thermal energies, K should be a function of temperature and the free energy of interfacial adsorption, ΔG^0, determined by the previously given variables.[5,17,18] Quantitatively, K can be given by an equation of form

$$K = \exp(\Delta G^0 / kT) \tag{2}$$

Although the Boltzmann relationship holds for solutes, Sharp and Brooks[18] and Clarke and Wilson[19] have shown it to be invalid for particles such as cells. The ΔG^0 can be calculated by measuring the equilibrium contact angle θ formed between the line tangent to the surface and the two-phase boundary. For a spherical particle of radius r in a system with an L-L interfacial tension of γ_{TB}, Young's equation holds, and

$$\Delta G^0 = -\gamma_{TB} r_c^2 (1 - \cos\theta)^2 \tag{3}$$

where K is the number of cells in PEG-rich phase/remaining number of cells in the system, γ_{TB} is the system interfacial tension, and r_c is the cell radius.

Theoretical and measured values of K can be compared for γ_{TB} and θ altered by varying phase system composition. Sharp et al.[18] have estimated the force needed to remove cells such as human red blood cells ($r_c = 3.5\ \mu m$) from the interface in a typical phase system such as system C, Table 1, to be on the order of 1 to 5×10^{-6} dyne (see also K. A. Sharp, PhD Dissertation, University of British Columbia, Vancouver, 1985). As predicted K depends exponentially on ΔG^0; however, the factor multiplying ΔG^0 is several orders of magnitude smaller than $1/kT$.[17-19] Nonthermal forces present during phase demixing evidently contribute to the observed particle partition. Therefore, cell partition in unit gravity has both equilibrium (i.e., thermodynamic) and kinetic (i.e., mechanistic) determinants. The nonthermal stochastic influences may be related to viscous drag forces to which cells adsorbed to the L-L interface are exposed during phase demixing (see the following paragraph).

On Earth partitioning is carried out by adding biologicals to an ATPPS that is then emulsified. Given the low interfacial tensions of the systems (Table 1), manual mixing is sufficient to create an emulsion of micron-sized droplets of one phase in the other. The PEG-rich phase typically forms the matrix. This emulsion rapidly demixes, with gravity accelerating the energetically favored reduction in L-L interfacial area. On Earth the process of phase emulsion demixing passes through a number of relatively discrete stages. These include an initial stage where residual fluid mixing flows accelerate the coalescence of proximal phase droplets, a subsequent stage involving organization of the system into rapidly moving convective fluid columns, a further stage involving formation of a macroscopic planar interface (Fig. 4; see also Fig. 10). This is followed by a long-duration final stage in which isolated droplets buoyantly clear from the complementary

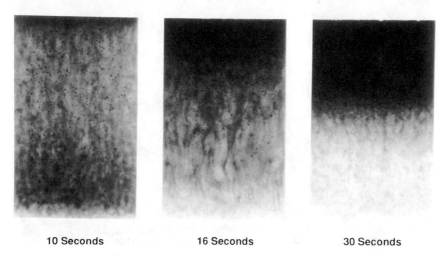

| 10 Seconds | 16 Seconds | 30 Seconds |

Fig. 4 Demixing of typical two-phase system, PEG-rich phase dyed with Trypan Blue.

phase. Concomitant with emulsion demixing, particles such as cells distribute between the phases and the L-L interface. The first two stages can result in 90% complete demixing within a minute (depending on system volume and chamber shape). However, the third stage can take from minutes to hours, depending on chamber geometry, chamber volume, phase volume ratio, viscosity, and density difference. The sampling time is therefore a compromise between emulsion demixing and cell sedimentation.

During the demixing process cells are localized in the bulk phases and at the interface, i.e., the surface of interfacial structures such as phase droplets (Fig. 5). Interfacially adsorbed cells can be exposed to viscous drag forces associated with system demixing. The measured distribution of cells would more closely represent (equilibrium) surface chemical interactions if it were not affected by sedimentation.[18] It may also be affected by convective fluid motions removing cells from the L-L interface.[19]

In many ways the environment of interfacially localized cells during the fluid column demixing stage (Fig. 4, middle) resembles a continuous flow flotation column in which there is an isotropic flow of liquid of viscosity (η) at net velocity (v) past an adsorbed spherical particle of radius r_c.[19] In the simplest case the upper limit viscous drag force (f) is given by the

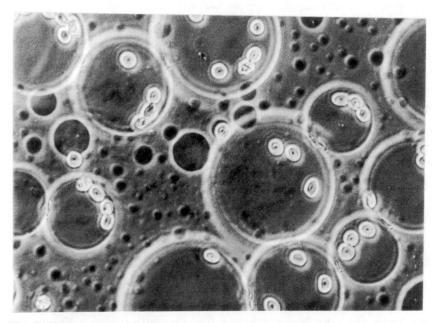

Fig. 5   Human red blood cells adsorbed at surface of dextran-rich phase droplets of NaCl-enriched (5 : 4 : 0) system.

Stokes equation

$$f = 6\pi\eta r_c v \qquad (4)$$

Under the condition $v = 0.5\,\text{cm-s}^{-1}$ (Fig. 4), $r_c = 3.5\,\mu\text{m}$ and matrix viscosity $\eta = 4\,\text{cP}$ (Table 1), then $f = 1.32 \times 10^{-4}$ dyne. This estimate is based on a number of assumptions[19]; nevertheless, due to the relatively fast fluid column velocities, it is two orders of magnitude larger than necessary to randomize the cell partition (K. A. Sharp, PhD Dissertation, University of British Columbia, Vancouver, 1984). Gravity may therefore play a significant role in both the emulsion demixing and particle distribution aspects of phase partitioning.

Given the previously stated low-g phase partitioning, experiments are being carried out to study the influence of gravity-independent variables, such as phase volume ratio and phase liquid-chamber interaction, on system demixing kinetics. Initial experiments, presented later, suggest that in microgravity ATPP systems demix via unique mechanism(s) in a manner amenable to operator control. These studies should allow the design of an unambiguous test of whether the lack of particle sedimentation and fluid column demixing in space will result in improved bioseparations of particles and/or molecules.

General Methodology

Two-Phase Systems

Preparation and physicochemical analysis (Table 1) of aqueous two-phase polymer systems has been detailed previously.[9,20,21] All chemicals were ACS grade or better. Phase systems were used following a 24-h period during which the phases came to equilibrium. Data tables are from Ref. 22 and represent typical mean values of an extensive series of experiments. Data vary with polymer lot, but individual system values exhibit standard deviations less than 10% of the mean. Polymer wall coatings were prepared and analyzed as described previously.[24,25] Equilibrium contact angles of $1\,\mu\text{l}$ to $5\,\mu\text{l}$ dextran-rich phase droplets on glass surfaces suspended in the corresponding PEG-rich phase were measured as described by Boyce et al.[25,26] Contact angles of 0, 90, and 180 deg represent complete, equal, and nonwetting by the dextran-rich phase.

Low-Gravity Phase Partition Experimentation

KC-135 experiments were carried in a manner similar to Shuttle experiments (see following) using prototype Shuttle experiment hardware and techniques. The KC-135, an aircraft used to obtain short-term, low-g conditions via parabolic flight trajectory, typically provided 10 s of $10^{-2}\,g$ and allowed for the development of simple but efficient Shuttle hardware.[21,23]

PPE-MODULE STS-26

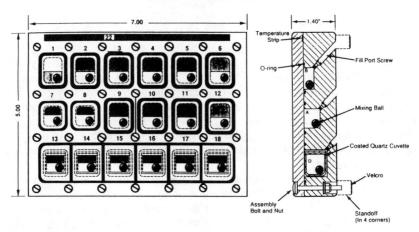

Fig. 6   Design of handheld phase partition experiment (PPE) flown on STS-26.

The handheld phase partition experiment (PPE) unit flown in October 1988 on Shuttle flight STS-26 is an improved version of the unit flown on STS-51D in April 1984.[17,35] It consists of a Plexiglas body plate into which 18 chambers were machined (Fig. 6). Chambers were sealed with O-rings that, together with offset fill ports, allowed for clear photovisualization of any interaction between the phases and the chamber wall. Each chamber contained a mixing ball, which appears as a dark spherical object in the photographs in Figs. 9 and 10. Mixing ball surfaces of D-rich phase wetting metal, (i.e. gold or stainless steel) or P-rich phase wetting teflon were chosen so that the mixing balls "wet" the phase which did not preferentially wet the container wall. Control of ball wetting was not possible in the fluorocarbon oil and water phase systems.

The PPE unit was mixed by manual agitation and mounted via Velcro-coated standoffs to a mid-deck fluorescent light box fitted with a light diffuser (Fig. 7). A $T = 0$ s picture was taken immediately following the mixing period. Table 2 indicates the experimental allocation of the chambers for STS-26, a Space Shuttle Mission flown in October 1988. Chambers 1–4, 6, and 9–12 had the same dimensions as those on STS-51D, a mission flown in April 1985 (1.90 cm high × 1.57 cm wide × 0.90 cm deep) and held 3.2 ml. Chamber 5 was 1.4 cm deep and held 4.2 ml. Chambers 7, 8 and 13–18 were 1.4-cm cubes to allow the nonwetting demixed phase to form a sphere. The latter six chambers held quartz cuvettes fully or half-coated with either D[23,25] or polyacrylamide (PA),[36] both of which preferentially wet the D-rich phase. Figures 10 and 11 represent the unit-g and low-g demixing appearance of the experimental unit.

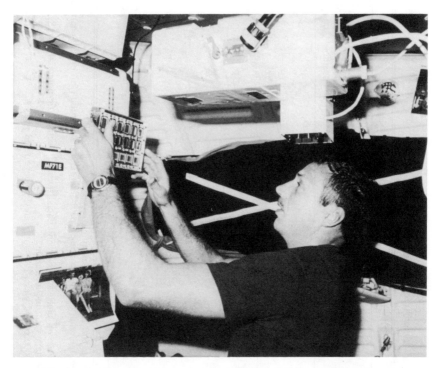

Fig. 7   Astronaut D. Hilmers mounts PPE on lightbox during STS-26.

Photograph Analysis of Phase System Demixing in Low Gravity

Quantitative analysis of the low-gravity demixing of two-phase polymer systems was introduced previously.[17,21] A photographic record was generated by taking a sequence of pictures of the backlit PPE using a Nikon 35-mm camera equipped with an hour : minute : second data back that imprints the time on the negative margin. Visualization of the phases is enhanced by the addition of a small concentration, e.g., 0.05 mg/ml, of Trypan Blue (Aldrich), which preferentially dyes the PEG-enriched phase and does not affect the physical properties of the system.[21] Dye concentrations were varied in order to compensate for differences in chamber dimensions, phase volume ratios, and dye partition coefficients.

Control studies supported the validity of the basic approach[21] which was subsequently modified to utilize two-dimensional Fourier transformation similar to the method of Beysens.[28] Photographs were digitized using a Panasonic video camera, Model WV1800, equipped with a 12.5/75-mm zoom lens, and a TV Vision Frame Grabber (Image Technology, Woburn, MA). Computer-aided manipulation and analysis of the digitized data was performed as discussed later in the chapter using a VAX 8300 (Digital

Table 2 STS-26 PPE unit chamber/experiment allocation

Chamber no.	Ball	System[b]	Volume ratio[b]	Chamber surface[c]	Experiment variation[a]							
					Magnetic control	Particle wetting and separation	Wall wetting	Chamber geometry and volume	Oil-water system	Phase volume ratio	Polymer system composition	STS-51D related
1	G	(4,6*,5*)I	1	PG	P	S	S				P	
2	G	(5,4)I	1	PG	P	S	S				P	P
3	G	(4,3,2*)I	1	PG	P	S	S				P	
4	G	(5,3.5)I	1	PG	P	S	S	P		P		P
5	G	(5,3.5)I	1	PG	P	S	S	P				
6	S	FC43 : 1 μm KCl	1	PG		P	P		P			P
7	S	(7*,5)I	1	PG	C	C	P				P	P
8	S	(5,3.5)I	1	PG	C	C	P	P				
9	S	(5.5 : 0.7 : 9.5)III	2/3	PG	C	S	S	P			S	
10	S	(5,3.5) I	3/2	PG		S	S			P		P
11	S	(5,3.5)I	1	PG		S	S			P		P
12	S	FC43 : 0.5 M KCl	1	PG		P	P		P			P
13	T	(5,3.5)I	1	D2000		P	P					P
14	T	(7*,5)I	1	PA		P	P					
15	T	(7*,5)I	1	G/PA		P	P					
16	T	(7*,5)I	1	G/D2000		P	P					
17	T	(7*,5)I	1	D2000		P	P					P
18	T	(53.5)I	1	D2000		P	P					P

Allocation set up so that related experiments are close together in unit to aid real-time observation.

[a]P, primary; S, secondary; C, control. Wall wetting can include chamber material, polymer coating type mol wt, and distribution (full or half-chamber). Influence on demixing rate and final disposition.

[b]System composition includes polymer type and composition and influence on phase viscosity, interfacial tension, electrostatic potential, etc. Systems chosen in relation to STS = 51D include reflights of same system or system with one alteration.

[c]Chamber inner surface; PG = Plexiglas; G = glass; D2000 = dextran T2000 coated glass; PA = poly(acrylamide) coated glass.

Equipment Corporation) and Interactive Data Language (IDL) software (Research Systems, Denver).

Computer Analysis of Photograph Images

Computer-Aided Analysis

Demixing kinetics in fluid phase systems is often studied using light scattering, holography, or photography. Light scattering allows the study of distributions of smaller droplets than photography (without the use of microscopy). It produces an intensity distribution corresponding to the two-dimensional power spectrum (see the following paragraphs). Holography provides a three-dimensional image and can follow the behavior of individual phase droplets, but is limited to much lower droplet concentrations (i.e., phase volume ratios and/or chamber volumes) than are needed for partition.[29] Light scattering would have required a more sophisticated experimental setup than was available with our Shuttle experiment. For these reasons we chose photography using 35-mm cameras and a light box which form part of the Shuttle's standard equipment. Photographs were labeled and digitized as discussed earlier. The following is a short discussion of methodology used in the photoanalysis of Shuttle partitioning experiments. This methodology may be applicable to a variety of fluid phase experiments involving high-volume-fraction emulsions.

A digital image typically consists of a two-dimensional array of elements or pixels, each of which can assume one of 256 gray levels. Several image manipulations have to be performed before an analysis of repetitive domains can be carried out.[30-32] Image data files often contain a header file that must be removed prior to mathematical manipulation. The image may have to be rotated by 90 deg or have its data rows and columns interchanged in order for it to be displayed correctly. Factors relating pixel distance to distance in the digitized object should be the same in both dimensions. However, in the present setup the 512×480 pixel image had a ratio of horizontal to vertical pixel per centimeter of 0.8. The easiest way to "square" such an image is to drop every fifth row, whereas a more accurate method involves the use of an algorithm that transforms each pixel via interpolation (Ref. 31, p. 115).

Shuttle data photographs frequently exhibit local variation in image intensities due to uneven illumination. Given a continuous gradient of average intensity across an image, the fast Fourier transformation (FFT) will exhibit a strong signal in the lowest channels. The following data adjustment corrects this situation by adding an intensity gradient to the image. The image is divided into four equal sections numbered 0–3 clockwise, the average intensities of which I_0, I_1, I_2, and I_3, can be thought of as being located in the center of each section. To construct an image corrected for uneven illumination, a bilinear plane of the same dimensions as the original image is fitted through all four points (Ref. 31, Chapter 8). Ignoring the intensity offset (i.e., lowest intensity value within the plane is

0) the new image has the intensity

$$G(x,y) = \frac{X(I_2 + I_3 - I_0 - I_1)}{N} + \frac{Y(I_2 - I_3 - I_0 + I_1)}{M} \tag{5}$$

with x and y being the image coordinates. This image is substracted from the original one. A constant is added to the result such that all intensities will be zero or positive (Ref. 31, Chapter 7).

"Smoothing" of an image reduces high frequency noise and enhances important data features. It is performed by replacing the individual intensity by the average intensity inside a square surrounding the individual point. The length of the square should be considereably smaller than the average characteristic length of the domains.

Strong local intensities (or shadows) in images tend to reduce the contrast over the rest of the image. This results in significant low-frequency peaks after FFT. If all values above (or below) an appropriate threshold are set to the threshold value, and the image is rescaled to maximum range, the image display and power spectrum are both improved.

Power Spectra

Repetitive patterns in data sequences are often analyzed by producing a power spectrum (autospectral density function). If $f(l)$ is such a data sequence, then the power spectrum is the frequency vs the intensity distribution, $S(u)$, where u is proportional to $1/l$, with t representing a repeated length. The method used most frequently to determine the power spectrum is to calculate the FFT and multiply the result by its own complex conjugate.[32,33] The FFT of a two-dimensional image is calculated by transforming first the columns and then the rows of the new array. Since in the present application, the density distribution should be independent of direction, the two-dimensional distribution can be reduced to one dimensional by averaging the frequencies within a given range over all angles. This is accomplished by calculating the distance from the origin (r) for each point, $S(u,v)$ in the power spectrum of an $N \times N$ image:

$$r = (u^2 + v^2)^{\frac{1}{2}} \tag{6}$$

A new one-dimensional array S_1 is created, and the intensity, $S(u,v)$ is added to the element in S_1, which has an index closest to the frequency r. All elements in S_1 are then divided by the number of additions performed (1 if none was performed). In an $N \times N$ image, the length of S_1 need only be $N/2$ (Nyquist frequency), since only intensities in elements $\leqslant N/2$ have physical meaning. Given a nonuniform domain size distribution, the resulting power spectrum should peak at a frequency proportional to the inverse of the average distance between adjacent domains (the mean wave number).

A FORTRAN listing for FFT can be found in Ref. 34, but many mainframe computers offer library routines that include FFT. Power

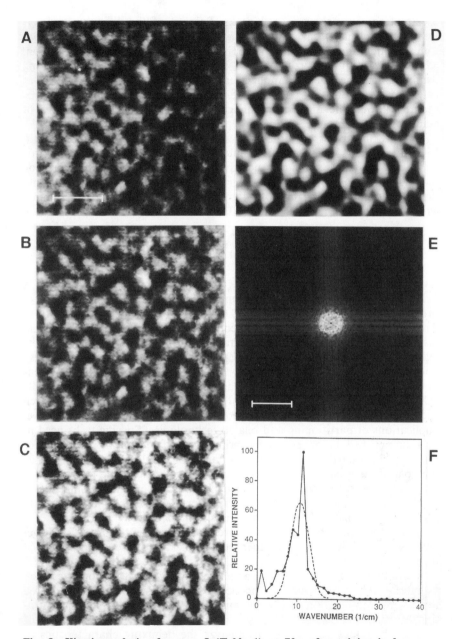

Fig. 8 Kinetic analysis of system I (Table 1) at 70 s after mixing in low g on STS-26. The photographic image a was corrected for uneven illumination b, its contrast was enhanced by setting an upper and lower intensity threshold c, and high-frequency noise was reduced by image smoothing d. Panel e indicates the two-dimensional intensity distribution of the FFT of the image in panel d. Panel f indicates the power spectrum (———) and its Gaussian approximation (— — —). The bars in panels a and e represent 0.2 and 50 cm^{-1}, respectively.

spectra frequently show ragged peaks. Because of the difficulty in determining appropriate peak values in such cases, the portion of the distribution surrounding the peak can be approximated by a Gaussian distribution, whose peak frequency value is used as a data point at each time during demixing. In the present application the range of approximation was extended to intensities one-fifth of the maximum intensity. This procedure allowed the analysis of images during more advanced stages of demixing.

Figure 8 illustrates the processing steps described earlier. The image in Fig. 8d highlights the structures of interest much more clearly than the raw image in Fig. 8a. It should be noted that, since the photographs provide records of domains rather than of individual droplets, domain size and kinetics should be proportional to that of the individual droplets assuming, as verified, that the droplets do not form discrete clusters.

Results and Discussion

Overview

NASA's interest in phase partitioning began in the late 1970s with a proposal from D. E. Brooks of the Oregon Health Sciences University to develop electrophoretic methods to control the rate of demixing and final disposition of the phases in microgravity. Initial research supported the feasibility of this and other lines of partition research.[9,36] Today, closely interacting groups at the University of British Columbia (UBC), Eisenhower Medical Center (EMC), University of Alabama at Huntsville (UAH), and Marshall Space Flight Center (MSFC) conduct research encompassing, theoretical, and applied studies on ATPPSs, as well as novel applications of partitioning and related technologies. Specific tasks need to be accomplished in order for the prime goal of evaluating the effect of gravity on cell partitions obtained on Earth to be achieved and for unique bioseparations in space to be carried out. These include the following:

1) development of control "isopycnic" phase systems possessing phases with equal densities and use of such systems to evaluate the ability of ATPPS emulsions to demix spontaneously in space[21];

2) development of specific hardware and experiments for KC-135, get away special (GAS, an STS experimental package) and sounding rocket flights (SRFs)[17,21];

3) analysis and understanding of phase system physical properties such as interfacial tension, viscosity, and electrostatic phase potentials, as related to system composition (polymer and buffer salt type and concentration)[9,20,22];

4) methodology to record and analyze the demixing characteristics of high- (i.e., 50%) volume-fraction, aqueous polymer emulsions in cubic chambers holding 1- to 5-ml samples (see above and Ref. 21);

5) development of materials and coatings capable of differentially interacting with the phases in order to "passively" control the demixing and final disposition of the phase systems[22,24-26,35];

6) understanding the effect of chamber shape on the demixing and final disposition of the phases[22];

7) development and testing of magnetic, electrophoretic, and other methods to "actively" control the demixing and final disposition of the phases[9,36];

8) understanding the effect of phase system volume ratios, and physical properties with regard to points 5, 6, and 7[17,22];

9) integration of the above to develop automated hardware capable of efficiently carrying out multisample, multistep partition separations in unit g and low g[36,37];

10) development of partition methods separating cells on the basis of quantifiable, physiologically important cell surface characteristics[10,12–14];

11) definition of particle separation challenges of medical and biotechnical importance amenable to point 10 and the unique facility/time constraints of space bioprocessing[10,11,37,38]; and

12) development of a theoretical understanding of the forces influencing phase demixing and the separations obtained via partitioning on Earth and in space.[9,17–19,21,22]

These tasks have all been accomplished or are being actively pursued and the reader is encouraged to consult the references provided. The following sections summarize recent results related to some of the preceding tasks. It should be noted that much of the work is of practical consequence for other separation methods such as electrophoresis[9,24,35,39] and chromatography.[37,40,41] It also impacts the creation and evaluation of biotechnical/biocompatible interfaces.[24–26,42] For example, polymer coatings developed to control chamber wall coatings have been shown to control the electroosmosis, which affects electrophoretic separations both on Earth and in space.[24,43] They also make excellent "spacer" coatings for use in anchoring proteins to surfaces without loss of protein activity.[42] PEG-derivatized fatty acyl molecules and immunoglobulins (antibodies) have been shown to be potentially useful for the separation of biologicals by hydrophobic affinity or immunoaffinity electrophoresis.[39,44] NASA-sponsored research into the basic physical properties of two-phase polymer systems (e.g., Table 1; Refs. 9 and 20) continues to be used by a variety of scientists working on theoretical as well as applied aspects of partition biotechnology.[6–8]

Ground-Based and KC-135 Research

Ground-based research has progressed in many areas. Isopycnic systems were created by adding small amounts of PEG to two-phase systems made using D and the poly(sucrose) Ficoll (F). Although D-F systems differ from D-PEG systems in their phase viscosities (Table 1) and usefulness for affinity cell separations,[5,6] they are similar in many ways.

Studies on isopycnic D-F-PEG systems suggested that in the absence of significant gravitational effects D-PEG phase systems would demix at a relatively slow rate, the equilibrium appearance of the system being one phase suspended "yolk-like" in the phase preferring to wet the container wall. Polymer coatings were developed to control the wetting of phases.

Contact angle and electro-osmosis measurements were used to characterize such coatings.[23-26] These studies suggested that phase volume ratio, interfacial tensions, phase viscosity, and chamber geometry would all influence the low-gravity phase demixing rate. They verified that spontaneous phase emulsion demixing would proceed slowly by a process that, over the time period of observation, is consistent with coalescence and therefore tractable to the kinetic analysis outlined in the previous section.[21,22,28]

KC-135 studies were undertaken concomitantly with isopycnic system research. Initial studies verified the ability of D-PEG systems used in bioseparations to spontaneously demix in low g. Most two polymer systems of biotechnical significance were found to demix significantly during the 10-15 s of low g (where low g is defined as $< 10^{-2} g$ for the KC-135, $10^{-4} g$ for the STS) available on the KC-135. The demixing behavior of these systems was shown to mimic the initial behavior of isopycnic systems.[21,23] KC-135 experimentation allows cost effective development of experimental systems, apparatus, and operator protocols for use on the Shuttle.

Shuttle-Based Experimentation

Two initial handheld PPEs were performed on STS-51D (April 1984) and STS-26 (October 1988). Separations of living cells were not attempted due to lack of a) detailed knowledge about and sufficient control of the demixing behavior of phase systems in low g and b) Shuttle facilities and crew time needed to carry out a meaningful cell separation experiment. In keeping with an approved flight proposal, the general plan has been to initially characterize the spontaneous low-g demixing behavior of candidate cell separation systems. Variables studied include those predicted by theory and verified via the experiments outlined in the previous subsection. The results of these initial studies should enable the design of systems and apparatus capable of providing an unambiguous test of cell partition in space (see the following paragraphs). The ability to control the demixing and final disposition of the phases is of key importance to such an evaluation.

On STS-51D a handheld PPE unit similar to that shown in Fig. 6 was flown.[17,22] The experiment consisted of shaking the unit to emulsify the phases, placing it on a fluorescent light box, and photographing the demixing process. The phase systems flown (Table 1) represented a broad range of physical properties. Two control systems contained fixed blood cells. An isopycnic system was also flown as a control. The experiment was performed once over a 2-h period. Results for $T > 10$ min were only partially useful because the flourescent light box unexpectedly warmed the samples and uncontrollably altered their physical properties. Nevertheless, in agreement with KC-135 and ground-based studies, the results indicated the following:

1) Spontaneous demixing occurs more slowly than in 1 g but at a rate compatible with space bioprocessing and similar to that of isopycnic systems in unit g (Fig. 10). Demixing was not inhibited by the addition of

cells at 10^8 per milliliter, and demixed systems were quite stable regarding random g-shock.

2) The typical demixed configuration is minimum interfacial area spheroid enclosed by the phase that, in agreement with contact angle studies, is energetically favored to interact with the container wall.

3) Interaction between the phases and a variety of surfaces such as mixing balls, biological cells, and chamber walls appeared to be as predicted from equilibrium contact-angle studies.

4) Systems not homogeneously mixed exhibited heterogeneous demixing within individual chambers.

5) Systems B, C, G, and H (Table 1) were chosen to cover a wide range of interfacial tension and phase viscosity. (In systems containing the same molecular weight polymer, both of these properties increase with polymer concentration (i.e., phase diagram tie-line length).[1,20] Quantitative results indicate that the low-tension, low-phase viscosity systems such as B and G (i.e., those closest to the phase diagram critical concentration point) demixed most rapidly (Fig. 11).

Demixing analysis of the results from STS-51D and related experiments[17,21,23] indicated that the systems studied appeared to demix via a process resembling the growth of evenly scattered phase regions, at a rate characteristic for each system. This observation is consistent with the low-g demixing behavior of other immiscible materials and alloys.[28,45,46] A notable exception is a 1975 Skylab experiment in which two-phase emulsions consisting of fluorocarbon oil and water were shown to be stable for hours in low g.[47,48]

The isopycnic system J (Table 1) demixed at comparable rates on Earth and in space, whereas systems such as G demixed similarly on the KC-135 and in space (Fig. 11). The discrepancy in low-g rate between systems G and J is most likely related to the greater phase viscosity and interfacial tension of the latter. The employed method of analysis is naturally limited to the observation period,[49] i.e., the period between the initial photograph and the point at which the systems achieve 3- to 4-mm-diam inclusions. Data are consistent with linear growth in average phase domain size during this period. As noted by Beysens et al.[28,49] and Brooks et al.,[17,21,22] such dependence is predicted for processes involving coalescence of proximal droplets facilitated by local droplet coalescence; or the coarsening of a extended, continuous, highly branched phase in a manner resembling spinodal decomposition. Either mechanism would be sensitive to phase volume ratio, phase viscosity, and interfacial tension.

Extrapolation of the three converging regression lines in Fig. 11 indicate initial droplet diameters of <250 μm. Laboratory data suggest that manual mixing can initially produce droplets of 1–10 μm diam with droplet growth to 50 μm occurring a few seconds even in viscous isopycnic systems.[21] There is obviously a need for further study of the initial demixing kinetics and the influence of short-term residual mixing flows.[17] The fourth regression line in Fig. 11, which indicates a larger initial droplet size, may have been offset a few seconds due to astronaut error.

The PPE flown on STS-26 (Figs. 6 and 7) was designed to improve on shortfalls of the STS-51D apparatus and experimental protocol. O-rings and offset fill ports allowed for greater visualization of liquid-chamber wall interactions, and temperature control and monitoring were provided by standoffs and a ±1°C liquid crystal temperature strip. Phase systems and chamber wall materials or coatings (Table 3) were chosen to augment data from STS-51D and support future PPEs. Isopycnic system I was flown as a control in chamber 9 (C-9), as were the fluorocarbon oil and water systems in C-6 and C-12.[22] The STS-26 experiment was performed four times and included still photography as well as video recording of the demixing process. Typical results are shown in Figs. 9, 10, and 12. Although the experiment was intended to be performed at 25°C, the Shuttle mid-deck temperature varied between 27 and 31°C during the mission. Fortunately, it only varied ±2°C during each experiment.

STS-26 PPE results have been discussed in more detail elsewhere.[22] As noted earlier, the demixing of phase systems on Earth involves a complicated process featuring relatively rapid movement of complementary fluid phase columns. This is in contrast to the behavior of the isopycnic system in C-9. In unit g the polymer systems demixed at individual rates slower than the fluorocarbon oil and water systems, which possess phase density differences on the order of 0.9 g/ml.[22] The ATPPS demixing rates (Fig. 9) were not strictly related to phase density differences. The most rapidly demixing D-PEG ATPP systems were those containing lower polymer concentrations, i.e., those closest to the critical concentration curve of the systems' isothermal phase diagram. Such systems exhibit lower phase density differences, interfacial tensions, and phase viscosities.[20]

It is expected that chamber geometry will affect gravity-driven demixing.[18] It is somewhat surprising that the unit-g demixing was slower in chambers coated in order to "wet" the D-rich phase (rather than the PEG-rich phase, which preferentially wets Plexiglas). For instance, compare C-7 with C-13 and C-18, and C-8 with C-14 and C-17. In chambers C-4, C-10, and C-11, system B demixed at a rate that varied with phase volume ratio (PEG-rich phase/D-rich phase) in the order 1.0 > 1.5 > 0.7. The 1.0 ratio may have afforded the most rapid demixing due to its

Table 3 Demixing rates in selected phase systems on STS-26

System	Buffer	Volume ratio	Demixing rate, mm/s
A (4 : 3.2 : 0)	I	1	0.127
B (5 : 3.5 : 0)	I	1	0.029
B (5 : 3.5 : 0)	I	3 : 2	0.067
C (5 : 4 : 0)	I	1	0.087
I (5.5 : 7 : 9.5)	III	1	0.0097

See Table 1 for details.

DEMIXING OF STS-26 PPE IN 1 g

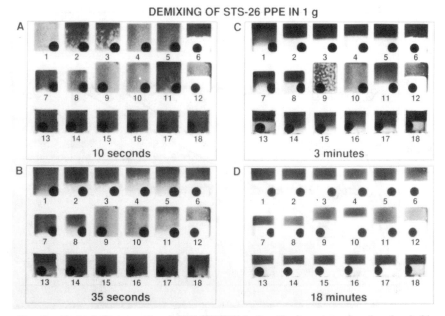

Fig. 9   Control photographs of STS-26 PPE in 1 *g*. Each rectangular chamber holds one phase system and one mixing ball. See Tables 1 and 2 for the compositions of the phase systems.

DEMIXING OF STS-26 PPE IN LOW g

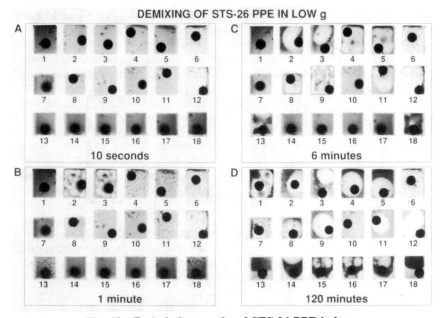

Fig. 10   Typical photographs of STS-26 PPE in low *g*.

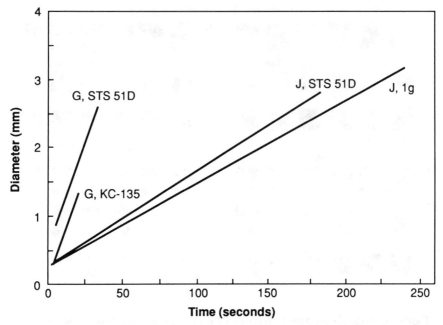

Fig. 11   Linear regression lines of mean droplet size vs time in phase system G (Table 1) on STS-51D and KC-135, and phase system J on STS-51D and on Earth.

favoring the formation and stability of fluid columns (Fig. 9) and/or the probability of contact and fusion between regions of each phase.

Typical low-*g* demixing of the STS-26 PPE is shown in Fig. 10. Systems again demixed, with the final disposition of phases and mixing balls being consistent with contact angle measurements.[22] Chambers 1 to 5 contained gold-coated magnetic iron mixing balls which were preferentially wet by the D-rich phase. In demixed systems it was possible to move these balls around using a handheld magnet. When this was done the wetting phase moved with the ball.

Demixing rates varied with polymer composition and chamber dimensions in a manner analogous to those of the 1-*g* controls. The fastest rates were observed for systems A, B, and C, whose lower phase viscosities and interfacial tensions would provide less hindrance to phase domain growth via mechanisms linked to rearrangement of interfacial structures.

The fluorocarbon oil and water system in C-6 did not appreciably demix over the 2-h duration of the study, confirming the Skylab observations of Lacy and Otto[47] (see the preceding paragraphs). This system contained only small amounts of salt and was expected to be electrostatically stabilized by droplet surface potentials. The fluorocarbon and saline system in C-12 contained enough added salt to significantly reduce such stabilization. Not surprisingly, it appeared to demix at a rate faster than the ATPPSs.

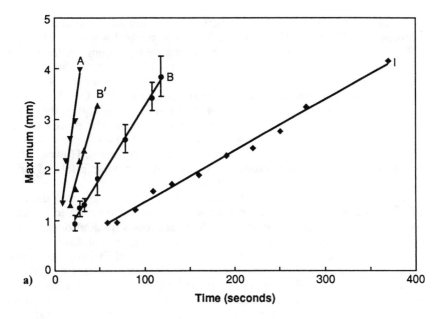

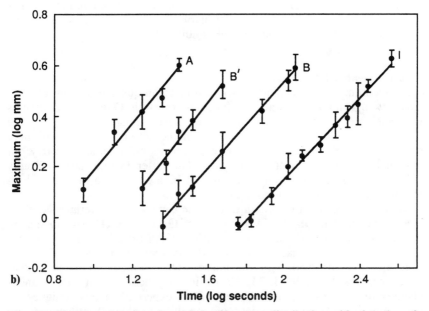

Fig. 12 Changes of peak values of the frequency distribution with time (see the section on computer analysis of photograph images) of systems A, B, and I (Table 1) on STS-26. The lines represent linear regression of time vs maximum plot a) and a double logarithmic plot of the same data b). See Table 3.

Plexiglas chambers C-4, C-10, and C-11 demixed in the phase volume ratio order $1.5 > 1.0 > 0.7$, which at 30°C was really $1.4 > 0.9 > 0.6$. This effect may be related to wall wetting altering phase volume ratios in the center of the chamber.[22] At unit bulk phase volume ratios, the importance of chamber wall wetting was confirmed on STS-26 for chambers of similar geometry (compare the demixing behavior of system B in C-7, C-13, and C-18, as well as system F in C-8, C-14, and C-17). As in unit g, the low-g demixing rates were decreased in chambers in which the more viscous D-rich phase wet the container wall. As a result, the two systems used in C-12 to C-18 did not completely demix over the experiment. However, the results support plans to influence low-g phase demixing rate and final disposition via control of phase wall wetting.

Although the results from STS-26 provided greater insight into the role of phase viscosity and interfacial tension, it is not yet possible to identify individual effects, since systems in which tension varied also exhibited differences in phase viscosities.[22] More data will be gained on future flights.

Quantitative analysis of the demixing results is still in progress. Some data are shown in Fig. 12 and Table 3. Multiple experiments on STS-26 yielded very reproducible results (see error bars). Figure 12a confirms that the low-g demixing systems exhibited, over the duration of photograph measurement, relatively uniform growth of phase domains. Figure 12b and Table 3 confirm the similarity of the results for relatively rapidly demixing systems such as A, slowly demixing systems such as I, and systems featuring variation in phase volume ratio. This supports a common demixing mechanism sensitive to and controllable via alteration of phase system physical properties and experimental conditions.

Future Research Directions

As a commercially significant separation technique affected by gravity, partitioning is a natural candidate for low-g research. Ongoing experiments will elucidate the effect(s) of gravity on partitioning. This will aid the development of improved apparatus and methods to carry out partitioning on Earth and in space. In addition, experiments are planned to carry out unique biological separations in space. The latter include large (rapidly sedimentating) cells and cell aggregates of biomedical importance such as megakaryocytes, tumor emboli, etc.

In conjunction with C. Lundquist of the Center for Materials Development in Space (CMDS) at UAH, a series of GAS and SRF experiments are planned to augment the data given earlier. GAS experiments will involve the influence of thermal variation and chamber geometry on demixing rate and final disposition of the phases. The chamber geometry studies have been instigated by J. M. Harris in conjunction with Paul Concus (University of California, Berkeley). SRF experiments are designed to study the influence of a variety of variables (Table 2) utilizing phase systems (Table 1, systems A, B, C, and G) that demix significantly in 6–10 min at low g (Fig. 11). The fully automated apparatus will allow better characterization of the kinetics of phase demixing during its initial stages. SRF PPE studies are also being undertaken by other research groups (see the chapter by P. Todd).

The Canadian Space Agency and National Research Council are sponsoring a PPE as part of the Space Physiology Experiments package that will fly on IML-1 (IML is the International Microgravity Laboratory). D. E. Brooks is the principal investigator. This PPE, designated PARLIQ after partition of liquids experiment, will study the ability of electric fields to "actively" control the demixing rate and final disposition of the phases as well as stabilize a planar interface. The latter is required in order to easily separate demixed phases, for sampling or carrying out multistep countercurrent separations.

J. M. Van Alstine, C. Lundquist, and J. M. Harris of the CMDS are planning multisample cell separation experiments for the Spacehab 1 Shuttle mission. The low-g separation behavior of various medically important fresh and fixed cell samples will be carried out. W. Hymer (Pennsylvania State University), L. Karr (NASA/MSFC), S. Bamberger (EMC), and D. E. Brooks will be coinvestigators. These experiments will utilize the results of previous experiments, particularly with regard to hardware design. A second set of PPEs involving cell separation is anticipated during a reflight of PARLIQ on IML-2.

Acknowledgments

The authors acknowledge the contribution of their colleagues R. S. Snyder, L. J. Karr, B. Hovanes, K. A. Sharp, and R. Cronise. We thank C. Chassay, R. Chassay, R. E. Shurney, and P. A. Curreri and Shuttle astronauts J. Garn, D. Hilmers, and G. Nelson for their help and encouragement. This work was supported by NASA-OSSA-Code E. Personal fellowship support from USRA (J.V.A.) and NRC (S.B.) is appreciated.

References

[1]NRC Committee on Separation Sciences and Technology, "Separation and Purification: Critical Needs and Opportunities," National Research Council, Washington, DC, 1987.

[2]Office of Technology Assessment, "Commercial Biotechnology: An International Analysis," U.S. Congress, Washington, DC, Rept. OTA-BA-218, 1984.

[3]Dwyer, J. L., *Bio/Technology* Vol. 2, 1984, pp. 957–964.

[4]Snyder, R. S., "Separation Techniques," *Materials Science in Space*, edited by B. Feuerbacher, H. Hamacher, and R. J. Naumann, Springer-Verlag, New York, 1986, pp. 191–222.

[5]Albertsson, P-Å., *Partition of Cell Particles and Macromolecules*, Wiley-Interscience, New York, 1986.

[6]Walter, H., Brooks, D. E., and Fisher, D. (eds.), *Partitioning in Aqueous Two-Phase Systems: Theory, Methods, Uses and Applications to Biotechnology*, Academic, New York, 1985.

[7]Hustedt, H., Kroner, K. H., and Kula, M.-R., "Applications of Phase Partitioning in Biotechnology," *Partitioning in Aqueous Two-Phase Systems: Theory, Methods, Uses and Application to Biotechnology*, edited by H. Walter, D. E. Brooks, and D. Fisher, Academic, New York, 1985, pp. 529–589.

[8]Fisher, D., and Sutherland, I. (eds.), *Separations Using Aqueous Phase Systems*, Plenum, New York, 1989.

[9]Brooks, D. E., Sharp, K. A., Bamberger, S., Tamblyn, C. H., Seaman, G. V. F., and Walter, H., *Journal of Colloid and Interface Science*, Vol. 102, 1984, pp. 1–13.

[10]Van Alstine, J. M., Sorensen, P., Webber, T. J., Greig, R., Poste, G., and Brooks, D. E., *Experimental Cell Research*, Vol. 164, 1986, pp. 366–378.

[11]Van Alstine, J. M., Trust, T. J., and Brooks, D. E., *Applied and Environ. Microbi.* Vol. 51, 1986, pp. 1309–1313.

[12]Karr, L. J., Shafer, S. G., Harris, J. M., Van Alstine, J. M., and Snyder, R. S., *Journal of Chromatography*, Vol. 354, 1986, pp. 269–282.

[13]Karr, L. J., Van Alstine, J. M., Snyder, R. S., Shafer, S. G., and Harris, J. M., *Journal of Chromatography*, Vol. 442, 1988, pp. 219–227.

[14]Sharp, K. A., Yalpani, M., Howard, S. J., and Brooks, D. E., *Analytical Biochemistry*, Vol. 154, 1986, pp. 110–118.

[15]Tumbs, M. P., and Harding, S. E., "Protein Transport Processes in the Water-Water Interface of Incompatible Two Phase Systems," *Separation Using Aqueous Phase Systems*, edited by D. Fisher and I. Sutherland, Plenum, New York, 1989, pp. 229–232.

[16]Baskir, J. W., Haton, T. A., and Suter, U. W., *Biotechnol. and Bioeng.* Vol 34, 1989, pp. 541–558.

[17]Brooks, D. E., Bamberger, S., Harris, J. M., Van Alstine, J. M., and Snyder, R. S., "Demixing Kinetics of Phase Separated Polymer Solutions in Micrgravity," *Proceedings of the 6th European Symposium on Materials Sciences Under Microgravity*, European Space Agency Publ. SP-256, 1987, pp. 131–138.

[18]Brooks, D. E., Sharp, K. A., and Fisher, D., "Theoretical Aspects of Partitioning," *Partitioning in Aqueous Two-Phase Systems: Theory, Methods, Uses and Applications to Biotechnology*, edited by H. Walter, D. E. Brooks, and D. Fisher, Academic, New York, 1985, pp. 11–84.

[19]Clarke, A. N., and Wilson, D. J., *Foam Flotation: Theory and Applications*, Dekker, New York, 1983, Chap. 3.

[20]Bamberger S., Brooks, D. E., Sharp, K. A., Van Alstine, J. M., and Webber, T. J., "Preparation of Phase Systems and Measurement of Their Physicochemical Properties," *Partitioning in Aqueous Two-Phase Systems: Theory, Methods, Uses and Applications to Biotechnology*, edited by H. Walter, D. E. Brooks, and D. Fisher, Academic, New York, 1985, pp. 86–131.

[21]Bamberger, S., Van Alstine, J. M., Harris, J. M., Baird, J. K., Snyder, R. S., Boyce, J., and Brooks, D. E., *Separation, Science and Technology*, Vol. 23, 1988, pp. 17–34.

[22]Van Alstine, J. M., Bamberger, S., Harris, J. M., Snyder, R. S., Boyce, J. F., and Brooks, D. E., "Phase Partitioning Experiments on Shuttle Flight STS-26," *Proceedings of the 7th European Symposium on Materials Sciences under Microgravity*, European Space Agency Publ. SP-295, 1989, pp. 399–477.

[23]Van Alstine, J. M., Harris, J. M., Snyder, R. S., Curreri, Bamberger, S., and Brooks, D. E., "Separation of Aqueous Two-Phase Polymer Systems in Microgravity," *Proceedings of the 5th European Symposium on Materials Sciences Under Microgravity*, European Space Agency Publ. SP-222, pp. 309–313.

[24]Herren, B. J., Shafer, S. G., Van Alstine, J. M., Harris, J. M., and Snyder, R. S., *Journal of Colloid and Interface Science*, Vol. 115, 1987, pp. 46–55.

[25]Harris, J. M., Brooks, D. E., Boyce, J. F., Snyder, R. S., and Van Alstine, J. M., "Hydrophilic Polymer Coatings for Control of Electroosmosis and Wetting," *Polymer Surface Dynamics*, edited by J. D. Andrade, ACS Monograph Series, Plenum, New York, 1988, pp. 111–118.

[26]Boyce, J. F., and Brooks, D. E., "Contact Angles as an Analytical Tool for Investigating Two-Phase Interactions with Biological Surfaces: A Review," *Separations Using Aqueous Phase Systems*, edited by D. Fisher and I. Sutherland, Plenum, New York, 1989, pp. 239–247.

[27]Schürch, S., Gerson, D. F., and McIver, D. J. L., *Biochemica et Biophysica Acta*, Vol. 640, 1981, pp. 557–571.

[28]Beysens, D., "Critical Phenomena," *Materials Science in Space*, edited by B. Feuerbacher, H. Hamacher, and R. Naumann, Springer-Verlag, New York, 1986, pp. 191–222.

[29]Lacy, L. L., Witherow, W. K., Facemire, B. R., and Nishioka, G. M., "Optical Studies of a Binary Miscibility Gap System," NASA TM-82494, 1982.

[30]Gonzalez, R. C., and Wintz, P., *Digital Image Processing*, 2nd ed., Addison-Wesley, Reading, MA, 1987.

[31]Castleman, K. R., *Digital Image Processing*, Prentice-Hall, Englewood Cliffs, NJ, 1979.

[32]Singleton, R. C., *Communications of the ACM*, Vol. 10, 1967, pp. 647–654.

[33]Bendat, J. S., and Piersol, A. G., *Random Data*, 2nd ed., Wiley, New York, 1986, pp. 391–400.

[34]Kay, S. M., *Modern Spectral Estimation: Theory and Applications*, Prentice-Hall, Englewood Cliffs, NJ, 1988, pp. 36–38.

[35]Van Alstine, J. M., Karr, L. K., Harris, J. M., Snyder, R. S., Bamberger, S., Matsos, H. C., Curreri, P. A., Boyce, J. F., and Brooks, D. E., "Phase Partitioning on Earth and in Space," *Immunobiology of Proteins and Peptides IV*, edited by M. Z. Atassi, Plenum, New York, 1987, pp. 305–326.

[36]Brooks D. E., and Bamberger, S., 1987, "Studies on Aqueous Two Phase Polymer Systems Useful for Partitioning of Biological Materials," *Materials Processing in the Reduced Gravity Environment of Space*, edited by G. Rindone, Academic, New York, 1982, pp. 233–240.

[37]Van Alstine, J. M., Snyder, R. S., Karr, L. J., and Harris, J. M., *Journal of Liquid Chromatography*, Vol. 8, 1985, pp. 2293–2313.

[38]Stocks, S. J., and Brooks, D. E., "Second Immunoaffinity Ligands for Cell Separation," *Separations Using Aqueous Phase Systems*, edited by D. Fisher and I. Sutherland, Plenum, New York, 1989, pp. 183–191.

[39]Van Alstine, J. M., Harris, J. M., Karr, L. J., Bamberger S., Matsos, H. C., and Snyder, R. S., "Application of Partition Technology to Particle Electrophoresis," *Separation Using Aqueous Phase Systems*, edited by D. Fisher and I. Sutherland, Plenum, New York, 1989, pp. 463–470.

[40]Skuse, D. R., and Brooks, D. E., *Journal of Chromatography*, Vol. 432, 1988, pp. 127–135.

[41]Müller, W., *European Journal of Biochemistry*, Vol. 155, 1986, pp. 213–224.

[42]Harris, J. M., and Yoshinaga, K. J., *Bioactive and Compatible Polymers*, Vol. 4, 1989, pp. 281–295.

[43]Van Alstine, J. M., Harris, J. M., Shafer, S., Snyder, R. S., and Herren, B., U.S. Patent 4,690,749, 1987.

[44]Van Alstine, J. M., U.S. Patent Application 1,376,487, 1989.

[45]Gelles, S. H., and Markworth, A. J., "Space Shuttle Experiments on Al-In Liquid Phase Miscibility Gap (LPMG) Alloys," *Proceedings of the 5th European Symposium on Materials Sciences Under Microgravity*, European Space Agency Publ. SP-222, 1984, pp. 417–422.

[46]Otto, H., "Stability of Metallic Dispersions," *Proceedings of the 5th European Symposium on Materials Sciences Under Microgravity*, European Space Agency Publ. SP-222, pp. 379–388.

[47]Lacy, L. L., and Otto, G. H., "Stability of Liquid Dispersions in Low Gravity." AIAA Paper 74-1242, Nov. 1974.

[48]Naumann, R. J., and Herring, H. W., "Materials Processing in Space; Early Experiments," NASA SP-443, 1980, pp. 82–83.

[49]Guenoun, P., Gastaud, R., Perrot, F., and Beysens, D., *Physical Review, A.*, Vol. 37, 1987, pp. 4876–4890.

Separation of Binary Alloys with Miscibility Gap in the Melt

Dieter Langbein*

Battelle, Frankfurt, Federal Republic of Germany

Introduction

THIS chapter is concerned with various attempts to produce finely dispersed mixtures of monotectic alloys under microgravity conditions as well as with the theoretical interpretation of the results obtained. There exist several hundreds of such metallic pairs, and many of them hold out a prospect of practical application. Microgravity experiments hitherto have been flown with aluminum-indium, aluminum-lead, lead-zinc, bismuth-zinc, and a few other systems.[1-11] Aluminum-lead, for instance, is a most promising material for producing bearings. Tests with sputtered samples have revealed excellent quality. However, sputtering is even more expensive than space production.

Figure 1a shows the finely dispersed mixture as expected from experiments under microgravity conditions, whereas Fig. 1b shows a real result from an experiment in a Space Application Rocket (SPAR),[2] a U.S. microgravity sounding rocket that was canceled after six successful flights in the early 1970s. The corresponding German program, Technologische Experimente unter Schwerelosigkeit (TEXUS), has, on the contrary, been expanded and has had about 20 very successful flights. Looking at Fig. 1 we have to ask ourselves: What went wrong with the expectations, and which mechanisms caused the obvious separation under microgravity conditions?

Figure 2 shows a phase diagram of a monotectic pair of metals, namely, the miscibility gap of aluminum-indium.[12] If these two metals are melted

*Leading Research Scientist, Electronic Systems Department.

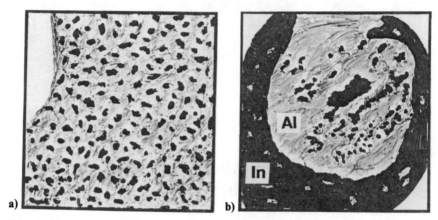

Fig. 1 Expected and obtained separation of monotectic systems under microgravity conditions: a) expected finely dispersed mixture; b) observed accumulation of indium at the container wall and of aluminum in the center.

and heated to a temperature above 830°C, their melts mix without any restriction on composition. On the other hand, when a mixture of 20% indium in 80% aluminum is cooled down below 810°C, the melt separates into two phases: an aluminum-rich phase and an indium-rich phase. The composition of these two phases during further cooling down follows the consolutal curve limiting the miscibility gap, until at 660°C aluminum solidifies from the aluminum-rich phase. The indium-rich phase, which still contains a small amount of aluminum, does not solidify until the temperature has come down to 156°C.

During cooling under terrestrial conditions, gravity usually dominates the separation. The heavier component, the indium-rich phase, sinks to the bottom. Figure 1b indicates, on the other hand, that cooling under microgravity conditions does not bring much advantage: Indium has been found outside and aluminum inside the flight sample. A similar separation has been found with the other systems mentioned earlier. (In most of the sounding rocket experiments performed with monotectic alloys, the samples have been heated to a temperature above the miscibility gap before launch and have been cooled down to solidification during the 6 min of microgravity available. Two experiments have been flown applying acoustic mixing.[13-15])

Figure 1b suggests that wetting effects play a dominant role under microgravity conditions. There are other mechanisms that strongly contribute to the separation as well. Table 1 presents a list of 16 relevant mechanisms.[16,17] Two of these, free convection and sedimentation, do not apply in microgravity. Some are important ones, and others less so. There are fast ones, such as wetting, and slow ones, such as Ostwald ripening. However, in order to produce finely dispersed mixtures (if this is possible at all), one has to be aware of all of the effects listed.

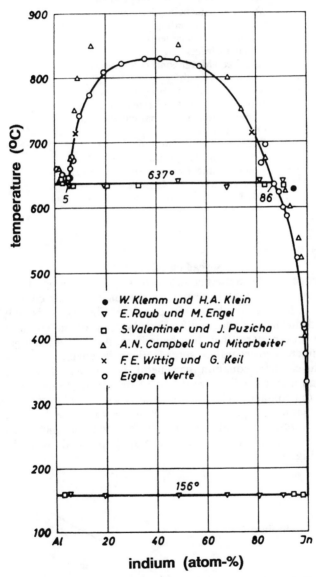

Fig. 2 **Phase diagram of the monotectic system aluminum-indium.**

Table 1 Mechanisms of separation

Free convection (gravity-dependent)
Sedimentation (gravity-dependent)
Differing tendencies of coagulation[a]
Differences in wetting[a]
Differing interface tensions[a]
Capillary forces[a]
Interface (Marangoni) convection[a]
Diffusion growth[a]
Momentum due to oriented growth
Momentum due to directed diffusion
Minimization of internal energy
Ostwald ripening
Repulsion by solidification front[a]
Volume shrinking during solidification[a]
Volume content of minority component[a]

[a]Mechanisms that are more important than others given in table.

Spreading Along Edges

In the edges of the container shown in Fig. 1b, there exist gas volumes. The contact angle between the indium-rich melt and the container, which consists of alumina, is about 150 deg (see Fig. 3). It is only roughly maintained during indium solidification. Wetting of the container wall by aluminum is even worse that than by indium. As a result, any gas enclosed in a container in microgravity spreads along its edges.

From the capillary equation the general law has been derived that a fluid must spread along a solid edge if half of the dihedral angle α of the solid edge plus the contact angle γ of the fluid is less than 90 deg[18] (by a fluid, one means either a liquid or a gas). With $\alpha = 45$ deg and $\gamma = 30$ deg, this condition is obviously satisfied for the gas volume enclosed. Capillary forces or wetting effects cause the general shape of the solidified sample shown in Fig. 1b.

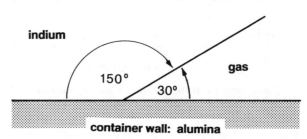

Fig. 3 Contact angle of indium vs gas at the container wall of alumina.

Capillary effects, in addition, are fast effects. For proving wetting effects it is sufficient to have a few seconds of microgravity; i.e., such effects can be studied during the parabolic flight of an aircraft or even during free fall in a drop tower. Figures 4 and 5 show the results of two wetting experiments performed during parabolic flights of the KC-135 aircraft over the Gulf of Mexico.[19,20] Figure 4 shows a transparent cell with 40 mm × 20 mm × 10 mm, which has also been used for studying separation in two TEXUS flights (see the section on sounding rocket experiments with cyclohexane-methanol). During the KC-135 flights in June 1985, this cell was filled with 2 : 1 methanol-cyclohexane and was observed by both a normal camera and a video camera. Methanol is slightly denser than cyclohexane (ρ-methanol = 0.79 g/cm^3, ρ-cyclohexane = 0.78 g/cm^3), such that during normal flight (1 g) and upward acceleration (1.8 g) of the aircraft, methanol rested on the bottom of the cell. However, methanol wets the container walls much better than cyclohexane does. Therefore, during reduced gravity ($10^{-2}\,g$) methanol rapidly spreads along the container's edges. From the stopwatch mounted next to the container, it is obvious that wetting takes less than 10 s.

Figure 5 shows a cylindrical 80 mm × 10 mm × 10 mm container of Plexiglas. It was filled to 1/8 with fluorinert, a liquid with density 1.8 g/cm^3 and a very low contact angle γ on glass, Plexiglas, and several other solids. The series of pictures taken during the KC-135 flights in May 1988 show

Fig. 4 **Spreading of methanol around cyclohexane under reduced gravity (during a KC-135 flight).**

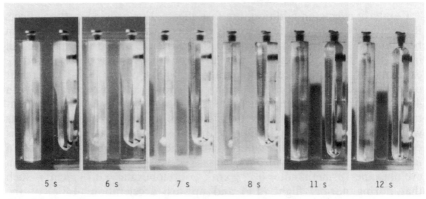

5 s   6 s   7 s   8 s   11 s   12 s

Fig. 5 Spreading of fluorinert in cylindrical containers with rhombic cross section under reduced gravity (during a KC-135 flight).

that spreading along the edges of the container is very fast.[21] After about 10 s the final configuration is reached.

Volume Balancing Systems

The message from these wetting experiments is that any gas volumes should be strictly avoided during experiments on the separation of monotectic alloys. On the other hand, there is volume expansion during heating and melting and volume shrinking during cooling and solidification, with a few exceptions (e.g., bismuth and antimony). Thus, in order to avoid gas volumes it is necessary to use a volume balancing system. Up to now various systems have been tested (see Table 2).

Figure 6 shows the volume balancing system based on capillary pressure that was used during melting and resolidification of a copper sample in the Spacelab D1 experiment WL-IHF-06.[22] The copper sample contained 1 vol% of molybdenum particles that were $2-3$ μm in size. The intention was to observe the displacement of the molybdenum particles by the melting and the solidification fronts of copper. The melting and solidification rates were chosen within the theoretical range for pushing particles. Because of time delays during setting up of the isothermal heating facility (IHF) and

Table 2 Systems Used or Considered for Volume Balance

Pistons squeezed by quartz, metal, or graphite springs
Nonwetting crucibles with conical ends
Nonwetting capillaries made from ceramics
Nonwetting conically or stepwise narrowing capillaries
Viscoelastic springs based on glass bulbs
Porous materials surrounded by flexible materials
Volume balancing on the basis of memory metals

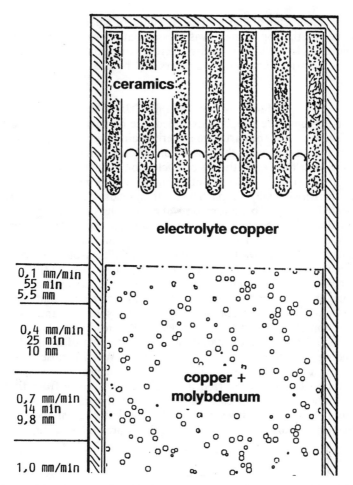

Fig. 6 Capillary volume balancing system used in the Spacelab D1 experiment WL-IHF-06.

again during installation of the gradient cooling block, the slow melting of the sample starting at the side of the volume balancing system had to be canceled. The sample became rapidly molten from the center to both ends instead.

The capillary volume balancing system consisted of a cylinder of magnesium silicate 10 mm in length, into which 19 holes, each 1 mm in diameter, had been drilled. Since molten copper has $\gamma > 90$ deg on magnesium silicate, it assumes a convex meniscus within the holes. It penetrates into the holes during melting, but is pushed back by capillary pressure during solidification. As a result of rapid melting of the sample from its center, the volume balancing system was not able to pick up the molten copper in the beginning, but perfectly fulfilled its tasks once the melting front had reached the upper end of the sample. Because of small differences in the hole's diameters, the copper melt was pressed out of the thinner holes prior to the thicker holes. Thus, in future microgravity experiments holes with conically or stepwise narrowing diameters will be used.

Influence of the Cartridge Material

Once gas volumes are properly excluded, the contact angle between the two metallic melts and the cartridge material gains primary importance. As a rule, the component with the lower contact angle nucleates at the wall first. This is what the Gibbs theory of nucleation claims. Once a droplet of the component with the lower contact angle has formed, it spreads out along the cartridge wall. In particular, it spreads along the edges. As already stated, in Fig. 1b indium wets the cartridge material better than aluminum does. The contact angle of indium on the cartridge material alumina is less than 90 deg if the second fluid is aluminum (see Fig. 7a).

The effect of the contact angle gains credibility from the complete wetting argument. There is theoretical reasoning that, close to the consolutal point of a pair of immiscible liquids, the interface tension between one component and the container wall becomes so small that Young's boundary condition on the contact angle has no solution, and the other component completely wets the container wall instead. This is even truer between a liquid and the corresponding solid: Spreading of a liquid on the corresponding solidification front is most likely. We will raise this argument again in the section on the TEXUS-9 experiment.

The wetting situation is reversed if the cartridge is made from titanium nitrate (see Fig. 7b). In that case aluminum wets the cartridge wall better than indium.[23] Systematic investigations on the influence of the cartridge material on the separation of monotectic alloys are still missing.

Coagulation

Another important class of mechanisms has to do with coagulation. Once a droplet of the minority component has been formed, it gives rise to a concentration gradient, a region of depletion, through which further material diffuses toward the droplet. The droplet grows. The question has

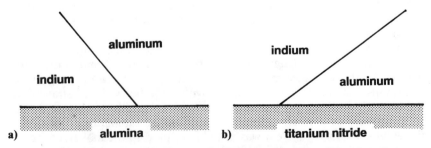

Fig. 7   Contact angles of indium vs aluminum at alumina and titanium nitrate.

to be raised: To what extent do two droplets of the minority component with a small spacing take notice of each other? They may mutually lose surface energy by coagulation and joining to form a single droplet. However, this either requires an external action (e.g., Brownian motion or a microscopic convective flow) or a direct interaction via an overlap of the depleted regions.

The effects of pressure and convection on coagulation are exhibited in Fig. 8. It shows the transparent cell, which was filled with methanol and cyclohexane during the KC-135 flights in June 1985. The cell was shaken by hand before one of the parabolas following the wetting experiment shown in Fig. 4, such that an emulsion of cyclohexane in methanol was caused. During the rest of the 1-g phase and the 1.8-g phase, the less dense cyclohexane droplets moved upward and rapidly coagulated. However, in reduced gravity, during the 20–30 s with 10^{-2} g shown in Fig. 8, both convection and coagulation fully came to rest.[19,20] The message from this experiment is that coagulation is strongly reduced under microgravity conditions. Two droplets of the same component may swim close to each other without any mutual notice.

The reservation must be stated that coagulation in a transparent system such as methanol/cyclohexan is not necessarily representative of coagulation in metallic systems such as aluminum/indium, etc. Transparent organic liquids consist of oriented molecules that, along a surface, may assume preferred orientations. It has been observed, for example, that methanol droplets in cyclohexane have a much stronger tendency to coagulate than cyclohexane droplets in methanol (see the section on the TEXUS-9 experiment). A methanol-cyclohexane interface may look different from

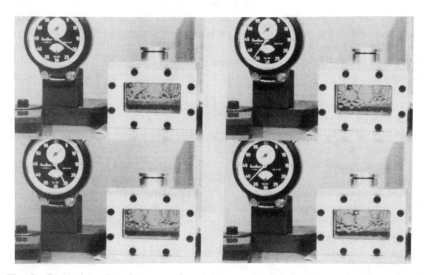

Fig. 8   Reduction of coalescence of cyclohexane droplets in methanol under reduced gravity (during a KC-135 flight).

D. LANGBEIN

each side. Metallic melts, on the other hand, are made up of atoms that are stripped of their electrons. The miscibility gap between two metallic melts usually arises from the differing electronic states, by the valence bands and conduction bands not fitting to each other, or by the differing atomic diameters. This hampers quantitative comparison of coagulation in transparent organic and metallic systems.

Effects Supporting Coagulation

There are several hypotheses on the effects caused by the mutual overlap of the depleted zones. On a molecular basis one may argue that a liquid mixture exhibiting concentration gradients assumes a state of lowest energy if there is maximum attraction between atoms of the same species; i.e., if the surfaces of equal concentration (the isopleths) assume a spherical shape. This hypothesis treats regions with different concentrations such as different liquids and the isopleths such as interfaces with infinitesimal interface energy. If the isopleths around two growing droplets contract to spheres, they bring the droplets enclosed into contact and thus favor coagulation.[24]

Another hypothesis makes use of Ostwald ripening and Marangoni convection. A small droplet generally has a higher solubility than a large one. The former tends to increase the concentration of the respective component in the outer liquid, whereas the latter tends to decrease it. A concentration gradient arises that is directed from the large to the small droplet. The growing of the latter at the cost of the former is known as "Ostwald ripening." It depends on diffusion and therefore generally is slow in comparison to the other mechanisms discussed in this chapter. However, it strongly gains weight if it is examined in connection with Marangoni convection: Once the large droplet notes the concentration gradient caused by the small droplet, it migrates toward the latter due to Marangoni convection. This is a fast mechanism. (Marangoni convection will be discussed in detail in the next section.) And, of course, any stirring in the liquid due to coagulation further favors coagulation.

Coagulation becomes more likely the higher the volume percentage of the minority component becomes.[17,25] This is a matter of statistics, of randomly filling a given volume with spherical droplets. Figure 9 shows cross sections of random distributions of drops with equal diameter. In the upper row, which corresponds to 5 vol% most droplets are singles; in the middle row, which corresponds to 10 vol% are a lot of pairs and triples; in the lower row, which corresponds to 20 vol%, singles form the minority. In Fig. 10 the percentage of droplets (particles) that exist as singles, pairs, triples, etc., is plotted vs the volume percentage. At 8 vol%, less than 50% of the droplets exist as singles.

In Figs. 9 and 10 random distributions of droplets were assumed without regard for the fact that droplets growing in demixing melts are more uniformly distributed than random. Close to a growing droplet, the probability of a new nucleus is strongly reduced.

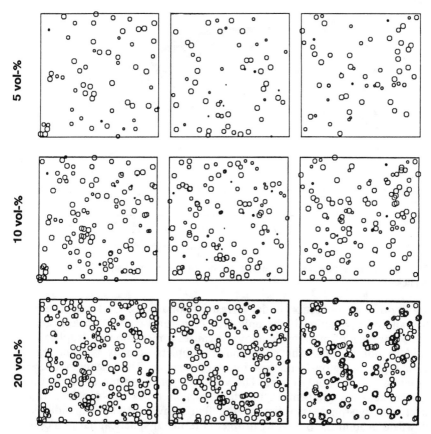

Fig. 9 Three-dimensional random distributions of spherical particles. Top: 5 vol%; center: 10 vol%; bottom: 20 vol%.

Marangoni Convection

One of the most important contributions to the separation of monotectic alloys, Marangoni convection, has been mentioned only marginally. It represents the main topic throughout the rest of this chapter.

Whenever a gradient in interface tension along a fluid interface arises, the regions with high interface tension pull more strongly than those with low interface tension, such that a shear force parallel to the gradient of the interface tension results. A gradient in interface tension may be caused by temperature differences, concentration differences, or electric or magnetic fields. The terms thermocapillarity, solutocapillarity, and electrocapillarity can be used, respectively. The original effect observed by Marangoni,[26] the formation of regular convection rolls, as when wine or cognac creeps up a glass and again slides down, is basically solutocapillarity.

Gradients in temperature or concentration usually give rise to density gradients, such that Marangoni convection always appears together with

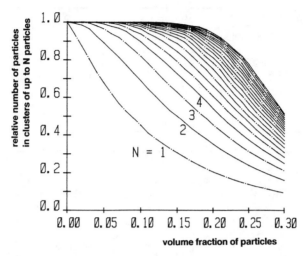

Fig. 10   Relative number of multiples arising in a random distribution of spherical particles.

gravity-driven convection under terrestrial conditions. Under microgravity Marangoni convection is the only type of natural convection left. During the separation of monotectic alloys, it causes a migration of droplets from cold to hot regions. The migration of droplets actually is from regions with high to regions with low interface tension. This means a decrease in energy of the fluid system. The velocity v of migration is proportional to the negative gradient in interface tension times the droplet's radius over the dynamic viscosities of the fluids involved. By expanding the flowfield around the migrating droplet in terms of spherical harmonics, one obtains[27]

$$v = (-\nabla\sigma)R/(2\eta_0 + 3\eta_1)$$

where η_0 and η_1 are the dynamic viscosities of the outer fluid 0 and the droplet 1, respectively. Figure 11 exhibits the flowfield in a coordinate system moving with the droplet. The gradient in interface tension is made up of both the thermal and the solutal gradient.

A cartridge containing melts of two monotectic metals necessarily has to be cooled from the outside. The first droplets of the minority component arise at or close to the container wall. They experience the temperature and concentration gradient from the wall to the center; i.e., Marangoni migration of the growing droplets is directed inward. The distribution of components shown in Fig. 1b (and that observed with all other samples solidified under microgravity conditions) clearly suggests that Marangoni migration together with wetting play a dominant role.

There is continuing controversy over whether it is more favorable to cool a sample fast or to cool it slowly. Fast cooling causes strong temperature and concentration gradients and strong Marangoni migration during a

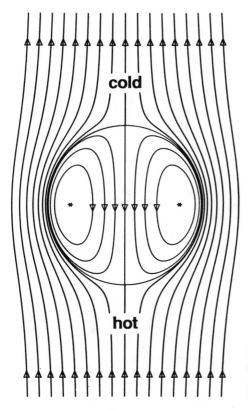

Fig. 11 Flowfield inside and outside a droplet undergoing Marangoni migration in a temperature field.

short time. During slow (so-called isothermal) cooling, Marangoni migration is slow as well, but solidification takes a long time. The considerable reduction of coagulation nevertheless is a strong argument in favor of the latter procedure.

Sounding Rocket Experiments with Cyclohexane-Methanol

In order to discriminate between the different mechanisms listed in Table 1 and in particular to demonstrate Marangoni migration of droplets, two TEXUS experiments with the transparent model system methanol-cyclohexane have been performed.[28,29] Figure 12a shows the phase diagram of this pair of liquids. The consolutal temperature is 46°C, which is easier to achieve than the 800–1200°C required for melting and mixing monotectic alloys. The disadvantage in doing model experiments with transparent systems is that the thermal diffusivity is lower than that of metals by two orders of magnitude. The Prandtl number is correspondingly higher. With the penetration of cooling into a transparent sample being slower, the droplets have more time for growing to sizes at which Marangoni convection becomes effective.

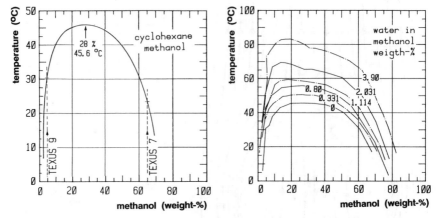

Fig. 12   Phase diagram of the transparent system cyclohexane-methanol: a) mixtures used in missions TEXUS-7 and TEXUS-9; b) dependence of the miscibility gap on the water content.

A great advantage of the system methanol-cyclohexane for terrestrial investigations is that the densities of these two liquids are very similar (see the section on spreading along edges). The densities may even be matched by adding deuterated cyclohexane, C_6D_{12}, with density 0.89 g/cm^3. With the temperature gradients of the densities being similar as well, it is possible to severely reduce gravity effects also during terrestrial investigations into separation.

Methanol (as are all alcohols) is very hydrophilic. Figure 12b shows the dependence of the miscibility gap on a possible water content. When methanol/cyclohexane is used, the water content has to be carefully watched.

Figure 13 exhibits the principle of the TEXUS experiments performed. Before launch of the rocket the liquid mixture is heated to a temperature

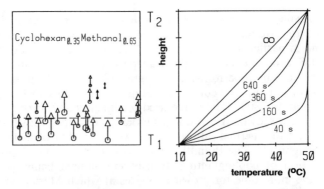

Fig. 13   Growth, migration, and shrinking of droplets during unidirectional cooling.

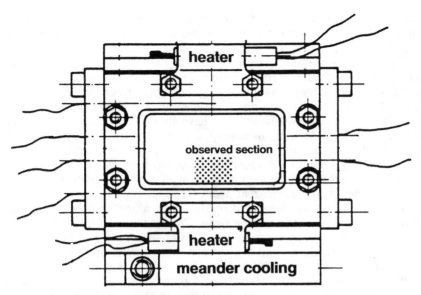

Fig. 14 TEXUS flight cell with the heaters and the cooling device.

above the consolutal point, to 50°C. After microgravity is begun it is unidirectionally cooled to 5 and −10°C in missions TEXUS-7 and TEXUS-9, respectively. The right-hand side shows the penetration of cooling within 40 s. Only close to the bottom, within a region of about 5 mm, does a temperature gradient exist. After 160 s the region of penetration has doubled, and the temperature gradient has fallen accordingly. Even after 360 s, at the end of the microgravity time, cooling has not really proceeded to the upper region of the cell. In the region between 15 and 20 mm, the temperature is still close to 50°C.

The left-hand side shows the growing of droplets of the minority component in the cooled region and their upward migration due to thermocapillary and solutocapillary convection. If they move faster than the cooling penetrates, they shrink in the upper, still-hot region.

Figure 14 shows the TEXUS flight cell, which was also used during the parabolic flights in June 1985. On the top and bottom, heaters are mounted for heating the liquid mixture to 50°C and to achieve complete mixing before launch. After microgravity is begun, the lower heater is switched off and the meander cooling system is used instead. The movie camera was directed toward the lower center part of the test cell. Magnification was 1 : 1; i.e., a section with 10 mm × 7 mm was observed.

TEXUS-7 Experiment

In mission TEXUS-7 the mixture used was 35% cyclohexane in 65% methanol. This composition was selected according to reference experiments on the ground, which showed only a very few cyclohexane droplets for lower concentrations. In the opposite case, for higher concentrations, the number of cyclohexane droplets became so large that transparency of the cell was lost.

The flight film shows an extending fog zone at the cell bottom during the first 30 s of cooling. The fog is formed by tiny cyclohexane droplets. The fog front moves upward as the cooling front penetrates (see Figs. 15 and 16). After 35 s cyclohexane droplets of about 50–200 μm appear, which rapidly migrate to the fog front. They have grown at the cooled cell bottom and now undergo Marangoni migration in the upward temperature gradient. At the cooling front, which is nearly identical to the fog front, migration stops due to the vanishing temperature gradient. Subsequently, the number and size of the migrating particles rapidly increase and the fog zone becomes partly transparent again (see Fig. 16).

The migrating cyclohexane droplets carry along coolness from the bottom of the cell to the cooling front. This additional transport of coolness causes the cooling front to move faster than predicted by the heat-conduction equation (a moving droplet front arises; see Fig. 15). The droplets move in the temperature gradient, which they carry along in a collective manner. The situation is even more complex since the droplets also bring along concentration. According to the consolutal curve, their cyclohexane equilibrium content at the cooled cell bottom is higher than that at the cooling front, where they dissolve cyclohexane or else take up further methanol. The droplets thus increase the cyclohexane concentration at the cooling front. The decrease in interface tension from bottom up is due to temperature and concentration as well.

Because of these transport effects, droplet migration in the TEXUS-7 experiment is faster than that in the 1-*g* reference experiments. In the latter experiments the cyclohexane droplets enter the upper hot region due to

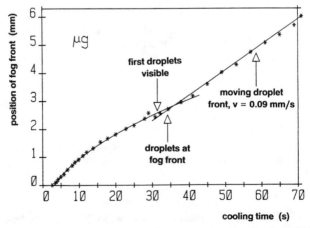

Fig. 15 Position of the fog front caused by cyclohexane droplets during mission TEXUS-7.

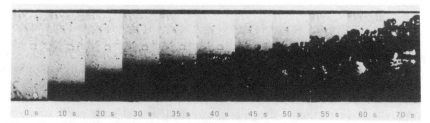

Fig. 16   Motion of the fog front and Marangoni migration of cyclohexane droplets during mission TEXUS-7.

buoyancy, where their dissolution does not further decrease the interface tension (it is zero anyway). The theory of the joint droplet migration in the temperature field has been tackled only recently.[30-32]

TEXUS-9 Experiment

In order to verify the results obtained in the TEXUS-7 mission and to study the additional effects of a solidification front (volume shrinking, wetting), the concentrations of methanol and cyclohexane were reversed in the mission TEXUS-9. The 1-g reference experiments showed that a reasonable but still not too high number of methanol droplets could be obtained by applying 5% methanol in 95% cyclohexane. The melting temperature of cyclohexane is 6°C. The cooling temperature thus was lowered to −10°C. Figure 17 shows the beginning of cooling in the TEXUS-9 experiment. A fog of methanol droplets arises (instead of the fog of cyclohexane droplets observed in the TEXUS-7 experiment) that moves upward with the speed of the cooling front. About 10 s after cooling has begun, a second, darker fog front arises. It moves upward faster than the first one and passes it after about 15 s. This fog front turns out to be formed by small methanol droplets undergoing collective Marangoni migration. After about 30 s larger methanol droplets migrate from the bottom to the cooling front. They may even pass it by 1−2 mm. This is particularly true for some large droplets, which migrate along the same trail. The transport of coolness and concentration causes local changes of the temperature and concentration field.

Fig. 17   Motion of the fog front and Marangoni migration of methanol droplets during mission TEXUS-9.

After 36 s, when the cell bottom assumes the solidification temperature of cyclohexane, the distribution of the migrating methanol droplets becomes more uniform. The average droplet size decreases. This can be ascribed to a change in the wetting conditions. The migration of many droplets along the same trail obviously has been caused by nucleation centers at the cell bottom, by small scratches, or even by dirt. At these positions large methanol droplets grow, detach, and undergo Marangoni migration. However, when the nucleation centers are blocked by the solidification front of cyclohexane, a more uniform growth of methanol droplets and a more uniform Marangoni migration result. As stated in the section on spreading along edges (see Fig. 4), methanol wets the cell bottom better than cyclohexane does. This means that methanol droplets generally must grow to a larger size in order to detach than cyclohexane droplets. This changes completely when the cell bottom gets covered with solid cyclohexane. Cyclohexane now wets the bottom much better than methanol. Even complete wetting of liquid cyclohexane at solid cyclohexane may happen, even though the solidification temperature and concentration of cyclohexane are far away from the critical point.

Because of the fog zone formed by methanol droplets, the solidification front of cyclohexane is not actually visable in the TEXUS-9 flight film. Its position was calculated theoretically and verified by the 1-g reference experiments. The final position of the solidification front became visible during re-entry of the payload, when there was strong stirring and the cell became transparent again. The position still agreed with the calculations.

The TEXUS-7 and TEXUS-9 films reveal several additional effects. There is a much faster coagulation of methanol droplets in the cyclohexane matrix than of cyclohexane droplets in the methanol matrix, despite the fact that in both cases the same interface tension is working. Another effect was the formation of regular convection rolls during the TEXUS-9 mission. They evolve more and more clearly after about 3 min and may be ascribed to a correlation between the growth front of cyclohexane and the nucleation and migration of the methanol droplets.

D1 Experiment WL-FPM-03

In order to reduce the observed contributions to separation, two approaches have been suggested:

1) Lower heterogeneous nucleation at the crucible by changing the material of the crucible, or preferably, by containerless processing. This procedure bears the risk of increasing capillary effects and Marangoni convection along the free surface.

2) Cool very slowly in order to achieve uniform nucleation within the sample and to minimize Marangoni convection. However, the latter may be jeopardized by the much longer cooling time required.

The D1 experiment WL-FPM-03 was a first step toward a free fluid surface and toward slow cooling as well.[33] A free liquid column between coaxial disks was established, and the mechanisms causing mixing during active heating and demixing during passive cooling were observed. A closed

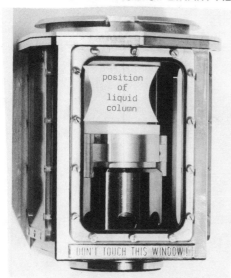

position
of
liquid
column

DON'T TOUCH THIS WINDOW!

Fig. 18 Flight cell of the experiment WL-FPM-03.

liquid container adapted to the fluid physics module (FPM) was developed (Fig. 18). The two circular disks holding the column are concave in order to avoid high flow velocities during liquid injection. Separation of the disks is achieved by rear plate rotation of the FPM. The liquid volume required for achieving a cylindrical column is simultaneously injected from the reservoir. The front disk contains the heater whose operation is redundantly controlled by two thermistors. Cooling of the liquid column is passive by heat radiation and conduction.

For safety reasons it was not possible to use the liquid system cyclohexane-methanol in Spacelab D1. Methanol is toxic, its flammability temperature is 11°C, and its boiling temperature is 65°C. Benzylbenzoate-paraffin oil was chosen instead. The miscibility diagram of this liquid pair, which was obtained by volumetric methods, is shown in Fig. 19.

The surface and interface tensions of the two liquids and their contact angles with aluminum and Teflon have been determined by the sessile drop method. The sum of the interface tensions benzylbenzoate-paraffin oil plus paraffin oil-air turns out to be much smaller than the interface tension benzylbenzoate-air, such that spreading of paraffin oil along the free fluid surface of the column was expected. In addition, benzylbenzoate wets the supporting aluminum disks better than paraffin oil. Therefore, this component would either form two separated spherical segments at the disks, or an inner liquid column in an outer liquid column made up from paraffiin oil. In view of the low contact angle of both liquids on aluminum, the supporting disks had sharp edges. They were equipped with Teflon rings, which also served as seals during storage, launch, and landing. Several wetting tests in parabolic flights with KC-135 were performed with this design.

Because the configuration of the two liquid components is unknown during heating of the column and because diffusion is very slow, Marangoni convection is a necessary requirement for mixing and unmixing.

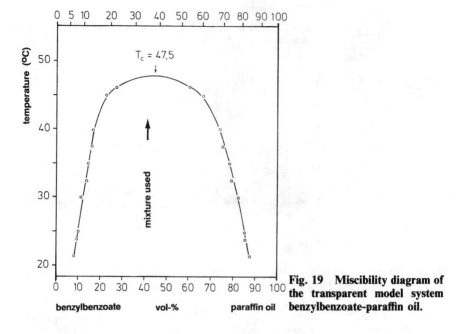

Fig. 19   Miscibility diagram of the transparent model system benzylbenzoate-paraffin oil.

The temperature coefficient of the surface tensions paraffin oil-air and benzylbenzoate-air is about 10^{-5} N/mK. Depending on the temperature profile, this may lead to flow velocities of up to 30 mm/s along the free fluid surface and to migration of bubbles up to 1 mm/s. This suggested using 20 min for heating and 40 min for cooling of the column.

Results of Run A of WL-FPM-03

Figure 20 shows the different steps of WL-FPM-03 and the actual position of run A in the D1 timeline. Real-time television and on air-to-ground loop were available for 46 min. A fruitful cooperation resulted between payload specialist Ernst Messerschmid in the Spacelab and the experimenter on the ground. Both sides pointed out important phases of the experiment and discussed steps to be taken.

The most important observations of the experiment consisted of the following:

1) Thermal and solutal Marangoni convection along the free fluid surface during heating. The fog regions resulting from mixing migrate to the lower, cold end of the column with speeds up to 3 mm/s. This speed, which is about 20% of the estimated value, further decreases in the course of heating, when the stationary temperature profile is reached (Fig. 21).

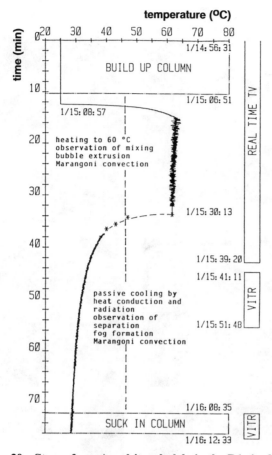

temperature (°C)

Fig. 20   Steps of run A and its schedule in the D1 timeline.

2) Downward Marangoni convection along the free fluid surface causes an upward counterconvection inside the column that takes along the bubbles. The bubbles eventually rupture when they reach the surface. The inclusion of bubbles was by no means planned, but turned out to be very helpful for observing the flow velocity and proving the effectiveness of bubble expulsion by Marangoni convection (Fig. 21).

3) Spreading of benzylbenzoate along the heated disk. This spreading, which is due to complete wetting of aluminum by this component, brings about the expected axisymmetric configuration of the liquid column. The speed of spreading and the dynamic contact angle clearly decrease with the approach of the symmetric configuration (Fig. 22).

4) Capillary effects together with a Rayleigh instability. The inner column of benzylbenzoate, which is caused by the preceding spreading, broke into two spherical segments at the supporting disks (Fig. 23).

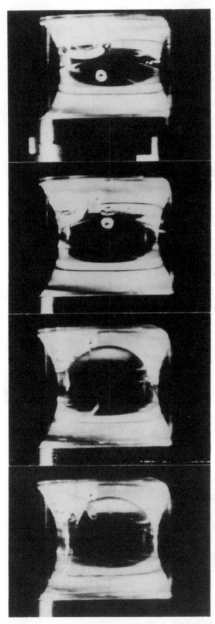

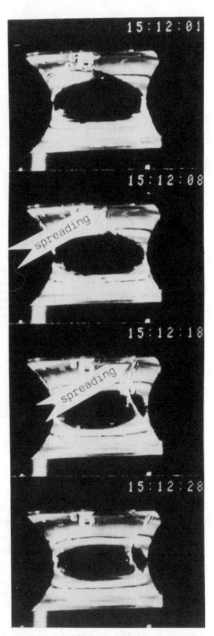

Fig. 21 Downward Maragoni convection of fog regions and upward migration of bubbles in the beginning of heating.

Fig. 22 Spreading of benzylbenzoate along the upper, heated disk.

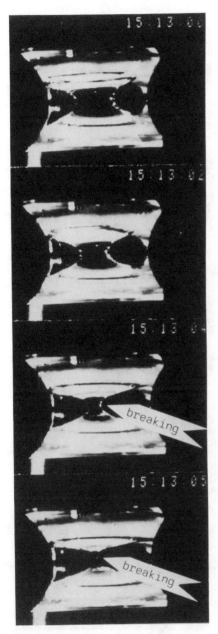

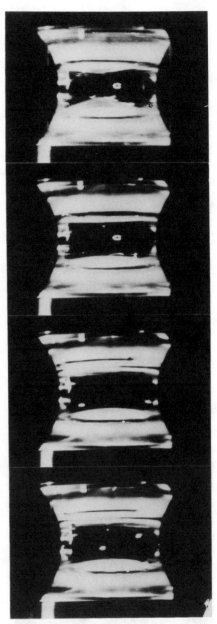

Fig. 23 Breakage of the inner column of benzylbenzoate in the outer column of paraffin oil due to Rayleigh instability.

Fig. 24 Continuing mixing and bubble migration in the end of the heating phase.

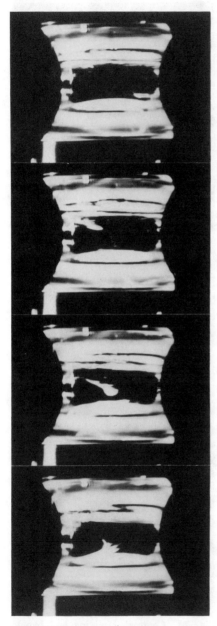

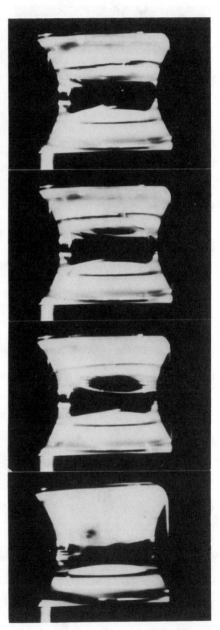

Fig. 25 Marangoni convection of fog regions in the beginning of the cooling phase.

Fig. 26 Progressive demixing in the upper half of the liquid column.

Breakage started at the relative height $H/2R = 0.8$ and the relative neck radius $R_n/R = 0.05$ of the inner column, a much larger neck radius and liquid volume than the values predicted by theory.

5) Continuing mixing due to diffusion and Marangoni convection (Fig. 24) until full dissolution of the spherical segment of benzylbenzoate at the upper disk.

6) Progressive demixing during the cooling phase. As in the beginning of the heating phase, there arose fog regions that moved downward due to Marangoni convection. The demixed region eventually extended over the upper half of the column (Figs. 25 and 26).

The final recovery of the column into the reservoir worked as intended.

Since the liquid mixture spread across the Teflon rings surrounding the supporting metallic disks, the liquid column formed an unduloid rather than a cylinder. The neck-to-disk ratio was $R_n/R = 0.7$. This shape of the liquid surface represents a convex-concave lens. Radially all distances are enlarged, whereas axially all distances shrink. Figure 27a shows the image of a quadratic grid in the meridian plane. The dashed-dotted lines in Fig. 27b are the images of circles in the meridian plane; the solid lines are images of circles on the supporting disks. Comparison with the video pictures reveals that the dark horizontal rings repeatedly visable within the column are images of the nonilluminated concave disks.

An analysis of the accelerometer data, which have been recorded within the Werkstofflabor, exhibits several obvious correlations. Some accelerations of about $5 \times 10^{-3}\,g$ cause vibrations of the video camera or, rather, of the video picture. No response at all can be detected of many other perturbations with the same order of magnitude. The frequency dependence obviously is important. The typical resonance frequencies of the column considered have the order of magnitude of 1 Hz, which means that they are not resolved by the peak detection method used in the Spacelab.

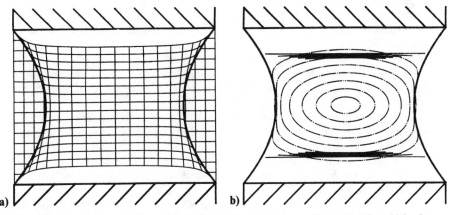

Fig. 27   Ray tracing in the unduloid obtained a) the image of a quadratic grid in the meridian plane and b) the image of circles in the meridian plane (dash-dot lines) and the image of circles on the supporting disks (full lines).

Results of Run B of WL-FPM-03

Following the successful performance of run A and the realization of many expected and unexpected effects, a second run was rendered possible by the science coordination team. The heater temperature was increased to 75°C in order to speed up Marangoni convection and to distinguish reproducible from irreproducible mechanisms. Because of insufficient coordination between JPL/Houston and GSOC/Oberpfaffenhofen, only the last seconds of the heating phase, the cooling phase, and the recovery of the column were recorded.

The first video pictures were rather surprising: Although the column was heated for 20 min and a stationary temperature profile resulted, there existed a clear interface between the two liquids. Paraffin oil formed a large asymmetric drop on one side of the column made up by benzylbenzoate. The contact line between this drop and the column is clearly visible despite gradual shrinking of the former. Since the temperature at the upper disk equalled 75°C, the column assumed the critical temperature of 47°C at about 40% of its height. Therefore, neither an interface tension nor an interface should exist in the upper half of the column.

Opposite to the drop's position, the column of benzylbenzoate shows a fog region extending between the two disks. In the course of cooling and demixing, this fog region becomes narrower and clearer and assumes a line structure caused by Marangoni convection. During recovery the column of benzylbenzoate breaks into two spherical sections at the supporting disks. Because of the volume reduction, once more a Rayleigh instability has arisen.

Conclusions

Both runs of WL-FPM-03 clearly demonstrated the importance of capillarity, stability, and spreading during mixing and unmixing of liquids exhibiting a miscibility gap. Despite being effective during short time intervals only, these mechanisms essentially affected the final distribution of the two liquids.

Marangoni convection along the free fluid surface turned out to be lower by at least one order of magnitude than had been expected from the reference experiments on the ground. This may be ascribed to the following:

1) A contamination of the liquid mixture due to its storage and transportation for several months (contamination regularly lowers surface and interface tensions); and

2) Opposite effects of temperature and concentration on the surface tension due to the component having the lower surface tension (paraffin oil) getting enriched on the cold side.

If the latter explanation is correct, the advantage of a free fluid surface, the suppression of heterogeneous nucleation, will not generally be balanced by the disadvantage of stronger Marangoni convection. In that case containerless processing appears commendable also for metallic alloys.

The liquid system benzylbenzoate-paraffin oil used in the Spacelab experiment WL-FPM-03 exhibits a much lower interface tension than the liquid system cyclohexane-methanol used in the TEXUS missions. This, together with the much lower cooling rate, strongly reduced Marangoni migration of the growing droplets. There was little contact and coagulation of these droplets, such that foggy regions lasted for the full cooling time. This suggests that slow cooling may reduce demixing also in the case of metallic alloys. Since Marangoni migration favors contact and coagulation and the speed of migration increases with droplet size, Marangoni migration exponentially speeds up demixing. It therefore appears advisable to reduce it by means of soluble additives.

Future Research

The realization of finely dispersed mixtures of monotectic alloys has recently led to a joint research program between German universities and industrial companies. The aim is to further elucidate the conduct of separation, and, by systematic variations, to render possible the production of compound materials.[34] In this context the intention is

1) to quantitatively study the different mechanisms of separation in terrestrial and microgravity experiments;

2) to improve the theories of the different mechanisms and to check them experimentally;

3) to develop a general theory of separation that fully integrates the hitherto isolated results; and

4) to produce samples with high technical potential, such as aluminum-lead for bearings, silver-nickel for electrical contacts, and aluminum-lead-zinc for superconductors.

A vital step will consist of measuring and collecting from literature (in dependence of temperature and concentration)

1) the surface tensions and interface tensions between different metallic melts and their contact angles with the crucible material;

2) dynamic viscosities, diffusion coefficients, and thermal diffusivities (thermal conductivity and heat capacity); and

3) phase diagrams and thermodynamic data such as enthalpy and entropy of mixing.

These basic studies will be paralleled by

1) further investigations into transparent model systems and their compatibility with metallic systems;

2) theoretical and experimental studies into homogeneous and heterogeneous nucleation and the effect of fast cooling;

3) studies into droplet growth due to coagulation, Ostwald ripening, and the mutual overlap of depleted zones;

4) the effect of the volume content of the minority component, which appears to render differing droplet sizes in different systems;

5) reconsideration and extension of existing theories of Marangoni migration inclusive of the inherent temperature and concentration transport;

6) TEXUS and Spacelab experiments on the particle transport by a solidification front; and

7) theoretical and experimental studies on the influence of additives, which by affecting the inherent interface tensions and contact angles, change the conduct of the separation.

References

[1] "The Effect of Gravity on the Solidification of Immiscible Alloys," European Space Agency Rept. ESA SP-219, 1984.

[2] Ahlborn, H., and Löhberg, K., "Ergebnisse von Raketenversuchen zur Entmischung Flüssiger Aluminium-Indium-Legierungen," Statusseminar Spacelab-Nutzung des BMFT, Paper 12.1, 1976.

[3] Ahlborn, H., and Löhberg, K., "Influences Affecting Separation of Monotectic Alloys Under Microgravity," European Space Agency Rept. ESA SP-222, 1984, pp. 55–62.

[4] Ahlborn, H., and Löhberg, K., "Separation of Immiscible Alloys Under Reduced Gravity," Scientific Results of the German Spacelab Mission D1, edited by P. R. Sahm, R. Jansen, and M. Keller, WPL, 1987, pp. 297–304.

[5] Ang, C. J., and Lacy, L. L., "Monotectic and Syntectic Alloys," NASA TM-58173, 1976.

[6] Carlberg, T., and Fredriksson, H., "The Influence of Microgravity on the Structure of Bi-Zn Immiscible Alloys," European Space Agency Rept. ESA SP-142, 1979, pp. 233–243.

[7] Gelles, S. H., and Markworth, A. J., "Agglomeration in Immiscible Liquids," Final Post-Flight Rept. on SPAR II, Experiment 74-30, NASA TM-78125, 1977.

[8] Gelles, S. H., and Markworth, A. J., "Agglomeration in Immiscible Liquids," Final Post-Flight Rept. on SPAR V, Experiment 74-30, NASA TM-78275, 1980.

[9] Hodes, E., and Steeg, M., "Herstellung einer Aluminium-Blei-Legierung unter Mikrogravität," Zeitschrift Flugwissenschaften Weltraumforsch., Vol. 2, 1978, pp. 337–341.

[10] Otto, G. H., and Lorenz, H., "Simulation of Low Gravity Conditions by Rotation," AIAA Paper 78-273, 1978.

[11] Walter, H. U., "Preparation of Dispersion Alloys—Component Separation During Cooling and Solidification of Dispersions of Immiscible Alloys," European Space Agency Rept. ESA SP-219, 1984, pp. 47–67.

[12] Predel, B., Z. Metallkde, Vol. 56, 1965, p. 791.

[13] Clancy, P. F., Heide, W., and Langbein, D., "Sounding-Rocket Flight Test of an Acoustic Mixer by Manufacture of a Lead-Zinc Emulsion Alloy in Microgravity," European Space Agency Rept. ESA SP-191, 1983, pp. 99–104.

[14] Clancy, P. F., and Heide, W., "Acoustic Mixing of an Immiscible Alloy (Pb-Zn) in Microgravity," European Space Agency Rept. ESA SP-219, 1984, pp. 73–77.

[15] Langbein, D., "Materialforschung unter Mikrogravitation," Spektrum der Wissenschaft, April 1984, pp. 28–42.

[16] Langbein, D., "On the Separation of Alloys Exhibiting a Miscibility Gap," European Space Agency Rept. ESA SP-219, 1984, pp. 3–12.

[17] Langbein, D., and Pötschke, J., "The Engulfment of Discrete Particles," Composites, both Artificial and In-Situ in the Earth's and the Space Laboratory, edited by D. Potard and P. R. Sahm, CEN, Grenoble, France, 1985, pp. 9–32.

[18] Finn, R., "Equilibrium Capillary Surfaces," Grundlehren der Mathematischen Wissenschaften, Vol. 284, Springer-Verlg, Berlin, 1986, pp. 1–244.

[19]Langbein, D., and Heide, W., "Fluid Physics Demonstration Experiments," *Science Demonstration Experiments During Parabolic Flights of KC-135 Aircraft,* ESTEC-WP 1457, 1986, pp. 47–54.

[20]Langbein, D., and Heide, W., "Study of Convective Mechanisms Under Microgravity Conditions," *Advances in Space Research,* Vol. 6/5, 1986, pp. 5–17.

[21]Langbein, D., Grobach, R., and Heide, W., "Parabolic Flight Experiments on Fluid Surfaces and Wetting," Applied Microgravity Technology, Vol. II, 1990, pp. 198–211.

[22]Langbein, D., and Roth, U., "Interactions of Bubbles, Particles and Unidirectional Solidification Under Microgravity," European Space Agency Rept. ESA SP-256, 1987, pp. 183–189; also *Scientific Results of the German Spacelab Mission D1,* edited by P. R. Sahm, R. Jansen, and M. Keller, WPL, 1987, pp. 309–315.

[23]Potard, C., "Etudes de Base Preparatoires de l'Experience de Solidification Divergée d'Alliages Immiscibles Al-In en Fusée Sonde," European Space Agency Rept. ESA SP-142, 1979, pp. 255–262.

[24]Langbein, D., "Theoretische Untersuchungen zur Entmischung nicht Mischbarer Legierungen, Battele Frankfurt: Schlußbericht für das BMFT," BMFT, Dec. 1980; also Forschungsbericht W 81-04 des BMFT, 1981, pp. 1–66.

[25]Walter, H. U., "Stability of Multicomponent Mixtures Under Microgravity Conditions," European Space Agency Rept. ESA SP-142, 1979, pp. 245–253.

[26]Marangoni, C. G. M., "Über die Austrietung einer Flüssigkeit auf der Oberfläche einer anderen," Annalen Physik (Paggendorf), Vol. 143, 1871, pp. 337– .

[27]Young, N. O., Goldstein, J. S., and Block, M. J., "The Motion of Bubbles in a Vertical Temperature Gradient," *Journal of Fluid Mechanics,* Vol. 6, 1959, pp. 350–356.

[28]Langbein, D., and Heide, W., "Entmischung von Flüssigkeiten Aufgrund von Grenzflächenkonvektion," *Zeitschrift Flugwissenschaften Weltraumforsch.,* Vol. 8, 1984, pp. 192–199.

[29]Langbein, D., and Heide, W., "The Separation of Liquids due to Marangoni Convection," *Advances in Space Research,* Vol. 4/5, 1984, pp. 27–36.

[30]Meyyappan, M., Wilcox, W. R., and Subramanian, R. S., "The Slow Axisymmetric Motion of Two Bubbles in a Thermal Gradient," *Journal of Colloid and Interface Science,* Vol. 94, 1983, pp. 243–257.

[31]Lasek, A., and Feuillebois, F., "Migration of Particles and Drops in a Microgravity Environment," *Composites, Both Artificial and In-Situ in the Earth's and the Space Laboratory,* edited by C. Potard and P. R. Sahm, Grenoble, France, 1985, pp. 204–209.

[32]Langbein, D., "On the Temperature and Flow Fields Caused by Marangoni Migration of Fluid Particles and the Inherent Particle Interactions," *Journal of Fluid Mechanics* (submitted for publication).

[33]Langbein, D., and Heide, W., "Mixing and Demixing of Transparent Liquids Under Microgravity," European Space Agency Rept. ESA SP-256, 1987, pp. 117–123; also *Scientific Results of the German Spacelab Mission D1,* edited by P. R. Sahm, R. Jansen, and M. Keller, WPL, 1987, pp. 321–327.

[34]Ahlborn, H., "Monotektische Legierungen; Verbundprojekt zur Untersuchung des Entmischungsverhaltens von Nicht Mischbaren Metallischen Schmelzen Unter Mikrogravitation im Hinblick auf Deren Einsatz für Technische Anwendungen," *Programmvorschlag für das BMFT,* 1987.

Ostwald Ripening in Liquids

L. Ratke*
*German Aerospace Research Establishment, Cologne,
Federal Republic of Germany*

Introduction

OSTWALD ripening, or competitive coarsening, is a process observed in a wide variety of multiphase materials. It occurs generally in mixtures of small solid particles or droplets that are dispersed in a solid or liquid environment. Provided there is enough atomic mobility, such a dispersion will coarsen by transfer of matter from small to large particles, reducing the free energy associated with the particle matrix interface area. In 1900, Ostwald[1] reported the first systematic study of the increased solubility of small HgO particles as it depends on their radii. The phenomenon has come to bear his name. The basic theory of particle coarsening, however, was developed 60 years later by Lifshitz and Slyozov[2] and Wagner.[3] For brevity we shall refer to it as the LSW theory.

Examples for solid-solid, solid-liquid, and liquid-liquid dispersions, where coarsening plays a role, are numerous, since, in general, any first-order transformation process results in a two-phase mixture of a dispersed second phase in a matrix. To mention only a few, being important in metallurgy, here some examples:

Age-hardened alloys are one of the classical examples of solid-solid dispersions. They are one of the most important metallic microstructures. The dispersoids are precipitated from a supersaturated solid solution by a thermal treatment. The strength of such an alloy depends strongly on the size and arrangement of the precipitates. Any process that increases the particle size leads to a reduction in strength. Examples for these alloys are the age-hardenable aluminum alloys (e.g., Al-Cu), maraging steels, and the superalloys.

*Senior Researcher, Institute for Space Simulation.

A dispersion of liquid particles in a liquid matrix can be obtained from all liquids that are immiscible in the liquid state either by mechanical stirring or by a first-order transition from a homogeneous one-phase state into a two-phase region. The process is similar to that of precipitation from a supersaturated solid solution. The stability of such a dispersion depends on Earth on buoyant forces leading to sedimentation or various types of convection and also on the size of the precipitated droplets. Coarsening can accelerate spatial phase separation even in space under reduced-gravity conditions.

One of the most important production methods in powder metallurgy is the so-called liquid phase sintering. Materials produced by this technique generally consist of particles of one or more phases dispersed in a binder phase, which during sintering is the liquid phase. The most widely known and used products of this technique are hard metals (e.g., WC-Co). The microstructural changes during sintering, the changes in spatial arrangement, and the size of the particles control the mechanical properties.

The deoxidation of steel melts is performed by the addition of iron-aluminum and iron-silicon alloys to a melt prior to casting. Immediately after dissolution of the alloys the oxygen dissolved in the steel bath reacts with aluminum to solid corundum particles and with silicon to quartz droplets. They grow and coarsen rapidly in size and are removed from the melt pool by buoyant forces. The grain refinement of cast aluminum alloys is done by the addition of TiB particles to the melt shortly before casting. The particles act as heterogeneous nucleation sites during solidification. Coarsening and sedimentation reduce the grain refinement ability.

A quite different example for the occurrence of Ostwald ripening even in one-component systems is dendritic solidification. The solid dendrites that grow in a liquid of near-equilibrium temperature and composition coarsen; i.e., they become rounded. The rate of dendrite coarsening determines the mechanical properties of cast multiphase alloys.

The following sections will outline the theoretical basis for Ostwald ripening, describe the specialities met if one deals with liquid matrices, and compare the theories developed so far with experiments performed on Earth and in space under microgravity conditions. Reviews on Ostwald ripening have been published by Greenwood,[4] Li and Oriani,[5] Ardell,[6] and Fischmeister and Grimvall.[7] New theoretical developments of the last years are summarized by Voorhees.[8]

Basic Theory

Consider a single spherical particle of radius R in an infinitely extended matrix. The free enthalpy of it is given by

$$G = G_0 + \sigma \cdot A \tag{1}$$

where σ is the interfacial tension, A is the surface area, and G_0 represents contributions to G not attributed to the particle/matrix interface. Assume that the particle and matrix are comprised of two types of atoms, A and B.

The particle shall be nearly pure B, and the matrix a dilute solution of B in A. Then the chemical potential of B in the spherical particle μ_B^P is given by

$$\mu_B^P = \frac{\partial G}{\partial n_B} = \sigma \cdot \Omega \frac{\partial A}{\partial V} + \mu_B^0 = \mu_B^0 + \frac{2\sigma\Omega}{R} \qquad (2)$$

where n_B is number of moles of B, $\mu_B^0 = \partial G_0/\partial n_B$, Ω is the mean atomic volume of B within the particle ($\Omega = \partial V/\partial n_B$) and V is the particle volume. Since the matrix is a dilute solution of B in A, the chemical potential of B in the matrix is

$$\mu_B^M = \mu_B^1 + k_B T \, \ell n c \qquad (3)$$

where k_B is the Boltzmann constant, T is the absolute temperature, c is the concentration of B in A, and μ_B^1 refers to a standard state.

From equilibrium of the two phases we have $\mu_B^M = \mu_B^P$; thus,

$$k_B T \, \ell n c = \frac{2\sigma\Omega}{R} + \mu_B^0 - \mu_B^1 \qquad (4)$$

If the interface between the two phases is flat, $R \to \infty$, we obtain

$$k_B T \, \ell n c_\infty = \mu_B^0 - \mu_B^1 \qquad (5)$$

From Eq. (5) we obtain the final relationship for solute enhancement in the matrix,

$$c(R) = c_\infty \exp\left(\frac{2\sigma\Omega}{k_B T R}\right) \cong c_\infty\left(1 + \frac{2\sigma\Omega}{k_B T R}\right) \qquad (6)$$

which is in equilibrium with the particle of radius R. This is the classical Gibbs-Thomson or Thomson-Freundlich[9] relationship, which was used by Ostwald in 1900 to explain his results and is one basic equation of competitive growth. Extensions of it to treat a nonideal solution have been given by Purdy[10] and Chaix et al.[11]

The linearization in Eq. (6) is always valid in systems with small interfacial tensions and for particles that are not too small (approximately greater than 0.1 μm). For very small particles ($\approx 1-10$ nm) the exponential form has to be used as shown by Kampmann and Wagner.[12]

Now consider a dispersion of particles with a certain size distribution in a matrix. From Eq. (6) we have that a particle is in equilibrium within the matrix if the concentration of dissolved atoms around it has a value given by Eq. (6). Since we have a large number of particles with various radii, the concentration of solute in the matrix adjusts itself to some mean value $\langle c \rangle$. This is schematically shown in Fig. 1. All particles whose radii are just the critical one, $R = R^*$, such that $\langle c \rangle$ is equal to $c(R^*)$, are in equilibrium with the matrix. Smaller particles have a higher concentration of solute in

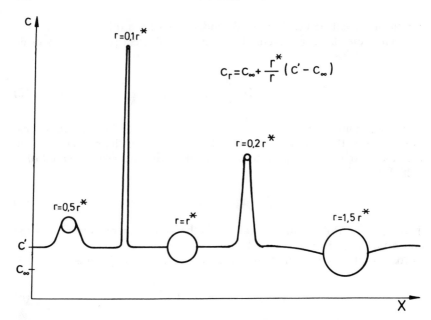

Fig. 1 **Concentration variation of solute in a matrix due to the Gibbs-Thomson effect in the case of a dispersion with particles of different size. The c' is the mean concentration in the matrix $\langle c \rangle$. Particles having a radius r^* are in equilibrium with the matrix of concentration $c' = \langle c \rangle$.**

their neighborhood and larger a lower compared to the mean value. Therefore, concentration gradients exist in the matrix. Diffusion currents arise that lead to a dissolution of the smaller particles at the expense of the larger ones. Inasmuch as some particle dissolve the situation continuously changes, i.e., the mean value decreases, the critical radius R^* changes, defining new conditions for particles being "smaller" or "larger."

In order for coarsening to occur solute atoms have to be exchanged from smaller to larger particles by some kind of transport within the matrix. It may be the diffusion of atoms between the particles that is the slowest and thus the rate-determining step. We refer to this as diffusion-controlled growth. When the deposition of atoms at the particle surface is the slowest step, we refer to it as reaction-controlled growth. The first case was first analysed by Lifshitz and Slyozov,[2] and the second was analyzed by Wagner.[3] If the particles or droplets are dispersed in a liquid matrix, it is possible that they move relative to it, e.g., by the Stokes motion due to different density. Depending on the particle/droplet velocity (strictly speaking, the Peclet number, as discussed later), there may be additional transport by convection. The theory for Ostwald ripening including particle motions of various origins has only recently been developed by Ratke and Thieringer,[13] Ratke,[14] and Ratke and Host.[15]

Diffusion-Controlled Growth

In contrast to the early stages of a first-order transformation, the relative supersaturation of solute $s = (c - c_\infty)/c_\infty$ is very low during Ostwald ripening. Therefore, a quasistationary approximation for the concentration profile around a particle may be employed. The diffusion field in the matrix is then governed by the Laplace equation

$$\nabla^2 c = 0 \tag{7}$$

along with the boundary conditions

$$c(R) = c_\infty\left(1 + \frac{2\sigma\Omega}{k_B T R}\right) = c_\infty\left(1 + \alpha\frac{1}{R}\right) \tag{8}$$

$$\lim_{r\to\infty} c(r) = \langle c \rangle \tag{9}$$

where $\langle c \rangle$ is the mean concentration of solute in the matrix. The growth rate of a single particle is given by

$$\dot{R} = \frac{dR}{dt} = \frac{\Omega \cdot I}{4\pi R^2} \tag{10}$$

where the transport current I through the interface is given by

$$I = \int j(r = R)\, df = j4\pi R^2 \tag{11}$$

and the diffusion current density j is determined from

$$j = -D\frac{\partial c}{\partial r}\bigg|_{r=R} \tag{12}$$

From Eqs. (7–12) one obtains for the growth rate

$$\dot{R} = \frac{D\Omega c_\infty}{R}\left(\frac{\langle c \rangle - c_\infty}{c_\infty} - \frac{\alpha}{R}\right) \tag{13}$$

From Eq. (13) its mean field nature is readily evident. A particle grows or shrinks only in relation to a mean field concentration set at infinity. From the definition of a particle of critical size R^*, which is in equlibrium with the mean solute concentration $\langle c \rangle$, we obtain from Eq. (6)

$$\frac{\alpha}{R^*} = \frac{\langle c \rangle - c_\infty}{c_\infty} \tag{14}$$

Thus, the growth law is

$$\dot{R} = \frac{D\Omega c_\infty \alpha}{R}\left(\frac{1}{R^*} - \frac{1}{R}\right) \tag{15}$$

Both growth rates, Eqs. (13) and (15), contain a yet unknown function of time, the supersaturation $s = (\langle c \rangle - c_\infty)/c_\infty$ or the critical radius. Since s and R^* depend on the size spectrum of the particles and its evolution with time, it is necessary to consider the size distribution in order to obtain knowledge on the kinetics of Ostwald ripening.

Let us introduce a size distribution function $f(R,t)$ in order to character- ize the morphology of the dispersed second phase. The f is defined as the number of particles per unit volume at time t in a size class R, $R + dR$. From the definition of f it is clear that $N(t) = m_0$, where $N(t)$ is the number of particles per unit volume and

$$m_n = \int_0^\infty R^n f(R,t)\, dR \tag{16}$$

is the nth moment. The flux of particles passing through a size class R to $R + dR$ is $f \cdot dR/dt$; therefore, f obeys the continuity equation

$$\frac{\partial f}{\partial t} + \frac{\partial}{\partial R}(\dot{R} \cdot f) = 0 \tag{17}$$

The final element of the LSW theory is mass of solute conservation, which is explicitly added to the theory because the concentration profile around a particle, and thus the growth rates in Eqs. (13) and (15), are solutions of Laplace's equation [Eq. (7)]. With no external solute sources, solute conservation demands that the total amount of dissolved atoms in the matrix and the particles is constant:

$$\langle c(t) \rangle = c_0 - \frac{4\pi}{3}\frac{m_3(t)}{\Omega} \tag{18}$$

where c_0 is the initial amount of solute. Equation (13) or (15), together with Eqs. (17) and (18), can be rewritten as a nonlinear partial integro- differential equation of the first kind.

A complete time-dependent solution has not yet been obtained analyti- cally, but has been obtained only with numerical methods (Voorhees,[30] Venzl,[16,17] Ratke,[14] and Ratke and Host[15]).

In the original LSW theory the system of Eqs. (13), (15), (17), and (18) is not solved for all times, but Lifshitz, Slyozov, and Wagner found so-called asymptotic solutions valid as t tends to infinity. Because their procedure has been employed since then in many publications on that topic, it will be outlined here in an abbreviated fashion.

Asymptotic LSW Analysis of Ostwald Ripening

Before we explain the asymptotic technique of the LSW theory, let us introduce dimensionless variables for all further treatment. We introduce dimensionless radii,

$$r = \frac{R}{\langle R_0 \rangle} \quad \text{and} \quad r^* = \frac{R^*}{\langle R_0 \rangle} \tag{19}$$

and a dimensionless time,

$$\tau = \frac{2D\sigma\Omega^2 c_\infty}{k_B T \langle R_0 \rangle^3} t \tag{20}$$

Then Eq. (15) is rewritten as

$$\frac{dr}{d\tau} = \frac{1}{r^2}\left(\frac{r}{r^*} - 1\right) \tag{21}$$

and Eq. (17) reads

$$\frac{\partial f}{\partial \tau} + \frac{\partial}{\partial r}(\dot{r}f) = 0 \tag{22}$$

The solution conservation is rewritten as

$$\langle s(t) \rangle = s_0 - p \cdot m_3 \tag{23}$$

where $p = 4\pi/3\Omega c_\infty \langle R_0 \rangle^3$, and m_3 is given by Eq. (16) replacing R by r. In the LSW theory an additional transformation of variables is performed. Introducing the variables

$$\rho = \frac{r}{r^*} \quad \text{and} \quad \Theta = \ell n\left(\frac{r^*}{r_0^*}\right) \tag{24}$$

a new growth law replacing Eq. (21) is obtained:

$$\frac{d\rho}{d\Theta} = \frac{1}{r^{*2}}\frac{d\tau}{dr^*}\frac{\rho-1}{\rho^2} - \rho = v \cdot \frac{\rho-1}{\rho^2} - \rho \tag{25}$$

and the continuity equation [Eq. (22)] preserves its form (one has simply to replace $\tau \rightarrow \Theta$ and $r \rightarrow \rho$). LSW were able to show that

$$v = \frac{1}{r^{*2}}\frac{d\tau}{dr^*} \tag{26}$$

is a constant if the dimensionless time Θ tends to infinity or $\Theta \gg 1$. They also showed that thus $r^* = <r>$ and the volume fraction of dispersed particles $\Phi = 4\pi/3\Omega/m_3$ is a constant. LSW also proved that under these conditions the particle size distribution function $f(\rho,\Theta)$ can be separated into a product $f(\rho, \Theta) = h(\rho) \cdot g(\Theta)$ and that $h(\rho)$ has a so-called cutoff radius ρ_m, i.e., $h(\rho \geqslant \rho_m) = 0$. The distribution functions $h(\rho)$ and $g(\Theta)$ are obtained from Eq. (22) in ρ, Θ variables and after rearranging yield (see White[18])

$$\frac{1}{g(\Theta)}\frac{dg(\Theta)}{d\Theta} = -\frac{1}{h(\rho)}\frac{dh}{d\rho}\frac{d\rho}{d\Theta} - \frac{d}{d\rho}\left(\frac{d\rho}{d\Theta}\right) = q \qquad (27)$$

Integration of both sides of Eq. (27) together with the condition of constant volume fraction determines q and both parts of the distribution function:

$$g(\Theta) = g_0 \exp(q \cdot \Theta) \qquad (28)$$

and

$$\ell n h(\rho) \sim -\int_0^\rho \frac{q\,d\rho'}{\dfrac{d\rho'}{d\Theta}} - \ell n\left(\frac{d\rho}{d\Theta}\right) \qquad (29)$$

Liftshitz and Slyozov showed that the constants v and ρ_m can be obtained from the conditions

$$\left.\frac{d\rho}{d\Theta}\right|_{\rho_m} = 0 \quad\text{and}\quad \left.\frac{d}{d\rho}\left(\frac{d\rho}{d\Theta}\right)\right|_{\rho_m} = 0 \qquad (30)$$

In the case of diffusional growth the result is $v = 2(3/2)^3$ and $\rho_m = 3/2$. One can then integrate Eq. (26), giving

$$r^{*3} - r_0^{*3} = \frac{4}{9}\cdot\tau \qquad (31)$$

or, since, $r^* = \langle r \rangle$ for diffusional growth,

$$\langle r^3 \rangle - 1 = \frac{4}{9}\tau = K_{LSW}\cdot\tau \qquad (32)$$

The distribution function $h(\rho)$ is obtained as

$$h(\rho) = \frac{81}{2^{5/3}}\frac{\rho^2}{(1.5 - \rho)^{11/3}}\frac{1}{(3 + \rho)^{7/3}}\exp\left(\frac{\rho}{1.5 - \rho}\right) \qquad (33)$$

This distribution function is shown in Fig. 2. It looks strongly skew symmetric to the right, decreases steeply after the maximum, and has a concave curvature at the origin (there the relation holds $\lim_{\rho\to0} d\rho/ d\Theta \cdot f(\rho,\Theta) = 0$).

There are some astonishing features of this LSW theory. First, it is predicted that, independently of the initial particle size distribution, an asymptotic state is reached such that a scaled particle size distribution $h(\rho)$

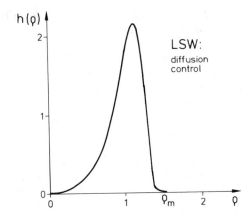

Fig. 2 Normalized distribution function of particle radius as predicted by the LSW theory for diffusion-controlled growth: $\rho = R/\langle R \rangle = r/\langle r \rangle$.

is independent of time. Thus, a universal self-similar nature of the coarsening process at long times is predicted. This self-similarity of the Ostwald ripening process makes it so promising to apply it to a wide variety of two-phase mixtures. (The self-similarity has nicely been shown theoretically in a recent paper by Mullins.[19])

Second, it has sometimes been asked how such a low cutoff radius of $r = 1.5 \langle r \rangle$ can be reconciled with the fact that according to Eq. (21) the maximum growth rate $dr/d\tau$ appears for $r = 2\langle r \rangle$. Note, however, that the distribution function is given in terms of the reduced variable $\rho = r/\langle r \rangle$. Because of the disappearance of smaller particles, $\langle r \rangle$ may increase relatively faster than r at the maximum growth rate. In fact, the cutoff corresponds to the size for which $d(r/\langle r \rangle)/d\Theta = 0$.

The now classical theoretical treatment of LSW does not explicitly use the solute conservation as stated in Eqs. (18) and (23) but uses the fact that in the asymptotic limit the volume fraction Φ of dispersed particles is a constant (this is especially used in the work of Wagner[2]). All applications of the LSW theory to other types of solute transport employed since then use the constancy of volume fraction as the additional constraint for solving the system of Eqs. (15) and (17) in a unique way. It was stated by Marqussee and Ross[20] that this is not entirely correct. Therefore, they employed a scaling ansatz in which, as τ tends to infinity, $\rho = r\tau^{-x}$ and $f = f_0\tau^{-y}$, where x and y are fixed by the conservation of solute. Their result is then in full accordance with the LSW theory.

The assumptions made in the derivation of the preceding results of the LSW theory for diffusional transport of solute are worth mentioning explicitly:

1) dilute solid solution of matrix and particles;

2) infinitely extended system;

3) near-zero volume fraction of dispersed particles, with no overlap of diffusion fields;

4) transport of matter occurring solely by volume diffusion (i.e., the particles are at rest and the interface reaction is a fast, high mobility interface); and

5) isotropic interfacial energy, spherical particles.

Assumption 1 is not a serious one. As mentioned earlier, deviations from Raoults's or Henry's law in the binary and ternary case have been treated by Purdy,[10] Oriani,[21] Feingold et al.,[22] and Chaix et al.[11] The factor c_∞ in Eq. (6) is always to be replaced by a more complicated one, including the thermodynamics of the solutions.

Assumption 2 is also not serious from a physical point of view. It only serves to handle the mathematics. Regarding assumption 5, it is difficult to deal with anisotropic interfacial energies and nonspherical particles. Both situations are often encountered in liquid phase sintering of cemented carbides. However, all treatments of Ostwald ripening in these systems have disregarded the asphericity of the particles and the highly anisotropic interfacial energy. It was only recently that Zwillinger[23] showed that even an ensemble of nonspherical particles follows the coarsening laws developed in the LSW theory and its extensions (see the following subsection).

Influence of Finite Volume Fraction

Assumption 3 is very often validated in experimental situations, especially in liquid phase sintering. There liquid fractions down to 5% per volume are used. In order to remove the zero volume fraction assumption it is essential to take into account interactions between the diffusion fields around the particles. There have been several attacks on this problem,[24-27] but only recently realistic models of the coarsening at finite volume fraction have been proposed.[28-33] In all of these treatments the particles are assumed to be spherical and fixed in space. Diffusion of species within the matrix determines the growth and shrinkage of the dispersoids. They all solve Laplace's equation for solute, placing point sources or sinks of solute at the center of each particle:

$$\nabla^2 s = 4\pi \sum_{i=1}^{N} B_i \delta(\mathbf{r} - \mathbf{r}_i) \qquad (34)$$

The source/sink strengths B_i are unknown and obtained together with the mean relative matrix concentrations s_m by requiring interfacial equilibrium and solute conservation. A general solution of Eq. (34) is

$$s = s_m - \sum_{i=1}^{N} B_i \frac{1}{|\mathbf{r} - \mathbf{r}_i|} \qquad (35)$$

From the boundary condition (Gibbs-Thomson effect) and solute conservation, a set of nonlinear equations for B_i and s is obtained. Each of the new theoretical developments employs a different procedure for determining the statistically averaged growth rate dR/dt or the averaged source/sink strength $B(R) = R^2\, dR/dt$ of a given particle at a fixed volume fraction Φ.

The differences between the various approaches are extensively discussed by Voorhees[8] and thus are not repeated here. Their common essential results consist of the following:

1) The temporal power law of the LSW theory is not a function of volume fraction. However, the rate constant is

$$\langle r \rangle^3 - 1 = K(\Phi)t \tag{36}$$

2) In the asymptotic limit scaled time invarant distribution functions exist.

3) As Φ increases, the distribution function becomes broader; the cutoff radius ρ_m is greater than 1.5.

4) The rate constant rises rapidly with Φ (see Fig. 3).

Two results from the work of Voorhees and Glicksman[29-31] are shown in Figs. 4 and 5. First it can be seen that the distribution function changes remarkably with increasing Φ; it becomes broader, and the maximum decreases inasmuch as the cutoff increases. Figure 5 shows that up to a volume fraction of 0.3 the new theories predict rate constants $K(\Phi)$ that are in close agreement.

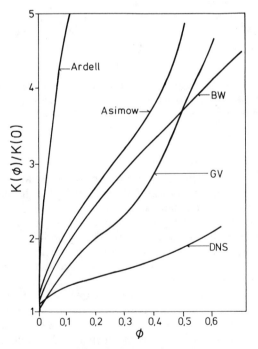

Fig. 3 **Variation of the rate constant in the growth law for the mean particle radius as a function of volume concentration predicted by various theories.**

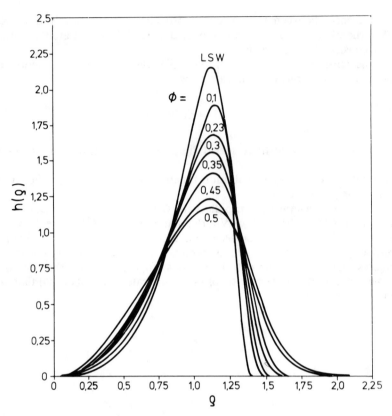

Fig. 4 Dependence of the shape of the time-invariant particle size distribution function $h(\rho)$ with the volume concentration of the second-phase particle for diffusive transport of solute as predicted by Voorhees and Glicksman.[29-31]

Convective Contributions to Ripening

Assumption 4, which assumes that particles are fixed at their position in space, is in most instances not valid if the matrix is a liquid. Convective fluid flows of various origins can exist that will give rise to particle motion and additional convective transport of solute to the particles. Therefore, a change of the growth kinetics is to be expected. In addition, particle motion and fluid flow may lead to an agglomeration of particles.

The concurrent action of diffusive and convective transport of matter on Ostwald ripening in a liquid matrix was only recently treated by Ratke and Thieringer.[13] They analyzed various cases of particles or droplets moving relative to the matrix (constant velocity, Stokes motion, or Marangoni motion). Their theoretical treatment was confirmed by experiments.[34] The RT analysis of the problem is only valid if the particles move with high velocities (U) or have large radii (R); i.e., the Peclet number, $Pe = UR/D$, is essentially larger than 1. Under this condition they

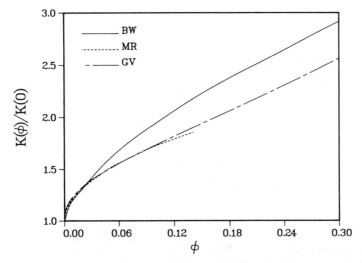

Fig. 5 Rate constant $K(\Phi)$ relative to $K(0)$ as a function of volume concentration as predicted by Marqusee and Ross,[19,28] Brailsford and Wynblatt,[27] and Voorhees and Glicksman.[29-31]

solved the problem of Ostwald ripening analogous to LSW and could show that stationary particle size distributions establish as in the case of pure diffusion. This treatment was extended by Ratke and Host to arbitrary Pe numbers.[15]

Since the main effects of additional convective transport to Ostwald ripening are most easily understandable from the RT theory, an abbreviated version of the theory will be given in this chapter.

In their treatment of convective contributions to Ostwald ripening, Ratke and Thieringer consider laminar flows in the matrix such that no hydrodynamic boundary layer exists around particles. At $Re \ll 1$ droplets retain their spherical shape. However, a diffusion boundary layer exists that is determined by the Peclet number $Pe = RU/D$ (Fig. 6). If $Pe \ll 1$, diffusion alone is important; if $Pe \gg 1$, there is outside the layer transport of matter by convection and inside by diffusion. From this consideration the general effects of convection are easily deduced: The concentration gradient at the particle/matrix interface increases with increasing Pe, thus leading to faster transport and broader size distributions. In order to calculate the flux I of solute from or to a particle, RT used Levich's[35] calculations of I for solid or liquid particles that are reasonably valid at high Peclet numbers.

For droplets moving with velocity U relative to a matrix, Levich obtained for the current I (atom/s),

$$I_L = 8\left[\frac{\pi}{3}\left(\frac{D}{2R}\frac{\eta}{\eta+\eta'}\right)\right]^{\frac{1}{2}} R^2(\langle c \rangle - c(R))U^{\frac{1}{2}} \tag{37}$$

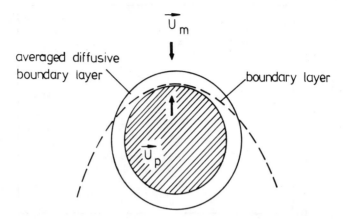

Fig. 6 Diffusion boundary layer at a particle moving with a velocity U_m relative to a liquid (dashed line) and mean diffusion boundary layer (solid line).

For solid particles he obtained

$$I_S = 7.98(\langle c \rangle - c(R))(D^2 U R^4)^{\frac{1}{3}} \tag{38}$$

With these expressions for I_S and I_L they solved the systems of Eqs. (8), (10), (14), and (17) within the asymptotic limit under the assumption of constant volume fraction of dispersed phase. They treated droplets moving with constant relative velocity, Marangoni motion as described by the Young et al. relationship,[36] and Stokes motion according to the Hadermard-Ribtsczinski equation.[37] Solid particles were considered moving with constant relative velocity or due to Stokes motion. In all cases the asymptotic analysis gave temporal power laws for the change of mean particle radius (Fig. 7):

$$\langle R \rangle^n - \langle R_0 \rangle^n = K_i \cdot t \tag{39}$$

where the K_i depend on various physical parameters (for further details cf. Ratke and Thieringer[13]). The size distributions obtained are shown in Figs. 8a and 8b, they differ remarkably from those for pure diffusional transport of matter. Table 1 summarizes some results of Ratke and Thieringer's paper.

From Table 1 it is obvious that, the more pronounced the particle velocity depends on particle radius, the faster the particles grow (the exponent n decreases from 3 to 3/2), the broader the particle size distribution, and the greater the cut off radius. The faster coarsening of droplets compared to solid particles is due to the continuity of the tangential velocity component of the matrix fluid flow at the particle surface. For solid particles the tangential velocity at the surface is zero; for liquids it is finite. Thus, transport of matter within the diffusion boundary layer of droplets does not only occur by diffusion but also by convection.

Table 1   Summary of results from Ratke and Thieringer's paper[13]

Coarsening exponent n	Width of the size distribution	Cut-off radius	Kind of flow	Relative velocity	Particle nature
3	narrow	3/2	No relative motion	Zero	Solid liquid
8/3	↑	8/5	Constant	$U = U$	Solid
5/2		5/3	Constant	$U = U$	Liquid
2		2	Stokes	$U \sim r$	Solid
2	↓	2	Marangoni	$U \sim r$	Liquid
3/2	broad	3	Stokes	$U \sim r$	Liquid

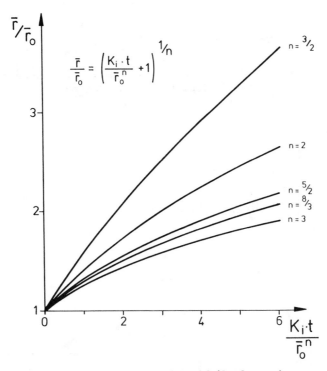

Fig. 7   Growth laws for various types of particle/droplet motion as predicted by Ratke and Thieringer.[13] In the designation used here $\bar{r} = \langle r \rangle$ and $\bar{r}_0 = 1$. Diffusion-controlled growth is given by $n = 3$, $n = 8/3$ is the growth exponent for solid particles moving with constant velocity, $n = 5/2$ is the growth exponent for droplets moving with constant velocity, $n = 2$ is equal for particles moving with Stokes velocity or droplets moving with Marangoni velocity, and $n = 3/2$ is for droplets moving with Stokes velocity.

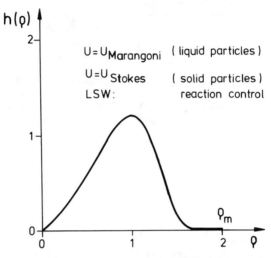

Fig. 8a   Size distribution function of particles or droplets moving relative to a matrix as indicated in the legend of the figure in Ratke and Thieringer.[13]

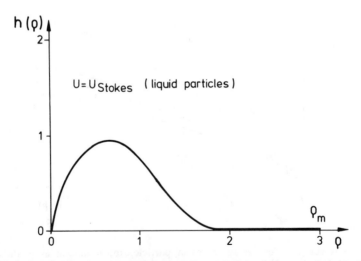

Fig. 8b   Size distribution for droplets moving relative to the matrix with Stokes velocity as described by the Hadarmard-Ribtsczinski relation.

From Fig. 7 it is obvious that for diffusional transport the slowest increase in $\langle r \rangle$ with time is obtained. The fastest increase should be observed for droplets moving with Stokes velocity ($\langle r \rangle \sim t^{2/3}$). There is another important difference to the purely diffusional LSW theory. The rate constant K_{LSW} depends on the product of σ and D. In all cases analyzed by Ratke and Thieringer the rate constants K_i depend on the product $\sigma \cdot D^q$, where q is 2/3 for solid and 1/2 for liquid particles.

The approach of Ratke and Thieringer has two shortcomings. First, the assumption of large Peclet numbers, $Pe \gg 1$, always breaks down at very small particles. For them the Peclet number is always around zero. Second, large Pe numbers are only attainable with high-speed particles or very large ones. If they are of large size, the effect of particle radius on solubility decreases and Ostwald ripening becomes extremely slow.

These shortcomings have been solved recently by Ratke and Host.[15] They used the complete solution for the transport current to a solid particle as given by Brian and Hales[70]:

$$ I = I_{\text{diff}} \left(1 + \frac{1.21}{4} Pe^{\frac{2}{3}} \right)^{\frac{1}{2}} \tag{40} $$

and employed a Pythagorean interpolation for the transport current to a droplet [squared addition to I_{diff} and I_L of Eq. (37) to cover the whole

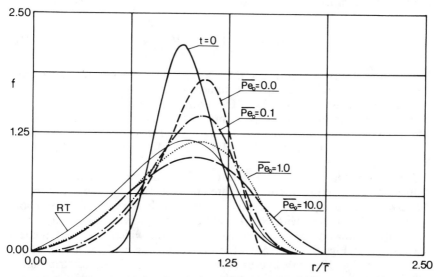

Fig. 9 Normalized particle size distribution after coarsening due to Ostwald ripening. The overall ripening time is $\tau = 100$. Distributions for various mean initial Peclet numbers $\overline{Pe_0} = \langle Pe_0 \rangle$ as calculated numerically are shown. The dispersed particles are either solid ones moving with Stokes velocity or liquid ones moving by Marangoni motion. Besides a constant factor in the growth rates, both yield the same particle size distribution function. Curve RT is the asymptotic result of Ratke and Thieringer for $Pe \gg 1$.

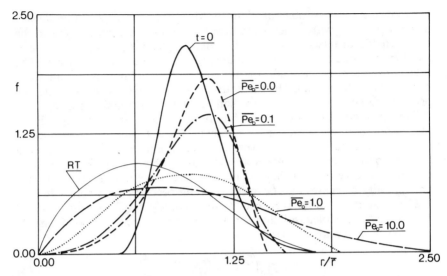

Fig. 10 Same as Fig. 9, but now for droplets moving with Stokes velocity relative to the matrix.

range of Pe numbers]:

$$I_L = I_{\text{diff}} \left[1 + \frac{2}{3\pi} \left(\frac{\eta}{\eta + \eta'} \right) Pe \right]^{\frac{1}{2}} \tag{41}$$

Using these transport currents they solved the problem of convective diffusion in Ostwald ripening first numerically, to obtain time-dependent results, and second analytically in the asymptotic limit.

From their numerical work two important results are shown in Figs. 9 and 10. They show how an initially log-normal distribution has changed after a reduced ripening time of $\tau = 100$ [Eq. (20)] for various initial mean Peclet numbers $\langle Pe_0 \rangle$.

The curves with $\langle Pe_0 \rangle = 0.0$ always reflect the evolution due to pure diffusional coarsening. Figure 9 represents the two cases of convective diffusive Ostwald ripening: droplets moving with Marangoni motion or solid particles moving with Stokes motion. The growth laws turned out to be identical with the exception of a constant, which does not render the results of Fig. 9. The case of droplets moving with Stokes motion is shown in Fig. 10. Both figures show that the particle size distributions depend at a given ripening time only on the mean initial Peclet number. For every $\langle Pe_0 \rangle$ there is a different distribution. With increasing $\langle Pe_0 \rangle$ the standard deviation becomes broader, and it seems that the greater the initial mean Peclet number the more the distribution function comes closer to the asymptotic solution of Ratke and Thieringer,[13] which is valid for large Peclet numbers and extremely long ripening times. Especially, Fig. 10 shows that with increasing $\langle Pe_0 \rangle$ the initially concave profile shape of the distribution function at the origin changes to a convex one as predicted by the RT theory. The numerical calculations revealed that the discrepancies to the asymptotic theories (LSW and RT) remain even after longer ripening times ($\tau = 1000$). The asymptotic solutions are really never achieved. This is not unexpected. Using the definition of the normalized time Θ of the asymptotic theory [Eq. (24)] it was estimated that Θ times are not greater than 10. Although this is a time scale never realized in metallurgical experiments,[7] it is not large enough to be in the asymptotic limit $\Theta \gg 1$. Therefore, it may be concluded that in experiments it is hardly to achieve the distributions calculated by asymptotic theory. Deviations should be usual and are usually observed (see below).

Ratke and Host[15] were also able to show that the evolution of the particle size distribution and the increase in mean particle radius are fully determined by the initial mean Peclet number, which in turn is determined by the initial particle size distribution function and the type of particle motion. Once this value is fixed, the evolution in time can be predicted exactly. The origin of this feature, which is astonishing on first sight, is the partitioning between convective and diffusive transport of matter. They both add in a Pythagorean way with a weight factor fully determined by the mean initial Peclet number.

Reaction-Controlled Growth

There are several mechanisms by which a surface reaction may become the decisive step in the growth rate: some hinderance for the transfer of atoms to the larger particles and site finding of arriving atoms, especially at surfaces of solid particles. Wagner[3] assumed in his treatment of reaction-controlled growth that the rate of transfer is proportional to the difference between the equilibrium concentration $c(R)$ at a particle of radius R and the actual surface concentration $c^*(R)$. If the diffusivity is very high, $c^*(R)$ will be equal to $\langle c \rangle$. Thus, one obtains for the growth rate (first-order kinetic equation)

$$\frac{dR}{dt} = -K_T \Omega [c(R) - \langle c \rangle] \tag{42}$$

where K_T is the temperature-dependent transfer constant. Using the outlined scheme to treat Ostwald ripening in the asymptotic limit, we have

$$\frac{dr}{dt} = \frac{1}{r^*} - \frac{1}{r} \tag{43}$$

with

$$\tau = \frac{2K_T \Omega^2 C_\infty \sigma}{k_B T \langle R_0 \rangle^2} \quad t \text{ and } r = R/\langle R_0 \rangle$$

The growth law of the critical particle then reads, after Wagner,[3]

$$r^{*2} - 1 = \tfrac{1}{2}\tau \tag{44}$$

There is a simple relation between r^* and the mean radius $\langle r \rangle$, $r^* = 8/9\langle r \rangle$. The stationary distribution function is particularly simple (Fig. 8b):

$$h(\rho) = \frac{8}{3} \frac{\rho}{(2-\rho)^5} \exp\left(\frac{-3\rho}{2-\rho}\right) \qquad (45)$$

if $\rho < 2$, and $h(\rho) = 0$ for $\rho \geqslant 2$. The cut off radius $\rho_m = 2$ is larger than in the case of diffusion-controlled growth.

It is worth mentioning that in the purely reaction-controlled growth the rate-determining mechanism is a strict surface phenomenon. Therefore, coarsening should be independent of volume fraction as long as the particles do not touch each other.

Experimental Techniques

The coarsening of two-phase mixtures has been studied fairly extensively since LSW obtained a complete theoretical solution for a system of isolated dispersed spherical particles (compare the reviews mentioned in the Introduction). Most of the work was done with solid-solid systems (e.g., Ni_3Al precipitates in a Ni matrix). A few results have been obtained with solid-liquid mixtures and only two with metallic liquid-liquid systems. Solid-solid systems are not proper candidates for validating the various theoretical approaches to the Ostwald ripening problem (see above). In most instances Ostwald ripening is complicated by the presence of stress fields generated by the second-phase particles and interfacial stresses (which are different from interfacial energies in liquid-liquid systems). Neither of these effects is accounted for in the theories. In addition, the diffusion coefficient in solid-solid systems is extremely low compared to systems with a liquid matrix (up to nine orders of magnitude). Therefore, coarsening of solid precipitates in a solid matrix needs very prolonged anneals for a considerable increase in particle diameter to be observed.[38] Systems with a liquid matrix are therefore most suited. This has often been done within the context of liquid phase sintering. Unfortunately, the use of liquid matrices has other severe limitations. First, often there is a large density difference between particles and matrix giving rise to sedimentation (buoyancy). This promotes coalescence of particles, i.e., direct particle contacts with the establishment of a grain boundary and a sinter neck.[39] It also leads to an inhomogeneous spatial distribution of the particles and thus a locally varying volume fraction. Second, temperature inhomogeneities will lead to natural convections, promoting coarsening by collisions. Third, temperature gradients will lead in liquid-liquid systems to Marangoni motion of drops.

There have been several techniques developed to overcome these difficulties. Hardy and Voorhees[40] used such a high volume of Sn particles in an eutectic Pb-Sn melt (and vice versa) that a skeleton formed stabilizing the position of the particles. Since the entectic liquid in the Pb-Sn system competely wets Sn-Sn and Pb-Pb grain boundaries, no coalescence occurred. In liquid phase sintering of, for instance, W-Ni alloys such a skeleton formation was also used (e.g., Kang and Yoon[41] and Exner[42]) to stabilize the dispersion, but coalescence could most often not be avoided. Kang and Yoon[63,64] minimized the effect of sedimentation using Cu-Co alloys, in which the phases have nearly the same density. Experiments with liquid-liquid dispersions of Pb drops in a Zn melt have been performed by Kneissl and Fischmeister under the reduced-gravity conditions of Spacelab in the first Spacelab mission in 1983.[43,44] Microgravity conditions in general are able to overcome all of the difficulties with liquids mentioned earlier, because the major driving force of sedimentation and convection, gravitational acceleration, is drastically reduced.

Another method was used successfully by Kneissl and Fischmeister[45] and Thieringer and Ratke[34] with Zn-Pb and Al-Pb suspensions. A temperature gradient that is vertical to the gravitational acceleration induces a transverse convection roll as shown in Fig. 11 called natural convection. The driving force is a horizontal density gradient of the liquid matrix (the hotter side has a lower density than the colder one). This natural convection will carry all particles whose settling velocity is smaller than the greatest velocity of the flowing matrix. Thus, the technique prevents a certain number of droplets from settling to the bottom of the crucible. The

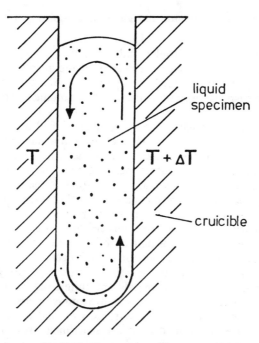

Fig. 11 Natural convection roll technique to stabilize a dispersion. A temperature gradient that is perpendicular to the gravity acceleration leads to a convection roll in the matrix. The velocity of the matrix fluid flow is controlled by the size of the crucible, the temperature gradient.

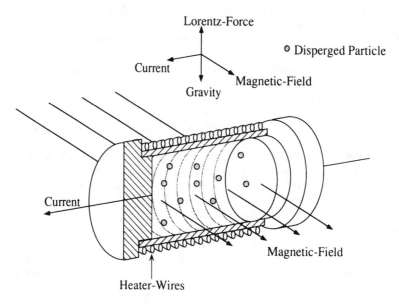

Fig. 12   Schematic arrangement to stabilize a dispersion by Lorentz forces.

relative velocity between a droplet and the matrix always is equal to the Stokes velocity of each particle, irrespective of the absolute velocity of the streaming matrix. By this forced convection technique both groups were able to stabilize suspensions with extremely low volume fractions of smaller than 0.1%.

Yet another technique has recently been developed and has proved its feasibility with Al-Pb suspensions. The new experimental technique is based on the well-known effect that in electrically conducting liquid the combined action of a magnetic field that is perpendicular to an electric current induces a Lorentz force that is a real volume force in the liquid-like gravity. Such an arrangement is shown schematically in Fig. 12. Proper adjustment of the electric current and the strength of the magnetic field induces a force field in the matrix such that the density difference between dispersed particles or droplets is compensated exactly. Therefore, sedimentation of droplets is suppressed and a dispersion is stabilized.[46]

Despite the problem of handling liquid matrices properly, experiments on Ostwald ripening with liquid metals are quite easily performed (simple heating of a sample to a definite temperature and isothermal annealing). In all cases the size distribution, mean particle size, and number of particles per unit volume are determined after an experiment from metallographic sections. Particle sizes are most suitably evaluated by the linear intercept method[47,48] or from intersect areas. Corrections must be made for the greater probability of intersecting larger particles and that of interesting spheres off center, which will broaden the distribution. Well-established procedures are available.[47,48] Conversion of intercept or intersection area distributions to true diameters is possible only by approximation, whereas the opposite transformation can be done stringently. Distribution shapes should therefore be checked by converting predicted distributions to intercepts or intersections and matching them against measured data. The procedure is outlined in full detail by Exner and Fischmeister[49] and Exner and Lukas.[50] As pointed out by Fischmeister and Grimvall,[7] the mean particle radius can be obtained more easily. It is sufficient to measure the mean free path between particles along a straight line, which is proportional to $\langle r \rangle$ for any distribution shape, provided it has become stationary. The total particle surface S and particle volume V in the system are proportional to the second and third moment of the distribution, $S = \beta_2 m_2$ and $V = \beta_3 m_3$. For any stationary distribution, $m_i(r) = \langle r \rangle(i + 1) \cdot m_i(\rho)$, where $\rho = r/\langle r \rangle$. Using the basic stereological relation[47] $\lambda = 4(1 - \Phi)V/S$ one obtains

$$\lambda = 4 \frac{(1 - \Phi)}{\Phi} \frac{\beta_3 m_3}{\beta_2 m_2} \cdot \langle r \rangle = \text{const} \langle r \rangle \tag{46}$$

because for any normalized stationary distribution the ratio of the moments m_3/m_2 is constant during growth.

Experimental Results

In contrast to the considerable progress the theoretical understanding of Ostwald ripening has gained since the now-classical papers of Lifshitz, Slyozov, and Wagner in 1961, there is generally no experimental research work that performed a complete quantitative test of even the LSW predictions. Only in the last few years have experiments been performed that may help in the future to close the gap between theory and experiment.[34,40,46,51]

Solid-Liquid Mixtures

Most of the experimental work with a liquid matrix has been performed within the context of powder metallurgical production techniques. There liquid phase sintering is routinely used in the fabrication of high-performance materials.[52] Solid-solid dispersions are heat treated such that the lower-melting-point material melts, thereby establishing a solid-liquid dispersion that rapidly becomes a pore-free compact body. The main problem then becomes the control of properties through grain size. The solid grains in a liquid matrix generally coarsen considerably by Ostwald ripening. A large variety of metal-metal and metal-nonmetal systems (including carbides and oxides) have been studied. Liquid phase sintered materials are applied, for example, as dental amalgams, cemented carbides (WC-Co), magnets, electrical contacts (Cu-W, Ni-Ag), automotive components (Fe-Cu), or high-density alloys (W-Ni-Fe). In liquid phase sintering the solid volume fraction is usually high, up to 95%.

Metal-Carbide Systems

Sarian and Weart[53] and Warren[54] investigated the growth of NbC in liquid cobalt, nickel, and iron. The growth exponent n for the increase in mean particle radius $\langle R \rangle$ as well as the rate constant were in general accordance with diffusion-controlled growth (compare with the subsection on asymptotic LSW analysis of Ostwald ripening). However, the variation of growth rate with temperature gave activation energies for the diffusion coefficient that were essentially too large for diffusion in a liquid phase. In addition, Sarian and Weart[53] could not detect a change of the rate constant with volume fraction. Exner et al.[55] and Santa-Marta[56] studied VC-Ni and VC-Co dispersions. Although they found activation energies that are consistent with liquid phase diffusion, their measurements of grain size distribution documented that the VC particles grow by an interface reaction (see the subsection on reaction-controlled growth). Figure 13 shows the square of the mean intercept length of VC in Co as a function of sintering time. The linear dependence is in agreement with Eq. (44) for reaction-controlled growth. Figure 14 shows that the Wagner function [Eq. (45)] for interface-controlled growth fits the observed distribution much better than either the LSW function for purely diffusion-controlled growth or a log-normal distribution. Application of a χ^2 test, as proposed by Exner and Lukas,[50] verified the validity of the reaction-controlled case. Similar findings have been made by Exner and Fischmeister[57] in the WC-Co system.

In light of the new theoretical developments made in the last decade to incorporate the effect of finite volume fraction (see the subsection on the influence of finite volume fraction), it can be argued that the agreement of

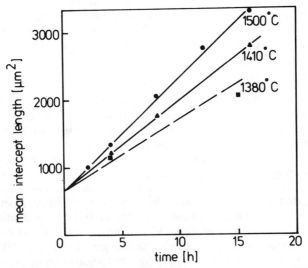

Fig. 13 Mean intercept length as a function of sinter time in VC-Co alloys as measured by Exner et al.[55] and Santa Marta.[56]

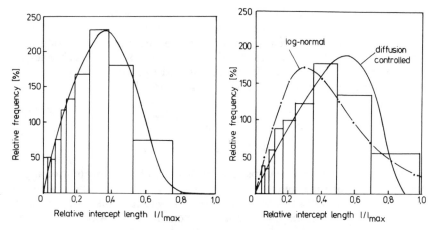

Fig. 14   Comparison of theoretical (LSW theory) and observed intercept distributions of VC grains in a Ni matrix after Exner et al.[55] and Santa-Marta.[56]

the particle size distribution measured by Exner et al.[55] with that predicted by Wagner[3] is fortuitous. At such high volume fractions of the dispersed phase (60%), even the distribution function for diffusion-controlled growth becomes very broad (see Fig. 4). However, the growth exponent is different. Therefore, it cannot be excluded that in WC-Co growth is interface controlled.

In contrast to the experimental findings of Exner et al.,[55] Waren[58] found that in VC-Co alloys the cube of the grain size varied linearly with time. In addition, he showed that with increasing VC content the growth rate varied by a factor of 4. The activation energy was estimated from an Arrhenius plot to 138 kJ/mole, which is in fair agreement with Exner's value of 109 kJ/mole. These apparent discrepancies between the different research groups have not been solved up to now. A reinvestigation taking into account the new theories of Ostwald ripening seems to be fruitful. The new techniques of microgravity research seems especially promising, since they allow a drastic reduction in solid volume fraction and therefore particle contacts.

Metal-Metal Systems

Watanable and Masuda[59] investigated particle coarsening in Fe-Cu, Cu-Ag, and W-Ni-Cu alloys. In all cases they found that the mean linear particle intercept varied proportionally to the cube root of time. Using different volume fractions of the solid phase (55–90%), they were able to show that the rate constant varied in the Fe-Cu system considerably (by a factor of 5). In the Cu-Ag mixtures only a small increase was measured (a factor of 2), as was also done for the W-Ni-Cu alloys. The particle size distributions deviated clearly from the LSW prediction for diffusion-controlled growth. They observed in all systems that the number of particle contacts and welds decreases inversely proportional with time. The size of the particle welds grows in such a manner that the proportion of particle diameter to weld diameter is a constant, independent of the sintering time. From this result they conclude that the coalescence of particles changes the growth rate in addition to diffusional Ostwald ripening.

Watanabe and Masuda[59] express this finding by a new growth rate equation

$$\frac{dR}{dt} = \left(\frac{dR}{dt}\right)_{OR} + \left(\frac{dR}{dt}\right)_{coalescence} \tag{47}$$

They conclude that the mean particle radius still changes proportionally to the cube root of time, but with a new rate constant. It is now a weighted sum of that for Ostwald ripening and coalescence. This idea was later adopted by Takajo et al.[60,61] in a study on concurrent Ostwald ripening and coalescence.

Kang and Yoon[62] investigated coarsening in a W-Ni mixture with 1.7, and 30 wt % liquid Ni. They observed that the linear particle intercept varied as predicted by diffusion-controlled Ostwald ripening in the case of higher Ni contents. The particle size distributions were only compared with the LSW results for diffusion and reaction control. They generally agree better with the function of Eq. (45), but this is meaningless because one could expect from the theoretical results presented in the subsection on the influence of finite volume fraction a strong deviation from the LSW curve. Kang and Yoon did not take into account the effect of particle coalescence.

The microstructural evolution in liquid phase sintered Fe-Cu alloys was also studied by Niemi and Courtney.[62] In agreement with Watanabe and Masuda, they observe a power law with $n = 3$ for the increase in particle size with time. In fact, they measured the interface area to volume ratio S_v, which is then proportional to $t^{-\frac{1}{3}}$. They also showed that, beyond a certain volume fraction and contiguity (see Underwood[47]), the microstructure makes a transition from a dispersion of particles with some interparticle contacts to a skeleton structure. The rate constants of coarsening varied considerably with contiguity, which is a measure of interfacial area relative to the sum of interfacial and grain boundary (interparticle) area. From this dependence they conclude that coalescence and Ostwald ripening concurrently determine the evolution of microstructure, at least for volume concentrations greater than 50%.

To overcome the difficulties of sedimentation in liquid phase sintering, Kang and Yoon[63] used Cu-Co alloys. The relative density difference of Cu and Co is smaller than 1% at room temperature. They used dispersions with 34, 42, and 55 vol % Co. The cube of the average intercept length increased linearly with time. The activation energy for diffusion within the liquid Cu matrix was evaluated to be 100 kJ/mole. The growth rates depend definitively on volume fraction (for a comparison with the new theories of volume fraction see below). Kang and Yoon[64] extended this study including Fe-Cu alloys with various volume fractions of Fe (59–90 vol %) and Co-Cu alloys and extended the range of Co content to

95 vol %. Their experimental findings are in general agreement with diffusional Ostwald ripening.

Takajo et al.[60,61] also obtained in their experimental study an increase of mean particle intercept length proportional to $t^{\frac{1}{3}}$. However, they include in their theoretical analysis both the effect of Ostwald ripening, following the approach of Brailsford and Wynblatt,[27] and coalescence. They were able to show that even then a power law $\langle R \rangle^3 \sim t$ holds. The particle size distributions measured agree reasonably well with a coarsening by a coalescence mechanism, slightly disturbed by Ostwald ripening. The authors assumed a collision rate of solid particles that is inversely dependent on the mean particle volume. This is equivalent (see Ratke[14]) to a constant collision volume as was often discussed in connection with the coarsening of aerosols.[65] The origin and mechanisms of coalescence are left open by Takajo et al.[61] Although slight deviations with the theoretical coalescence distribution are observed, especially for particles near zero radius and at the tail of the distribution, they argue that these are due to the aspherical shape of the Fe particles. The interpretation of Takajo et al. is generally in accordance with the findings of Niemi and Courtney.[62] The relevance of both papers is that they clearly demonstrate the difficulty of separating the diffusion controlled Ostwald ripening from coalescence. The experimental observation of $\langle R \rangle^3$ proportional to time is not sufficient, and the comparison of particle size distributions may not be accurate enough.

An important step for solving these problems can be found in the work of Hardy and Voorhees.[40] They investigated two-phase mixtures of Pb-Sn alloys with various volume fractions of solid phase, 57–96% (Pb particles in a eutectic melt or vice versa). At solid volume fractions above approximately 50%, a solid skeletal structure inhibits sedimentation. In contrast to the Fe-Cu and Cu-Ag alloys investigated by Takajo et al., no real solid-state interparticle contacts develop in this system. The eutectic liquid completely wets the solid Sn or Pb such that a liquid film is always in between the dispersoids. Hardy and Voorhees clearly demonstrated the self-similarity of the microstructures. Figure 15 shows micrographs of such a solid-liquid mixture. In the top row three micrographs of the same alloy after different ripening times demonstrate that the particles have grown in size. Using the cube root dependence of time for the mean particle size, one can magnify the micrographs so much that the mean size becomes equal in each picture. The result is shown in the bottom row. This series of pictures demonstrates most clearly that the process of competitive coarsening is self-similar in nature.

Hardy and Voorhees measured the mean particle intercept and the particle size distribution. The average linear intercept depends on time, as predicted by theory (Voorhees and Glicksman[30,31]). The distribution functions could be described almost exactly with the distribution functions predicted theoretically (Fig. 16; see also the subsection on the influence of finite volume fraction). In contrast to other solid-liquid mixtures, the material parameters for the Pb-Sn system are well known. Therefore, an exact quantitative comparison between the rate constant (for the growth of the mean particle radius) as predicted by theory and the experiment could

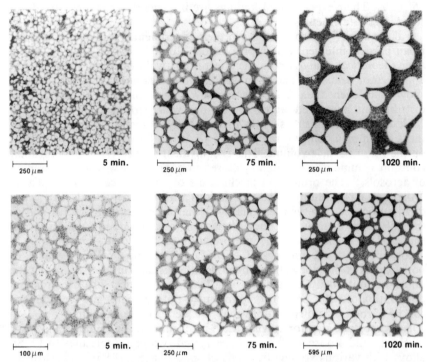

Fig. 15 **The microstructure of samples with Sn-rich coarsening phase at a volume fraction of $\Phi = 0.64$ as a function of time. The top row is at constant magnification; the bottom row is at variable magnification and illustrates the self-similarity of coarsening. (This picture is reproduced with permission of P. W. Voorhees.)**

be performed for the first time. The result is shown in Fig. 17. In general, there is a discrepancy in both by at least a factor of two. The origin of this disagreement is not yet clear. Several possibilities, including convective effects, are discussed by Hardy and Voorhees.

Figure 18 summarizes many results on solid-liquid mixtures. It gives a normalized presentation of the rate constants of the various systems mentioned earlier as a function of volume fraction. The solid line represents the result of the mean field calculation of diffusional Ostwald ripening as developed by Voorhees and Glicksman.[30,31] The agreement with the experimental data is promising.

Liquid-Liquid Mixtures

The investigation of Ostwald ripening in immiscible liquid metals is theoretically most promising because liquids best fulfill the assumptions made theoretically. Experimentally they are even more difficult to handle than solid-liquid mixtures. Coalescence of droplets will generally lead to formation of new, larger ones within a short time (an order of milliseconds[66]).

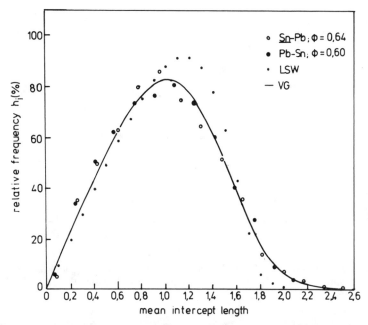

Fig. 16 **The relative intercept probability as a function of scaled intercept length in Sn-Pb alloys as measured by Hardy and Voorhees.[40] The dotted curve is the prediction of the LSW theory, and the solid curve is due to the Voorhees and Glicksmann theory.[29-31]**

   The two different techniques for dealing with this problem are described in the section on experimental techniques. The first successful investigation using Zn-Pb alloys was reported by Kneissl et al.[45] They used an alloy of Zn with 2 wt % Pb. The starting material had a fine dispersed microstructure due to rapid solidification from a temperature in the one-phase homogeneous region of the phase diagram. Pb particles of around 1 μm were produced. Heating the samples right into the miscibility gap retains the two phase structures and avoids nucleation. They included natural convection rolls in the samples and could thus prevent a certain part of the Pb droplets in the liquid Zn matrix. Depending on the size of the specimens, the speed of the convection rolls varies, giving different times for droplets to grow by Ostwald ripening. If they are so big that their Stokes velocity is larger than the velocity of the convection roll, they leave the dispersion. Figure 19 shows one result of this investigation. It is clearly seen that the time law for the increase in particle size follows the LSW law

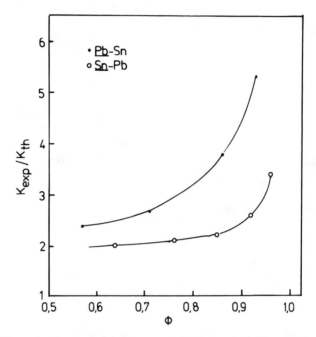

Fig. 17 The ratio of experimentally measured rate constant K_{exp} and the theoretical predicted rate constant K_{th} as a function of volume fraction Φ after Hardy and Voorhees.[40]

for pure diffusional transport of the solute (note that the volume fraction of Pb is around 0.3 vol %). From the rate constants measured they estimated an activation energy for Pb diffusion in liqud Zn of 23 kJ/mole.

The process of Ostwald ripening with immiscible liquids was studied by Kneissl and Fischmeister[43] in a Spacelab 1 experiment and was analyzed later by Ratke et al.[44,67] They used Zn-Pb alloys with various Pb concentrations (2–5 wt %Pb). Alloys with a very fine microstructure of the dispersion were produced on Earth by rapid quenching of homogeneous alloys ($dT/dt > 100$ K/s) and then heated the samples in Spacelab 1 right into the miscibility gap to avoid nucleation and held them isothermally (1 h 475°C) followed by a free cooling until solidification. Metallographic sections of the samples prior to and after flight were analyzed in a SEM connected on-line to an automatic image analyzing system so that the section radii and their distribution could be evaluated with a large number of particles.

All samples showed no massive segregation of the lead droplets. They were really homogeneously distributed throughout the samples (besides small-scale inhomogeneities[44]). It was expected that, after the temperature of isothermal anneal was reached, the dispersion would coarsen by diffusional Ostwald ripening. Since Kneissl and Fischmeister had only one

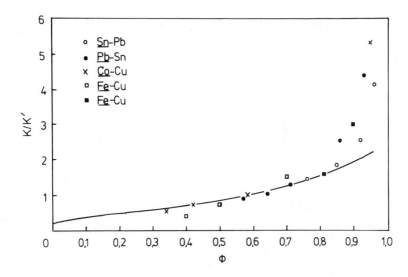

Fig. 18 The relative rate constant K/K', where K' is the rate constant at $\Phi = 0.6$ as a function of Φ after Hardy and Voorhees.[40] The solid curve is the theoretical prediction by Voorhees and Glicksman.[29-31]

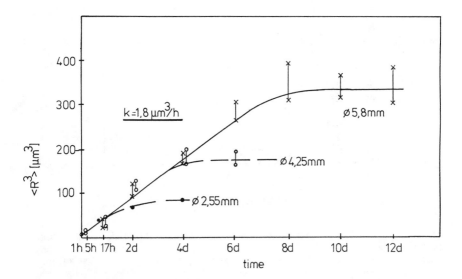

Fig. 19 Growth of Pb droplets in a liquid Zn matrix as measured by Kneissl et al.[45] with the natural convection roll technique. Different specimen sizes Φ define maximum particles sizes that can be preserved in the stirred liquid. The larger the sample size, the larger the maximum droplet diameter.

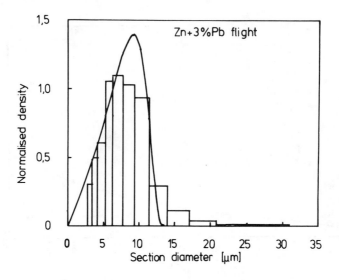

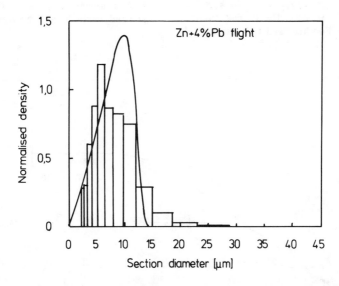

Fig. 20   Comparison of the measured (bars) and theoretical particle size distribution (curve) as predicted by LSW in Zn-Pb suspensions after Ostwald ripening under microgravity conditions in Spacelab 1: a) Zn + 3% Pb flight; b) Zn + 4% Pb flight.

experimental run with one time for growth, a comparison between experiment and theoretical prediction can only be achieved via the particle distribution functions. The experiment fulfilled the requirements of the LSW theory presented in the basic theory section almost perfectly. The dimensionless time Θ defined in Eq. (24) is approximately 8. Therefore, it should be possible to compare the experimental results with the theoretical predictions (Figs. 20).

The droplet diameter distributions of both samples after flight show some similarities with the LSW distribution. For the smaller diameters the agreement between experimental and calculated LSW distribution is satisfactory. The modal diameters of the curves coincide, but the measured frequency is lower than predicted. At larger diameters the experimental curves tail off more slowly than predicted. It is worth noting that the deviation from theory increases from the samples with 3 wt % to that with 4 wt % Pb.

The comparison shows that some additional coarsening process must have been superimposed on the Ostwald ripening. The disturbance increases with the volume fraction of Pb droplets so that in specimens with more than 4 wt % Pb the observed distributions (not shown here; see Refs. 43 and 44) become so broad that no similarity with the LSW distribution is left. Since the nucleation of droplets is excluded by the experimental procedure used, the only way to take into account such a deviation from diffusional Ostwald ripening is by the collision of droplets with subsequent coagulation. However, collisions can only occur if the droplets move with different velocities driven by some force field.

Ratke et al.[67] calculated the effect of collisions due to Marangoni motion in the samples which occurs concurrently with Ostwald ripening. The results are shown in Figs. 21. The agreement between theory and experiment is remarkably good. Therefore, in this microgravity experiment diffusional Ostwald ripening in a liquid dispersion was observed, disturbed by collision processes due to droplet motion in a temperature gradient.

There is only one investigation that sought the effect of convective transport of solute on Ostwald ripening. Thieringer and Ratke[34] and Thieringer et al.[68] used liquid Al-Pb dispersions and induced via a temperature gradient a natural convection roll in the samples as described in the section on experimental techniques. The samples were heated into the liquid two-phase field so that Pb-rich droplets were immersed in the Al matrix with a total volume content of 0.75%. After 1 h the volume content was down to 0.04% and was approximately constant over times up to 20 h. By variation of the temperature gradient (15–30 K/cm), the mean speed of the natural convection roll could be varied between 2 and 4 mm. Because of normal-gravity acceleration the lead Pb droplets moved relative to the Al matrix. Their Peclet numbers have been calculated to be between $Pe = 1$ and 40, depending on size and sedimentation velocity. Thus, the requirements of the theory of Ostwald ripening by convective diffusion were fulfilled. From the numerous experimental results two extreme cases are shown in Figs. 22 and 23. In Fig. 22 a temperature gradient of 15 K/cm

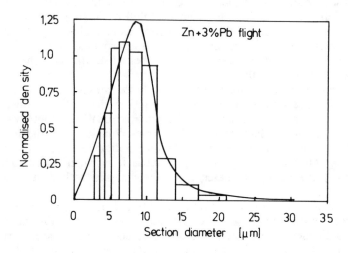

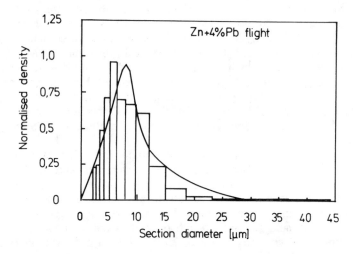

Fig. 21   Comparison of the measured (bars) and theoretical size distribution of the samples but now including concurrent Ostwald ripening and coarsening by collisions due to Marangoni motion after Ratke et al.[67]: a) Zn + 3% Pb flight; b) Zn + 3% Pb flight.

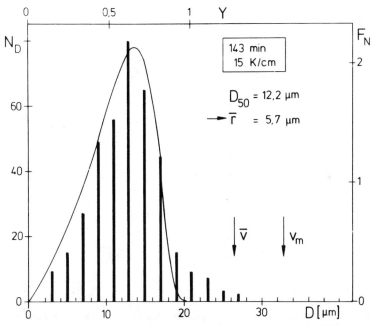

Fig. 22 Normalized particle size distribution function of Pb droplets dispersed in a liquid Al matrix after Thieringer and Ratke.[34] They used the natural convection roll technique to stabilize a suspension. The bars are the experimental distribution, and the solid curve is the prediction from LSW theory for diffusion-controlled growth.

yielded a maximum Pe number of 18. Thus, the experimental distribution (bars) fits better to the LSW distribution for diffusional transport. In Fig. 23 the maximum Pe number was nearly 40; therefore, convective contributions to solute transport are not negligible. The experimental distribution and the theoretical for Ostwald ripening of liquid droplets moving with Stokes velocity (Fig. 8b) agree almost perfectly.

Conclusions

The progress in the theory of Ostwald ripening within the last five years has been enormous. Dispersions with finite volume fractions of second phase can now be treated satisfactorily. The effect of additional convective transport on the coarsening is treated for several cases of particle motion. Deviations from ideal solution behavior can easily be incorporated. Concurrent transport of solute and heat (or only heat) can be taken into account.[69] In contrast to this theoretical progress, the experimental results with solid-liquid or liquid-liquid dispersions are lagging behind. There is a great need for conclusive experiments even to validate the classical LSW theory for dilute dispersions. The availability of space laboratories with

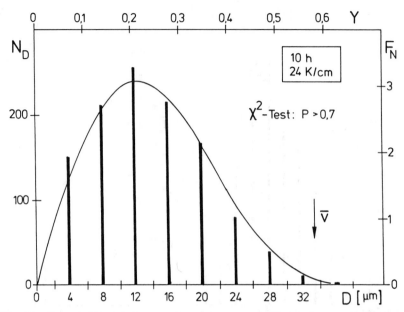

**Fig. 23   Same in Fig. 24, but now the temperature gradient is increased and thus the
fluid flow velocity is increased. The Pb droplets move relative to the matrix by Stokes
motion. The solid curve is the theoretical prediction for Ostwald ripening with
convective diffusion[13,34] (cf. Fig. 8b).**

their reduced-gravity environment will fertilize experiments on Ostwald
ripening with liquids and hopefully close the gap between theory and
experiment.

References

[1]Ostwald, W., *Zeitschift fuer Physikalische Chemie*, Vol. 34, 1900, pp. 495–503.

[2]Lifshitz, I. M., and Slyozov, V. V., "The Kinetics of Precipitation from
Supersaturated Solid Solutions," *Journal of Physics and Chemistry of Solids*, Vol. 19,
1961, pp. 35–50.

[3]Wagner, C., *Zeitschrift fuer Elektrochemie*, Vol. 65, 1961, pp. 581–591.

[4]Greenwood, G. W., *1969 Proceedings of the Conference on the Mechanism of Phase
Transformations in Crystalline Solids*, Institute of Metals, London, 1969, p. 103.

[5]Li C.-Y., and Oriani, R. A., *Proceedings of Symposium on Oxide Dispersion
Strenthening*, edited by G. S. Ansell, T. D. Cooper, and T. V. Lenel, New York, 1966,
p. 431.

[6]Ardell, A. J., *1969 Proceedings of the Conference on the Mechanism of Phase
Transformations in Crystalline Solids*, Institute of Metals, London, 1969, p. 111.

[7]Fischmeister, H. F., and Grimvall, G., *Proceedings 3rd Conference on Sintering and
Related Phenomena*, edited by G. C. Kuczynski, Plenum, New York, 1972, p. 119.

[8]Voorhees, P. W., "The Theory of Ostwald Ripening," *Journal of Statistical Physics*,
Vol. 38, 1985, pp. 231–252.

[9]Freundlich, H., *Kapillarchemie*, Akademie Verlagsges., Leipzig, 1922.

[10]Purdy, G. R., *Metal Science*, Vol. 5, 1971, pp. 81–85.

[11]Chaix, J. M., Eustathopoulos, N., and Alibert, C. H., *Acta Metallurgica*, Vol. 34, 1986, pp. 1589–1592.

[12]Kampmann, R., and Wagner, R., "Decomposition of Alloys—the Early Stages," *Proceedings of the 2nd Acta-Scripta Metallurgica Conference*, edited by P. Haasen, V. Gerold, R. Wagner, and M. F. Ashby, Pergamon, Oxford, UK, 1984, p. 91.

[13]Ratke, L., and Thieringer, W., "The Influence of Particle Motion on Ostwald Ripening in Liquids," *Acta Metallurgica*, Vol. 33, 1985, pp. 1793–1802.

[14]Ratke, L., PhD Dissertation, University of Stuttgart, Stuttgart, FRG, 1987.

[15]Ratke, L., and Host, M., *Journal of Physicochemical Hydrodynamics* (submitted for publication).

[16]Venzl, G., *Berichte der Bunsengellschaft fuer Physikalische Chemie*, Vol. 87, 1983, pp. 318–324.

[17]Enomoto, Y., and Kawasaki, K., "Computer Simulation of Ostwald Ripening with Elastic Field Interactions," *Acta Metallurgica*, Vol. 37, pp. 1399–1406.

[18]White, R. J., *Material Science and Engineering*, Vol. 40, 1979, pp. 15–20.

[19]Mullins, W. W., *Journal of Applied Physics*, Vol. 59, 1986, pp. 1341–1349.

[20]Marqusee, J. A., and Ross, J., *Journal of Chemical Physics*, Vol. 79, 1983, pp. 373–378.

[21]Oriani, R. A., *Acta Metallurgica*, Vol. 12, 1964, pp. 1399–1409.

[22]Li, C.-Y., Blakely, J. M. and Feingold, A. H., "Mass Transport Analysis for Ostwald Ripening and Related Phenomena," *Acta Metallurgica*, Vol. 14, 1966, pp. 1397–1402.

[23]Zwillinger, D., "Coarsening of Non-Spherical Particles," *Journal of Crystal Growth*, Vol. 94, 1989, pp. 159–165.

[24]Asimov, R., "Clustering Kinetics in Binary Alloys," *Acta Metallurgica*, Vol. 11, 1963, pp. 72–73.

[25]Ardell, A. J., *Acta Metallurgica*, Vol. 20, 1972, pp. 61–71.

[26]Davies, C. K. L., Nash, P., and Stevens, R. N., *Acta Metallurgica*, Vol. 28, 1980, pp. 179–189.

[27]Brailsford, A. D., and Wynblatt, P., *Acta Metallurgica*, Vol. 27, 1979, pp. 489–497.

[28]Marqusee, J. A., and Ross, J., "Theory of Ostwald Ripening: Competitive Growth and Its Dependence on Volume Fraction," *Journal of Chemical Physics*, Vol. 80, 1984, pp. 536–543.

[29]Voorhees, P. W., and Glicksman, M. E., *Metallurgical Transactions*, Vol. 15A, 1984, pp. 1081–1088.

[30]Voorhees, P. W., and Glicksman, M. E., *Acta Metallurgica*, Vol. 32, 1984, pp. 2001–2012.

[31]Voorhees, P. W., and Glicksman, M. E., *Acta Metallurgica*, Vol. 32, 1984, pp. 2013–2030.

[32]Tokuyama, M., and Kawasaki, K., "Statistical-Mechanical Theory of Coarsening of Spherical Droplets," *Physica*, Vol. 13A, 1984, pp. 386–411.

[33]Tokuyama, M., Enomoto, Y., and Kawasaki, K., *Acta Metallurgica*, Vol. 35, 1987, pp. 907–913.

[34]Thieringer, W. K., and Ratke, L., *Acta Metallurgica*, Vol. 35, 1987, pp. 1237–1244.

[35]Levich, V. G., *Physicochemical Hydrodynamics*, Prentice-Hall, Englewood Cliffs, NJ, 1962.

[36]Young, N. O., Goldstein, J. S., and Block, M. J., *Journal of Fluid Mechanics*, Vol. 6, 1959, pp. 350–356.

[37]Landau, L. D., and Lifshitz, I. M., *Hydrodynamik*, Akademie Verlag, Berlin, 1981.

[38]Ratke, L., *Proceedings of the Nordeney Symposium on Scientific Results German Spacelab Mission D1*, edited by P. R. Sahm, R. Jansen, and M. H. Keller, DLR, Cologne, 1985, p. 332.

[39]Takajo, S., PhD Dissertation, University of Stuttgart, Stuttgart, FRG, 1981.

[40]Hardy, S. C., and Voorhees, P. W., "Ostwald Ripening in a System with a High Volume Fraction of Coarsening Phase," *Metallurgical Transactions*, Vol. 197, 1988, pp. 2713–2721.

[41]Kang, T.-K., and Yoon, D. N., "Coarsening of Tungsten Grain in Liquid Nickel-Tungsten Matrix," *Metallurgical Transactions*, Vol. 9A, 1978, pp. 433–438.

[42]Exner, H. E., *Zeitschrift fuer Metallkunde*, Vol. 64, 1973, 273–280.

[43]Kneissl, A., and Fischmeister H., *Proceedings of the 5th European Symposium on Material Science Under Microgravity*, European Space Agency Rept. SP-222, 1984, p. 63.

[44]Ratke, L., Fischmeister, H., and Kneissl, A., *Proceedings of the 6th European Symposium on Material Sciences Under Microgravity Conditions*, European Space Agency Rept. SP-256, 1986, p. 161.

[45]Kneissl, A., Pfefferkorn, P., and Fischmeister, H., *Proceedings of the 4th European Symposium on Material Sciences Under Microgravity*, European Space Agency Rept. SP-191, 1984, p. 55.

[46]Ratke, L., and Uffelmann, D., *Proceedings of the 7th European Symposium on Material Sciences Under Microgravity*, European Space Agency Rept., 1989 (to be published).

[47]Underwood, E. E., *Quantitative Stereology*, Addison-Wesley, Reading, MA, 1970, p. 109.

[48]Exner, H. E., "Analysis of Grain and Particle-Size Distributions in Metallic Materials," *International Metallurgical Reviews*, Vol. 17, 1972, pp. 25–42.

[49]Exner, H. E., *Zeitschrift fuer Metallkunde*, Vol. 61, 1970, pp. 218–225.

[50]Exner, H. E., and Lukas, H. L., *Metallography*, Vol. 4, 1971, pp. 325–338.

[51]Voorhees, P. W., and Schafer, R. J., *Acta Metallurgica*, Vol. 35, 1987, pp. 327–339.

[52]German, R. M., *Liquid Phase Sintering*, Plenum, New York, 1985.

[53]Sarian, S., and Weart, H. W., "Kinetics of Coarsening of Spherical Particles in a Liquid Matrix," *Journal of Applied Physics*, Vol. 37, 1966, pp. 1675–1681.

[54]Warren, R., *Journal of Materials Science*, Vol. 3, 1968, pp. 471–485.

[55]Exner, H. E., Santa-Marta, E., and Petzow, G., *Modern Developments in Powder Metallurgy: Processes*, Vol. 4, Plenum, New York, 1971, p. 315.

[56]Santa-Marta, E.A.G.M., PhD Dissertation, University of Stuttgart, Stuttgart, FRG, 1970.

[57]Exner, H. E., and Fischmeister, H., *Archiv Eisenhüttenw*, Vol. 35, 1966, pp. 417–426.

[58]Warren, R., "Microstructural Development During the Liquid-Phase Sintering of VC-Co Alloys," *Journal of Materials Science*, Vol. 7, 1972, pp. 1434–1442.

[59]Watanabe, R., and Masuda, Y., *Sintering and Catalysis*, edited by G. C. Kuczynski, Plenum, New York, 1975.

[60]Takajo, S., Kaysser, W. A., and Petzow, G., *Acta Metallurgica*, Vol. 32, 1984, pp. 107–113.

[61]Kaysser, W. A., Takajo, S., and Petzow, G., *Acta Metallurgica*, Vol. 32, 1984, pp. 115–122.

[62]Niemi, A. N., and Courtney, T. H., "Microstructural Development and Evolu-

tion in Liquid-Phase Sintered Fe-Cu Alloys," *Journal of Materials Science*, Vol. 16, 1981, pp. 226–236.

[63]Kang, C. H., and Yoon, D. N., "Coarsening of Cobalt Grains Dispersed in Liquid Copper Matrix," *Metallurgical Transactions*, Vol. 12A, 1981, pp. 65–69.

[64]Kang, S. S., and Yoon, D. N., *Metallurgical Transactions*, Vol. 13A, 1982, pp. 1405–1411.

[65]Drake, R. L., "A General Mathematical Survey of the Coagulation Equation," *Topics in Current Aerosol Research*, Vol. II, edited by G. M. Hidy and J. R. Brock, Pergamon, Oxford, UK, 1970, p. 203.

[66]Bisch, C., and Lasek, A., *Proceedings of the 6th European Symposium on Material Science Under Microgravity Conditions*, European Space Agency Rept. SP-256, 1986, p. 209.

[67]Ratke L., Fischmeister, H., and Kneissl, A., *Proceedings of the 7th European Symposium on Material Science Under Microgravity Conditions*, European Space Agency Rept., 1989 (to be published).

[68]Thieringer, W., Ratke, L., Fischmeister, H., *Proceedings of the 6th European Symposium on Material Science Under Microgravity Conditions*, European Space Agency Rept., SP-256, 1986, p. 169.

[69]Voorhees, P. W., "Coarsening in Binary Solid-Liquid Mixtures," *Metallurgical Transactions*, Vol. 21A, 1990, pp. 27–37.

[70]Brian, P. L. T., and Hales, H. B., "Effects of Transpiration and Changing Diameter on Heat and Mass Transfer to Spheres," *AIChE Journal*, Vol. 15, 1969, pp. 419–425.

Chapter 7. Combustion

Particle Cloud Combustion in Reduced Gravity

A. L. Berlad*

University of California, San Diego, La Jolla, California

Introduction and Background

THE characteristic flame propagation and flame extinction processes sustained by a uniformly premixed, quiescent fuel and oxidizer are of central interest in the fundamental and applied combustion sciences. Treatises on fire safety for gaseous systems feature experimental data describing the range of fuel-oxidizer ratios within which quasisteady flame propagation can (or cannot) occur.[1] Corresponding combustion theory attempts to describe the flame's structure, its propagation speed, the extinction process, and other conditions that may limit quasisteady flame propagation.[2] Fundamental to this experimental and theoretical correspondence is the requirement that the unburned combustible medium is initially quasisteady. For premixed combustible gaseous systems initial spatial and temporal uniformity of a combustible system's chemical, transport, and thermophysical properties is easily achieved prior to combustion experimentation. At normal gravitational conditions, the corresponding quasisteady requirements for particulate fuel clouds is not achieved.[3-6] Accordingly, the extensive body of experimental observations characteristic for premixed gaseous combustion (autoignition, ignition, laminar flame propagation, extinction, oscillatory oxidation, and other phenomena) is not matched by a corresponding body of combustion data characteristic for initially quasisteady premixed particle clouds. At normal gravitational conditions, uniform, quiescent particle clouds cannot be established for the following reasons:

1) Gravity causes the sedimentation of particles of significant size, thereby rendering spatial and temporal uniformity of a quiescent fuel particle cloud impossible.

*Professor, Department of Applied Mechanics and Engineering Science, Center for Energy and Combustion Research.

2) Mixing-induced turbulence and secondary flows, which are used to suppress sedimentation and to create uniform particle clouds, imply ill-defined transport properties. Such mixtures sustain the problems of item 1 when particle-uniformity-promoting stirring ceases.

3) Fuel-oxidizer ratio fluxes through freely propagating flame fronts are functions of the gravity vector.

The systematic experimental study of freely propagating flames through clouds of uniform, quiescent particulates is thus not feasible at normal gravity. However, premixed particle cloud flames have been stabilized on burners and studied at normal gravity.

Premixed coal-air flames have been stabilized on burners, and their properties in the upward propagation mode have been observed[7-9] at normal gravity. More recent experiments with burner-stabilized premixed lycopodium-air flames have been carried out in upward propagation ($g = +1$), in downward propagation ($g = -1$), and in reduced gravity ($g \simeq 0$). These latter studies[10,11] show that the stability and structures of these three flame propagation modes are substantially different. Premixed, stabilized particle cloud flame characteristics are related to those for freely propagating flames. However, burner-stabilized flame propagation rates and existence limits depend importantly on experimentally predetermined flow and burner conditions. Freely propagating flames (sustained by initially quiescent, uniform combustible media) display characteristic flame speeds and existence limits. Thus, it is the limiting, fuel-lean concentrations of freely propagating flames that are associated with the "lean flammability limit" for any given premixed fuel-oxidizer system.[1,2,12]

In the neighborhood of lean flammability limit fuel concentrations, normal-gravity buoyancy effects on the slowly propagating flame structures are most pronounced. This is widely observed for premixed, freely propagating gaseous flames.[1,12-14] It is observed for both stabilized and for freely propagating[4] particle cloud flames. Oddly enough, stabilized premixed gaseous flame studies analogous to those done recently (at $g = 0, \pm 1$) for particle clouds[10,11] have not been reported. Nevertheless, the experimental combustion literature shows that normal-gravity buoyancy effects are most pronounced for the cases of near-lean-limit fuel concentrations. This is observed for premixed gaseous systems as well as for premixed particle cloud flames. It is observed for burner-stabilized as well as for freely propagating flame systems.

Fundamental flame theory seeks to describe flame structure, flame propagation speeds, and flame existence limits.[2,15-21] Currently, tractable fundamental flame theory generally neglects gravitational (and other body force) effects of flame propagation characteristics.[2,15-21] Heavily truncated phenomenological theories are generally used to characterize experimental flame propagation and extinction data where buoyancy plays a substantial role.[12,13]

For premixed flame systems in general, the available $g \simeq 0$ theoretical formulations do not correspond satisfactorily to the available $g = \pm 1$ experimental observations for freely propagating flames sustained by near-lean-limit fuel concentrations. For premixed particle cloud flames sedimentation as well as buoyancy effects further degrade the correspondence between $g = \pm 1$ experimentation and $g \simeq 0$ theory.

The studies described in this chapter are concerned with understanding the experimental behavior of particle cloud flame propagation and extinction processes under reduced-gravity conditions where sedimentation and buoyancy effects do not significantly modify the underlying $g \simeq 0$ combustion processes. Such experimental observations are associated with initially quasisteady, defined combustible particle clouds. These experimental data may then be utilized, together with existing fundamental flame theory, to help provide an understanding of these underlying $g \simeq 0$ flame processes of interest. Understanding of these underlying $g \simeq 0$ flame propagation and extinction characteristics is needed as a basis for understanding general particle cloud flame processes, including the effects of gravitational, transport, compositional, and other combustion parameters.

Gravitational Effects and Particle Cloud Combustion Experiments

The principal objective of particle cloud combustion experiments (PCCEs) is to provide flame propagation rate and extinction condition data. At normal gravity, particle sedimentation processes compromise our ability to properly prepare a combustable system for meaningful study of freely propagating flames and their limiting conditions for propagation. During combustion experimentation, buoyancy effects (as well as continuing sedimentation processes) further compromise our ability to understand the experimental observations.[3-6,10-14] An examination of the character and magnitude of these gravitational effects is useful. Normal-gravity difficulties as well as the unique research opportunities afforded by reduced gravity are thereby characterized.

Particles of interest in most investigations have maximum densities (~ 1.35) and maximum diameters ($\sim 70 \, \mu$m). Thus, their settling speeds at normal gravity are in the Stokes regime. In this regime the particle settling speed (in air) is given by

$$V_t = gr^2\left(\frac{2}{9}\right)\left(\frac{\rho_p}{\mu_g}\right) \tag{1}$$

Where g is the acceleration due to gravity, r the particle diameter, ρ_p the particle density, and μ_g the gas (air) viscosity. Lycopodium spores may serve as prototypical fuel particulates, due to their well-defined shape and size and high volatility. The lycopodium spore has a mean diameter of about $27 \, \mu$m and a density very close to unity. Some properties of lycopodium are given in Table 1. Pocahontas coal particles have a density that is about one-third higher than that for lycopodium.

At normal gravity, lycopodium has a settling speed of about 3 cm/s. If one attempts to mix lycopodium with air in a 5-cm-i.d. tube (classically, the inside diameter of a flame tube selected for flammability limit measurements[1]) to create a uniformly mixed cloud, the quiescent uniformity criterion for the combustion experiment cannot be met. Mixing-induced

Table 1 Some properties of lycopodium particles

Lycopodium spore diam	Major axis	$30 \pm 1.5 \, \mu m$
	Minor axis	$25 \pm 1.5 \, \mu m$
Stoichiometric ratio		125 mg/l
Combustion enthalpy		30.25 MJ/kg
Density	Single particle	1015 kg/m³
	Bulk density	400 kg/m³
Adiabatic flame temperature		1975 K
Analysis	Carbon	65.8%
	Oxygen	21.9%
	Hydrogen	9.6%
	Nitrogen	1.2%
	Sulfur	0.2%

turbulence and secondary flows must be allowed to decay prior to flame initiation. The unburned combustible medium must display time invariant properties. Near fuel-lean flammability limits (the minimum fuel-to-air ratio capable of supporting quasisteady flame propagation), flame speeds of the order of 10 cm/s or less may be characteristic. Any reasonable criterion of both quiescence and uniformity for the unburned medium (for $g \simeq 1$) generally cannot be met. Postmixing periods needed to achieve quiescence have been estimated to be of the order of 10 s for energetic acoustically induced mixing processes. Minimum time requirements for maintenance of a quiescent cloud, prior to flame front arrival, is of the order of 10 s or more. If the cloud of vigorously mixed particles is to be allowed to settle (or drift) by no more than 10% of a tube diameter during a total experimental time of some 50 s, utilization of Eq. (1) implies the need for a reduced-gravity environment (g^*) of about $10^{-3}g_0$ to $5 \times 10^{-4}g_0$, where g_0 is the acceleration due to normal gravity. It is clear that mixing of lycopodium (to uniformity) in reduced gravity is an easier and shorter task than mixing attempts at g_0. This requirement for a reduced-gravity environment is one of three shown in Table 2.

Another normal-gravity obstacle to the achievement of the experimental objectives relates to the effective number of particulates consumed by a flame front that propagates upward ($g_0 = +1$) or downward ($g_0 = -1$) at normal gravity. Even if the particle cloud were uniform and quiescent (at normal gravity), the particle number swept out per unit time by a given quasisteady flame front (propagating upward or downward) is different. The effective concentration of such a flame front is given by[4]

$$c^* \simeq c_0 \left(1 \pm \frac{V_t}{U_f} \right) \qquad (2)$$

where c^* is the effective concentration, c_0 the actual concentration, V_t the settling speed, and U_f the flame speed. The positive sign corresponds to

Table 2   Experimentally required g values

1) Reaction zone's effective concentration is to be kept close to actual, where $C^* = C_0(1 \pm V_t/U_f)$ and $V_t/U_f \leqslant 0.02$	$g^* \approx 10^{-2}g_0 - 10^{-3}g_0$
2) Dispersed cloud is to be restricted to small drift during a 50-s combustion process, where $\delta \simeq 0.05$ cm and $\rho = 1.0$ g/cm^3	$g^* \approx 10^{-3}g_0 - 5.0 \times 10^{-4}g_0$
3) Buoyancy-induced flows due to postreaction zone products are to be inhibited during the combustion process (see Ref. 12) $[t_s^3 g^2/\nu] \approx \text{const}$	$g^* \approx 10^{-2}g_0$

upward propagation, and the negative sign corresponds to downward propagation. The upward propagating front enjoys an enriched fuel concentration. The downward propagating front experiences a depleted fuel concentration. This effect is particularly troublesome for large, dense particles and for the very slow flames anticipated in the neighborhood of flammability limits. For example, a lycopodium-air flame speed of 3 cm/s observed in upward propagation would correspond to an enriched particle concentration of $c_+^* \sim 2c_0$. The same flame speed in downward propagation is a physical impossibility. In the latter case $c_-^* \sim 0$. Here again, use of the reduced-gravity environment allows the imposition of effective limits on the difference between c^* and c_0, regardless of the direction of the gravity vector. If we require that $(V_t/U_f) \leqslant 0.02$, we find that the required range of reduced gravity conditions is of the order $g \simeq 10^{-2}g_0 - 10^{-3}g_0$. This requirement for a reduced-gravity environment is one of three shown in Table 2.

Another normal-gravity obstacle to the achievement of an important experimental objective relates to the effects of buoyancy in flames. These effects are well known for purely gaseous flames[1,12-14] and are also observed for particle cloud flames. Lovachev[12] has discussed the relation between a characteristic time and other parameters for hot combustion product buoyancy effects on premixed flame propagation rates. His studies lead to the relation

$$\frac{t_k^3 g^2}{\nu} \simeq \text{const} \tag{3}$$

where t_k is the time (after ignition) necessary for the development of significant buoyancy effects on flame propagation, g the gravitational constant, and ν the kinematic viscosity. The characteristic group of Eq. (3) was developed through observation of hot product buoyancy effects of premixed gaseous flames as a function of pressure, at normal gravity. Lovachev finds that an ambient pressure of one-tenth atmospheric is adequate for virtual suppression of this buoyancy effect. Based on the

Lovachev data[12] and a conservative estimate of some 50–100 s needed for a particle cloud combustion experiment at reduced gravity, it follows that a value of $g^* \simeq 10^{-2}g_0$ is needed to achieve about the same buoyancy-suppression effects for particle cloud flames at reduced gravity. This requirement for a reduced-gravity environment is one of three shown in Table 2.

Finally, it is to be noted[14] that the reaction zone ratio of the gravitational to pressure terms for a flat freely propagating gaseous flame is given by the ratio

$$\frac{(\rho_1 + \rho_2)gh}{2(p_2 - p_1)} = \sigma \tag{4}$$

where ρ_1 and ρ_2 are the densities downstream and upstream of the reaction zone, respectively, h is the reaction zone depth, and $(p_2 - p_1)$ is the pressure drop across the reaction zone. The reaction zone thickness varies inversely with flame speed. Near the flammability limits it is expected that (h) may become a relatively large value. Inasmuch as the near-lean-limit flame speed for experiments of interest (in this study) are not known, the value of σ is not known, and its possible significance cannot be fully assesed. However, the observations[12] that led to Eq. (3) suggest that the reaction zone buoyancy effects are no more significant than the combustion product buoyancy effects for the range of stoichiometries studied and reported in Ref. 12. A plot of (σ) vs the equivalence ratio for methane-air flames is shown in Ref. 14.

Based on the previously mentioned considerations, it is concluded that gravitational conditions of the order of $g^* = 10^{-3}g_0$ are adequate to fulfill virtually all requirements for suppression of sedimentation and buoyancy effects. Furthermore, gravitational conditions of the order of $g^* = 10^{-2}g_0$ may suffice if the experimental time period is of the order of 10–20 s. Experimental times of the order of 2–5 s are available in drop tower facilities. Experimental times of the order of 20 s are available in suitably equipped (NASA) aircraft, piloted to sustain Keplerian trajectories.

Particle Cloud Combustion Experiments

The principal objective of reduced-gravity particle cloud combustion experiments is to obtain flame property and flame extinction limit data for a variety of premixed, quiescent two-phase combustion systems under near-zero-gravity conditions. Several of the NASA reduced-gravity facilities must be employed to help gather the scientific data required. The study of extinction phenomena and flame propagation rates of fuel-lean systems requires both the long reduced-gravity time and the low values of the gravity levels offered by space transportation system (STS). Near the lean flammability limits flame speeds are at their slowest (perhaps significantly less than 10 cm/s) and the *rate* of change of flame speed with fuel concentration is at its highest.[9-14] Accordingly, mixing times and flame propagation times associated with a 75-cm-long flame tube are expected to require STS conditions. However, the higher flame propagation rates of most other mixtures do permit the collection of necessary scientific data in ground-based facilities (e.g., airplane flights in Keplerian trajectories).

Particle Cloud Combustion Experiments: Stabilized Flames in Low Gravity

A combustible system's experimentally determined flammability limits, quenching limits, and pressure limits are not completely independent of one another.[1-4,18-21] Comprehensive theoretical consideration of premixed flame existence limits is complicated by the substantial effects that gravitational conditions may impose.[22-24] Tractable combustion theory is rarely more than one dimensional. Free convective flows are generally three dimensional. Upward ($g = +1$) flame propagation limits are generally wider than downward ($g = -1$) propagation limits. The findings to date show that this general combustion limit behavior obtains for both premixed gaseous and premixed particle cloud flames.[10] This behavior also obtains for burner-stabilized[10] as well as for freely propagating flames.[1-4,10-14]

Recent studies[10] of burner-stabilized lycopodium-air flames have been carried out under all three conditions of interest ($g = 0, \pm 1$). A single, fully self-contained apparatus was used for all three gravitational conditions. The $g = 0$ data[10,11] were obtained at the NASA Lewis Research Center 2.2-s drop tower facility. It may be surprising that no similar studies of other $g = 0, \pm 1$ stabilized flames have been reported, *either for premixed gaseous systems or for the premixed particle cloud systems.* These recent findings[10,11] are derived from a burner apparatus capable of measuring heat-transfer rates to the burner lip and capable of thermocouple probing of flame structure. Figure 1 is a schematic of the burner apparatus. Figures 2–4 show the measured heat-transfer rates from stabilized lycopodium-air flames to the burner lip. Figure 5 shows three flame temperature structures for the three conditions ($g = 0, \pm 1$). The finding of Refs. 10 and 11 for stabilized lycopodium-air flames appear to describe stabilized flame properties expected for both premixed particle cloud flames *as well as for* many premixed gaseous flame systems. These findings include the following:

1) Omnidirectional heat losses sustained by the upstream flame structure at $g = 0$ are smaller than those for the other two modes ($g = \pm 1$) and result in substantially higher peak temperatures for the $g = 0$ case.

2) Omnidirectional heat losses sustained by the upstream flame structure at $g = +1$ are smaller than those for the $g = -1$ case. This helps to account for the wider stability limits at $g = +1$ than those observed at $g = -1$.

3) Buoyancy effects move the $g = +1$ flame closer to the cold boundary (than is the case for $g = 0$). Buoyancy effects move the $g = -1$ flame farther away from the cold boundary. This helps to account for the larger cold boundary heat losses at $g = +1$ than are observed for $g = -1$.

4) From findings 2 and 3 it follows that upstream transverse heat losses at $g = -1$ are larger than those at $g = +1$.

5) From findings 2–4 it follows that large omnidirectional heat losses (rather than simply cold boundary heat losses) lead to narrowing of these flame stability limits.

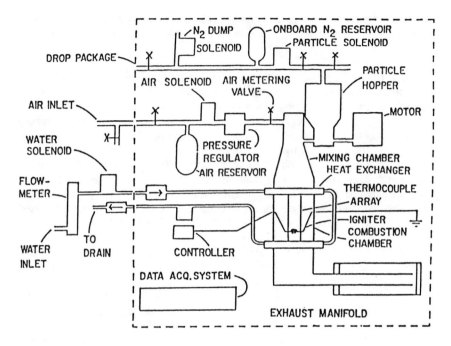

Fig. 1 Schematic of the burner apparatus used for the study of stabilized lycopodium-air flames at $g = 0, \pm 1$.

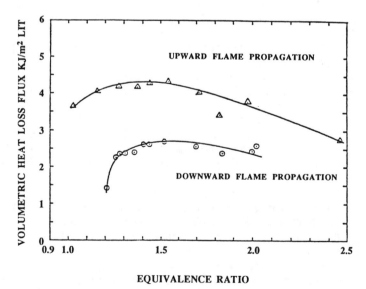

EQUIVALENCE RATIO

Fig. 2 Volumetric heat loss flux for stabilized lycopodium-air flames for a flame velocity of 17.1 cm/s.

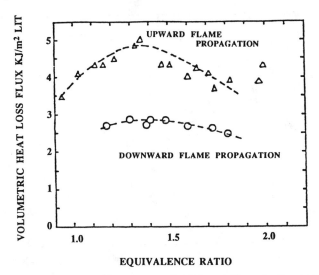

Fig. 3 Volumetric heat loss flux for stabilized lycopodium-air flames for a flame velocity of 13.9 cm/s.

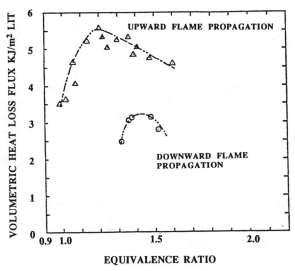

Fig. 4 Volumetric heat loss flux for stabilized lycopodium-air flames for a flame velocity of 11.4 cm/s.

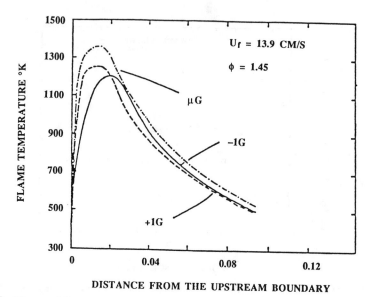

Fig. 5 Measured flame temperature profiles for stabilized lycopodium-air flames for upward $(+1\,g)$, downward $(-1\,g)$, and microgravity flames.

6) From findings 1–5 it follows that flame stability limit theory requires inclusion of omnidirectional heat loss rates. This general observation applies to both purely gaseous flames as well as to particle cloud flames.

7) The data show that cold boundary heat losses due to molecular conductive processes are a small fraction of the total heat loss rate to the cold boundary. These data, together with the observed temperature structures, show that radiative omnidirectional losses are large and that gravitationally induced transverse losses can be significant.

Commonly employed nonadiabatic flame theory[2] (for stabilized flames) is generally one dimensional and does not take into account gravitational effects. Recent theoretical efforts to consider gravitational effects[25] (in one dimension) and to consider two-dimensional flows[18,26] (without gravitational effects) represent promising starts to more general representations. Flame propagation and extinction theory recently employed for particle cloud combustion generally fails to account for transverse radiative losses.[9,19,20,27]

An interesting aspect of other $(g \simeq 1)$ studies of quasisteady, burner-stabilized particle cloud flames concerns the effects of particle sedimentation. In general, the experimenter fixes the particle number flux that is to support the steady, stabilized flame. Cold gas particle concentrations are gravitationally influenced and are determined from

$$c_0 = \frac{\dot{n}''}{V_t \pm V_g} \qquad (5)$$

where $\dot{n}''$ is the number flux per unit area, V_t the settling speed, V_g the experimenter-imposed gas speed, and c_0 the volumetric number concentration of particulates. The positive sign is used where settling velocities and gas velocities are parallel. The gaseous flux is given by

$$\dot{m}''_g = V_g \cdot \rho_g \tag{6}$$

and the equivalence ratio (defined to be the fuel/oxidizer mass fluxes divided by the stoichiometric fuel/oxidizer mass flux ratio) is given by

$$\phi = \frac{c_0(V_g \pm V_t)}{\rho_g V_g M_p} \tag{7}$$

where M_p is a (stoichiometric factor) constant. We may also define ϕ^* as the ratio of fuel to oxidizer mass densities divided by the stoichiometric fuel/oxidizer mass density ratio:

$$\phi^* = \frac{c_0}{\rho_g M_p} \tag{8}$$

For purely gas phase systems ϕ and ϕ^* are identical. This is generally not the case for flowing particle cloud systems. The magnitudes and directions of V_t and V_g become exceedingly important as experiments are carried out near flammability limits (low values of V_g) and for large particle sizes (high values of V_t). These effects are generally important but are not analyzed in the body of data provided by $g = \pm 1$ burner-stabilized studies. Consider, for example, the data derived from the very careful experimental studies reported in Ref. 8. Those data[8] are shown in Fig. 6 and show flame speed vs particle concentration for burner-stabilized Pocahontas coal dust-air flames. Flow was downward, and flame propagation was upward. The apparatus used[8] was not capable of performing downward flame propagation studies. Left unresolved are the following experimental (and theoretical) issues: What downward propagating flame speeds would be measured if the apparatus could accomodate the observations? What are the effects of sedimentation, particularly where flame speed values are lowest (and flame stability is marginal)? As one investigates larger and larger particle sizes, a particle size regime is reached where no flame propagation is observable in downward flame propagation $(V_g - V_t) \simeq 0$. To what extent is flame extinction due to buoyancy effects? To what extent is this due to sedimentation effects? To what extent is this due to low volumetric vaporization-pyrolysis rates associated with increased particle size heat-transfer effects? Clearly, mixed particle size, burner-stabilized flames are impossible to study at $g = \pm 1$, where particle settling speeds are a significant fraction of the fundamental flame speeds.

a) I(X) for uniform distribution of particles

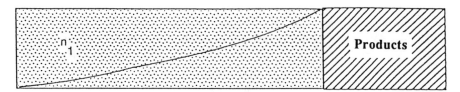

b) I(X) for nonuniform distribution of particles

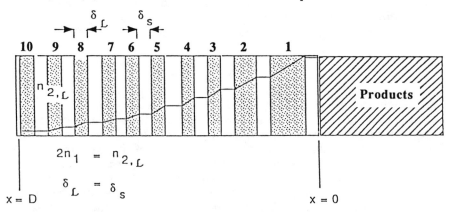

c) I(X) for random distribution of particles

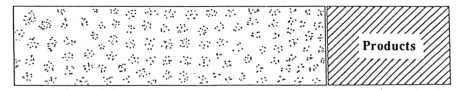

Fig. 6 Radiative flux density vs distance from reaction products for several modes of particle cloud flame propagation: a) uniform particle cloud, b) acoustically segregated particle cloud, and c) nonuniformly mixed particle cloud.

Particle Cloud Combustion Experiments:
Freely Propagating Flames in Low Gravity

Fuel Particulates in Air

The Lewis Research Center drop tower as well as Learjet research aircraft have been employed to achieve low-gravity experimental conditions. Freely propagating flames (in 5-cm-i.d. flame tubes) have been studied for uniform particle clouds of lycopodium in air at constant pressure.[28-31] The experimental arrangement is essentially that described in Ref. 32. A long flame tube is mounted on a Learjet and contains a premeasured lycopodium particulate charge. During the some 20 s of low g ($g^* \simeq 10^{-2}g_0$) available on a Learjet in a Keplerian trajectory, the following occurs: 1) an acoustic driver is turned on (for ~ 0.5 s) to mix the particles into a suitably uniform cloud; 2) the driver is turned off, and secondary airflows are given time (~ 7 s) to decay; 3) a nitrocellulose igniter is powered, burning through a mylar diaphragm and igniting the particle cloud; and 4) flame propagation proceeds axially down the tube length.

Four LED optical detectors measure cloud uniformity. High-speed (200 frames/s) photography provided details of flame propagation. The initial results of these studies have been cited elsewhere.[33,34] Several interesting flame propagation modes have been observed. Each mode is strongly dependent on the radiative fields peculiar to its unique flame structure. Figure 6 shows the attenuation patterns for the radiative fields associated with the several different observed modes.[33,34] The first case of Fig. 6 corresponds to a uniform cloud of (lycopodium) particulates in air, subject to an initial radiative flux density, I_0 (incident from the right), originating from hot combustion products. For the same initial equivalence ratio, ϕ_0, the second case of Fig. 6 corresponds to an acoustically segregated system of laminae,[33] where the equivalence ratio prior to acoustic segregation is also ϕ_0. Particle segregation derives from flame-induced acoustic excitation of the Kundt's tube phenomenon[33,34] The third case of Fig. 6 corresponds to a microscopically nonuniform mixture that is macroscopically uniform at the same equivalence ratio, ϕ_0. Experimental data pertaining to modes 1 and 3 are currently being analyzed. The novel mode 2 flames have been designated as *chattering flames*. These mode 2 flames progress through a series of radiatively induced autoignitions where

$$u_f = (\delta_1 + \delta_2)/\tau$$

and where δ_1 is the (cold) width of the particle-concentrating laminae and δ_2 of the particle-depleted neighboring space. Based on $\bar{u}_f$ and δ values observed, the autoignition time τ is about 30 ms for $\phi_0 = 1.2$. Use of the autoignition data of Conti and Hertzberg[35] along with a theory for radiatively driven particle cloud autoignition yield calculated autoignition times that agree well with the observations.[33,34]

These novel flame structures observed for freely propagating flames ($g^* \simeq 10^{-2}g_0$) owe their features to a number of special properties. First, partial confinement encourages flame-acoustic interactions. Second, clouds of particles are segregated in laminae, in a manner prescribed by a complex acoustic field. Third, the flame's radiative flux density penetrates far into the unburned particle cloud regimes. These special features are characteristic for clouds of particulates as well as for some clouds of drops.

Substantial safety implications derive directly from these observations. In low gravity, sedimentation and buoyancy effects are generally not capable of destroying the conditions needed for a flame-induced Kundt's tube segregation of particulates. Accordingly, for partially/wholly confined systems of combustible particulates, average values of fuel concentrations may not characterize the flammability (or explosive danger) of a combustible system. For particles in confined spaces uncontrolled fire and explosion may be a threat even where ϕ_0 values (averaged over a large regime) are below some apparent lean limit. For small particles such threats appear real even for normal gravitational conditions (e.g., flour mills) and for large containment vessels.

Particle segregation processes that may give rise to highly nonuniform fuel (or other particle) concentrations have no simple parallelism to premixed gaseous systems. Particle-particle agglomerative forces and complex fluid flowfields (acoustically influenced, in some cases) can create large particle nonuniformities, even in initially uniform concentrations of particulates. [32,36,37] Where particle-segregation effects occur, flame spread characteristics and safety limits may be substantially modified. [37]

Inert Particulates in Uniformly Premixed Gaseous Fuel and Oxidizer

The lycopodium-air system sustains a variety of combustion processes characteristic of segregated (and unsegregated) fuel particulates in an oxidizing gas. It is to be recognized that related particle cloud effects may occur for combustible systems other than those for which the only fuel elements are combustible particulates. In general several cases may be considered: 1) fuel particulates in an oxidizing atmosphere, 2) inert particulates in premixed gaseous fuel and oxidizer combination, and 3) hybrid systems of the two preceding subsections.

The lycopodium-air system is illustrative of case 1. The theoretical studies of Joulin and Eudier[38] of uniform clouds of inert particulates in methane-air is illustrative of case 2. Unfortunately, reduced-gravity experimental studies of inert particulates in the methane-air system are not yet available. Nevertheless, the Joulin studies indicate, for a cloud of uniformly mixed inert particulates in methane-air, that flame structure, flame propagation rates, and flammability limits are different from those for the pure gaseous system. Moreover, it appears that flammability limits may be *widened* by the addition of inert particulates to a premixed gaseous methane-air system. [38]

We may consider theoretically the consequences of inert particle segregation for a system such as that considered by Joulin. Such a particle-segregated system, for inert particulates + methane + air, may be associated with a number of limiting cases:

Case 1: Laminae spacings are such that $T_p \ll \delta_2/U_g$, where T_p is the autoignition time for the radiatively driven autoignition of the particle-enriched regimes, and U_g is the burning velocity of the radiatively transparent purely gaseous regime. The variable δ_2 is the distance between particle-enriched laminae. Flame propagation for such a system resembles that discussed previously for chattering flames. That is, the time for ordinary flame propagation through the purely gaseous regime is very long:

$$T_p \ll (\delta_2/U_g) = T_g$$

Thus, the controlling overall flame propagation rate is prescribed by

$$\bar{U}_f = \frac{\delta_1 + \delta_2}{T_p}$$

The flame propagates via a series of radiatively driven autoignitions, with the combustion of the purely gaseous regimes occurring in the wake of the leading portions of the sequence of autoignitions.

Case 2: Laminae spacings are such that $T_p \gg \delta_2/U_g$. Here

$$\bar{U}_f \doteq \frac{\delta_1 + \delta_2}{T_p + T_g}$$

For such a case the radiative preheating of the particle-enriched laminae does *not* lead to an autoignition that is largely independent of the purely gaseous system. Rather, the particle-enriched regimes (though somewhat preheated, radiatively) are ignited by the arrival of a flame that has traversed the δ_2 regime.

Case 3: Laminae spacings are such that $T_p \simeq \delta_2/U_g$. Here, the strong interaction between combustion processes in the δ_1 and δ_2 regimes makes it generally impossible for simple expressions (such as those for the previous two cases) to apply.

Case 1 corresponds most closely to the chattering flame system observed for lycopodium-air. For inert particulates in premixed methane-air, the radiatively induced autoignition process is characterized by $T_p = (\delta_1 + \delta_2)\bar{U}_f$. Here, radiative heating of inert particulates brings the methane-air mixture (within a particle-laden lamina) to autoignition. The overall methane oxidation rate expression may be taken to be[39]

$$\frac{d}{dt}[CH_4] = -6.7(10^{12})[CH_4]^{0.2}[CO_2]^{1.3} \exp\left[-\frac{48,400}{RT} \right]$$

where units are in terms of cubic centimeters, moles, Kelvin, and calories. For case 1 conditions ($T_p \ll \delta_2/U_g$) we have calculated[37] the space-time structure of the autoignition phenomenon for a segregated cloud of inert particles in a methane-air mixture. Figure 7 shows the laminae temperature structures for $j = 4,2,3,1$, where autoignition occurs within the $j = 1$ lamina. Radiative heating (flux is incident from the combustion products, situated

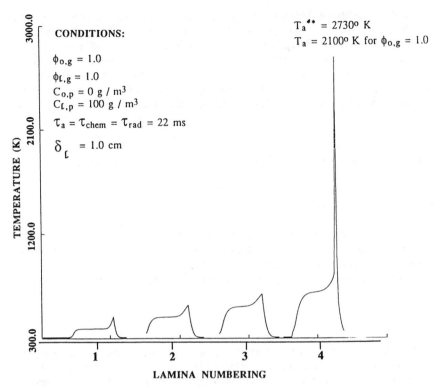

CONDITIONS:

$\phi_{o,g} = 1.0$
$\phi_{\ell,g} = 1.0$
$C_{o,p} = 0 \text{ g } / \text{m}^3$
$C_{\ell,p} = 100 \text{ g } / \text{m}^3$

$T_a = T_{chem} = T_{rad} = 22 \text{ ms}$

$\delta_\ell = 1.0 \text{ cm}$

$T_a^{**} = 2730^\circ \text{ K}$
$T_a = 2100^\circ \text{ K for } \phi_{o,g} = 1.0$

LAMINA NUMBERING

Fig. 7 Segregated inert dust cloud in a premixed methane-air-system; case 1) for $T_p \ll T_g$.

to the right of the irradiated particle-laden regimes) leads to superadiabatic temperatures, based on an adiabatic flame temperature of some 2100 K for an equivalence ratio of unity. The effect of molecular transport of heat and mass is to suppress the superadiabatic temperature actually achieved.

Concluding Remarks

Flames supported by premixed clouds of particulates in a gas may be of several types: fuel particulates in an oxidizing gas, inert particulates in a premixed gaseous fuel and gaseous oxidizer, or hybrids of these two limiting cases. Unlike purely gaseous systems, it is difficult or impossible to create uniform, quiescent systems (e.g., clouds of particulates of lycopodium in air) at normal gravity for the purposes of studying flame propagation and extinction. In order to deal with this problem of sedimentation (as well as the problem of buoyancy effects on flame propagation and extinction), reduced-gravity experimental facilities are required. It is found that NASA drop tower and aircraft low-g facilities are extremely useful in the study of both burner-stabilized particle cloud flames and freely propagating particle cloud flames.

Experiments to date include premixed fuel particulates in air for the cases of burner-stabilized and of freely propagating flames. Novel flame structures have been observed, which derive from flame-excited acoustic phenomena in a confined environment. The complex three-dimensional flows induced by the flame-acoustic interaction lead to particle segregation effects, i.e., the creation of particle concentration nonuniformities. Accordingly, initially uniform fuel particle clouds that are initially nonflammable may be rendered flammable where particle segregation effects adequately enrich the concentrations of local particle-containing regimes. These (fuel segregation) characteristics of two-phase combustible systems are quite unlike those for premixed gaseous systems. These characteristics imply that initially uniform nonflammable, fuel-lean systems may be transformed to locally flammable stoichiometries by agglomerative and fluid mechanical processes commonly encountered. Hybrid systems such as those involving inert dusts plus gaseous fuels in air (not yet studied in reduced gravity) have been of great interest to combustion safety specialists for many years.

Many "dusty" flame systems of fundamental and practical interest remain to be examined. A particularly interesting case of a dusty flame system involves hydrogen-air in which inert particulates (or small water droplets) are suspended. Studies of hydrogen combustion safety in nuclear containment buildings generally assume that radiative fields are not important to flame structure. In the absence of suspended dusts (and/or fine water droplets), this appears to be a valid assumption. The reality of nuclear containment building environments[37,40] suggests that they are necessarily "dusty" and that radiative transport *and* particle-segregation effects on flame structure may be important.

Finally, it is necessary to observe that the literature is rich in data related to limiting flammable concentrations (of particulates mixed with gases) associated with "flammability limits" of two-phase systems. The issues of particle sedimentation, buoyancy-induced flows, particle segregation effects, and unknown mixing-induced transport processes are currently not adequately resolved to permit safety specialists the straightforward use of such two-phase "flammability limit" data.[33,37]

References
[1]Coward, H. F., and Jones, G. W., "Limits of Flammability of Gases and Vapors," U.S. Bureau of Mines Bulletin 503, 1952.

[2]Williams. F. A., *Combustion Theory*, 2nd ed., Cummings, 1985.

[3]Berlad, A. L., Huggett, C., Kaufman, F., Markstein, G., Palmer, H. B., and Yang, C. H., "Study of Combustion Experiments in Space," NASA CR-134744, Nov. 1974.

[4]Berlad, A. L., "Combustion of Particle Clouds," *Progress in Aeronautics and Astronautics: Combustion Experiments in a Zero-Gravity Laboratory*, Vol. 73, edited by T. H. Cochran, AIAA, New York, pp. 91–127.

[5]Hertzberg, M., Conti, R. S., and Cashdollar, K. L., "Spark Ignition Energies for Dust Air Mixtures: Temperature and Concentration Dependences," *Twentieth Symposium (International) on Combustion*, The Combustion Institute, Pittsburgh, PA, 1984, p. 1681.

[6] Palmer, K. N., *Dust Explosions and Fires*, Chapman & Hall, London, 1973.

[7] Smoot, L. D., Horton, M. D., and Williams, G. A., "Propagation of Laminar Pulverized Coal-Air Flames," *Sixteenth Symposium (International) on Combustion*, The Combustion Institute, Pittsburgh, PA, 1976 p. 375.

[8] Horton, M. D., Goodson, F. P., and Smoot, L. D., "Characteristics of Flat, Laminar Coal Dust Flames," *Combustion and Flame*, Vol. 28, 1977, pp. 187–195.

[9] Smoot, D., and Horton, M., "Propagation of Laminar Coal-Air Flames," *Progress in Energy and Combustion Science*, Vol. 1, 1977, pp. 235–248.

[10] Berlad, A. L., and Joshi, N. D., "Gravitational Effects on the Extinction Conditions for Premixed Flames," *Acta Astronautica*, Vol. 12, 1985, pp. 539–545.

[11] Joshi, N. D., and Berlad, A. L., "Gravitational Effects on Stabilized, Premixed, Lycopodium-Air Flames," *Combustion Science and Technology*, Vol. 47 1986, pp. 55–68.

[12] Lovachev, L. A., Babkin, V. S., Bunev, V. A., V'yun, A. K., Krivulin, V. H., and Baratov, A. N., "Flammability Limits: An Invited Review," *Combustion and Flame*, Vol. 20, 1973 pp. 259–289.

[13] Levy, A., "An Optical Study of Flammability Limits," *Proceedings of the Royal Society of London, Series A*, Vol. A283, 1965, pp. 134–145.

[14] Hamins, A., Heitor, M., and Libby, P. A., "Gravitational Effects on the Structure and Propagation of Premixed Flames," International Astronautical Federation Paper 86-279, 1986.

[15] Spalding, D. B., "A Theory of Inflammability Limits and Flame Quenching," *Proceedings of the Royal Society of London, Series A*. Vol. A20, 1956, pp. 83–98.

[16] Berlad, A. L., and Yang, C. H., "On the Existence of Steady State Flames," *Combustion and Flame*, Vol. 3, 1959, pp. 447–452.

[17] Essenhigh, R. H., and Csaba, J., "The Thermal Radiation Theory for Flame Propagation in Coal Dust Clouds," *Ninth Symposium (International) on Combustion*, The Combustion Institute, Pittsburgh, PA, 1963, p. 111.

[18] Aly, S. L., and Hermance, C. E., "A Two-Dimensional Theory of Laminar Flame Quenching," *Combustion and Flame*, Vol. 40, 1981, pp. 173–185.

[19] Mitani, T., "A Flame Inhibition Theory by Inert Dust and Spray," *Combustion and Flame*, Vol. 43, 1981, pp. 243–253.

[20] Joulin, G., "Asymptotic Analysis of Non-Adiabatic Flames. Heat Losses Towards Small Inert Particles," *Eighteenth Symposium (International) on Combustion*, The Combustion Institute, Pittsburgh, PA, 1981 p. 1395.

[21] Kailasnath, K., and Oran, E., "Effects of Curvature and Dilution on Unsteady Premixed Laminar Flame Propagation," NRL Rept. 5659, 1985.

[22] Berlad, A. L., Joshi, N., Ross, H., and Klimek, R., "Particle Cloud Kinetics in Microgravity," AIAA Paper 87-0577, 1987.

[23] Gat, N., and Kropp, J. L., "Summary of Activities on the Zero-Gravity Particle Cloud Combustion Experiment," Summary Rept. on NASA Contract NAS3-23254, 1983.

[24] Berlad, A. L., Burns, R. J., and Ross, H. D., "Science Requirements for the Particle Cloud Combustion Experiment (Revision A)," NASA Document, Jan. 1987.

[25] Clavin, P., and Nicoli, C., "Effects of Heat Losses on the Limits of Stability of Premixed Flames Propagating Downwards," *Combustion and Flame*, Vol. 60, 1985, pp. 1–14.

[26] Oran, E. S., Young, T. R., Boris, J. P., Picone, J. M., and Edwards, D. H., "A Study of Detonation Structure: The Formation of Unreacted Gas Pockets," *Nineteenth Symposium (International) on Combustion*, The Combustion Institute, Pittsburgh, PA, 1982, pp. 573–582.

[27] Smoot, L. D., Horton, M. D., and Williams, G. A., "Propagation of Laminar Pulverized Coal-Air Flames," *Sixteenth Symposium (International) on Combustion*, The Combustion Institute, Pittsburgh, PA, 1976, pp. 375–387.

[28] Oran, E. S., and Boris, J. P., *Numerical Simulation of Reactive Flows*, Elsevier, New York, 1987.

[29] Nagy, J., Cooper, A. R., and Dorsett, H. G., Jr., "Explosibility of Miscellaneous Dusts," U.S. Bureau of Mines Rept. 2708, Dec. 1968.

[30] Hertzberg, M., Cashdollar, K. L., Zlochower, I., and Ng, D. L., "Inhibition and Extinction of Explosions in Heterogeneous Mixtures," *Twentieth Symposium (International) on Combustion*, The Combustion Institute, Pittsburgh, PA, 1984, pp. 1691–1708.

[31] Berlad, A. L., "On Characterization and Mitigation of Combustion Hazards Involved in the Handling of Particulate Materials," *Drying Technology*, Vol. 3, 1985, pp. 123–147.

[32] Ross, H., Facca, L., Tangirala, V., and Berlad, A. L., "Particle Cloud Mixing in Microgravity," AIAA Paper 88-0453, Jan. 1988.

[33] Berlad, A. L., Ross, H., Facca, L., and Tangirala, V., "Particle Cloud Flames in Acoustic Fields," *Combustion and Flame* (to be published).

[34] Berlad, A. L., Ross, H., Facca, L., and Tangirala, V., "Radiative Structures of Lycopodium-Air Flames in Reduced Gravity," AIAA Paper 89-0500, Jan. 1989; also *Journal of Propulsion and Power* (to be published).

[35] Conti, R. S., and Hertzberg, M., "Industrial Dust Explosions," edited by K. L. Cashdollar and M. Hertzberg, ASTM-PCN-04-958000-31, Philadelphia, PA, 1986, pp. 45–59.

[36] Rayleigh, Lord, "On the Circulation of Air Observed in Kundt's Tubes and on Some Allied Acoustical Problems," *Phil. Trans.*, Vol. 175, 1883, p. 1.

[37] Berlad, A. L., Tangirala, V., Ross, H., and Facca, L., "Particle Segregation Effects on the Combustion Safety of Dust-Containing Systems," *12th International Colloquium on the Dynamics of Explosions and Reactive Systems*, Paper 034, 1989.

[38] Joulin, G., and Eudier, M., "Radiation Dominated Propagation and Extinction of Slow Particle-Laden Gaseous Flames," *Twenty-Second Symposium (International) on Combustion*, The Combustion Institute, Pittsburgh, PA, 1989, pp. 1579–1585.

[39] Westbrook, C. K., and Dryer, F. L., "Simplified Reaction Mechanism for the Oxidation of Hydrocarbon Fuels in Flames," Spring Technical Mtg., The Combustion Institute, 1981.

[40] Berman, M. (ed.), "Proceedings of the Workshop on the Impact of Hydrogen on Water Reactor Safety," Vols. I–IV, Scandia National Lab. Rept. NUREG-CR-2017, 1981.

Opposed-Flow Flame Spread with Implications for Combustion at Microgravity

Robert A. Altenkirch* and Subrata Bhattacharjee†
Mississippi State University, Mississippi State, Mississippi

Nomenclature

a_P = Planck mean absorption coefficient, m^{-1}
C = specific heat, J/kg · K
C = dimensionless conduction ratio defined in Eq. (6)
f = fraction of gas-phase radiation directed to solid surface
L_g = gas-phase length, α_g/V_r, m
L_{gx} = gas-phase length in x direction, m
L_{gy} = gas-phase length in y direction, m
L_s = solid-phase length, α_s/V_f, m
L_{sx} = solid-phase length in x direction, m
L_{sy} = solid-phase length in y direction, m
Q' = heat transfer into or out of region I in Fig. 1, W/m
R_g = dimensionless number for gas-phase radiation
R_s = dimensionless number for surface radiation loss
T_f = scale for the gas temperature at the flame tip, K
T_v = scale for the evaporating surface temperature, K
T_∞ = ambient temperature, K
t = characteristic time, s
V_f = flame spread rate, m/s
V_{f0} = spread rate with no surface radiation, m/s
V_g = opposing flow velocity due to natural and/or forced convection, m/s
V_r = reference velocity, $V_f + V_g$, m/s
x = coordinate parallel to solid surface (see Fig. 1), m

*Dean, College of Engineering.
†Post-Doctoral Fellow, Department of Mechanical and Nuclear Engineering.

y = coordinate normal to solid surface, m
α = thermal diffusivity, m^2/s; surface absorptance
ϵ = surface emittance
λ = thermal conductivity, $W/m \cdot K$
ρ = density, kg/m^3
σ = Stefan-Boltzman constant, $W/m^2 \cdot K^4$
τ = fuel half-thickness, m

Subscripts
comb = combustion
evap = evaporation
g = gas phase
gx = x direction in gas phase
gy = y direction in gas phase
gsc = gas-to-surface conduction
gsr = gas-to-surface radiation
l = loss to environment
s = solid phase
sx = x direction in solid phase
sy = y direction in solid phase
ser = surface-to-environment radiation
sfc = solid forward conduction
sic = solid interior conduction

Introduction

STEADY laminar flame spread against an opposing flow of oxidizer constitutes a classic combustion phenomenon including interaction among fluid mechanics, heat transfer, and chemical processes. A considerable number of experimental and theoretical studies have been performed to investigate this apparently simple phenomenon that is critical to our understanding of fire spread. The literature is rich with descriptions of works on different aspects of this basic flame spread problem. An exhaustive discussion of the existing literature is beyond the scope of this article; a few references, mostly theoretical in nature, will be cited with the intent to highlight the complex physics of what appears to be a simple problem.

Many factors contribute to making a particular flame spread problem different from others. Notable among them are fuel type, fuel thickness, nature of the opposing flow, environmental conditions, and gravitational acceleration. Two kinds of fuel, namely polymethylmethacrylate (PMMA) and thin cellulosic materials (paper) have been mostly used in flame spread studies, since they generally exhibit two distinctly different kinds of mechanistic behavior.[1,2] Flames spreading over PMMA have been found to exhibit different sensitivity to gas-phase chemistry than those spreading over cellulosic fuels. The response of the spread rate to different ambient conditions has been found to depend strongly on the thickness of the fuel.[3,4] For example, the spread rate does not vary (or slightly decreases) with an increase in the opposing velocity for a thin fuel, whereas it may exhibit quite an opposite trend for a thick fuel. An opposing flow can be due to forced, natural, or mixed convection, and the character of the flow has a dominant effect in determining the flame shape and spread rate.[5]

At high opposing flow velocities, the flame may extinguish due to blow off.[6] The environmental oxygen level and pressure strongly influence the spread rate,[7-9] and flame spreading in low oxygen levels near extinction has been given special attention for the study of flammability limits.[7,10] The effect of various diluents in the oxidizer also has been considered.[11] Past studies aimed at determining the role of buoyancy in flame spreading[8,9] complement current ones in which flame spreading in a microgravity environment, where any opposing flow may be almost entirely absent due to a lack of buoyancy (gravity), is being investigated.[5,10-14] These latter studies show that flame behavior may be completely different from that which occurs in normal gravity, and they ultimately have implications for fire safety onboard spacecraft.

Our objective here is to classify various regimes of the flame spread phenomenon in an opposing flow environment depending on the mechanism of flame spread. Simple analyses of energy balances at the tip of the flame will be used for this purpose and will complement more sophisticated numerical analysis of the field problem associated with flame spreading in microgravity. The microgravity flames are shown to constitute a separate regime or class by themselves and are discussed in the most detail.

Energy Analysis

Figure 1 is a schematic of the flame spreading over a solid fuel into an opposing flow. With respect to the flame the oxidizer has a relative velocity of $V_r = V_f + V_g$, where V_g is the oxidizer velocity due to forced, natural, or mixed convection. In this flame-fixed coordinate the fuel approaches the flame with velocity V_f, the flame spread rate, and its surface temperature rises from T_∞ to T_v, the evaporation or vaporization temperature, near the flame tip. Evaporated fuel reacts with the ambient oxygen to form the laminar diffusion flame anchored near the surface at the flame tip.

The flame spreads by transferring heat to the unburned fuel ahead of it,[15] and this transfer takes place in the vicinity of the flame tip. Therefore, the physics of the spread process can be discussed by confining attention to the flame-tip region.

Consider the length scales indicated by an L with a subscript in Fig. 1. Because diffusion is important in both the x and y directions in the gas, a balance between convection and diffusion produces the following relation for a gas-phase length scale, L_g:

$$L_g = L_{gx} = L_{gy} = \frac{\alpha_g}{V_r} \qquad (1)$$

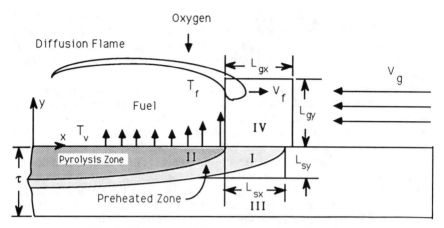

Fig. 1 Schematic of the flame spread configuration.

Under many situations, as will become clear later, conduction from the gas to the solid overwhelms any other heat-transfer pathway, and L_g becomes the only length scale of importance. With L_g being inversely proportional to V_r, flame size is expected to decrease with an increase in opposing flow speed.

For the flame to propagate three different processes, i.e., heating of the fuel in region I (see Fig. 1) from T_∞ to T_v, evaporation of the fuel, and gas-phase reaction releasing the enthalpy of combustion, must take place simultaneously. A quick comparison of the characteristic times involved shows that evaporation and gas-phase reaction are generally relatively much faster than the remaining process such that the spread rate is often heat-transfer-limited.

The net heat transfer to region I is due to the following: 1) conduction from the gas, Q'_{gsc} (region IV to region I); 2) conduction through the heated solid, Q'_{sfc} (region II to region I); 3) conduction from region I, Q'_{sic} (region I to region III); 4) radiation from the surface of region I escaping to the environment, Q'_{ser}; and 5) radiation from the gas onto the surface of region I, Q'_{gsr}. This net heat transfer can be equated to the enthalpy rise of region I, which yields

$$\rho_s C_s L_{sy} V_f (T_v - T_\infty) = Q'_{gsc} + Q'_{sfc} - Q'_{sic} - Q'_{ser} + \alpha Q'_{gsr} \qquad (2)$$

where Q' is the total heat transfer, per unit width, to region I.

V_f explicitly appears in Eq. (2) and is determined by the competition among various terms arising from the heat transfer from different sources. By scaling each term, it is possible to explore various regimes under which the flame spread problem simplifies, and the behavior of the spread rate can be predicted in a qualitative sense without detailed analysis.

Nonradiative Flames

Propagation Mechanisms

We desire to be able eventually to determine the dominant heat-transfer mechanism(s) for flames spreading in a microgravity environment where V_g is zero. As we shall see, this regime is one of the most complicated in which radiation is important; thus, we approach it by first discussing the more simple regimes.

First we neglect both the surface and gas-phase radiation. This will later be shown to be justifiable under many circumstances. We are then left with two driving mechanisms, i.e., Q'_{gsc} and Q'_{sfc}, to supply heat for the enthalpy rise of the solid and the heat lost to the solid interior, Q'_{sic}. We take the oxidizer flow to be uniform and parallel to the surface, i.e., the Oseen approximation, and the heat transfer from the gas to the solid becomes purely conductive. The Q' of Eq. (2) can then be approximated as

$$Q'_{gsc} \cong \lambda_g(T_f - T_v)L_g/L_g = \lambda_g(T_f - T_v) \tag{3a}$$

$$Q'_{sfc} \cong \lambda_s(T_v - T_\infty)L_{sy}/L_{sx} \tag{3b}$$

$$Q'_{sic} \cong \lambda_s(T_v - T_\infty)L_{sx}/L_{sy} \tag{3c}$$

In order to estimate the solid-phase length scales, note that the gas-phase conduction term, Q'_{gsc}, in Eq. (3) is independent of the solid-phase length scales, whereas the solid-phase forward conduction, Q'_{sfc}, is proportional to L_{sy}, the depth of the heated layer in the solid. For the situation where forward solid conduction is the driving force for flame propagation, i.e., $Q'_{sfc} \gg Q'_{gsc}$, a balance between the remaining terms of Eq. (2) yields

$$L_s = L_{sx} = L_{sy} = \frac{\alpha_s}{V_f} \tag{4}$$

Because L_{sy} cannot be larger than τ,

$$L_{sy} = \min(\tau, L_s) \tag{5}$$

Using these length scales in Eqs. (3), the ratio of gas-phase to solid-phase conduction heat transfer is

$$Q'_{gsc}/Q'_{sfc} \cong \frac{CL_s}{\min(\tau, L_s)} \tag{6}$$

where

$$C = \frac{\lambda_g(T_f - T_v)}{\lambda_s(T_v - T_\infty)} \tag{7}$$

The preceding estimate for the heat-transfer ratio Q'_{gsc}/Q'_{sfc} is based on the solid-phase length scales given by Eq. (4), which are correct only under conditions in which gas-phase conduction is relatively unimportant. Under other circumstances solid-phase conduction may be overestimated by using Eq. (6). Because L_s as given in Eq. (4) is usually small compared to τ for most flame spread situations, it is usually the value of C that determines whether gas- or solid-phase conduction plays the dominant role, as Williams[15] has previously pointed out.

Gas-Phase Conduction Dominance

For flames spreading over cellulosic fuels, C is larger than 5 in most usual situations. Therefore, conduction through the gas phase is the only significant means of heat transfer ahead of the flame in the absence of radiation. However, it must be realized that fuel type alone does not determine C. Environmental conditions affect T_f and T_v and can therefore alter C. For example, flames spreading near extinction exhibit a lowered T_f at relatively fixed T_v, thereby assuming a smaller value of C.

For high-C flames, the solid-phase length scale must be estimated again because the previous estimates, i.e., Eqs. (4) and (5), only apply to low-C flames. With solid-phase forward conduction negligible, the streamwise length over which solid heating occurs, L_{sx}, must be entirely determined by the gas-phase length scale. Hence, $L_{sx} = L_g$. To determine L_{sy}, Q'_{gsc} in Eq. (2) is balanced with Q'_{sic} to yield

$$L_{sy} = L_g/C \tag{8}$$

$$L_{sx} = L_g = \alpha_g/V_r \tag{9}$$

Because Q'_{sic} is always less than Q'_{gsc}, the expression for L_{sy} given in Eq. (9) constitutes a lower bound for the actual depth of heating. Accounting for the fact that L_{sy} cannot be larger than τ, Eq. (8) can be rewritten to give

$$L_{sy} = \min(\tau, L_g/C) \tag{10}$$

When $\tau < L_g/C$, the fuel is uniformly heated across its thickness, and Q'_{sic} can be dropped from Eq. (2), which results in the simple expression for V_f:

$$V_f \cong \frac{\lambda_g(T_f - T_v)}{\rho_s \tau C_s(T_v - T_\infty)} \tag{11}$$

for $CL_s/\min(\tau, L_s) \gg 1$, and $\tau < L_g/C$. Except for a factor of $\sqrt{2}$ in the denominator, this expression is the same as de Ris's analytical result for thin fuels.[3]

The conditions of Eq. (11) clearly identify a thin fuel. For cellulosic fuels where C is usually larger than 5, τ must be about an order of magnitude smaller than L_g for the fuel to be thin. This is not a particularly stringent criterion at low opposing flow velocities, since L_g can be quite large. L_s is usually much smaller than L_g such that $\tau \ll L_s$ as the criterion that determines whether or not a fuel is thin is a stringent one that will meet both the conditions of Eq. (11) for any realistic type of fuel.

For a fuel of sufficient thickness, τ in Eq. (11) must be replaced by the depth of heating, L_{sy}. A little manipulation results in the following formula for V_f:

$$V_f \cong \frac{\rho_g C_g \lambda_g V_r}{\rho_s C_s \lambda_s} \left| \frac{T_f - T_v}{T_v - T_\infty} \right|^2 \tag{12}$$

for $CL_s/\min(\tau, L_s) \gg 1$ and $\tau \gg L_g/C$. This result is the same as de Ris's analytical formula for thick fuels.[3] The determination of whether or not a fuel is thick or thin is intimately connected to the environmental conditions; C depends on T_f, which depends on the environmental conditions and the opposing flow velocity due to finite-rate chemistry effects, and L_g depends on the opposing flow velocity. This connection will be pronounced for flame spreading at microgravity, where L_g increases drastically and, as we shall see, T_f decreases. An attempt is made in Fig. 2 to define graphically the various flame spread regimes discussed thus far and the ones to be identified later. The qualitative behavior of C^{-1} as a function of V_r is based on arguments given in the following.

With an increasing opposing flow velocity the residence time, α_g/V_r^2, quickly decreases and eventually is insufficient to support the finite-rate

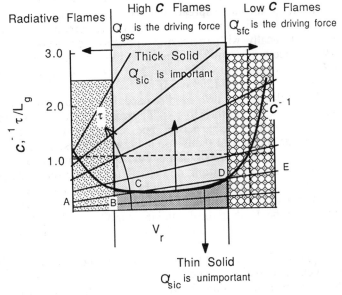

Fig. 2 Different flame spread regimes.

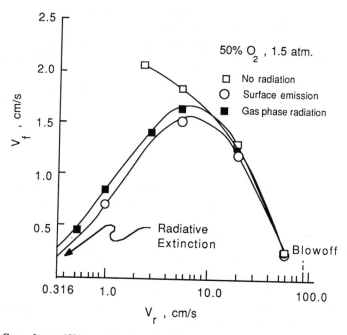

Fig. 3   Spread rate (V_f) as a function of reference velocity ($V_r = V_f + V_g$) calculated with different models. (From Ref. 14. Published with permission of The Combustion Institute.)

chemical reactions that occur in the gas. The result is flame blow off as illustrated in Fig. 3, for flame spreading over ashless filter paper at 50% O_2 and 50% N_2 by volume and 1.5 atm, where the spread rate is seen to decrease monotonically with increasing opposing flow velocity. Without including the effects of finite-rate chemistry, the curve of Fig. 3 would follow a line of constant V_f as demanded by Eq. (11). As T_f decreases with increasing V_r due to finite-rate chemistry effects, so does C. For low opposing flow velocity, as will be discussed later, radiative effects may cause a reduction in T_f and therefore C. Thus, at low and high V_r, the C^{-1} curve of Fig. 2 exhibits a sharp rise. The results shown in Fig. 3 were obtained from the theoretical flame spreading model described in detail in Refs. 12–14. The model consists of the partial differential equations that describe mass, energy, momentum, and species conservation in the gas; the ordinary differential equations that describe mass and energy conservation in the solid; appropriate boundary conditions; and the eigencondition needed to determine V_f that the boundary conditions along the fuel surface must be consistent with both the gas- and solid-phase differential equations.

The resulting C^{-1} curve qualitatively shown in Fig. 2 separates the thin from the thick solid regime. Superimposed on Fig. 2 are $\tau/L_g = (\tau/\alpha_g)V_r$ lines, which are lines of constant τ. Given a V_r and τ, on which side of the C^{-1} line the coordinate is located can be obtained, thus what is the governing flame spread mechanism can be determined. For example, consider a fuel of thickness τ that corresponds to line ABCDE in Fig. 2. Between A and B, the flame is strongly radiative, which will be discussed later; between B and C, τ/L_g is less than C^{-1} and $C > 1$, which meets the conditions of Eq. (11), and thus the fuel can be considered thin. With similar arguments it is easy to see that the same fuel becomes thermally thick between C and D; beyond the point D, C^{-1} quickly rises, and flame propagation is no longer solely governed by gas-phase conduction. Because T_f depends strongly on the ambient oxygen level, the C^{-1} curve shown in Fig. 2 corresponds to a particular ambient oxygen level. Figure 2, then, is a plane of constant O_2 concentration cut from a three-dimensional figure for which the third axis is ambient oxygen level.

Solid-Phase Conduction Dominance

Flames spreading over PMMA under most normal situations exhibit a low value of C. We have already established that under such a situation the solid-phase lengths are given by Eqs. (4) and (5). However, if $\tau < CL_s$, it has been shown that any flames including low-C PMMA flames will be similar to flames spreading over thin cellulosic fuels. For most situations involving PMMA, though, $\tau > CL_s$, and solid-phase forward conduction is responsible for flame propagation. Experimental observation[2] in this regard shows that Q'_{sfc} is almost 10 times as high as Q'_{gsc}.

There are no analytical solutions for spread rate similar to de Ris's formula[3] for low-C flames. An attempt to find an expression for V_f from Eq. (2) with Q'_{sfc} and Q'_{sic} as the only terms on the right-hand side results in $V_f = \alpha_s/L_s$, which simply expresses one unknown in terms of another. The spread phenomena are similar to those of a premixed flame, as Williams[15] points out. As such, the length scale is established by the flame, and heat transfer is no longer a limiting factor. The residence time in the solid phase, L_s/V_f, cannot be less than the t_{evap}, whereas in the gas the residence time, L_s/V_r, cannot be less than the gas-phase reaction time, t_{comb}. Available kinetic parameters suggest that $t_{evap} \ll t_{comb}$. Hence, gas-phase reaction controls the spread of such flames, and the time of reaction must be of the same order of the smaller of the available residence times in the solid and gas phase. V_r being always $\geqslant V_f$, we obtain

$$L_s = V_r t_{comb} \qquad (13a)$$

and

$$V_f \cong \alpha_s/(V_r t_{comb}) \qquad \text{for} \qquad \tau > CL_s \qquad \text{and} \qquad C \ll 1 \qquad (13b)$$

Altenkirch et al.[9] measured the downward spread rate of PMMA flames as a function of different gravity levels at various ambient conditions by creating artificial gravity using a centrifuge. At low oxygen levels, and therefore low levels of T_f and C, the spread rate increased with increasing gravity. At the higher oxygen levels, however, the spread rate decreased with increasing gravity. This result is consistent with Eq. (13).

In the absence of any forced flow, the opposing flow velocity is produced by buoyancy with a characteristic velocity of $V_g = (\alpha_g g\{(T_f/T_\infty)-1\})^{\frac{1}{3}}$. It can be expected that the opposing flow created by gravity level g would have a similar effect on the flame spread rate as that produced by a forced flow of velocity V_g, provided the flame is embedded well within the boundary layer. Theoretical support for such a conclusion is shown in Fig. 4, in which numerically computed spread rates from the model presented in Refs. 5, 12, and 13 for opposing flows of characteristic velocity V_r, caused by both forced and natural convection are presented. It is evident from Fig. 4 that an increase in g is equivalent to an increase in V_r, with both causing a reduction in V_f for thin cellulosic fuel.

For the normal ambient oxygen level, C for PMMA flames is less than unity, and Eqs. (13) suggest a drop in spread rate with increasing V_r as seen in the experiments.[7,9] At high oxygen levels, C is high enough for the PMMA flames to behave like the high-C flames discussed earlier, where V_f increases with V_r in Eq. (12). In fact, Altenkirch et al.[9] treated PMMA flames with a high value of C and obtained an excellent correlation for a dimensionless V_f with an appropriate reduced Damkohler number for the high oxygen levels.

Just as PMMA flames may at times behave like the cellulosic flames, the cellulosic flames may behave as the PMMA flames normally do. At low oxygen levels, C may be quite low, even for cellulosic flames. Solid-phase conduction should therefore play an increasingly important role in near-limit flames. Chen[17] reached a similar conclusion while modeling near-limit flame spread over thin cellulosic fuels in a microgravity environment.

Radiative Flames

In the discussion thus far, radiation from the gas and the fuel surface, Q'_{ser} and Q'_{gsr}, have been neglected without justification. An examination of

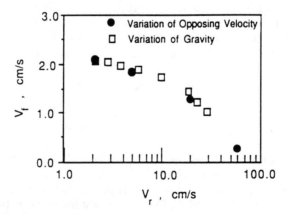

Fig. 4   Spread rate as a function of reference velocity. Similarity between forced and natural convection (variable gravity).

these terms will be carried out to determine under what circumstances they become important.

For the sake of simplification we will limit the discussion to flames spreading over cellulosic fuels where the gas-phase length scale establishes the extent of the preheat zone in the solid, i.e., $L_{sx} = L_g$. The radiation loss from the surface can be easily approximated as

$$Q'_{ser} \cong \epsilon\sigma(T_v^4 - T_\infty^4)L_g \tag{14}$$

To estimate the radiative transfer from the gas to the preheat zone, consider the gas mixture of dimensions L_g by L_g of unit width at the tip of the flame. A fraction f of the total emission from this volume is taken to be directed back to the surface, and the rest escapes to the environment. Using the emission approximation, Q'_{gsr} can be approximated as

$$Q'_{gsr} \cong f4a_P\sigma(T_f^4 - T_\infty^4)L_g^2 \tag{15}$$

where a_P is the Planck mean absorption coefficient for the emitting gas mixture. These radiative heat-transfer rates can be compared to the gas-phase conduction rate, Q'_{gsc}, to produce the following dimensionless numbers:

$$R_g \equiv \frac{\alpha Q'_{gsr}}{Q'_{gsc}} = \frac{\alpha 4a_P f \sigma T_\infty^3}{\lambda_g}\left|\frac{(T_f/T_\infty)^4-1}{T_f/T_\infty - T_v/T_\infty}\right|L_g^2 \tag{16}$$

$$R_s \equiv \frac{Q'_{ser}}{Q'_{gsc}} = \frac{\epsilon\sigma T_\infty^3}{\lambda_g}\left|\frac{(T_v/T_\infty)^4-1}{T_f/T_\infty - T_v/T_\infty}\right|L_g \tag{17}$$

The magnitude of R_g and R_s should indicate the level of involvement of radiation in establishing the spread rate. High values of R_g indicate augmentation of the driving force responsible for propagation, whereas high values of R_s signify more surface radiative loss and hence lower net heat flux to the solid. In addition to these two effects, an indirect influence of radiation comes from the fact that the flame radiates heat to the environment, causing the flame to become cooler. The total loss due to radiation can be approximately written as

$$Q'_l = Q'_{gsr}(1-f)/f + Q'_{gsr}(1-\alpha) + Q'_{ser} \tag{18}$$

where the first two terms on the right-hand side represent the gas-phase radiation directly escaping to the environment and the radiation reflected from the surface, respectively. As a result of this radiative cooling, T_f is reduced, which causes a drop in the conductive heat flux from the gas.

Although the length scale in the gas phase will be modified by radiation, the expression given in Eq. (1) for L_g can be used to determine approximately the importance of radiation. Because L_g is directly proportional to

V_r, all of the quantities R_g, R_s, and Q'_l can be expected to be small for high opposing flow velocities. This is the rationale for neglecting radiation in most of the analyses available in the literature.

In a normal-gravity environment in downward flame spread, buoyancy produces an opposing flow. This induced flow may not be strong enough to justify a sweeping assumption that radiation effects are not important, an assumption that is pervasive in the literature. Little work has been done in which radiation has been included in theoretical modeling. By arbitrarily varying the radiation parameters, Sibulkin et al.[18] conclude that surface radiation is important whereas gas radiation is not for vertically burning thick PMMA slabs. Models that include radiation in spreading flames include those of Fakheri and Olson[19] and Bhattacharjee and Altenkirch.[13,14,16]

In Refs. 13 and 19 only surface radiation is considered for flame spread over thin cellulosic fuels. In Ref. 13 attention is focused on a quiescent, microgravity environment such that V_r is at its minimum, i.e., $= V_f$. Surface radiation was found to be significant, and the spread rate normalized when the spread rate in the absence of surface radiation was found to decrease linearly (see Fig. 5) with an increase in the radiation parameter $\epsilon\sigma T_\infty^3/(\rho_g C_g V_r)$. Such behavior is understandable. With only Q'_{gsc} and Q'_{ser} on the right-hand side of Eq. (2), V_f can be expressed as

$$\frac{V_f}{V_{f0}} \cong 1 - \frac{\epsilon\sigma T_\infty^3}{\rho_g C_g V_r} \left| \frac{(T_v/T_\infty)^4 - 1}{T_f/T_\infty - 1} \right| \qquad (19)$$

where V_{f0} is the spread rate expression of Eq. (11).

In a companion study Bhattacharjee and Altenkirch,[14] in addition to surface radiation, coupled CO_2 and H_2O emission to the fluid dynamics and obtained spread rates for spread over thin cellulosic fuels for different environmental conditions. Spread rates obtained from the different models are shown as a function of V_r in Fig. 2. As V_r decreases, both surface- and gas-phase radiation eventually become important as the spread rate decreases due to increasing radiative effects. This type of behavior has been observed experimentally in drop tower studies of flame spreading in microgravity.[10] In the absence of radiative effects, the spread rate is expected to decrease with increasing V_r as T_f in Eq. (11) is reduced by finite-rate chemistry effects in the gas. The merging of the different models at high V_r can be anticipated from earlier scaling.

It is evident from Fig. 2 that at low V_r radiation effects in flame spreading cannot be neglected. In a quiescent, microgravity environment, where the induced flow is completely absent, V_r is as low as V_f, and radiative effects may completely control the spreading of such flames.

The radiation loss Q'_l is also expected to become more significant at low V_r, which results in an altering of the flame shape and structure. For example, for a particular environmental condition, Figs. 6a and 6b show temperature contours with and without radiation in the mathematical models.[12,14,16] Flame size is sharply reduced due to radiative loss. But it is gas-phase radiation that is mainly responsible because, when only surface

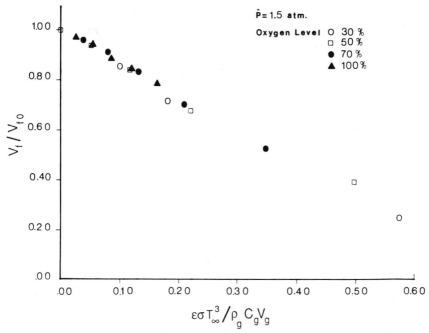

Fig. 5 Normalized spread rate as a function of the radiation parameter $\epsilon \sigma T_\infty^3/(\rho_g C_g V_r)$. **(From Ref. 13. Published with permission of The Combustion Institute.)**

radiation is included in the model, flame size does not undergo such a drastic change, as can be seen in Fig. 6c.

The optical properties of the fuel over which a flame is spreading are not known accurately. However, with $\alpha = \epsilon$, both R_g and R_s increase with surface emittance. Whereas an increase in R_g tends to increase the spread rate, a simultaneous increase in R_s and Q_l' favors a decrease as the net flux to the surface is reduced. These competing effects appear to result in the spread rate increasing with increasing ϵ as shown in Fig. 7 for a particular environmental condition.[14] The results in Fig. 7 imply that the increase in R_g more than compensates for the increase in R_s and Q_l'.

The role of gas-phase radiation is also demonstrated in Fig. 8, where the maximum flame temperature for a spreading flame is shown for two different models: one that considers both surface and gas-phase radiation and another that neglects gas-phase radiation. The flame temperature decreases by almost 1000 K when gas-phase radiation is included. In the studies cited in which gas-phase radiation has been included,[14,16] only CO_2 and H_2O radiation was considered. For sooty flames this may be inadequate, and for such a situation, gas-phase radiation is expected to assume an even bigger role. The consideration of soot radiation is a complexity

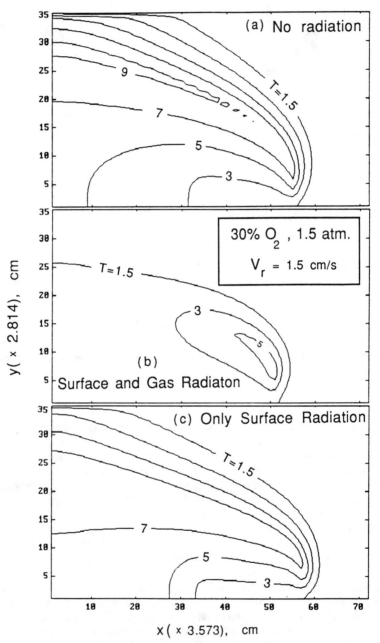

Fig. 6 Temperature contours (normalized with 298 K) obtained with different treatments of radiation in the mathematical model. (From Ref. 14. Published with permission of The Combustion Institute.)

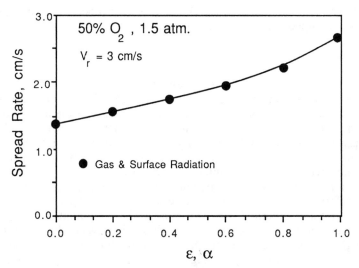

Fig. 7   Variation of V_f with surface optical properties. (From Ref. 14. Published with permission of The Combustion Institute.)

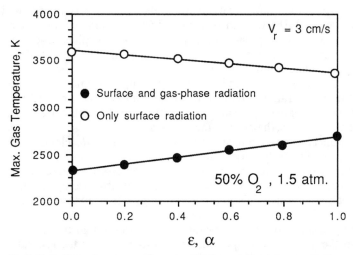

Fig. 8   Variation of maximum gas-phase temperature with surface optical properties. (From Ref. 14. Published with permission of The Combustion Institute.)

that has not yet been addressed in flame spreading models that are already inherently complex without its consideration.

Conclusions

Flame spreading into an opposing flow has been divided into three basic regimes based on simple analysis of the energy balance at the flame tip. The three important mechanisms of heat transfer to the unburned fuel, i.e., gas-phase conduction, solid-phase forward conduction, and gas-phase radiation, may all interact to supply the radiative loss from the surface, the heat conducted into the solid interior, and the enthalpy rise of the fuel. A balance among these form a fundamental scaling equation, Eq. (2), from which different asymptotic cases can be investigated.

An important criterion for determining the dominant driving force for propagation is the magnitude of the dimensionless conduction ratio C. For high values of C (the middle region of Fig. 2), which are usually observed for flames spreading over cellulosic fuels, gas-phase conduction is the dominant driving force, provided the opposing velocity is not too extreme in magnitude. Such flames may again be divided into two categories depending on whether or not the heat absorbed by the solid interior is important. For flames such that $\tau < L_g/C$, i.e., for a given V_r, if the location of the coordinate (τ, V_r) is below the C^{-1} line in Fig. 2, the fuel is thermally thin, and interior conduction can be neglected. For $\tau \gg L_g/C$, i.e., for locations above the C^{-1} line, interior heating is important, and the fuel is thermally thick. Although analytical expressions for spread rate of such thick and thin fuels have been available for some time,[3] the demarcation between the two situations has not always been clear.

For nonradiative flames with a low value of C, as is found in flames spreading over PMMA, solid-phase conduction assumes a dominant role, provided $\tau > L_s$. The spread rate depends on the gas-phase chemistry and is not heat-transfer-limited. However, for $\tau \ll L_s$, flame spread is dominated by gas-phase conduction once again.

The magnitude of C depends strongly on environmental conditions; hence, the spread rate mechanism is not determined solely by fuel type. For a given fuel the value of C can vary widely with the ambient oxygen level, causing a change in the spread mechanism.

Radiation becomes significant for low opposing flow velocities. Because buoyancy is effectively absent in a microgravity situation, the opposing flow velocity may be completely absent in a quiescent, spacecraft environment. These flames become strongly radiative irrespective of their size under such situations. This is a class of flames that apparently cannot be observed in the Earth's normal gravity environment because of the flows created by natural convection. Radiation has been found to have a striking effect on the flame size, shape, and spread rate of these flames. Additional research is necessary to characterize this relatively new class of spreading flame.

Acknowledgment

We thank Sandra Olson for serving as Contract Monitor and Kurt Sacksteder for serving in that capacity during the early phases of the project. This work was supported by NASA through Contract NAS3-23901.

References

[1] Frey, A. E., and T'ien, J. S., "A Theory of Flame Spread over a Solid Fuel Including Finite-Rate Chemical Kinetics," *Combustion and Flame*, Vol. 36, No. 3. 1979, pp. 263–289.

[2] Fernandez-Pello, A., and Williams, F. A., "Laminar Flame Spread over PMMA Surfaces," *Fifteenth Symposium (International) on Combustion*, The Combustion Institute, Pittsburgh, PA, 1975, pp. 217–231.

[3] de Ris, J. N., "Spread of a Laminar Diffusion Flame," *Twelfth Symposium (International) on Combustion*, The Combustion Institute, Pittsburgh, PA, 1969, pp. 241–252.

[4] Di Blasi, C. D., Crescitelli, S., Russo, G., and Fernandez-Pello, A. C., "Predictions of the Dependence on the Opposed Flow Characteristics of the Flame Spread Rate over Thick Solid Fuel," *Second International Symposium on Fire Safety Science*, IAFSS, Tokyo, June 1988.

[5] West, J., Bhattacharjee, S., and Altenkirch, R. A., "Buoyancy in Flame Spreading: A Comparison of the Role Played by Natural and Forced Convection" Central States Section, The Combustion Institute Meeting, Cincinnati, OH, May 1990.

[6] Altenkirch, R. A., and Vedha-Nayagam, M., "Opposed-Flow Flame Spread and Extinction in Mixed Convection Boundary Layers," *Twenty-Second Symposium (International) on Combustion*, The Combustion Institute, Pittsburgh, PA, 1989, pp. 1495–1500.

[7] Fernandez-Pello, A. C., Ray, S. R., and Glassman, I., "Flame Spread in an Opposed Forced Flow: The Effect of Ambient Oxygen Concentration," *Eighteenth Symposium (International) on Combustion*, The Combustion Institute, Pittsburgh, PA, 1981, pp. 579–589.

[8] Altenkirch, R. A., Eichhorn, R., and Shang, P. C., "Buoyancy Effects on Flames Spreading Down Thermally Thin Fuels," *Combustion and Flame*, Vol. 37, No. 1, 1980, pp. 71–83.

[9] Altenkirch, R. A., Eichhorn, R., and Rizvi, A. R., "Correlating Downward Flame Spread Rates for Thick Fuel Beds," *Combustion Science Technology*, Vol. 32, June 1983, pp. 49–66.

[10] Olson, S. L., Ferkul, P. V., and T'ien, J. S., "Near-Limit Flame Spread over a Thin Solid Fuel in Microgravity," *Twenty-Second Symposium (International) on Combustion*, The Combustion Institute, Pittsburgh, PA, 1989, pp. 1213–1222.

[11] Olson, S. L., Stouffer, S. C., and Grady, T., "Diluent Effects on Quiescent Microgravity Flame Spread over a Thin Solid Fuel," *Eastern Section/ The Combustion Institute Meeting*, Albany, NY, 1989.

[12] Bhattacharjee, S., Altenkirch, R. A., Srikantaiah, N., and Vedha-Nayagam, M., "A Theoretical Description of Flame Spreading over Solid Combustibles in a Quiescent Environment at Zero-Gravity," *Combustion Science Technology*, Vol. 69, Jan. 1990, pp. 1–15.

[13] Bhattacharjee, S., and Altenkirch, R. A., "The Effect of Surface Radiation on Flame Spread in a Quiescent, Microgravity Environment," *Combustion and Flame* (to be published).

[14]Battacharjee, S., and Altenkirch, R. A., "Radiation Controlled Flame Spread in a Microgravity Environment," *Twenty-Third Symposium (International) on Combustion*, The Combustion Institute, Pittsburgh, PA (to be published).

[15]Williams, F. A., "Mechanisms of Fire Spread," *Sixteenth Symposium (International) on Combustion*, The Combustion Institute, Pittsburgh, PA, 1977, pp. 1281–1294.

[16]Bhattacharjee, S., and Altenkirch, R. A., "Radiative Effects in Opposed-Flow Flame Spread over Thin Fuels," *Eastern Section/ The Combustion Institute Meeting*, Albany, NY, 1989.

[17]Chen, C. H., "Flame Propagation: Effect of Solid-Phase Heat Conduction," *Eastern Section/ The Combustion Institute Meeting*, San Juan, Puerto Rico, 1986.

[18]Sibulkin, M., Kulkarni, A. K., and Annamalai, K., "Effects of Radiation on the Burning of Vertical Fuel Surfaces," *Eighteenth Symposium (International) on Combustion*, The Combustion Institute, Pittsburgh, PA, 1981, pp. 611–617.

[19]Fakheri, A., and Olson, S., "The Effects of Radiative Heat Loss on Microgravity Flame Spread," AIAA Paper 89-0504, Jan. 1989.

Author Index

PROGRESS IN ASTRONAUTICS AND AERONAUTICS
SERIES VOLUMES

*1.   Solid Propellant
Rocket Research (1960)
Martin Summerfield
Princeton University

*2.   Liquid Rockets
and Propellants (1960)
Loren E. Bollinger
Ohio State University
Martin Goldsmith
The Rand Corp.
Alexis W. Lemmon Jr.
Battelle Memorial Institute

*3.   Energy Conversion
for Space Power (1961)
Nathan W. Snyder
*Institute for Defense
Analyses*

*4.   Space Power
Systems (1961)
Nathan W. Snyder
*Institute for Defense
Analyses*

*5.   Electrostatic
Propulsion (1961)
David B. Langmuir
*Space Technology
Laboratories, Inc.*
Ernst Stuhlinger
*NASA George C. Marshall
Space Flight Center*
J.M. Sellen Jr.
*Space Technology
Laboratories, Inc.*

*6.   Detonation and
Two-Phase Flow (1962)
S.S. Penner
*California Institute
of Technology*
F.A. Williams
Harvard University

_____

*Out of print.

*7.   Hypersonic Flow
Research (1962)
Frederick R. Riddell
AVCO Corp.

*8.   Guidance and
Control (1962)
Robert E. Roberson,
Consultant
James S. Farrior
*Lockheed Missiles
and Space Co.*

*9.   Electric Propulsion
Development (1963)
Ernst Stuhlinger
*NASA George C. Marshall
Space Flight Center*

*10.   Technology of
Lunar Exploration (1963)
Clifford I. Cummings
Harold R. Lawrence
Jet Propulsion Laboratory

*11.   Power Systems
for Space Flight (1963)
Morris A. Zipkin
Russell N. Edwards
General Electric Co.

*12.   Ionization in High-
Temperature Gases (1963)
Kurt E. Shuler, Editor
*National Bureau of
Standards*
John B. Fenn,
Associate Editor
Princeton University

*13.   Guidance and
Control – II (1964)
Robert C. Langford
General Precision Inc.
Charles J. Mundo
Institute of Naval Studies

*14.   Celestial Mechanics
and Astrodynamics (1964)
Victor G. Szebehely
*Yale University
Observatory*

*15.   Heterogeneous
Combustion (1964)
Hans G. Wolfhard
*Institute for Defense
Analyses*
Irvin Glassman
Princeton University
Leon Green Jr.
*Air Force Systems
Command*

16.   Space Power Systems
Engineering (1966)
George C. Szego
*Institute for Defense
Analyses*
J. Edward Taylor
TRW Inc.

17.   Methods in
Astrodynamics and
Celestial Mechanics (1966)
Raynor L. Duncombe
U.S. Naval Observatory
Victor G. Szebehely
*Yale University
Observatory*

18.   Thermophysics and
Temperature Control
of Spacecraft and
Entry Vehicles (1966)
Gerhard B. Heller
*NASA George C. Marshall
Space Flight Center*

102. **Numerical Methods for Engine-Airframe Integration** (1986)
S.N.B. Murthy
Purdue University
Gerald C. Paynter
Boeing Airplane Co.
ISBN 0-930403-09-6

103. **Thermophysical Aspects of Re-Entry Flows** (1986)
James N. Moss
NASA Langley Research Center
Carl D. Scott
NASA Johnson Space Center
ISBN 0-930403-10-X

104. **Tactical Missile Aerodynamics** (1986)
M.J. Hemsch
PRC Kentron, Inc.
J.N. Nielsen
NASA Ames Research Center
ISBN 0-930403-13-4

105. **Dynamics of Reactive Systems Part I: Flames and Configurations; Part II: Modeling and Heterogeneous Combustion** (1986)
J.R. Bowen
University of Washington
J.-C. Leyer
Université de Poitiers
R.I. Soloukhin
Institute of Heat and Mass Transfer, BSSR Academy of Sciences
ISBN 0-930403-14-2

106. **Dynamics of Explosions** (1986)
J.R. Bowen
University of Washington
J.-C. Leyer
Université de Poitiers
R.I. Soloukhin
Institute of Heat and Mass Transfer, BSSR Academy of Sciences
ISBN 0-930403-15-0

107. **Spacecraft Dielectric Material Properties and Spacecraft Charging** (1986)
A.R. Frederickson
U.S. Air Force Rome Air Development Center
D.B. Cotts
SRI International
J.A. Wall
U.S. Air Force Rome Air Development Center
F.L. Bouquet
Jet Propulsion Laboratory, California Institute of Technology
ISBN 0-930403-17-7

108. **Opportunities for Academic Research in a Low-Gravity Environment** (1986)
George A. Hazelrigg
National Science Foundation
Joseph M. Reynolds
Louisiana State University
ISBN 0-930403-18-5

109. **Gun Propulsion Technology** (1988)
Ludwig Stiefel
U.S. Army Armament Research, Development and Engineering Center
ISBN 0-930403-20-7

110. **Commercial Opportunities in Space** (1988)
F. Shahrokhi
K.E. Harwell
University of Tennessee Space Institute
C.C. Chao
National Cheng Kung University
ISBN 0-930403-39-8

111. **Liquid-Metal Flows: Magnetohydrodynamics and Applications** (1988)
Herman Branover,
Michael Mond, and
Yeshajahu Unger
Ben-Gurion University of the Negev
ISBN 0-930403-43-6

112. **Current Trends in Turbulence Research** (1988)
Herman Branover,
Michael Mond, and
Yeshajahu Unger
Ben-Gurion University of the Negev
ISBN 0-930403-44-4

113. **Dynamics of Reactive Systems Part I: Flames; Part II: Heterogeneous Combustion and Applications** (1988)
A.L. Kuhl
R & D Associates
J.R. Bowen
University of Washington
J.-C. Leyer
Université de Poitiers
A. Borisov
USSR Academy of Sciences
ISBN 0-930403-46-0

114. **Dynamics of Explosions** (1988)
A.L. Kuhl
R & D Associates
J.R. Bowen
University of Washington
J.-C. Leyer
Université de Poitiers
A. Borisov
USSR Academy of Sciences
ISBN 0-930403-47-9

115. **Machine Intelligence and Autonomy for Aerospace** (1988)
E. Heer
Heer Associates, Inc.
H. Lum
NASA Ames Research Center
ISBN 0-930403-48-7